PHYSICS AND NUCLEAR ARMS TODAY

Edited by
David Hafemeister

Readings from *Physics Today*

Number Four

American Institute of Physics
New York, New York
1991

Readings from *Physics Today*

PHYSICS TODAY, a publication of the American Institute of Physics, provides news coverage of national and international research activities in physics as well as government and institutional activities that affect physics. Both technical and nontechnical developments are covered by scientific articles, news stories, book reviews, letters to the editor, calendars of meetings, and editorial opinion.

Articles in PHYSICS TODAY are intended to be of interest to—and understandable by—a broad audience of professionals from all subfields of physics as well as people with a general interest in physical science.

Physics and Nuclear Arms Today is the fourth book in a series of volumes that contains reprinted articles and news material from PHYSICS TODAY in other areas and subfields of physics.

Cover and title design by Armen Kojoyian

Printed in the United States of America

Library of Congress Cataloging-in-Publication Data

Physics and nuclear arms today / edited by David Hafemeister.
    p. cm.—(Readings from Physics Today; no. 4)
    Includes bibliographical references.
    ISBN 0-88318-626-8.   ISBN 0-88318-640-3 (pbk.)
    1. Nuclear weapons.   2. Nuclear arms control.   I. Hafemeister, David W.   II. Series.
    U264. P48 1990
    355.8 25119—dc20   90-110
                                                                                          CIP

INTRODUCTION

The Early Years

This collection of reprints from PHYSICS TODAY contains articles written by physicists about nuclear weapons, their effects, their delivery systems, and the agreements that are intended to control these weapons. For good or bad, the history of nuclear weapons has been inexorably linked with physics. As physicists we have always been primarily committed to "the advancement and diffusion of the knowledge of physics," but indirectly the spin-offs of our research have, of course, been crucial to the development of nuclear weapons. Early on in 1919, Rutherford carried out the dream of the alchemists by transmuting oxygen to nitrogen, commenting,

> "Talk softly please. I have been engaged in experiments which suggest that the atom can be artificially disintegrated. If it is true, it is of far greater importance than a war."

Many events paved the way for the establishment of the Manhattan Project in the US and the smaller projects elsewhere. These included: the discoveries of deuterium by Urey in 1931 and of the neutron by Chadwick in 1932, the neutron experiments with uranium by Fermi in 1934, the discovery of fission by Hahn and Strassman in 1938, and the multiplication of neutrons from the fissioning of uranium by Fermi, Szilard, and Joliot in 1939. The physicists' domination of the field of nuclear weapons ended with the explosions of atomic bombs in 1945 under Oppenheimer, and the hydrogen bomb in 1954, using the Teller–Ulam compression designs. Thereafter, the dominance shifted as engineers also developed more sophisticated bombs and delivery systems of missiles, based on land, sea, and air.

The Role of Physicists

Prediction of the future is not easy. Perhaps the deterrence of nuclear war by the macabre threat of porcupine-quilled nuclear weapons will continue to give us another 40, and possibly many more years of peace between the industrialized states. If this were to be so, it would be very good. However, because of accidents, fear, or malice, it is possible that the weapons of deterrence could become the weapons of mass destruction. If this were to happen, it would be very bad. The question of which of these two possibilities is the more important, the one to act on, split the physics community the past 40 years, and will, no doubt, continue to polarize us. As honest academics, we must acknowledge that we cannot know what the future will bring, so we cannot speak with great assurance on the future outcomes of what will be writ large on the firm stone in the moveable sands of time. However, as in some form of rite of passage, we seem to have adopted an ethical code, a sort of Hippocratic oath. This code of ethics implies a theorem that we are obligated to examine the scientific facts, calculate what we can, and tell the truth with all of the assumptions, limitations, errors and caveats, laid out for easy reading, and not hidden away. It is our duty then to inform our colleagues, our students, and the public in general what we have learned from our analysis (on an unclassified basis). A corollary to this code of ethics is that we distinguish in our pronouncements between statements based on our technical expertise and those based on our humanity. And that is just what PHYSICS TODAY has done, working with and for the physicists in telling this "truth," particularly over the past 20 years of its 40-year history. When an issue is very controversial, such as the Comprehensive Test Ban Treaty, civil defense, nuclear waste disposal, SALT II, the "nuclear freeze," or the Strategic Defense Initiative (SDI, or "Star Wars"), PHYSICS TODAY has found experts on both sides of the issue to debate the details so that we can determine the technical truths. When an issue is less controversial, then purely technical articles have been written. The American Physical Society itself has contributed very positively to the scholarly literature on nuclear arms. The APS studies on "The Nuclear Fuel Cycle" and on "Directed Energy Weapons Systems" have been among the best resources available on these topics. In particular, I know of no document on directed-energy weapons comparable in scientific quality to the DEWS study, either inside or outside the government. These independent studies have been particularly valuable since governments often find it difficult to assemble qualified experts on controversial topics, such as nuclear arms, without some pressure, from within the government or from outside, to bias the results in some particular way.

Who is ahead? What is stabilizing? What is predictable?

The public is often concerned with "which side is ahead in the arms race." One can make numerical comparisons of the many parameters, systems, and scenarios such as:

Parameters: Number of RVs per launcher, number of warheads in each system and in total, explosive yield, accuracy of warheads, hardness of silos, reliability of warheads and missiles, availability and survivability of warheads, penetrability of warheads, rapid reload, etc.

Systems: ICBMs (silo-, road-, rail-based), SLBMs, bombers, ALCMs, SLCMs, tactical weapons, INF weapons, NAVSTAR/GLONASS, ABM/BMD and ASAT weapons, 3rd generation nuclear weapons, etc.

Scenarios: First strike ("bolt out of the blue," or time to flush mobiles), second strike (counterforce, or countervalue), launch-on-warning, launch-under-attack, nuclear-war fighting, proportional and flexible responses, nuclear-free zones, ASW-free zones, nuclear proliferation, etc.

When one tries to compare the various scenarios, considering the wide ranges of values for the quantifiable parameters and the number of different systems that have been or may be built, one is quickly overwhelmed with the possible outcomes of the arms race. And, because nuclear weapons are so powerful, and there are so many, it may indeed be impossible to say who is "ahead" overall. One may reasonably argue that the two competing sides have reached a point of "approximate parity," and proceed from there.

The technical aspects of nuclear arms are more predictable than the political aspects. For example, the evolution in offensive systems to higher accuracy with more re-entry vehicles per missile was predicted and has come to fruition with the American MX strategic missile and its Soviet counterparts, the SS-18 and SS-24. Various predictions have been made for defensive systems; we shall have to watch to see which are correct. On the other hand, predictions in the political arena seem much more difficult. Who would have predicted President Reagan's shift from seeing the Soviet Union as the "evil empire" to signing the INF Treaty ("trust, but verify")? Who would have thought that General Secretary Gorbachev would allow limited elections to the Congress of Deputies, invite challenge inspectors to go almost anywhere in the Soviet Union, shower the West with various arms control proposals, and encourage the destruction of the Berlin Wall? Who would have believed as recently as three years ago that US and Soviet scientists would monitor nuclear explosions at the test sites of their rivals? And, who would have predicted the collapse of Communism in eastern Europe?

PHYSICS TODAY on the Nuclear Arms Race

PHYSICS TODAY began publishing four decades ago, in 1948. During the first two decades between 1948 and 1968, PHYSICS TODAY published very few articles on the arms race. Of course, physicists did influence public policy during that period, but physics journalism was much more silent in those days. During that period J. Robert Oppenheimer was president of The American Physical Society (1948), six years before the AEC began an inquiry into his security clearance. Members of the APS were on both political sides of the big issues: how to contain the Soviet Union and the concomitant growth in nuclear weaponry. The arms race spiraled in agreement with the predictions of Niels Bohr and Leo Szilard. Physicists of various persuasions agreed that a system that controls 50 000 nuclear weapons could get out of control, but disagreed over a sensible political path of action. An example is the decision in 1949 of the General Advisory Committee to the Atomic Energy Commission to recommend building multi-stage, boosted uranium bombs rather than the hydrogen bomb, a decision in which Oppenheimer's opinion was influential. Part of the drama of the Oppenheimer hearings was Oppenheimer the physicist, but the larger landscape behind Oppenheimer was the political decision whether to build the "super," the hydrogen bomb. PHYSICS TODAY covered these issues (July and August 1954) in a responsible, but subdued fashion, as the journalism of the 1950s was not up to today's standard of asking the more penetrating questions.

In the second pair of decades, between 1968 and 1988, the debates in the physics community on nuclear arms policies became more focused. Spurred by the first ABM debates of the late 1960s, the physics community recognized both the benefits and the responsibilities of performing good physics analysis of public-policy issues. Beginning with the Washington meeting of 1969, physics and society sessions at the APS meetings became more lively as the debates over nuclear arms, energy policy, and environmental problems filled the evening sessions. In parallel, PHYSICS TODAY increased its coverage of nuclear-arms issues in both quantity and quality. This volume is a collection of the best of these articles of the past decade. Because events move rapidly in the national security area, we have made a special effort to update some of the articles by asking the authors for their comments on the changing circumstances.

The PHYSICS TODAY articles have been divided into eight sections. Each section begins with a brief editorial review of the articles, followed by the authors' reflections on the impact of changing events on their contributions. The eight sections are:

> Section 1: The Effects of Nuclear Weapons (pages 1 to 43)
> Section 2: Nuclear Testing (pages 45 to 70)
> Section 3: The Offense (pages 71 to 111)
> Section 4: The Defense (pages 113 to 185)
> Section 5: Nuclear Proliferation (pages 187 to 239)
> Section 6: History of Nuclear Weapons (pages 241 to 334)
> Section 7: Commentary on the Era of Nuclear Weapons (pages 335 to 388)
> Section 8: For Further Reading (pages 389 to 390)

The last section, "For Further Reading," lists additional books and compilations of references for those who would like to delve into the subject more deeply. I would like to thank the individual authors for taking time to provide updates on their articles and to Dietrich Schroeer for a critical reading of the manuscript. I especially would like to acknowledge Tom von Foerster and Michael Hennelly of AIP, who created the idea for this book and were extremely helpful on this project. Lastly, I would like to thank the Center for Energy and Environmental Studies at Princeton University for pleasant hospitality during the summer of 1989 while I completed this project.

David Hafemeister
San Luis Obispo/Princeton
September 1990

TABLE OF CONTENTS ══════

AUTHOR AFFILIATIONS

Robert B. Barker is the assistant to the Secretary of Defense for Atomic Energy Matters and a former assistant associate director of arms control at the Lawrence Livermore National Laboratory.

Henry H. Barschall is professor emeritus of physics at the University of Wisconsin, Madison.

Barton J. Bernstein is professor of history and Mellon Professor of Interdisciplinary Studies as well as director of the international relations program at Stanford University.

Nicolaas Bloembergen, winner of the 1981 Nobel Prize in physics, is the Gerhard Gade University Professor in the Division of Applied Sciences, Harvard University.

Arthur A. Broyles is professor of physics and physical science at the University of Florida, Gainesville.

Gregory H. Canavan is assistant leader of the physics division at Los Alamos National Laboratory.

Timothy Coffey is director of research at the Naval Research Laboratory in Washington, D.C.

Jean Coonan was an editor at *Physics Today*.

Dale Corson is a physicist and former president of Cornell University.

Hugh DeWitt is a physicist at the Lawrence Livermore National Laboratory, V Division.

Fred A. Donath is president of a consulting firm, CGS, in Urbana, Illinois.

Sidney D. Drell is deputy director of the Stanford Linear Accelerator Center and is a past-president of the American Physical Society.

Freeman J. Dyson is professor of physics at the Institute for Advanced Study, Princeton, New Jersey.

Jack Evernden is a research geophysicist at the U. S. Geological Survey in Menlo Park, California.

Harold A. Feiveson is senior research policy analyst at the Center for Energy and Environmental Studies at Princeton University.

Daniel Gladstone was an editor at *Physics Today*.

Irwin Goodwin is a senior associate editor at *Physics Today*.

David Hafemeister (editor) is a professor of physics at California Polytechnic State University at San Luis Obispo and is a staff member of the Senate Foreign Relations Committee.

Michael Jacobs was an editor at *Physics Today*.

Paul Josephson teaches science policy and the history of science at Sarah Lawrence College in New York.

Mark Kuchment is a science historian at Harvard University's Russian Research Center and at the Fletcher School of Law and Diplomacy, Tufts University.

Harold W. Lewis is professor of physics at the University of California, Santa Barbara.

Barbara Goss Levi is the senior associate editor at *Physics Today*.

Gloria Lubkin is the editor of *Physics Today*.

Paul N. McCloskey, Jr., former Congressman from California, is currently in the private practice of law with Brobeck, Phleger & Harrison in Palo Alto.

Gerald E. Marsh is a defense analyst at Argonne National Laboratory.

Ernest Moniz is a professor of physics at the Massachusetts Institute of Technology.

Thomas Neff is a research affiliate in the Center for International Studies at the Massachusetts Institute of Technology.

Wolfgang K. H. Panofsky is the former Director of the Stanford Linear Accelerator Center.

Kumar Patel is executive director of the division of research, materials science, engineering and academic affairs at AT&T Bell Laboratories.

Robert O. Pohl is professor of physics at Cornell University.

Corey S. Powell is an assistant editor at *Scientific American*.

Tony Rothman is a consultant at the Center for Relativity at the University of Texas, Austin.

Andrei Sakharov was a member of the Academy of Sciences of the USSR and the winner of the 1975 Nobel Peace Prize.

Leo Sartori is a professor of physics at the University of Nebraska, Lincoln.

Dietrich Schroeer is a professor of physics at the University of North Carolina, Chapel Hill.

Charles Schwartz is a professor of physics at the University of California, Berkeley.

Robert W. Seidel is a research historian at the Laser History Project in Los Alamos, New Mexico.

Jeffrey D. Schmidt is an associate editor of *Physics Today*.

Bertram M. Schwarzschild is an associate editor of *Physics Today*.

Philip Sprangle is head of the plasma theory branch of the Naval Research Laboratory.

Fritz Stern is Seth Low Professor of History at Columbia University.

Roger H. Stuewer is professor of Physics at the University of Minnesota.

William Sweet is an associate editor at *Physics Today*.

Edward Teller is professor of physics emeritus at the University of California and is also a fellow at the Hoover Institution at Stanford University.

Yevgeni Velikhov is a vice president of the USSR Academy of Sciences and director of the Kurchatov Institute.

Frank von Hippel is professor in the Woodrow Wilson School of International Affairs and senior resident physicist at the Center for Energy and Environmental Studies, both at Princeton University.

Craig Waff was an assistant editor at *Physics Today*.

Spencer R. Weart is the director of the Center for the History of Physics at the American Institute of Physics.

Victor Weisskopf is Institute Professor of Physics at MIT and formerly Director General of CERN.

Eugene Wigner is professor of physics emeritus at Princeton University and winner of the 1963 Nobel Prize.

Gerold Yonas is the Director of Laboratory Development at Sandia National Laboratory and former chief scientist of the Strategic Defense Initiative.

Herbert F. York, former director of research and engineering at the Defense Department, former chief U. S. negotiator for the Comprehensive Test Ban Treaty, and former director of Lawrence Livermore National Laboratory, is professor emeritus of physics at the University of California, San Diego.

SECTION 1

THE EFFECTS OF NUCLEAR WEAPONS

The articles in this section deal with the many, diverse effects of nuclear weapons. Because a megaton explosion is almost a million times more energetic than a World War II blockbuster, the weapons effects for a single nuclear weapon have grown from part of a "block" to a large city. A barrage on the nation's cities with 400 effective megatons is far beyond any precedent in destruction. In his article, Leo Sartori describes the destruction of cities by thermal radiation, blast, and radioactivity. Sartori also discusses the effects on the populace of counterforce attacks aimed at the missile fields in the center of the US. Because these effects are so devastating, a vigorous civil defense program is recommended by Arthur Broyles and Eugene Wigner to save lives. Sidney Drell counters that such programs will not protect most of the populace from nuclear war, and might indirectly enhance a decision to attack by one side, knowing that his own population is somewhat protected.

The leukemia data for Hiroshima and Nagasaki have been re-evaluated to account for new estimates of the radiation from these weapons. Bertram Schwarzschild tells us that the neutron levels from the two weapons are now thought to have been quite similar, giving similar leukemia mortality rates of about 4 deaths per year per million person rads.

In 1982 Paul Crutzen and John Birks showed that small soot particles from burning cities and forests, lifted into the stratosphere, could cause severe global cooling. Jeffrey Schmidt describes the chemical kinetics of this first paper on "nuclear winter." This work stimulated further computer calculations by several groups. Barbara Levi and Tony Rothman reviewed these results, concluding that the nuclear winter effects would be real, but that it is a matter of degrees, that is, depending on the choice of parameters, how severe the temperature of the winter would be.

Because of concerns raised by the Chernobyl accident in 1986, the Department of Energy and the National Academy of Sciences examined the US production reactors. These reactors lack containment structures, and one of these, the Hanford N-reactor, used graphite as a moderator. Irwin Goodwin describes the similarities and differences between the N-reactor and Chernobyl Unit 4, and he describes the decision to shut down the N-reactor. Goodwin then describes DOE's plans for new production reactors to produce additional plutonium and tritium. Lastly, Goodwin describes the current environmental cleanup problems at various DOE nuclear weapons facilities, a task which could cost $100 to $150 billion.

CONTENTS

Effects of nuclear weapons

**An understanding of the technical details—
thermal radiation, shock or blast wave, nuclear radiation—
makes for more effective participation in the debate.**

PHYSICS TODAY/MARCH 1983

Leo Sartori

The intensity of public debate on issues involving nuclear weapons and strategic policy is currently at an all-time high. This is surely a welcome development. Although the issues are largely political, they cannot be addressed without some knowledge of the properties of nuclear weapons and of the

Leo Sartori is professor of physics at the University of Nebraska, Lincoln. He served with the US Arms Control and Disarmament Agency in Washington and Geneva, 1978–1981.

destruction that their use could bring about. It is our particular responsibility as technically trained citizens to be informed of the basic facts concerning nuclear weapons and to help our fellow citizens understand them so that they can contribute more effectively to the debate. It is my hope that this article—a presentation of the fundamental principles governing nuclear explosions and their effects—will be useful to those interested in carrying out this responsibility for education on nuclear war.

I begin by summarizing the nature of the three major components of a nu-

clear explosion—thermal radiation, shock wave and nuclear radiation—and by describing how the effects of these components depend on the yield of the weapon, its height of burst and the distance from the point of detonation. I follow this with a description of the destruction that we could expect if nuclear weapons were actually employed in each of three contexts frequently discussed: a single weapon detonated over a major city, a large-scale "counterforce" exchange limited to strategic targets, and, finally, the extreme case—an all-out nuclear war.

Direct information on nuclear-weapons effects comes from study of the two explosions that actually took place in a combat setting—the bombs dropped on Hiroshima and Nagasaki in 1945 (see figure 1)—and from numerous tests conducted since then. All testing by the United States and the Soviet Union since 1963 has been underground, atmospheric tests having been banned by treaty.

Prediction of nuclear-weapons effects is difficult because the weapons have changed a great deal since 1945. Present-day weapons have yields one to two orders of magnitude greater than the ones dropped on Japan. A thermonuclear weapon has never been used in combat, nor has a ballistic missile ever been tested with a live warhead. Moreover, most scenarios for nuclear war entail the use of nuclear weapons on a scale far beyond that of an individual explosion. It is obviously impossible to predict the outcome of such exchanges with high confidence. One has to make many assumptions about how technical systems would work under conditions in which they have never been tested, and, perhaps more important, about how human beings would react to a disaster of unprecedented proportions. The best one can do in making models is to try to include as many dimensions of the problem as possible, make the most informed guesses as to the values of the relevant parameters, and attach large probable errors to the results. In the case of all-out nuclear war, the scale of destruction is so great that the specific numbers almost don't matter. The only firm conclusion one can draw—that humankind must find a way to prevent this disaster from coming about—is independent of any model.

Most of my data on weapons effects come from the authoritative source, *The Effects of Nuclear Weapons*, by Samuel Glasstone and Phillip J. Dolan.[1] The results of government studies on nuclear war scenarios appear in another valuable publication, *The Effects of Nuclear War*, by the Office of Technology Assessment.[2]

Thermal radiation

Some 70 to 80 percent of the energy yield of a nuclear weapon is emitted as thermal radiation within about a microsecond of the explosion. Because the temperature during this phase is in the tens of millions of degrees, the primary radiation is principally soft x rays. Unless the burst is at a very high altitude, the x rays are largely absorbed within a few feet, heating the surrounding air to form a fireball. As the fireball expands, about half its thermal energy goes into a shock wave; the rest is ultimately reradiated at much lower temperatures, mostly between 6000 K and 7000 K. The duration of the thermal pulse increases with yield from about 1 second for a 10-kiloton burst to 10 seconds for explosions of about 1 megaton.

As the thermal pulse propagates, it undergoes geometrical spreading and is attenuated by absorption and scattering. We can write the integrated thermal flux Q as

$$Q = f\tau Y/4\pi D^2$$

where Y is the yield, D is the slant range, or distance from the point of detonation, and τ is a transmission factor whose value depends on atmospheric conditions as well as on the slant range. (The attenuation is generally not a simple exponential.) The thermal energy fraction f is the fraction of the total yield that remains as thermal energy in the fireball after the shock wave forms. For airbursts, f is between 0.35 and 0.40, and for contact surface bursts it is about 0.18. Because 1 kiloton of TNT releases 10^{12} calories, we can write the integrated thermal flux in hybrid units as

$$Q \approx 3000(f\tau Y/D^2) \text{ cal/cm}^2$$

with Y in megatons and D in miles. On a clear day the attenuation is quite gradual: Typical transmission factors for low-altitude bursts are 60%–70% at 5 miles, and 5%–10% at 40 miles. At high altitudes the absorption is even weaker. In haze or fog, thermal radiation is more strongly absorbed. Figure 2 shows the radiant exposures as a function of slant range and yield for low-altitude bursts on a clear day.

The harmful effects of thermal radiation on people are of two kinds: "flash" burns, caused by radiation striking exposed skin directly, and secondary, or "contact" burns, caused either by ignited clothing or by a general fire started by the radiation. The severity of a flash burn depends on the duration of the exposure and on the individual's skin pigmentation. Five to six calories per square centimeter received in ten seconds will cause second-degree burns, and 8–10 cal/cm² will cause third-degree burns. As figure 2 shows, such exposures would be experienced on a clear day out to 7 or 8 miles from a one-megaton burst. Even someone indoors near a window would be in danger. The pulse from a high-yield burst lasts long enough, however, to enable people indoors or with nearby shelter to take cover and thus reduce their exposure. Ignition of clothing, which requires 20–25 cal/cm², would occur to about 5 miles from ground zero. A person whose eyes were actually focused on the fireball would suffer serious retinal burning that could cause permanent blindness.

Because any opaque material is an effective shield against thermal radiation, people not in the direct path of the thermal pulse would be safe from flash burns. However, the danger of fire would be high over a large area. Wood-frame houses are not likely to be set afire directly; in tests, such structures charred heavily and emitted dense smoke but did not burn. Ten calories per cm² in ten seconds suffices to ignite light combustibles such as paper or dry leaves; ignition of household items like overstuffed chairs or beds requires 20–30 cal/cm². Interior fires, once started, are more likely to be sustained than outdoor ones. The strong wind associated with the blast wave, which arrives a few seconds after the thermal pulse, might extinguish some fires; test evidence on this effect is inconclusive. On the other hand, the blast wave is likely to start many fires by breaking gas lines, by overturning stoves and furnaces, or by causing electrical short circuits.

Under certain circumstances, individual fires could merge into a single mass fire that consumes a large area. Such "fire storms" occurred during World War II, in Hiroshima after the atomic attack (but not in Nagasaki) as well as in several German and Japanese cities after massive incendiary bomb raids. If a fire storm does occur, people in underground shelters might be killed by heat, by asphyxiation or by carbon monoxide entering through the ventilation system.

Care of burn victims would be one of the most taxing medical problems in case of nuclear war. Third-degree burns over 25% of a person's body or second-degree burns over 30% generally produce serious shock and are likely to be fatal unless promptly treated. Burn treatment is complicated, requiring specialized facilities and intensive care. In the entire United States there are facilities to treat only a few thousand severe burn cases, fewer than the number likely to result from a single explosion in a populated area.

Blast

Blast is the effect that would generally be counted on to accomplish the military objectives assigned to nuclear weapons in a conflict. (A noteworthy exception is the neutron bomb, which relies on prompt radiation.) The shock or blast wave produced by a nuclear explosion is characterized by a sharp increase in air pressure, accompanied by extremely strong winds. Sudden overpressure crushes objects and collapses structures, while the wind can turn any fairly light object into a lethal projectile.

Under ideal hydrodynamic conditions the shock velocity U in air is given[1] by

$$U = c_0(1 + 6p/7P_0)^{1/2}$$

and the wind velocity u by

$$u = {}^5\!/_7 (p/P_0)c_0(1 + 6p/7P_0)^{-1/2}$$

Here p is the peak overpressure, c_0 the ambient velocity of sound ahead of the shock front, and P_0 the ambient pressure. Table 1 gives the relation between the peak overpressure, the peak dynamic pressure of the wind, and the maximum wind velocity, obtained from the last equation. The dynamic pressure is given by $\frac{1}{2}\rho u^2$, where ρ is the density of air.

An important hydrodynamic scaling law simplifies the analysis of shock effects: The peak overpressure is a function only of the parameter z, defined as $Y^{1/3}/D$. This scaling law has been verified experimentally up to and including the megaton range, for all but very-high-altitude explosions. It follows that the area over which a given overpressure is produced is proportional to $Y^{2/3}$; this quantity is defined as the equivalent megatonnage of the weapon. Notice that the combined equivalent megatonnage of several low-yield weapons is greater than that of a single weapon with the same total yield.

Figure 3 is a log–log plot of overpressure p against $1/z$ for a free-air burst. In the high-overpressure range, p varies approximately as z^3; for low overpressures the fall-off is less steep. A fairly good approximate formula is[3]

$$p = 22.4\ z^3 + 15.8\ z^{3/2}$$

where p is measured in lb/in^2 and z in megatons$^{1/3}$/mile.

When the shock wave from an air burst strikes the ground, it is reflected. Because the reflected shock passes through an already shocked region in which density and temperature are higher than ambient, it propagates faster than the primary shock. For certain geometries, the reflected front catches up to the primary and the two fronts merge, creating an overpressure typically twice that of the primary shock alone. The merged fronts are known as the Mach stem and the region in which the phenomenon occurs is called the Mach reflection region. In

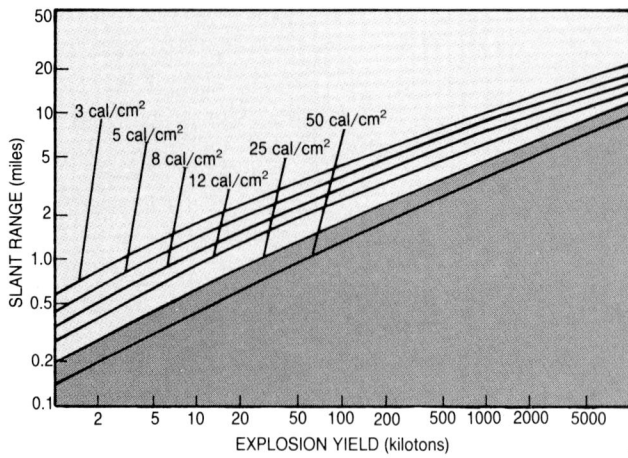

Slant ranges for specified radiant exposures on the ground, as a function of yield. These data, valid for airbursts at altitudes up to 15 000 feet, assume 12-mile visibility. (From reference 1.)   Figure 2

the "ordinary" reflection region the reflected shock front arrives at a given point after the primary front has already passed, and two distinct pressure peaks are experienced.

Because of the Mach effect, the peak overpressure at a given distance from ground zero can be greater for a weapon detonated at a substantial altitude than for a ground burst, even though the slant range is obviously less for the ground burst. This effect is most pronounced in the low- and medium-overpressure regions, that is, fairly far from ground zero. Figure 4 is a contour plot of peak overpressure against altitude and distance from ground zero. Due to the scaling law, the overpressure is a function of the "scaled height of burst" $h/Y^{1/3}$ and the scaled distance from ground zero $d/Y^{1/3}$; a single plot provides the data for any yield. By convention, all lengths are scaled to a 1 kiloton yield. Thus, a 200-foot scaled height of burst corresponds to 200 feet for 1 kiloton, 2000 feet for 1 megaton, and so on.

As the bulges in the curves of figure 4 indicate, for relatively low overpressures there is a well-defined "optimum height of burst" that maximizes the overpressure at a given distance from ground zero or, alternatively, that maximizes the distance to which a specified overpressure extends. For 15 lb/in^2, for example, the "optimal" scaled height of burst is 650 feet. For this height of burst the 15-lb/in^2 contour extends to 1200 scaled feet, as compared to only 800 feet for a contact burst. The area exposed to at least 15 lb/in^2 is more than doubled.

In the high-overpressure region, closer to ground zero, the Mach effect occurs only for very low heights of burst and the peak overpressure is nearly independent of height of burst in the Mach reflection region.

The height of burst likely to be employed depends on the nature of the target. In an attack on an extended "soft" target, such as an airfield or

industrial plant, the objective would be to create overpressures of 10–30 lb/in^2 over as large an area as possible. For such an attack, an air burst at a scaled height of 600 to 800 feet would be the indicated choice. On the other hand, a "hard" target, such as an ICBM silo, can withstand overpressures as high as 1000 lb/in^2 or more. (Hardness figures much higher than 1000 lb/in^2 have been cited in recent discussions, principally in connection with the proposed closely-spaced basing mode for the MX missile.) To attack such a target successfully, the weapon must detonate at very close range. Although an optimum height of burst exists in this case, it is quite low, and a burst at this height is only slightly more effective than a ground burst.

As an example, consider a 1-megaton weapon with a 600-foot circular error probable (a measure of the delivery vehicle's median miss distance) directed against a target able to withstand overpressures of 2000 lb/in^2. A burst at the optimum height, about 1000 feet, gives a 93% probability of destroying the target, compared to 89% for a ground burst.[4]

Because the overpressure at the target drops off quite steeply with height of burst above the optimum, a small fuzing error could lead to substantial degradation in effectiveness. This consideration favors a lower-altitude burst. On the other hand, if two or more weapons are aimed at the same target the attacker would wish to avoid raising large amounts of dust that could cause "fratricide," the destruction or impairment of a warhead by the effects of a prior explosion. To avoid fratricide, the burst must be as high as possible.

Table 2 summarizes the effects of various overpressures on common types of structures, and gives the distance to which each overpressure would be produced by a one-megaton weapon detonated at 6000 feet. The analogous distances for any other yield are easily obtained with the help of the scaling law.

The human body can withstand quite high overpressures without sustaining serious injury. The chief dangers from sudden compression are hemorrhage and possible rupture of abdominal and thoracic walls. The lungs are particularly prone to hemorrhage and edema; their threshold for injury is 12 lb/in² and severe damage occurs at 25 lb/in². Lethal effects begin at about 40 lb/in². The principal danger to people from blast, however, is indirect—it comes from the collapse of buildings, from flying glass and other fragments, and from winds strong enough to hurl people against hard surfaces.

Of all the effects of a nuclear explosion, blast is the most difficult to protect against. Underground facilities such as subway stations provide adequate protection. A home blast shelter requires special construction and is more expensive than a fallout shelter. Most blast shelters are designed to protect against overpressures of 40 lb/in² or less. There is little point in trying to do better, because an airburst near the optimum height of burst exposes little or no area to more than 40 lb/in². Against a ground burst, the incremental protection provided by a shelter able to withstand 60 lb/in², say, would be slight and the incremental cost considerable. Naturally, people would have to have advance warning to make any use of a blast shelter.

Nuclear radiation

The important nuclear radiations associated with nuclear explosions are neutrons, gamma rays, and to a much smaller extent, beta particles. Essentially all the neutrons and most of the gammas are produced in fission and fusion reactions during the explosion proper. The capture of neutrons in the weapon debris and in the surrounding air, earth or water produces additional gamma rays and a large variety of radioisotopes, which make up the radioactive fallout.

The harmful biological effects of nuclear radiation are attributable to damaging ionization produced in the cell bodies of exposed organisms. The mean free paths for both gamma rays and neutrons in animal tissue are on the order of 20 cm—just in the range that causes maximum harm to the organism. If the mean free path were a few millimeters or less, the radiation would be absorbed in the surface layers of the skin and would not reach vital organs; if it were several meters or more, most of the radiation would simply pass through the organism without interacting. As it is, both neutrons and gamma rays are strongly absorbed throughout the body, affecting all organs.

Radiation dosage is measured in *rads*; one rad is defined as the deposition of 100 ergs of ionizing radiation per gram of material. Because equal amounts of energy from different kinds of radiation do not necessarily have the same biological effects, it is necessary to define for each radiation and each effect a parameter called the "relative biological effectiveness." The RBE is the ratio of the absorbed dose of gamma radiation at a specified reference energy to the absorbed dose of the radiation in question that produces the same effect. The biological dose in *rems* is the energy dose in rads, multiplied by the relative biological effectiveness.

The relative biological effectiveness of gamma rays is unity, by definition, at the reference energy, and is only slightly energy-dependent. The RBE of neutrons varies both with energy and with type of injury. For the energy spectrum of nuclear weapons, however, the relative biological effectiveness of the neutrons that cause acute radiation injury is close to unity; hence one may use rems and rads interchangeably in most instances with little error. (The RBE of neutrons for certain injuries, such as cataracts, leukemia and genetic damage, is substantially greater than one.)

Table 3 summarizes the medical effects of radiation on human beings as a function of the total, or whole-body, dose received. As the table indicates, serious illness begins to appear at about 200 rems exposure, and above 600 rems the effects are fatal to most people.

It is useful to distinguish between initial or prompt radiation, which is released in the first minute or so after the explosion, and residual radiation, which is due entirely to induced radioactivity. The initial radiation produces huge doses at close range: The prompt gamma-ray dose one kilometer from a one-megaton burst is on the order of a million rads. The gamma rays and neutrons are, however, strongly attenuated by interactions with air molecules as they propagate.

Their mean free paths in sea-level air are a few hundred meters. As a result, even for a weapon of several megatons, the prompt dose a few kilometers from the point of detonation is already below the harmful range. Although the prompt exposure in the immediate vicinity of an explosion is many times the lethal dose, its impact is blunted by the fact that most individuals exposed to so great a dose are likely to have been already killed by the effects of blast and fire.

The characteristics of the residual radiation depend on how much of the explosive yield is due to fission and, most sensitively, on whether or not the fireball touches the ground. Fission reactions form more than 300 different radioactive isotopes, whereas fusion produces negligible activity. Hence, the higher the fission yield the more radioactivity is produced directly in the explosion.

Dangers of fallout. In an air burst, the residual radioactivity comes primarily from fission products, with some coming from weapon residues activated by neutron capture. A surface burst, on the other hand, produces radioisotopes copiously as neutrons are captured by a variety of nuclei in the earth. Much rock and soil vaporized by the intense heat become part of the fireball. As the fireball cools, the vaporized matter condenses into particles in which radioactive nuclei are imbedded. Radioactive nuclei lodge also on particles of dirt and dust sucked up by strong afterwinds as the fireball rises. The contaminated particles eventually descend to Earth, a phenomenon commonly known as fallout.

Other things being equal, there is much more fallout if the fireball touches the ground than if it doesn't. The height of burst below which significant fallout is generated is given[1] by the following approximate formula, in which Y is the yield in megatons

$$h = 2900 \, Y^{0.4} \text{ feet}$$

The probable error here is ± 30%; Glasstone warns[1] that one should not assume that a burst above this altitude will necessarily result in negligible fallout.

The radioactive nuclei in the fallout decay, each isotope according to its own halflife, emitting both beta and gamma radiation. The gamma rays can produce whole-body doses high enough to cause acute radiation sickness; this is the principal hazard from fallout. Beta rays have a range of only a few millimeters in body tissue and cannot penetrate the skin; they are absorbed by a thin layer of clothing. Fallout particles can, however, cause painful beta burns if they land directly on the skin, and there can be serious consequences if the particles are inhaled or if

Table 1. Shock-front relationships

Peak over-pressure (lb/in²)	Peak dynamic wind pressure (lb/in²)	Maximum wind velocity (mi/hr)
200	330	2078
150	222	1777
100	123	1415
50	41	934
30	17	669
20	8.1	502
10	2.2	294
5	0.6	163
2	0.1	70

For sea-level air. (Reference 1.)

they find their way into the food chain, ultimately to be ingested and reach sensitive organs. Cancers and genetic abnormalities, the most serious effects of this nature, can be caused by extremely low doses.

After the deposition of fallout particles has ceased, the activity at a given location declines quite slowly, as the sum of exponentials with many different half-lives. A $t^{-1.2}$ dependence provides a good empirical fit to the measured activity for t between about a half hour and six months; the subsequent decline is somewhat steeper. About 80% of the total dose is received in the first day, and about 90% during the first week. The remaining 10%, however, is distributed over a long time; in a high-exposure area the activity can remain at a dangerous level for many weeks or even months.

The shape, extent and location of the fallout pattern from a single surface burst are determined primarily by the height of the troposphere, the distribution of winds and the occurrence of rain or snow. The size of the fallout particles also affects their rate of deposition. The particles range in diameter from less than a micron to several millimeters. The larger ones begin falling back to Earth even before the radioactive cloud has reached its maximum height, so they are all deposited locally. The smallest particles may remain suspended in the atmosphere for long periods and may be carried great distances by winds. Although the radioactivity is very much diluted, low-level

effects of long-lived isotopes can extend worldwide.

Figure 5 is a contour map of the total fallout from a surface burst of one-megaton yield, 50% of which is due to fission. Calculated with the assumptions of a steady 15 mph wind and no precipitation, the distribution is sharply peaked in the downwind direction. Under these conditions, a region roughly a thousand square miles in area would be subjected to a total dose over 900 rems, and an area some four times greater would receive more than 100 rems. With shifting winds the distribution would be less asymmetric, but the qualitative features of the fallout pattern would be similar.

Because all matter absorbs nuclear radiation, anything between an individual and the source reduces the dose received. A rough rule of thumb is that a foot of concrete or 18 inches of earth reduces the intensity of gamma rays by a factor of ten; the effect on neutrons is similar, although neutron absorption is somewhat more complicated. Unlike a blast shelter, a fallout shelter does not require special construction; even an ordinary basement provides considerable protection. With earth piled over windows and against walls, radiation may be reduced by a factor of about 20.

Effects of a single explosion

The preceding summary of the behavior of the three major components of a nuclear explosion—thermal radiation, shock wave and nuclear radiation—gives us enough information to

estimate the extent of destruction that would occur if a single large nuclear weapon were detonated over a major city. For concreteness we consider a one-megaton weapon exploded at 6000 feet.

According to table 2, the overpressure would exceed 10 lb/in² within a circle of 2.7-mile radius centered on ground zero. Virtually every structure in this region would be destroyed and there would be almost no survivors except for people in blast shelters. In a ring between 2.7 and 4 miles from ground zero, overpressures ranging between 5 and 10 lb/in² would destroy individual residences, leaving only some foundations and basements. Stronger commercial buildings might remain standing, but with walls blown out, while some industrial buildings might remain nearly functional. Debris would pile in the streets to depths up to several feet. About half the people in this ring would be killed, mostly as the result of buildings collapsing on them. Almost all the survivors would be injured.

In the ring with 2–5 lb/in² peak overpressure—4 to 7 miles from ground zero—there would be extensive building damage and many injuries, although the number of immediate fatalities would be small. The danger of fires would be most severe in this region, because fires are more likely to ignite and spread if buildings are left standing than if they are demolished. According to the Office of Technology Assessment,[2] perhaps five percent of the buildings in this region would ignite initially, and it is highly likely that the fire would spread where the separation between buildings is less than 50 feet. Fires would continue to spread for at least 24 hours, and with little opportunity for fire-fighting, perhaps half the buildings would ultimately be consumed.

Finally, in a ring extending to ten miles from ground zero, overpressures of 1–2 lb/in² would cause only light damage to commercial structures and moderate damage to residences. Fatalities would be few, but about 25% percent of the residents would be injured.

The total number of casualties would depend on the population density, time of day, atmospheric conditions and numerous other factors that are unpredictable. An Office of Technology Assessment study of an assumed explosion at night over downtown Detroit estimated that blast effects would cause 470 000 immediate deaths and 630 000 injuries. Casualties due to flash burns and fires are even more unpredictable because they are sensitive to weather and to how many people happen to be outdoors. The study found that as few as 1000 or as many as

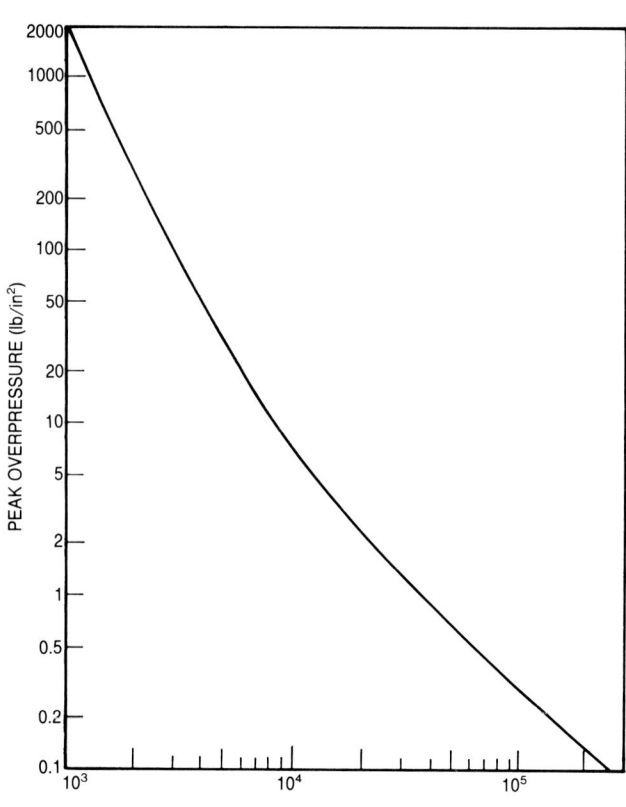

Pressure amplitude of the shock wave from a free-air burst. The horizontal axis is the distance to the blast, scaled by the yield of the blast. No reflected wave is included here. (Adapted from reference 1.) Figure 3

Table 2. Effects of blast wave from a nuclear explosion

Peak overpressure (lb/in^2)	Effects	Distance* to which effects are felt (miles)
20	Multistory reinforced-concrete buildings demolished. 500 mph wind	1.8
10	Most factories and commercial buildings collapsed; small wood and brick residences destroyed. 300 mph wind	2.7
5	Unreinforced brick and wood houses destroyed; heavier construction severely damaged. 160 mph wind	4
2	Moderate damage to houses—wall frame cracked, interior walls knocked down, severe damage to roofs. People injured by flying glass and debris. Wind about 70 mph	7
1	Light damage to commercial structures; moderate damage to residences.	10

*Based on a one-megaton burst at 6000 ft altitude.

190 000 burn fatalities could be expected, depending on the conditions assumed.

Counterforce exchange

Analysts generally consider two types of scenarios for large-scale nuclear exchanges. One is a so-called counterforce exchange, in which each side strikes only the strategic forces of the other—ICBM silos and launch-control facilities, nuclear-submarine bases and strategic-bomber bases. The second scenario is an all-out exchange in which the targets include additional military facilities as well as major urban-industrial targets—factories, oil refineries, communication centers and the like. It is generally assumed that cities and civilian populations would not be targeted *per se*. However, because of the proximity of military and industrial targets to major population centers, most large cities would inevitably be hit. Such an all-out exchange would involve thousands of high-yield warheads.

A pure counterforce exchange would have relatively little direct effect on civilian populations. In fact, because ICBM silos are situated in sparsely populated regions, an attack confined strictly to ICBMs would cause virtually no direct damage to civilians, provided none of the attacking missiles were grossly mistargeted. If submarine bases and strategic airfields were included the toll would be greater, because some of those facilities are close to large cities. But most of the civilian populations would still escape direct attack.

For most of the country, the major impact of a counterforce attack would be due to fallout. As we saw earlier, attacks on hardened silos require low-altitude bursts. According to the equation on page 35, the minimum height of burst to avoid significant fallout in a one-megaton explosion is 2900 feet. But figure 4 shows that to produce 1000 lb/in^2 overpressure on the ground requires a height of burst less than 2400

feet, even if the detonation is directly overhead. Because the critical height of burst for avoiding high fallout scales as $Y^{0.4}$ and the height of burst for producing a given overpressure scales as $Y^{0.33}$, the comparison is essentially yield-independent.

Hence it is impossible to create an overpressure of 1000 lb/in^2 or more at ground level with an explosion high enough to keep the fireball from touching the ground, and extensive fallout is unavoidable in any attack on hardened silos. The harder the silo, the lower the height of burst required, the more the fireball touches the ground, and the greater the amount of fallout produced. It is ironic that any attempt to improve the survivability of fixed silos by superhardening would ensure that if the silos are attacked, the detonations would create the most fallout. The proposed dense-pack deployment is designed to *force* the attacker to detonate at low altitude, so as to maximize

"fratricide" effects.

A Soviet attack against US strategic forces, including two warheads directed at each of the Minuteman and Titan ICBMs, would entail some 2000 explosions, either ground bursts or low-altitude air bursts, in the 1-megaton range. One can estimate the distribution of fallout resulting from such an attack by superposing the fallout patterns for individual bursts. Figure 5 shows the fallout pattern of an individual burst. Inevitably, a substantial fraction of the country would be exposed to high levels of radiation. With the silos concentrated in the West, prevailing westerly winds would drive much of the fallout into densely populated areas in the Midwest and East.

The Department of Defense, the Arms Control and Disarmament Agency and the intelligence community have made detailed analyses of counterforce exchanges, and the Office of Technology Assessment has summarized[2] the results. Even though the locations of all the targets are known precisely, the predicted numbers of civilian casualties vary over a broad range because of uncertainties in many input variables and assumptions. The seasonal variation in wind patterns, for example, can by itself cause the outcome to change by as much as a factor of three. The most critical variable, however, is the degree of fallout protection assumed for the population.

The government studies indicate that between 2 million and 20 million Americans would die from radiation effects within 30 days after an attack confined to ICBM silos. (If other strategic targets were attacked as well, the results would be altered only slightly,

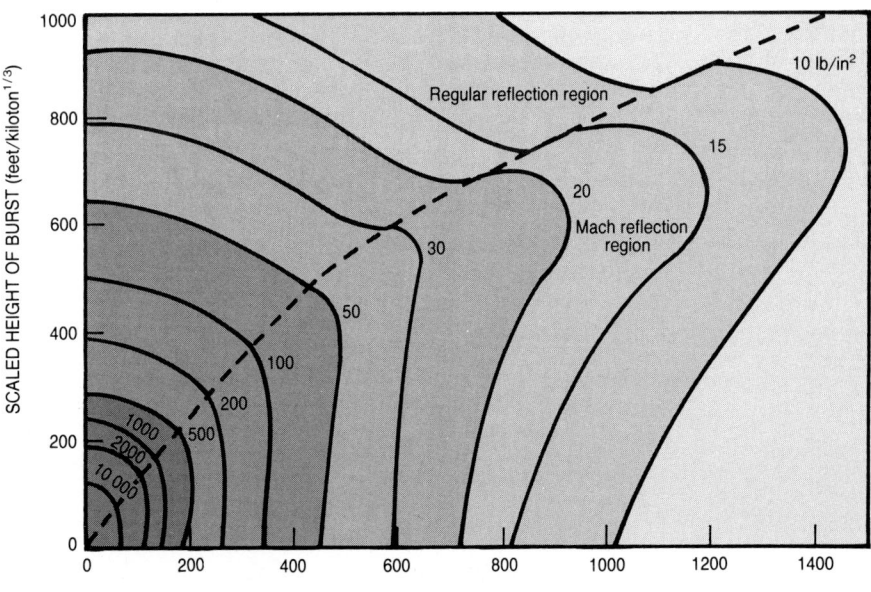

Peak overpressures on the ground for designated heights of burst and distances from ground zero. Lengths are scaled to a 1-kiloton yield. (Adapted from reference 1.)   Figure 4

because those targets would probably receive only airbursts.) Moreover, the low end of the range of estimates—fewer than 8–10 million deaths—requires quite optimistic assumptions not likely to be satisfied. The estimated number of injuries is about equal to the number of fatalities. One could also expect extensive radiation damage to crops and livestock. For a US counterforce attack on the Soviet Union the estimated number of casualties is comparable to the above but somewhat smaller, principally because US missile warheads have lower yields than those of the Soviet Union.

These results cast considerable doubt on the validity of an argument that plays a prominent role in the current strategic debate. According to this argument, a Soviet "surgical strike," that is, a strike strictly limited to US ICBMs, could destroy almost all of these missiles while inflicting only light casualties on the civilian population. Although the US would still have a formidable force of surviving bombers and submarine-launched ballistic missiles, a retaliatory strike against Soviet urban-industrial targets would cause heavy civilian casualties and would almost surely elicit a massive Soviet counterstrike in which US cities would be devastated. Fear of such an outcome, it is claimed, would inhibit the US from retaliating. Lacking a strong counterforce capability of its own, the US could make no response to the destruction of its ICBMs. In effect, then, the US deterrent is no longer credible. This is the essence of the "window of vulnerability" argument invoked in support of proposals to increase US counterforce capability and to develop more-survivable basing for US ICBMs.

By conservative estimates, a Soviet "surgical strike" would probably result in at least 7–10 million Americans dead, a comparable number injured, and large-scale disruption of US society, with many millions confined to

fallout shelters for weeks. One has to question whether a Soviet leader could confidently assume the US would absorb a blow of such magnitude without retaliating, regardless of the consequences.

All-out exchange

The executive-branch studies referred to above considered also the consequences of a massive all-out exchange. In such an exchange the effects rise to truly catastrophic proportions. Most large cities in both the US and the USSR would be subjected to the kind of devastation that I described earlier (see table 2). All the studies agree that very large numbers of civilians would be killed by blast and thermal effects and that a high percentage of the economic and industrial capacity of both countries would be destroyed, no matter which one struck first.

As in the case of a counterforce exchange, the quantitative results depend on the assumptions concerning the nature of the attacks and the protective postures of the populations, and are subject to large uncertainties. Assuming the US population remains in place and utilizes only locally available shelter, the estimated numbers of prompt fatalities after a Soviet first strike range between about 80 and 170 million. A US retaliatory strike would cause between 50 and 100 million deaths in the USSR.

If we assume that urban populations are evacuated, the estimated casualty levels drop significantly. One Defense Department study predicted 40–55 million prompt fatalities in the United States, while a study by the Defense Civil Preparedness Agency, which assumed substantially greater fallout protection for the evacuees, estimated "only" 20 million deaths. Soviet fatalities after a US retaliation were estimated to be between 23 and 37 million.

The potential saving of significant numbers of lives through civil defense,

and particularly through evacuation, has occasioned a heated debate. It is reported that the Soviets devote substantial resources to civil defense preparations, which are said to comprise extensive shelters, hardening and dispersal of industrial facilities, and an evacuation plan. Some have argued that the "civil defense gap" would give the Soviets a decisive advantage in case of nuclear war.

The US public has never taken civil defense very seriously. At present, the Federal Emergency Management Agency is developing a large-scale "crisis relocation" plan, in which host communities in lower-risk areas would be designated to receive evacuees from large cities. There are many obvious problems with such a plan, not least being the feasibility of carrying out such massive evacuations even if the required warning time were available. Many have also questioned the efficacy of Soviet civil defense.

I must emphasize that the casualty figures cited here refer to the first 30 days only, and take no account of the aftermath or the long-term effects. Formidable problems would face the survivors of a large-scale nuclear exchange. Devastated areas could count on little help from the outside, because neighboring communities are likely to have been hit themselves.

For a period of perhaps a month, radiation levels due to fallout would be higher in much of the country, particularly in the eastern half. Figure 6, from a study[5] by the Arms Control and Disarmament Agency, shows the areas that would receive a total dose in excess of 1000 rems. Survivors in those areas would have to spend most of their time in shelters, many of which would be overcrowded. Supplies of food, water and medicine might not be adequate. Maintaining even minimal hygiene and sanitation would be a taxing problem. The level of stress in the shelters is likely to be high. Not knowing how much radiation they had received, peo-

Table 3. Effects of acute radiation on human beings

Dose (rems)	Symptoms	Treatment	Prognosis
0 to 100	Little or no visible sign. Some blood changes are detectable above 25 rems.	No treatment required.	Excellent.
100 to 200	Vomiting, headache, dizziness. Moderate leukopenia (loss of white blood cells).	Hematologic surveillance; reassurance. Hospitalization not required	Full recovery in a few weeks.
200 to 600	Severe leukopenia; internal bleeding; ulceration; hair loss above 300 rems; infection likely.	Blood transfusions; antibiotics. Hospitalization required.	Guarded. Probability of death near 0 at low end, 90% at high end.
600 to 1000	Same as above but more severe.	Consider bone marrow transplant.	Poor. Probability of death 90–100%. Long convalescent period for survivors.
1000 to 5000	Diarrhea, fever, disturbance of electrolyte balance.	Maintain electrolyte balance.	No chance of recovery, death occurs in 2–14 days.
over 5000	Convulsions, tremor, ataxia.	Sedatives.	Death in 1–2 days or sooner.

Adapted from reference 1.

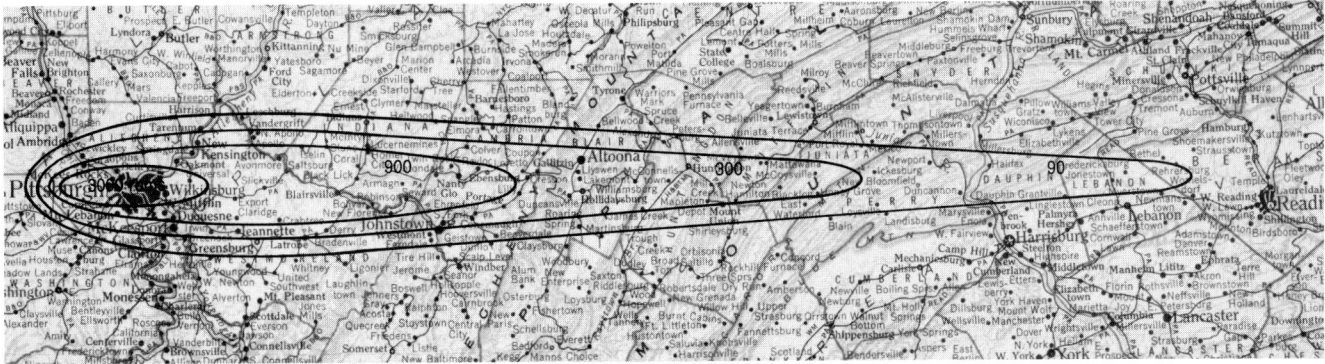

Radiation-dose contour map showing accumulated 7-day exposure following a 1-megaton ground burst. The map assumes a steady wind of 15 mph. The contours indicate the areas receiving 3000, 900, 300 and 90 rems. (Adapted from reference 2.)   Figure 5

ple might panic at any symptom of radiation sickness.

After radiation levels had subsided sufficiently, people could leave their shelters and begin efforts at recovery. However, they would be working under extreme duress. Millions of people would be homeless. Many essential commodities, including food, could be in short supply. Fallout would have killed much of the nation's livestock. If the attack had come early in the growing season, much of the crop would have been killed by radiation; if it had come at harvest time, the crop could have been lost because of farmers' inability to harvest it.

Doctors have drawn attention[6] to the high danger of infection and communicable diseases during the recovery period. Resistance to infection would be sharply lowered because of radiation, malnutrition and dehydration during the shelter period, and general exposure and hardship. Poor sanitation, lack of refrigeration, and inadequate waste disposal would encourage the spread of disease. Under such conditions epidemic diseases that have long been under control—cholera, plague, typhoid fever—could reemerge as dangerous threats, and hepatitis, salmonellosis, and other diseases of the intestinal tract are likely to become more prevalent and more deadly. Tuberculosis could also increase dramatically.

Under such hardships the survivors would have to set about rebuilding and restoring the industrial plant, the communications and transportation networks, and the commercial, medical-scientific, and cultural systems—in short, practically every aspect of the complex US society. Both material and psychological obstacles would confront them. Deficiencies in each component of society would hamper the recovery of the others. The government might be hard put to maintain law and order and to establish and enforce priorities in the recovery effort. Significant changes in the political and social systems might come about. There could be a serious effort to restrict democratic procedures and perhaps even to establish some kind of autocratic regime.

One can only speculate how the populace would react to such an unprecedented catastrophe. People might roll up their sleeves and set about vigorously to rebuild, or a general depression might set in, with people concerned only with survival. Some predict a return to a society like that of the Middle Ages. Any estimate of how long it might take to restore conditions to something resembling their prewar state cannot be much better than a guess. Some highly optimistic estimates predict recovery in as little as four years. Others think society would never recover, at least not for many decades or even centuries.

Long-term effects

Some consequences of nuclear radiation become apparent only months or even years after exposure. These include cataracts, leukemia and other forms of cancer, and genetic effects, such as deformed births, various mutations, and abortion due to chromosomal damage. Because the low-level radiation from late fallout can induce all these effects, the medical consequences of nuclear war would be manifested worldwide. The numbers of cancers and genetic abnormalities that would be caused by a large-scale nuclear war are highly uncertain, but are generally estimated to be in the millions—far fewer than the direct effects, but far from negligible.

A 1975 study by the National Academy of Sciences[7] called attention to the danger that nitrogen oxides from high-yield nuclear explosions might deplete the ozone layer in the stratosphere and increase the ultraviolet radiation reaching the Earth's surface by a factor estimated at between 2 and

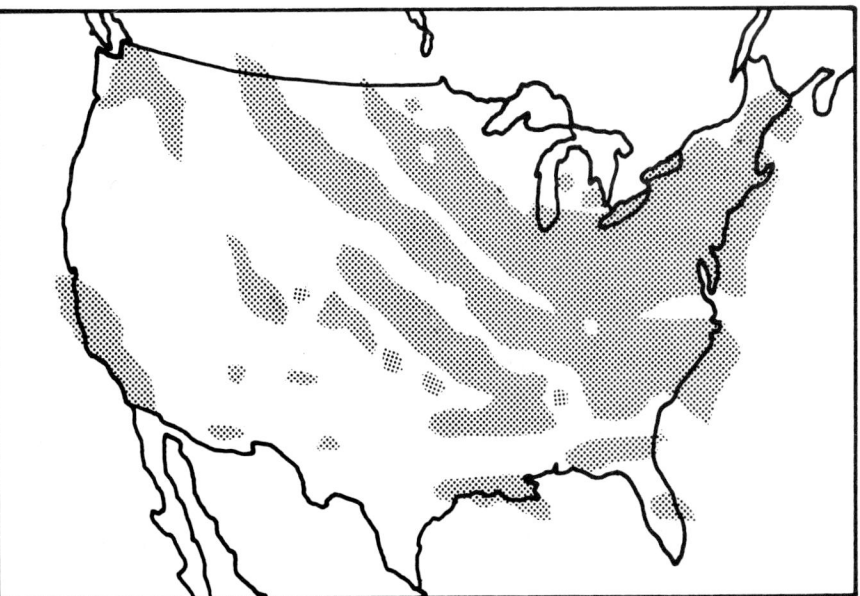

Fallout distribution. Map shows areas in the United States that would receive fallout doses of 1000 rems or more in an all-out exchange of strategic nuclear weapons, according to a study by the US Arms Control and Disarmament Agency. Table 3 outlines the effects on human beings. (From reference 5.)   Figure 6

100, causing increased skin cancer, severe sunburn, and a variety of potentially harmful ecological effects. The chemistry of the upper atmosphere is complex, and the likely extent of the ozone depletion remains controversial.

Another potential disaster is the deposition of large amounts of smoke in the atmosphere from fires in cities, forests, agricultural lands, and oil and gas fields. A recent study estimates[8] that smoke would reduce the average amount of sunlight reaching the ground in the Northern Hemisphere by a factor of between 2 and 150 for many weeks, perhaps months. This would strongly reduce and perhaps totally eliminate the possibility of growing agricultural crops for an entire season, leading to widespread famine as well as causing a variety of other harmful ecological changes. (See PHYSICS TODAY, October 1982, page 17.)

It is quite possible that there exist other effects of nuclear war, as yet unidentified, that could bring about significant changes, either temporary or irreversible, in the planet's ecology. The ecological system is fragile; it would not take very large changes, for example, to diminish substantially the world food supply. Obviously, one cannot estimate the magnitude of such unknown effects. The likelihood of large but inestimable effects is but another manifestation of the inherent uncertainty in trying to assess the consequences of nuclear war. About the only thing one can predict with certainty is that it would be a disaster of unparalleled dimensions.

References

1. S. Glasstone, P. J. Dolan, *The Effects of Nuclear Weapons*, third edition, US Departments of Defense and Energy, Washington, DC (1977). Available from US Government Printing Office.

2. *The Effects of Nuclear War*, US Office of Technology Assessment, Washington, DC (1979).

3. H. L. Brode, Ann. Rev. Nuc. Sci. **18**, 153 (1968).

4. B. W. Bennett, *How to Assess the Survivability of US ICBMs*, Rand Corp. Reports R-2577-FF and R-2578-FF (Appendices), Santa Monica, Calif. (1980).

5. *The Effects of Nuclear War*, US Arms Control and Disarmament Agency, Office of Operations Analysis, Washington, DC (1979).

6. H. Abrams, W. von Kaenel, New England Journal of Medicine **305**, 1226 (1981).

7. National Academy of Sciences, *Long-Term Worldwide Effects of Multiple Nuclear-Weapons Detonations*, National Academy of Sciences, Washington, DC (1975). Available from the Committee to Study Long-term Worldwide Effects of Multiple Nuclear-Weapons Detonations, National Research Council, 2101 Constitution Avenue, Washington DC 20418.

8. P. J. Crutzen, J. W. Birks, Ambio **11**, No. 2–3 (1982).   □

Studies revise dose estimates of A-bomb survivors

PHYSICS TODAY/SEPTEMBER 1981

We know very little about the long-term health hazards of human exposure to neutron irradiation—certainly less than we thought we knew a year ago. The painstaking follow-up of the Hiroshima and Nagaski survivors has provided our only extensive data on human neutron exposure. A joint Japanese–American effort has for more than three decades attempted to keep track of the medical history of every person within a few kilometers of Ground Zero who survived the cataclysm and its immediate aftermath in these two ill-fated cities.

The interpretation of the epidemiological data—mostly the occurrence and death rates for various kinds of cancer—in terms of absorbed neutron and gamma doses, has until recently been based on a 1965 estimate of the radiation doses as a function of distance from Ground Zero in the two cities. The dose–response curves derived from this generally accepted "T-65 dosimetry" (tentative 1965 dose estimates, producd by John Auxier at Oak Ridge) showed a striking difference between the two cities. Leukemia mortality rates in response to a given dose, for example, appeared to be much higher in Hiroshima than in Nagasaki. The difference was generally attributed to the much higher neutron component in the T-65 estimate of the Hiroshima radiation field, resulting from the different detonation mechanisms and casings of the two atomic bombs.

Last fall, however, William Loewe and Edgar Mendelson at Livermore made public a sharply revised estimate of the A-bomb dosimetries[1] that has thrown all conclusions based on T-65 into turmoil. Their transport calculations, starting from a 1976 Los Alamos calculation of the gamma and neutron spectra emerging from the bombs, conclude that the neutron doses at Hiroshima were an order of magnitude lower than previously believed. Gamma dose estimates for Hiroshima, on the other hand, are increased markedly in the Livermore paper.

Replotting the leukemia mortality dose–response curves, Loewe and Mendelsohn now find no difference between Hiroshima and Nagasaki. The new

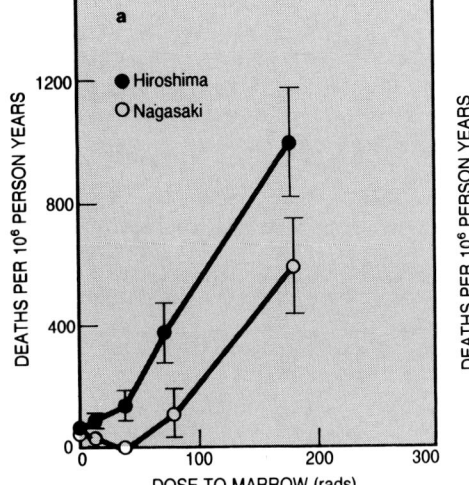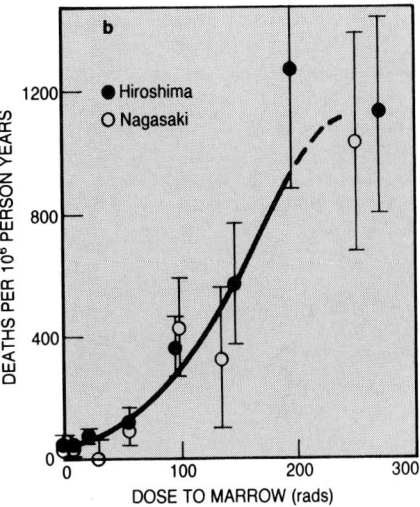

Leukemia mortality dose-response curves for Hiroshima and Nagasaki, **(a)** using the old T-65 dose estimates (adapted from Rossi and Mays, 1978), and **(b)** using the new Livermore dosimetry (Straume and Dobson, 1981). Striking difference between cities in **(a)** had been attributed to the higher neutron component at Hiroshima and very high relative biological effectiveness of neutrons. With the lower neutron dose estimates in **(b)**, the difference between cities appears to go away. Both fit well to a single linear-quadratic function out to 200 rads.

neutron dose estimates are now so low in both cities, they conclude, that one can derive only very limited information (from the leukemia data) about the relative biological effectiveness of neutron and gamma irradiation.

Tore Straume and Lowrie Dobson, also at Livermore, have however drawn some conclusions[2] about the carcinogenic effectiveness of neutrons by applying the new Loewe–Mendelsohn dose estimates to a more extensive body of cancer data for the Japanese survivors. Their conclusion, that neutrons may be orders of magnitude more carcinogenic than gammas at low doses for some forms of cancer, is disputed by those who contend that one can no longer draw any plausible inferences from so small a neutron component.

At a Munich symposium on neutron dosimetry, held in June, George Kerr reported[3] the results of his work with Joseph Pace at Oak Ridge, also reestimating the Hiroshima–Nagasaki dosimetry. Their technique and conclusions are quite similar to those of Loewe and Mendelsohn, but they argue that the Livermore group has overestimated somewhat the gamma radiation

field at Hiroshima. This conclusion, based on a recent recalculation for gammas from the post-explosion fission clouds and on a disagreement about the total yield of the Hiroshima bomb, would spoil to some extent the identity of the dose–response curves for the two cities that Loewe and Mendelsohn had found for leukemia.

A week earlier Kerr had given a preliminary version of his results at a Minneapolis meeting sponsored by the National Council for Radiation Protection and the Radiation Research Society, at which Loewe and Auxier also spoke. Warren Sinclair, president of NCRP and chairman of the meeting, told us that most participants were agreed that it was unwise to set about revising radiation risk estimates before the new dosimetry was well tied down. "We've lost information about neutrons," he told us, "but we've gained information about the gamma risks, because the reduction of the neutron doses has brought the two cities into better agreement."

Last year the Biological Effects of Ionizing Radiation (BEIR) committee of the National Academy of Sciences finally released a much disputed report (BEIR III) on the health risks of gamma radiation (PHYSICS TODAY, July 1979, page 78). The BEIR III committee did not have at its disposal the new A-bomb dosimetry estimates from Livermore or Oak Ridge. Although these new calculations have not yet laid to rest the hot contention over low-gamma-dose risks that split the BEIR III committee into three factions, it is widely agreed that the new gamma dose estimates have now brought the A-bomb data into better agreement with gamma exposure data from epidemiological studies of medically irradiated patients.

The Hiroshima bomb, whose unlikely code name "Little Boy" is still in use, was a uranium device whose detonation system was a massive steel gun assembly; a subcritical mass of U^{235} was fired down a gun barrel into another subcritical mass to initiate the fission blast. The Nagasaki plutonium bomb, code-named "Fat Man," was implosively detonated by a spherically symmetrical surrounding shell of high explosive. The high hydrogen content of the Fat Man implosion shell served to attenuate the fission neutron flux to such an extent that very few neutrons managed to emerge from the bomb. In Little Boy, by contrast, the fission neutron spectrum was softened (shifted to lower energy) by the steel assembly, but many more neutrons were able to escape with sufficient energy to survive more than a kilometer in air.

For purposes of studying long-term effects on survivors, one is concerned primarily with gamma and neutron fluxes at distances between one and two kilometers from the blast. Those unfortunate enough to be much closer in did not in general survive the *prompt* effects of the bomb; for distances beyond about two kilometers, long-term effects, if any, are difficult to ascertain. When Auxier was preparing his dose estimates for Hiroshima and Nagasaki in the early 1960's, computer techniques were not yet available to give reasonably reliable estimates of the neutron and gamma spectra emerging from the bombs. A number of devices of the Fat Man variety had been test-fired in the Nevada desert, but Little Boy was the only bomb of its type ever detonated.

Unlike the recent dosimetry estimates from Livermore and Oak Ridge, Auxier's empirical 1965 estimates were based neither on calculations of the bomb outputs nor transport calculations of the resulting radiation fields at distant points. In addition to limited weapons-test data, Auxier used a reactor mockup of Little Boy at Los Alamos to simulate the Hiroshima bomb's output, and he determined attenuation in air by measuring the distant radiation fields emanating from an unshielded

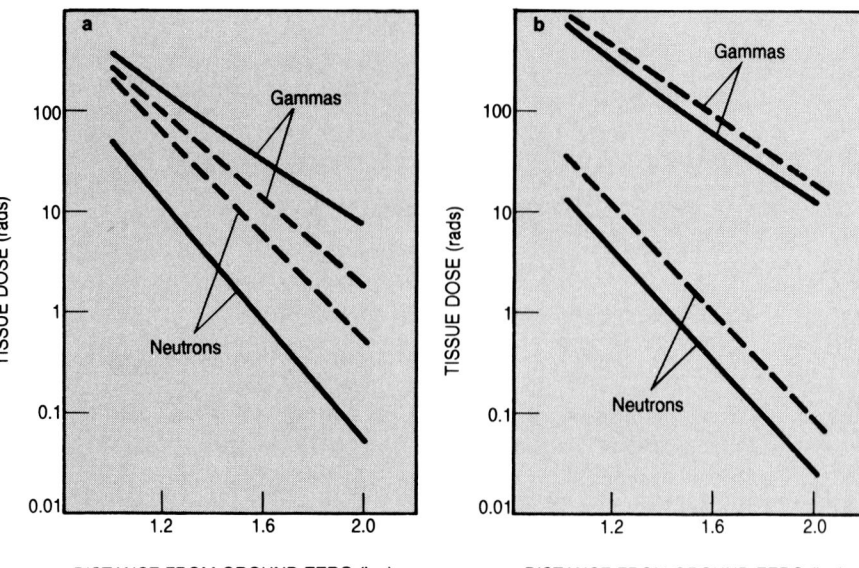

Revised estimates of neutron and gamma doses absorbed by unshielded persons in Hiroshima **(a)** and Nagasaki **(b)**, as a function of distance from Ground Zero at the moment of the A-bomb explosion. Solid lines are 1981 estimates calculated by Livermore group. Dashed lines are previous standard (T-65) estimates produced by John Auxier in 1965. The most striking changes are reduced neutron estimate and increased gamma estimate for Hiroshima.

uranium metal reactor sitting on a tower at the Nevada Test Site. Because the bare reactor output spectrum was not softened by a massive steel enclosure, Loewe told us, the Nevada measurements overestimated the fraction of neutrons surviving to large distances. Furthermore, neutrons traversing a kilometer or two of dry Nevada air suffer considerably less degradation than those passing through the humid atmosphere of the Japanese cities. These, Loewe contends, are the principal reasons for the overestimate of neutron doses for Hiroshima in the T-65 dosimetry.

The immediate impetus for the Livermore reevaluation of the Hiroshima and Nagaski dosimetries was a disquieting 1978 paper by Harald Rossi (Columbia) and Charles Mays (University of Utah). Using the T-65 dose estimates, corrected for structural and body shielding to give radiation doses absorbed by bone marrow, Rossi and Mays produced dose–response curves for leukemia mortality among the survivors in both cities. (A dose–response curve is a plot of the rate of occurrence of some biological effect as a function of absorbed radiation dose). From the striking difference between the Hiroshima and Nagasaki curves thus produced, they sought to deduce the relative biological effectiveness (RBE) of neutrons as a function of dose. Particles such as neutrons and alphas, which produce densely ionizing tracks in tissue, can cause more cell damage than gammas, and the neutron RBE coefficient is the neutron dose divided *into* the gamma dose that would be required to elicit the same biological response.

Assuming that the small Nagasaki neutron component had a negligible effect on leukemia mortality, Rossi and Mays calculated from the difference between the two dose–response curves that the neutron RBE was as high as *sixty* at low doses. That is, a 1-rad dose of neutrons would cause as much leukemia as 60 rads of gamma irradiation. Although this result was consistent with the "dual action theory" of radiation effects developed by Rossi and Albrecht Kellerer in the early 1970's, it was the first human epidemiological study to present evidence for such enormous neutron RBEs.

As far as environmental risks to the general public are concerned, gamma radiation is much more important than neutrons. But the high RBE estimates of Rossi and Mays were troubling for people working around reactors or nuclear-weapons production facilities. Because of Livermore's involvement with the weapons program, Loewe and Mendelsohn were asked to review the T-65 dosimetry on which the Rossi–Mays work had been based. Although Kerr and Pace at Oak Ridge had begun a similar review in 1977, shortly after the new Los Alamos bomb-output calculations became available, neither group was aware of the other's effort until shortly before the results of Loewe and Mendelsohn were made public a year ago.

Transport calculations. Both groups took the Los Alamos output spectra for Fat Man and Little Boy as the starting points for their transport calculations. Only about a tenth of the gammas encountered a kilometer or so from Ground Zero came directly from the bombs. The bulk of the gammas come in roughly equal measure from neutron capture reactions in air and from the decay of fission products in the fireball rising from the blast. Loewe and Mendelsohn used a Monte Carlo computer code to follow individual hypothetical neutrons and prompt and neutron-capture gammas as they were scattered and attenuated in the air and ground. Because of the high attenuation rate, the Monte Carlo calculation yields good statistics out to only about a kilometer if one starts with a manageable number of hypothetical particles at Ground Zero. To solve the Boltzmann transport equation for distances beyond 1 km, the Livermore group used a "discrete-ordinate transport" (DOT) program developed at Oak Ridge. The DOT calculations treat the transported flux in discrete bins of energy and angle, instead of following each individual particle through thousands of scatterings. The good agreement between their Monte Carlo and DOT results, Loewe told us, contributed much to his confidence in the correctness of the Livermore results when it became clear they were in strong disagreement with the standard T-65 dosimetry.

Having concluded that the results of their transport calculations were also in good agreement with neutron-activation and thermoluminescence measurements of building materials from the Japanese cities, and with recent dose measurements in liquid air using a neutron source at Livermore, Loewe and Mendelsohn had sufficient confidence in their surprising new dosimetry to make it public last September. The results of the Oak Ridge transport calculations, which used only the DOT program, were confined to private reports (to DOE and the National Council on Radiation Protection) until Kerr's presentations in Minneapolis and Munich a few months ago.

The new dosimetry arrived at by both groups gives only about one-tenth the neutron dose estimated by T-65 at 2 km from the Hiroshima Ground Zero. At 1 km, it deflates the T-65 neutron dose by a factor of only five. For Nagasaki, the new estimates reduce the already much smaller neutron doses by about a factor of three. Thus the only two nuclear weapons ever fired in earnest no longer appear to be so generous a source of information about the long-term health hazards of neutron irradiation.

Livermore and Oak Ridge are in agreement that Auxier's estimate of the Nagasaki gamma doses must be reduced by about 30%—"not a very significant change," Loewe told us. A bigger revision is the Livermore quadrupling of the T-65 gamma dose estimates for Hiroshima at 2 km; at 1 km it is only doubled. Although Kerr agrees that Auxier's Hiroshima gamma doses were much too low, he contends that Livermore has now gone too far—overestimating the gamma doses by about 25%.

Both groups have relied on outside sources for estimates of the very significant decay gamma component from the fission fireball. The gamma flux from this maelstrom of radioactive fission products rising out of the blast is very difficult to calculate. Livermore used a 1966 phenomenological model (based on earlier work of Loewe) confined to classified Defense Department manuals. Kerr's lower gamma dose estimate results from a recent calculation of the fireball component by William Scott of Science Applications Inc (San Diego), and from a lower estimate of Little Boy's total yield. Scott is continuing to refine this fireball calculation, and Loewe agrees that a thorough new look at the fireball output has been long overdue.

Radiation risk implications. The new Livermore dosimetry wipes out the difference between the Hiroshima and Nagasaki dose–response curves for leukemia mortality—previously attributed to a significant neutron component at Hiroshima, with very high neutron RBE at low doses. The best estimate of Loewe and Mendelsohn for the neutron RBE (for leukemia) is close to unity, but with a large uncertainty because the new neutron dose estimates are so small. With their somewhat lower estimates for gammas at Hiroshima, the Oak Ridge group does not find that the leukemia data for the two cities fall so nicely on a single dose–response curve. Kerr argues that it is too early to speculate on neutron RBEs from the still tentative data.

A serious three-way dispute arose in the BEIR III committee over the form of the dose–response curves for cancer risk at low gamma doses. A linear response curve without threshold, as advocated by Chairman Edward Radford, a University of Pittsburgh epidemiologist, enhances the risk estimate for low gamma exposure. Rossi argued for a quadratic fit to the existing dose–response data (prior to the new Livermore and Oak Ridge dosimetries), implying smaller low-dose risk. The final BEIR III report last year, from which both Radford and Rossi dissented, used a compromise linear–quadratic fit (of the form $ax^2 + bx + c$).

From the new Livermore dosimetry, Dobson and Straume have now produced new dose–response curves for leukemia, breast cancer, total cancer mortality and chromosomal aberrations. The result is a somewhat perplexing mixed bag. For leukemia and breast cancer mortality, they find that the data for the two cities coincide, fitting well to linear–quadratic dose–response curves that assume no threshold "safe" dosages. Because the Hiroshima neutron doses, albeit small, are still much larger than those for Nagasaki, the coincidence of the dose–response curves for the two cities lead Dobson and Straume to conclude that neutron RBEs for leukemia and breast cancer are close to unity.

For total cancer mortality and chromosome aberrations, on the other hand, Dobson and Straume find that the Hiroshima responses lie well above those for Nagasaki at a given dose level. If one ascribes the remaining differences between the two cities to neutrons, they calculate, much as Rossi and Mays had done earlier, that neutron RBEs would be of the order of 100 for chromosome aberrations and malignancies other than leukemia and breast cancer at low neutron dose levels. Radford argues that one is not justified in calculating RBEs from the low neutron doses and death rates involved here; there were only 20 Nagasaki leukemia deaths in the low-dose bins, he stresses. Nonetheless, the human RBE estimates produced by Dobson and Straume do in fact agree well with a number of studies of animals and human tissue (*in vitro*) irradiated with neutrons.

Mortimer Mendelsohn (no relation to Edgar), associate director of Livermore for environmental and biomedical research, told us that Dobson and Straume have suggested a possible explanation for the striking differences observed for different malignancies. There is evidence, he points out, that leukemia and breast cancer are of viral origin. For virally induced cancers, it is plausible that malignancies are triggered by a "one-hit mechanism," the interaction of a single ionizing particle with a virus. For such a mechanism, he explained, a given dose of gammas and neutrons would be about equally effective—giving an RBE near unity. On the other hand, forms of cancer induced by chromosomal aberrations would require a number of adjacent hits. It is known that single ionizing lesions do not break chromosomes; both DNA strands must be severed. Whereas this would require several gamma hits, a single neutron, generating multiple ionizing tracks, could do the trick by itself. Thus cancers produced by radiation-induced chromosome damage could be expected to exhibit very high neutron RBEs at low dose levels, he argues.

There is a considerable body of data on human cancer induction by gamma and x-ray sources other than the atomic bombs—mostly from studies of patients exposed in various medical procedures. With the old Auxier dosimetry, Mendelsohn told us, the response rates of the Japanese survivors had appeared anomalously low. The new dosimetries appear to have closed the gap between the two data sets. Risk coefficients (death rates per rad) calculated by Dobson and Straume for gamma-induced leukemia and breast cancer are consistent at low doses with those promulgated in 1977 by the International Commission on Radiological Protection. At high gamma doses, however, the leukemia risk coefficients calculated at Livermore are four times as large as the dose-independent ICRP risk factor. For total cancer mortality, on the other hand, the new gamma radiation risk coefficients are lower than the ICRP value.

It is generally agreed that more work needs to be done before the new atomic-bomb dosimetries can provide a firm basis for a revised set of radiation-risk standards. Uncertainties remain with respect to the total yield of the Hiroshima bomb and the gamma doses coming from fission products in the fireballs. New estimates of structural and body shielding are yet to be calculated for the revised radiation spectra.

Dissenting from this cautious consensus, Radford argues that too much emphasis has been put on the leukemia and other mortality data, leading to underestimated gamma risks. The "more reliable" data on total cancer incidence (as distinguished from mortality), he contends, remove all differences between the two cities (without reference to neutron RBEs) and lend strong support to his belief that the dose–response functions are indeed linear. —BMS

References

1. W. E. Loewe, E. Mendelsohn, Lawrence Livermore Lab preprint UCRL-85446 (1980).
2. T. Straume, R. L. Dobson, Lawrence Livermore Lab preprint UCRL-85697 (1981).
3. G. D. Kerr in *Proc. Fourth Symposium on Neutron Dosimetry*, Gesellschaft für Strahlen und Umweltforschung, Munich–Neuherberg (1981), to be published.

Global atmospheric effects of nuclear-war fires

PHYSICS TODAY/OCTOBER 1982

When Yves Laulan says, "We would be returning to the Dark Ages," the chief economist of the largest bank in France is talking about the economy after a nuclear war. With half the Northern Hemisphere's urban population wiped out by the war's direct blasts alone, those remaining would find themselves in a barter economy trading in nothing more than the basic elements of survival: food, shelter and medical care.[1]

Now, some physical scientists, too, are saying that the months following a nuclear exchange may be a dark age—although in a more literal and more lethal sense. They estimate that massive secondary fires could inject enough smoke into the atmosphere to block 50 to 99 percent of the sunlight that would otherwise reach the surface of the Earth in the agricultural latitudes of the Northern Hemisphere. Research

on this little-studied atmospheric effect of nuclear war indicates that the blockage could persist for several months, stopping much of the world's food production, and subjecting all but a small fraction of the survivors of the initial nuclear effects to famine and disease. These survivors would see a darkened world much like the one some believe the dinosaurs saw as they perished 65 million years ago (PHYSICS TODAY, May 1982, page 19). This time, however, even after the particulate matter clears, a dangerous photochemical smog might take its place for several months.

These are some of the results of a new study[2] of the atmospheric effects of nuclear war by Paul J. Crutzen (Max Planck Institute for Chemistry, Mainz, West Germany) and John W. Birks (University of Colorado, Boulder).

They evaluated many effects that they say have been overlooked or not carefully examined in previous considerations of the postwar atmosphere.

The Crutzen–Birks study of atmospheric effects appears with papers on other consequences of nuclear war in a special issue of the Royal Swedish Academy of Sciences publication, *Ambio*. The studies are the result of a two-year project, which the academy initiated "in the belief that a realistic assessment of the possible human and ecological consequences of a nuclear war may help to deter such a catastrophe." All the studies use the same reference scenario, which gives a detailed distribution of megatonnage over targets, under specific weather and seasonal conditions. It is 11 am in New York, 6 pm in Moscow, on a weekday in early June 1985 when the nuclear

Blowout on an offshore well injects large quantities of aerosol and gases into the atmosphere. When special firefighting equipment and transportation are available, a team can extinguish a well fire in less than a month. If relief wells are necessary, the process may take several months because of the highly accurate drilling required. (Photograph courtesy of the American Petroleum Institute.)

exchange is assumed to begin. Most cities in the Northern Hemisphere with populations over 100 000 are targeted, as are military installations, command posts, communication and transportation centers, industrial complexes, energy supplies and other economic or military targets. These include oil fields and refineries, chemical plants, hydroelectric stations and nuclear reactors.

With these things destroyed, the authors of the *Ambio* scenario call an end to the hostilities, even though they doubt that an actual war would end at that point. After exploding fewer than one-quarter of the total US and Soviet supply of warheads, they have run out of targets. In the final tally, then, the scenario uses
▶ about 14 700 warheads out of a total of about 59 000 available
▶ about 5700 megatons—almost all exploded in the Northern Hemisphere—out of about 13 000 megatons available.

The energy of a warhead determines in large part how it affects the atmosphere. In explosions that have an energy equivalent of more than 1 megaton of TNT, material rises beyond the top of the troposphere—12 km above the Earth's surface—and penetrates the stratosphere. Material from smaller explosions is confined to the troposphere. Earlier studies of atmospheric effects of nuclear war, by the US National Academy of Sciences[3] and others, focused mainly on ozone depletion in the stratosphere, which would allow increased levels of harmful ultraviolet radiation to reach the Earth's surface. Stratospheric ozone depletion is the result of the action of oxides of nitrogen, which are produced in the high temperatures of nuclear explosions in much the same way they are produced in automobile and airplane engines.

Fire and smoke. The *Ambio* scenario is based on the size-distribution of weapons in the nuclear arsenals, Birks told us. It assumes the use of relatively small weapons, only a few of energy greater than 1 megaton. In such a war, little NO_x enters the stratosphere. Crutzen and Birks, then, concentrate on the tropospheric effects of nuclear war. They calculate that the troposphere would be profoundly affected by the aftermath of the thousands of lower-yield explosions, which would touch off fires in cities, fossil-fuel stockpiles, industrial centers and refineries where huge quantities of combustible chemicals are stored, fields of natural gas wells and oil wells, and forests and grasslands close to targets. In the United States, but especially in the Soviet Union and Canada, there are vast forests close to important urban stra-

Ozone mixing ratios in parts per billion by volume (ppb) 50 days after the beginning of the nuclear war. Inputs are from forest fires and oil- and gas-well fires. (From ref. 2.)

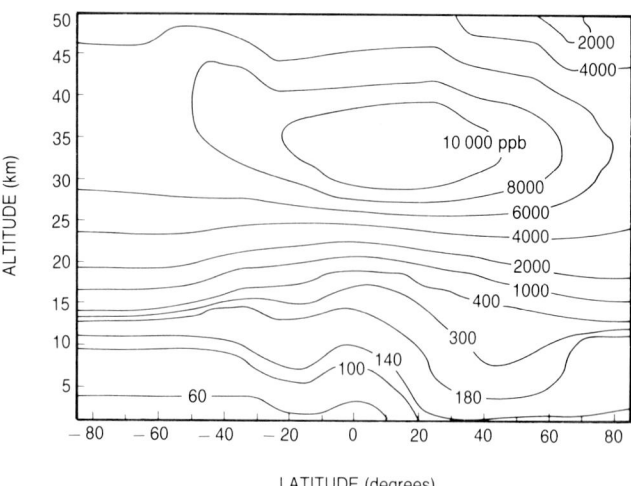

LATITUDE (degrees)

tegic targets, say Crutzen and Birks.

For their calculations, Crutzen and Birks limit themselves to the forests and the gas and oil wells, making what they consider to be underestimates of the quantities of combustible material from those sources that would be involved. They assume the blasts would
▶ start fires that would consume 10^6 km² of forest, an area twenty times that presently consumed by wildfires each year
▶ break production wells so as to release gas and oil at the current rate of worldwide usage
They assume further that 25% of the gas, which is mainly methane (CH_4), ethane (C_2H_6) and propane (C_3H_8), would escape without burning.

An area of 10^6 km² of forest contains about 10^{16} g of carbon and 10^{14} g of fixed nitrogen, not counting organic material in peat bogs and the soil. Assuming that fires initiated by nuclear war behave like ordinary wildfires, the atmosphere would take up $1.3–2.5\times10^{15}$ g of carbon in various gases, and $2–4\times10^{14}$ g of particulate matter.

Measurements show that particulate matter injected into the atmosphere by forest fires is 55% tar, 25% soot and 20% ash, and has a maximum particle-number density at a diameter of 0.1 micron. It strongly absorbs sunlight and infrared radiation, say Crutzen and Birks. Assuming that submicron-sized particles reside in the atmosphere for 5 to 10 days, that the forests take two months to burn and that the aerosol spreads out over half the Northern Hemisphere, Crutzen and Birks calculate that a vertical column of air will maintain an integrated particle density of 0.1–0.5 g/m². (For comparison, gold leaf has a mass per unit area of about 0.5 g/m².) Even at noontime during the summer, this would reduce the amount of sunlight reaching the ground by a factor between 2 and 150, they say. Much of the Northern Hemisphere would be darkened for a

number of weeks.

Not everyone is convinced that the darkening effect predicted by Crutzen and Birks would materialize. James Hansen, head of the Goddard Institute for Space Studies in New York, has studied minor constituents in the atmosphere, such as volcanic dust. "Any efforts to point out added dangers of nuclear war are meritorious," Hansen told us, "but I believe that any aerosol or climate impact would be negligible compared to the horrifying direct effects." Hansen questions the numbers Crutzen and Birks use to estimate the reduction of sunlight—"They are only guesses," he says. Even if the particulate mass injected into the atmosphere is $2–4\times10^{14}$ g as Crutzen and Birks say, Hansen questions its effect, pointing out that this is only comparable to that injected by a very large volcanic eruption—less than one cubic kilometer of material. Furthermore, Hansen notes that one cannot be sure the fires would actually burn for two months, continuously replenishing the tropospheric aerosol as it is rained out. Thus, Hansen believes that "the inference of prolonged extreme darkness is dubious."

However, Siegfried Gerstl (Los Alamos National Laboratory) told us that the tar and soot produced by fires is at least an order of magnitude more absorbing than volcanic aerosol, so that the attenuation factors calculated by Crutzen and Birks appear reasonable. Working with Andrew Zardecki of Los Alamos, Gerstl has done similar calculations[4] to test the hypothesis that the mass extinctions of species at the boundary between the Cretaceous and Tertiary periods were caused by an asteroid-induced aerosol that blocked enough sunlight to curtail photosynthesis and reduce the food supply. Gerstl notes that Crutzen and Birks do not include the effects of the "nonnegligible amounts of dust" that nuclear explosions near the ground loft into the atmosphere, something Gerstl has studied and considers significant. If Crutzen and Birks are otherwise correct in their aerosol estimates, says Gerstl, their calculated attenuation factors "look more like underestimates than overestimates."

Smog. Although most of the gas released by forest fires is CO_2, the already-large atmospheric content of this gas would not increase significantly. Carbon monoxide, however, would increase by up to a factor of four in the midlatitudes of the Northern Hemisphere and there would be a significant injection—tens of teragrams (10^{12} g)—of reactive hydrocarbons, mostly ethylene (C_2H_4) and propylene (C_3H_6), which contribute to the formation of photochemical smog. Even more important, forest fires would put 15 to 30 Tg of nitrogen into the atmosphere as oxides— a quantity greater than the 12 Tg produced in the nuclear fireballs and comparable to the 20 Tg emitted annually by industrial processes. Whereas nitric oxide (NO) catalyzes the destruction of ozone in the stratosphere, it catalyzes the *production* of ozone in the troposphere, where the chemical composition and the spectrum of transmitted light are quite different. Because of this, Crutzen and Birks say ozone could accumulate in sufficient quantities to constitute a global photochemical smog.

Although there are more than 800 000 oil and gas wells in the world, much of the total production is vulnerable to a relatively small number of nuclear weapons because production varies dramatically from one region to another. Saudi Arabia's 800 oil wells, for example, outproduce the United States' 590 000 oil wells. About 3.5 percent of the world's petroleum comes from about 40 drilling platforms in the North Sea. Just 4 megatons of explosive yield would uncap the wells there, Crutzen and Birks estimate. It would take fewer than 300 breakages of the magnitude of the gaswell blowout at Gassi Touil in the Sahara to release natural gas at the current rate of global consumption. Known as "The Devil's Cigarette Lighter," this well powered a 200-meter-high flame with 15×10^6 m³ of gas per day until it was capped.

The liberated oil and gas that burns would put copious amounts of particulate matter into the atmosphere, and would produce oxides of nitrogen as well, in an amount Crutzen and Birks estimate at the current annual industrial NO_x production. The oil and gas that escapes unburned, together with the gases injected into the atmosphere by the forest, gas and oil fires, would take part in a complex series of photochemical reactions, some of which have been studied previously because of their role in the production of smog. These reactions would commence as

soon as the atmosphere was sufficiently clear of light-absorbing particulates.

Crutzen and Birks looked in detail at the reaction sequences that would involve the NO, CO, CH_4 and C_2H_6 produced in a nuclear war. They looked in particular at the photochemical production of ozone, which is of interest because of its effect on human health and because it inhibits the growth and yield of plants. According to Crutzen and Birks, when the concentration of NO in the troposphere exceeds 1/4000 that of ozone, the oxidation of CO increases the concentration of ozone:

$$CO + 2O_2 + h\nu \rightarrow CO_2 + O_3$$

The oxidation of CH_4 and C_2H_6 also produces ozone:

$$CH_4 + 4O_2 + 3h\nu \rightarrow$$
$$CO + H_2 + H_2O + 2O_3$$

$$C_2H_6 + 10O_2 + 6h\nu \rightarrow$$
$$CO + H_2 + 2H_2O + CO_2 + 5O_3$$

The CH_4 reaction takes place in 10 steps (see box). The C_2H_6 reaction, which takes place in 24 steps, produces as an intermediate product peroxyacetylnitrate (PAN), a strong plant toxin. (PAN is now most familiar as an eye-irritating component of smog.)

Crutzen and Birks use a two-dimensional computer model to calculate the distribution of ozone in the postwar atmosphere as a function of latitude and altitude. Their model couples atmospheric photochemistry—including nearly 100 reactions thought to be important in global air chemistry—with atmospheric dynamics covering altitudes from the ground to 55 km and latitudes from the South Pole to the North Pole. When it is tested with war scenarios used in previous studies of the postwar stratosphere, the model reproduces the predicted ozone distributions, say Crutzen and Birks. Using the *Ambio* scenario, the calculations

---

Reaction cycle

The oxidation of methane in the atmosphere leads to formation of ozone as follows.[2] (M, which can be any molecule, serves to carry away energy.)

$$CH_4 + OH \rightarrow CH_3 + H_2O$$
$$CH_3 + O_2 + M \rightarrow CH_3O_2 + M$$
$$CH_3O_2 + NO \rightarrow CH_3O + NO_2$$
$$CH_3O + O_2 \rightarrow CH_2O + HO_2$$
$$HO_2 + NO \rightarrow OH + NO_2$$
$$NO_2 + h\nu \rightarrow NO + O \text{ (twice)}$$
$$O + O_2 + M \rightarrow O_3 + M \text{ (twice)}$$
$$CH_2O + h\nu \rightarrow CO + H_2$$

$$CH_4 + 4O_2 + 3h\nu \rightarrow$$
$$CO + H_2 + H_2O + 2O_3$$

(some results of which are shown in the figure on page 18) show the possibility of "severe world-wide smog conditions" in the troposphere, they say, with high concentrations of ozone (160 parts per billion by volume) and large accumulations of ethane (50–100 ppb) and PAN (1–10 ppb). Concentrations over land, where the gases are generated, would exceed these latitudinal averages. However, any aerosol remaining in the atmosphere may act to reduce the concentrations through adsorption.

The food supply. Agricultural crops, in addition to suffering from radioactive contamination and the blockage of sunlight, would be subject to other atmospheric effects. Dark aerosol deposits and photochemical smog would severely limit plant productivity, and oxides of nitrogen could make rainwater highly acidic, with an average pH less than 4. Crucial ocean life would also be affected. According to some estimates,[5] the reduction of sunlight by a factor of 100, which Crutzen and Birks think quite probable in an all-out nuclear war, would lead to the death of much of the ocean's phytoplankton and therefore much of its herbivorus zooplankton.

How harmful is several months of exposure to ozone at 160 ppb and PAN at 1–10 ppb? When ozone reaches 200 ppb in the Los Angeles basin, the South Coast Air Quality Management District issues an alert characterizing the situation as "unhealthful for everyone," Armando Zumaya of the district tells us. At 120 ppb, ozone is officially "unhealthful for sensitive people." So it would seem that most survivors of the blasts would survive at least the direct effects of the 160 ppb ozone predicted by Crutzen and Birks.

How serious, then, is the threat to the food supply? Plants vary in their sensitivity. Crutzen and Birks say the effect on agricultural crops may be particularly severe, but there seems to be room for further research here. Birks told us of a recent study[6] by the EPA-initiated National Crop Loss Assessment Network, a group investigating the agricultural impact of air pollution. Their open-field experiments indicate that crops grown in an atmosphere that is polluted with 160 ppb ozone for just 7 hours each day show losses in yield as follows: corn, 15%; winter wheat, 42%; soybeans, 64%; spinach, 72%; head lettuce, 88%; peanuts, 99%; turnips, 100%.

Paul Miller of the United States Forest Service (Statewide Air Pollution Research Center, Riverside, California) tells us that a month of exposure to 160 ppb ozone, by itself, "wouldn't have a devastating effect." Even if we ignore the effects of the smoke, radiation and fallout, there might be some scarcity of food because of the ozone, says Miller, but some of the more tolerant crops—generally the less leafy ones—would survive. Edgar Stephens (University of California, Riverside) tells us that during the Los Angeles "smog season"—June through mid-October—PAN levels comparable to or exceeding those predicted by Crutzen and Birks occur quite often. Although some plants exposed to these levels for just a few hours show markings on their leaves, the postwar PAN alone would probably not kill them, says Stephens.

It is difficult to estimate the combined effect on the food supply of reduced sunlight, aerosol deposits, acid rain, ozone, PAN and other air pollutants, radioactive contamination, and the disruption of cultivation and distribution systems. A further complication, say Crutzen and Birks, is that meteorological patterns of wind, temperature and rain would be altered for several weeks following a nuclear war. The absorption of most solar radiation in the atmosphere rather than at the ground would stagnate the flow of air below 10 km, and slow the removal of pollutants. The excess cloud condensation nuclei—up to 6×10^{10} per gram of wood burned—could narrow the range of sizes of cloud droplets and decrease the efficiency with which they coalesce and precipitate.

Crutzen and Birks conclude that the atmospheric effects of the explosions and fires would probably result in agricultural yields sufficient to feed only a small fraction of the initial survivors; many would die of starvation during the first year after the war.

Whatever the fate of life on Earth, the atmosphere and meteorology may return to prewar conditions after a few years, although one cannot rule out irreversible changes. One question, for example, is whether melting due to soot settling on glaciers or arctic ice and snow would cause permanent changes in important climatic parameters. We do have some data on the effect of atmospheric aerosol in nuclear-war quantities. The 1815 volcanic eruption of Mount Tambora in the Dutch East Indies (now Indonesia) was followed in the summer of 1816 by four waves of crop-killing snow and frost in New England and food shortages in parts of Europe. The volcanic eruption at Krakatoa in 1883 lowered global mean temperatures, but for only a few years. One must be careful in using volcano data to generalize, however, because the physical characteristics of volcanic aerosol are not the same as those of fire-produced aerosol.   —JDS

References

1. Y. Laulan, Ambio **11**, 149 (1982); H. Middleton, Ambio **11**, 100 (1982).
2. P. J. Crutzen, J. W. Birks, Ambio **11**, 114 (1982).
3. National Academy of Sciences, *Long Term World-wide Effects of Multiple Nuclear-Weapon Detonations*, Washington, DC (1975).
4. S. A. W. Gerstl, A. Zardecki, *Reduction of Photosynthetically Active Radiation under Extreme Stratospheric Aerosol Loads*, Los Alamos National Laboratory report LA-8938-MS (August 1981).
5. D. H. Milne, C. P. McKay, in *Geological Implications of Impacts of Large Asteroids and Comets on the Earth*, L. Silver, *et al.*, eds., Special Paper 190, Geological Society of America (1982, in press).
6. W. W. Heck, *et al.*, J. Air Pollution Control Assoc. **32**, 353 (1982).

Nuclear winter: A matter of degrees

The major climatic effects from nuclear war would come from soot generated by urban fires; much research will be needed to clarify the uncertainties.

Barbara G. Levi and Tony Rothman

PHYSICS TODAY/SEPTEMBER 1985

Although climatic impacts have long been on the list of potential consequences of nuclear war, two years ago they were predicted to be so devastating as to earn the label, "nuclear winter." That term describes a world that is plunged into cold and darkness because the sunlight is blocked by smoke from fires ignited by nuclear explosions and by the dust from nuclear groundbursts. This vision is considerably more severe than any painted earlier because the calculations for the first time included massive injections of soot, the carbonaceous component of smoke. The soot strongly absorbs the incoming sunlight but transmits most of the outgoing infrared radiation, creating an inverse greenhouse effect.

The dramatic prediction of nuclear winter naturally caught the public spotlight. It reinforced calls for sharp arms reductions and stimulated debate over possible policy implications. It also spurred review of the predictions by other scientists. The calculations underlying nuclear winter involve a wide variety of disciplines, and the theory rests on numerous assumptions for which the data are sparse or even nonexistent. As a result, the uncertainties in virtually every parameter are wide. The calculations predict such massive injections of smoke and dust into the atmosphere that they greatly

challenge atmospheric scientists who are already struggling to model more modest perturbations to the normal climate patterns. Over the past two years, the input assumptions have been scrutinized and the models modified (see figure 1). While no one can yet either prove or disprove that a nuclear winter might result from a large-scale war, it does seem that some climatic impact would occur. Much of the debate now centers on the extent of the cooling and on the magnitude and type of the nuclear exchange that might cause it.

History of the problem

One of the first questions to be asked about nuclear winter was why no one had predicted it sooner. Actually, although the recent quantitative treatment required a convergence of knowledge from widely different fields, ingredients to this calculation had been at least qualitatively recognized over the years. The first substantial concern over the global atmospheric impacts of nuclear weapons involved the possible effects of the nitrous oxides produced by nuclear explosions on the stratospheric ozone layer that now protects Earth from harmful ultraviolet radiation. One study in the mid 1970s investigated[1] the suggestion that the nuclear fireballs might produce such a large quantity of nitrous oxides that the reactions they catalyze would seriously deplete the ozone layer. In a 1984 study of the climatic effects of nuclear war, the National Academy of Sciences reiterated this concern[2] and cautioned the world to worry not only about nuclear winter but also about "ultra-

violet summer."

Another concern over atmospheric impacts has been that the dust from groundbursts might block out the solar radiation and cause Earth to cool. The earliest treatments of dust from nuclear bombs concluded that a single explosion couldn't produce enough dust for a noticeable effect on Earth's climate. However, by the mid 1960s the arsenals were sufficiently large that a massive exchange had become a possibility. The quantity of aerosols potentially raised by a war involving nuclear explosions equivalent to 5000–10 000 Mt of TNT might be comparable to what was spewed out by large volcanic eruptions in the past. Volcanoes are only a weak analogy, but historical evidence of their impact tends to suggest what might result from the dust ejected by a large-scale nuclear war. For example, the 1815 eruption of Tambora in Indonesia drove so many particles into the stratosphere that it apparently affected weather patterns around the globe. Although the annual mean temperature in the Northern Hemisphere for the subsequent year may have dropped[3] by less than one degree Celsius, 1816 was marked by anomalous weather events such as New England snowfalls in June and has been dubbed[4] "the year without summer." It has also been hypothesized[5] that the extinction of the dinosaurs occurred after an asteroid, perhaps 10 km in diameter, struck the Earth and raised a dust cloud that obscured sunlight (see PHYSICS TODAY, May 1982, page 19). (The energy of that asteroid impact would have been ten thousand times that released in a 5000–Mt nu-

Barbara G. Levi is a member of the research staff of the Center for Energy and Environmental Studies at Princeton University; she is a contributing editor at PHYSICS TODAY. Tony Rothman is a postdoctoral fellow in the applied mathematics department at the University of Capetown in South Africa.

Global smoke distributions. The maps show distributions 1, 5, 10, and 20 days after injections of smoke at five points: two in the US, one in Europe and two in the USSR. A total of 150 Tg of smoke is assumed to be injected; each dot represents 4000 tons of smoke spread throughout a volume of 550 000 km^3. (Lawrence Livermore National Lab.)   Figure 1

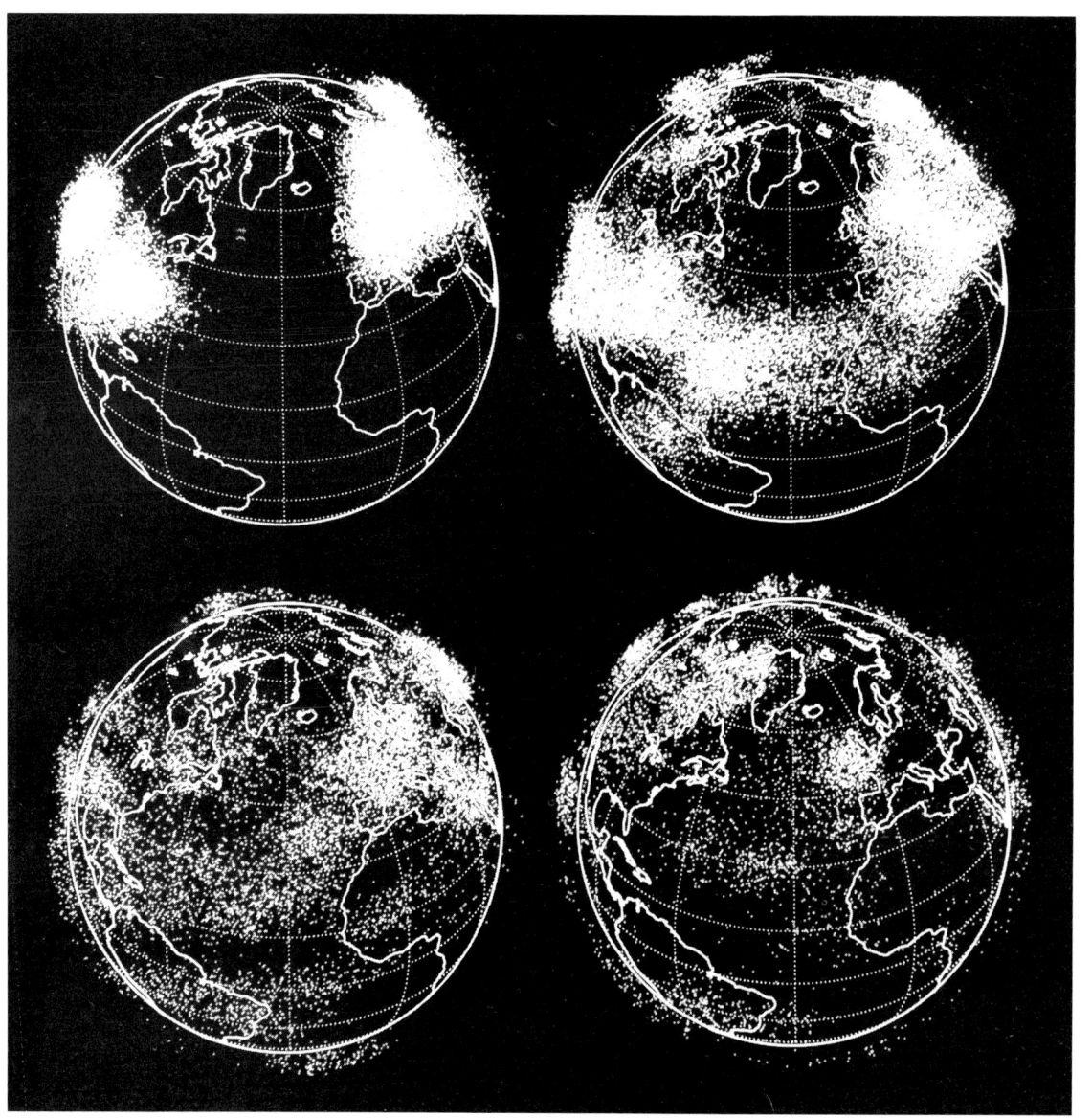

Convection column established over a 1600-acre, helicopter-ignited prescribed burn near Chapleau, Ontario, 3 August 1985. The burn served to provide information on smoke emissions from fires simulating those created in nuclear explosions; it was carried out in tramped, dead forest fuels to reduce surface fuel loadings and prepare the site for planting. Figure 2

clear war.)

By rough analogy with the Tambora eruption, the dust from a large-scale war might be expected to lower the average yearly northern hemispheric temperatures by about one Celsius degree. The first nuclear-winter calculations, however, estimated far greater temperature drops—with a maximum of about 35 °C. The key element is the soot. A relevant analogy from nature in this case is that of plumes from intense forest fires: In 1950, the smoke pall from a fire in western Canada spread across the continent, lowering the predicted maximum temperatures[6] in Washington, D. C. a few days afterward by an estimated several Celsius degrees. Some early studies of nuclear weapons effects done in the mid 1960s loosely mentioned[7] the possibility that smoke from fires set by nuclear war might severely affect the climate. However, no one treated the effect quantitatively until 1982, when Paul Crutzen and John Birks estimated[8] the solar attenuation that might result if a nuclear war set ablaze one million square kilometers of forest land. They suggested that fires in urban areas, especially in regions where quantities of petroleum are stored, might produce even more smoke (see PHYSICS TODAY, October 1982, page 17).

Predictions of the TTAPS model

The quantitative predictions that first inspired the term "nuclear win-

ter" were done[9] by Richard Turco, Brian Toon, Thomas Ackerman, Jim Pollack and Carl Sagan (see PHYSICS TODAY, February 1984, page 17). Contrary to the acronym TTAPS formed from their names, the group may have sounded reveille to a potentially important climatic effect. They all had backgrounds in analyzing such phenomena as asteroid impacts, volcanic eruptions and Martian dust storms and applied these backgrounds to an analysis of Crutzen and Birks's idea. The TTAPS group took as their baseline a nuclear war involving 5000 Mt. They estimated the quantities, heights of injection, lifetimes and optical properties of both the dust and the smoke, including soot. They assumed the aerosols were instantaneously and uniformly spread over the Northern Hemisphere; and they fed this information into a one-dimensional calculation that simulated the convective and radiative processes within a globally averaged vertical column of the atmosphere. The model necessarily ignored horizontal transport and the buffering effect of the oceans.

Turco and his colleagues predicted that for their baseline scenario the temperature would drop by approximately 30 °C in the first 20 days and would still remain below normal after one year. These temperatures are relevant over land masses of a planet assumed to have no oceans. Based on model calculations with the surface

treated as ocean as well as land, the group estimated that the inclusion of the moderating effect of oceans should cause temperature drops in mid-continental regions to be about 30% smaller, and in coastal regions to be 70% smaller.

In another scenario TTAPS considered an attack involving only airbursts so that no dust was generated. In this case, the temperature returns to prewar values more quickly. In still another scenario, where dust and no smoke is produced, the temperature drops by at most 7 °C but still remains several degrees below normal after 300 days. The temperature over land surfaces for a land-only planet has an average drop of about 3 °C during the first year, in the "dust only" scenario, which seems consistent with an average temperature drop over the Northern Hemisphere, including oceans, of 1 °C. Hence, the smoke causes the severe temperature drops and the dust prolongs the cooling in the TTAPS model. The reason for the difference in the longevity of effect is that the smoke is generally injected by fires into the troposphere from which it is normally removed within a few weeks, while a significant fraction of the dust is carried by the fireball into the stratosphere, where it may remain for over a year. It must be noted, however, that the conditions of nuclear winter may greatly alter "normal" atmospheric patterns that promote the scavenging of smoke by rain.

Another TTAPS scenario, termed "100-Mt city attack," deserves special comment, as it is really an excursion from the baseline scenario. Although it involves only 2% of the total yield of the baseline case, this attack produces as sharp a cooling because the very particular circumstances of this scenario create 60% as much smoke as the baseline. This scenario is often cited in debates as evidence that the threshold for the onset of nuclear winter—as

Table 1. TTAPS baseline scenario

Weapon yield (Mt)	Target type	Type of burst	Atmospheric emissions Dust (Tg)	Soot (Tg)
1000	Urban, industrial	Airburst	0	149
4000*	Military	Airburst and groundburst	960*	80
5000		Total	960	229

From reference 9.
*Includes 1150 Mt airburst, which produce the 960 Tg dust.

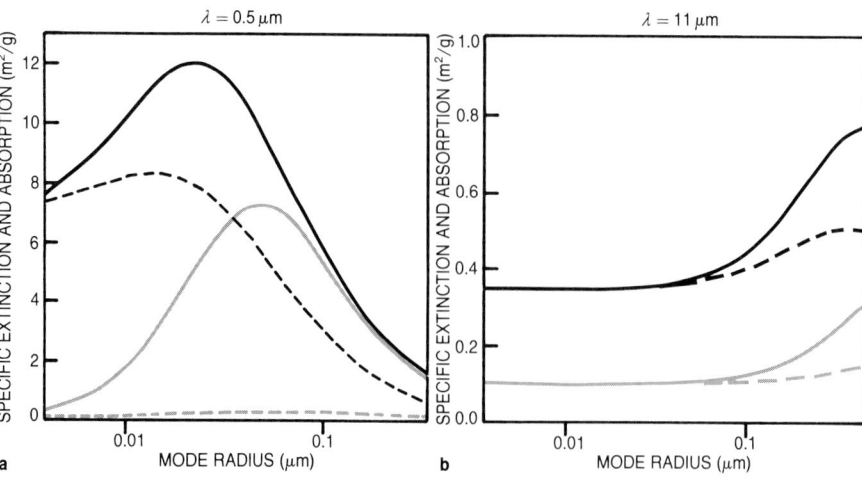

Specific cross sections for the absorption (dotted lines) and extinction (solid lines) of light by log-normal distributions of smoke (black) and dust (red) particles, at both visible wavelengths (**a**) and long wavelengths (**b**), as a function of their mode radii. For the size range of particles shown, the smoke absorbs visible light far more strongly than infrared, leading to an inverse greenhouse effect. The dust particles largely act to scatter light; much of the sunlight is scattered forward and still reaches the surface.
(See reference 19.)   Figure 3

measured by the megatons expended—is very low. However, this scenario is not a threshold as measured by its direct consequences on the population: To produce this much smoke with 100 Mt, the bombs must explode only on the built-up centers of cities, where the density of burnable materials is highest. In our analysis we estimate that the area burned in this scenario is equal to approximately 10% of the urban area of about 1000 cities, or virtually all the cities in the more developed countries with populations over 100 000 each[10]—in all, a population of more than 500 million people. Moreover, for this scenario, the TTAPS calculations doubled the parameters assumed for both the fuel loading and for the percentage of urban areas occupied by city centers, and also more than doubled the value assumed for the net smoke emission factor. (They essentially reduced their estimates of the degree of incineration and of the amount of smoke rained out.) Thus they increased the smoke predicted from these urban-center fires by a factor of nearly ten over what they would have calculated using the same parameters as in the baseline scenarios.

The TTAPS estimates of temperature change were reasonably consistent with results from several two- and three-dimensional climate models that appeared at about the same time. All examined the climatic effects of approximately the same, or greater, quantities of soot (see PHYSICS TODAY, March 1984, page 17). Even so, these multidimensional climate models were not fully realistic. Two of the models constrained the soot to a fixed region and calculated the impact on circulation patterns; the third used unperturbed circulation patterns and calculated how the soot might spread. Because they included ocean moderation, these studies generally predicted lower

drops than the original TTAPS study. None of the early attempts simulated feedbacks, such as changes in precipitation patterns caused by smoke, which might in turn prolong its residence time in the atmosphere. Some results are beginning to emerge from more interactive models. They have shown that the climatic impact may be far greater for a summer war and that the smoke may be lofted to high altitudes by self-heating. Those working on such models stress[11-14] the complexity of the problem and the (fortunate) absence of real experience against which to test their models in this new regime.

Key questions

The smoke produced in a nuclear exchange is far more instrumental in producing the extreme temperature drops than the dust particles and hence will be the focus of our attention. To estimate how much smoke might be generated requires one to ask many more questions. We begin with the targeting scenario: How many nuclear weapons are aimed at which categories of targets? Table 1 gives the baseline scenario assumed by TTAPS. (A subsequent study by the National Academy of Sciences[2] made similar assumptions.) The total exchange involves 5000 Mt—a significant fraction of the yield in the US and Soviet strategic arsenals and an amount that could conceivably be unleashed in an all-out nuclear war. Just over half of those weapons are assumed to be groundbursts against military targets—presumably against such hardened targets as ICBM silos. About one-fifth of the total yield is assumed to be airbursts on other military targets (such as air bases) and the remaining one-fifth is targeted on urban and industrial targets. Only groundbursts and very low airbursts produce dust; in the model, urban and industrial targets produce about 60% of the total smoke although

they account for only 20% of the total yield. Only a fraction of the total 960 Tg of dust estimated to be emitted in this scenario is submicron, stratospheric dust (1 Tg = 10^{12} g).

The quantity of smoke produced by the TTAPS baseline scenario is detailed in table 2. The amount of smoke produced in a fire is the product of: the area burned, the burnable fuel loading (the amount of combustible material per unit area times the fraction of the fuel that actually burns), the fraction of mass that is released as smoke, and the fraction of smoke that is not immediately rained out. Turco and his colleagues assumed that their hypothetical attack would burn 500 000 km^2 of forests and 250 000 km^2 of cities. Although the area of forests assumed to burn was twice that of cities, the fraction of the fuel in forests that actually burns is much lower. The equivalent dry biomass content of forests is about 2 g/cm^2, but typically only a fraction of that is consumed in a fire. The group estimated the burnable fuel loading to be 0.5 g/cm^2. By contrast, in urban areas both the fuel loading and fraction burned would be considerably higher. They estimated that firestorms would develop in the inner 5% of the city area, consuming all the fuel, which they assumed to have an areal density of 10 g/cm^2. The fires in the remaining 95% of the city were taken to burn 50% of the fuel, which has an areal density of 3 g/cm^2. Thus the assumed average fuel loading of material that burns in urban areas corresponds to 1.9 g/cm^2.

In any fire, only a few percent of the fuel by mass is converted to smoke, which is composed partly of soot. Some of the soot may be almost immediately removed in "black rain," which is the rainfall from the cloud that may form over the hot, rising fire plume, as occurred in Hiroshima. Turco and his colleagues assumed that the black rain does not occur at all for forest fires, but

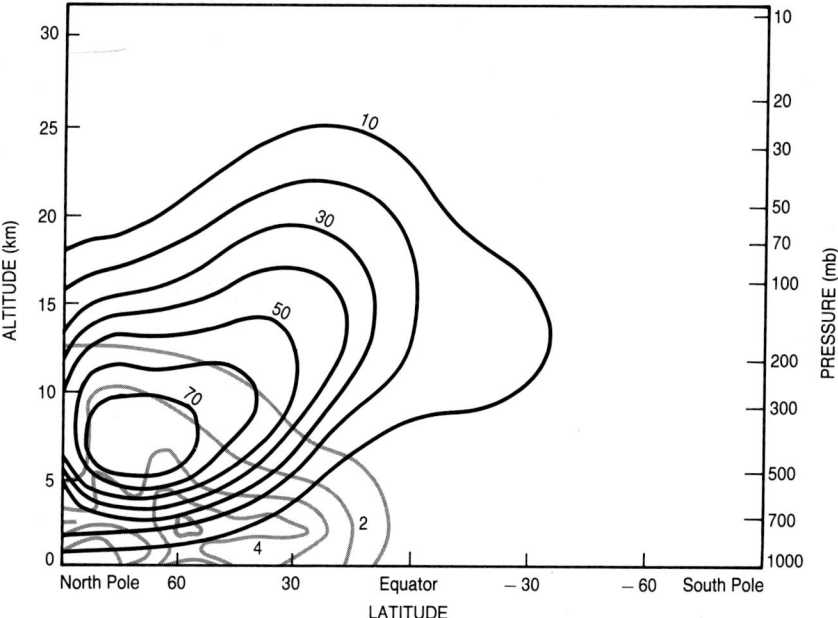

Smoke that absorbs sunlight (black contours) is heated and rises to very high altitudes compared to a passive tracer that does not absorb at all (red). Contours are labeled by the mixing ratio of smoke to air, in units of 10^{-9}. Curves show distributions 20 days after the July injection of 170 Tg of smoke in the height range 2–5 km, as simulated in a three-dimensional global atmospheric circulation model. Because smoke has been lofted above the level where precipitation occurs, ten times as much smoke as tracer remains at this point. (See reference 12.)   Figure 4

that rain removes 25% of the soot in urban fires and 50% in firestorms. They further assumed that the smoke emission in firestorms is only half that of general conflagrations, and that in both cases, the percentage of smoke particles less than one micron in radius is 90%. The total mass of smoke particles shown in table 2 is less than the 225 Tg predicted by TTAPS because their model also included a contribution from long-term, or smoldering, fires.

The ranges given in table 3 for the parameter values reflect, in our personal judgment,[15] the current state of knowledge. By comparison, the TTAPS values fall within these ranges, except for their assumptions about forest areas burned. The original TTAPS number was a rough estimate and the area of forest fires continues to be debated. One detailed study estimated[16] that no more than 70 000 km^2 of forests or 190 000 km^2 of vegetation of all types would burn in a 4100-Mt attack on nonurban targets at the most susceptible time of the year. The National Academy study of nuclear winter assumed that 250 000 km^2 of forests might go up in flames if 5000 Mt were targeted outside cities. The discrepancies in the various estimates stem from differing judgments as to the nature of the vegetation surrounding the targets, the susceptibility of the targets to combustion in different seasons (and whether that susceptibility is appreciably different for nuclear explosions), the probability and extent of fire spread and the likelihood that a second burst on the same target would increase the fire-start probability.

Despite the controversy over areas of forests burned, the more critical term in the calculation is the area of urban

fires. We have estimated[15] the upper limit for the area of urban fires in table 3 as being equal to the TTAPS value of 240 000 km^2 and close to the National Academy value of 250 000 km^2—but recognize it as a very improbable case. This area corresponds, in the TTAPS baseline scenario, to spacing the 1000 Mt evenly over the areas of virtually every city in the more developed nations having a population exceeding 100 000. An urban area this size might conceivably be burned in a nuclear war, especially if more megatonnage were aimed at cities; however, such an attack would probably require a deliberate policy by both alliances to burn virtually all the adversary's cities.

The lower limit in our range accounts for a possible overlap by a factor of 4 among warheads aimed at targets within the same city. For example, 60 warheads may be targeted[17] at Moscow alone.

Table 3 suggests that fires in outer-city areas are the source of most of the smoke. Current estimates of the urban fuel loadings rely on a few surveys[18] of the "typical" content of American buildings, coupled with assumptions about patterns of building height and density. More data—for Soviet cities as well—would help narrow the uncertainties here. A third factor, smoke emissions, is very difficult to determine for the conditions that might prevail in fires set by nuclear explosions. Current estimates rely either on laboratory-scale fires in controlled conditions or on samples from deliberate burnings of forest material (see figure 2). The range shown in table 3 is the same as that cited by the National Academy study of nuclear winter. The final factor—the percentage of smoke that might be immediate-

ly washed out by rain—is also a key factor for which little evidence now exists. The unknowns include the extent to which the soot particles may act as cloud-condensation nucleii.

The net result of all these uncertainties is that estimates of the total amount of smoke resulting from nuclear war vary over a wide range. The National Academy study estimated an "excursion" range from 20–650 Tg about their baseline figure of 180 Tg. To combine the ranges of uncertainties for each parameter given in table 3, we have assumed that all values between the given limits are equally probable. We then computed the mean and standard deviation for the product of these parameters. The amount of smoke could thus vary over a two-standard-deviation range of 4–318 Tg.

The optical properties of smoke greatly differ from those of dust. The dust particles kicked up by nuclear explosions scatter—but only weakly absorb—the incoming sunlight, and much of that scattered light is still directed forward. By contrast, smoke particles strongly absorb and scatter light in the visible region. Both types of particles scatter and absorb infrared radiation weakly. The effective cross section for scattering or absorption by an individual particle is given by the geometric cross section of that particle—πr^2 for a spherical particle—multiplied by the scattering or absorption efficiency Q_{scat} or Q_{abs}. The sum of these two efficiencies is the extinction efficiency. These efficiencies in turn depend on the complex index of refraction, m, of the particle. They indicate how much sunlight is removed from the collimated beam, although some of that light may still reach the ground. The attenuation of sunlight in a unit length of its path

Table 2. Key parameters—TTAPS values

Parameters	Forests	Inner city	Outer city	Total
Area burned (1000 km^2)	500	12	228	740
Burnable fuel loading (g/cm^2)	0.5	10	1.5	
Quantity of submicron smoke particles emitted (g/g)	0.032	0.023	0.045	
Fraction of smoke remaining after "black rain"	1.0	0.5	0.75	
Total smoke (Tg)	81	14	115	210
Absorption efficiency (m^2/g)	2	2	2	
Absorption optical depth	0.63	0.11	0.89	1.6

Based on reference 9.

by a collection of individual particles is determined by integrating the effective cross sections, weighted by the density of particles of each size, over the particle radius:

$$b_i(l) = \int \pi r^2 \, Q_i(r,m) \, [dn(r)/dr] \, dr$$

where $n(r)$ is the number density of particles of radius r at a position l along a path normal to the horizontal. The subscript i denotes scattering, absorption or extinction. Integrating $b_i(l)$ over the path length then gives the dimensionless parameter, τ, called the optical depth:

$$\tau_i = \int b_i(l) \, dl$$

The optical depth determines the attenuation of a beam of light (with initial intensity I_0) along a path at a fixed zenith angle ϑ, according to the relationship:

$$I = I_0 e^{-\tau/\cos\vartheta}$$

If one divides the attenuation parameter $b_i(l)$ by the mass of particles per unit volume of air, the resulting parameter, Ψ_i, called the specific scattering (or absorption or extinction) expresses the effective cross section per mass of aerosol present (m^2/g). If $b_i(l)$ is constant along the path length, then once the quantity of dust or soot is known, one can combine the specific scattering (or absorption) with the atmospheric loading of particles per unit area, σ, to calculate the optical depth, τ.

$$\tau_i = \Psi_i \sigma$$

Sample values[19] of specific absorption and extinction parameters are shown in figure 3 for dust and soot particles at both visible and infrared wavelengths.

The extinction of visible light by dust and soot particles is very sensitive to particle size for particles with radii from 0.01–0.1 microns. Most nuclear winter studies have assumed that the smoke particles lie within this size range. The studies represent the size distribution by assuming that the number distribution of smoke-particle radii is what is called a "log-normal" distribution (the mode radius, which is the center of that distribution, is the horizontal axis in figure 3). Dynamical processes will, of course, affect the evolution of any real distributions, but particles in the 0.1–1-micron size range tend to remain longest in the atmosphere.

Figure 3 indicates that at the longer wavelength of 11 microns, the extinction is lower by a factor of 10. Because submicron smoke particles absorb short-wavelength light but are nearly transparent to long-wavelength light, a layer of smoke can create an inverse greenhouse effect. For larger, agglomerated particles, however, the absorption at longer wavelengths increases relative to that at shorter wavelengths.

Smoke particles vary a great deal with fuel type and burning conditions. The greater the content of elemental carbon, the stronger will be its absorption. The percent of elemental carbon in smoke aerosols may vary from about 10% for burning vegetation up to 80% or so for oil and gas fires.[20] Smoke from well-ventilated fires tends to be highly oxidized and hence less absorbing. The absorption and extinction efficiencies for smoke particles shown in figure 3 fall in the middle of the range of measured values. The absorption efficiency may be anywhere from 2–5 m^2/g for particle distributions with mode radii ranging from 0.05–0.2 mi-

crons. If the smoke comes from fuel with a high carbonaceous content, such as petroleum, the upper limit could be higher. If the smoke is from forest fires, the lower limit could be smaller. Table 3 indicates our assumed range of absorption optical depths.

The height of injection for micron-sized particles of both dust and soot greatly influences their lifetimes in the atmosphere. Small dust particles injected into the stratosphere (above roughly 13 km for northern mid latitudes) may remain there for over a year. Small soot particles might remain[21] in the upper troposphere for one to two weeks or in the lower troposphere for only a few days to one week. These estimates are based on rather sparse data on aerosols in the normal atmosphere. In the conditions of nuclear winter, however, the cooling of the surface relative to the troposphere is expected to slow or stop convection, and hence may inhibit precipitation. Thus the aerosols may prolong their own impact. The scavenging rate for aerosols remains a critical unknown.

The smoke from fires set either by air- or ground-burst bombs rises to heights that depend primarily upon the intensity of the fire. Typical plume heights might be 1–5 km for forest fires (see figure 2) and 1–7 km for urban fires. Firestorms, which occurred in about 5% of the fire-bomb raids in World War II, might lift plumes into the stratosphere. (The plume over Hamburg is estimated to have been 13 km high.)[22]

The dust generated is lifted as high as the mushroom cloud—to a height roughly proportional to a fractional power of the weapon's yield. The bottoms of clouds from warheads of about 500 kt—typical of the warheads on

Table 3. Ranges of key parameters

Parameters	Forests		Inner city		Outer city		Total	
	From	To	From	To	From	To	From	To
Area burned (1000 km^2)	100	250	3	12	57	228	160	490
Fuel loading (g/cm^2)	0.1	0.5	4	20	1	4		
Quantity of submicron particles emitted (g/g)	0.01	0.04	0.01	0.04	0.01	0.04		
Fraction of smoke remaining after "black rain"	0.5	1	0.5	1	0.5	1		
Total smoke (Tg)	0.5	33	0.6	57	3	228	4	318
Absorption efficiency (m^2/g)	1	3	2	6	2	6		
Absorption optical depth	0.0	0.27	0.0	0.93	0.0	3.6	0.0	4.8

(The "Type of region" spans the Forests, Inner city, and Outer city column groups.)

modern Soviet ICBMs—would rise above 10 km, while only the tops of clouds from 40–50-kt bombs—about the size of warheads on the US Poseidon missiles—would reach this height.

Climate modelers have explored various distributions of the soot with height, but the appropriate representation remains a subject of debate. Most of the early work assumed that the smoke was initially distributed uniformly with height within a certain range. Some critics argued, however, that the smoke might be well mixed with air and thus, more realistically, its density might vary in direct ratio with the air density, which falls exponentially with height. The scale height (the distance at which the air density is $1/e$ times its value at the Earth's surface) is about 8 km. Perhaps a sum of distributions with different scale heights is the best way to represent the smoke from many fires. The recent National Academy study, in fact, argued[2] that a constant density distribution would result if the number of city fires of a given area (and hence their intensities) were inversely proportional to that area. The importance of height distribution is further illustrated[19] by one sensitivity study: In a one-dimensional model, the surface cooled by 32 °C when smoke was distributed with a constant density in the height range 0–10 km. The drop was only 22 °C when the smoke density varied with a scale height of 3 km.

Whatever the initial distribution of smoke in the atmosphere, it may well be altered by subsequent dynamic effects. For example, two recent climate models suggest[12,13] that the upper part of the smoke cloud may be lofted to much greater heights due to heating by

the sunlight it absorbs. One of these reports comes from a very detailed climate model that assumes 20 vertical levels in the atmosphere. As shown in figure 4, for a July simulation, with 170 Tg of smoke initially injected between 2–5 km, some fraction of the smoke gets lofted to altitudes of 15–25 km, from which it is removed very slowly. This effect is estimated to increase the amount of smoke remaining after 3 weeks by a factor of 10 in July, and by a factor of 3 in January, over the noninteractive simulation.

The climatic impact of smoke is influenced by its height of injection. Figure 5 illustrates this, using a very simple model of the radiation balance of Earth and its atmosphere. This model ignores convection and surface reflection and treats atmospheric gases, such as water vapor and carbon dioxide, as transparent to sunlight but opaque to infrared radiation. All atmospheric scattering is assumed to occur at the layer of greenhouse gases. Figure 5a shows a two-layer model representing Earth's surface and atmosphere in radiative equilibrium. The surface temperature is 303 K, 15 K higher than the global average of 288 K. In figures 5b and 5c, smoke layers are added and it is assumed that the optical properties of the smoke are the reverse of those of the atmospheric gases—that is, opaque to sunlight but transparent to infrared. In reality, the smoke cannot be transparent to infrared and still radiate it. The case shown is really the limit as the infrared absorption goes to zero and the smoke layer becomes very hot.

When the smoke is above the atmospheric gases (figure 5b), the surface

temperature drops dramatically, to 234 K. When it is below those gases (figure 5c), the temperature drop is much less because the atmospheric gases trap and reradiate the infrared from the soot layer and from Earth's surface. Although the numbers represent only very rough estimates of temperatures, the model does illustrate the importance of smoke height.

If the smoke layer is not too thick and not too high, its effect may actually be to warm the surface slightly rather than to cool it: The slowing of surface cooling by convection and the increased infrared radiation may then compensate for the decrease in the incoming sunlight. This possibility is relevant because the amount of smoke produced could conceivably be low. Any such warming would not be as severe or as prolonged as the cooling now predicted.

A uniform distribution of soot would have a greater overall impact on climate than a patchy distribution, especially because the attenuation varies exponentially with the optical depth. One three-dimensional model that addresses the question of patchiness has been developed[11] at Lawrence Livermore Laboratory. In that model, a quantity of smoke equal to that from city fires in the TTAPS baseline scenario is injected at discrete points in North America, Europe, and the Soviet Union. For a July simulation, the smoke cools the areas under the smoke by as much as 30–35 °C, but large areas of the globe are relatively unaffected. The soot concentrations at various times after injection are shown in figure 1. The patchy effect seems rather short-lived, however: After about a month, the distribution

Oversimplified diagram of Earth's radiation balance qualitatively illustrates the impact of smoke height on the climate. The model considers three layers (**a**): Earth's surface, with convective and sensible heat transport ignored; an atmospheric layer of greenhouse gases assumed to transmit all visible light, to absorb all infrared and to reflect with an albedo *w* having an average value of 0.3 for the Earth–atmosphere system; and a smoke layer assumed to absorb all visible light and to transmit all infrared. The incoming radiation I_0 is equal to 0.34 kW/m². Temperatures are considerably lower when the smoke is above the greenhouse gases (**b**) than when it is below them (**c**).   Figure 5

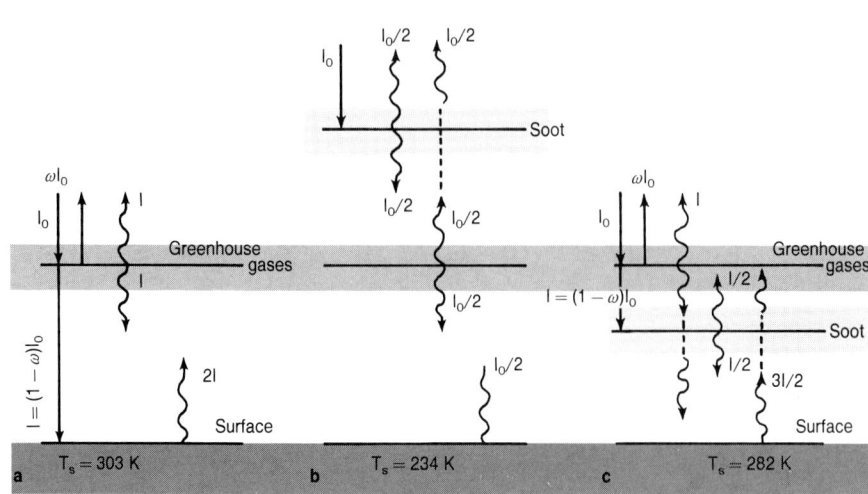

of smoke originating from discrete sources is similar to that from a uniform distribution. The simulation with one-tenth as much smoke indicated very little impact: Only a few spots on the globe had temperature drops of a few degrees. The spatial resolution of this model is quite coarse (the grid size is 4°×5°) and it includes only two tropospheric layers. Many effects—especially the appropriate smoke removal rate—still warrant further exploration.

The Southern Hemisphere might conceivably be affected by nuclear winter. If so, noncombatants may express greater concern over the size of the superpowers' nuclear arsenals because the countries not directly involved in a nuclear exchange might now appear to have more to risk from a conflict between those two nations. In the first climate model that addressed this question, the dust and soot were constrained[23] within a fixed region of latitude and altitude. The model then tracked the circulation patterns that resulted from this perturbation. Normally the so-called Hadley cells at the equator keep atmospheric circulations in the Northern Hemisphere apart from those in the Southern Hemisphere. In a simulation of nuclear-winter conditions for April, the model indicated that perturbation would cause these separate circulations to merge, so that dust and soot might be carried into the Southern Hemisphere if the aerosols were allowed to move in the model. The same effect was not apparent for a January simulation. The recent Livermore model, which is more interactive than the above model, does indicate some transport of smoke

into the Southern Hemisphere in July. The results from two other recent models[12,13] also suggest that as it is lofted, some heated smoke is transported to the Southern Hemisphere. How this smoke might affect the Southern Hemisphere has not been determined.

References

1. National Research Council, "Long-Term Worldwide Effects of Multiple Nuclear Weapons Detonations," NAS (1975).

2. National Research Council, "The Effects on the Atmosphere of a Major Nuclear Exchange," NAS (1985).

3. R. B. Stothers, Science **224**, 1191 (1984).

4. H. Stommel, E. Stommel, Sci. Am., June 1979, p. 176.

5. L. W. Alvarez, W. Alvarez, F. Asaro, H. W. Michel, Science **208**, 1095 (1980).

6. H. Wexler, Weatherwise, December 1950, p. 129.

7. R. U. Ayres, *Environmental Effects of Nuclear Weapons*, Vol.2, Hudson Inst., (1965) report number HI 518; E. S. Batten, "The Effects of Nuclear War on the Atmosphere and Climate," Rand Corp. Study RM-4989-TAB (1966).

8. P. J. Crutzen, J.W. Birks, Ambio **11**, 114 (1982); reprinted in *The Aftermath*, Pantheon, New York (1983).

9. R. P. Turco, O. B. Toon, T. P. Ackerman, J. B. Pollack, C. Sagan, Science **222**, 1283 (1983).

10. *Concise Report on The World Population Situation in 1979*, UN Dept. Int. Economic and Social Affairs, Population Studies, ST/ESA/SER.A/72 (1980).

11. M. C. McCracken, J. J. Walton, "The effects of interactive transport and scavenging of smoke on the calculated temperature change resulting from large amounts of smoke," Lawrence Livermore preprint UCRL-91446, December

1984.

12. R. C. Malone, L. H. Auer, G. A. Glatzmaier, M. C. Wood, O. B. Toon, "The influence of solar heating and precipitation scavenging on the lifetime of smoke from a nuclear war," to be published in Science.

13. S. L. Thompson, "Global interactive transport simulations of nuclear war smoke," to be published in Nature.

14. R. D. Cess, G. L. Potter, S. J. Ghan, W. L. Gates, "The climatic effects of large injections of atmospheric smoke and dust: A study of climate feedback mechanisms with one- and three-dimensional climate models," Lawrence Livermore preprint UCRL-92504, April 1985.

15. B. G. Levi, T. Rothman, *Nuclear Winter: A Review of the Key Factors*, CEES 197, Princeton Univ., in preparation.

16. R. D. Small, B. W. Bush, Science, **229**, 465 (1985).

17. "An analysis of civil defense in nuclear war," ACDA, December 1978.

18. C. G. Culver, "Survey results for fire loads and live loads in office buildings," NBS report NBS-BSS-85, PB-253 226, May 1976; *Attack Environment Manual*, CPG 2-1A3, FEMA (1982); D. A. Larson, R. D. Small, "Analysis of the large scale urban fire environment part II: Parametric analysis and model city simulations," Pacific Sierra Research Corp. report 1210, November 1982.

19. V. Ramaswamy, J. T. Kiehl, J. Geophys. Res. **90**, July 1985.

20. P. J. Crutzen, I. E. Galbally, C. Brühl, Climatic Change **6**, 323 (1984).

21. H. R. Pruppacher, J. D. Klett, *Microphysics of Clouds and Precipitation*, Reidel, Dordrecht, (1978).

22. G. F. Carrier, F. E. Fendell, P. S. Feldman, *Firestorms*, Defense Nuclear Agency, DNA-TR-81-102 (1982).

23. C. Covey, S. H. Schneider, S. L. Thompson, Nature **308**, 21 (1984).

Doe shuts N-reactor for safety repairs but fears persist

PHYSICS TODAY/FEBRUARY 1987

Halfway around the world from Chernobyl, the Soviet nuclear reactor that blew up last 26 April has claimed its most famous victim—the N-Reactor at the Hanford nuclear reservation, located on an arid and (relatively) isolated site of 570 square miles along the Columbia River near Richland, Washington. A graphite-moderated plant structured nearly like the Chernobyl unit 4 power plant, the Hanford reactor produced about 20% of the plutonium that winds up in US nuclear arms. One of the ironies of the similarity is that Chernobyl's explosion has forced the Department of Energy to shut down for the next six months the principal facility producing plutonium for the weapons the US points at the USSR.

The dependence on the N-Reactor as the nation's largest producer of weapons-grade plutonium is reason enough why the Department of Energy's announcement on 12 December took many people by surprise. Possibly just as surprising is DOE's decision to spend $50 million on extensive improvements of the safety systems and operating procedures but not to shut the reactor permanently. Still, with the memory of Three Mile Island unit 2 and Chernobyl unit 4 so clear in the public mind, the department's under secretary, Joseph F. Salgado, said the N-Reactor would be closed for good by the mid-1990s.

Even after the overhaul, say members of the independent panel that reviewed the N-Reactor, it is likely to remain vulnerable to a runaway critical excursion, a steam surge and a potentially catastrophic hydrogen explosion, perhaps worse than the sequence of events that led to the Chernobyl disaster (PHYSICS TODAY, December, page 17). Two out of the six reviewers urged DOE to close the N-Reactor immediately and permanently unless to do so would threaten national security. Worries about the reactor's safety come only weeks after DOE reversed the position it had steadfastly held throughout the Chernobyl crisis—that the department's 11 reactors, located in five states, are entirely safe to operate

N-Reactor site at Hanford, Washington, produces much of the plutonium used in US nuclear weapons. Its structure is much like that of the Chernobyl reactor.

as they stand.

Comparisons. Built 23 years ago at the Hanford Engineering Works, the N-Reactor was designed in the 1950s to have a life expectancy of 20 years. Like the RBMK-1000 reactors in the Soviet Union, including the four at Chernobyl, the Hanford plant is not only graphite moderated and water cooled, it also lacks a concrete containment dome capable of withstanding steam and hydrogen explosions that might occur in a runaway accident. Unlike the RBMK design, the N-Reactor has 365 metric tons of uranium metal in its core along with 1800 tons of graphite (Chernobyl unit 4 had 2000 tons). Besides plutonium, the Hanford plant produces 4000 MW of electricity, by far the highest power in the US.

After conducting three of its own audits of the Hanford reactor and finding that nothing drastic had to be done to restore it to relative fitness, considering its age, DOE appointed six nuclear plant experts to perform separate examinations. The consultants agreed that it is either "impossible" or

"not plausible" that a Chernobyl-type catastrophe could occur, given the design differences and additional operational safeguards of the N-Reactor. Even so, they each concluded that radioactivity might be released following a minor accident because of uncertainties about the reactor's safety system, which is designed to lower the pressure of steam or hydrogen when cold water is sprayed on the core.

The N-Reactor has never been tested under conditions involving rupture of many of its 1000 metal process tubes and accumulation of large amounts of hydrogen. It is like no other reactor in the world. Thus, experience in running it grows at the rate of one reactor-year per year. The reactor has no hydrogen control system or hydrogen monitors around the graphite stack, though such instrumentation has been recommended in the past.

Reviewers. The strongest statements about the reactor came from the chairman of the review group, Louis H. Roddis Jr, a nuclear engineer who was once president of Consolidated Edison of New York. The other reviewers, performing independently to avoid the public meetings required of an advisory committee, were Miles C. Leverett, a consultant with 25 years of experience with the N-Reactor; Harold Lewis, a physicist at the University of California at Santa Barbara with a long record of advising the government about nuclear safety; Thomas A. Pigford, chairman of the nuclear engineering department at the University of California at Berkeley, who has experience in nuclear plant studies for the National Research Council; Gerald F. Tape, a physicist who was once deputy director of Brookhaven National Laboratory, served on the Atomic Energy Commission and headed Associated Universities Inc, which functions as a board of directors for Fermilab; and retired Admiral Eugene P. Wilkinson, former president of the industry's Institute for Nuclear Power Operations.

Roddis begins his report by quoting the last independent report on the N-Reactor, completed in 1966 by the Advisory Committee on Reactor Safeguards. In the unlikely event of severe accident, the committee said, the N-Reactor would release more radioactivity than a commercial power reactor. Writing only three years after the plant's start-up, ACRS warned that operating the N-Reactor was riskier than running a civilian reactor and that it was justified only by military requirements for plutonium.

Two decades later Roddis recommends that DOE "shut down the N-Reactor unless a positive judgment is made that the requirements for defense material warrant accepting public hazards exceeding those of commercial reactors." Lewis agrees. Prudent policy, he asserts, requires DOE to close the plant in "the very near future . . . concomitantly forcing a decision on a new production facility." The other reviewers call for remedial work but do not believe it is dangerous to continue running the reactor for another three to five years.

In his review, Lewis attempts to answer those who defend the N-Reactor because it operated for 23 years without mishap and is likely to run safely for many more years. This was the argument used to defeat an amendment put before the House of Representatives last July by James Weaver, an Oregon Democrat, who sought to shut down the plant until the safety reviews were completed. Some members of Congress praised the reactor's redundant safety systems and reinforced concrete structure, which would minimize the risk of a radioactive release. DOE and the White House opposed the amendment on the grounds that the reactor had a superb safety record and that its plutonium output was necessary to meet current goals for nuclear weapons.

Lewis, however, regards what many see as virtues to be defects. He notes that the statistical record of the N-Reactor is specious and irrelevant. "The probability of a major accident can simply not be inferred from such short operating experience. TMI-2 happened after several hundred reactor-years of US commercial experience and Chernobyl after thousands of reactor-years of worldwide experience." He also recalls, somewhat sadly, that NASA boasted that 24 successful shuttle flights could justify another mission no matter what objections were raised (PHYSICS TODAY, August, page 41). In addition, because the reactor is unique, "it benefits only in part from operating experience, and I found little in the way of systematic effort to derive even those benefits." The problem, according to Lewis, is that management of the reactor "resembles a family operation" within DOE, with "no external peer pressure encouraging excellence of the entire structure," unlike what prevails in the commercial domain of reactors. All six outside reviews agreed that the facility's management was somewhat lax and that workers were unmindful of safety rules.

Complacency. Some problems, say Hanford's critics, involve what they call "widespread complacency" at the plant, which led to no less than 2800 pounds of plutonium unaccounted for more than a decade ago. Hanford's antagonists say it is already the world's biggest radioactive waste dump, and DOE would like to make it the permanent graveyard for 77 000 tons of waste with a halflife longer than human history on the planet.

According to a report issued last August by Congress's General Accounting Office, the N-Reactor's continued operation beyond the 1990s would require spending at least $1.2 billion for changes and repairs. One trouble is that equipment used to sample water in the plant for levels of radioactivity is inoperable, so that manual sampling of high-pressure and high-temperature steam is required, causing delays in obtaining readings. Another involves the primary coolant pumps, designed to function five years between major overhauls. These are so old that the work must be done every two years. What's more, electrical wiring often fails, causing reactor outages, and motors for running various valves often burn out but such motors are no longer made.

Operators of the plant say the N-Reactor encounters 20 to 25 "trips," or unplanned outages, each year. Commercial reactors average about seven trips per year, and the Nuclear Regulatory Commission wants to lower the number as a way of reducing wear and tear on operating parts and safety equipment.

Even routine operation is becoming more difficult. The N-Reactor building was designed to withstand only 5 psi above atmospheric pressure, compared with about 50 psi for a modern US nuclear plant and 26 at Chernobyl. Examinations reveal that neutrons are loosening the structure of the graphite blocks at the heart of the reactor with the result that the pile of blocks has been expanding vertically and is almost certain to "hit the roof" of shielding by 1990 or 1991, Roddis asserts, causing a shutdown. In addition, neutrons from the fission reaction bombard the metal tubes that isolate the fuel rods and high-pressure hot water from the graphite, causing them to become so brittle they are prone to rupture. If that were to happen, steam would strike the hot graphite, resulting in a catastrophic accident.

After Salgado announced DOE's plan to repair the N-Reactor, attorneys for the Natural Resources Defense Council asked for a delay in all work until a full environmental impact statement is completed. If the government is unwilling to do this, NRDC threatens to take the issue to court. Meanwhile, Washington's Governor Booth Gardner, a frequent critic of the Hanford nuclear facility, commended DOE for its action, while observing that the various reviews "read like a script from a disaster movie."

—IRWIN GOODWIN

DOE WANTS NEW WEAPONS REACTORS TO REPLACE AGING, TROUBLED ONES

PHYSICS TODAY/SEPTEMBER 1988

The political fallout from the explosion and meltdown at Chernobyl's Unit 4 in April 1986 led to reappraisals of reactor safety around the world—nowhere with such swift effects as on the materials production program for US nuclear weapons. Chernobyl led the Secretary of Energy, John S. Herrington, to order several independent studies of reactors producing plutonium and tritium. Before the end of 1986, one panel called for major modifications or a complete shutdown of the large N-Reactor at the Hanford Nuclear Reservation near Richland, Washington, which is like Chernobyl's four RBMK-1000 plants in having a graphite core. A few months later, the department turned off the N-Reactor to make mechanical changes (PHYSICS TODAY, February 1987, page 63). Since then, DOE has decided not to restart the reactor.

Other studies of DOE reactors, conducted by the National Research Council and by the department's independent Advisory Committee for Nuclear Facility Safety, found some either rapidly nearing or already past the ends of their expected lifetimes. This was especially true for defense production reactors at Savannah River, not far from Aiken, South Carolina, where deteriorating components and design changes since the reactors were turned on in the 1950s have put increased demands on control-room staffs and their supervisors. Indeed, for the past seven years the Government Accounting Office, Congress's watchdog over executive agencies, has criticized operating practices at Savannah River and designated the reactors as "high-hazard facilities."

So it came as no surprise when Herrington called in the news media on 3 August to announce that DOE proposed to replace the three tritium production plants at Savannah River with a single reactor that is similar in concept to those now there but would provide 100 percent of current military requirements. Experts have warned that the P, K and L reactors, all more than 30 years old, may not last the ten years it will take to design and build the successor, which would

Savannah River facilities include three tritium production reactors and processing plants at which the isotope is separated from irradiated lithium-aluminum targets.

be capable of yielding plutonium as well as tritium.

Herrington's statement also contained a surprise: A smaller reactor, based on a novel high-temperature gas-cooled technology, would be constructed at the Idaho National Engineering Laboratory near Idaho Falls. This reactor would also take ten years to complete and would yield another 50 percent of the tritium the Pentagon currently uses.

The reason the Energy Department wants the capability of producing so much tritium is that this isotope, the basic fuel of thermonuclear bombs and useful in upgrading the power of fission warheads, must be replenished periodically. Tritium degrades at a rate of 5.5 percent each year. Plutonium, by contrast, is a relatively stable material, with a halflife of about 23 000 years.

"As long as this nation relies on the nuclear deterrent," explained Herrington, "we must have the capability for a steady, reliable supply of tritium and plutonium." The need for two production plants at widely dispersed locations, he said, will "minimize the

technical risks to national security." This "two-reactor strategy," he declared, "involves proceeding on an urgent schedule" to make the US less vulnerable to operating interruptions. DOE estimates the price tag for the two reactors would be $6.8 billion if built today. Critics say the final cost is likely to be more.

The reactor for Savannah River would be based on proven technology for making weapons-grade material, using heavy water or deuterium. The Idaho reactor would be based on technology developed in the US and the Federal Republic of Germany. The concept originated in Britain in 1956 with the Magnox-class power reactors, which use carbon dioxide as a coolant and natural uranium as fuel. In the US, GA Technologies of San Diego pioneered the development of a reactor using helium as the coolant and graphite as the moderator. In 1967, the 40-MW Peach Bottom No. 1 plant, designed by GA Technologies (then known as General Atomics) went on line in the Philadelphia Electric Co system as part of the Atomic Energy Commission's power reactor demonstration program. The reactor was shut down in 1984 when scheduled tests were completed. The Fort St. Vrain plant was another experimental reactor built by GA Technologies for Colorado's Public Service Co. It went critical in 1974 and continues to operate. But as a commercial demonstration it was not a good advertisement for the concept. It produced only about 10 percent of the power it was designed to provide if it could have run at full capacity. Although GA Technologies claims the plant showed the benefits of helium instead of ordinary water as a coolant, frequent breakdowns of the helium circulator raise questions about its reliability.

High-temperature gas-cooled reactors have been developed in West Germany for a 15-MW power plant that has been operating since 1968 and a 300-MW plant that is set to start up soon.

The use of graphite is uncommon among nearly all reactors. Graphite was used in Hanford's N-reactor, which produced plutonium, and in Chernobyl's RBMK-1000 types. The GA design departs markedly from those reactors. A major difference from the Chernobyl reactors is that as the core temperature rises in the high-temperature, gas-cooled reactor, the nuclear reaction is choked off. At Chernobyl the opposite was true.

Unfortunately, six days after Herrington's announcement, an incident at Savannah River called into question the safety of the aging weapons-material reactors. A DOE source called the episode a "complete collapse" of safety procedures that, in worst circumstances, could have resulted in a Chernobyl-type calamity.

The history of the problem has its origin in power cutbacks of the Savannah River defense reactors to 45 percent capacity because of safety concerns. According to DOE officials, the P reactor, which had been shut down since early April for safety modifications and routine maintenance, was being restarted on 7 August when operators found that the position of its control rods apparently prevented a sustained reaction. When they attempted to restart the reactor on 9 August, its temperature and pressure surged unexpectedly in what is termed a "power spike."

Instead of trying to control the surge, however, operators not only continued to run the reactor but did exactly the wrong thing: They turned up the power. It seems that when the operators had trouble getting the reactor to sustain a chain reaction, they did not do the customary thing,

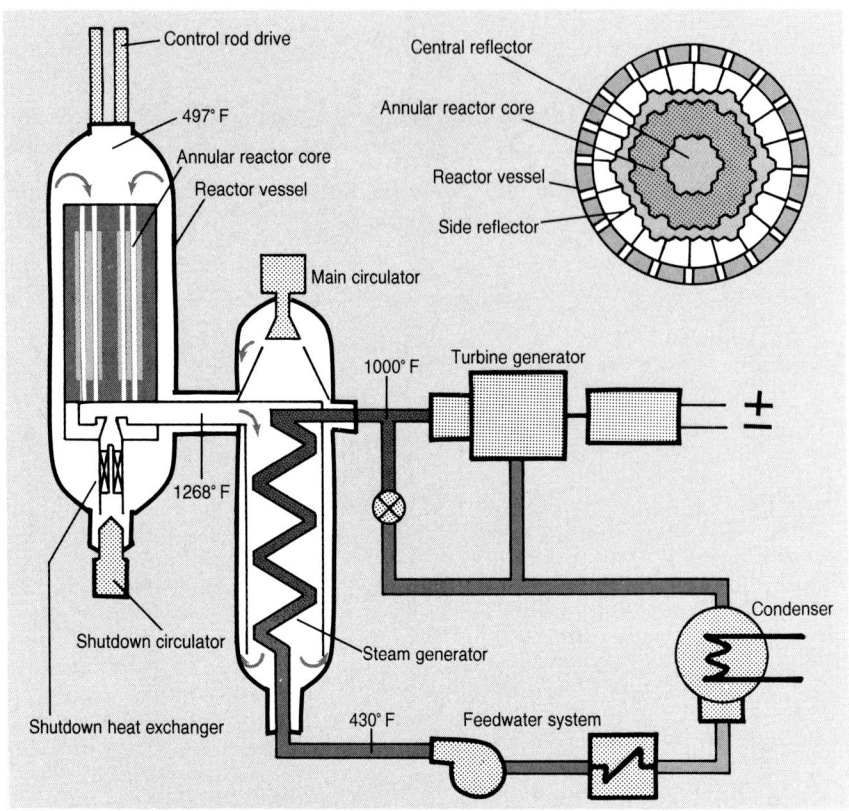

High-temperature, gas-cooled reactor, shown schematically, is being developed to satisfy worries about nuclear safety. Based on technology demonstrated in the US and West Germany, the system used refractory coated nuclear fuel, helium gas as an inert, noncorrosive coolant and graphite as a core moderator that remains stable at extremely high temperatures. The Department of Energy has selected the HTGR to produce tritium and plutonium at the Idaho reactor test station. DOE proposes to build a more conventional reactor at Savannah River to replace three old tritium production reactors.

pushing in the control rods to suppress the reaction, but instead pulled the rods further out. If the reactor had been running at higher power, a tragedy might have resulted, said an official at Du Pont, which operates the Savannah River complex.

John F. Ahearne, the former chairman of the Nuclear Regulatory Commission, who heads a special DOE safety advisory committee named by Herrington last year (see page 38 for an article by Ahearne) expressed anger that the problems at the P reactor were not reported instantly. In an electronic mail message to plant managers, Ahearne admonished operators for not informing him of the problem. When Ahearne was asked what he would do in the circumstance, he recommended an immediate shutdown. That was done.

DOE's assistant secretary for environment, safety and health, Ernest C. Baynard III, said that at one point the operators increased the reactor's power to 60 percent, one-third higher than the 45-percent maximum approved by the board. The operators

detected more decay products—primarily helium-3, which acts to absorb neutrons and suppress the reaction. The presence of helium-3 made the reactor more difficult to start. Puzzled by this, the operators pulled more control rods in an effort to boost the reaction. Each time they pulled the rods, the reactor surged briefly and then subsided. "You can't have people operating a nuclear reactor acting as if it is business as usual when something unusual occurs," said Baynard.

After interviewing the operators, DOE issued a report explaining that the reactor did not "exhibit uncontrollable behavior." The surge had never been more than 1 percent of the authorized power level. But the DOE criticized the operators for neglecting the "checks and balances that would prevent a recurrence of the events." Still, members of Congress have expressed concern that this worrisome episode happened at the very time that DOE is seeking their support for the new reactors.

—Irwin Goodwin

PERILS OF AGING US WEAPONS PLANTS STIR OUTRAGE AND FEAR OF A 'TIME BOMB'

PHYSICS TODAY/NOVEMBER 1988

The irony is inescapable: It is open season on US defense materials plants, which have been virtually off–limits to public scrutiny since they began producing isotopes for nuclear weapons in strict secrecy in the early 1950s. Suddenly the old and new problems at the reactors making bomb-grade chemicals and the factories processing the stuff are being exposed in Congressional hearings, front-page stories and television news shows. To many in Washington and across the country, the revelations of technical mishaps and radioactive spills almost defy belief.

The scope and seriousness of the troubles, which, it now appears, have plagued the plants virtually from their start, caused members of Congress who saw the reports and heard the testimony of investigators in late September to warn of disasters waiting to happen. If the defense reactors had been commercial power plants, said Senator John Glenn, an Ohio Democrat, they would have been closed years ago. Senator Ernest F. Hollings, a Democrat from South Carolina, where the Savannah River plutonium and tritium production complex is located, demanded an independent inquiry into all weapons materials facilities. "The plants must be run in ways that protect both the nation's defense and the workers' safety," said Hollings.

The intense scrutiny now being given the nation's weapons production plants began soon after the disastrous fire half a world away in a reactor at the Soviet Union's Chernobyl complex in 1986. To his credit, Energy Secretary John S. Herrington responded promptly to Chernobyl. He asked the National Research Council to study DOE's defense materials reactors and mobilized the department's own safety experts, known as "junkyard dogs," to examine the problems of all 14 nuclear installations. Herrington's most immediate worry was the N-Reactor at the Hanford Reservation, located near Richland, Washington. Built in the early

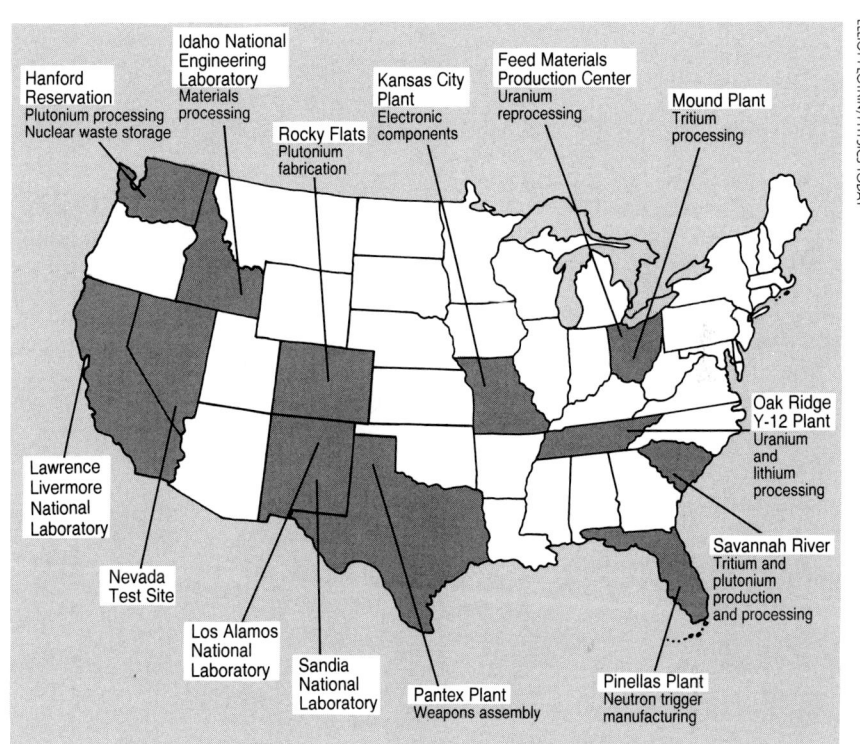

Major facilities of DOE's nuclear weapons complex.

1960s with a life expectancy of 20 years, the N-Reactor was most nearly like the RBMK-1000s at Chernobyl. All are graphite-moderated, water-cooled and without a concrete containment dome capable of withstanding steam and hydrogen explosions that might occur in a runaway accident.

Although a DOE team had looked at the N-Reactor and decided nothing drastic had to be done, Herrington appointed six nuclear plant specialists to perform separate reappraisals. On the basis of the team's reports citing faulty equipment, "widespread complacency" about safety among workers and managers, and practices that left Hanford the world's biggest waste dump, he ordered the reactor shut down (PHYSICS TODAY, February 1987, page 63.)

The most recent incident occurred on 29 September at the Rocky Flats plutonium-processing center near Golden, Colorado. A DOE inspector and two employees of Rockwell International Corp, which manages the installation, received small doses of plutonium radiation when they walked into an unmarked room of Building 771 where workers were cleaning contaminated equipment. The building, which is central to reprocessing plutonium, was closed on 8 October. Three days later DOE was under assault for numerous instances of accidental leaks and deliberate dumps of uranium waste at a processing plant outside Fernald, Ohio. The state's governor, Richard Celeste, demanded that the facility be permanently closed. "If terrorists had buried all that uranium there, there would be strong action taken immediately," he declared. "In this case, it was our own government that left the time bomb."

Over the past 33 years, though, the most perturbing problems took place at the five reactors of the Savannah River complex, near Aiken, South Carolina, operated from the start by E. I. du Pont de Nemours Inc. In April the P-Reactor was shut down for improvements to its safety systems. When technicians attempted to restart it in August, they encountered a power lag that prevented a sustained reaction. When they tried again two days later, the reactor suddenly experienced an unexpected neutron surge. The operators stopped it within minutes by reinserting the control rods. DOE safety teams rushed to the Savannah River facility to investigate the trouble and, to their horror, discovered that the reactor operators had neither understood the situation nor followed the proper procedures. The inspectors recommended that the P-Reactor be shut down until DOE was satisfied that the operators could run it safely (see PHYSICS TODAY, September, page 47).

'Prelude to disaster'

Meanwhile, accounts of the incident began appearing in newspapers in South Carolina and in Washington, DC. Nervous about the whole episode, DOE shut the reactor down on 17 August and initiated a full safety review of the plant. A memo written by Richard Starostechi, whom Herrington had brought in from the Nuclear Regulatory Agency to be DOE deputy assistant secretary for safety, health and quality assurance, criticized the attitude of the plant operators as being "a prelude to disaster."

The P-, K- and L-reactors at Savannah River are currently the sole suppliers of plutonium and tritium for the nation's nuclear arsenal. Because plutonium has a halflife of about 24 400 years, there are abundant quantities for US nuclear weapons. In fact, Herrington told a Congressional committee last February, using an apt if somewhat disconcerting metaphor, the nation is "awash in plutonium." The real cause for concern is tritium, a hydrogen isotope that boosts the explosive power of nearly all 22 000 US nuclear warheads. Because it has a relatively short halflife of 12.3 years, tritium is in constant demand for topping up warheads. Pentagon sources say that a critical shortage of tritium could compromise national security by next summer and that without production the present stockpile would need to be cannibalized.

While incidents like the one at the P-Reactor have been alarmingly frequent for decades, they have previously been shrouded in secrecy. Indeed,

John S. Herrington: Candor for bureaucrats.

since the Savannah River facility went on line in 1953, top officials in the weapons agencies were routinely kept in the dark about its troubles. Apparently only local and regional managers were informed. In recent months, however, investigators at the Government Accounting Office made public dozens of memorandums and reports that have shattered the silence.

On 30 September, at a joint hearing of the Senate Governmental Affairs Committee and the House Government Operations Subcommittee, Senator Glenn voiced the worries of many members of Congress, saying that "there seems to be no end to the problems uncovered." Congress decided in the early 1950s that the Atomic Energy Commission should keep civilian nuclear plants scrupulously separate from military ones. As a consequence, DOE answers only to itself for safety at its nuclear operations.

Critics have long assailed DOE's lack of external oversight as a formula for failure. Some have urged that all reactors and processing facilities be routinely inspected by the Nuclear Regulatory Commission, which has rigorous procedures and trained staff to do the job. After Glenn and Representative Mike Synar, Democrat of Oklahoma, released memorandums and reports documenting the failures at Savannah River and other facilities, Herrington was forced to concede DOE's haphazard oversight. He now admits that safety was frequently overlooked in pursuit of maintaining production schedules for weapons. "Things got too cozy" between DOE managers and plant operators, says Herrington. In the event, DOE made public some unsettling accounts of years of chaotic and complacent oper-

Richard E. Heckert: "Bum rap" for du Pont.

ations.

Among the most damaging evidence is a 1985 memorandum, by du Pont plant supervisor G. C. Ridgely, listing 30 "reactor incidents of greatest significance" at Savannah River between 1957 and 1985. The specific problems included an accident in 1970 in which two fuel rods were inadvertently allowed to melt, resulting in radioactive contamination inside the reactor core; in 1960, as a reactor was being restarted, operators allowed it to run wild, causing a volatile power burst more than 12 times faster than what is considered safe. Many of the reactor incidents were attributed by Ridgely to "gross procedural violations."

Another memo, prepared by du Pont engineer Frederick Christensen when he retired in 1981, stated that an incident in 1965 could have turned into a catastrophe when a foreman attempted to stop a coolant leak by closing off the flow of water to the reactor. That act was prevented by a senior supervisor, who realized that a steam build-up might follow and possibly cause an explosion. Christensen wrote at the time, "One trained man stood between us and disaster."

Synar released a report by the NUS Corp, dated May 1988, reviewing the operating history of the Savannah River complex through 1987. The NUS document cites dozens of radioactive spills and worker exposures. Another report, written by an official at DOE's Idaho Operations Office, discloses as many as 43 unplanned reactor shutdowns per year at Savannah River—far greater than the number of similar shutdowns at commercial power plants.

DOE officials now admit that the department and its predecessor agencies, AEC and the Energy Research and Development Administration, withheld reports about serious accidents from the public. Glenn Seaborg, a former AEC chairman, does not recall being informed of many of the incidents now coming to light. Robert C. Seamans Jr, the onetime head of ERDA, speculates that local managers and agency officials did not pass on accounts of accidents for several reasons: They might worry administrators unnecessarily, draw blame to plant operators, panic local citizens if the incidents were announced publicly, and suggest that the system was incapable of producing adequate quantities of plutonium and tritium for the nation's nuclear arms. Equally important, an obsession with secrecy in defense matters, a legacy of World War II and the cold war, has been the enemy of free-flowing information even within defense-related agencies. Whatever the reason or reasons, by not passing on reports of incidents at the plants, operators and managers were covering their critical assets.

'A bum rap'

For its part, du Pont, which has been involved in nuclear weapons work virtually since Fermi's first chain reaction in 1942, claimed it had dutifully notified regional Federal offices of the many mishaps at Savannah River. The company's chairman, Richard E. Heckert, held a press conference in Washington on 11 October to defend its record and employees. "Things are fine down there if the government will let us go on with our business," he declared. "It's a bum rap." Du Pont decided last April to give up running Savannah River, which it had operated from the beginning. Westinghouse will take over next April.

Though the troubles at the nuclear defense facilities stunned Congress and the public, most of the incidents in fact should have been familiar. A National Research Council study headed by Richard A. Meserve, who has a PhD in physics from Stanford and a JD from the Harvard Law School, found that safety was being compromised at the plants for decades. The Meserve report, issued in October 1987, chastized DOE on three main counts: failure to set clear safety guidelines; skimping on technical and hardware upgrades that would improve safety and compensate for aging; and neglecting to manage and review the operations of its contractors, with the result that "safety oversight of the production reactors is ingrown and largely outside the scrutiny of the public." The Meserve report also was critical of the backlog of approximately 200 unresolved reactor incidents—mishaps whose causes remained unidentified and whose solutions were unknown.

A new safety oversight board

Anticipating the Meserve report, DOE closed the Hanford N-Reactor and ordered power levels reduced to 45% of full capacity at Savannah River to reduce stress on the aging reactors. Herrington appointed John Ahearne, formerly chairman of the Nuclear Regulatory Commission and now at Resources for the Future, to head an independent safety oversight board within DOE.

Although DOE and its predecessor agencies had claimed most of the defense reactors could be operated indefinitely, it is now obvious that they either have exceeded their designed lifetimes of 20 years or so or are certainly in need of major overhauls. Cracks in the reactor vessel forced the shutdown of the Savannah River C-Reactor in late 1986. The Meserve report noted that "all of the Savannah River reactors may eventually have to be retired from service due to stress corrosion cracking."

A recent report by the Government Accounting Office warns of deterior- ating defense production plants, and it singles out Savannah River as being "less than marginal." According to GAO, operating the defense production plants safely would take between $15 billion and $25 billion. But to make improvements at all the facilities, install modern waste disposal equipment and clean up the environment in and around all the plants would run to at least $100 billion, according to GAO estimates.

Herrington has already announced plans for the construction of two new production plants, which would cost $6.8 billion in 1988 dollars and, if construction starts in 1990, would be on line in the year 2000.

A cheaper alternative

Meanwhile, DOE is considering a cheaper alternative that may be ready much sooner: the WNP-1, a conventional light-water reactor at Hanford that was 63% completed before financial problems and uncertainties about future electricity demand forced the owner, Washington Public Power Supply System, to stop construction. GAO has calculated that WNP-1 could be converted into a tritium-producing facility for $2.6 billion plus an undetermined amount that DOE would pay WPPSS for the reactor. The conversion would take about six years.

Congress has taken action in the 1989 Defense Appropriations Act to make the defense reactors safer. One of the provisions of the act amends the Atomic Energy Act of 1954, in order to create the Defense Nuclear Facilities Safety Oversight Board, an independent organization of the sort that DOE has lacked all these years. The new board would operate like the Nuclear Regulatory Commission, monitoring defense reactors just as the NRC now watches commercial ones.

—IRWIN GOODWIN, WITH REPORTING BY COREY S. POWELL

Civil defense in limited war—

Have recent developments in strategic weapons
given us reason to look at civil defense in a new context?

PHYSICS TODAY/APRIL 1976

Civil defense, once a hotly debated issue of
the 1960's, has again surfaced as a topic of
controversy.   It reappears amid the
discussions of possible new strategies being
proposed by the Defense Department.   In
January 1974, the then Secretary of Defense
James R. Schlesinger announced the
intention of the US to develop long-range
ballistic missiles of unprecedented accuracy.
Because such weapons would have a
relatively small error radius their yield would
not have to be as large to be effective against
military targets such as land-based offensive
missiles.   Hence the Defense Department
has raised the possibility of a limited nuclear
war with counterforce strikes (that is, against
the opponent's offensive force) coupled with
a program of civil defense to ensure a
minimal level of civilian casualties.

   We present here two different viewpoints
regarding civil defense in this context.   Arthur
Broyles and Eugene Wigner will argue that
civil defense can be effective as a defense
against a nuclear attack.   Sidney Drell will
argue that the price of civil defense is too
high in relation to the degree of
protection it buys.

a debate

In favor:

Arthur A. Broyles and Eugene P. Wigner

Should the American people be protected from the effects of nuclear war? Let us first narrow that intensely studied question[1] to one that lies within the realm of physics to answer—namely, can such protection be effective? Evaluations of various evacuation and shelter systems show that they can greatly reduce the number of casualties in a nuclear encounter. Our response thus agrees entirely with the statement by V. Chuykov in the *Civil Defense Handbook* of the USSR: "Although the discussed means of destruction are called mass means, with knowledge and skillful use of modern protective measures, they will not destroy masses of people, but only those who neglect the study, mastery and use of these measures."[2]

The question then broadens into one with psychological and political aspects and cannot be answered precisely or completely. Nevertheless we feel that our nation's civil-defense preparations may determine the balance of power in some future nuclear crisis. Civil defense is more important than ever at a time when other nations have extensive civil-defense plans and when the balance of terror that has reigned to date is being upset by the development of new types of weapons.

The protective measures against nuclear explosions and their effectiveness can be evaluated on the basis of a wealth of data gathered by the Atomic Energy Commission in its nuclear testing program. Besides making quantitative measurements of such phenomena as blast-wave pressures, fallout intensity patterns and heat-ray intensities, the AEC constructed buildings and other structures in the vicinity of nuclear explosions and observed the resulting damage.[3] This information has been used by the AEC (now ERDA) laboratories, Stanford Research Institute, RAND Corporation, the Hudson Institute, the National Research Council and other institutions to devise and determine the effectiveness of methods for protecting people. Their results are in surprisingly close agreement.

Unfortunately the general public is not well informed about such studies, probably because a large fraction of the physics community as a whole is not aware of them. And yet so much physics is involved that physicists bear a responsibility to understand it themselves and to pass on the information through the classroom and other contacts. A clear presentation of the facts is essential because it is possible, as we shall see, that a nation's civil-defense preparedness may determine the balance of power in some future nuclear crisis.

continued on next page

Opposed:

Sidney D. Drell

The strategic doctrine of "limited nuclear counterforce strikes" has been revived in the United States during the past few years. This return to a policy that was discarded more than a decade ago is accompanied by a renewed interest in extensive and organized civil-defense programs, which would require massive relocation and evacuation of populations during crises. Official government statements during the past two years allege that this combination offers the prospect of low levels of fatalities and casualties resulting from the immediate blast, thermal, radiation and subsequent radioactive fallout effects. In particular the former Secretary of Defense, James R. Schlesinger, in the Annual Defense Department Report for FY 1976 stated that "Relocation of the population from high risk areas near key military installations and the protection of the rest of the population against fallout could reduce nationwide fatalities due to fallout from a limited Soviet counterforce attack to relatively low levels well under 1 million—provided that the people in the communities that would be most exposed by fallout from such an attack make effective use of the shelters available."

The conclusion drawn from these claims and analyses is that limited nuclear war may be palatable and need not escalate to the level of an all-out nuclear exchange, which would cause unimaginable horror. In fact, on 11 September 1974, Secretary Schlesinger testified[8] to a subcommittee of the US Senate Committee on Foreign Relations that "the likelihood of limited nuclear attacks cannot be challenged on the assumption that massive civilian fatalities and injuries would result."

Because the basis for this change in strategic doctrine is the relatively low fatality level, we must examine not only the total civil-defense implications of this doctrine but also the assumptions about the nature and effectiveness of the weapons used in the attack.

Civil defense in the larger context of an all-out nuclear strike against population centers will not concern us here, not only because it is not being proposed at present but also because most who have studied the financial and societal costs, not to mention the technical challenges, of such a program have concluded that it is not practical. But how practical and how effective is civil defense in a limited counterforce context?

The resurgence of the doctrine of limited nuclear counterforce has been spurred by progress in weapons technology—in particular, the development of accurate and reliable

continued on page 39

The principal sources of danger and the most effective measures against them are listed in the table on this page. (Of course a far more convincing display of the data requires something like the elaborate descriptions in the USSR handbook.) Because of the short time available for action to protect against effects of nuclear weapons, survival depends very heavily on previous planning and preparation. The effectiveness of all the protective measures would be much increased if the population were familiar with them well before the attack. The stockpiling of relatively simple tools can also help in the long-term recovery effort. Because this subject is complicated and requires extensive considerations, we shall limit our discussion to the problems of survival of the initial effects of the attack that are listed in the table.

The most obvious way of protecting against all these effects is to prevent the bombs from exploding. For example, the US might attack the enemy launch site before the missile leaves it. Such an attack is the purpose of the "smart bombs" bemoaned by Bernard T. Feld in the July 1975 issue of PHYSICS TODAY. Or, the US might destroy the incoming missile with its own missile— the Anti-Ballistic Missile. Despite extensive debate over the ABM, it cannot be generally implemented now. As a result of the SALT I treaty, the ABM is restricted, as far as nonmilitary defense is concerned, to Moscow (with a population of 4.5 million) and Washington, D.C. (population of 1.5 million). Nevertheless, even a small ABM system could be very effective. By destroying the first wave of incoming missiles, it can give time to the people to enter shelters or to protect themselves in other, although less effective, ways.

Once a bomb does strike, the first effect is the electromagnetic pulse. This pulse threatens electric power transmission rather than human lives, although the disruption of radio transmission is of concern during an emergency.

The protection against the other effects of nuclear explosions can be pro-

vided in two ways—evacuation and shelter. Evacuation takes very much longer than the missile flight time and hence can not be considered to be a truly defensive measure. If evacuation is undertaken during a crisis, it will greatly aggravate the situation. It can be effected before provoking a showdown and serve as an aggressive move. Hence, since the advent of missiles, our country did not seriously propose it until the elaborate evacuation preparations of the USSR became known. Now it is being seriously planned as a "counterevacuation," that is, as a response to a possible evacuation of the cities of the USSR. The Ponast study, which was organized by the National Security Council,[4] considered a nuclear attack in which the USSR aimed two thirds of its destructive force at civilian targets. This attack would destroy 45% of the US population under present circumstances. The preparation for the "counterevacuation" would cost about $500 million—one day's welfare expenditure—and would reduce the popula-

H-Bomb major immediate effects			
EFFECT	**CAUSE**	**DAMAGE**	**DEFENSE**
ELECTROMAGNETIC PULSE	Expanding charged particles from bomb explosion	Damage to electronic equipment up to hundreds of miles; power stations at shorter ranges	Special protective equipment related to lightning security devices; no effects on humans
PROMPT NUCLEAR RADIATION	Nuclear reactions during bomb explosion	Normally less than blast	(Normally negligible compared to blast)
HEAT RADIATION	Radiation from the hot fireball generated by the explosion	Fires ignited a few tens of miles but greatly reduced by clouds or smog and dampness	Eliminating exposed inflammable material; shelters including large public buildings
BLAST WAVE	Expansion of hot bomb material pushes air into a wave of wind and high pressure	Destruction of buildings as well as serious injuries to people from flying objects and falling buildings from five to ten miles	Evacuation blast shelters; reinforced public buildings
FALLOUT	Radioactive products of nuclear fission mixed with vaporized earth	Heavily wind dependent; can be the order of one hundred miles	Sheltering by large public buildings or special shelters for a few days or weeks until the radiation level has died down

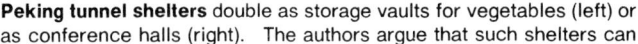

Peking tunnel shelters double as storage vaults for vegetables (left) or as conference halls (right). The authors argue that such shelters can be effective in reducing US casualties in the event of a nuclear attack to about 5.5% of the total population.   Figure 1

tion loss to 11%. Because the USSR population is less crowded into cities than ours, their losses would be smaller yet—less than 5% according to our calculations.[5] This loss is half of that experienced by the Soviets in World War II.

Shelter design

The defense measure advocated in the US, and installed by the Chinese, is the provision of shelters. The technical problem is to design a shelter with maximum blast resistance, minimum access time and minimum cost. The Chinese appear to have conquered the problem, as shown in figure 1. US scientists, during a 1970 study at the Oak Ridge Civil Defense Project,[6] estimated that effective shelters could be built at a cost of $23 billion. In similar conclusions four years later, the Ponast study found that a $35-billion investment—very much larger than that needed for preparation for counterevacuation and one tenth of one year's federal expenditures—would reduce the casualties caused by an attack by the USSR to 5.5%.[4] For this reason we can not possibly accept Feld's conclusion in PHYSICS TODAY that "there is no defense against nuclear weapons, now or in the forseeable future." Actually, as we have just described, the effectiveness of shelters should not be surprising: If shelters were ineffective, the expenditure on their construction by the government of China, the government of a nation much poorer than ours, would be entirely unjustifiable.

A third intermediate arrangement for

Arthur A. Broyles is professor of physics and physical science at the University of Florida, Gainesville, and Eugene Wigner is professor emeritus in the department of physics at Princeton University.

defense, also indicated already in the Soviet handbooks on civil defense,[2] is to move most city dwellers away from densely populated areas but not as far as the pure counterevacuation proposes. Instead, the Soviets would build "expedient shelters" using materials at hand. Rather ingenious designs, which can be built by untrained prospective occupants, give a blast resistance of 30 pounds per square inch. A sample plan is shown in figure 2. Such a system, not significantly more expensive than the simple evacuation plan (not much over $500 million, according to the Ponast study) could reduce the fatalities as well as does the elaborate and rather expensive shelter system referred to above. However neither one can provide protection against a sudden attack.

In the design of shelters, prompt nuclear radiation can generally be ignored in comparison with the blast wave unless the blast protection is very good or the weapon is very small. The reason is that prompt-radiation effects decrease much more rapidly with distance than do blast effects. To see this, note that the blast pressure in pounds per square inch from a W kiloton explosion at a distance r in kilometers is given approximately by

$$p = \frac{1.6\ W^{2/3}}{r^2}$$

The intensity of the prompt radiation decreases more rapidly than $1/r^2$ because of the absorption by air. Thus, according to the equation, blast shelters designed for 100 psi will be effective against a 1-megaton weapon for distances greater than about $1\frac{1}{4}$ km. The area within which the pressure exceeds a given amount is inversely proportional to this pressure. Thus the area where the pressure exceeds 5 psi—the pressure often considered as the survival pressure for unprotected people—

is twenty times the area for 100 psi.

The effects of blast decrease more rapidly with bomb yield than do those from prompt nuclear radiation. For very small nuclear weapons, prompt radiation can be more harmful than the blast. Thus for a 1-kiloton bomb, neutron and gamma radiation at 750 meters are 700 and 400 R if no protection is provided. The blast pressure at that distance is 5 psi—quite tolerable. Indeed the mid-lethal blast pressure for a well instructed person, who knows how to protect himself from flying objects, is well in excess of 30 psi.

Blast shelters are designed not only to diminish the air pressure to which a person is subject, but also to protect him from flying objects. A properly designed blast shelter will also place sufficient mass between a person and the outside fallout particles to shield him adequately from the radiation. One foot of earth cover reduces radiation perpendicular to it by a factor around ten, and more than that for slanting rays. Shelters also provide cover against heat radiation and external fires. Two feet of earth will provide adequate protection from actively burning fire.

Global consequences

Worldwide effects from the detonation of a nuclear explosion naturally demand as much concern as the immediate effects. Many wonder whether the global consequences such as fallout might not be so severe as to deter any nation from even precipitating an attack. The most recent investigation of this question, the Nier report by the National Academy of Sciences,[7] verified previous conclusions that world-wide fallout produced in a nuclear attack would not be sufficient to deter the attack. It found, however, that the de-

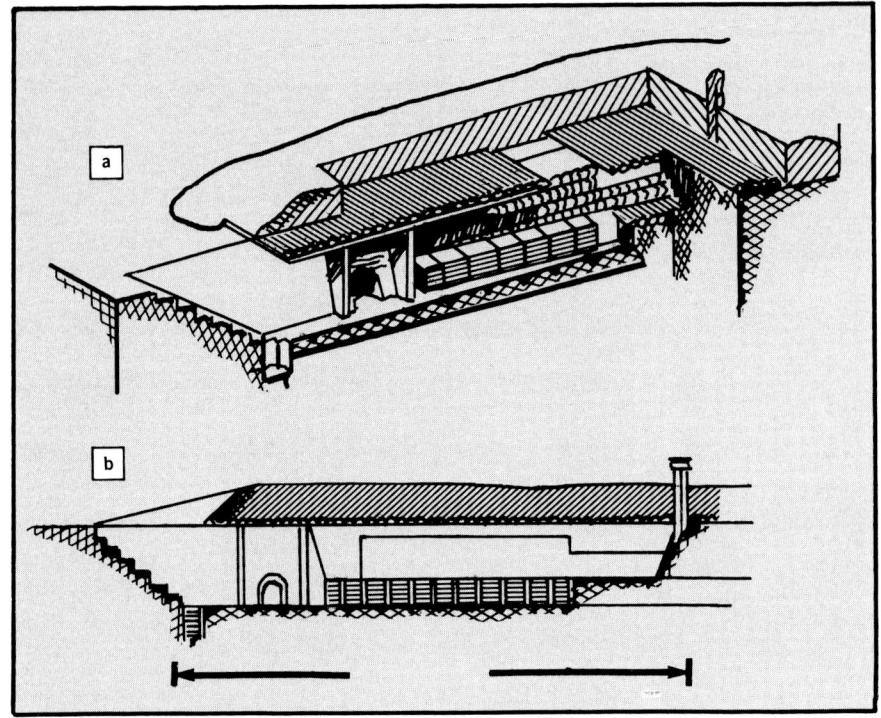

Hasty shelter plan of the Soviets is a dugout in dense soil with a ceiling of pine poles. The plan shows the general view (a) and cross section (b). From reference 2.   Figure 2

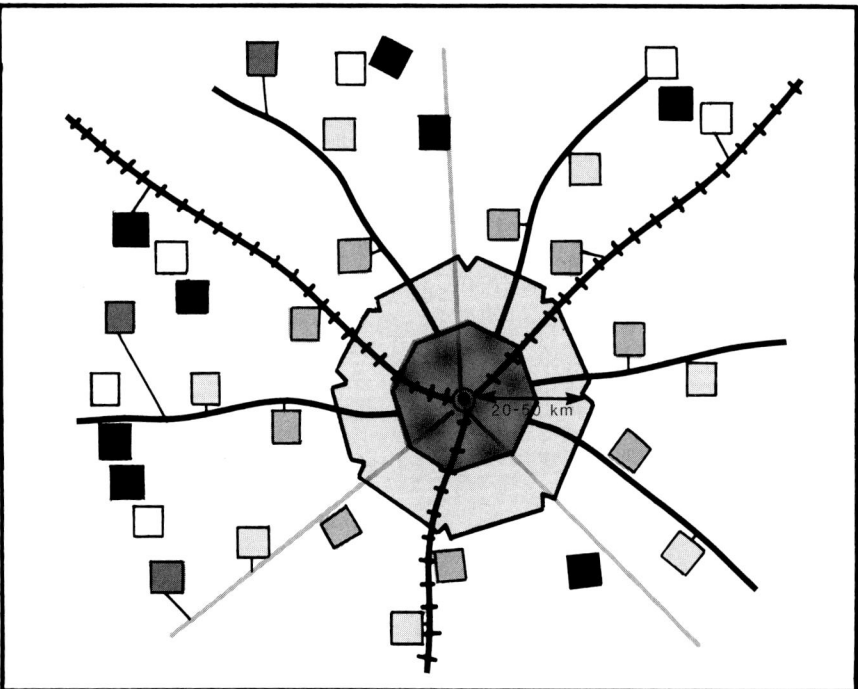

Soviet evacuation scheme illustrates their detailed planning. Safe zone is outside light colored region surrounding populated district of city (dark color). Map shows districts for relocation of workers of plants that do not stop their operation (dark colored squares) and for those that temporarily suspend operation (light color). Also shown are relocation sites for evacuees (light gray) and for plants and organizations (open squares). Black squares are existing communities. Colored lines denote operational control limits.   Figure 3

"At the same time, the governments of the United States and of other major nuclear powers should be alert to the possibility that a geographically distant, populous other nation might determine that the degree of short-term damage to itself in this report, would be 'acceptable' and that, since long-term recovery would be highly likely, might conclude that its own self-interest is compatible with a major nuclear exchange between other powers."

In other words, we cannot count on global effects in themselves as deterrents.

Even though civil-defense measures can be effective as population protection, the US lags behind many nations of the world in building such systems. The Chinese have installed extensive blast shelter systems; the Russians have preferred an evacuation procedure that removes the city population to outlying areas where hasty shelters are to be constructed from materials at hand. A sample evacuation plan from the USSR handbook is shown in figure 3. Admittedly, this system would lose effectiveness if another nation initiated the war: It takes two or three days to evacuate cities and to build emergency shelters. However, if such time is available, the USSR system is cheaper and probably more effective than the Chinese blast shelters. The Chinese, however, can occupy their shelters in a very short time and thus be prepared for an attack with very little warning. Evidently the Chinese are afraid that someone will attack them with little notice, while the Russians believe that they are in a position to determine when the nuclear exchange will come and that they can carry out their evacuation and construction in time.

Political aspects

The United States, on the other hand, has essentially no civil-defense system. This lack is deliberate, and the reasoning behind it is clearly evident in the hearings before Congress on military matters.[8,9] Our leaders recognize that, if the nuclear powers have the capability of destroying the opposing nuclear attack forces, they will be tempted to strike first. If they wait, their own weapons may be destroyed first and they would be defenseless. Thus the US, until quite recently, carefully designed its nuclear strike force to be effective against the population of an opponent but ineffective against his weapons. We also did not protect our people. This inaction assured him that we would not attack first and therefore, that he need not strike a preventive blow.

The trouble with our strategy was that the Soviets, and more recently the Chinese, have not accepted this "balance of terror." The Soviets' large mis-

pletion of the ozone layer could be more serious. Increased radiation might force people to adopt special protection against sunburn, and it would lead to an increase in the skin-cancer rate by a factor of almost two. The depletion of ozone would also upset some ecological systems in important ways. Although this study calls for additional research to answer some remaining questions regarding world-wide effects, Philip Handler, President of the National Academy, makes the following statement in his letter accompanying the Nier report:

siles are effective against our land-based missiles and their killer submarines can attack our Polaris submarines. In addition, our population is so exposed that it is doubtful we would accept the casualties required to participate in any stage of nuclear war through a second, third, or any strike with our missiles. Perhaps such considerations led Secretary of Defense James R. Schlesinger to propose the addition to our arsenal of missiles that would be effective against sheltered enemy ICBM's.[8] However we are disappointed that Washington has not given strong support for measures that will protect the US population from the effects of a nuclear war.

As a final remark we wish to add that it disturbs us greatly that passionate opponents of the protection of our own civilians against nuclear attack do not oppose, and do not even mention, the elaborate preparations of the USSR in this direction. The Soviet handbook on civil defense is circulated in millions of copies. (It has been carefully studied at the Oak Ridge National Laboratory.) The USSR gives instruction on civil defense in the high schools, they carry out exercises in their factories and, most distressingly, they have made elaborate preparation to evacuate their cities preceding a confrontation. If the opponents of the civil defense feel that these preparation are not even worth mentioning, why do they consider the protection of our own civilians objectionable and even provocative?

Drell: continued from page 35

MIRV's (multiple independently targetable reentry vehicles), which enable a single missile to attack several different targets with high accuracy. These MIRV's can selectively attack hardened military targets such as underground silos containing the fixed land-based ICBM forces and at the same time can cause relatively low casualty levels. Indeed this combination of factors forms the basis for the military value and strategic credibility that are claimed for such an attack.

Of course the effect of weapons against both military targets and civilians depends critically on such factors as the numbers and yields of incoming warheads, their height of burst and the level and extent of civil-defense protection. One example described by Secretary Schlesinger in his Senate testimony envisioned an attack against all the fixed ICBM's—1000 Minutemen and 54 Titan missiles—with a single one-megaton warhead incident on each silo and with the warhead fuzed to detonate in air at the optimum height of burst. The attack would result, he claimed, in fewer than 800 000 dead and 800 000 injured or ill from radioactive fallout.

The fatality levels for such an attack are calculated by making certain assumptions about the civil-defense protection provided in terms of the protection factors of various shelters. These numbers are the reciprocals of the fraction of radiation that penetrates the shelter. Thus the existing civil-defense program requires that, for a shelter space to be identified as such and stocked, it must have a protection factor of 50–100. That is, it must shield against all but 1–2% of the radioactive fallout. This factor is equivalent to a dirt cover of approximately two feet or a concrete wall of about 16 inches. By comparison,[3] a single-story residence has a protection factor of three, and a residential basement, a factor of 25.

In the attack described by the Secretary, the Department of Defense assumed that for 30 days roughly 35% of the US population remained in designated shelters with protection factors of 50–100, that 20% sought residential-basement protection and that the remaining 45% were protected by the average residential protection factor of 3. These calculations were stopped after this thirty-day period and thus do not include the final 6% of the fallout nor the long-range effects.

However, the Secretary did not describe the military effects of this attack, which was designed to cause such low civilian casualty levels. Straightforward calculations show that the nuclear attack assumed in the above calculations would destroy well under one half of our fixed ICBM force if carried out by missiles with the targeting accuracies projected for the Soviet ICBM force. This conclusion follows even if we assume that the Soviet missile systems have a perfect 100% reliability, which is surely a gross overestimate, particularly when you recall that we are talking of a massive attack coordinated in time so that all 1054 US ICBM silos are hit essentially simultaneously. I can see no practical military value to such an attack. On the contrary it would surely invite lethal retaliation.

In response to these and other DOD calculations on collateral civilian damage related to counterforce attacks, the Senate Foreign Relations Committee in September 1974 asked Congress's Office of Technology Assessment to review the DOD analyses. A panel convened by OTA for this purpose raised questions about the sensitivity of the DOD analyses to various assumptions, including a range of possible weather conditions, civilian protection factors and parameters of the incoming attack.[8] The DOD responded with more calculations, which showed that the expected fatalities are indeed very sensitive to the nature of the attack and can vary by large factors. In particular, the DOD now finds that fatalities in the range of 10 to 20 million will result from prompt effects and fallout alone if the attack is delivered by the nuclear weaponry of today or of the near future and is designed to destroy the majority of the attacked ICBM force.[8] Figure 1, which is based on DOD calculations, illustrates the fatalities as a function of the percentage of ICBM silos destroyed. (Note that the DOD reduced the civil-defense protection factors assumed for the last two attacks by 25% relative to that described earlier; otherwise, with identical protection factors, one would expect the one-megaton ground burst to cause more fatalities than two 550-kiloton bursts—one in air and one on the ground.) Even at the highest level in figure 1 a healthy retaliatory force of some 210 ICBM's would remain as well as all the SAC bombers and missile submarines.

Naturally the predictions of figure 1 are subject to such uncertainties as the weather and winds at the time of attack, and are sensitive to the degree of civil-defense protection and to the ability to provide medical care to the ill or injured. Nevertheless, one can clearly not contemplate an effective strategic attack designed to decimate our ICBM force in terms of casualty levels of one million civilians, but rather must consider it in terms of upwards of tens of millions, even assuming extensive protection of the population.

The price of civil defense

The most recent DOD reports also make clear that civil defense would be a central element of our policy of flexible response, with emphasis on limited nuclear counterforce. Indeed the justification for the civil-defense budget was expressed in the report for FY 1976 largely in terms of its role as a necessary adjunct of our policy emphasis on flexible response. The DOD report also argues that we must have the same population-evacuation options as the Soviet Union for two reasons:

Sidney D. Drell is deputy director of the Stanford Linear Accelerator Center. This text is adapted from his testimony presented on 18 September 1975 to the Subcommittee on Arms Control, International Law and Organization of the US Senate Foreign Relations Committee.

▶ "to be able to respond in kind if the Soviet Union attempts to intimidate us in time of crisis by evacuating population from its cities," and

▶ "to reduce fatalities if an attack on our cities appears imminent."

This position marks a major shift in emphasis of the civil-defense program since the 1974 Annual DOD Report, when it was largely justified by Secretary of Defense Elliott Richardson to help recovery from peacetime disasters. I personally endorse this previous objective and furthermore I support the existing program of identifying and stocking shelters as a prudent insurance program against a wide range of incidents, including the accidental launch of nuclear weapons, a severe nuclear-reactor accident or natural disasters such as hurricanes. However, a comprehensive civil-defense program involving both sheltering and evacuating the population on a very large scale is a different thing. Undoubtedly it can be demonstrated to have a great lifesaving potential in the event of a nuclear attack against specific military targets. But the issue is in essence an issue of the price one has to pay for a civil-defense program in relation to the degree of protection one buys against specified attacks: What price in our priorities, values and style as a society? What price in dollar costs?

Investment in a civil-defense program could, as one function, protect the population from the blast, thermal and radiation effects in the immediate vicinity of a nuclear explosion—roughly within a radius of four miles for a blast of one megaton. Such protection against the close-in effects is either impossible or tremendously costly.

Another function of civil defense is to reduce casualties from fallout generated at distances well beyond several miles. This effect of dangerous fallout levels, extending many hundreds of miles downwind from nuclear explosions, plus the long-range effects of radioactive contamination to extensive areas, differentiates nuclear war from all other previous experience. The range and extent of the threat to life of radioactive fallout depends critically on many factors including the height of burst (that is, whether or not the fireball from an explosion near Earth's surface scoops up and spreads an enormous cloud of radioactive debris); the fraction of fission yield in the bomb design and the weather.

The biological effect of fallout is measured in terms of the standard dosage unit of the roentgen-equivalent mammal (the rem). Whole-body exposures to less than 100 rems cause blood changes but no disabling illness. Experience following the Hiroshima and Nagasaki blasts shows that doses of 100 to 200 rems cause a certain amount of ill-

ness including fatigue and perhaps some nausea, but are rarely fatal. However, levels of about 450 rems of whole-body exposure can cause severe illness and produce a 50% fatality rate. This scale is the basis for assessing how much protection must be provided for an effective civil defense. As is shown in figure 2, an unsheltered person as far away as several thousand miles downwind from an attacked missile field or military base would be exposed to an expected 600 rems.

The time scale of the radioactive fallout is also of great importance in considering protection. For how long a period of time after an explosion must one be sheltered from fallout in order to survive? For typical burst altitudes in the atmosphere a human body totally and completely shielded from fallout during the first hour immediately following a nuclear explosion will still receive 45%, or almost half, of the total fallout if exposed thereafter. Twenty percent of the total dose is deposited after the first day, and a person emerging after four weeks of complete protection from fallout will still be subject to 6% of the total dosage. The decrease in rate of fallout follows a $1/T^{1.2}$ law, and evidently the required time scale for protection is measured in weeks.

This discussion of fallout effects shows the required physical parameters of civil-defense shelters. Few dispute the technical facts concerning the means to protect large populations for one to two weeks after an attack from the physical effects of blast, fire, radiation and fallout. However, major social parameters and costs are also involved because identified shelter spaces and

evacuation plans do not by themselves make an effective civil-defense program, in my judgment. A total system must be organized and interwoven extensively into civilian life through training programs, rehearsals, and volunteer activities. The pre-attack shelter organization envisioned by the 1962 Office of Civil Defense Guide planned that a shelter accommodating 100 civilians would require an operating cadre of 25, of which 10–12 would need prior training. This number constitutes 10% of the sheltered or 20% of the adult population.

To recruit the required large cadre of trained personnel the government would have to look beyond existing community safety personnel such as policemen and firemen. Perhaps the military reservists and National Guard units could play a central role in organization and training, but they would still have to rely on a large functioning organization involving a much larger number of trained civilians.

One task of trained personnel would be to operate communications systems over large distances in order to deal with shortages of food, water and medical supplies. They would also have to know how to use radiation dosimeters, because in the immediate post-attack period the fallout levels can vary greatly from one locale to another. Like the snow, radioactive debris accumulates where driven, depending on wind and weather conditions as well as on the location and shadows of tall buildings. Local pockets of relative safety may exist amid areas with lethal levels of radioactivity. Finally the trained cadre would have to provide leadership in the

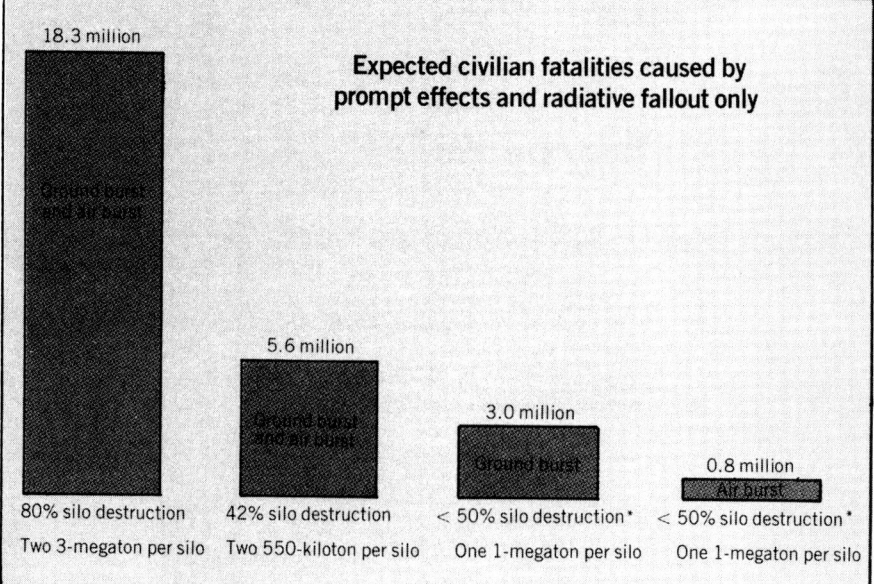

Casualty toll varies with the type of nuclear attack, among other parameters. All the calculations were done by the DOD in its analysis "Sensitivity of Collateral Damage Calculations to Limited Nuclear Scenarios," sent to the Senate Foreign Relations Committee on 11 July 1975, except for the two with asterisks, which are by the author.   Figure 1

long period of extreme social duress after the attack and would have to reestablish requisite services for a society with a large proportion of ill and injured citizens.

Beyond the training of these special leaders, the plans for massive population relocation and evacuation out of high-risk areas near the possible counterforce target system require a heightened level of public awareness and concern, and a willingness to rehearse the evacuation plans. Without them, surely a chaos spawned by panic will ensue at the time of implementation. How can one draw public attention, much less commitment, to such plans without "overselling" them by a sustained escalation of apprehensions from the mood of today *vis-a-vis* the dangers of nuclear exchange between the US and the Soviet Union? Is not such an escalation of apprehensions more to be feared than desired as the US and Soviet Union move further from the brink of a nuclear conflict due to misunderstanding, misapprehension or mistake and strive mutually at SALT for a more stable nuclear balance at lower levels of nuclear armaments? Indeed one of the lessons of the civil-defense shelter exercises in 1961 and 1962 was that the large expenditures for civil defense and the general dislocations accompanying a major shelter program could only be sold to the American public by presenting the very real threat of nuclear war.

Strategy

Consideration of civil defense as an element of strategy has been given renewed importance by the new emphasis on fighting a limited nuclear war. This policy changes our nuclear doctrine of the past decade, which has been dominated by the recognition that once a nuclear weapon is detonated on US or Soviet territory there would be substantial probability that nuclear exchanges could not be terminated before both nations were destroyed and the casualties numbered hundreds of millions. The new strategic doctrine raises the issue of whether this unpleasant "balance of terror" and mutual hostage relationship might be changed by the adoption of new tactics and the development and purchase of new weapons for fighting limited nuclear wars at acceptably low casualty levels. I believe such a policy would cause the following deleterious effects:

▶ Harm to strategic stability. The development of a new missile force designed specifically as hard-silo killers would fuel concern on both sides about the vulnerability of the fixed ICBM's to a preemptive first strike. It would emphasize the importance of striking first and could thereby destabilize a crisis situation. Furthermore the development and rehearsal of civil-defense

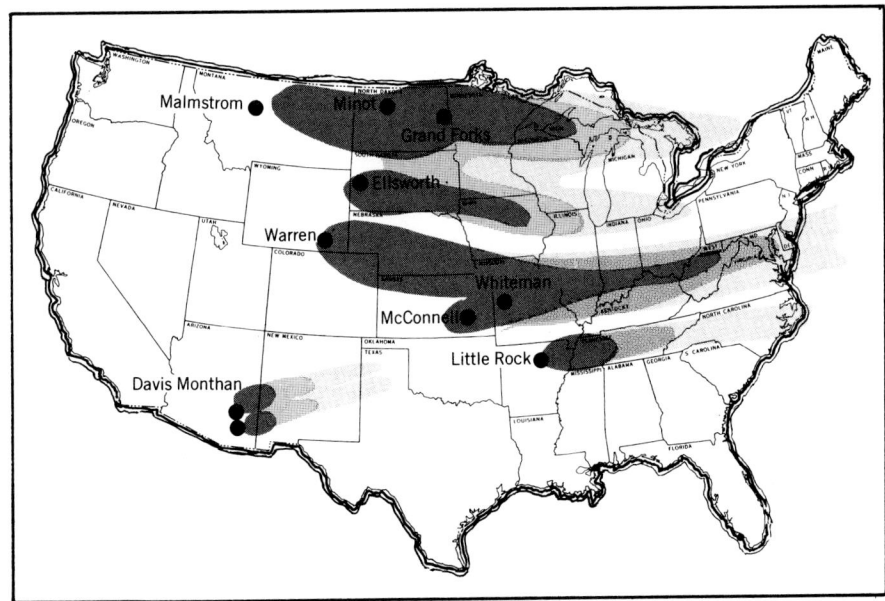

Fallout patterns for an attack on US ICBM silos (black dots). Two inner contours show radiation doses of 450 and 200 rems for a person with a protection factor of 3. In the lightest colored regions strontium-90 contamination exceeds 2 microcuries/meter^2. Data are for a winter day and will vary with wind patterns. (From R. L. Garwin, reference 8.)   Figure 2

plans involving evacuation and relocation of large populations could be viewed with alarm by an opponent as preparation for executing a first strike.

▶ Harm to SALT talks. The development and testing of the required new missiles will create pressures against quantitative reductions in the numbers of strategic forces and against such verifiable qualitative restraints as missile test-flight quotas and limits on the rate of deployment of new systems that would slow down the pace of progress in the arms race.

▶ Waste of resources. The plans justified by this year's rhetoric may materialize into the multibillion-dollar weapons systems of the next decade unless the rationale behind them is rejected.

▶ Shift of values. Implementation of an extensive civil-defense system through massive training will affect the priorities of our society and will require heightened concern about nuclear war,

which would counter the progress that has been made toward reduced international tensions.

Finally, what will prevent the eventual escalation of an initially limited nuclear war to an all-out nuclear holocaust? Once nuclear weapons are used in war at all it will be very difficult, if not impossible, to verify yields, sizes, numbers and types of the nuclear explosions on both sides. However, the one technically unambiguous fact is whether or not nuclear weapons have been used at all. Therefore it is wisest for the US to adopt as a national policy the highest possible nuclear threshold. We should maintain a gap between nuclear and non-nuclear warfare that is as clear and wide as possible, and resist the temptation to develop doctrines and civil-defense programs that understate, on dubious technical and strategic premises, the collateral damage and the casualty levels of nuclear conflict.

Broyles and Wigner reply to Drell

Our own discussion is principally concerned with the technical question of whether defense against nuclear weapons is possible. We feel that as physicists we should be able to judge the extent to which such defense is possible and we also feel that the physics community at large should have a degree of familiarity with this problem. Sidney Drell's article is less concerned with the physical problem than with the more important but less precisely ascertainable one concerning the political implications and consequences of a vigorous civil-defense effort—a subject to which only the last section of our own article

refers. Nevertheless, we would like to comment, first, on a problem of physics concerning which our opinions differ.

We differ with Drell in our estimation of the radiation danger from fallout after a reasonably long sojourn in shelter, let us say two weeks. First of all we calculate that the total radiation dose from the fallout after two weeks amounts to less than 7% of the total radiation of the fission products from 1 minute on to infinity. In addition, the radiation becomes softer as time goes on, so that it becomes easier to protect against it. More importantly, the radiation after two weeks is stretched out over a rather long period—six months or so. Although the damage done to Man by 10% of this radiation is not reversible the damage done by the remaining 90% appears to decrease by $2\frac{1}{2}$% per day. As a result, by the end of the half year, the effect of the radiation received in the early period after emergence from the shelter has decreased to 11% of its initial magnitude. Altogether, the damage caused by the radiation received after the two-week sheltering period hardly exceeds 4% of the damage that a person outside would receive in the initial two-week period. Even more importantly, because the radiation intensity after two weeks is only one thousandth of its intensity at one hour after the explosion, after two weeks the shelter can be abandoned for reasonably long periods. Thus survivors can possibly clean up surroundings or, in extreme cases, move to a less contaminated location. We conclude that the danger from the fallout radiation can be easily guarded against after a period of two weeks from the time of the explosion and that the emergence from the shelter after that period produces much less difficulty than indicated in Drell's article. We do not wish to deny, of course, that it is even better if no nuclear explosion takes place.

The second, still somewhat technical, point to which we wish to take exception is the statement that "Protection against the close-in effects (blast and heat) is either impossible or tremendously costly." The gross national product per person of China is a small fraction of ours, yet most visitors to their land return greatly impressed by the very effective and easily accessible civil-defense shelters that were proudly shown to them. More concretely, the implementation of the counter-evacuation plan would cost $2.50 per person and the Chinese-type shelters $175 per person (or $35 per person per year, because their construction may take about five years). Surely, neither of these figures can be called "tremendous;" yet they would really buy each of us a great deal of security and would discourage attacks or threats of attacks—an equally important accomplishment. In fact, the Swiss civil-defense book says that the most important accomplishment of civil-defense preparations is that they will never have to be used.

On the other hand, we agree with Drell that an unlimited nuclear exchange between the USSR and the US would result in more than one million casualties on both sides. But in our opinion, we must strive for an approxi-

mately equal casualty rate—not 2 or 3% in the USSR and about 45% here. We also note that as Drell points out, the US Secretary of Defense believes that nuclear attacks on military targets may be feasible. Unfortunately the Soviet government may share this view.

Our last objection to Drell's statement is nontechnical and is in the spirit of his own article. He says "Furthermore, the development and rehearsing of civil-defense plans involving evacuation and relocation of large populations could be viewed with alarm by an opponent as preparation for executing a first strike." If that is so (and we believe it is) we do not understand the failure of his article to mention the USSR development and rehearsal of civil-defense plans involving evacuation and relocation of large populations. Evidently, he is not concerned by these plans and does not view them with alarm; he does not even think that they are worth mentioning. What he sees with alarm is that we may duplicate these efforts, that we put an end to the situation in which we may have to face an enemy who can destroy fifteen times more citizens in the US than we can destroy of his. Frankly, this current situation is what alarms us and is what we wish to terminate.

Drell replies to Broyles and Wigner

Although Arthur Broyles and Eugene Wigner frequently allege that the Soviets have extensively interwoven a civil-defense program into their society, to the best of my information no evidence exists that they have in fact exercised a civil-defense system capable of massive population relocation or evacuation. A large number of emigrés from many parts of the Soviet Union have been received in the West; had there been any widespread civil-defense rehearsals in the Soviet Union we surely would have heard about them by now. The Soviets have indeed written much on the subject and have given their population a more intensive exposure to civil defense. Apparently they have spent much more money on plans and organizations and involved small numbers of people with key skills in exercises. However, I believe that in view of the unprecedentedly large scale of the nationwide disaster we are considering, an effective civil-defense program must also include, as one of its essential components, full-scale rehearsals and survival living exercises involving the population.

Selective quotations from civil-defense manuals are not reliable guides to the effectiveness of a civil-defense program. If it were, we might cite from their manuals the removal of anti-Western polemics in the 1974 edition. We

might also cite the fact that their civil-defense manuals for 1970 and 1974 (see reference 2 for the former and ORNL-tr-2845, 1975, for the latter) contain elementary substantive errors such as the translation, from US sources,[3] of miles directly to kilometers without the conversion factor of 1.6 in giving ranges of destruction from given bomb yields. Furthermore, the Soviet analysis of

minimum requirement for air supply in shelters has not changed from old manuals. Thus the US editor of the translation is led to comment, in the preface, that "The Soviet Union has not conducted mass shelter living experiments or even simulated ones as has been done in the US." The editor then comments further: "We believe that this is the most serious flaw in the whole Soviet Civil Defense planning." In my judgment, plans and manuals, on one hand, and an effective operating system, on the other, are very different things!

In referring to the Nier report Broyles and Wigner stated that it "verified previous conclusions that worldwide fallout produced in a nuclear attack would not be sufficient to deter the attack." In fact the report contains no such conclusion, nor does it address questions as to what will or will not deter war. Its task was the much more narrow one of considering the consequences of a nuclear conflict "by examining, independently, possible effects upon, respectively, the atmosphere and climate, natural terrestrial ecosystems, agriculture and animal husbandry, the aquatic environment and both somatic and genetic effects upon humans," as remarked by Handler in his letter of transmittal. In my reading of the Nier report I was more impressed by how extensive are the unknowns that will determine the scale of the disaster resulting from a major nuclear conflict and by how little can be predicted with confidence.

I believe there is no basis in fact for the statement by Broyles and Wigner that "the Soviets' large missiles are effective against our land-based missiles and their killer submarines can attack our Polaris submarines." This allegation is also at variance with assessments given by our civilian and military leaders. To quote Secretary Schlesinger, for example, in the Annual Defense Department Report for FY 1976, "Our sea-launched ballistic-missile force provides us, for the foreseeable future, with a high confidence capability to withhold weapons in reserve."

References

1. *Civil Defense,* a Report to the Atomic Energy Commission by a Committee of the National Academy of Sciences, Washington, D.C., 1968. Available as TID-24690 from Division of Technical Information Extension, ERDA, P.O. Box 62, Oak Ridge, Tennessee 37830; T. L. Martin, D. C. Latham, *Strategy for Survival,* University of Arizona Press, Tucson (1963); C. M. Haaland, *Systems Analysis of US Civil Defense Via National Blast Shelter Systems,* Oak Ridge National Laboratory, Oak Ridge, Tennessee, Report ORNL-TM-2457 (1970).

2. *Civil Defense,* (N. I. Akimov, ed.), Moscow, 1969. Translated by S. J. Rimshaw, ORNL-tr-2306 (1971).

3. S. Glasstone, *The Effects of Nuclear Weapons* (revised edition), US Government Printing Office, Washington, D.C. (1974).

4. R. H. Sandwina, "Ponast II," *Proceedings of the Radiological Defense Officers Conference,* South Lake Tahoe, 23–25 October 1974, State of California Governor's Office of Emergency Services.

5. E. P. Wigner, "The Myth of Assured Destruction," in *The Journal of Civil Defense (Survive),* July–August 1970, P.O. Box 910, Starke, Florida.

6. D. L. Narver Jr, D. T. Robbins, *Engineering and Cost Considerations for Tunnel Grid Blast Shelter Complex,* ORNL-tm-1183 (1965); D. T. Robbins, D. L. Narver Jr, *Engineering Study for Tunnel Grid Blast Shelter Concept for Portion of City of Detroit, Michigan,* ORNL-tm-1223 (1975).

7. *Long-Term World Wide Effects of Multiple Nuclear-Weapons Detonations,* The National Research Council (Committee Chairman, Alfred O. C. Nier); The National Academy of Sciences, Washington, D.C. (1975).

8. *Analyses of Effects of Limited Nuclear Warfare,* prepared for the Subcommittee on Arms Control, International Organizations and Security Agreements, of the Committee on Foreign Relations, US Senate, September 1975.

9. Hearings before the Subcommittee on International Organization and Disarmament Affairs of the Committee on Foreign Relations, US Senate, Ninety-first Congress. □

SECTION 2 _____

NUCLEAR TESTING

The articles in this section deal with issues related to tests of nuclear weapons. In their debate on the Comprehensive Test Ban Treaty, Hugh DeWitt and Robert Barker discuss the issues of reliability, verification, modernization, and arms stability. DeWitt says that a comprehensive test ban is feasible and that the weapons labs themselves bear a heavy responsibility for the failure of the negotiations because of their desires for third-generation weapons and because of their failure to design fewer and more robust systems. DeWitt comments that the new designs for "defensive" nuclear weapons, such as the x-ray laser pumped with a nuclear weapon, would ultimately need to be tested in space, which would be a violation of the Outer Space Treaty of 1967. DeWitt says that for a comprehensive test ban one must compare the small disadvantages of possible clandestine tests of a few kilotons in large cavities with the advantage of the curtailment of newer generations of nuclear weapons. Barker argues that nuclear testing is necessary for ensuring the reliability of weapons and for modernizing nuclear forces. He recently commented on his article that "The debate that I called for in the concluding sentence of my article has indeed occurred—it continues. There has been a substantial swing on the debate away from a CTB to lower-yield-threshold treaties. I would like to believe that the arguments that were fully laid down in the article have, in part, been responsible for the increased recognition that a CTB would pose great risks for the US deterrent. The potential damage of lower thresholds now needs to be examined more thoroughly. Certainly, the proposed benefits of lower yield thresholds have not yet been clearly articulated."

The question of whether the Soviets have violated the Threshold Test Ban Treaty hinges on the geophysical differences between the US and Soviet testing sites. Because the US test site in Nevada is on a much younger geological plate than the Soviet site at Semipalatinsk, the seismic signals from nuclear explosions at the Nevada test site are smaller than explosions of the same yield at Semipalatinsk. This seismic magnitude correction, called the bias factor, is not controversial, but there have been differences of opinion as to the numerical value of the bias factor. Jack Evernden and Gerald Marsh have estimated that the value is between 0.35 and 0.45; from this they conclude that the Soviets have not violated the TTBT. The government has raised its value for the bias factor, resulting in a split decision within the Executive Branch on the charge of a "likely" violation, but the National Security Council maintains the "likely" verdict. Since this article was written there has been some progress toward reducing the uncertainties. In the summer of 1988 US and Soviet scientists made independent measurements of yield of a explosion at the other's test site. By comparing CORRTEX data and seismic data from about five explosions, one should be able to determine the bias factor with enough accuracy to substantially reduce the systematic errors, improving the quality of future seismic data. We now await the results. In addition, the two nations have exchanged seismic data on past explosions, as well as geological information. These Joint Verification Experiments are encouraging, since they have shown that cooperation is possible.

Before the Intermediate Nuclear Force Treaty, most observers felt that the Soviets would never allow foreigners on their soil to monitor compliance of their military programs to terms of arms-control treaties. William Sweet describes how the Natural Resource Defense Council bypassed the official negotiations to show that the Soviets were willing to have seismographs placed on Soviet soil, attended by American scientists. The agreement that was signed in May 1986 between the NRDC and the Soviet Academy of Sciences was not only historic, it also paved the way for the Joint Verification Experiments mentioned above, and the acceptance of on-site inspections in the INF treaty.

CONTENTS

Debate on a comprehensive

Pro

A critic on the inside says that a treaty is vital to world security, that it is technically feasible, and that the nuclear weapons laboratories should end their opposition to the idea.

PHYSICS TODAY/AUGUST 1983

Hugh E. DeWitt

The 1963 Limited Test Ban Treaty prohibits nuclear explosions in the atmosphere, the oceans and space. Nevertheless, nuclear tests continue at an alarming rate—almost one explosion per week somewhere in the world—although now the testing is largely underground.

In the 1963 treaty, the United States committed itself to negotiate toward a comprehensive test ban, which would end the testing of nuclear weapons altogether. Many people in this country and in other countries strongly believe that a comprehensive test-ban agreement between the major nuclear powers would put a brake on the current runaway development of new nuclear weapons, and reduce the possibility of nuclear war. As a physicist for 26 years on the staff of the Lawrence Livermore National Laboratory, I have observed the development of nuclear weapons from the inside of the weapons establishment. During this time I have reached some possibly heretical conclusions for a weapons-lab employee, and the rest of this article should be understood in their light:

▶ Some form of comprehensive test ban treaty that ends nuclear testing is both feasible and vital to the security of the world.

▶ The weapons labs themselves bear a heavy responsibility for our present situation in which the two superpowers compete to obtain an illusory nuclear superiority.

Both of these personal convictions contrast sharply with positions taken by the weapons laboratories. The labs' stance on the first point was well summarized by Livermore director Roger Batzel when he stated:[1]

　...I believe the continued credibility of the US nuclear weapon deterrent cannot be assured for long without nuclear testing.

continued on page 51

Subsidence craters. This aerial photograph shows part of the Nevada test site, where US nuclear weapons are tested underground. The surface is marked by craters caused by the

nuclear weapons test ban Con

In the interest of maintaining a reliable nuclear deterrent capable of surviving Soviet attack, the United States should reexamine its commitment to achieving a comprehensive test ban.

Robert B. Barker

PHYSICS TODAY/AUGUST 1983

collapse of explosion-created cavities. Above-ground testing was banned by treaty in 1963. (Lawrence Livermore National Laboratory photograph, Mercury, Nevada).

We are seeing today an unprecedented public interest in the control of nuclear arms. The resulting plethora of books and articles on all aspects of nuclear weapons and nuclear-weapon strategies is providing the basis for a new, informed discussion of arms-control objectives and priorities. While there may be a general desire for the elimination of all nuclear weapons, there is also an appreciation that nuclear weapons will be with us for the foreseeable future, and that we will achieve bilateral and verifiable reductions only over time. The public understands that even as we work for reductions in nuclear arms, those nuclear weapons that do exist should continue to preserve deterrence and stability between the two nuclear-weapons superpowers, the Soviet Union and the United States. Hence we should measure all proposals for arms control or arms reductions against their ability to reduce the numbers of nuclear weapons while preserving deterrence and stability.

Just as it is appropriate to assess each new arms-control proposal against this criterion, it is appropriate to use this same measure to reassess past proposals. A comprehensive test ban has been a stated goal of both the United States and the Soviet Union since 1958. (Herbert York discusses the subsequent history in his article in PHYSICS TODAY, March, page 24.) The Reagan Administration has reaffirmed that goal in the context of deep and verifiable arms reductions, expanded confidence-building measures and improved verification capabilities. There has, however, been little recent public discussion of the significance and advisability of a comprehensive test ban. Should a comprehensive test ban be a current goal because it has been one since 1958? Should a comprehensive

continued on next page

BARKER *continued from previous page*

test ban be a current goal because, according to the 1982 edition of *Arms Control and Disarmament Agreements*,[1] the non-nuclear weapons signatories of the 1968 Non-Proliferation Treaty view it as a *sine qua non* for the prevention of nuclear proliferation?

Recognizing that this is not 1958, or 1968, or even 1978, we should reexamine and debate the desirability of a comprehensive test ban. The executive and legislative branches of the government will ultimately determine US policy on this issue, assessing in the process international political as well as technical issues. This article is a personal assessment of the technical issues. I will examine the objectives of nuclear testing and ask the reader to evaluate their compatibility with nuclear-arms reduction and the maintenance of deterrence and stability.

Goals of testing

For too many people, the objectives of continued nuclear weapons tests are unnecessarily a mystery. While classification rules prevent discussion of some specifics, one can discuss the rationale for current testing.

The current US nuclear testing program has several goals: ensuring the reliability of existing nuclear weapons; providing new designs for nuclear weapons intended to replace aging and ineffective weapons in the national stockpile; developing nuclear weapons with better safety and security features; guaranteeing against technology surprise through research on new weapons concepts; improving the fundamental understanding of nuclear-weapon performance; and maintaining the competence of nuclear-weapons scientists and engineers. All nuclear tests are carried out for the Department of Energy by this country's two nuclear-weapons laboratories, the Los Alamos National Laboratory in Los Alamos, New Mexico, and the Lawrence Livermore National Laboratory, in Livermore, California. The actual testing takes place at the Nevada test site, north of Las Vegas. (See the photograph on the previous page.)

Reliability. The number of nuclear weapons in the US stockpile is often cited as approximately 30 000. The weapons range from those delivered by missiles and aircraft to those carried by artillery shells. Whatever their delivery system, these weapons have no place in the stockpile if the nation does not have confidence that they are capable of fulfilling their role if called upon to do so. As we will see, the

current nuclear weapons testing program plays a critical role in maintaining confidence in the country's nuclear deterrent.

Exactly how is this "confidence" established and maintained? Periodically, the Department of Energy must certify to the Department of Defense that weapons in the stockpile continue to meet the criteria established for them. While nuclear weapons are designed to last for the lifetimes of the systems that carry them—10, 20 or even 30 years—age does take its toll and not always in ways that were anticipated. Therefore, DOE periodically samples non-nuclear components of nuclear weapons in the stockpile and tests those components to ensure that age has not impaired their function. Scientists and engineers of the weapons laboratory responsible for the initial design disassemble and examine the nuclear components, and make an assessment as to whether any age-related changes will degrade the performance of the weapon.

In general, the Department of Energy performance certification is based upon non-nuclear testing and the judgment of experienced personnel. However, when these inspectors find unacceptable deterioration, nuclear testing may be necessary to determine whether performance is truly unacceptably degraded. Acceptable nuclear test performance results in DOE certification; unacceptable performance requires that weapons in the stockpile be replaced through a new production of identical weapons or by weapons of a new design not subject to the same deterioration. The decision on which course to take is based on the age of the weapon. It makes little sense to rebuild if several rebuilds will be required during the life of the weapon system. If a new design is required, nuclear testing will be necessary to certify that it performs as predicted.

The weapons designers have done a good job throughout the history of the US nuclear stockpile, but perfection has escaped them. Only infrequently has nuclear testing been required to verify adequate performance or to develop a new weapon to replace one that no longer works. However, in those few cases, until a solution was found and the testing was done, the weapon systems involved were suspect—confidence was lost—and the effect was the same as if they had been unilaterally removed from the stockpile.

Modernization of US nuclear-weapons delivery systems has been an ongoing process. Weapons systems based on newer technology replace those that have lost effectiveness because of obsolescence; for example, air-launched cruise missile carriers and B-1 bombers are to replace the B-52s. Weapons systems whose survivability may be threatened are replaced with less vulnerable systems; thus the Trident missile system is replacing the Polaris and

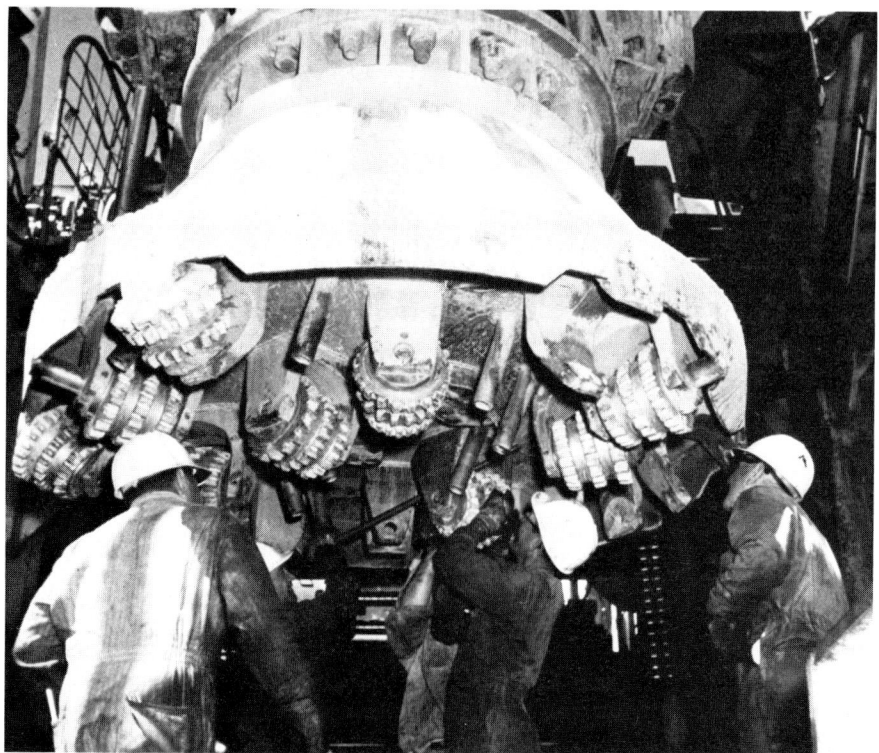

Drill bit. Since 1963, all US and Soviet nuclear tests have been conducted in deep underground holes. Drillers are shown here working on one of the large-diameter bits used to drill the holes that hold nuclear devices and associated test instruments such as those shown on the opposite page. Each hole is backfilled and plugged to contain the radioactive debris.

Robert B. Barker is assistant associate director for arms control, Lawrence Livermore National Laboratory.

Poseidon systems.

In every case to date, the replacement system has required a nuclear weapon different from that of the system it replaces. In some cases, physical dimensions alone preclude use of the older weapon. In other cases, existing warheads cannot survive the heat, acceleration, vibration and environmental extremes that a nuclear weapon will meet in the stockpile or during delivery. Even the yield requirement of the new system may be different from that of the system it replaces. As J. Carson Mark, retired head of the Theoretical Division of Los Alamos, noted[2] recently in the *Bulletin of the Atomic Scientists*,

> The nuclear explosive and its carrier consistute a "weapon system" of which neither part is of much use without the other... The weapon, tailored for...[its] particular delivery mode, cannot easily be used in any other way.

A substantial fraction of current nuclear testing is directed toward providing new weapons for new delivery systems. The Navy designed its C-4 missile to have a longer flight range, thereby permiting the Poseidon and Trident submarines to operate in larger ocean areas. No existing Navy reentry body could survive the harsher reentry environments associated with the greater missile range; a new reentry body with a new warhead made the C-4 system possible. The B-1 bomber will replace the B-52 in its role as a penetrating bomber. For the new bomber to fulfill its mission of deterrence, it must credibly be able to penetrate Soviet air defenses, deliver its weapons and escape. Accordingly, bombs delivered by the B-1 must be able to withstand release at greater speed, survive ground impact, and delay detonation while the aircraft flies out of range of the bomb's explosion. The criteria are very different from those for bombs designed for delivery by the B-52. The weapons labs have developed new nuclear designs to enable the B-1 to fulfill its mission.

In the area of tactical nuclear weapons, new development work has established the survivability of nuclear weapons in long-range artillery. The original nuclear artillery shells were designed to withstand the acceleration associated with the range of the 8-inch and 155-mm howitzers of the 1960s. In the following decades, US and Soviet artillery doubled in range. Without new nuclear shells, capable of withstanding the acceleration associated with the longer range, US nuclear artillery would be "outranged" and therefore vulnerable to destruction by conventional weapon fire.

While concern for survivability is the primary motivation for modernizing

Instrumentation canister with signal cables attached, being lifted into an assembly tower prior to going "downhole." The effort to extract large amounts of information about device performance had led to diagnostics instrumentation canisters that are complex underground "physics laboratories." (Lawrence Livermore National Laboratory photograph, Mercury, Nevada.) ▶

Instrumentation trailers. Electrical signals from underground measurements of nuclear devices are transmitted via coaxial cables to recording instruments located in trailers placed outside of the surface subsidence region. The trailers move from one test location to another. ▼

BARKER *continued*

nuclear weapons systems, there are other important reasons for doing so. The military effectiveness of established systems declines as the hardness of intended targets increases. To reestablish past destructive capability requires new nuclear weapons systems. Another motivation for modernization comes directly from developments in the area of nuclear-weapons design: In the last decade the nuclear-weapons laboratories have developed the technology to increase dramatically the safety and security of nuclear weapons.

Improved safety and security. In the laboratories' work on nuclear-weapons safety, the concern is not that of an accidental nuclear explosion. As Mark has stated,[2]

> The high explosives which have been mostly used in connection with nuclear weapons . . . can reliably withstand the jolts and impacts encountered in normal handling, even if they should be dropped from modest height; but they might detonate on falling on to a hard surface from a plane, for example. The concern is not that a full-scale nuclear explosion would result, since that requires a thoroughly symmetric detonation of the explosive which could not be induced by impact at one point.

In fact, two aircraft accidents have caused the high explosives in nuclear weapons to detonate: in 1966 at Palomares, Spain, and in 1968 in Thule, Greenland. In both cases there was no nuclear chain reaction, but the explosions dispersed plutonium, requiring extensive clean-up operations to eliminate the hazard to health.

As a result of developments at the nuclear weapons laboratories, it is now possible to preclude accidents that disperse plutonium. There are some relatively insensitive high-explosive mixtures that can survive quite violent impacts. The laboratories are now in the process of incorporating such explosives in new weapons systems as they are modernized. Due to the number of different nuclear-weapons designs in the US stockpile, it will be many years before all the weapons incorporate this improved safety feature. Because the weapons with insensitive explosives are based on new designs that differ substantially from those using older explosives, testing will play a critical role in the conversion to safer nuclear weapons.

Security is another area where recent developments in design are leading to dramatic improvements. Again, as weapons systems are modernized, features are being included that make it impossible for unauthorized persons to make use of a nuclear weapon.

The B-1 bomber. This high-speed plane, which will replace the B-52 in its role as a penetrating bomber, is designed for low-altitude weapon delivery. The nuclear-weapons laboratories have developed nuclear bombs that can withstand the rigors of this delivery mode.

These features are an intimate part of the nuclear design and require nuclear tests to ensure that only authorized use would result in the expected performance.

Technical surprise. One long-standing mission of the nuclear-weapons laboratories is to understand all means by which a nuclear explosion might be of military use. In part, this represents a desire to understand all the ways in which the US might employ such explosives to enhance its security. It also represents a desire to avoid surprise from the advantages others might obtain from nuclear-weapons developments.

The evolution of nuclear-weapons design is not a one-dimensional process; there is no unique path that a nuclear-weapons state must follow from its first nuclear explosion to subsequent developments. One cannot be confident that findings by the United States match those of the Soviet Union. With the maturity of the US nuclear program, new concepts are less frequent, but they do occur. Nuclear testing is critical to determining whether a new concept will work.

Verification that a concept is feasible does not imply that it will be incorporated into weapons in the US stockpile. Far from it. But establishment of feasibility does permit the evaluation of the threat to this country should the Soviet Union have already incorporated it into their nuclear arsenal.

Fundamental understanding. Despite the lengthy history of nuclear-weapons

testing in the United States, weapons scientists do not fully understand some fundamental phenomena that bear on the performance of nuclear explosives. The nuclear-weapons laboratories possess the country's most impressive computer resources and a very impressive cadre of theoretical physicists. Yet, there are sometimes substantial discrepancies between calculation and experimental results; the mathematical models are just not yet adequate to predict reality. Economic considerations alone motivate the nuclear-weapons laboratories to maximize the role of calculations so as to husband the scarce and expensive resource of nuclear tests. Thus, the objective of some nuclear tests is to improve calculations by exploring fundamental phenomena that are not yet understood, and which may be the cause of the discrepancies between calculation and experiment.

A further very real consideration since 1958 has been the recognition that a comprehensive test ban may someday preclude testing, leaving the laboratories with calculation as the sole tool for meeting their obligation to maintain confidence in the US nuclear-weapons stockpile. We are not now at the point where we can maintain current confidence requirements with calculation alone.

Experienced judgment. Nuclear testing also plays an essential role in developing and maintaining the competence of the scientists and engineers at Livermore and Los Alamos who are responsible for the nuclear weapons' reliability. The same personnel who are charged with assessing the reliability of stockpiled nuclear weapons are involved in developing nuclear weapons. In their development work they are continuously having their judgments, based upon calculation and experience, tested against the reality of nuclear tests. Discrepancies between expectations and results are a constant reminder of the fallibility of computer calculation and "experience." Without testing, confidence in the judges of reliability will justifiably erode even in the unlikely event that the weapons laboratories can retain experienced personnel under such a circumstance.

Reopen the debate

As we have seen, nuclear weapons testing plays a major role in maintaining confidence in the country's nuclear deterrent. Some have argued that the United States understands fundamental nuclear-weapons phenomena well enough. Some have, as an act of faith—not through hard evidence—asserted that the US understanding of the application of nuclear explosions for military purposes encompasses all Soviet developments. They have said that US weapons are already safe enough and secure enough. They have argued that modernization without nuclear testing can maintain the survivability of weapons systems even in the face of as-yet-unknown threats; and they assert that the United States can, without testing, maintain confidence in its stockpile, or that, in any case, for some unspecified reason, confidence in the nuclear-weapons stockpiles of the United States and the Soviet Union will erode at the same rate.

There should be a debate on the subject. The United States in 1983 should reevaluate the desirability of a comprehensive test ban as a national goal. Indeed, it should be more than a national debate, it should be an international debate, because US confidence in its nuclear deterrent has international implications.

In 1983 we find nuclear weapons parity between the United States and the Soviet Union. It is a delicate balance. Will a comprehensive test ban increase chances of maintaining that stability or will it detract? Will a comprehensive test ban allow the US to maintain the reliable deterrent that a majority of its citizenry wants? A real discussion is called for. An informed debate should begin.

References

1. *Arms Control and Disarmament Agreements*, 1982 edition, United States Arms Control and Disarmament Agency, Washington, D.C. (1982).
2. J. C. Mark, Bulletin of the Atomic Scientists, March 1983, page 45. □

DEWITT *continued from page 46*

On the second point, the weapons laboratories maintain that they do not make national policy, but carry it out, and that when they advise policymakers they simply present the scientific truth. In this article I want to take a close look at these claims as a way of addressing the question of why even now, 38 years after the end of World War II, the labs feel that it is so important to continue nuclear testing.

Labs are an active lobby

Glenn T. Seaborg, as chairman of the Atomic Energy Commission during the Kennedy Administration, was deeply involved in the US–Soviet negotiations that culminated in the Limited Test Ban Treaty and moved nuclear testing underground. He has recently written a very significant book,[2] *Kennedy, Khrushchev and the Test Ban*, in which he describes the intricate negotiations that began in 1958 and culminated five years later in the first serious nuclear weapons treaty between the rival superpowers. At the time, Seaborg himself was strongly in favor of a treaty to end all nuclear weapons tests, and he still holds this position. Furthermore, as he explains in his book, both Kennedy and Krushchev had a deep commitment to a total ban on nuclear testing. Indeed, both men felt that such a ban would be a major step in the direction of world peace.

With this kind of commitment, coupled with the scare of the 1962 Cuban missile crisis, one may ask why Kennedy and Krushchev were not able to attain their goal of a comprehensive test ban in 1963. Obviously, each man had to contend in his own country with powerful forces opposed to a nuclear test ban. Seaborg discusses some of the opposition and how it affected the treaty negotiations. A recurrent theme in his book is the role of the American nuclear weapons labs and the efforts of leading weapons scientists to block the treaty; Krushchev evidently had similar troubles with the Soviet nuclear-weapons establishment. A few examples from the American side illustrate the influence of the US weapons labs:

▶ In 1957, Edward Teller and Ernest Lawrence met with President Eisenhower to argue against a moratorium on testing. They told Eisenhower that the Soviets could cheat with clandestine tests, and that US testing must continue anyway, to develop "clean" bombs, which they foresaw coming within seven years. These fallout-free devices would be deployed as tactical nuclear weapons in Europe. It has now been 26 years since the meeting with Eisenhower, and we have no "clean" bombs.

▶ As the test-ban negotiations proceeded, verification became a major issue because of studies from the weapons labs suggesting that nuclear bombs could be exploded in large cavities deep underground and decoupled sufficiently to look like much smaller explosions. One such study suggested that a 300-kiloton bomb might look like a one-kiloton explosion. There were also arguments about testing in space, on the other side of the Sun, for example. By 1963 the weapons labs prevailed and nuclear testing was allowed to continue underground.

I don't have the space to detail the numerous later examples of the influence of the weapons labs on US policy, but a recent example is important.

▶ President Carter began his four-year term with a determination to complete the comprehensive test-ban treaty negotiations. In the summer of 1978, Department of Energy secretary James Schlesigner took Harold Agnew, director of Los Alamos, and Batzel to see Carter to argue against United States participation in a comprehensive test-ban treaty. At that time, the Soviet Government was in favor of the treaty, and agreement seemed to be very near.[3] Yet the arguments of the weapons-labs leaders were apparently persuasive, and progress toward a comprehensive test-ban treaty stopped after that visit. Agnew later said[4] concerning that meeting,

> No question about it.... We influenced Carter with facts so that he did not introduce the [treaty] which, we subsequently learned, he had planned to do.

At this point, one can only speculate as to what alarming facts caused Carter to change his mind on the need to complete the comprehensive test-ban treaty. The Reagan Administration is far more inclined to see things the same way as the nuclear-weapons establishment, and on 19 July 1982 Reagan announced[5] an end to negotiations toward a comprehensive test-ban treaty, and thus a change in a 20-year-old US policy.

Corporate survival. The nuclear-weapons establishment occupies a very secure place in the American government. This is illustrated by the revealing testimony[6] of Major General William W. Hoover, director of the Department of Energy's Office of Military Application, before the Procurement and Military Nuclear Systems Subcommittee of the House Armed Services Committee. Hoover, speaking to a friendly Congressional subcommittee, indulged in a bit of humor and

Hugh DeWitt, a theorist in the H division of Lawrence Livermore National Laboratory's physics department, works on the statistical mechanics of strongly coupled plasmas.

DEWITT *continued*

likened the weapons establishment to a large corporation:

> We are something unique in the US Government—that is, a totally government-owned, integrated industry. A corporation, if you will, for which we are responsible.
>
> I would like the committee to consider themselves as the board of directors of that corporation. My remarks are in essence a prospectus of our corporation, and the record of this hearing will serve as our stockholders report.
>
> Let me touch briefly on the assets of our corporation.... The total number of employees is about 35 000. That includes production plants, test facilities, and the laboratories—those people who work for the weapons program.
>
> ... The results of our R&D activities lead to our product line ... warheads supporting weapons systems of the Department of Defense.

Hoover goes on to describe the weapons laboratories' "product line," which includes nine different types of mainly strategic nuclear warheads, such as the W76 for the Trident I missile, the W87 for the MX missile, and the B83, a "modern strategic bomb" for high-speed low-altitude delivery. He also talks of the "theater nuclear product line," meaning smaller nuclear bombs for fighting tactical nuclear wars.

I must comment at this point that many Livermore staff members who are committed to their work on weapons design believe sincerely in the idea of deterrence, and they will say that their nuclear bombs are designed and made for the purpose of never being used. One can only ask: Does the General's "theater nuclear product line" really mean only deterrence? I should also mention here that the Livermore Lab, which designs and develops many of these devices, is a large establishment with over 7000 employees and a proposed FY 1984 budget of 584 million dollars. Nuclear weapons work is big business!

Later in his testimony, Hoover makes a significant statement about the weapons laboratories' stake in testing:

> Like any good corporation, we have an investment strategy which we have been pursuing for the last couple of years and we intend to pursue it in the decade of the eighties.... We think we need to increase our manpower in research, development, and technology by about 15% above what it was a couple of years ago. We think we need to increase the level

of underground testing.

This kind of direct statement to Congress from a high Department of Energy military official provides one clear answer as to why we have no comprehensive test-ban treaty now and are not likely to have one in the near future. The nuclear-weapons establishment is a very powerful "corporation," staffed with intelligent and dedicated people whose livelihoods are tied to never-ending nuclear-weapons work. This establishment will not remain neutral and quietly allow elected representatives to curtail their enjoyable and profitable weapons work through limitations such as a comprehensive test-ban treaty.

Objections to a test ban

There are more serious reasons given for the ongoing nuclear testing. Two suggested reasons come from an unlikely source, Jack Anderson's nationally syndicated newspaper column. In a column[7] titled "Test Ban Folly," Anderson refers to classified White House documents presented at a secret Pentagon technical briefing and shown to him. From this information Anderson makes essentially two points:

▶ The Soviet Union is believed to have cheated extensively on the Threshold Test Ban Treaty of 1974, and is supposed to have exploded as many as 11 underground shots above the agreed-on 150-kiloton limit since 1978. (Although the United States has not ratified this

treaty, the US and the USSR have said they will comply with the 150-kt limit.)

▶ Nuclear weapons testing must continue indefinitely because the weapons labs are not confident that new bombs manufactured from proven designs will actually explode to design specifications. In other words, without continued proof testing, America's nuclear stockpile cannot be relied on the future. On both points it is my impression that Anderson was taken advantage of and shown the supposedly sensitive documents to spread ideas that cannot withstand scrutiny.

Let me first dispose of the question of Soviet cheating on the 150-kt limit. Reputable seismologists outside the weapons establishment have not confirmed the claim of Soviet cheating. In their recent article[8] in *Scientific American* on the verification of a comprehensive nuclear test ban, Lynn Sykes of Columbia University and Jack Evernden of the United States Geological Survey state that

> When the correct calibration is employed, it is apparent that none of the Russian weapons tests exceed 150 kilotons, although several come close to it.

From inside the weapons establishment we have a statement[9] by Michael May, associate director-at-large of Livermore, that classified documents "conclude that there was no evidence that the Soviets had cheated on the Threshold Test Ban Treaty...." Ger-

Cannon. This Army photograph shows the M100E2 8-inch self-propelled cannon, typical of modern long-range artillery. The nuclear-weapons laboratories developed atomic projectiles that can survive the acceleration associated with long-range delivery. The survival of the warheads is asssured by testing.

ald E. Marsh gives more details on this subject in his commentary in the March issue of the *Bulletin of the Atomic Scientists*. While people in the nuclear-weapons labs believe in their work, they are honest and don't believe in the story of Soviet cheating. That story emanates from officials in Washington, and it has the appearance of an attempt to justify American renunciation of the unratified Threshold Test Ban Treaty so that the US can once again test at above 150 kilotons.

Anderson's second point is far more serious. From the documents shown to him, he states that leaders of the US nuclear-weapons labs believe that they must have the ability to test up to five kilotons to guarantee the performance of weapons in the US stockpile. He quotes from one of the unspecified White House documents:

> In the continued non-nuclear testing of weapons components, it turns out with some regularity that individual components fail or degrade. Even acceptable components may become unavailable as manufacturers shift product lines or go out of business.

Note that this reason for the necessity of continued testing has nothing to do with developing new designs or even modifying proven old designs. The documents that were shown to Anderson claim that even to maintain a dependable stockpile of nuclear weapons manufactured from well-tested designs, it is necessary to test the bombs occasionally. This would preclude a comprehensive test-ban treaty forever, because no US president is likely to sign a such a treaty knowing that the US stockpile of nuclear weapons may degrade to the point of unreliability.

The suggestion that certain necessary materials might become unavailable as manufacturers change their line of products is a startling excuse for reserving the right to continue to set off nuclear bombs. Surely the Department of Energy can somehow solve this problem, given the money and resources available to it! Furthermore, if present-day proven bomb designs are that sensitive to slight changes in materials, then one must ask why the weapons labs have produced such designs. I think the answer is simply that the weapons labs have never had to contend seriously with the prospect of cessation of nuclear tests, and thus felt no need to design bombs that could be dependably manufactured in the distant future. I will come back to this problem later to argue that the weapons labs could solve it quickly if they felt the need to do so.

Exciting new weapons. Another reason why the labs want to avoid the restrictions of a comprehensive test-ban treaty is the exciting prospect of developing a whole new class of nuclear weapons. These are described rather vaguely as directed-energy weapons or third-generation weapons, and they have been widely promoted by Edward Teller since last summer. (See the news story on page 17.) The nature of these new devices is hidden behind walls of secrecy, and I may say very little about them. Teller claims that it is imperative for the US to develop these weapons because they would be "defensive" in nature and would provide a reliable defense against a Soviet nuclear attack. One of the ideas is the bomb-pumped x-ray laser described[10] a couple of years ago in *Aviation Week and Space Technology*. This marvelous device would supposedly send a burst of x rays at a Russian missile high above the Earth's atmosphere, and destroy it long before it reached the US. Teller and his colleague Lowell Wood from Livermore are reported[11] to have met with President Reagan last summer to promote the new weapons ideas and to propose a major increase in funding—$200 million per year—for the x-ray laser and related systems.

These ideas for new weapons provide excitement and challenge for the weapons laboratories. Regardless of whether they will ever work as weapons systems, they have their own dangers, I think, and should be examined carefully. The promise of a new nuclear defense against Soviet missile attack, as described by Teller, is misleading and dangerous if accepted uncritically by the American public and ill-informed officials. I see a number of serious consequences:

▶ For the x-ray laser to be developed into a weapon, it would have to be tested in space. This would probably violate the Outer-Space Treaty of 1967, which prohibits the placement of nuclear weapons in space. A US abrogation of this treaty could lead to the unraveling of all the arms-control agreements negotiated with such difficulty since 1963.

▶ Any complicated and expensive system, such as the x-ray laser, would be subject to a variety of countermeasures. For example, pieces of metal chaff near the target missile may give the same radar image as the missile itself.

▶ Reliance on new "defensive" nuclear weapons could lead to a false sense of security for the nation. Maybe these new ideas could be made to work after a few decades of expensive development, but for now they strike me as high-technology fixes that belong in "Star Wars" stories.

Finally, I want to point out that the new weapons systems give the weapons labs an additional strong argument against a test-ban treaty. If the third-generation weapons ideas are sold to the Reagan Administration and the Congress, then the weapons labs will need many years if not decades to develop them, and during that time a comprehensive test-ban treaty would obviously be impossible. This, in my opinion, is the main danger of Teller's new third-generation weapons.

Detectability

Sykes and Evernden conclude[8] that seismological monitoring techniques have become so good in recent years that compliance with a comprehensive test-ban treaty could be effectively verified. They state that present-day seismic monitoring methods are capable of detecting and identifying underground explosions in the Soviet Union down to yields of one or two kilotons. If an array of 15 unmanned seismic monitors were placed in the Soviet Union by treaty agreement—something the Soviets have already agreed to in principle[12]—then the detectability limit would be reduced to a fraction of a kiloton. Sykes and Evernden base these estimates on explosions in rock.

Decoupling by conducting explosions in cavities complicates the issue considerably. Sykes and Evernden say that with conceivable cavities in rock or in salt domes, the largest blast that the Soviet Union could mask in the presence of 15 seismic monitors is still only two or three kilotons. As one might expect, the weapons labs are quite disturbed by Sykes and Evernden's conclusions, and they dispute them. Milo Nordyke, who is in charge of treaty verification work at Livermore, says[13] that it is possible to decouple relatively small explosions. He suggests that in a large cavity, a 10-kt explosion may give the seismic signal of a 0.2-kt explosion. Such a decoupling by a factor of 50 would be a serious matter for treaty verification purposes—but note that the yields being discussed in 1983 are far smaller than the 300 kilotons that Teller in 1957 told Eisenhower could be hidden. There seems to be a healthy technical debate going now among seismologists in the nuclear-weapons labs and outside, and the Defense Advanced Research Projects Agency is making every effort to dispute and discredit the Sykes and Evernden work.[14] Whatever the outcome of this debate, it now seems clear that seismologists can detect quite small nuclear explosions and that this represents a serious threat to the weapons labs. If explosions above one kiloton were prohibited by a new treaty, the labs would be effectively out of business.

Clearly, the present 150-kiloton limit

of the Threshold Test Ban Treaty is unrealistically high. Some future US Administration may resume test-ban negotiations with the Soviet Union. If the direction of future negotiations is simply to modify the 150-kt limit, then it will be necessary to consider a treaty based on either a yield limit or a detectability limit. A yield limit would have to be determined by what both sides agree is a yield large enough to be detected in spite of decoupling. This might be considerably more than 10 kilotons in salt-dome cavities, for example. Conversely, the negotiating countries could try for a treaty based on a seismic detectability limit that might, for example, correspond to a 0.2-kiloton explosion in rock. It is important to note that a yield limit would be much more favorable in the eyes of the weapons laboratories, because to be realistic such a limit would have to be at least 10 kt. The labs could then continue their work with explosions below that limit. A detectability limit of a fraction of a kiloton, however, would seriously restrict the weapons labs.

One problem in negotiating a limit based on seismic detectability is that it would require much discussion of the masking of larger explosions in decoupling cavities. One would expect that in such negotiations the American weapons labs will argue strongly that the Soviets might clandestinely cheat occasionally with explosions of a few kilotons in expensive cavities. At some point, the political leaders of the US and the USSR would have to come to some understanding as to whether there is any advantage to be gained from small-scale clandestine weapons programs that risk detection as seismological methods improve.

Proof testing

As I see it, the nuclear-weapons establishment likes the protracted argument over verification of a comprehensive test ban or a low-yield threshold test ban because it focuses attention on the possible cheating capabilities of the Soviets. The seismic verification questions, as long as they sound alarming, serve the purpose of distracting attention from the really serious argument against a test ban that Anderson publicized. If the seismologists can make their case that verification of a test ban is really no longer a problem, then the labs will have to face the real question: Can dependable working bombs be manufactured in the future from today's proven designs? People at the top of the weapons-labs hierarchy say that bombs manufactured in years to come will not be dependable without continued nuclear testing. Other weapons experts deny this assertion. In August of 1978, when it seemed that

Treaty signing and toast, 5 August 1963. Top photo shows the signing of the Limited Test Ban Treaty in St. Catherine's Hall in the Kremlin. Seated are US Secretary of State Dean Rusk, Soviet Foreign Minister Andrei Gromyko and British Foreign Secretary Lord Home. Those standing include, from the left, US Senators George Aiken, William Fulbright and Hubert Humphrey, US Ambassador to the UN Adlai Stevenson, UN Secretary General U Thant and Soviet Premier Nikita Khruschev. Rusk, Home and Gromyko (below) are among the celebrants after the signing. (United Press International photographs.)

negotiations with the Soviets on a comprehensive test-ban treaty were close to success, three men from Los Alamos wrote President Carter a very significant letter concerning testing and the reliability of the stockpile. They were Norris Bradbury, director of Los Alamos from 1945 to 1970, J. Carson Mark, head of the Theoretical Division of Los Alamos for 26 years, and Richard Garwin, a consultant at Los Alamos since 1950. In their letter they argue that it is possible to have a reliable stockpile even with a comprehensive test-ban treaty. They pose the question,

> Can the continued operability of our stockpile of nuclear weapons be assured without future nuclear testing? That is, without attempting or allowing *improvement* in performance, reductions in maintenance cost, and the like, are there non-nuclear inspection and correction programs which will prevent the degradation of the reliability of stockpiled weapons?

Their answer is "yes," and they go on to address several problems that must be solved to maintain and manufacture reliable bombs, including the problem of materials acquisition mentioned by Anderson. They further point out that

> It has also been rare to the point of non-existence for a problem revealed by the sampling and inspection program to *require* a nuclear test for its resolution.

Livermore personnel disagree with the assessment of Bradbury, Mark and Garwin. Thus, as I quoted earlier, Livermore's director Batzel stated in September 1978 that continued nuclear testing is necessary to keep the US stockpile credible.

The disagreement among weapons scientists in 1978 may have provoked some further thinking about the problems of a potential comprehensive test ban treaty. In February 1980, Joseph Landauer, then assistant associate director for arms control at Livermore, wrote a classified report[15] titled *National Security and the Comprehensive Test Ban Treaty*. Later that year he released a declassified version[16] that is fairly close in content to the classified document. This report gives reasons for and against a comprehensive test ban treaty—primarily against—and is a rare example of what amounts to a publicly available policy statement from a weapons laboratory. As one argument against a comprehensive test ban treaty, Landauer raises the question of materials replacement, saying:

> We expect that all nuclear weapons will have to be replaced or remanufactured within a few decades of their original manufacture. More and more of our stockpiled weapons are approaching

retirement age. No amount of good intentions or executive decisions will ensure the availability of exact replacement materials or prevent subtle changes in manufacturing processes.

The implication is clear that nuclear testing is required in the future to make sure that replacements or newly manufactured bombs actually work. There are also some classified aspects of current weapons designs that persuade people at Livermore that standard bombs built in the future must be tested to make sure they work. These design questions need to be examined by qualified scientists from outside the weapons laboratories.

Clearly, what has happened in the 20 years since the signing of the Limited Test Ban Treaty is that the weapons laboratories have produced sophisticated designs that are very efficient but so delicate that the labs seem to have no confidence that they can be manufactured reliably in the future. This raises two questions:

▶ Why have the laboratories been allowed to produce weapon designs that effectively preclude the US from ever signing a comprehensive test-ban treaty? Surely this is strange considering that such a treaty has been a US policy goal for 20 years.

▶ Can this situation be changed? That is, can the weapons laboratories quickly modify some of their designs so that bombs can be built reliably in years to come in the event of a comprehensive test-ban treaty?

Landauer makes another point that we have to consider seriously. He notes that Russian nuclear bombs are generally heavier, possibly less sophisticated, and possibly more dependable for manufacturing in the future. Thus he fears a serious degradation gap favoring the Soviets after a few years of a comprehensive test ban. He says

> We cannot assume that stockpile degradation will be symmetrical in respect to US and Soviet weapons. We do not know how Soviet weapons are made, what their remanufacturing problems are, or by what means the Soviets can maintain the skills of their weapon scientists.

It would be ironic indeed if the cruder and more robust Soviet bomb designs allow the Soviets to be better prepared for a comprehensive test ban. I believe—for whatever it's worth—that the Livermore and Los Alamos Laboratories are full of clever weapons scientists who can in a short time meet the technical challenge posed by a test ban, and can produce bomb designs that avoid stockpile degradation problems. Certainly this should be one of the

duties of the weapons laboratories.

Nominally, the University of California manages both the Livermore and Los Alamos laboratories under contract with the Department of Energy. The University of California obviously cannot interfere with the nuclear weapons design work done at the two labs, but it does have some oversight role. Part of this role is handled by a committee, the Livermore and Los Alamos Scientific and Academic Advisory Committee, which reports to the president of the university and occasionally to the university's regents.

What should be done?

In February 1982, Ray Kidder, one of my colleagues at Livermore, made a formal presentation to the Advisory Committee concerning the question of the necessity of continued nuclear testing to assure stockpile reliability. He asked the committee to look into this question as a technical scientific matter and to try to resolve the conflicting claims of weapons experts. In July of last year, Kidder sent a letter to David Saxon, then president of the university, stating that

> The purpose of the report [of the Advisory Committee] would be to provide government policy makers with information that is of fundamental importance in the formulation of national policy concerning nuclear weapons, and that directly concerns the statements and activities of the two weapons laboratories under the stewardship of the University.

After some months of prodding, the answer finally came. It was *no*, the Scientific Advisory Committee would not be authorized to take up the question. Evidently this question impinges on national policy and is simply too difficult for the university to study.

To conclude, I will give my own opinions on what should be done.

▶ I think Congress or some part of the US Government should appoint a high-level committee of competent scientists, with members from outside the weapons establishment, to examine carefully the problem of bomb replacement and to figure out what needs to be done so that a comprehensive test-ban treaty will be possible.

▶ Congress, with the help of a group of qualified scientists from outside the weapons establishment, should examine carefully all the ideas for third-generation weapons, and make sure that they don't instantly become mammoth secret projects that attain their own momentum and destabilize the present precarious arms-limitation agreements.

▶ Technology for seismic verification of a comprehensive test-ban treaty appears to be sufficient already, and I am not convinced by the labs' arguments about the need for indefinite nuclear testing. The world needs to stop nuclear testing even more now than it did in 1963. Thus I hope that a more enlightened Administration in a few years will approach the Soviets again and complete the agreement that Kennedy and Krushchev tried to attain, namely a comprehensive test-ban treaty. Not everybody agrees that stopping continued nuclear-bomb development will reduce the possibility of nuclear war, but I think so.

Seaborg concludes his book with a strong recommendation for a renewed effort to reach agreement on a comprehensive test-ban treaty. The final words in his book reflect the urgency of this task: "The hour is late. Let us hope not too late."

References

1. This September 1978 statement by Roger Batzel is quoted in the abstract of J. K. Landauer, *National Security and the Comprehensive Test Ban Treaty*, Lawrence Livermore National Laboratory report number UCRL-52911(SRD), 29 February 1980. The abstract has been declassified; the report is classified as Secret Restricted Data.

2. G. T. Seaborg, *Kennedy, Khruschev, and the Test Ban*, U. Calif. P., Berkeley (1981).

3. Science **201**, 1105 (22 September 1978).

4. Interview with Harold Agnew, Los Alamos Science, volume 2, number 2 (Summer/Fall 1981).

5. Widely reported in national newspapers on 20 July 1982. Implications of this announcement are discussed at length in The Defense Monitor, volume XI, number 8, Center for Defense Information, Washington, D.C. (1982).

6. Excerpts from General Hoover's testimony given in Public Interest Report, Federation of American Scientists, Washington, D.C. (October 1982), page 8.

7. See, for example, the *San Francisco Chronicle*, 10 August 1982.

8. L. Sykes, J. Evernden, Sci. Am., October 1982, page 47.

9. Letter from Michael May to Gerald E. Marsh, 17 December 1982.

10. Aviation Week and Space Technology, 23 February 1981.

11. Aviation Week and Space Technology, 20 September 1982.

12. See Herbert York's article, PHYSICS TODAY, March 1983, page 24.

13. Quoted in *The New York Times*, 8 March 1983, page 13.

14. Unpublished notes from the office of Thomas C. Bache Jr., Geophysical Sciences Division, Defense Advanced Research Projects Agency, January 1983.

15. J. K. Landauer, UCRL-52911(SRD), see reference 1.

16. J. K. Landauer, *National Security and the Comprehensive Test Ban Treaty*, Lawrence Livermore National Laboratory report number UCRL-84848, August 1980 (unclassified).   □

NUCLEAR TEST BAN VERIFICATION AGREEMENTS YIELD NEW SEISMIC DATA

PHYSICS TODAY/NOVEMBER 1987

New ground in the field of nuclear test ban verification will be broken next year under an agreement concluded in June between the Natural Resources Defense Council, an American advocacy organization that works primarily on environmental and arms control issues, and the Soviet Academy of Sciences. The agreement, the successor to another agreement between the two parties reached in May 1986 (PHYSICS TODAY, July 1986, page 63, and August 1986, page 57), provides for the establishment of five seismic stations in the Soviet Union at distances greater than 1000 kilometers from the Soviet test site in Kazakhstan. In contrast to the three existing NRDC–Academy stations, which are between 100 and 200 km from the test site, the five new stations will be permitted to record signals from Soviet nuclear weapon tests.

Implementing the agreement will involve moving the three existing seismic stations to locations more than 1000 km from the test site, establishing two entirely new stations and—starting on 15 July 1988—training academy personnel to operate the equipment with support, as needed, from the NRDC. "The academy will staff and operate the new and relocated stations with occasional assistance as required and requested from the NRDC," the agreement says. "All seismic data, including recordings of American and Soviet nuclear tests [made] at these five stations, will be available for both sides."

The purposes of the new five-station network will be, first, to test whether a low-threshold (on the order of 1 kiloton) test ban treaty could be verified with confidence by relying on a comprehensive network of stations spaced at intervals of about 1000 km and, second, to support an intergovernmental exchange of seismic data that is to take place next year under the auspices of the Ad Hoc Group of Scientific Experts of the disarmament conference in Geneva. The first such

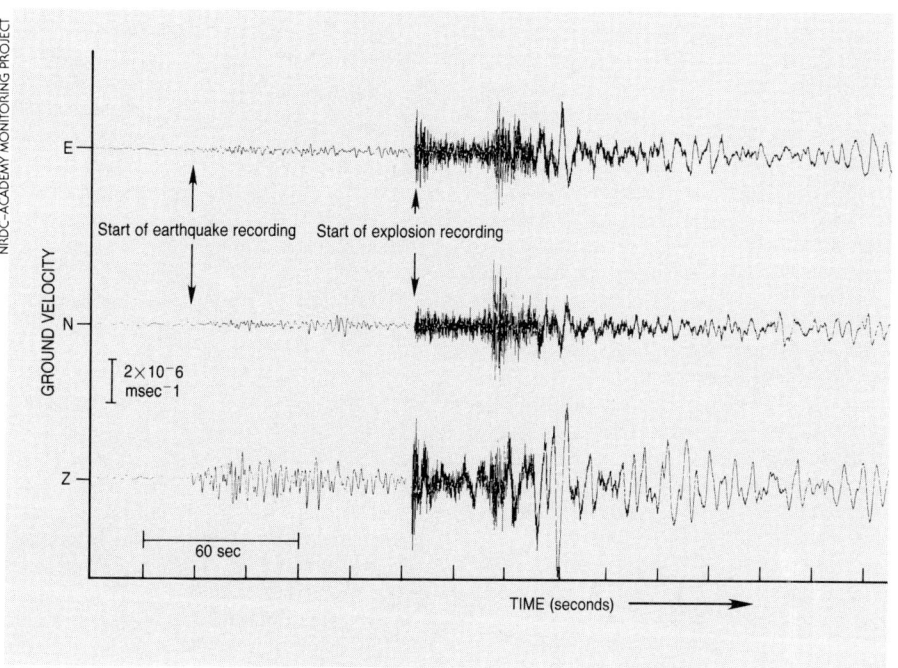

Seismometer recording from detonation of a 10-ton chemical explosive near the Soviet test site on 3 September. The first signals from the explosive arrived barely a minute after the first signals from an earthquake in the McQuarrie Islands, south of New Zealand. The higher frequency of the signals from the explosion distinguishes its signature from that of an earthquake. Three components of ground motion are shown: vertical (Z), east–west (E) and north–south (N).

exchange, starting in 1984, involved basic information on matters such as arrivals of pulses and estimated event locations. Next year, binary waveform data—the actual seismograms—are to be traded among participating government agencies.

The NRDC–Academy agreement also provides for an experiment to be performed in which the Soviet Union will detonate chemical explosives of known yield near the test site so that equipment at the three current seismic stations can be calibrated. Three such tests took place in early September, two with 10-ton charges and one with a 20-ton charge. According to Thomas B. Cochran, the NRDC physicist who first suggested establishment of the stations to the Soviets, the

experiment was "successful beyond our wildest dreams."

The experiment was designed to see whether the stations would be able to pick up signals from a "decoupled" test of a low-yield nuclear weapon—that is, a test in which the explosive is placed inside a large cavern so that shock waves are muffled. In all three experiments, the stations picked up a rich mix of strong signals, including the high-frequency signals that are considered crucial to detecting decoupled tests and distinguishing weapons explosions from other seismic events such as earthquakes or industrial explosions.

By happy chance, an earthquake occurred moments before one of the tests, and the signals from the earthquake and the test proved to be readily distinguishable on seismometer recordings.

Summing up the results of the tests, Charles Archambeau of the University of Colorado has written: "It is apparent that high-frequency seismic signals could be easily detected to distances well beyond 1000 km from such small explosions. Since the explosions were designed to be seismically comparable to decoupled 1-kt nuclear explosions, this means that a decoupled 1-kt test, at or near the Soviet test area, could be detected over a very wide area (i.e., over an area exceeding 3 million square kilometers)."

Archambeau has overall responsibility for seismic research in the NRDC–Academy project. Jonathan Berger of the Scripps Institution of Oceanography is director of the US field team. Igor Nersesov, chief of seismology at the USSR's Institute of Physics of the Earth, heads the Soviet team.

Following the chemical explosive tests in September, the Soviets expressed interest in leaving the three existing stations in place and adding five entirely new ones at distances of more than 1000 kilometers. Berger has prepared a new plan and budget for the five new stations and has submitted them to the Soviets.

Bumpy background

When the Soviet academy and the NRDC negotiated the original test monitoring agreement in May 1986, the stated objectives were "to demonstrate that in-country nuclear weapons test verification is not an obstacle to a comprehensive test ban or a moratorium on testing, to demonstrate that scientists of the United States and the Soviet Union are prepared to cooperate to work toward a common goal of a CTB, and to obtain baseline seismic data that would be useful in designing and operating a seismic verification network."

Despite the Reagan Administration's unenthusiastic attitude toward the project, the NRDC succeeded in quickly obtaining Commerce Department approval for the export of seismic equipment to the Soviet Union, and with support from a number of private foundations, US and Soviet scientists managed within a few months to set up stations in the Soviet Union that were able to start gathering data on ambient ground noise and propagation of various wave types. This work confirmed that compression body waves (P waves) suffer less attenuation at the Kazakhstan test

site than at the US test site in Nevada.

Efforts to establish similar stations in the United States were less successful. The Reagan Administration took the position that it would not grant visas to Soviet scientists unless they agreed to observe a demonstration at the Nevada test site of CORRTEX, a monitoring technique involving the placement of coaxial cable in a bore hole near the test explosive. When the Soviet government refused to authorize participation in such a demonstration, which it has regarded as irrelevant to the problem of verifying compliance with a comprehensive test ban treaty, its scientists were permitted to come to the United States only for a week, and in that time they were only able to select proposed sites for US monitoring stations. When they applied for visas a second time in February, the same conditions were imposed, and they remain without visas.

A further setback came in February when the Soviet Union resumed nuclear testing after an 18-month moratorium and ordered the NRDC to turn off its monitoring equipment during tests. Negotiations between the NRDC and the academy had left unresolved the issue of whether the stations would be allowed to continue gathering data during tests, but the two parties had agreed to a memorandum of understanding saying that "recording of tests of nuclear weapons is not necessary to the success of the joint research being undertaken." The Soviet side took the position that the point of the project was to verify compliance with a test ban, not to estimate yields of actual tests.

Achievements to date

With the resumption of Soviet testing and the reorientation of the bilateral monitoring project toward verification of low-threshold test bans, the original goal of demonstrating the feasibility of a comprehensive test ban has receded somewhat. Nonetheless, the project has demonstrated the political feasibility of establishing in-country monitoring stations. In addition, the project is providing—or will soon provide—a variety of information that until now was unavailable outside the Soviet Union:

▷ Ironically, as a result of the periodic orders to the NRDC to turn off its equipment, for the first time there is official notification of Soviet tests.

▷ From the September experiments with chemical explosives near the Soviet test site, the first local measurements are publicly available of how seismic waves propagate from

well-defined and scheduled Soviet explosions.

▷ As a result of US tests of known yield in Nevada being monitored by the NRDC–Academy stations in the Soviet Union, scientists have obtained the first unclassified data on attenuation of compressional waves between the two test areas.

▷ With the establishment next year of the expanded network, the first local estimates of Soviet test yields will be publicly available.

Finally, for better or worse, the project may have helped contribute to the resumption of test ban negotiations and to a small but notable shift in the Reagan Administration's position on the ultimate goal of a comprehensive test ban. During Soviet Foreign Minister Eduard A. Shevardnadze's visit to Washington on 17 September, the two superpowers agreed to launch a multistage negotiation in which the goals would be first to improve verification of existing threshold test ban treaties—a step the USSR had previously resisted—and then to discuss intermediate, and presumably lower, test limits. The ultimate goal, according to an announcement the two sides released, would be the complete cessation of all nuclear weapon testing. Previously, the Reagan Administration's position had been that a complete test ban would not be worth discussing as long as the United States continued to rely on nuclear weapons for its security.

A fringe benefit?

In mid-September, Representatives Thomas J. Downey of New York, Bob Carr of Michigan and Jim Moody of Wisconsin visited the NRDC–Academy monitoring stations in the Soviet Union in the company of a larger party that included reporter William J. Broad of *The New York Times*. After arriving in the USSR, the party received permission—partly as a result of NRDC mediation—to visit the Soviet Union's controversial radar at Krasnoyarsk, exhibit A in the Reagan Administration's case that the USSR has not abided by the letter and spirit of the 1972 ABM treaty.

The highly publicized visit produced an ambiguous result and shed some light on the risks political leaders run when they dare to permit closer inspection of their most sensitive military installations. The US party concluded that the radar would indeed be a violation of the ABM treaty if it is ever brought into operation, but it also concluded that the installation was ill suited to be used for battle management of a missile-defense system, as the Pentagon had contended. Members of the party described the facility as "shoddy," and one congressman said that the generals who built it should be court-martialed.

In the estimation of Jeremy Stone, staff director of the Federation of American Scientists, the radar visit "put a human face on a situation that was abstract before and dramatized that the Soviets are not denying the radar is there." Stone thinks the visit also conveyed that the Soviets are trying to be helpful to groups in the United States that favor arms control but that are handicapped by reports of alleged Soviet treaty violations.

Such efforts are not without pitfalls. Stone points out that Soviet scientists were quite annoyed about the denigrating language some of the American visitors used to describe the radar. He says one leading Soviet scientist remarked: "It's as though you invite guests to your home, and when they leave they report that you have dirty toilets. Naturally your wife tells you not to invite them back."　　　—WILLIAM SWEET

Yields of US and Soviet nuclear tests

Failure to account properly for geological and seismological differences between the US and Soviet test sites has led to overestimates of the yields of Soviet tests and to incorrect claims of Soviet cheating on the treaty limit of 150 kilotons.

Jack F. Evernden and Gerald E. Marsh

PHYSICS TODAY/AUGUST 1987

The likelihood that the United States will negotiate a comprehensive or low-threshold test ban treaty with the Soviet Union in the relatively near future depends not only on the ability of the US to monitor such an agreement but also on US perception of past Soviet compliance with treaties that limit nuclear testing. Of particular importance is the 1974 Threshold Test Ban Treaty, which prohibits tests of nuclear weapons exceeding 150 kilotons in yield. This treaty is unratified, but both the United States and the Soviet Union have avowed their compliance since 1976, when the treaty was scheduled to go into effect. (For an annotated list of treaties, see Herbert York's article in PHYSICS TODAY, March 1983, page 24.)

The US needs an accurate method for estimating the yields of Soviet nuclear tests, not only to assess compliance with existing or future treaties but also to estimate the effectiveness of Soviet strategic forces. Although several systems employing a variety of methods are currently available for detecting Soviet tests, only signals recorded by

seismic networks give accurate estimates of the yields of individual underground explosions. Other techniques give estimates of the yields only of sets of Soviet tests.

We begin this article by reviewing the several seismological procedures for estimating the yields of Soviet explosions, our purpose being to demonstrate that yield estimates made by direct seismological measurements are accurate and do imply that the Soviets have abided by the 150-kiloton limit of the Threshold Test Ban Treaty.

Seismic magnitudes

The discussion centers on two seismic "magnitudes," m_b and M_S. The magnitude m_b is calculated from measurements of P waves, which are compressional seismic waves in the body of the Earth. The magnitude M_S is calculated from measurements of Rayleigh waves, which are surface seismic waves. These magnitudes are simple logarithmic functions of the amplitudes of the pertinent seismic waves. The amplitudes are normalized for distance from the epicenter and for path of propagation, so that magnitude estimates made at arbitrary distances and locations are internally consistent. The normalizations for distance and propagation path are based upon extensive empirical data obtained over decades by many seismologists.

A seismic event does not have a single magnitude—it has several. When one uses different seismic waves

and frequencies to estimate the magnitude of an event, one gets markedly different numerical values for the magnitude. For instrumental and seismological reasons, routine magnitude estimates are based on P waves of 1-second period and Rayleigh waves of 20-second period. The magnitudes m_b and M_S assigned a seismic event are averages of the magnitudes determined at many seismological stations. By applying empirically observed corrections based on data from many earthquakes and explosions, one can reduce the standard deviation of the magnitude estimates for any given event to 0.20–0.25 units of magnitude. The standard deviation of the mean depends on the number of stations used to determine the magnitude. The relationships between the magnitudes and the yields of nuclear explosions are also derived empirically: Yields of an adequate number of US explosions have been declassified and published to allow accurate estimates of the relationship between magnitude and yield at the Nevada test site.[1]

Magnitude bias. Because of geological

Jack Evernden is a research geophysicist at the US Geological Survey in Menlo Park, California. Gerald Marsh is a defense analyst at Argonne National Laboratory, in Argonne, Illinois. The views they express in this article are not to be construed as those of the US Geological Survey, Argonne National Laboratory, the Department of Energy or the US government.

Test of a nuclear device with an energy yield equal to that of 10 kilotons of TNT. The test, code-named Baneberry, took place at the Nevada test site, 110 miles north of Las Vegas, on 18 December 1970. The venting of radioactive dust was accidental. Releases of radioactive material in underground explosions at the site since this test have been limited to minor seepage. **Figure 1**

variations within the crust and upper mantle of the Earth, an explosion at the Soviet test site at Semipalatinsk or Novaya Zemlya produces seismic P waves of much greater amplitude at distances than does an explosion of the same yield at the US test site in Nevada, shown in figure 1. The difference is determinable quantitatively; to obtain it one solves the so-called P-wave magnitude bias or m_b bias problem. The large regional variations in the properties of the Earth's crust and upper mantle lead to different attenuations of body waves, thus causing amplitude differences and magnitude bias along different paths between epicenters and seismic stations.

Quantitative determination of the P-wave magnitude bias has been a major issue in estimating the yields of Soviet underground explosions. Studies done in the early 1970s demonstrated the existence of the bias.[2,3] The US government, however, refused to accept the conclusions of those studies, even though they were conducted within the Defense Department and cleared by that department for publication in the open scientific literature. Finally in 1977 the government admitted that seismic amplitudes of P waves do vary with the sites of the sources and receivers, but chose for reasons never specified to apply correction factors that were too small. This choice formed the basis for claims by the Reagan Administration of Soviet noncompliance with the 150-kiloton test limit. Only now is the government moving toward acceptance of the proper correction factor, bringing its position into conformance with long demonstrated fact.

Something still not appreciated by many is that one can estimate the yields of large Soviet explosions through seismological procedures that do not require estimates of the magnitude bias between the US and Soviet test sites, and that these procedures confirm the validity of the method used to calibrate the P-wave magnitude bias estimates made by totally separate seismological procedures. One such technique is use of Rayleigh wave amplitudes, or magnitudes M_S.

Following our review of seismological procedures for estimating the yields of Soviet tests, we introduce a method for estimating the yields of US tests and show that this method gives estimates that agree with previously published yields of individual events and with recently declassified data on the US test pattern for 1980–84. We then investigate the Soviet testing program by placing the calculated pattern of Soviet tests within the constraints of physical law and military requirement—that is, to compare the US test

program, a program with known logic, with that of the Soviet Union while assuming various values of P-wave magnitude bias. We also investigate the differences in the two test programs as a function of time. We show that if US and Soviet nuclear weapons are similar in basic design, meaning that most are two-stage thermonuclear weapons having primaries of comparable yields (we believe that military requirements and the physics of nuclear weapons demand this), then we would expect the testing programs of the two countries to show similar distributions of yields for a given level of weapons sophistication. We show that the level of the P-wave magnitude bias needed for this to be the case is consistent with that determined from seismological considerations. Conversely, if we accept the magnitude bias as determined by seismology, a comparison of the two test programs shows them to be strikingly similar. The comparison we make of the complete test programs of the two countries is the first to appear in the unclassified literature.

Explosion seismology

An underground explosion creates a radially symmetric shock wave that propagates outward with a velocity that depends on the yield. If the rock containing the explosion has little or no initial stress and if the properties of the volume of rock experiencing the shock wave are uniform, then the shock wave will consist almost completely of P waves and Rayleigh waves. The former are compressional body waves that propagate throughout the entire volume of the Earth, while the latter are surface waves whose amplitudes decrease exponentially with depth. Two types of surface waves are possible: Rayleigh waves and Love waves. Rayleigh waves are the elastic analog of waves on the surface of water; in addition to vertical motion, particles undergo horizontal motion in the plane of propagation, resulting in elliptical particle motion. Love waves are shear surface waves displaying horizontal particle motion perpendicular to the direction of propagation.

Simple conditions of very low initial stress, or pre-stress, are not general throughout the shallow zones of the Earth in which explosions occur. There is very little or no pre-stress in rocks overlying some subduction zones (the Aleutian Islands, for example), salt deposits and unconsolidated alluvium, but most other rock is under some pre-stress (10–100 bars in seismic areas), which arises from the never ending deformational processes going within the Earth. An explosion produces a sphere of shattered rock of zero

strength, allowing a readjustment of stress in the surrounding rock. This readjustment produces radiation of the full complement of possible types of elastic waves—shear body waves and Love surface waves in addition to the P body waves and Rayleigh surface waves. The waves resulting from this tectonic release can complicate identification criteria and can require calibration of Rayleigh-wave amplitudes when those amplitudes are used to estimate yields. However, the contribution of stress release to P-wave amplitudes has never been sufficient to alter the P-wave magnitude bias.

Estimating the bias

Figure 2a is a plot of the P-wave magnitudes m_b of Soviet nuclear tests as a function of time. The points of immediate interest are the cessation of high-magnitude tests by the Soviets in 1976 because of the 150-kt limit, and the rise in magnitude from 5.8 to 6.2 in the years following 1976. All the large Soviet tests before 1976 were at Novaya Zemlya, although most Soviet tests have been at Semipalatinsk.

As we will see below, an explosion at the Nevada test site with a magnitude m_b of 6.2 would have a yield of 600–700 kilotons. However, there is a large P-wave magnitude bias between Semipalatinsk and the Nevada test site, and one must take this into account if one is to use yield-vs-magnitude curves based on the Nevada test site to make accurate estimates of the yields of Soviet tests of such magnitude. Let us review the seismological procedures that demonstrate the magnitude bias.

Figure 3a is based on data for 33 Soviet explosions for which Lynn Sykes and Inés Cifuentes at Columbia University have published carefully calibrated mean magnitudes m_b and M_S calculated from data from many seismic stations.[4] Yields were calculated from magnitudes M_S using the following formula:[5]

$$\log_{10}(\text{yield in kt}) = 0.762M_s - 1 \quad (1)$$

This equation, which is based on explosions of known yield from numerous sites worldwide, applies to explosions in hard rock anywhere. The existence of a universally applicable relationship between magnitude M_S and yield for explosions in hard rock when Rayleigh waves are little affected by the release of tectonic stress was first shown[2] in 1971, and the bases for it were first summarized[6] in 1977.

In figure 3a, the yields based on the magnitudes m_b of the 33 Soviet explosions were calculated using the relationship that is correct for the Nevada test site (see figure 4b), only the high-

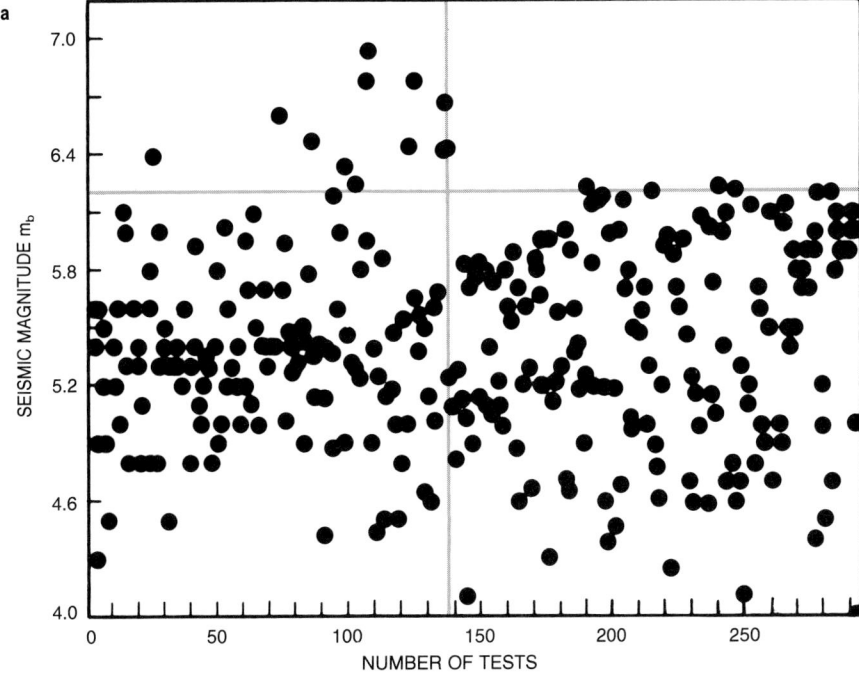

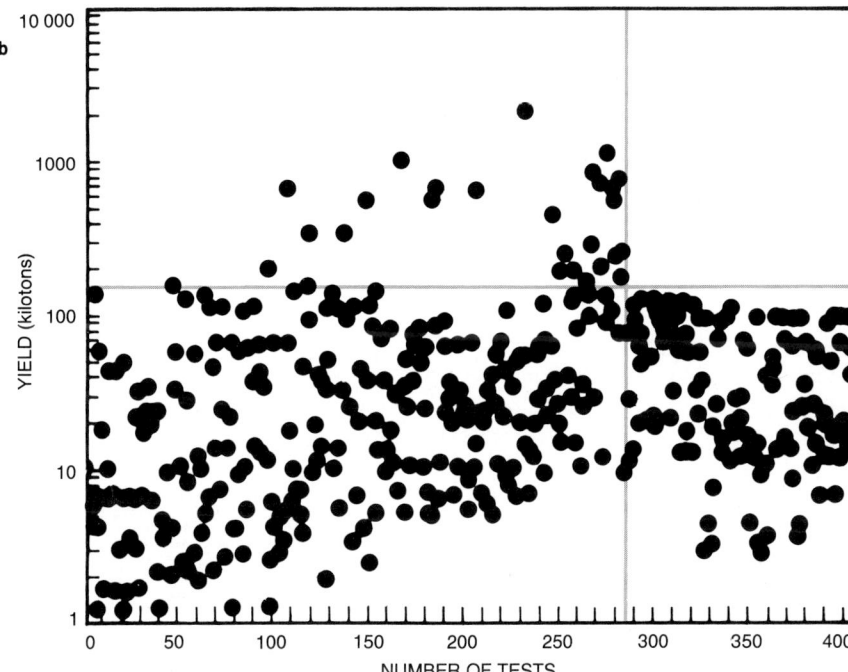

Soviet and US testing patterns. a: Seismic magnitudes m_b as a function of time for Soviet tests, 1964–85. The time axis is not uniform, as each tick mark indicates 10 tests. **b:** Yields of US tests as a function of time, 1963–84. The vertical lines mark the effective date of the Threshold Test Ban Treaty.　　　　　　　　　　　　Figure 2

US test site. To resolve this problem, one must determine whether the problem is with the magnitudes m_b or the magnitudes M_S.

Evernden and John Filson in the early 1970s were the first to demonstrate[2] the existence of this problem, and Evernden explained[6] it in the mid-1970s. Simply put, there are significant variations in anelastic processes and in the velocity structure of seismic waves at depths of 0–200 km in different regions of the Earth. Tectonically active terrains, such as that found at the Nevada test site, are characterized by high mean elevation, high heat flow, low P-wave velocities at the Mohorovičić discontinuity (the boundary between the Earth's crust and mantle), marked travel-time delays of both P and S body waves, and significant anelastic attenuation of all seismic body waves. The P-wave attenuation leads to lower measured amplitudes and thus to lower measured magnitudes m_b than in terrains such as Canada, the eastern United States and most of the Soviet Union. None of these Earth structures significantly perturb the amplitudes of 20-second Rayleigh waves, the waves used to determine the magnitudes M_S.

Several procedures

The contrast in properties between active and stable terrains leads to several distinct procedures for estimating the nature and size of the magnitude bias illustrated by figure 3a:
▶ Both empirical data and theoretical analysis establish that 20-second Rayleigh waves do not experience significantly different propagation effects in different terrains, and also establish the uniformity of explosion-generated low-frequency source amplitudes, and thus Rayleigh-wave amplitudes, for explosions of the same yield in a wide variety of "hard rock."[6] Therefore, if tectonic release does not perturb the magnitudes M_S, no correction need be applied to these magnitudes, and one can estimate[7] directly from figure 3a the bias in the magnitudes m_b.
▶ If it becomes necessary to correct the observed magnitudes M_S for the release of tectonic energy, one can use procedures for calculating both the explosion- and earthquakelike components of the 20-second Rayleigh waves from multi-azimuth multiperiod observations of both Rayleigh and Love waves. Use of the corrected magnitudes M_S then permits estimation of the bias in the magnitudes m_b. The assumptions required, the sensitivity of the calculations to the details of the earthquakelike energy release, the size of the necessary correction for some of the Semipalatinsk events and the limit-

yield portion of the relationship being needed:

$$\log_{10}(\text{yield in kt}) = 1.25m_b - 4.95 \quad (2)$$

Note that using equations 1 and 2 gives pairs of values for the Soviet explosions, that is, for log yield(M_S) and log yield(m_b), and that these pairs result in a distribution of points (figure 3a) that scatter around a straight line with a slope of essentially 1. This

implies, as expected, that the fundamental physics controlling the slopes of equations 1 and 2 is identical at hard rock sites in the US and USSR. Note also that the yield-versus-magnitude curves appropriate to the Nevada test site give m_b-based estimates of yields of Soviet tests that are 3 to 4 times greater than the M_S-based estimates. Thus there is a problem of "magnitude bias" between the Soviet test sites and the

ed data base may well prevent determination of the proper value of the magnitude M_S within 0.1 magnitude units for an explosion with high tectonic release. Some have thought that this problem of marked contamination of the M_S value by the release of tectonic stress seriously perturbed the M_S values obtained for the large post-treaty explosions at Semipalatinsk. It is now generally accepted, however, that there is no evidence for such a perturbation. Contamination of the magnitude M_S by tectonic release is not a problem for any explosions at the Nevada test site if those explosions are observed over a broad range of azimuths.

▶ The velocities P_n of the horizontally traveling seismic waves along the Mohorovičić discontinuity are strongly correlated with the amplitudes of the vertically traveling P waves.[8] The contrast in the velocities P_n between the Soviet test sites and the US test site (8.2–8.3 km/sec versus 7.9 km/sec) confirms a bias in the magnitude m_b of a few tenths between the Soviet and US sites. However, this technique is inadequate to determine the bias in the magnitude m_b at a particular site to the 0.1–0.2 level.

▶ There is an extremely close correlation between the travel times and the amplitude attenuations of P waves through the crust and upper mantle of the Earth. This correlation is associated with variations as great as 0.2 in m_b bias for stations on similar rock in terrains with indistinguishable velocities P_n, and with variations of greater than 0.5 in m_b bias between such stations in different-P_n terrains.[3] To apply this technique of travel time versus amplitude to evaluation of the P-wave bias at Novaya Zemlya and Semipalatinsk relative to the Nevada test site one needs the P-wave arrival times at seismic stations throughout western North America and Eurasia and data from a station at or very near each test site. The Soviets have for several decades published in their seismological bulletins such P-wave arrival times for several standard stations, including one at Semipalatinsk. One can get a detailed estimate of the m_b bias at Semipalatinsk relative to Nevada by this procedure[6]; the resulting value is 0.45 ± 0.02. Unfortunately, there were no Soviet stations near enough to Novaya Zemlya to allow an equally accurate estimate of P-wave magnitude bias there by this technique. Note that two totally separate procedures for estimating the m_b bias between Semipalatinsk and Nevada—M_S and P-wave travel time—agree on a value of about 0.45.

▶ If one knows the shape of the source

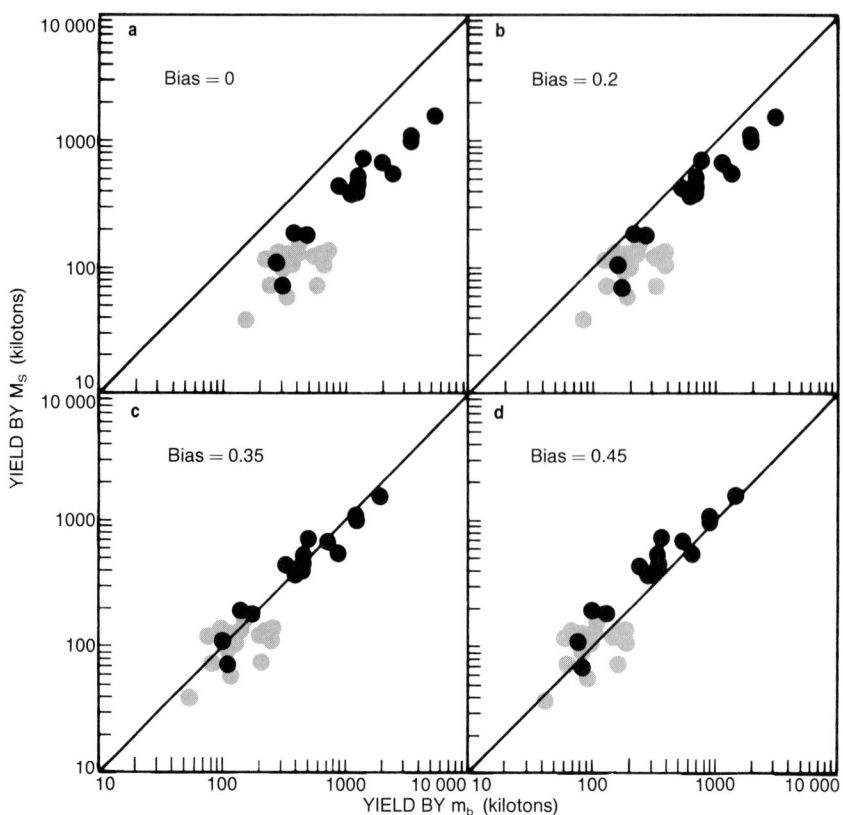

Calculated yields for 33 Soviet tests with accurately determined seismic magnitudes m_b. The yield of each test was calculated two different ways—by its seismic magnitude m_b and by its seismic magnitude M_S. The two methods give the same results when one assumes a bias to correct for seismological differences between US and Soviet test sites. Black points represent tests at Novaya Zemlya; red points, tests at Semipalatinsk.[4]   Figure 3

spectrum of an explosion, one can observe the broadband spectrum at a distance, calculate the attenuation parameter required to reshape the observed spectrum to the source spectrum and then apply this attenuation parameter[9] to the passband used in estimating the magnitude m_b. The problems in this procedure are the assumptions one must make about the details of the source spectrum and the difficulty of uniquely separating elastic and anelastic propagation effects. This technique has reportedly given estimates of an m_b bias of 0.35 for Semipalatinsk. The assumptions in the technique may lead to errors of as much as 0.1 in estimating the correct bias value.

▶ The ideal and direct technique, of course, would be to have amplitude data for P waves of distant earthquakes or explosions as recorded at or very near the Soviet test sites. The Soviets have not made such data available in their seismological bulletins. However, a group from the University of California at San Diego is working in the Semipalatinsk region on a project funded by the Natural Resources Defense Council (see PHYSICS TODAY, August 1986, page 57); one hopes this group will obtain such data.

Recalculating yields. Figures 3b–d present the yields for the 33 events used in figure 3a, recalculated assuming various values of P-wave magnitude or m_b bias for the Soviet test sites relative to the US test site. A value of 0.2 (figure 3b), decreed by bureaucratic fiat within the US government in 1977 in spite of evidence presented at the time that the proper bias value for Semipalatinsk was at least 0.40, is seen to be inadequate to achieve agreement of yield estimates calculated from the magnitudes m_b and M_S. Figure 3c indicates that an m_b bias value of 0.35 quite nicely fits the Novaya Zemlya data, in agreement with Sykes and Cifuentes and the analyses given earlier.[4] A value of 0.45 (figure 3d) gives a better fit to the Semipalatinsk explosions, in agreement with the analysis of P-wave amplitudes versus P-wave travel times discussed above. As pointed out earlier, the appearance of different bias values for the two sites is not surprising, as greater differences in m_b bias were demonstrated many years ago within geotectonic terrains of the US comparable to those at Novaya Zemlya and Semipalatinsk.[3]

Figure 3d indicates that even when we use an m_b bias of 0.45 for Semipala-

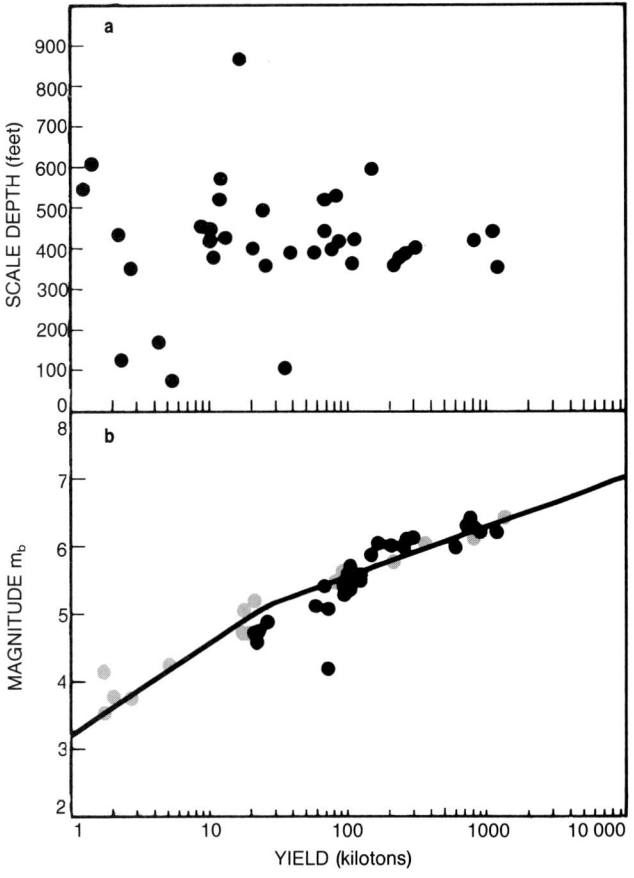

Scale depth and magnitude m_b as functions of yield. **a:** Plot of calculated scale depth versus yield for US explosions with announced yields and depths, 1963–76. **b:** Plot of magnitude m_b versus yield for Nevada test site explosions in mesa tuff or granite. Red points denote explosions during the period 1963–76 with announced yields,[13] while black points denote explosions during the period 1971–83 with yields calculated on the basis of announced depths and a scale depth of 450 feet. The linear segment in the high-yield region is given by equation 2.   Figure 4

for the quantitative evaluation of the P-wave magnitude bias problem, Richards first quoted from a 1977 report by Evernden[6] and then stated that "Although in some details the scientific argument has required slight revision in light of later work, the main conclusion is [essentially that?] adopted recently when the US government's [position?] was revised." The phrases in brackets are our guesses of the words that the government censored before releasing Richards's testimony in unclassified form.

US testing program

To use an indirect procedure to estimate the yields of Soviet nuclear tests, it is necessary to have a reasonably accurate picture of the entire US test program. The yields of only a few US explosions at the Nevada test site have been declassified. (For a complete list of these events, see *Nuclear Weapons Databook*, Natural Resources Defense Council, Washington, DC, 1984.) However, the government has announced the depths of nearly all the Nevada explosions. Figure 4a, a plot of data for Nevada events with announced yields and depths, indicates a strong tendency to detonate explosions at a scale depth D_s of 400–450 feet, based on the relation

$$\text{Depth(feet)} = D_s \, [\text{yield(kt)}]^{1/3} \quad (3)$$

Figure 4b also suggests use of a scale depth of 450 feet. The red points on Figure 4b are data for explosions in mesa tuff or granite for which the government announced yields; the m_b values for these explosions are as published by Evernden and Archambeau[5] or in the bulletins of the National Earthquake Information Service.[13] The line drawn through these points is somewhat different from that published in Evernden and Archambeau, because the high-yield curve of figure 4b has the added constraint imposed by the two explosions near 90 kilotons, code-named Miniata and Starwort. The bend in the curve indicates[5] a change in the frequency spectrum for yields exceeding about 50 kilotons. The black points in figure 4b are based upon m_b values published by NEIS for explosions in mesa tuff or granite between 1971 and 1983, with the "yields" for these events being based on their announced depths, use of equation 3 and a scale depth of 450 feet. The agreement of the black points with the curve based on the red points indicates that usual US practice in recent years certainly has been to use a scale depth of close to 450 feet. The single black point far below the other data implies either a greatly over-buried explosion—one much deeper than the 450-foot scale

tinsk, the several high-yield events since 1978 have calculated yields with a mean value of about 175 kilotons and maximum and minimum values of about 195 and 155 kilotons. The figure suggests that either the set of five large explosions with calculated yields greater than 150 kt (the five rightmost crosses in each frame of figure 3) were actually each somewhat greater than 150 kt, that a greater m_b bias value is required or that some other factors such as rock properties or slight errrors in the yield-vs-magnitude curves slightly perturbed the m_b value. That the calculated yields are slightly above the 150-kt limit therefore does not necessarily mean cheating by the Soviets. (The standard deviation of the mean m_b value of these explosions is 0.018, meaning that the standard deviation of the yield estimates is about 10 kt, assuming correctness of the yield-vs-mgnitude curves; if all are assumed to be of the same yield, the mean calculated yield is about 175 kt with a standard deviation of 5–10 kt.)

Lawrence Livermore National Laboratory reached a similar interpretation of the seismological data of the Semipalatinsk explosions some years ago. In a report in *The Bulletin of the Atomic Scientists*, Marsh presents quotations from letters by Warren Heckrotte and Michael May of Livermore; in these

May states explicitly that internal Livermore documents "did conclude that there was no evidence that the Soviets had cheated on the Threshold Test Ban Treaty."[10] Roger Batzel,[11] the present director of Livermore, reiterated that evaluation in testimony to Congress in 1985.

The analysis of the US and USSR testing programs presented below indicates that the Soviets would accrue no technical advantage by testing at 175 kilotons rather than at 150 kilotons. Although it has not yet been demonstrated, we expect that the explanation for these higher-yield tests will probably be found in seismological details such as those suggested above.

Thus several simple, direct and totally independent seismological procedures for estimating yields of Soviet explosions agree on a positive P-wave magnitude bias value of 0.35 or greater for both Novaya Zemlya and Semipalatinsk relative to Nevada, while the procedures we deem most accurate indicate a P-wave magnitude bias of 0.45 for Semipalatinsk. In this regard it is worth quoting from a presentation to Congress by Paul G. Richards, a geophysicist at Columbia University's Lamont–Doherty Geological Observatory, who had reviewed all available information.[12] Discussing the adequacy of then available data and analyses

depth—or an explosion that did not go off at the expected yield—one with a magnitude m_b lower than expected.

Figures 5a and 5b are yet another demonstration that the United States uses a scale depth of approximately 450 feet. The black curve of figure 5a is the plot of $df(Y)/d\log(Y)$ published by Ray Kidder[14] of Livermore for test yields at the Nevada test site during 1980–84. Here $f(Y)$ is the fraction of tests whose yields were Y kilotons or less. Kidder's curve is based on the actual yields. The red curve in figure 5a is the function $f(Y)$ derived by integrating Kidder's curve. The red histogram of figure 5b is derived from the function $f(Y)$ of figure 5a, while the black histogram is derived from US tests during the same period by using published depths of explosions, a scale depth of 450 feet and equation 3. The vertical scale of the red histogram is adjusted to imply the same number of tests as the black histogram. The two patterns of figure 5b are different in detail but basically very similar. Though the curves have peaks of somewhat different shape, they do show nearly the same number of high- and low-yield tests. The agreement is adequate to support the conclusion that explosions at the Nevada test site are routinely detonated at a scale depth very near 450 feet. In the following discussion, interpretation of the US testing pattern is based on a scale depth of 450 feet for all explosions. We do not mean to imply that one can accurately estimate the yield of each US test in this manner; it does, however, yield a close estimate of the pattern of US testing.

USSR testing program

Non-seismological arguments based on the above discussion of the US testing program and on various other considerations all indicate a high bias in the magnitudes m_b for Semipalatinsk and Novaya Zemlya relative to the Nevada test site, with an associated lack of regional variation of the magnitudes M_S. Statements both old and recent by people connected with the US weapons laboratories provide the bases for these arguments.

An argument for a large m_b bias between Semipalatinsk and Nevada was presented by weapons designers 20 or so years ago—before there was a seismological understanding of the bias problem. Recent statements in the open literature permit us to present that argument, which is as valid today as it was originally.

To begin with, it has often been stated that nuclear warheads are characterized by having "primary" and "secondary" explosive devices, both of them nuclear. This is confirmed in a

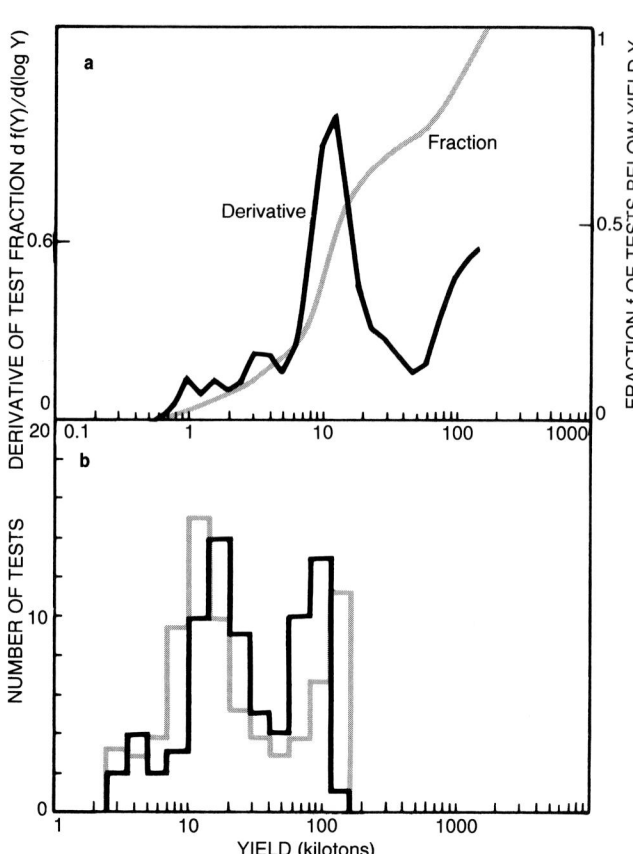

Yield distribution of explosions[4] at the Nevada test site, 1980–84. The black curve in **a** is the published function $df(Y)/d(\log Y)$. The red curve, obtained by integration, is the implied fraction $f(Y)$ of tests with yields at or below Y kilotons. The black curve in **b** is a histogram of yields calculated from announced depths when one assumes a scale depth of 450 feet for all explosions. The red curve in **b** is a histogram based on the red curve in **a**. This histogram and those in subsequent figures are terminated at 2.5 kilotons with yield-bin boundaries increasing by a factor of $2^{1/2}$.   Figure 5

1983 Lawrence Livermore bulletin, which states that "x-rays produced during the nuclear explosion of the primary transfer energy to compress and ignite thermonuclear fuel contained in the secondary."[15] It is the testing of the primary that is critical to the design of a nuclear weapon, as Richard L. Wagner Jr, then assistant to the Secretary of Defense for atomic energy, stated in 1986: "The primary design is the sort of bellwether of whether it will work … we can test all primary designs which are at a lower yield, thereby giving us continued reasonable assurance that the design will work at full yield."[16] Wagner was speaking within the context of the Threshold Test Ban Treaty, which, he noted, allows one to test "the primary or a combination of the primary and an altered secondary, which would not exceed the threshold."

One can infer from a comment by Harold Agnew in a letter to Congressman Jack Kemp that testing at a yield of about 10 kilotons or somewhat more is of particular importance to a rational test program[17]: "I don't believe testing below five or ten kilotons can do much to improve (as compared to maintaining) strategic posture." As we will see, the distinction between "improve" and "maintaining" corresponds to that between testing secondaries or only primaries. Discussing the effect that a comprehensive test ban treaty would

have on weapons design, Agnew said, "The military significance to either the USSR or the USA of conducting clandestine tests below 5 or 10 kilotons is *per se* of relatively little importance today."

Figure 6a shows the yields of US tests during 1963–65, the first three years of underground testing, calculated assuming a scale depth of 450 feet. The histogram has peaks at a few kilotons and indicates an almost uniform level of testing from 10 to 56 kilotons. The testing at low yields is almost certainly related to the development of tactical nuclear weapons and primaries; note in figure 2b the concentration of low-yield tests in the very early years of the underground program.

During the years 1965–70, as US seismologists were establishing that it is possible to identify the seismic waves of underground explosions down to magnitudes m_b of around 4.75 (corresponding to yields of 10–15 kilotons in hard rock, as figure 4b indicates), representatives of the weapons laboratories repeatedly depreciated these accomplishments as being of no significance to a test-ban-monitoring environment: They stated that it is possible to scale up to full yield without testing above 10–20 kilotons. The implications of that repeated statement agree with the supposition that primaries have a maximum yield of 10–20 kilotons and with Wagner's statement about the

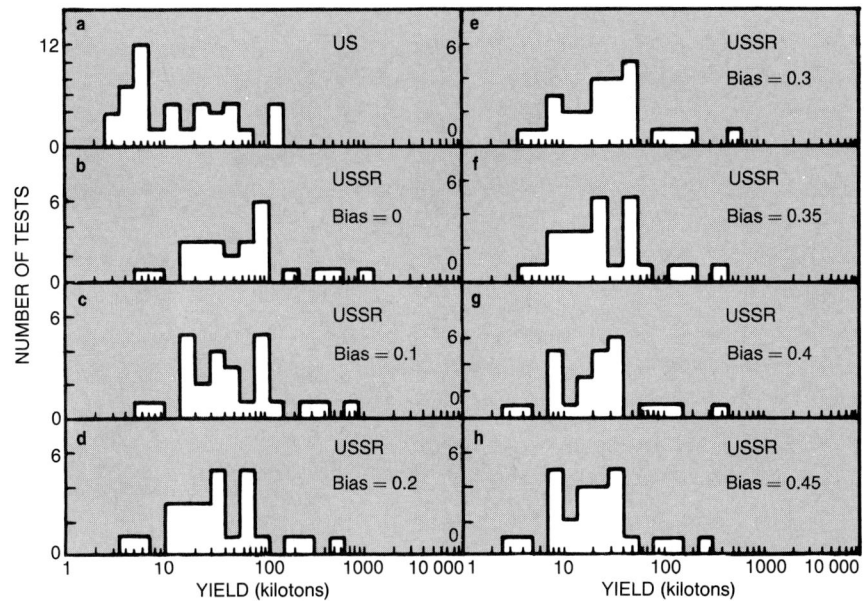

Histograms of yields for the US and Soviet test programs during the first three years of underground tests. Yields for US tests during 1963–65 were calculated using announced depths and a scale depth of 450 feet. Yields for Soviet tests during 1964–66 were calculated using National Earthquake Information Service magnitude values, equation 2 and the indicated biases in the magnitudes m_b.   Figure 6

possibility of scaling up from the testing of primaries or of primaries and degraded secondaries.[16]

Figures 6b–h are based on the first three years of Soviet underground testing. Figure 6b, which assumes no bias in the magnitude m_b, indicates that there were no Soviet tests between 10 and 14 kilotons and that the highest peak in activity was at 80–112 kilotons. This pattern is drastically different from the US pattern in figure 6a and implies the unlikely conclusion that the Soviets were testing secondaries and relatively few primaries or that they were using 100-kt primaries. However, as the P-wave magnitude bias increases from figure 6b to figure 6h, the calculated pattern of Soviet testing becomes very similar to the US pattern. Under the obvious constraint that physical and military requirements dictate the character of testing patterns, figures 6f–h imply an m_b bias distinctly greater than 0.35 at Semipalatinsk.

Further study of the patterns of testing by the US and the USSR strengthens these conclusions as well as leading to some interesting interpretations. Figure 2b presents the US testing pattern for 1963–84, based on a scale depth of 450 feet and the use of all events for which a depth was announced. Each tick mark along the horizontal axis indicates ten tests. The vertical line demarcates 31 March

1976, the acceptance date of the 150-kt limit of the Threshold Test Ban Treaty. Note the burst of high-yield testing just prior to the treaty and the absence of tests exceeding 150 kt since the treaty. (The magnitudes m_b of these events confirm them to be less than 150 kt.) The concentration of low-yield tests in the mid- to late 1960s is obvious.

Figure 2a gives the pattern of Soviet testing during 1964–85, expressed in magnitudes m_b as published by the National Earthquake Information Service. The maximum magnitude m_b of 5.8 immediately after the treaty and the subsequent rise to 6.2 are clearly visible. This rise at Semipalatinsk and confusion about its proper interpretation triggered the now largely resolved intragovernmental disputes about the magnitude bias in 1976.

Sykes has pointed out that study of the Soviet pattern of high-yield tests is yet another means of demonstrating a large bias in the magnitudes m_b at Novaya Zemlya.[18] He notes that between 1973 and 1976 the USSR deployed five strategic systems, including the MIRVed versions of the SS-17, SS-18 and SS-19, with warhead yields beteen 300 and 600 kilotons according to CIA National Intelligence Estimates quoted by Peter Samuel.[19] Samuel's statements make it clear that these yield estimates are based on non-seismological criteria. Samuel says that if one uses the official value of the m_b

bias, presumably 0.2, then no Soviet tests prior to 1976 have had yields in the 300–600-kt range. He concludes that either the methods of determining yields from seismic data are incorrect or the yields of Soviet warheads given by the National Intelligence Estimates are wrong.

Let us assume that the Soviets tested these warheads at full yield. Figure 2a shows there to have been five explosions prior to 1976 in the m_b range 6.4–6.5, all of them at Novaya Zemlya and of very similar yield. Use of an assumed m_b bias of 0.2 leads to the conclusion that there was little testing in the range 300–600 kt, but that there were five tests in the range 600–800 kt, a yield range with no known purpose for Soviet warheads. Using an m_b bias of 0.35, however, one finds the set of five explosions to have calculated yields of 400–500 kt, while a bias of 0.4 gives yields of 300–400 kt for these explosions. This adjustment of calculated yields by use of the seismologically correct m_b bias brings the seismological estimates and the National Intelligence Estimates into full agreement. The National Intelligence Estimates thus give another demonstration of a P-wave magnitude bias at Novaya Zemlya of 0.35–0.40 as shown in figure 3.

The impact of using the correct m_b bias to evaluate the Soviet test program is succinctly indicated by comparing the aggregate explosive yields of that program obtained assuming no magnitude bias and using the proper biases at Semipalatinsk and Novaya Zemlya. Use of zero m_b bias leads to a calculated aggregate yield of 59 megatons for all Soviet underground tests, while use of the proper m_b biases lowers the estimate to 19 megatons.

Historical comparison

Figure 7 shows histograms of the patterns of testing by the US and USSR over the years. The figure separates the testing programs into four time intervals. In the US program, the years 1965–74 were characterized by development of sophisticated warheads, with little further advance in design in later years. One can interpret the nearly uniform level of testing between 20 and 1000 kilotons during this period as a reflection of the development of sophisticated secondaries. Contrast this US pattern with that of the USSR during the same years. With the assumed bias of 0.45 at Semipalatinsk, the site of all these tests, the Soviet testing pattern shows marked peaking at 15–30 kilotons. This is the expected pattern if primaries were of those yields and if the test program included mostly primaries, with full-yield weapons requiring little further

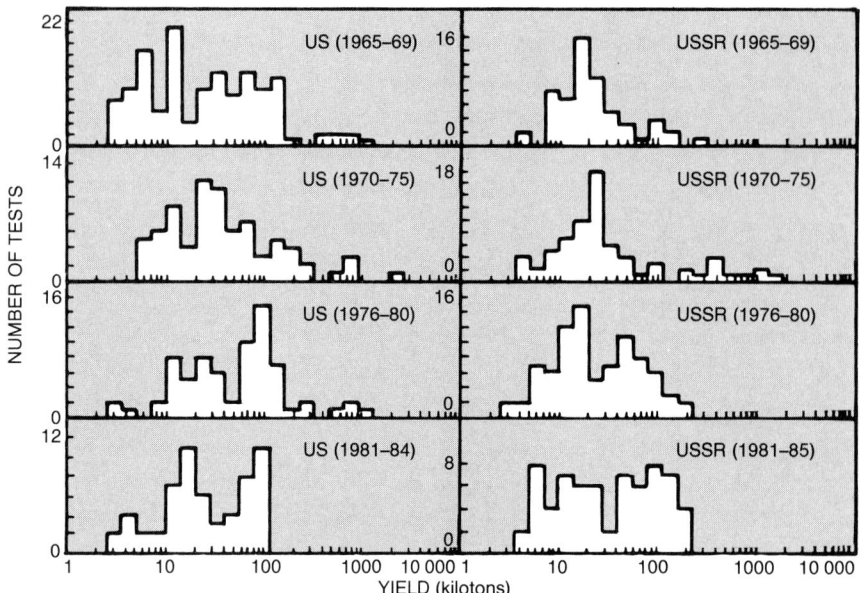

Evolution of US and Soviet underground testing programs. Some US test yields in the period 1976–80 exceed 150 kilotons because that period brackets the acceptance of the treaty limit. The absence of US test yields exceeding 113 kilotons in the period 1981–84 may reflect use of a scale depth of slightly less than 450 feet for the larger explosions of this period. As discussed in the text, yields of the largest Soviet tests since the treaty are calculated as 170 ± 20 kilotons; most of these explosions thus fall in the yield bin 160–226 kt. The histograms assume a 450-ft scale depth for US tests and a magnitude bias of 0.45 for Soviet tests.   Figure 7

testing as discussed above.

Comparison of the US and Soviet testing patterns for 1981–85 suggests that both countries do indeed use primaries with maximum yields of 10–30 kilotons. Further, inspection of the periods 1976–80 and 1981–85 indicates a marked change in Soviet testing; these patterns look very much like the US 1965–69 and 1976–80 patterns, respectively, lacking the high-yield events barred by treaty. Thus it seems reasonable to conclude that the Soviet testing program and stockpile designs have been proceeding toward greater sophistication along the lines pursued by the US 5 to 10 years earlier. Arguments for the greater reliability and slower degradation of Soviet weapons under a comprehensive or low-threshold test ban treaty, based as they are on the presumed simplicity of Soviet warheads, may not be as solidly based as frequently implied.[20] It is also not necessarily true that any degradation of the nuclear weapons stockpile need occur under such treaties.[21]

The multiplicity of seismological arguments demonstrating a large bias in the P-wave magnitude m_b between the Soviet test sites and the Nevada test site—arguments extant in the seismological literature since the early 1970s—should finally be accepted as definitive. The only remaining issue is the exact value of the bias at any particular point, as there is marked variation in bias even within terrain that appears uniform by superficial analysis.

References

1. For an introduction to seismological definitions relevant to nuclear tests, see L. Sykes, J. Evernden, Sci. Am., October 1982, p. 47; J. Evernden, Bull. At. Sci., March 1985, p. 9.

2. J. Evernden, J. Filson, J. Geophy. Res. **76**, 3303 (1971).

3. J. Evernden, D. Clark, Phys. Earth Planet. Inter. **4**, 1, 24 (1970).

4. L. Sykes, I. Cifuentes, Proc. Natl. Acad. Sci. USA **81**, 1922 (1984).

5. J. Evernden, C. Archambeau, in *Technical Means of Verification of Compliance with Arms Control Treaties*, K. Tsipis, ed., Pergamon, New York (1986), ch. 16.

6. J. Evernden, *Regional Bias in Magnitude Versus Yield Measurements: Its Explanation and Modes of Evaluation*, unclassified report submitted to Arms Control and Disarmament Agency, Washington, DC (1977). This report contains the most expansive discussion and supporting list of references available on the numerous geophysical parameters that correlate with bias in the magnitudes m_b and that explain its fundamental cause; the report also discusses the lack of regional bias in the magnitudes M_S.

7. C. Archambeau, Geophys. J. R. Astron. Soc. **29**, 329 (1972). C. Archambeau, C. Sammis, Rev. Geophys. **8**, 473 (1970). J. Evernden, C. Archambeau, E. Cranswick, Rev. Geophys. **24**, 143 (1986).

8. P. Marshall, D. Springer, H. Rodean, Geophys. J. R. Astron. Soc. **57**, 609 (1979).

9. Z. A. Der, T. W. McElfresh, Bull. Seismol. Soc. Am. **70**, 921 (1980). Z. A. Der, A. C. Lees, M. R. Marshall, EOS Trans. **67**, 302 (1986).

10. G. E. Marsh, Bull. At. Sci., March 1983, p. 4.

11. R. Batzel, in *Department of Energy Fiscal Year 1986 Authorization for Defense Programs*, testimony before the Subcommittee on Strategic and Theatre Nuclear Forces of the Senate Armed Services Committee, 99th Congress, 1st session, 14 March 1985; available in Senate Hearing 99–485, US Govt. Print. Off. (1986), p. 169.

12. P. Richards, presentation to the House Permanent Select Committee on Intelligence, Subcommittee on Oversight and Evaluation, 6 May 1986.

13. Depths of US explosions as announced by AEC or DOE are included in the seismological bulletins issued by the National Earthquake Information Service of the US Geological Survey. This article uses the magnitudes m_b calculated by the NEIS and published in its bulletins.

14. R. Kidder, oral presentation at DOE-sponsored Workshop on Cavity Decoupling, 29 July 1985, document no. UCRL-93194, Lawrence Livermore National Laboratory, Livermore, Calif. (August 1985); reprinted in Public Interest Report (Federation of American Scientists, Washington, DC), September 1985, p. 1.

15. *LLNL Classification Bulletin WNP-45A* (supplement), Lawrence Livermore National Laboratory, Livermore, Calif. (28 June 1983).

16. R. L. Wagner Jr, in *Review of Arms Control and Disarmament Activities, 99th Congress, 1st Session*, House Armed Services Committee report no. 99-18, US Govt. Print. Off. (1986), p. 115.

17. H. M. Agnew, letter to J. Kemp, 19 April 1977, in *Effects of a Comprehensive Test Ban Treaty on United States National Security Interests*, House Armed Services Committee report no. 95-89, US Govt. Print. Off. (1978), p. 192.

18. L. Sykes, letter to D. B. Fascell, chairman of House Foreign Affairs Committee, 30 August 1985.

19. P. Samuel, Defense Week, 5 August 1985, p. 1.

20. J. K. Landauer, *National Security and the Comprehensive Test Ban*, document no. UCRL-84848, Lawrence Livermore National Laboratory, Livermore, Calif. (August 1980).

21. R. E. Kidder, *Evaluation of the 1983 Rosengren Report from the Point of View of a Comprehensive Test Ban*, document no. UCID-20804, Lawrence Livermore National Laboratory, Livermore, Calif. (1986). R. E. Kidder, *Stockpile Reliability and Nuclear Test Bans: Response to J. W. Rosengren's Defense of His 1983 Report*, document no. UCIS-20990, Lawrence Livermore National Laboratory, Livermore, Calif. (February 1987).   □

NRDC and Soviet Academy sign unusual test-verification pact

PHYSICS TODAY/JULY 1986

In a dramatic move that some might characterize as taking US foreign policy into its own hands, the Natural Resources Defense Council concluded an agreement on 28 May with the Soviet Academy of Sciences providing for establishment of seismological stations in the USSR and United States to collect data on nuclear-weapons tests. The agreement was signed in Moscow by NRDC Chairman Adrian W. Dewind, a lawyer with Paul, Weiss, Rifkind, Wharton and Garrison in New York, and Evgeny P. Velikhov, vice-president for physical and mathematical affairs of the Soviet Academy.

Under the agreement, NRDC would set up three stations in the Soviet Union, each within 200 kilometers of the main Soviet nuclear-weapons test site near Semipalatinsk in the south-central part of the country. In exchange, three similar stations would be located in the United States near the Department of Energy's Nevada Test Site. The stations would be staffed jointly by scientists sponsored by NRDC and the Soviet Academy. Participation by US government scientists is not precluded by the agreement.

The idea for the agreement was hatched at NRDC by Thomas B. Cochran, a senior staff scientist known for his critical study of the liquid-metal fast-breeder reactor and, more recently, the data books he coedits on nuclear weapons. NRDC is a private organization with headquarters in New York City and offices in San Francisco and Washington, DC, that lobbies and takes legal actions on a wide range of matters affecting the environment, including the proliferation of nuclear weapons.

Cochran says he had an opportunity in the early spring, during a Federation of American Scientists conference at Airlie House near Washington, DC, to discuss his idea for a monitoring agreement with a Soviet delegation headed by Roald Z. Sagdeev, director of the Soviet Academy's Institute of Space Research in Moscow. In April, Frank von Hippel of Princeton University had another chance to discuss the idea with Velikhov during a trip to Moscow, and they arranged for a workshop to take

place at the Soviet Academy at the end of May. Dewind and Cochran presented their idea at the workshop on 22–23 May, and five days later the Soviets signed the agreement.

Cochran emphasizes that the mission of the seismic stations would be to provide improved data, not to make independent determinations of yields or of whether tests in fact have taken place. Cochran notes that Charles Archambeau, a seismologist at the University of Colorado at Boulder, is heading the NRDC technical team.

Lynn Sykes, a geophysicist at the Lamont–Doherty Geological Laboratory of Columbia University, describes Archambeau as one of the few university people who have worked continuously on the seismology of test verification for 25 years. Sykes says Archambeau is "regarded as one of the very best people working in that area" and notes that Archambeau is the coauthor with Jack F. Evernden and Ed Cranswick of a paper in the May 1986 *Reviews of Geophysics* on the propagation of high-frequency waves—the aspect of verification seismology in which "the biggest advances in maybe ten years have taken place."

Sykes sees several technical advantages in the NRDC–Academy agreement. First, if the stations were in place and operating, one could tell down to a very low level whether the Soviets are observing—at least at Semipalatinsk—the moratorium on nuclear tests that they claim to have initiated last August. Sykes thinks that the stations could detect blasts "well under 1 kiloton, assuming there were no attempts at muffling."

Second, if the Soviets resumed testing, then the stations would provide some valuable information on high-frequency waves, Sykes observes. Third, the stations would pick up ongoing information about seismic events such as chemical explosions, earth-quakes and so on—data useful in discrimination of nuclear tests from other events.

Finally, Sykes believes that the agreement establishes the principle that the three stations in each country

could become the prototypes for a larger network if a comprehensive test ban were negotiated.

State Department reaction. In a press directive issued to members of the State Department, foreign-service officers have been instructed to respond to questions about the NRDC agreement by focusing on the asymmetry in the representation of scientists—official Soviet Academy versus NRDC–nongovernmental—and on the implications of the agreement for verification of the Threshold Test Ban Treaty and the Peaceful Nuclear Explosives Treaty. The press directive says:

We're taking a careful look at this development in the context of improved TTBT and PNET. To the extent that it may reflect a change in Soviet attitudes toward the need to share additional verification data on nuclear testing it appears to be encouraging.

While the agreement between the NRDC and the USSR Academy of Sciences has no official US government status, we are interested in learning more about the discussions. Obviously issues with such strong national security implications as nuclear testing ultimately must be discussed in a government-to-government context.

The Soviet position, as we understand it was conveyed to the NRDC, provides for the participation of Soviet government scientists in improved in-country monitoring of underground nuclear explosions. We therefore see no reason why the Soviet Union should not agree to the President's standing offer to have a meeting of US and Soviet government scientists and experts on improving TTBT and PNET verification with on-site monitoring with the CORR-TEX technique [a method of estimating yield by means of a coaxial cable placed in or near the test hole, with an accuracy of $\pm 30\%$].

Because effective verification is a matter that can only be resolved at the government-to-government level, one would expect the Soviets—if indeed they are serious regarding verification improvements—to accept the standing US invitation for a meeting of US and Soviet government experts to discuss verification improvements for the TTBT and PNET.

Reacting to the State Department's directive, NRDC Chairman Dewind says: "NRDC is not involved and would not want to be involved in the discussion of official US–Soviet relations. And that, I think, is one of the merits of what we are doing here. This is purely a scientific exchange of basic data, and what it will demonstrate is that the Soviet Union and the United States will not make in-country verification any sort of obstacle to a test ban. That's our objective here. Much of the State Department directive relates to other policy relationships and negotiations—that's not our bag."

Regarding the alleged asymmetry between NRDC-sponsored and Academy scientists, NRDC officers point out that US government scientists are welcome to participate and that the Academy is one of the more independent Soviet institutions. The Academy's Institute of Earth Physics, the participating unit, is thought to be primarily research oriented and according to a CIA guide is a "leading center for the study of earthquakes and other seismic phenomena."

Test-ban negotiations. In theory at least, it was official US policy from 1963 to 1981 that a verifiable comprehensive test-ban treaty was an immediate objective. The partial test-ban treaty and the nonproliferation treaty, both ratified by the US, contain clauses committing the parties to the pursuit of a complete test ban.

Until the end of 1977, it was the Soviet position that any test-ban treaty had to make allowance for so-called peaceful nuclear explosives, which the Soviets hoped to use in civil-engineering projects. The Threshold Test Ban Treaty of 1974 and the Peaceful Nuclear Explosives Treaty of 1976 therefore allowed nuclear explosives of under 150 kilotons per explosive.

Both the House and Senate have passed resolutions requesting that the two treaties be submitted for ratification, but neither has been submitted. Hawks disliked the treaties because they were seen as an opening wedge for a comprehensive ban, and doves disliked them because they fell short of a total ban and provided legitimation for the idea of peaceful nuclear explosives, an idea that India exploited in 1974 to justify a nuclear-weapon test.

On 2 November 1977, in a landmark speech to the Supreme Soviet, Soviet leader Leonid Brezhnev called for a full moratorium on nuclear explosions. Cyrus R. Vance, US Secretary of State at the time, hailed Brezhnev's offer as "a major step." Vance said the Soviet decision to forgo "so-called peaceful nuclear explosions is in the direction of what we have been talking about for several months."

The two superpowers opened negotiations on a comprehensive test ban, and President Jimmy Carter officially endorsed the idea of a five-year ban. Under pressure from the Pentagon and the nuclear-weapons laboratories, represented in the Carter Administration by Energy Secretary James Schlesinger, the President backed off from a five-year ban in August 1978 and instead gave his blessing to the idea of a three-year ban.

At that point, confronted with a US demand that they accept placement of ten national seismic stations in each signatory country, the Soviets raised the question of why they should accept so much surveillance for such a short treaty. Nonetheless they agreed in principle to ten stations in early 1979, on condition that the United Kingdom accept ten as well, despite its size.

By that time it had become apparent that Carter was not going to win ratification of the newly negotiated SALT II treaty without a big Senate fight, and so the President put the CTB on the back burner. Test-ban talks were suspended in November 1980, and President Reagan, upon taking office, decided not to resume with them. The Reagan Administration proceeded to shift attention back to TTBT and PNET, raising questions about their verifiability (PHYSICS TODAY, November 1983, page 70).

The intent behind the NRDC agreement with the Soviet Academy plainly is to focus national attention back on the possibility of a comprehensive test ban. "The NRDC proposal is a sort of pilot program to demonstrate that you can put in place in-country monitoring for a comprehensive test-ban treaty," Cochran says.

NRDC has submitted a request to the Commerce Department for transfer of US seismic equipment to the Soviet Union. Commerce routinely handles such requests and presumably will solicit the views of interested departments including State and Defense. The Office of Political and Military Affairs at State expects the matter to receive extensive review, as any such request would.

—WILLIAM SWEET

SECTION 3 _____

THE OFFENSE

To analyze nuclear arms race issues, one must have strategic-arms data bases. Barbara Levi lists various weapons parameters and discusses the differences. The strategic concepts have not changed greatly in the intervening six years, except for an increased emphasis on land-based mobile missiles and cruise missiles. The table on strategic nuclear arsenals in Levi's paper has been updated to display the changes. Of course, in the intervening years much has happened in the diplomatic arena, as the INF treaty and expanded on-site inspections have come to fruition and SALT II has been allowed to lapse. Lines were clearly drawn in the 1979 "Debate on SALT II" between Wolfgang Panofsky and Edward Teller. Panofsky argued that SALT II did limit the number of launchers, and the number of new systems that each side could have. SALT also limited the number of warheads by creating the counting rules for warheads, and other "rules of the road." Panofsky and Teller disagreed as to which side "was ahead": Panofsky believed in an approximate parity of forces, while Teller believed that the Soviets had an advantage. They also disagreed on the verifiability of the SALT II Treaty: Panofsky believed that national technical means were adequate, providing enough timely warning to protect our national security, while Teller believed that potential break-outs of 40 RVs on the SS-18's were worrisome.

The Freeze movement in the early 1980s was born in response to the Reagan buildup of counterforce, high-accuracy ICBMs, capable of destroying hard targets. It was to be a "mutual freeze on the testing, production and deployment of nuclear weapons and of missiles and new aircraft designed primarily to deliver nuclear weapons." The Freeze would have stopped the "modernization" program, including the MX, Trident II, B1, B2 (Stealth bomber), Pershing II, various cruise missiles, and the corresponding Soviet systems. Harold Feiveson and Frank von Hippel wrote the position in favor of the Freeze, and their current perspective is that:

> Although the freeze movement is now history, it created a public much more knowledgeable and skeptical about the 'counterforce race.' Also, the new Gorbachev leadership has made extraordinary demonstrations of

its genuine interest in ending this wasteful and dangerous competition and negotiating many of the elements of a freeze. But the arms race is not easy to stop. For example, the US is about to deploy the accurate Trident II submarine-launched ballistic missile to threaten Soviet missile silos and underground command posts. And the super-expensive B-2 bomber is being promoted for the complementary mission of hunting down and destroying mobile missiles and command posts.

The position in opposition to the Freeze was written by Harold Lewis, who recently concluded that:

> Time has confirmed that the Freeze movement was a minor historical anomaly. Contrary to the not-so-hidden assumption that all the blame and coercion should be directed at the United States government, it required instead a major change in the Soviet government to make real progress possible. To borrow from Macbeth, the Freeze movement was a 'a poor player, Who struts and frets his hour upon the stage, And then is heard no more.' RIP.

In his essay, "Arms-Limitation Strategies," Herbert York reviews the various approaches to arms control, from the Baruch plan to the Freeze. During the late 1970s and 1980s, the Executive Branch attempted to enhance the effectiveness of national security export controls to prevent technology transfer to the Eastern bloc. These government restrictions collided with the academic process in universities, with the dissemination of results on unclassified research at conferences, and with industrial exports. Two National Academy Panels were established to examine the broader costs of the policy. The first one under Dale Corson examined the effects on research in universities and research laboratories, which is summarized in this section. The second study, lead by Lewis Allen, examined effects on industrial laboratories, and it appeared as *Balancing the National Interest: U.S. National Security Export Controls and Global Economic Competition* (National Academy Press).

CONTENTS

Debate on SALT II

W. K. H. Panofsky, in favor . . .

PHYSICS TODAY/JUNE 1979

Wolfgang Panofsky is director of the Stanford Linear Accelerator Center.

SALT is a product of *negotiation* aimed at limiting strategic nuclear weapons spanning four administrations since 1967. Negotiations are not only across the table between the United States and the Soviet Union, but also involve resolutions of diverse positions at home on each side. Specifically, on the US side, SALT positions are the result of decisions by the President faced with inputs from the Defense Department, the Joint Chiefs of Staff, the Arms Control & Disarmament Agency, the CIA and the State Department. These decisions are also affected by consultation with Congressional leaders and with our Allies. One should be aware of this complex pattern to realize that the SALT outcome cannot make everyone happy. The negotiating history should be kept in mind when judging the sincerity of criticism that claims: "I support *a* SALT treaty but not this particular one." Little purpose is served in discussing an "ideal" treaty that would be optimal only to a particular set of interests.

Since SALT II is the product of negotiation, its very structure reflects the present status of agreement and disagreement between the two nations. The *Treaty,* running until 1985, represents agreement between the US and the Soviets on numerical limitations and minor qualitative constraints and on the means to assure verification. The *Protocol* (expiring by the end of 1981) represents items on which definitive agreement could not be reached on a long-range basis, but which are restrained for a shorter time while still preserving freedom of action after the expiration of the Protocol. For instance, the Protocol prohibits deployment (but not research, test and development) of ground-launched and sea-launched cruise missiles reaching beyond 600 kilometers. This provision offers no constraint as such, since development of these weapons cannot have led to deployment before expiration of the Protocol. On the other hand, inclusion in the Protocol will automatically place these items on the agenda for discussion for the next round of SALT. The Protocol prohibits the test and deployment of mobile ICBM's. The Soviets have developed such systems while we have not, and the US could not possibly test or deploy these before the Protocol expires. Accordingly, this item constrains the Soviets but not the United States. Again, this subject will clearly be a matter for discussion in an ongoing SALT process.

Finally, there is a *Statement of Principles,* which promises more incisive arms control and touches on those subjects that were not seriously considered, let alone resolved, in SALT I and II, but which urgently need consideration if the strategic arms race is to be limited in its burdensome dangers.

Achievements

Let me list some of the achievements of SALT II that are both positive and important:

1. While SALT II will not limit the evolution of any of the US strategic weapons systems that are now under development or definitely planned for deployment, it does limit both US and USSR expansion in the next generation of strategic nuclear delivery systems. As a result, the projections that forecast US and Soviet future weapons systems under a variety of assumptions are facing a cap beyond which the numbers of weapons cannot grow. This, in turn, places a limit on the justifiable demands for new weapons that a "worst case" defense planner would make on the governments of each country.

2. In the past the pattern of buildup of strategic weapons has generally been

continued on page 76

At PHYSICS TODAY's invitation, two physicists with opposing views on the proposed Strategic Arms Limitation Talks discuss the arguments for and against US approval.

Edward Teller, opposed . . .

For many years the United States enjoyed unchallenged military superiority. This secure position was due to our wealth and to the high level of our technology.

Today our wealth is being dissipated. Many young people consider technology, and even science, irrelevant. The result is that our superiority disappeared and even turned into a position of inferiority vis-à-vis the Soviet Union.

To some this statement may appear exaggerated. However, hardly anybody will deny that the continuing trend is shifting the balance of power in favor of the Soviet Union.

The situation should be considered not only from the point of the interest of the United States. We should put the greatest emphasis on the question of how the best chances for peace can be achieved. I believe that stability depends on power being in the hands of those who are determined to prevent war.

The question of SALT II must be viewed in this context. Is it appropriate at this time to rely on an agreement? Let us remember that in 1961 the test moratorium was broken by Russia on a 48-hour notice.

If a SALT II agreement is signed and ratified, this will constitute a signal to the American people that they need not worry—all's right with the world. Unfortunately, all will be right with the world only if we are prepared to defend such a happy condition. Events in Angola, Sudan, Yemen, Afghanistan indicate that "Pax Americana" no longer exists. What we have today is no longer Americana, and it is pax, at best, in a shaky manner.

It has been argued that, in the absence of a SALT treaty, an unlimited and disastrous arms race will result. Some fear the arms race because they believe that arms have a dynamism of their own; if they are produced they will be used. This view is supported by the history leading up to World War I. On the other hand, World War II was preceded by a race in disarmament—a race the western democracies won with ease. As a result, a single fanatic had a chance to initiate Armageddon. He almost succeeded in permanently subjugating all of Europe.

One should also note that we are involved not so much in an expensive arms race but rather in an intricate race of technology. The question is not primarily quantity and expenditure, but rather determination and ingenuity. As far as military affairs and innovations are concerned, the Russians are beating us at our own game.

Alternatives to SALT

In the absence of a SALT agreement we could, and probably would, pursue some essential and not too expensive developments.

Civil defense. The Russians have made considerable progress in plans to evacuate if nuclear war is impending. Indeed, the Russians might disperse the population of their cities before giving us an ultimatum, with nuclear attack as an alternative. They have deployed shelters and food storage. Our affluent and motorized society has done next to nothing. We should, as a minimum, develop and implement plans for counter-evacuation.

Tightened alliances. We should and could tighten our alliance. The burden for defense and the responsibility for decisions could be divided in a more equal manner. For instance, cruise missiles might be jointly developed. The industrialized countries of the free world

Edward Teller is Professor of Physics Emeritus at the University of California.

Teller continued

have a common interest in a secure peace. The issues that divided us are few and relatively unimportant.

Electronics and electronic computers is one area of technology in which our consumerist society enjoys a clear advantage. This capability has been applied to defense to a limited extent. The cruise missiles represent a small but significant step in this direction. One should note that the Russians are attempting to use SALT II for the purpose of limiting our cruise-missile development. In the end we might move toward highly sophisticated unmanned weapons with two-way communications using lasers and microwaves. These would be less expensive and more flexible than our present instruments and still could perform better than manned planes, tanks or small ships. They will not function without sophisticated electronics.

The Triad. At present we rely on the "Triad," a three-fold system of nuclear retaliation: bombers, nuclear submarines, and land-based missiles. None of these three are completely reliable. One of them, the land-based missiles, could be wiped out by a Russian first strike. Thorough modernization of this system is essential. In this connection, some mobility of the missiles is essential.

The list above is incomplete. The most urgent and least expensive items are mentioned first. This sketchy program may illustrate to the reader that if we become aware of the deficiency in our defenses soon enough, we shall be able to improve our situation at an acceptable cost.

A SALT agreement will be presented to the American public as a harbinger of peace between the two superpowers. In that case none of the programs mentioned above are apt to be pursued with vigor. By the time the danger of our situation is apparent it may be too late to catch up.

A position of inferiority

Previously treaties for arms limitation have been sought under conditions of American superiority. Today the United States negotiates from a position of inferiority. The attempt to agree with the Russians on restrictions of armaments is dangerous for two additional reasons: the US is an open society, while the USSR is not; and the US has no plans to dominate the world, while the USSR has a clearly stated program to extend their philosophy and their rule around the globe.

I believe that a SALT agreement is dangerous for the United States at this time under any conditions. There are even further dangers which are inherent in the plan that has been evolved for SALT II.

SALT II attempts to limit delivery vehicles while allowing nuclear weapons to proliferate. Under this arrangement the

Russians could build exceedingly great numbers of nuclear weapons and missiles to carry these weapons. These systems could be stored in warehouses and not deployed in silos we can see. Even at this time Russia has a four or five-fold advantage in "throw-weight,"—that is, the weight that their nuclear-tipped missiles can carry from Russia to the United States. The Russians could double or triple their advantage. A limit is not in sight. They could develop a superiority so great that resistance would become folly. All this could be done in complete secrecy and without violating the letter of SALT II. Indeed, they could not be limited in fabricating the missiles; the treaty would restrain them only from deployment in silos. Therefore, there does not seem to be any method by which the proliferation of nuclear weapons could be controlled in a society such as Russia's. However, in the open society of the US such proliferation could not occur.

The Russians consider their modern

Backfire bombers as non-strategic. Therefore, these are not included in the planned SALT treaty. Yet these bombers could take off from Russia, deliver their bombs on the US without the planes having refueled, and then land in Cuba.

The Russians count their SS-20 missiles as non-strategic. These missiles could wipe out all European defenses in a single blow. A signing of SALT II would, therefore, give rise to fully justified worries for our NATO allies. To weaken the NATO alliance is one of Russia's main near-term objectives.

The Russians consider our cruise missiles as strategic. SALT II would limit our cruise missiles, which represent a step toward sophisticated unmanned vehicles. Since we honor treaties, in the spirit as well as in the letter, the treaty would discourage the one development that we are now actively pursuing and that might lead to a more equal balance between the superpowers in the 1980's.

If we do not sign SALT II we have a

SALT II—the terms of the Treaty

The Strategic Arms Limitation Talks aim to limit the escalation both in numbers and in types of strategic nuclear delivery systems. Reflecting the view that SALT is an ongoing process, the proposed SALT II agreement is organized in three parts, each corresponding to a different duration of enforcement. The Treaty would remain in effect until 1986; the Protocol would last about three years and the Joint Statement of Principles would establish guidelines for future negotiations.

The SALT II Treaty would subject both parties equally to quantitative and qualitative

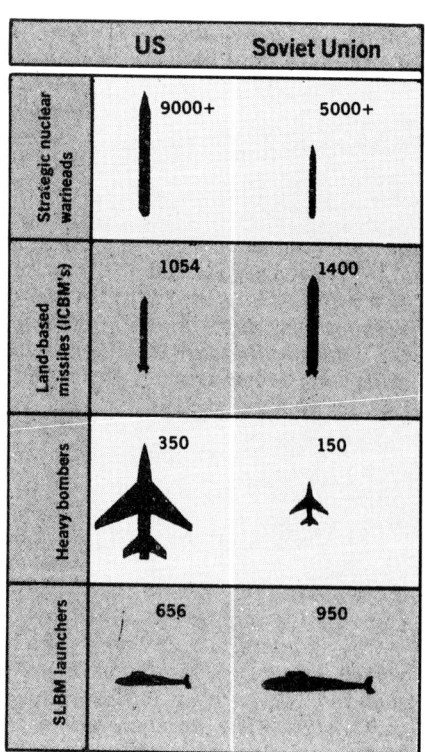

restrictions, including the following:
▶ A ceiling of 2400 on the aggregate number of strategic nuclear delivery vehicles, to be lowered to 2250 by 1982
▶ A limit of 1320 on launchers of ballistic missiles equipped with multiple, independently targetable reentry vehicles (MIRV's) and on heavy bombers equipped for long-range cruise missiles
▶ A lid of 1200 on the number of MIRV'd ballistic missiles alone
▶ A ceiling of 820 on MIRV'd, land-based intercontinental ballistic missile (ICBM) launchers
▶ A prohibition against any increase in the number of reentry vehicles (RV's) carried by existing types of ICBM's. The number of warheads on existing type of ICBM's will be frozen at the maximum with which that missile type has been tested. New types of ICBM's will be limited to 10 RV's and new types of SLBM's to 14.
▶ A ban on the construction of additional fixed ICBM launchers and on any increase in the number of heavy ICBM launchers, defined as those larger than the Soviet SS-19 missile
▶ A restriction to one new type of ICBM during the lifetime of the Treaty
▶ A ban on certain new types of strategic offensive systems such as ballistic missiles on surface ships
▶ A ban on the development of the Soviet SS-16 missile for the Treaty's duration
▶ An exchange of data on the weapons systems that are limited by the Treaty

The SALT II Treaty would continue the verification provisions of SALT I. (The SALT I Interim Agreement expired in October 1977, but the US and USSR have unilaterally stated their intents not to take any actions contrary to that accord as long as SALT II negotiations continue in good faith.) Under the verification arrangements of that earlier treaty, each side is to confirm by national

chance to recover from the dangers of military inferiority. The power that we acquire would not be used for nuclear aggression but might well serve as a nuclear deterrent.

For more than two decades we have pursued a policy aimed at arms limitation. In this period we have gone from superiority to parity, from parity to essential equivalence, and from equivalence to a situation that we hopefully call "sufficiency." SALT II may complete the process and introduce an obvious lack of balance. Such a situation would encourage Russian imperialism and is bound to endanger peace. This is why this is not the proper time to sign or to ratify SALT II.

Panofsky's rebuttal

Edward Teller's article clearly states his philosophy: The Soviets are now superior militarily, they are bent on world domination while we are not, they are capable of very large-scale, high-technology, military undertakings that remain hidden from us over many years, the US is determined to avoid war, the USSR is not. Therefore we need a "pax Americana."

These are highly debatable assertions in their own right, but they are hardly related to SALT as drafted. *None* of the programs Teller strongly advocates for the United States—expanded civil defense, tightened alliances, increased emphasis on electronic warfare, cruise-missile development, strengthening of the strategic Triad—are inhibited by SALT. Teller states: "If a SALT II agreement is signed and ratified, this will constitute a signal to the American people that they need not worry—all's right with the world." Teller thus considers any SALT agreement with the Soviets dangerous to the United States. In other words, Teller suggests that under our democratic processes no arms-limitation agreement can be signed without endangering our security through neglect of our defense. Only continuing superiority will do.

Not only is such a position destructive to the search for an alternative to the "race for oblivion," but it is also unsupported by history. Past arms-control agreements with the Soviets have not caused us to ignore our defense. The Limited Test Ban Treaty has actually accelerated our rate of testing nuclear weapons underground, as permitted by that Treaty. While SALT I has assured the penetration of each one of our missiles by restricting Soviet ABM's to militarily insignificant numbers, it has not reduced the number of missiles the US has deemed necessary. In the current budgetary debates I have heard no voices pleading that SALT I, the Nuclear Test Ban Treaties, or other arms-control agreements obviate the need for an adequate defense. On the contrary, there have been successful arguments that we should *accelerate* weapons acquisitions in order to "bargain from strength" at SALT; the price for support for Senate ratification of SALT II by military spokesmen may well be a commitment to *increased* military effort where not constrained by SALT. Despite these escalatory forces SALT II will achieve significant arms limitation on both sides. Moreover, basic technological innovation in military weapons, which Teller so strongly advocates, is much more hindered by institutional inertia than by arms control.

Teller hardly addresses himself to the actual provisions of SALT II. However, his references to the Treaty lead to misinterpretation of the provisions as drafted.

Teller states: "SALT II would limit our cruise missiles . . . the treaty would discourage the one development that we are now actively pursuing, and that might lead to a more equal balance . . ." In the strategic area this program is the ALCM (Air Launched Cruise Missile), *permitted and not limited* in range by either the SALT II Protocol or Treaty.

Teller states correctly that the Soviets could stockpile missiles and warheads *without violating the letter of the Treaty.* However, he states that these could be fired (presumably from transportable launchers) since the treaty would only restrain them from deployment in silos. This is incorrect; the Treaty limits launchers, not silos. Any such augmentation in throw weight would have to be clandestine, and I doubt that this is possible on a significant scale.

Teller states: "The Russians count their SS-20 missiles as non-strategic." These are intermediate-range missiles threatening Western Europe. It was the United States that refused to include the "Forward Based Systems" such as the US-manned, European-based aircraft as strategic weapons in SALT I and II.

technical means (NTM)—that is, by observations not requiring the active cooperation of the other nation—that the other party is complying with the treaty. Interference with the NTM is forbidden, as is the use of deliberate concealment measures.

Adding to these verification measures, the proposed Treaty contains type rules to assist in counting MIRV'd launchers. For example, if an ICBM or SLBM of a certain type has ever been flight tested with MIRV's, it is considered to be MIRV'd, even if it has also been tested with a single RV. All missiles of that type are considered to be MIRV'd. The new Treaty would add the requirement for advance notification of certain ICBM test launches but would not limit their number.

The Protocol would allow testing, development and deployment of air-launched cruise missiles (ALCM's) of unlimited range. (A cruise missile is an unmanned, guided-weapon delivery vehicle that flies by means of aerodynamic lift in the manner of an airplane. They generally fly at low altitude to elude detection.) The Protocol forbids the deployment of land- and sea-based cruise missiles with ranges beyond 600 km but permits their development and testing. It would ban the deployment (but not the development) of mobile ICBM's, as well as the testing and deployment of long-range, air-launched strategic ballistic missiles. A separate package of constraints will deal with the Soviet Backfire bomber, an aircraft with capabilities somewhat less than those of heavy bombers. This aircraft can reach the US unrefueled, but only on high-altitude, one-way trips.

Under the Joint Statement of Principles, the US and USSR would commit themselves to pursue further reductions in the numerical ceilings, to extend the qualitative limitations, and to resolve the issues covered by Protocol.

Barbara G. Levi

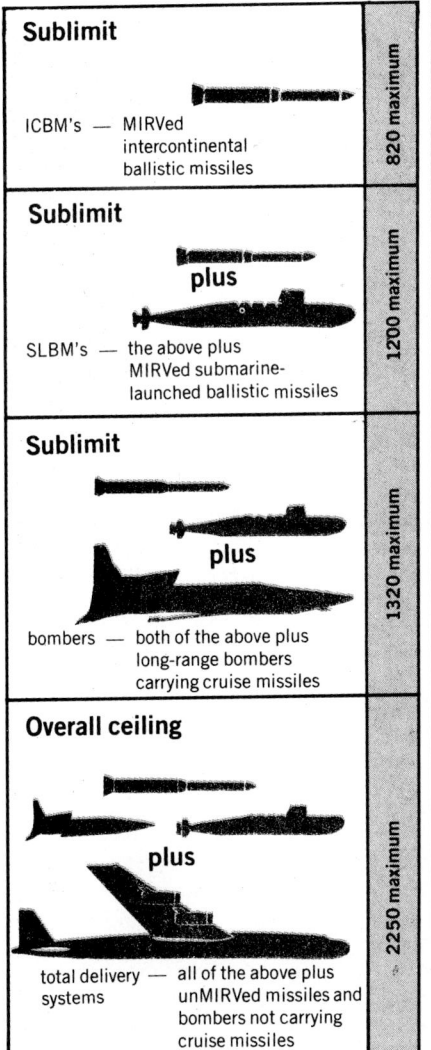

Sublimit

ICBM's — MIRVed intercontinental ballistic missiles

820 maximum

Sublimit

plus

SLBM's — the above plus MIRVed submarine-launched ballistic missiles

1200 maximum

Sublimit

plus

bombers — both of the above plus long-range bombers carrying cruise missiles

1320 maximum

Overall ceiling

plus

total delivery systems — all of the above plus unMIRVed missiles and bombers not carrying cruise missiles

2250 maximum

These can strike western Russia with nuclear weapons. As a result, the Soviets and the US agreed to omit consideration of these and other systems specifically affecting the European balance.

Teller states: "The signing of SALT II would, therefore, give rise to fully justified worries for our NATO allies." Ratification of SALT II has been supported by Chancellor Schmidt of Germany, Prime Minister Callaghan of England, President Giscard d'Estaing of France and other West European leaders.

Teller states correctly in his concluding paragraph that the US and USSR world situation has changed substantially for more than two decades, with the Soviets growing in military power. Yet this evolution can hardly be blamed on the ongoing arms-control effort. On the contrary, analyses of the actual provisions of SALT II lead to the conclusion that enactment of this treaty would increase US security.

Teller states: "... the agreement will be presented to the American Public as a harbinger of peace between the two superpowers." This I do not claim; all I maintain is that SALT II is a positive step in Man's efforts to limit the increasing dangers and burden of nuclear arms.

Panofsky

continued from page 72

that the US has led in a given technology (for example, nuclear weapons, ICBM's, SLBM's, ABM's, MIRV's and now strategic cruise missiles) and has proceeded to rapid buildup. The USSR then followed after a lag of several years while the US leveled off its own deployments. Thus, the Soviets often passed the US in terms of number and size of weapons deployed, but rarely in quality. We are currently in a phase where the US has leveled off its numbers of ICBM's and SLBM's, while the USSR is still building up. As a result, SALT does a great deal more to arrest Soviet strategic arms momentum than it does to impede planned US systems. SALT II actually requires the Soviets to destroy some, albeit older, strategic systems.

3. SALT II limits the number of warheads that can be placed on MIRV'd missiles. For instance, the SS-18 (the largest Soviet ICBM) will be constrained to carry no more than ten warheads, while its size theoretically permits carrying twice that many. As a result the US planner, in evaluating countermeasures to the emerging vulnerability of the land-based ICBM force, faces a threat with a specific upper limit.

4. The SALT process places the technical national surveillance systems (satellites, radars and so on) of each side under legal protection. In other words, short of actual war, we are assured of the continuity of information flow so neces-

"Strategic Arms." Clockwise from top: An airborne command post, ready to assume command of SAC missiles; a Titan-II ICBM launched from an underground silo; an FB-111 strategic bomber, a variation of the F-111 fighter that can carry nuclear or "conventional" weapons; the *Ohio,* a Trident-class submarine, at Groton, Conn.

sary for verifying compliance with SALT and for assessing the military status of other nations.

5. SALT II continues the Standing Consultative Commission that was created in SALT I to resolve controversial matters relating to SALT compliance. It thereby preserves the forum in which military matters (whose discussion had previously been taboo) can be aired. This Commission has been invaluable in clearing up suspicions of violations on both sides, resulting in the conclusion that *no* violations of significance of SALT I have occurred.

6. SALT is not an expression of mutual trust between the US and the USSR. Both parties must be persuaded that their intelligence apparatus is adequate to ensure that evasion of the SALT provisions to a militarily significant extent should not go undetected. Much has been written about the adequacy of US satellites, radar and other means to collect information in "verifying" Soviet compliance

with SALT II. Let me simply state here that the numerical provisions of SALT II can indeed be very well policed. Checking the restraints on modernization is somewhat more difficult but still surprisingly good.

7. There are real economic benefits of SALT, at least in the long run, because purchases for strategic hardware need not be as large with SALT as without. However, the cost of such weapons is not a dominant factor in the overall economic burden of armaments.

Misconceptions

After having recited what SALT II will accomplish in dampening the US–USSR nuclear arms capabilities, let me list and discuss some frequent misconceptions:

1. SALT is not a major factor in shifting the relative defense posture of the United States versus the Soviet Union. Competition both in the military and economic arenas of both nations will not be diminished substantially by SALT II

in itself, although certain limits are imposed. The United States must still establish defense policies to do whatever is believed necessary in its own security interest. Military procurements must still be gauged by the internal priorities as seen by government. In other words, the "guns vs. butter" debate will not be silenced by SALT.

2. SALT is not a "zero sum game" in which one party's gain must be achieved at the expense of the other party's loss. Therefore, the frequently heard opinion that since the Soviets seem to want SALT so much it must be bad for the US, and that therefore we should be able to charge the Soviets a high price for our agreement, makes little sense. The SALT process does reflect the conviction of *both* parties that reducing the dangers and burdens of nuclear weapons is a matter of overriding interest, and that the security of both nations can be increased through a SALT agreement. As Andrei Sakharov said recently: "I believe that the problem of

lessening the danger of annihilating humanity in a nuclear war carries an absolute priority over all other considerations."

3. SALT is not a reward for good Soviet behavior. Although successful negotiation of a SALT agreement requires, of course, a minimum of civil relations between the United States and the Soviet Union, it does *not* signify approval of Soviet ideology, of Soviet moves in Africa, or of their conduct vis-à-vis internal human rights any more than it reflects Soviet approval of United States mediation in the Israeli–Egyptian treaty. In fact, SALT I negotiations continued while the US was bombing Haiphong harbor with Soviet ships in that port.

The much-discussed problem of the vulnerability of US land-based ICBM's (the Minuteman force) is real but is not significantly affected by SALT; as a matter of fact, with SALT II the problem is somewhat more tractable than without such a treaty. This is because of the limit

that SALT II places on the fractionation of ICBM warheads, and thus on the total number of reentry vehicles carried by the Soviet ICBM force. The options now before the US Government in replacing Minuteman by a more survivable weapons system are not limited by SALT.

The problem of ICBM vulnerability should be put in proper perspective. It is indeed true that by early in the next decade the number and accuracy of Soviet warheads will be adequate to destroy a large fraction of US ICBM's in their silos; conversely, US weapons will be able to destroy not quite one-half of Soviet ICBM's. However, US land-based ICBM's carry only about 25% of US retaliatory power while Soviet ICBM's carry about 75% of the total weight of nuclear weapons. Thus the vulnerability of these land-based, fixed ICBM's, while certainly undesirable and contributing to instability, does not even approach giving a first-strike potential to either side. On the contrary, the ability to retaliate in a devastating manner after absorbing a first strike, which is a necessary consideration for stability, is preserved for the foreseeable future, and this stability is significantly enhanced by SALT.

Objections

If the achievements of SALT II are indeed positive, then why is there any criticism at all? The objections fall in three basic classes:

Criticism based on issues totally unrelated to SALT, objecting to military decisions taken either by this or prior administrations. Among such items, not constrained by SALT, are the cancellation of the B-1 Bomber, the deferred decision on the neutron bomb, the choice of small versus large missiles, the level of defense spending and so on. Generally such criticism focuses on areas of Soviet strength and US weakness, while omitting matching areas of US strength and Soviet weakness. Obviously, one should expect disagreement on the wisdom of past decisions in the strategic military arena. Yet blaming whatever dissatisfaction the critic may have on SALT is clearly wrong.

SALT II will generate an atmosphere of false security which will prevent the United States from providing adequately for its military needs. This argument in essence pleads that a move toward a more stable and peaceful world is dangerous. I have confidence that the wisdom of our institutions in providing for the needs of national security in the broadest sense will be preserved.

SALT II does not achieve enough. It is indeed true that many people interested in arms control would have wished that SALT II had achieved deeper cuts and more stringent controls. In fact, one can maintain that technological progress during the time in which the SALT treaties have been negotiated has outpaced the achievement of that process.

Therefore, for SALT actually to reverse the growth of nuclear armaments, future treaties must achieve more. Yet there is no question that nuclear strategic weapons buildup on both sides projected for the future would be larger without SALT II than with SALT II in force. Moreover, defeat of SALT II would be a major setback towards attaining more incisive arms control in the future. The leaders of all Western European nations have expressed apprehensions about the security of their nations and the future of the Alliance if SALT II is not ratified. There is no question that US security and the hope for a peaceful world will be strengthened by SALT II.

It is important to keep the awesome reality of nuclear explosions in focus. The current inventory of the world is around 30 000 nuclear weapons, most of which are much larger than the two weapons which each killed about 100 000 Japanese when detonated over Nagasaki and Hiroshima. Many of the arguments swirling around SALT are couched in such terms as *perceptions* of strength, *perceptions* of resolve and national will, and other phrases that make nuclear weapons symbols of power rather than objects of physical reality. As physicists we must keep reminding ourselves and our fellow citizens of the real nature of these weapons and that any use of nuclear weapons for whatever purpose in war can cause dangers to our civilization that are impossible to quantify.

Teller's rebuttal

In a political debate such as the one on SALT II, there is not—and cannot be—a definitive answer. Some remarks on Wolfgang Panofsky's well-reasoned paper may, however, be in order.

Indeed, the "Protocol" will not be binding at all on the Soviet Union after expiration of a short period. In the United States, having raised hopes for more agreement, the Protocol will exercise great pressure, even after it has expired.

The Protocol may prohibit only deployment, but not research and development. But it is hard in the United States to appropriate money for research and development when deployment is not expected. It may also be hard to obtain the enthusiastic cooperation of scientists when the Protocol seems to have placed a military instrument outside the pale.

The Protocol prohibits the deployment of mobile ICBM's. This is apt to interfere with a viable option designed to ensure the survival of our Intercontinental Ballistic Missiles.

Panofsky lists six positive achievements of SALT II.

1. He expects a "cap" beyond which the numbers of weapons cannot go. The reality of such a cap depends on our surveillance, and also depends on the absence of future systems for which surveillance is more difficult. It may have been a serious mistake that in SALT I we overemphasized limitations on silos for the obvious reason that for these, surveillance is possible.

2. Panofsky states ". . . the Soviets often passed the US in terms of number and size of weapons deployed, but rarely in quality." Of the quantities we might be aware; to check quality is almost impossible. It is an uncomfortable situation when we have to admit that the Russians are ahead of us in those respects that might be measured, but we claim to be ahead in those fields where guesses must suffice.

3. How shall we check that an SS-18 carries fewer than ten warheads? The usual answer is that we shall find out when a missile carrying 40 warheads is tested. But, could the Russians not fire a missile carrying 40 warheads, release only seven, and be confident that those

not released would have worked as well as those released and observed?

4. SALT placed national surveillance systems under legal protection. These are words. Our foremost surveillance system consists of our satellites. We know that the Russians are perfecting satellite-killers.

5. The Standing Consultative Commission created in SALT I seemed to me, at the time, like real progress. In spite of its existence, bitter public debates ensued concerning violations—for instance, between Secretary Kissinger and Admiral Zumwalt. Due to secrecy, the public has no way to judge.

6. SALT tends to discourage proliferation of numbers. I agree with Panofsky that this may be advantageous to both sides. The United States should catch up by producing better quality, not by outdoing the Russians in quantity.

Concerning misconceptions related to SALT:

1. The "guns vs. butter" debate would, indeed, not be silenced. It would, however, be deeply influenced.

2. To avoid a nuclear war is truly the interest of everyone. However, to speak of annihilating humanity is an exaggeration.

3. Secretary Kissinger suggested that SALT should be linked to moderation in Russian expansionist policy. This seems to have some merit.

Panofsky mentions that only 25 percent of US retaliatory power is carried by ICBM's. But the great proportion of our weapons is carried by aircraft, and these aircraft are facing truly ample Russian air defenses.

In conclusion, Panofsky argues that we physicists must keep reminding everyone of the real nature of nuclear weapons. This point hardly needs more emphasis. Our proper role should be to point out the great and increasing role that new technologies play in national defense and survival.   □

The nuclear arsenals of the US and USSR

A survey of the particulars of the arms race—what kinds of weapons are there and how do US and Soviet arsenals compare?

PHYSICS TODAY/MARCH 1983

Barbara G. Levi

In the course of the nuclear arms race, both the US and the Soviet Union have filled their arsenals with tens of thousands of nuclear warheads and with thousands of vehicles to deliver those warheads. The nuclear weapons now deployed vary in size and design from sub-kiloton artillery shells to 20-megaton warheads, and their delivery vehicles include land- and sea-based ballistic missiles as well as long-range aircraft and the new cruise missiles. Not only have the numbers and types of nuclear weapons grown, so, too, have their capabilities. The newer ballistic missiles are tipped with multiple reentry vehicles. They have longer ranges, greater accuracies, and increasingly sophisticated guidance systems. These weapons are further augmented by ever-more-able systems for land, air and sea defense and for command, communication and control. These many dimensions of the US and Soviet nuclear arsenals require that one assess their respective strengths on a multiparameter scale. Even then, any inherently quantitative comparison is incomplete, for it ignores the equally important, qualitative features of the arsenals where, in fact, much of the recent development has occurred.

All nuclear weapons may be roughly grouped into two categories—strategic and tactical. The strategic weapons are those that each party targets against the territory of the other. The tactical weapons are deployed for a conflict on foreign soil or at sea. This distinction is useful although admittedly fuzzy: Some weapons systems on each side that are usually classified as

Barbara Levi is a member of the technical staff at Bell Laboratories and is a contributing editor at PHYSICS TODAY.

Cruise missile being launched from a truck during testing. Such missiles are being added to the air leg of the US strategic arsenal. (US Air Force photo)

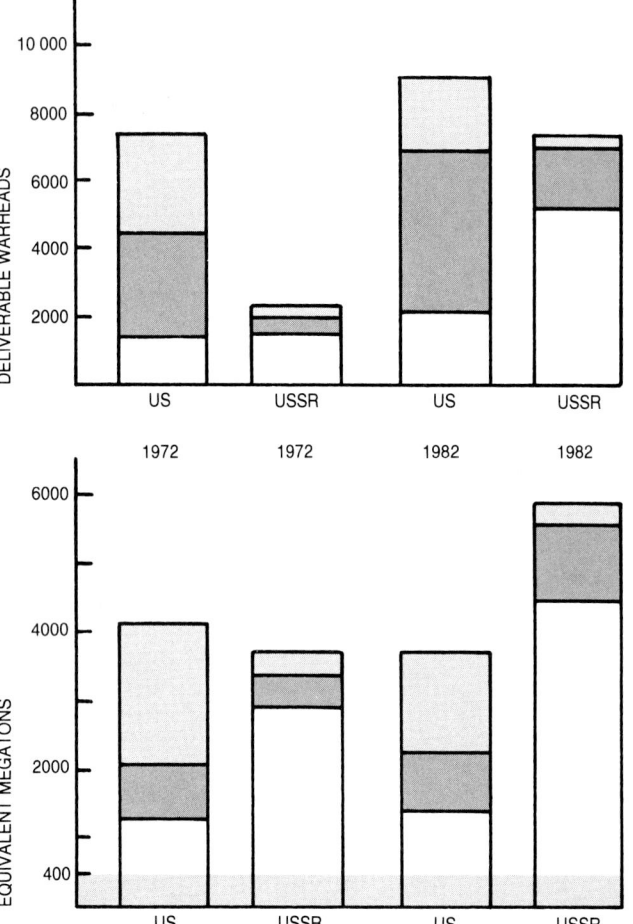

Aggregate arsenals of the US and USSR. The charts show the number of warheads and their destructive power (given as the equivalent number of single-megaton bombs) in air-based (dark color), sea-based (light color) and land-based (white) strategic forces. The level of 400 equivalent megatons indicated in the lower chart is the size of retaliatory force estimated to achieve the "mutual assured destruction" required for deterrence.

"tactical" can actually perform strategic missions as well.

Strategic forces

The US has designed its strategic nuclear force on the concept of a triad, building three strong legs—on land, in air and at sea. Substantial numbers of delivery vehicles in each of these areas would be expected to survive a Soviet attack. Although some are now questioning the survivability of the ICBMs, the current thought is that the submarines when at sea are nearly invulnerable and are expected to remain so for some time to come.

The US has about 9300 strategic warheads, with a total destructive power equivalent to that of about 3800 one-megaton bombs.[1] (The measure of equivalent megatons—EMT—is used here rather than yield to characterize the size of a weapon. An EMT is the yield in megatons raised to the two-thirds power. The area of blast damage caused by a nuclear explosion is directly proportional to the weapon's EMT; see the article by Leo Sartori on page 32). Because the three legs of the nuclear triad have unique strengths, these totals are not divided equally among them. Both the long-range aircraft and the intercontinental ballistic missiles (ICBMs) carry larger payloads than the submarine-launched ballistic

missiles (SLBMs). Hence the bombers account for just over 40%, and the ICBMs for just under 40%, of the US destructive power in EMT. By contrast, the SLBMs have greater numbers of multiple, independently-targetable reentry vehicles (MIRVs) per launcher, and so they carry roughly half of the total US warheads. The remaining half is split about equally between the air and land legs.

For tactical use, the US has deployed[2] about 15 000 warheads in Western Europe, in Asia and the US, as well as in the Atlantic and Pacific fleets. These weapons include antisubmarine missiles, battlefield weapons and bombs for strike aircraft.

The Soviet Union has concentrated its strategic nuclear strength in its ICBM force, although it has dramatically increased its deployment of SLBMs in the past ten years. The Soviet ICBMs carry nearly eighty percent of both the total destructive power (of about 6700 EMT) and of the total number (approximately 7300) of deliverable warheads.[1] The submarine launchers account for about 15% of these totals, and long-range bombers for roughly 5%. Experts estimate the tactical arsenal of the USSR to include about 7000 warheads[2] located in Eastern Europe, and in the USSR both east and west of the Urals.

The US has currently deployed three types of ICBM launchers: 52 Titan IIs, 450 Minuteman IIs and 550 Minuteman IIIs. The table on page 46 summarizes some of the quantitative features of these missiles, which are listed roughly in order of their dates of deployment. The table clearly shows the trend away from large, single warheads and towards smaller, multiple, independently-targetable reentry vehicles (MIRVs). It further illustrates the increase in the accuracy of these delivery vehicles, as measured by their "circular error probable." The CEP is the radius of a circle around the target within which the warhead has a 50% chance of landing.

Some 300 of the existing Minuteman III ICBMs have been fitted with the new Mark-12A warhead. This new warhead permits greater accuracy and higher yield (335 kilotons per reentry vehicle compared to 170 kt for the older versions of the Minuteman III). The low targeting error of the "improved" Minuteman IIIs has been enhanced by upgrading to the computers of the NS-20 inertial guidance system, for example, by providing more precise prelaunch calibration of the gyroscopes and accelerometers, and better mathematical descriptions of their in-flight performance.[3]

The US is developing one additional ICBM system, designated the "missile experimental," or MX. Designers want the MX to be both lethal in an attack against Soviet ICBMs and invulnerable to an attack by them. The Carter Administration proposed that this ICBM be based in multiple shelters—the so-called "shell-game" scheme. The Reagan Administration rejected that idea and, in November 1982, proposed the "dense pack" scheme. The theory behind the plan was that of fratricide: Siting the US ICBMs close together would mean that the detonation of one Soviet warhead would cause other Soviet warheads aimed at nearby silos to malfunction. This proposal failed to win approval of the Congress in December, and the basing mode remains undetermined.

Much less uncertainty surrounds the design of the MX missile itself. The ICBM will be MIRVed with ten warheads; the warhead will be[4] the Advanced Ballistic Reentry Vehicle, now called the Mark 21, and will have a yield of about 300 kt per warhead. The Air Force has been developing this reentry vehicle as part of its program on maneuverable and terminally guided reentry vehicles. The MX may well have[5] a CEP of less than 100 m.

The MX program has raised some questions regarding its compliance with the provisions of the SALT II Treaty. Although the US Senate never ratified that treaty, both parties to it appear to be abiding by its restrictions. The major terms of that document are summarized in the box on page 49. Among other provisions, SALT II allows the US to develop one new ICBM system but bans the addition of any new fixed ICBM silos. Whether the basing mode eventually selected entails "fixed" silos may be a future topic of debate.

In the sea leg of its nuclear triad, the US has[1] 304 Poseidon C-3 and 216 Trident C-4 SLBMs. The Poseidon launchers are carried on 19 Lafayette-class nuclear ballistic submarines (SSBNs), each of which has 16 launch tubes. Since October 1979 the Navy has modified 12 Poseidon subs to hold the higher-yield and longer-range Trident I missiles. Because the Trident I has a range of 7400 km compared to Poseidon's range of 4600 km, the modified SSBNs can patrol wide areas of the ocean and still remain within strike distances of their targets.

Trident I missiles are now being loaded onto a new class of submarines. The lead vessel in this class—the USS Ohio—made its first operational patrol in 1982. Eight other Ohio-class SSBNs are under contract and are expected to be built at the rate of one per year. The Navy has 6 more in its current five-year plan and could build up to 20.

The mammoth Trident submarine measures 560 feet long and displaces 18 700 tons of water when submerged. It will hold 24 Trident SLBMs. Compared to earlier vintage SSBNs, the Ohio is faster, quieter and deeper diving. The cost of each one is estimated to be more than $1.2 billion, excluding the cost of the ballistic missiles and the nuclear reactors.[4]

The US intends for the Trident submarine to carry the next-generation SLBM, the Trident II D-5 missile. The Trident II will have some combination of improved range, accuracy and yield. If equipped with the Navy's Maneuverable Reentry Vehicle (MARV), it could have terminal guidance. Both the Trident II and the MX missiles may use a US satellite system, scheduled for completion in the late 1980s, to make midcourse corrections. This NAVSTAR Global Positioning System consists of six satellites in each of three circular orbits. A ballistic missile could determine its position with very small error by receiving signals from several satellites and comparing the differences in their times of arrival.

The air leg of the US nuclear triad rests mainly on its B-52 long-range bombers. The Air Force has over 300 such aircraft, armed with variable-yield (0.1 to 1 Mt) gravity bombs. The newer B-52G and B-52H planes may in addition carry short-range air missiles (SRAMs), with yields of 200 kt apiece.[1] The FB-111A aircraft is a medium-range bomber that can play a strategic role if refuelled sufficiently. (The ranges given in the table are theoretical maximum distances a plane could fly, unrefuelled, at optimal altitude and speed.) The short range of the FB-111A excludes it from the provisions of the SALT II Treaty. Any inclusion of the FB-111A in such pacts would necessitate the inclusion of the Soviet Backfire bomber as well, and hence meets with Soviet resistance.

The latest addition to the air-based nuclear arsenal is the air-launched cruise missile (ALCM). The vehicle is small enough to be easily transported by bombers yet difficult to detect by radar. (It is about 20 feet long and weighs around 3100 pounds). It flies unmanned at subsonic speeds (about 500 mph) for distances up to 1500 km, often skimming just 100 feet above the earth, further confounding efforts at detection.

When fitted with ALCMs, the B-52s can stand further out from Soviet air defenses to strike their targets. Twelve of these craft may be mounted externally under the wings of a B-52G bomber while eight more fit in the bays. The first squadron (16 planes) of B-52Gs were equipped only with external cruise missiles in December 1982, and the Air Force plans to continue deploying the cruise missile on its B-52s. Each ALCM carries a 200-kt nuclear bomb.[4]

The capabilities of the B-52 bombers are being enhanced not only by the cruise missiles but also by the modernization of its electronic countermeasures, sensors, communications and so forth. Military planners are pushing for a successor (or successors) to the B-52 aircraft, however; they anticipate that by the late 1980s the USSR will be able to begin to send long-range interceptor aircraft to attack cruise-missile carriers and will soon thereafter deploy an airborne warning and control system.

Because the B-52 continued to be a useful aircraft, especially when fitted with ALCMs, President Carter cancelled production of the B-1 bomber in 1977. President Reagan restarted the program in 1981 as the B-1B, an interim replacement for the B-52 to serve in the second half of this decade and to be followed by an advanced-technology "Stealth" bomber in the early 1990s. The Stealth would be designed to be nearly invisible to radar. This low radar and infrared cross section might be acheived by such features as non-metallic material construction, lighter weight and better propulsion, lowered engine exhaust temperatures, and treated fuels that have reduced infrared signatures.[4]

USSR

Over the past decade, the Soviet Union has replaced many of its large-yield, single-warhead ICBMs with small-yield, more accurate, MIRVed ballistic missiles. Approximately 200 SS-7 and SS-8 launchers with 5-Mt yields were taken out of service to comply with SALT I ceilings as new sea-based missiles were deployed. The 20- to 25-Mt SS-9s have been superceded by about 300 SS-18s, most of which have eight warheads of about 0.9 Mt apiece.[1] Some 500 SS-11 ICBMs with relatively large CEPs are being replaced with the more precise SS-17 and SS-19 vehicles, although about 570 SS-11s and 60 SS-13s still remain. The total number of Soviet ICBMs stands at 1398.

The characteristics of the USSR's land-based nuclear forces are summarized by the estimates[1] in the table. These numbers are of course impossible to know precisely and are intended only to indicate general trends and relative sizes. In many cases, the

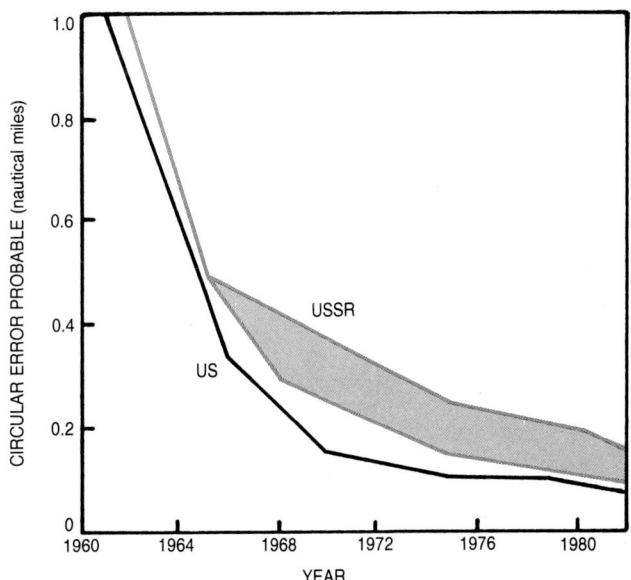

Estimated accuracies of ICBMs. The circular error probable is the radius of a circle around the target within which the missile has a 50% probability of landing. The band indicates the uncertainty in the values for Soviet missiles.

Strategic nuclear arsenals   (updated, 1990)

Delivery vehicle	Number of vehicles	Warheads per vehicle	Yield per warhead (Mt)	Total warheads N	Equivalent megatons $NY^{2/3}$	Range (km)	CEP (m)
US							
ICBMs							
Minuteman II	450	1	1.2	450	508	11 300	370
Minuteman III-Mk12	200	3	0.17	600	184	14 800	220
Minuteman III-Mk12A	300	3	0.335	900	434	12 900	220
MX/Peacekeeper	50	10	.3–.4	500	248‡	11 000	100
SLBMs							
Poseidon C-3	224	10	0.04	2240	262	4 600	450
Trident D-4	384	8	0.1	3072	662	7 400	450
SLCMS							
Tomahawk	300*	1	0.2	300	103	2 500	75⁺
Aircraft						Radius of Action (km)	Max. Speed (Mach)
B-52G	69			828	~828	4 600	0.95
Internal bombs or SRAMs'		⩽12	~1				
B-52G W/ALCM	98			⩽2352	⩽1578	4 600	0.95
Internal bombs or SRAMs		⩽12	~1				
External ALCMs		12	.2				
B-52H	99			⩽2304	⩽1546	6 140	0.95
Internal bombs, SRAMs, ALCMs		⩽12	~1				
External ALCMs		12	.2				
B-1B	97			⩽3686	1261–3686	4 580	
Internal bombs, SRAMs, ALCMs		⩽24	.2 − 1				
External bombs, SRAMs, ALCMs		⩽14	.2 − 1				
II-USSR							
II-ICBMs							
SS-11 Mod 2		1	1.00			13 000	1 100
3	400	3	0.1 to 0.3	400#	400#	10 600	1 100
SS-13 Mod 1	60	1	0.60	60	43	9 400	1 800
SS-17 Mod 3	138	4	0.50	552	348	10 000	400
SS-18 Mod 4	308	10	0.50	3080	1940	11 000	250
Mod 5		10	0.75			9 000	250
SS-19 Mod 3	350	6	0.55	2100	1410	10 000	300
SS-24	30⁺	10	0.10	300⁺	65⁺	10 000	200
SS-25	165⁺	1	0.55	1650⁺	111⁺	10 500	200
II-SLBMs							
SS-N-5	18	1	1.00	18	18	1 400	2 800
SS-N-6 Mod 1		1	0.5–1.00			2 400	1 300
3	240	2	0.50	360**	227–271**	3 000	13 000
SS-N-8 Mod 1		1	0.5–1.0			7 800	1 500
2	286	1	0.8	286	246$	9 100	900
SS-NX-17	12	1	0.50	12	7.6	3 900	1 400
SS-N-18 Mod 1		3	0.50			6 500	1 400
2	224	1	0.50–1.0	667⁺⁺	420–448	8 000	900
3		5	0.50			6 500	900
SS-N-20	100	6	0.10	600	129	8 300	500
SS-N-23	80	10	0.10	800	172	8 300	<900
II-SLCMs							
SS-N-21		1				3 000	
II-Aircraft						Radius of Action (km)	Max. Speed (Mach)
TU-95 (Bear)	175					5 690	0.9
Bombs	40	2–3		80–120			
AS-3–4 ALCMs	60	1–2		60–120			
AS-15 ALCMs	75	⩽8		600			

Data from *The Military Balance 1989-90*, The International Institute for Strategic Studies, London (1990).
*V. Thomas, *Science and Global Security*, Summer, 1989, p. 27.
⁺House Armed Services Committee, DoD Authorization Hearings FY85, part 2, p. 392.
† Short Range Air Missile.
‡Assuming an average yield of .35 Mt for the MX/Peacekeeper.
Assuming all SS-11s are mod 2 and that all SS-18s are mod 4.
**Assuming 120 SS–N–6s are mod 1, with .75 Mt yield, and 120 are mod 3.
$ Assuming all SS-N-8s are mod 2.
⁺⁺Assuming 74 SS-N-18s are mod 1, 75 are mod 2 and 74 are mod 3.

Soviets have tested several versions of a given ICBM and, while the number of each general type is known, the number of each model that is actually deployed is uncertain. The estimates for both the SS-18 and the SS-19 indicate that they may exceed the yield but not the accuracy of the Minuteman III. The SS-17, with fewer warheads per launcher and slower rate of deployment, is believed to have been developed in competition with the SS-19.[6]

The USSR has developed a potential mobile, solid-fuel ICBM designated the SS-16; that system is banned entirely by SALT II. Some controversy now surrounds the question of whether the Soviets have deployed the SS-16 in violation of that ban. The USSR may currently be developing several other new ICBM systems. A recent test is thought to be the first test of a new solid-fueled missile different from the SS-16.

The Soviet's two modern classes of SSBNs—the Yankee and the Delta classes—carry the vast majority of its submarine-launched ballistic missiles, although some of the older Golf and Hotel boats still have SLBMs in their launch tubes. About 25 Yankee-class subs are currently loaded with the SS-N-6 missile, which has a single reentry vehicle, in their 16 launch tubes. The 12 launch tubes of each of the 18 Delta I vessels and the 16 tubes of each of the 4 Delta II boats are fitted with the non-MIRVed, long-range SS-N-8 system. Newer Delta III submarines are replacing the Yankee-class SSBNs on a one-for-one basis. Each Delta III carries 16 MIRVed SS-N-18 missiles. Both the SS-N-8 and the SS-N-18 more than double the range and considerably reduce the CEP of the SS-N-6. The Delta III submarine is somewhat smaller, less capable and noisier than the Trident vessel.

One can make a rough quantitative comparison of the naval nuclear capabilities with the numbers in the table. By those measures, the USSR has more submarines and missiles and higher total EMT, while the US has greater numbers of warheads due to more extensive MIRVing. The table does not show the fact that only about 15% of the Soviet SSBNs are at sea at any one time while that percentage is more than 50% for the US.

The antisubmarine-warfare capability of the US is widely acknowledged to be superior and is expected to remain so. Furthermore, the American submarines are quieter and more capable of evading Soviet submarine seekers. The Soviets have sought to offset at least partly the US advantage in antisubmarine warfare by installing SLBMs capable of reaching their targets when launched from a submarine not far from home port.

The latest acquisition to the fleet of Soviet SSBNs is the Typhoon submarine. This vessel, which is now being produced, is comparable in size to the Trident submarine and has 20 launch tubes. It will carry the SS-NX-20 missile, which may have greater range and accuracy than the SS-N-18. The Soviets have been testing the SS-NX-20 for about two years.

The weakest component of the strategic force of the USSR is its fleet of long-range aircraft. This leg consists of approximately 105 Tu-95 Bear and 49 Mya-45 Bison bombers, which typically transport two, one-megaton gravity bombs or air-to-surface missiles (ASM) and have ranges of 11 000 to 13 000 km. The Tupolev Tu-22M Backfire bomber is the latest addition to the Soviet's long-range aircraft, and the USSR may be producing them at the rate of about 30 per year.[7] The range of the Backfire is about 8000 km[1] so that it could conceivably fly on intercontinental missions. However, the USSR has refused to classify the Backfire as a strategic aircraft. No limitations concerning it were included in the SALT II agreement, although President Brezhnev signed a letter at the time promising certain limitations on production and deployment of the craft.

Summary of trends

In both the American and Soviet arsenals, the numbers of launchers have remained relatively constant while the numbers of warheads per launcher have increased and the yield per warhead has decreased. The graph on page 44 depicts a few such trends by comparing the current numbers of warheads and the total destructive power of the two arsenals to those of ten years ago. The US retains its advantage in total warheads while the Soviet Union has moved ahead in total destructive power. The breakdown of these figures into the three components—land, sea and air—illustrates the US triad, the Soviet reliance on and increased strength in its land-based forces, as well as the growth in the sea-based forces of both powers.

To put these comparisons into perspective, the lower graph indicates the size of the retaliatory force—400 EMT—that was defined in the 1960s as that necessary to achieve "mutual assured destruction." The Defense Department under then-Secretary Robert McNamara calculated that the equivalent of 400 one-megaton bombs would destroy 25% of the Soviet population and 50 to 60% of its industry.[8] Although this estimate may be somewhat arbitrary, it provides an order-of-magnitude measure of the extent of overkill now contained in both superpower arsenals.

Another very important trend is the dramatic increase in targeting accuracy of the ballistic missiles, as illustrated in the graph on page 45.[9] The CEPs of the newest ICBMs are now so small that they threaten destruction of hardened targets such as missile silos, as well as population centers and "soft" military installations such as naval bases or airfields. The ability of a missile to damage a hardened target is related to its yield and its CEP. For a given yield, the missile must land close enough to its intended target so that the blast produced at the target exceeds the peak overpressure the silo was designed to withstand. (Most US silos are hardened to 2000 psi.) A Soviet SS-18 or SS-19 with a yield of about 0.5 Mt and a CEP of 300 m has greater than a 50% probability of destroying a Minuteman in its silo,[10] while a Minuteman III could threaten a Soviet land-based launcher.

This "countersilo" capability of the Soviet Union has prompted concern about a "window of vulnerability" for the US during which the USSR could initiate a first strike that could incapacitate nearly all the American ICBMs. Some advocates of the MX contend that it is needed to close this gap. Those who dismiss the concern about a "window of vulnerability" point out that the Russians would not be likely to attempt such an attack, which requires a high degree of simultaneity and reliability that is hardly attainable. Furthermore the US would be left with substantial retaliatory ability in its aircraft and submarines.

The "countersilo" capabilities of both superpowers are alarming to some analysts who view this aspect of the arms race as inherently destabilizing: The more threatening the weapons of one side become towards those of the other side, the greater will be the motivation for each to undertake a preemptive first strike.

Another trend in the arms race is the growing sophistication of electronics used in surveillance and in the command, communications and control (C^3) structures. For example, as part of its antisubmarine warfare, the US has laid an extensive array of passive sonar detectors along its own coast and that of its allies in the North Atlantic and North Pacific; this system is the Sound Surveillance System (SOSUS). For patrolling smaller areas for tactical purposes, the Navy relies on aircraft equipped with magnetic anomaly detectors and with sonobuoys that can be dropped into the ocean, and on the nuclear attack submarines that can both seek and destroy. The US is working to coordinate these various efforts and is implementing higher-speed data processing for more effective signal processing.

Both the US and USSR are considering some form of ballistic missile defense (BMD) system. The Anti-Ballistic Missile Treaty of 1972 allowed each of the two parties to have one ABM system with no more than 100 launchers. The Soviet Union has its Galosh ABM system of some 64 launchers stationed around Moscow. They are reportedly developing a rapidly deployable BMD system that would consist of a phased-array radar, a missile-tracking radar and an interceptor with a high-altitude interception capability.[11] The US has largely dismantled its earlier Safeguard system and is now discussing for the future a possible BMD system for its MX ICBMs. One scheme that might be considered is a layered one, with exo- and endo-atmospheric detectors and interceptors. The first layer would consist of an infrared telescope and non-nuclear interceptor launched to destroy a target above the atmosphere. The second layer would be the Low Altitude Defense System (LOADS) composed of a small phased-array radar and 100 hypersonic missiles armed with nuclear warheads. Depending on their specific implementation, these systems would almost certainly violate the ABM Treaty.

On a more exotic note, both nations are conducting research on high-energy particle beams and laser weapons, despite considerable controversy surrounding their potential. Work is also proceeding on antisatellite technology,

and the USSR has already deployed a "killer" satellite for low-orbit attacks.

Tactical forces

Nuclear explosives play a variety of roles in the tactical arena. Thousands of bombs are carried by strike aircraft based on land and on aircraft carriers. Other nuclear warheads arm the hundreds of short-range ballistic missiles and howitzers designed for battlefield use. Many more are transported by submarines, surface ships or aircraft in support of antisubmarine missions.

Of all these types of tactical nuclear armaments, the ones receiving the most attention are those in the European theater. The US and the Soviet Union are currently holding talks on the reduction of intermediate-range nuclear weapons in Europe. Here the distinction between strategic and tactical becomes truly fuzzy. Some American weapons stationed in Europe for defense of the Western Allies can reach the Soviet Union and are perceived by the Soviets as a strategic threat. Similarly, the Europeans feel strategically threatened by "tactical" missiles on Russian territory that can reach their cities and industries.

The major American weapons on the negotiating table in the current talks are the ground-launched cruise missile (GLCM) and the Pershing II missile. Both are scheduled for deployment later this year. NATO voted in 1979 to replace 108 of its Pershing 1A missiles in West Germany with the new Per-

shing weapon, and to install 464 GLCMs in five countries, although some of those nations have not yet made firm commitments to accept the missiles.

The Pershing II will have a range of 1770 km, more than twice that of its predecessor. The radar-assisted, terminal-guidance system of the Pershing II should allow it to have a CEP of about 30 m.[1] Each Pershing II would deliver a yield of about 250 kt.[1]

The GLCM will be the Tomahawk cruise missile developed by the Navy. It is similar to the air-launched cruise missile developed independently by the Air Force for its B-52 bombers. The GLCM will be deployed in combat flights of 16 missiles each. The range of the GLCM is estimated to be about 2250 km, enabling it to reach Moscow if fired from West Germany as the map on this page shows. The Navy also plans to install Tomahawks on its surface ships and attack submarines. Of the three sea-based versions of this cruise missile, one will be a nuclear land-attack missile.

A third new weapon to debut in the European theater is the "enhanced radiation warhead," or neutron bomb. President Reagan intends to proceed with their production, aimed at supplying about 380 warheads for the Lance short-range ballistic missile and about 800 eight-inch artillery shells.[4] The neutron bombs are to be stored in the US for rapid deployment to Europe.

The medium and intermediate range ballistic missiles on the Soviet side are its SS-4, SS-5 and SS-20 weapons. The USSR began deploying the first two types in 1959. It currently has about 275 SS-4s, with ranges of 2000 km and around 16 SS-5s, with ranges of 4100 km.[1] Much more recently the Soviet Union has deployed about 300 of the SS-20 ballistic missiles, with an estimated two thirds of these capable of striking Western Europe. The SS-20 is a mobile missile tipped with three reentry vehicles of 150 kt each. Its range is around 5000 km, and its CEP close to 400 m.[1]

Besides the intermediate-range nuclear missiles, the Soviet Union has more than 1000 short-range ballistic missiles. The latest three additions to this force are the SS-21, 22 and 23 mobile missile systems. Perhaps 100 or so of these missiles have been deployed. The SS-22 SRBM has a range of about 1000 km and could pose a threat to Western Europe if moved into positions within Eastern Europe.

In sum, the American and Soviet arsenals are large, diverse, complex, and changing all the time. This survey merely scratches the surface. At the very least, it may illustrate the many dimensions with which arms-control negotiators must wrestle.

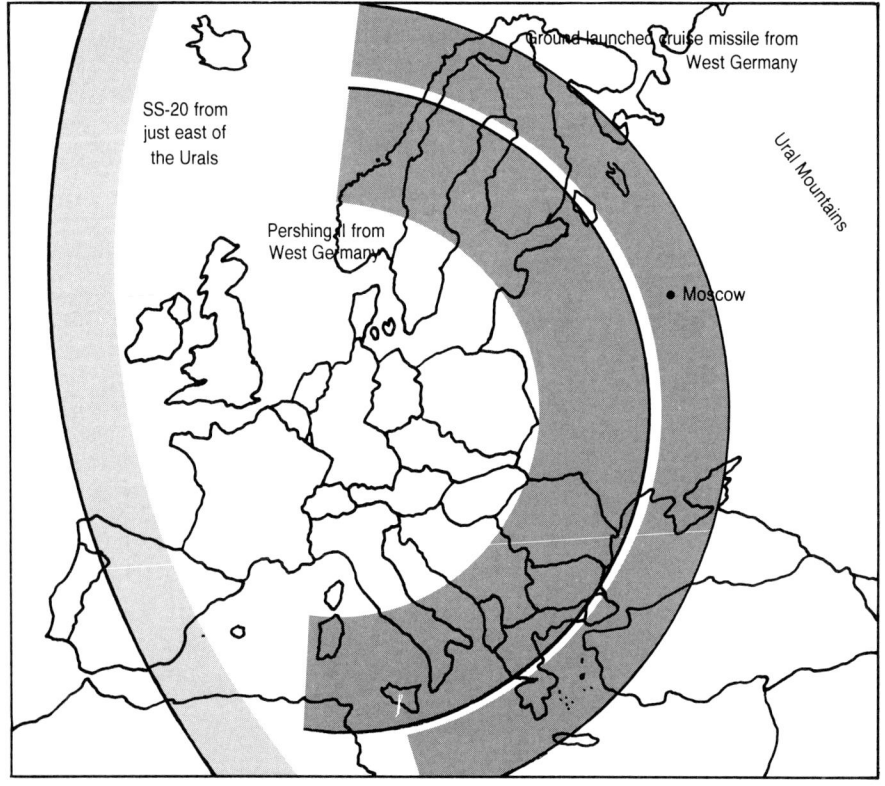

Ranges of tactical nuclear weapons. Some nominally "tactical" weapons—that is, weapons intended for use in the theater of a war—have ranges large enough to threaten, the home territory of the opponents, who therefore consider these weapons to be "strategic."

Salt II—terms of the treaty

▶ A ceiling of 2400 on the aggregate number of strategic nuclear delivery vehicles, to be lowered to 2250 by 1982

▶ A limit of 1320 on launchers of ballistic missiles equipped with multiple, independently targetable reentry vehicles (MIRVs) and on heavy bombers equipped for long-range cruise missiles

▶ A lid of 1200 on the number of MIRV'd ballistic missiles alone

▶ A ceiling of 820 on MIRV'd, land-based intercontinental ballistic missile (ICBM) launchers
▶ A prohibition against any increase in the number of reentry vehicles (RVs) carried by existing types of ICBMs. The number of warheads on existing type of ICBMs will be frozen at the maximum with which that missile type has been tested. New types of ICBMs will be limited to 10 RVs and new types of SLBMs to 14.
▶ A ban on the construction of additional fixed ICBM launchers and on any increase

in the number of heavy ICBM launchers, defined as those larger than the Soviet SS-19 missile
▶ A restriction to one new type of ICBM during the lifetime of the Treaty
▶ A ban on certain new types of strategic offensive systems such as ballistic missiles on surface ships
▶ A ban on the development of the Soviet SS-16 missile for the Treaty's duration
▶ An exchange of data on the weapons systems that are limited by the Treaty

References

1. *The Military Balance 1981–1982*, The International Institute for Strategic Studies, London (1982).

2. R. Forsberg, Sci. Am., November 1982, page 54.

3. F. Barnaby, AMBIO **11**, Nos. 2–3 (1982), page 76.

4. "Preparing for Nuclear War: President Reagan's Program," The Defense Monitor **10**, No. 8, Center for Defense Information (1982).

5. A. A. Tinajero, "US/USSR Strategic Offensive Weapons: Projected Inventories Based on Carter's Policies," Congressional Research Service Report 81-238F (1981).

6. R. Pretty, ed., *Janes' Weapons Systems 1981*, Key Book Service, Bridgeport, Conn. (1982).

7. *Soviet Military Power*, Pub. by the US Department of Defense (1982).

8. "US Strategic Offensive Forces in the 1960's," in, Commission on the Organization of the Government for the Conduct of Foreign Policies, Appendices, **4**, page 139.

9. R. Speed, *Strategic Deterrence in the 1980s*, Hoover Institution, Stanford, Calif. (1979) p. 38.

10. H. Feiveson, F. von Hippel, PHYSICS TODAY, January 1983, page 40.

11. *World Armaments and Disarmament*, SIPRI *Yearbook 1981*, the Stockholm International Peace Research Institute, London (1981).　□

Freeze on nuclear-weapons

Pro

The freeze and the counterforce race

The deployment of nuclear weapons that contribute to first-strike capabilities is destabilizing and must stop now

PHYSICS TODAY/JANUARY 1983

Harold Feiveson and Frank von Hippel

President Carter, in a 1980 report to Congress, speculated on how the Soviet Union might respond to the deployment of US weapons capable of destroying Soviet missiles in their silos:[1]

> ... adopting a launch-on-warning posture is perhaps the least expensive but the most potentially destabilizing and dangerous response option available to Soviet leaders.

Despite this risk, by 1980 the United States was already embarked on a massive effort to threaten Soviet land-based missiles. Similarly, despite the obvious danger that the US would adopt a policy of launching its missiles on warning of a Soviet attack, the Soviet Union had several years earlier initiated a massive deployment of missiles equipped with accurate multiple independently targetable reentry vehicles—"MIRVs"—capable of threatening US land-based missiles. (See the figure on page 40.)

This reckless superpower competition to develop "counterforce" weapons—that is, weapons designed to destroy the nuclear weapons of the adversary—has finally provoked, in the United States, a popular demand to "freeze" the nuclear arms race. In the words of the Nuclear Weapons Freeze Campaign's *Call to Halt the Nuclear Arms Race*,[2] this would be

> a mutual freeze on the testing, production and deployment of nuclear weapons and of missiles and new aircraft designed primarily to

deliver nuclear weapons.

In the last election, voters in states and cities representing a third of the country's population passed resolutions similar to this.

While a freeze would catch in its net many nuclear weapons systems, it was to counterforce systems that President Reagan referred when he rejected the idea of a bilateral freeze on the nuclear-arms race. He stated[3] that such a freeze would "only codify existing Soviet advantages," and he has made clear that the Administration is determined to deploy a new generation of counterforce missiles. This sets the stage for a sustained national debate on US policy toward counterforce weapons.

It is on the issues these weapons raise that we focus our discussion in this article. We argue that a freeze on the counterforce race at this time would be in the interests of both the US and the USSR, above all because it would fore-

continued on page 88

Six reentry vehicles streak toward targets on Kwajalein Atoll in the Western Pacific. Two Minuteman-III ICBMs, launched from Vandenberg Air Force Base in California, delivered the Mark-12 reentry vehicles—unarmed in this test. The deployment pattern evident in the

Harold Feiveson and Frank von Hippel are members of the research faculty at Princeton University's Center for Energy and Environmental Studies. This article was initiated in collaboration with Frank Barnaby, currently Guest Professor of Peace Studies at The Free University of Amsterdam.

development and deployment:

Con

The freeze—deep or shallow?

The freeze proposals are misdirected and will not help do what needs to be done to prevent nuclear war

Harold W. Lewis

PHYSICS TODAY/JANUARY 1983

The trouble with trite and banal sayings is that they are sometimes painfully to the point. In the case of the freeze, the observation that comes to mind is that to every complex problem there exists a solution that is simple, appealing, and wrong. Wrong may be too strong a term for the freeze proposal—it is wrong only in the sense that it is wrong to give laetrile to a cancer patient. In both cases there is little intrinsic harm done, *unless* the patient really believes the treatment will contribute to the cure of his disease, and thereby substitutes wishful thinking for therapy.

And the disease is all too real. Andrei Sakharov was right when he said that the prevention of nuclear war is the central problem for mankind. Yet it is equally true that no one wants it. That is the dilemma—how to forestall the occurrence of something no one wishes to occur, but that cannot be prevented by oversimplifying the issues. Some of the freeze advocates seem to think that there is a back-burner constituency for nuclear war, that there is a military–industrial complex that lusts after destruction, and that all that is necessary is to "send them a message" that we feel differently. Would that it were so—that would be an easy problem. Any serious discussion of these matters has to begin with the recognition that nuclear war is dreaded by everyone—hawks and doves, Russians and Americans, French and British, Japanese and Germans, and so on—and is yet possible.

What causes war?

What on Earth has this to do with the freeze? Simply that the connection between a freeze on nuclear weapons (not a reduction to zero by all nations, which *would* help but would probably make conventional war more likely) and the prevention of nuclear war is tenuous indeed. It appears to rest on the assumption that it is somehow the availability of weapons that leads to war, rather than international conflict over national interests, perceived as important by at least one side to the dispute. To prevent wars, we need a peaceful means of resolving genuine and difficult international questions, including questions that are regarded as threatening the existence or integrity of a nation. We are inching our way toward such a capability through international organization, but it is whimsical to believe we are yet there. The inventory of nuclear weapons has nothing to do with that. In fact, distasteful

continued on page 89

photograph is that of a counterforce attack in which two warheads from different missiles are directed against each hardened target. A Minuteman-III missile carries three warheads, each of which has a yield of either 170 kilotons or 335 kilotons.

Harold W. Lewis is professor of physics at the University of California, Santa Barbara.

FEIVESON/VON HIPPEL *continued*
stall a destabilizing enhancement of each side's first-strike capabilities. In the process of making this argument we provide some background information for scientists who may wish to become involved in the debate.

Why counterforce?

Because, to a considerable degree, opposition to a freeze and support for counterforce programs are two sides of the same coin, we must ask, "Why build up counterforce capabilities?" The answer appears to have many facets: the symbolic value of nuclear weapons as the "big sticks" upon which the superpowers depend as their ultimate recourse if they get into serious trouble abroad, the image held by many nuclear-weapons decision-makers of a zero-sum competition between the two superpowers, and the almost inevitable progress in the areas of technology that are critical for missile accuracy.

A key factor spurring the US side of the counterforce race has been an effort by the US to make credible its willingness to use nuclear weapons in areas where US conventional forces alone might be insufficient to deter Soviet aggression. During the 1950s the United States promised all-out nuclear attacks against the Soviet Union if the US decided that Soviet actions threatened US vital interests. Since the USSR developed a nuclear arsenal comparable to that of the US, however, such threats of "massive retaliation" against the Soviet population for anything other than an all-out Soviet nuclear attack on the US have become less and less credible.

Furthermore, as the US reached a level of nuclear plenty in which it had many times the number of nuclear weapons required to hold Soviet cities hostage, it became possible to think of using the extra weapons to develop more credible threats against targets whose destruction would not quite be the equivalent of the destruction of Soviet society. The obvious targets for such threats were the Soviet military and in particular their nuclear weapons aimed at the US.

The arms race ensued, in which each side has tried to threaten the other and foil the threats against its own nuclear weapons. In the case of the ICBMs, it has until now been a race between the hardening of underground silos on one side and the number and accuracy of the ICBM warheads on the other side. It is in this race that some see the Soviet Union as being ahead and the US thereby weakened in its ability to affect the decisions of the Soviet and other governments.

Thus, in 1980, the Carter Administration in its justification of the counterforce capabilities of the proposed MX land-based missile claimed that[1]

An asymmetry in hard-target-kill capability could lead to perceptions of Soviet advantage that could have adverse political and military implications including: (1) greater Soviet and less US freedom of action in the employment of conventional forces...

More recently, General Lew Allen Jr, then Chief of the Air Force, in a closed hearing before the Senate Armed Services Committee, stated that even

continued on page 90

Estimated size and destructive power of strategic arsenals, 1982

Delivery Vehicle	Warheads per delivery vehicle	Total number of warheads	Yield per warhead	Total blast area, all warheads, airbursts	Total fallout area, all warheads, groundbursts
US arsenal			(kilotons)	(10^3 km^2)	(10^3 km^2)
Intercontinental ballistic missiles					
Titan II	1	52	9000	60	400
Minuteman II	1	450	1200	120	370
Minuteman III					
Mark 12 warhead	3	750	170	60	50
Mark 12A warhead	3	900	335	110	150
Subtotal		2512		350	970
Submarine-launched ballistic missiles					
Poseidon	avg. 9	2736	40	90	130
Trident I	avg. 8	1920	100	100	230
Subtotal		4656		190	360
Bombers	avg. 7.5				
Bombs		1264	1200	330	1000
Short-range attack missiles		1114	170	80	70
Cruise missiles		192	200	15	15
Subtotal		2570		425	1095
Total		**9378**		**965**	**2425**
Soviet arsenal					
Intercontinental ballistic missiles					
SS-11	1	518	950	130	340
SS-13	1	60	600	10	25
SS-17	1	32	6000	30	170
	4	480	750	110	230
SS-18	1	58	20 000	120	930
	8	1400	900	350	770
	10	750	500	120	220
SS-19	1	60	10 000	80	520
	6	1800	550	320	580
Subtotal		5158		1270	3785
Submarine-launched ballistic missiles					
SS-N-5	1	18	1000	5	10
SS-N-6	1	356	700	70	160
SS-N-8	1	292	800	70	140
SS-N-17	1	12	750	5	5
SS-N-18	avg. 7	1680	200	150	130
SS-N-20	avg. 10	200	200	20	15
Subtotal		2558		320	460
Bombers	avg. 2				
Bombs		152	1000	40	90
Short-range attack missiles		161	1000	40	90
Subtotal		313		80	180
Total		**8029**		**1670**	**4425**

LEWIS *continued*

though the thought may be to some, these appalling weapons have probably contributed mightily to preserving the peace among the great powers for the last 35 years. The realistic course for the prevention of nuclear war lies first and foremost in the prevention of war among the nuclear powers, and, failing that, making the nuclear threshold high enough to deter a resort to nuclear weapons by a losing side. The freeze proposals are not directed to either of these.

In fact, as I listen to arguments for the freeze, the common theme I find is revulsion against nuclear weapons *per se*, with a strong undercurrent of anti-technology sentiment. Rarely is there anything resembling a considered effort to assess the probable impact of a freeze on the prospects for world peace. It is not too unlike the arguments against nuclear power (indeed, many of the same people are among the leaders of the freeze movement), which are often derived from an emotional and Luddite base, not from any expert assessment of whether nuclear power is or is not the cleanest, safest and cheapest way we know to make electricity. In both cases we find ourselves dealing with issues of symbolism, where the substance has become subordinated to the "message." In both cases that makes it possible to concentrate on the horrors of failure, rather than on the tools of success. Is that a good way to run a country?

Command center of the Strategic Air Command, Offutt Air Force Base, near Omaha, Nebraska. (Photograph courtesy of United States Air Force.)

But let's get back to what the various freeze initiatives say, and what they would really do if they were to pass. They vary somewhat among themselves, but the common theme (for example, in the California initiative and in the Senate Joint Resolution introduced by Senators Edward Kennedy and Mark Hatfield) is to call on the US government to propose to the Soviets a mutually verifiable freeze on the testing, producton and further deployment of nuclear "warheads, missiles and other delivery systems." They do not call for any of these actions

Trident missile rises from the nuclear-powered submarine USS John C. Calhoun in a "shakedown operation" launch off the east coast of Florida, 28 October 1980 (US Navy photograph.)

to be taken unilaterally (though there seems to be an underlying assumption that we are the impediments to progress), but view a freeze as a step toward halting "nuclear madness." (I have to express a particular resentment here about the tendency to brand anyone who may hold a different view as "mad." It makes a reasoned debate difficult. Of course two can play at that game. In an article in Business Week a year or so ago, a respected practicing psychiatrist assessed some public attitudes toward nuclear power as having the classic clinical features of a phobia. According to Webster, a phobia is an "irrational, persistent fear.") In any case, the apparent objective is solid—the two major nuclear powers do have in their arsenals enough firepower to destroy each other many times over, in part at least because each wants to have enough reserve to provide a credible deterrent in the event of an attack by the other.

Soviet buildup. We (the US) have actually not increased our firepower in many years, though the same can not be said of the Soviets. We are frozen at a little over a thousand land-based missiles, somewhat over five hundred sea-based missiles, and some aircraft of debatable penetration capability. We have, on the other hand, been converting to multiple independently targeted reentry vehicles. To the extent that one "MIRVs" a missile, the number of warheads goes up while the total firepower goes down; the destructive capability is a relatively complicated function of those two and of the missile's accuracy and reliability. It is, in particular, not true that we have been madly building missiles. The Soviets *have* been increasing their arsenal, for

continued on page 91

FEIVESON/VON HIPPEL *continued*
though the developing vulnerability of
US ICBMs is[6]

> perhaps not dangerous in that it
> will incite them [the Soviets] to
> first strike, it nevertheless gives
> them confidence in their nuclear
> forces. That confidence means
> that we will find the threshold of
> nuclear war much higher than in
> the past, and we will see greater
> Soviet confidence in their ability to
> be adventuresome and provocative
> to the United States across a broad
> range of areas.

General Allen is widely considered a
moderate in matters of nuclear-weapons policy. Yet here he was arguing in
favor of keeping the threshold of nuclear war low!

And still more recently, Richard
DeLauer, the Reagan Administration's
Undersecretary of Defense for Research and Engineering indicated the
same priorities when he worried that[7]

> increases in nuclear hardness of
> Soviet ICBM silos and other impor-
> tant facilities have reduced our
> ability to put those targets at risk.
> Knowing this the Soviets feel less
> constrained from adventurism
> around the world . . .

Another purpose of the Reagan Administration in pursuing increased
counter-silo capabilities is to undermine the economy of the Soviet Union
by forcing it to initiate costly programs
of military investments in new mobile
strategic systems or even active missile
defense. Thus, in the Reagan Administration's first 5-year defense guidance
document, the Defense Department
was advised to develop weapons that[8]

> are difficult for the Soviets to
> counter, impose disproportionate
> costs, open up new areas of major
> military competition and obsolesce
> previous Soviet investments.

Unfortunately, if either superpower
decides that its fixed land-based missiles are obsolete and deploys a mobile
land-based missile or deceptively-based
missile less vulnerable to attack by
accurate warheads, these missiles will
also be more difficult for the other side
to count by its "national technical
means" (primarily satellites) and
therefore to eliminate by agreement.
And, if either side decides to deploy a
defense of its ICBM silos, the result
could be the abrogation of the Treaty
on the Limitation of Anti-Ballistic Missile Systems.

A final reason for the drive by both
sides toward counterforce capabilities
is that it is the path of least resistance.
With major laboratories working continuously on more accurate systems as
well as new warhead designs and new
delivery vehicles, techniques for improvement will be found. Once new
technology is available—and it is often
available at relatively modest cost—
the defense establishments usually find
it irresistible. This is especially so for
counterforce weapons, which both the
US and Soviet military see as more
usable and appropriate to traditional
military roles than "city-busting" deterrent forces.

The chimera of limited war. An important element in the analyses used to
justify the counterforce race is the idea
that it might be possible to fight a
nuclear counterforce war in a carefully
controlled manner. However, because
the means of command and control are
inevitably vulnerable to nuclear destruction, it is extremely doubtful that
a nuclear war could be limited and
prevented from escalating into an all-
out civilization-shattering exchange.
Moreover, even if a nuclear exchange
could be strictly limited to military
targets, a strategically significant
counterforce attack would probably
cause tens of millions of civilian deaths.

Command and control systems can
be "hardened" to some extent against
nuclear attack, and the Reagan Administration proposes to spend about $20
billion over five years for that purpose.
But these systems will remain inherently more vulnerable than nuclear-
weapons systems. As John Stein-

Weapon capability curves showing the overpressures produced by warheads of various yields
as a function of distance from ground zero. Such curves indicate weapons' ability to destroy hard
targets. Cratering effects may knock out silos that are hard enough to escape destruction by
overpressure;[10] silos in the shaded region would be covered by debris to a depth of at least 4 m.
The points shown for various US and Soviet warheads indicate estimated yields and median
miss distances.[13,15]   Figure 1

continued on page 92

LEWIS *continued*

Strategic-missile submarine USS Ohio, underway off the coast of Connecticut. This US Navy photo, a starboard-bow view of the nuclear-powered submarine, was taken 4 September 1981.

reasons I have yet to understand—maybe their military–industrial complex is responsible, or whatever. It is true that, whatever the reason, they spend nearly twice as large a fraction of their substance on defense as we. Just as in our case, of course, the actual expenditures for strategic offensive weapons are only a small fraction of defense expenditures, the vast bulk of the budget going to maintain conventional forces. However, in their case, there is an expansion of the strategic forces, particularly those directed against our European friends. To negotiate a "mutual and verifiable" freeze, we have to deal with all that.

We also have to deal with the last point—verifiability—which has been a persistent roadblock for decades of negotiations about arms control and nuclear testing. The ABM and SALT negotiations finessed that question by agreeing that each nation was free to use its own intelligence resources, the so-called "national technical means," to verify compliance with the agreements, while each side undertook not to deliberately interfere with the process. This has come to mean satellite and other forms of remote surveillance, which are reasonably comprehensive with respect to deployment and testing, except for low-yield underground testing of nuclear weapons. Production is another matter.

Pressure our own government?

Presumably the intent of a freeze—in this case a ban—on the testing of

continued on page 93

US policy: strategic modernization and arms control

Robert W. Dean

The public discussion of the so-called warfighting strategy of this Administration completely misconstrues the basic objectives of US and alliance policies. We do not seek strategic superiority over the Soviet Union and the nations of the Warsaw Pact. We do seek equality in the form of a strategic balance that provides greater security for all nations.

The unrestrained growth of Soviet military power is the basic factor that motivates this Administration's modernization programs. This growth has forced us to reevaluate our force requirements and to make the necessary adjustments with determination and resolve.

The Soviet military buildup over the past two decades has been sustained and impressive. In most significant measures used to judge strategic forces—total number of systems, total number of ballistic missiles, total destructive power—the Soviets now surpass the United States. Soon they could equal and surpass us in number of warheads, the one area in which the United States has traditionally had an advantage. In nonstrategic nuclear forces, the Soviet buildup has been equally impressive. The Soviets now have an overwhelming superiority in numbers and capability of nuclear forces deployed for theater use.

The Administration's response to the challenge of the Soviet force buildup has been twofold: The first step was a commitment by the members of the NATO alliance to modernize both strategic and intermediate nuclear forces, to ensure that our deterrent remains strong. The second step is a commitment to pursue vigorously arms-control measures designed to increase stability and reduce the number of these formidable weapons.

The comprehensive modernization program announced by the President in October 1981 is designed to rectify the vulnerabilities and weaknesses in our strategic forces. The purpose of this program is to restore the eroding nuclear balance and to sustain the credibility of the United States deterrent.

The programs authorized as part of this modernization—the MX and Trident II missiles, the B-1 and Stealth bombers, air- and sea-launched cruise missiles—combined with improved air defenses and enhanced command-and-control capabilities will serve to counter many, if not all, of the advantages that have accrued to the Soviets as a result of their own deployments. The reliability, survivability and effectiveness of these US forces will do much to strengthen the deterrent posture of the entire NATO alliance.

Modernization is also moving forward at the nonstrategic level. As the result of a 1979 decision by the NATO alliance, deployments of Pershing II and ground-launched cruise-missile systems will begin at the end of 1983 unless there is a concrete agreement with the Soviet Union on intermediate-range nuclear forces. The presence of Pershing II and ground-launched cruise missiles in Europe will force the Soviets to recognize that to strike NATO Europe would engage US forces that can strike Soviet territory.

Such a clear United States commitment to the nuclear defense of the alliance will convey to Soviet leaders that they cannot use their territory as a sanctuary from which to launch nuclear attacks against the NATO allies. The willingness of the United States to take such a clearcut risk, to identify its security with that of its European allies in this way, ensures that the Soviets will not see any advantage in striking Europe because they will know they are subject to sure retaliation by the central force of the United States.

Arms-control agreements. Modernization of US nuclear forces is only one of the two essential elements of our program to restore the stability of the nuclear balance and thus guarantee our nuclear deterrent. The search for sound arms-control agreements is the other key feature of our program. The President has outlined the general principles that guide our arms-control policies:

▶ Arms control must be an instrument of, and not a surrogate for, a coherent security policy. We will work for agreements that truly enhance security by reinforcing deterrence.

▶ We seek balanced agreements that involve meaningful reductions on both sides. Balanced agreements are necessary for a relationship based on reciprocity with the Soviet Union, and are essential to maintaining the security of both sides. Quantitative parity is important, but balance is more than a matter of numbers. Of greater significance is the capacity of either side to make decisive gains through military operations or the threat of military operations. Agreements that do not effectively reduce the incentive to use force,

continued on page 93

FEIVESON/VON HIPPEL *continued*

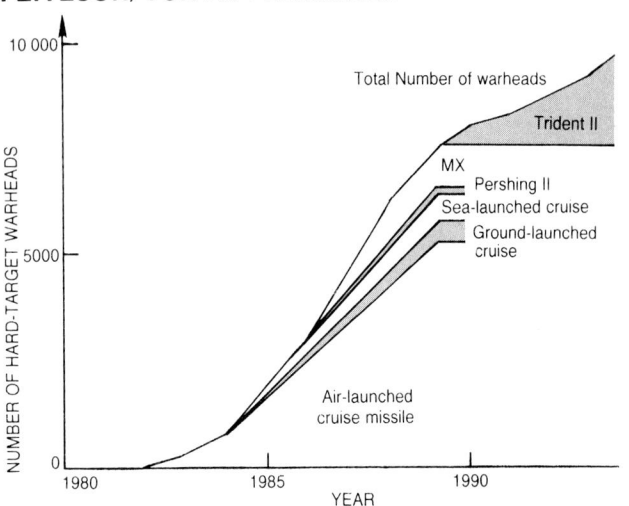

Proposed new US counterforce weapons. The proposed nuclear-weapons modernization program includes more than 10 000 new warheads, each of which is expected to be capable of destroying a Soviet ICBM silo.[6,14]   Figure 2

bruner, an expert in command and control, recently pointed out[9]

> ... once the use of as many as 10 or more nuclear weapons directly against the USSR is seriously contemplated, US strategic commanders will likely insist on attacking the full array of Soviet military targets. Political motives for engaging in limited strategic attacks will not likely prevail against the risks of leaving a vulnerable command system exposed to counterattack from a severely provoked enemy.

In the strategic literature, nuclear war often seems like a long-distance version of the artillery duels of World War I. The side-effects of the missile exchanges are labeled "collateral damage" and are seldom discussed. They are far from unimportant, however.

Hidden in the Defense Department's scenarios for limited strategic nuclear war, for example, are Soviet "barrage attacks" on US airbases that house bombers and refueling aircraft. In these scenarios, warheads of half-megaton size explode over and around the bases to destroy aircraft caught on the ground and aircraft that have just become airborne. The blast and heat from a single 0.5-megaton warhead exploded in the air over a B-52 base would kill the population in an area of the order of 100 square kilometers.[10] A number of urban areas in the US would be destroyed or partially destroyed by such barrage attacks on bomber bases. Such attacks would be still more damaging if, in a time of tension preceding the war, bombers were dispersed to major civilian airfields, as occurred during the 1962 Cuban missile crisis.

In the United States, ICBM bases are generally more isolated from nearby populations than are bomber and submarine bases. However, the Minuteman bases contain so many separate targets—150 to 200 silos, each of which is ordinarily assumed to be targeted by

two half-megaton warheads—that the lethal radiation field from the overlapping fallout patterns would extend for many hundreds of miles downwind.[11]

As a result of all these effects, the Department of Defense was forced in 1975 by the Senate Foreign Relations Committee to admit[11] that a full-scale Soviet attack on US ICBM, bomber and

missile-submarine bases would kill 3 to 16 million Americans. More recent US government estimates[12] have raised this range to 24 to 45 million. Estimates[12] of the consequences of a US counterforce attack on Soviet strategic nuclear forces are of the same order of magnitude.

Dangers of counterforce

Figure 1 shows some estimated yields and median miss distances, ordinarily termed "circular errors probable," or CEPs, of the various US and Soviet counterforce warheads. These points are superimposed on a graph showing the peak overpressure felt by a silo as a function of the explosion's horizontal distance from the silo. One can see that warheads on modern ICBMs—the US Minuteman II and III equipped with Mark 12A warheads, and the Soviet SS-18 and SS-19—are expected to produce within their CEPs an overpressure equal to or beyond that which US silos are designed to withstand. The labels at the top of the figure indicate that the next generation of US warheads—to be carried by the MX, Trident II and cruise missiles—are expected to be so

continued on page 94

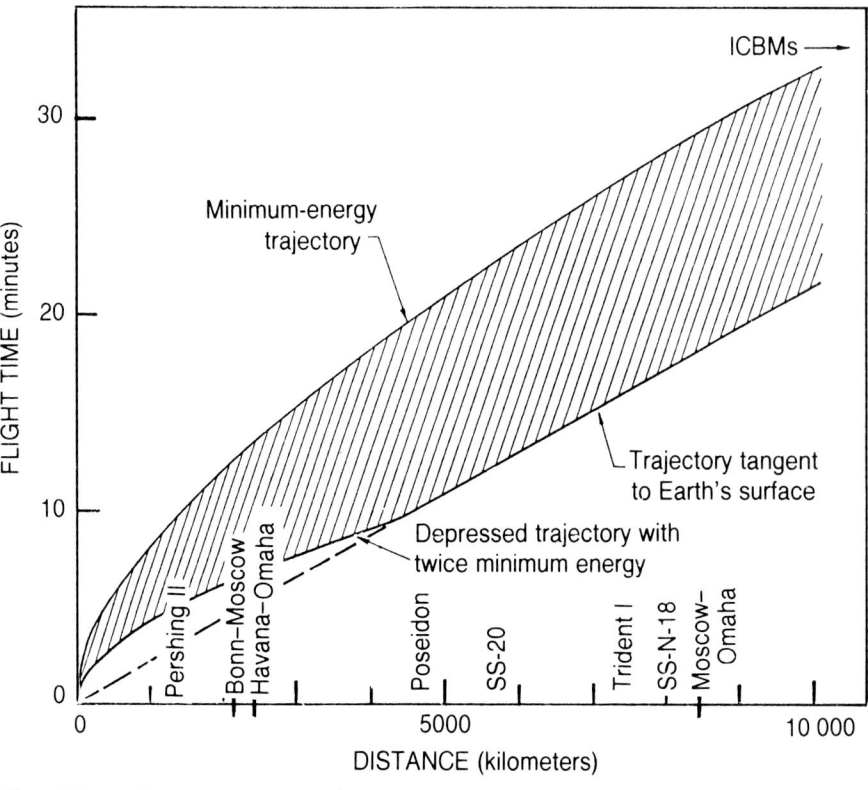

Travel times. The range of times required for warheads to follow ballistic trajectories between two points on the Earth's surface are shown here as a function of the great-circle distance between the two points. These times are upper limits on the times available for making decisions in launch-on-warning systems. The top of the band gives travel times for elliptical trajectories, which require the minimum energy for a given range. The bottom of the band gives times for circular orbits just above the Earth's surface. For short distances, the graph shows flight times along "depressed" ballistic trajectories of twice the minimum energy. Actual flight times would be 1–2 minutes longer than shown, principally because of the slower average speed of the warhead during the "boost" phase, when the average acceleration is on the order of a few times that of gravity. Along the bottom of the graph are a few relevant distances and the estimated full-load ranges of various US and Soviet ballistic missiles.[4,16]   Figure 3

LEWIS *continued*

nuclear weapons is to prevent a technology "breakthrough" that might disturb the rough parity that now exists between the Soviets and us. It is, in fact, hard to imagine either country depending upon untested weapons, so that a testing freeze is indeed very likely a technology freeze, and stabilizing. However, a ban on nuclear testing already exists, with the exception of low-yield underground testing, and the exception is there precisely because the verification problem has turned out to be technically very difficult in that regime. I don't want to sound overly pessimistic on this point, but one can always conduct a test too small to be detected by *any* system. The real question is whether the threshold for detection can be made low enough to make the rewards for the violator unequal to the risk. Perhaps the freeze advocates know how. Failing that, the only solution is relatively unconstrained on-site inspection, something we have offered and the Soviets have consistently rejected for decades. The leader of the California freeze movement, a real-estate developer, was

Minuteman III intercontinental ballistic missile. This US Air Force photograph shows a test launch in Florida.

quoted recently as having said that he expects the Soviets to eventually change their minds on this point, but he didn't explain why "sending a message" to *our* Government, or "changing *our* political climate" (his words, my emphasis) will accomplish this. This low-yield underground testing is the only nuclear testing we or the Soviets have done in years, so this is all the proposed testing ban refers to.

From all the above, it would probably be possible to conclude that I am opposed to the freeze proposals, but that is not so. Nor am I in favor. Nor do I feel that I even care enough to take position on an issue that is so disjoint from the prospects for avoiding nuclear war, an objective I regard as paramount. (Of course there are far too many nuclear weapons in the world, but that is a symptom of the disease, not the disease itself. The use of one-tenth the number in a nuclear war

continued on page 95

DEAN *continued*

especially in crisis situations, do nothing at all to enhance security.

▶ Arms control must include effective means of verification.

▶ Arms-control policies must take into account the totality of the national-security posture, not simply those elements that are the subject of a particular negotiation.

▶ Our efforts based on these principles are and will be guided by a seriousness of purpose reflected in our willingness to accept reductions to the lowest possible equal levels of nuclear forces.

The US proposals at the strategic-arms-reduction talks are the centerpiece of our arms-control efforts and are based on the five principles listed above. The START talks, which began 29 June 1982 in Geneva and resumed on 6 October, provide the opportunity to enhance world security and peace through a carefully constructed agreement to reduce strategic nuclear arsenals.

The United States proposal would limit to 850 the number of ballistic missiles that each side may have. Beyond this, both sides would be limited to a maximum of 5000 deployed ballistic-missile warheads, of which no more than 2500 could be on ICBMs.

This proposal breaks important new ground in strategic-arms control in several ways. It directly addresses, for the first time, the most pressing strategic problems that threaten world security. It suggests direct limits on ballistic-missile warheads rather than on missile launchers, a unit of account that was used in the past. The total number of warheads is a much more relevant measure of strategic capability than launchers alone, because a limit on the latter would treat equally weapons

systems that are not equivalent. For example, the single-warhead US Minuteman II is treated in SALT II no differently from the multiple-warhead Soviet SS-18.

The US proposal calls for major reductions in strategic armaments on both sides. About a year ago, a prominent American scholar urged a 50 percent reduction in the strategic arsenals of both sides. Many praised this goal, but this type of reduction was viewed as unlikely to win any support in this Administration.

But we have in fact challenged the Soviet Union to demonstrate its professed desire for strategic-arms reduction, calling on both sides to reduce levels of deployed ballistic missiles by almost 50 percent. The US proposal fully embraces the principle of equality. It is not motivated by a desire for strategic superiority, because it calls for equal levels of ballistic missiles and their warheads for both sides.

Because of the asymmetry in Soviet and US forces, particulary Soviet reliance on a larger number of land-based ICBMs, it will be necessary for the USSR to undertake greater reductions within ICBM forces to reach equal warhead levels. The US, on the other hand, would have to dismantle a larger number of submarine-launched missile warheads.

The quantitative increase in Soviet strategic nuclear forces, their enormous delivery vehicles, multiple warheads and deliverable throw weight, combined with qualitative improvements in accuracy, make it possible, in theory, for the Soviets to destroy the large majority of US land-based ICBMs. This is a highly destabilizing situation.

The point here is not that the Soviets would undertake an unprovoked preemptive first strike against US strategic forces.

Nevertheless, the danger exists that during a period of extreme international tension, when the advantage of striking first could be persuasive, the Kremlin leadership might be tempted to undertake such a strike. This is a recipe for potential catastrophe.

Therefore, to strengthen deterrence, we must modernize our forces to make a preemptive first strike an unthinkable option for the Soviets. We must remember that Soviet actions will be determined not only by their perception of American determination, but by their dispassionate assessment of objective American capabilities.

The US proposal recognizes what defense planners and arms-control specialists have known for years, that some types of strategic nuclear weapons pose greater threats to strategic stability and thus pose a greater risk of igniting a nuclear war than do others. Multiple-warhead ICBMs are among the most destabilizing strategic offensive arms. The Soviet SS-18, for example, with its ten large, independently targetable warheads, has done the most to render vulnerable our land-based ICBMs and thus undermine strategic stability. The United States proposal to establish a sublimit of 2500 warheads on land-based ICBMs is intended to address this threat to stability.

Submarine-launched ballistic missiles are included in the overall 5000-warhead limit and in the 850-missile limit, but they are not subject to a special sublimit because they are less accurate than land-based ICBMs and thus are less of a threat to stability.

Bombers, because they are not effective first-strike weapons, are generally recog-

continued on page 95

FEIVESON/VON HIPPEL *continued*

accurate that their target would ordinarily be within the radius of cratering effects produced by a nuclear ground burst.

The principal threat to US landbased missiles today is the large number of accurate warheads carried by two types of large Soviet ICBMs. These Soviet missiles are designated the "SS-18" and "SS-19" by the US Defense Department. The Defense Department believes[13] the Soviets soon will have (or may already have) roughly 4000 silolethal warheads on a total of almost 700 SS-18 and SS-19 ICBMs.

Each warhead on the 1000 US Minuteman missiles is believed[13] to have a similar ability to destroy silos, but there are only 2100 Minuteman warheads for 1400 Soviet ICBM silos of all types, versus about four SS-18/19 warheads per US silo. Therefore, given roughly equal destructive capability per warhead, a larger fraction of Soviet ICBMs might be expected to survive a US first strike than *vice versa*.

The Reagan Administration's plans[6,14] to increase the US threat to Soviet missile silos includes the following, as shown in figure 2: at least 1000 accurate high-yield warheads on a force of 100 MX missiles; thousands of silo-killing warheads on a force of new submarine-launched Trident II, or "D-5," ballistic missiles; several thousand warheads on slower but highly accurate air-, sea- and ground-launched cruise missiles; and 108 accurate Pershing II missiles, which could hit key targets in the western USSR within ten minutes of being launched from West Germany. The deployment of all these "hard-target killers" would make the Soviet land-based missiles at least as vulnerable as those of the US.

The Soviet Union should find the prospective vulnerability of land-based missiles even more disturbing than does the United States at present. Whereas only one quarter of all US strategic warheads are on ICBMs, two thirds of the Soviet Union's strategic warheads are based in silos. Furthermore, the other two legs of the Soviet "triad" are already somewhat vulnerable: The Soviet bomber force is much less capable than that of the US and is not ordinarily on alert status; and the security of the small percentage of Soviet ballistic-missile submarines that are at sea at any one time is being eroded by enormous US investments in large ocean sound-surveillance systems, nuclear attack submarines and antisubmarine aircraft.[1] (There is no comparable Soviet threat to US ballistic-missile submarines.)

Under these circumstances, as the US threat to Soviet missile silos grows, the Soviet Union may become tempted to put its ICBMs on a launch-on-warning status and, during periods of crisis, entertain ideas of preemptive attack.

An indication that the Soviet Union is at least considering launching its missiles on warning of US attack recently appeared in an article by Soviet Defense Minister, Dmitri Ustinov:[18]

> With modern detection systems and the combat readiness of the Soviet Union's strategic nuclear forces, the United States would not be able to deal a crippling blow to the socialist countries. The aggressor will not be able to evade an all-crushing retaliatory strike.

The warning times involved in an

Vulnerability of US silos

How does one estimate the percentage of the 1052 US ICBM silos that would survive a Soviet counterforce attack? In 1979, the Department of Defense released[5] the 1978–88 vulnerability projection shown below. In the unclassified version of the graph, the DOD deleted the numbers on the vertical axis—but not the ticks. Let us calculate the deleted numbers.

In calculating "silo kill probabilities," one ordinarily assumes that the distribution of warheads as a function of distance R from their targets will be proportional to

$$\exp[-0.7\ (R/\mathrm{CEP})^2\,]$$

a gaussian probability distribution in which CEP is the "circular error probable," or median miss distance. One assumes that a silo is destroyed if a warhead lands within a "lethal radius" R_L. For hardness higher than about 100 lb/in^2,

$$R_L \approx 460\ (Y/H)^{1/3}\ \text{meters}$$

Here the yield Y of the warhead is measured in megatons and the hardness H of the silo is measured in thousands of pounds per square inch.[10]

In 1978, as today, the principal threat to US missile silos was the large number of accurate warheads carried by the most modern Soviet "SS-18" and "SS-19" intercontinental ballistic missiles. In 1978 there were about 200 of each of these missiles, many of which were believed to carry payloads of 6–10 warheads. These warheads have estimated yields of 500–750 kilotons and estimated accuracies such that half would land within 315–425 meters of their targets.[13,16,17] We assume that in a counterforce attack two of these warheads would be aimed at each US missile silo; a Soviet attack using a larger number per silo would be much more disarming of the Soviet Union than the US because there are an average of only two warheads available to be destroyed in each US missile silo.

Silos in the US are designed[14] to protect their missiles against blast effects associated with peak overpressures up to 2000 lb/in^2. This corresponds to a lethal radius of 300–340 meters for the Soviet warheads. Assuming that each Soviet warhead had an arrival reliability of 80–100 percent, one can estimate that 225–640 US missile silos would have survived the postulated Soviet attack in 1978. We have, therefore, labeled the vertical axis in the figure as shown.   —HF & FvH

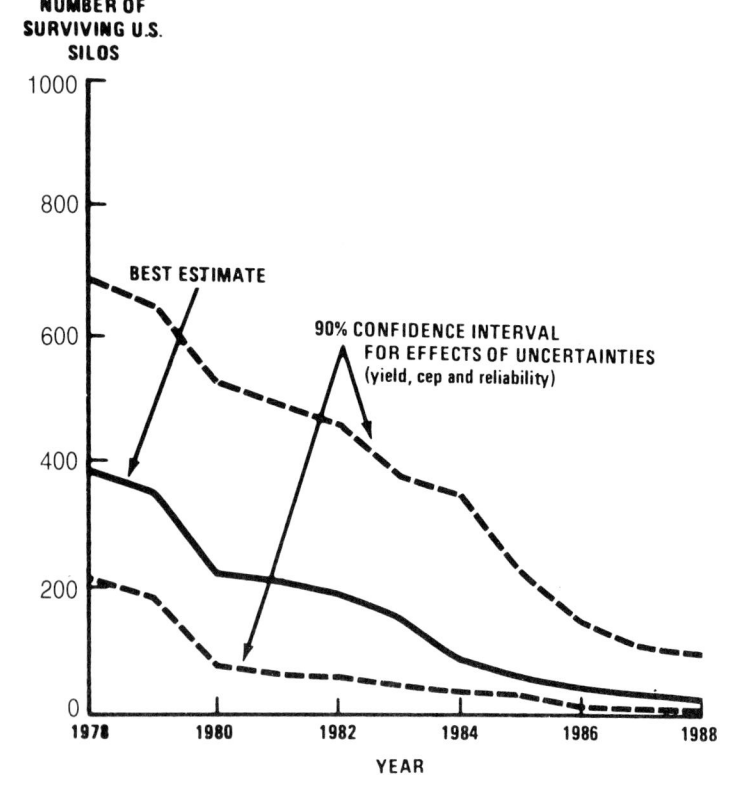

continued on page 96

LEWIS *continued*

would mean the end of western civilization, as we know it.) To be sure, the freeze movement is likely to provide somewhat more incentive for the negotiators in our Government, and that is good. It cannot provide any guidance in a formal sense, because a "mutual, verifiable freeze" is just one of many possible objectives for arms-control negotiations, and, while the desire for arms control and reduction is a proper subject for political influence, the form of any putative agreement is not a matter for bumper-sticker or rock-concert politics. Finally, one can ask whether political pressure on our Government by dissenting people on our side is likely to increase the Soviet incentive to strike a mutually satisfactory bargain.

It may not be fair, but it is common, to ask the question "What would you do?" First and foremost, I would like to prevent nuclear war, which I believe is far more likely to occur through the inexorable proliferation of nuclear weapons to parties less responsible than either we or the Soviets. I applaud the Israeli bombing of the Iraqi reactor, and believe we could and should put a great deal more effort and attention into controlling the proliferation of nuclear weapons. I believe we should do this by encouraging, and participating in, the international development of nuclear power, not by rejecting one of the few weapons we have against the oil sheiks. I am not averse to the use of strong measures,

Test of the MX cold-launch system at a site north of Las Vegas, Nevada, 26 January 1982. In this system a gas generator ejects the missile from a launch canister. The rocket motor would ignite when the missile is 100 feet from the canister. The dummy missile used in this test has the same dimensions, weight and center of gravity as the MX. (US Air Force photograph.)

preferably diplomatic, against those who lie about their efforts to acquire nuclear weapons. If that be elitism, so be it.

Given that, I would recognize that a much lower level of nuclear weaponry is possible, while maintaining a rough parity between us and the Soviets, provided the security of the deterrent against preemptive attack is assured. I believe this can be accomplished (on our side) by non-nuclear defense of a subset of the Minuteman silos, or even new silos if necessary, which is technically feasible. (Ballistic-missile defense got a bad name some years ago, when people were discussing the defense of cities, which is both technically infeasi-

ble and destabilizing. Non-nuclear hard-point defense is not easy, but it is feasible, and it is stabilizing. This is, incidentally, an alternate track for resolving the MX siting problem.) With reasonable security of the deterrent, the road would be open to reduction in nuclear weaponry, not because it would save money or provide symbolism, but because the excess numbers would no longer make a substantial contribution to either nation's security. The cart would then be behind the horse.

Many good songs end by repeating the refrain. I can't think of a better final paragraph for this article than the first. Please reread it. □

DEAN *continued*
nized as the least destabilizing systems and thus are not restrained as severely as ballistic missiles, though they are included in the US position at Geneva.

This emphasis on enhancing strategic stability is fundamental to the United States proposal. It ensures that the reductions that are achieved in START serve the essential objective of these negotiations: the enhancement of international security by reducing the risk of nuclear war.

The US START proposal points the way to a more stable strategic environment at equal and reduced levels of strategic forces. It is fair and equitable. It is internally consistent, and its limits are in the mutual interests of both East and West.

Through our START proposal and in conjunction with our strategic modernization programs, the United States will continue to be able to deter the Soviet Union while reducing the risk of war. We will thereby continue to support both our commitments to NATO and our obligations to the world as a whole to maintain the peace and security that we all seek.

The no-first-use and freeze proposals, which are currently receiving wide attention, are, in the Administration's view, mis-

steps on the road to effective arms control. The no-first-use policy, enunciated in the West and brought up again by Soviet Foreign Minister Andrei Gromyko at the UN special session on disarmament last June, is a superficially attractive suggestion, but it has numerous defects that would result in an erosion of deterrence.

In the first place, it is not credible. Simple declaratory statements have no meaning when the capability to violate the declaration is retained. Such a commitment would place NATO at the mercy of superior Soviet conventional forces. It would remove any further recourse NATO would have in the face of imminent conventional defeat. In essence, it would make Europe safe for conventional aggression.

Another arms-control proposal that has attracted considerable attention is a mutual freeze on the testing, production and deployment of nuclear weapons and their delivery systems. The drawbacks of this proposal are considerable.

A freeze at existing levels would codify US military disadvantages, especially in the strategic area, and it would lock us into a situation of dangerous instability. I have already mentioned the areas in which these vulnerabilities exist.

The nuclear-freeze proposal ignores the fact that some modernization will be required, along with arms controls, to ensure lasting stability and effective deterrence. A freeze is simply not good enough. Arms control, properly pursued, can and should result in lower numbers of nuclear weapons on both sides. The US START proposal and the US proposal for the reduction of intermediate-range forces in Europe are based on this premise.

A freeze on all testing, production and deployment of nuclear weapons would include important elements that cannot be verified. The practical result would be that we would live up to a freeze in all its aspects, while there would be considerable doubt that the Soviets were equally faithful to it. This would result in a highly unstable situation.

★ ★ ★

Robert W. Dean is deputy director of the Bureau of Politico-Military Affairs, United States Department of State. The above is based on Dean's testimony at a September 1982 Federation of American Scientists' hearing on the nuclear freeze. A complete transcript of the two-day hearing is being published by Brickhouse Press, Andover, Massachusetts. □

FEIVESON/VON HIPPEL *continued*

attack could be less than ten minutes, as figure 3 shows. Herbert York points out that with a US launch-on-warning system, these short times mean that[19]

the determination of whether or not doomsday has arrived will be made either by an automatic device designed for the purpose or by a preprogrammed President who, whether he knows it or not, will be carrying out orders written years before by some operations analyst.

The danger of preemptive attack is inherent in counterforce weapons. These weapons are presented by their proponents not as first-strike weapons, but as weapons that would only be used to destroy any enemy missiles held back in a first strike. Unfortunately, such a strategy would almost assuredly be futile, for, if there were ever a time when a nation would be prepared to launch its nuclear weapons on warning it would be after it had struck first. The only chance—a very small one— for a successful use of counterforce weapons would be in a "preemptive" first strike.

The Carter Administration recog-

nized the possibility of the Soviets being driven to a preemptive attack as one of the risks associated with the US counterforce development program:[1]

Under extreme crisis conditions Soviet leaders who had little confidence in the deterrent value of their own air-breathing, submarine and residual ICBM forces might perceive advantages in launching a first strike. In this context, such Soviet leaders might view the threat to their silo-based ICBM force as being of major concern since currently about 75 percent of Soviet strategic weapons . . . are in its fixed-silo ICBM force.

As of today neither side has been pushed into adopting a launch-on-warning system, for despite improving counterforce capabilities, each side maintains an overwhelming deterrence capability, sufficient beyond question to withstand a preemptive attack. The box below outlines this current, still relatively stable, balance of weapons. A new round of deployments dominated by counterforce weapons is therefore likely to be either

sufficiently threatening to each side's deterrent to provoke desperate measures such as launch-on-warning, or merely futile and wasteful. This is the fundamental rationale for the nuclear-weapons-freeze proposal.

Why not a bilateral freeze?

The Reagan Administration has cited several objections to a bilateral freeze that go beyond its concern to protect US counterforce programs. Its spokesmen have argued[20] that a freeze would be unverifiable, would lead to a loss of deterrent by eroding the ability of the US to respond to a Soviet nuclear attack, and would reduce Soviet interest in negotiating arms reductions. While such claims need detailed attention, as we hope independent scientists will provide, we can raise a few points of skepticism immediately.

Although it is not surprising that opponents of a freeze focus on the weakest points of its verifiability, this should not obscure the fact that methods of verifying the most important elements of a freeze have already been worked out in considerable detail. For example, methods to verify compliance

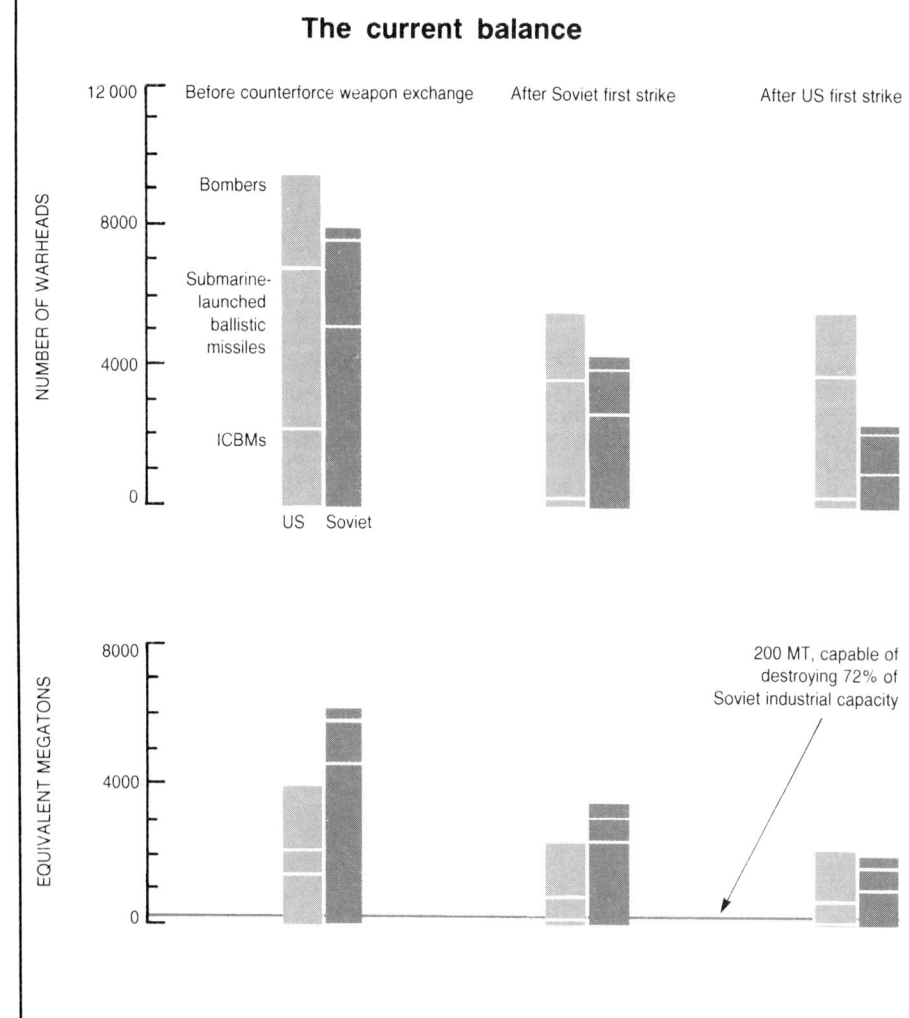

The current balance

Before counterforce weapon exchange · After Soviet first strike · After US first strike

NUMBER OF WARHEADS

12 000 — 8000 — 4000 — 0

Bombers
Submarine-launched ballistic missiles
ICBMs
US   Soviet

EQUIVALENT MEGATONS

8000 — 4000 — 0

200 MT, capable of destroying 72% of Soviet industrial capacity

Both superpowers now possess enormous strategic arsenals with vast destructive capability, as the table on page 38 shows. The blast areas given are those that would experience peak overpressures greater than 0.25 atmospheres, or 3.5 lb/in[2]. The destruction of ordinary buildings within these areas would be nearly total.[1] Fallout areas given in the table are those in which unsheltered people would be subject to a dose greater than 600 rads. Above this level, people sheltered above ground in ordinary housing would begin to die. (We use the methodology of reference 2 and assume the following: bursts on the ground, an average wind speed of 40 km/hr and a transverse wind shear of 1 km/hr for each km of altitude. We also assume that for weapons with yields over 100 kilotons, 50 percent of the yield is from fusion.)

If forces were frozen at their present levels, how much incentive would there be for a preemptive first strike? That is, how much firepower would one side have left after an all-out counterforce attack by the other? Defense Department reports contain charts[3] like the one shown at left, giving estimates in terms of numbers of warheads and "equivalent megatonnage." (Equivalent megatonnage takes into account the fact that the area a warhead subjects to a given peak overpressure increases as the two-thirds power of the warhead yield. A one-eighth-megaton warhead, for example, would contribute one quarter of an equivalent megaton.) However, in unclassified versions of such reports, the *absolute* numbers of surviving warheads and megatonnage are suppressed. Also, the results of a first strike by the United States are never shown.

with prohibitions on flight testing and on deployment of new ballistic missiles were incorporated in SALT II and subjected to intensive study. Similarly, the United States and the Soviet Union have over several years worked out verification procedures to monitor a comprehensive nuclear test ban, including underground tests. Verification of certain other elements of a freeze, such as a ban on building cruise missiles, would be more difficult, but not impossible when the numbers of missiles involved are large. Given our vast intelligence-gathering capabilities and the comprehensive character of a freeze, it would be very difficult for the Soviet Union to conceal cheating on a scale sufficient to create a threat anywhere nearly as serious as that posed by the current nuclear arms race.

In addition, we should not assume that verification beyond "national technical means" will remain impractical. While the Soviet Union is not about to convert itself into an open society, it did agree under the virtually complete Comprehensive Test Ban Treaty to allow the US to emplace sealed "black boxes" containing sensi-

tive seismometers in strategic spots around the Soviet Union. And, under the SALT I Treaty, the Soviet Union has participated constructively in the "standing consultative commission" in which each side has agreed to explain questionable activities detected by any of the multitude of telescopic "eyes" and electronic "ears" that continually monitor surface activities from outer space.

The Reagan Administration claims[20] that in the long run a freeze could erode US deterrent capability by stopping the development of offensive nuclear systems while allowing Soviet nonnuclear air defense and antisubmarine-warfare capabilities to develop unimpeded. In fact, under a freeze, a country's confidence in its ability to make a first strike will erode far more rapidly than its ability to deter a first strike by the other side. A first strike must be virtually perfect, with thousands of warheads coordinated in time and space to high precision. An effective deterrence force need be neither perfectly reliable nor highly accurate. For instance, the possibility that even a few bombers can get through an air defense

system would represent a substantial deterrent to any rational political leader. As McGeorge Bundy, national security advisor to President Kennedy, has written,[21]

...a decision that would bring even one hydrogen bomb on one city of one's own country would be recognized in advance as a catastrophic blunder; ten bombs on ten cities would be a disaster beyond history; and a hundred bombs on a hundred cities are unthinkable.

That said, it is true that freezing bomber- and air-launched cruise-missile technology while allowing continued development of air-defense technology will erode the capabilities of both US and Soviet bombers to deliver their weapons on target.

By contrast, there is a consensus in the US defense establishment that no foreseeable development in Soviet antisubmarine technology will be able to threaten US ballistic-missile submarines.[14] In any case, the contest between antisubmarine-warfare systems and ballistic-missile submarines would be largely unaffected by a freeze. For under a freeze, as usually defined,

---

Nevertheless, by making a few simple assumptions, one can reproduce the pattern of results roughly, as we have done in the chart at left.

We have assumed force structures such as those that would exist if a freeze on the deployment of new nuclear weapons occurs during the next few years. The inventories of strategic forces are those of 1982, but the missile accuracies are based on test results and might not be fully achieved in deployed missiles until a later date.

We have assumed also that the counterforce exchanges occur during a crisis, not as a "bolt out of the blue." That is, we assume the forces of both sides are on "generated alert," during which
▶ The Soviet Union increases the fraction of its ballistic-missile submarines at sea from 15 to 50 percent,[4] while the United States increases its corresponding fraction from 50 to 75 percent[3]
▶ The Soviet Union increases the fraction of its bombers on alert from 0 to 50 percent (75 percent in case of a Soviet first strike) while the United States increases its corresponding fraction from 30 to 75 percent.[3] We assume that all bombers not on alert and submarines not at sea will be destroyed unused.

With respect to attacks on ICBM silos, we have assumed that the Soviet Union in a first strike would assign two ICBM warheads to each US silo and that the United States in a first strike would assign two Minuteman III warheads to each silo containing a MIRVed Soviet ICBM. We assume that only 20 percent of the silos so attacked would survive; however, if we had credited a two-warhead attack with a 100 percent probability of destroying an ICBM silo, the overall results would not change significantly. We have further assumed

that, regardless of which side attacked first, each side would expend an additional 200 ICBM warheads attacking bomber and submarine bases and command and communications facilities.

It is evident from the figure that, under conditions of a generated alert, neither side has, at present, a strong incentive to strike first. Doing so would not significantly change the relative positions of the two sides. The Soviets, for example, in an attack on US Minuteman silos, would have to expend two high-yield warheads to destroy 1–3 mostly lower-yield warheads. Furthermore, even after a Soviet first strike, the US would still have about 2000 equivalent megatons of nuclear explosive power. Even a Soviet "bolt out of the blue" attack against US forces not on a crisis alert would leave the US with more than 1000 equivalent megatons. In 1969 the Department of Defense estimated[5] that an attack with 100 equivalent megatons could destroy 59 percent of Soviet industrial capacity and that an attack with 200 equivalent megatons would raise this figure to 72 percent.

There are, of course, those who argue that even thousands of equivalent megatons would not be enough to deter the Soviet Union. Paul Nitze, now chief US negotiator in the US–Soviet talks on intermediate-range nuclear forces in Europe, advanced such an argument in 1979. Nitze presented the Senate with a calculation in which it was assumed that[6]

...approximately 80 percent of the Soviet urban and industrial population will have been evacuated, distributed in ... an equal density over a million square miles ..., equipped with shelters with a PF [protection factor against radiation] greater than 200,

prepared to stay in those shelters for 2 weeks if necessary, and prepared to act with some prudence when the residual radiation levels call for prudence.

He then argued that the United States could only cover at most 3.5 percent of a million-square-mile area with fallout great enough to give the sheltered population lethal radiation doses. He did not discuss how the evacuated population would survive in the longer term with the Soviet economy destroyed.

More recently, the *Los Angeles Times* reported[7] that Thomas K. Jones, US Deputy Undersecretary of Defense for Strategic and Theater Nuclear Forces, believes that with a proper civil defense it would take only two to four years for the United States to fully recover after and all-out nuclear war with the Soviet Union.   HF & FvH

References

1. S. Glasstone, P. J. Dolan, eds., *The Effects of Nuclear Weapons*, 3rd ed., US Departments of Defense and Energy (1977), pages 111, 115.
2. L. A. Schmidt Jr, Methodology of fallout-risk Assessment, Institute for Defense Analyses, Arlington, Virginia (1975).
3. See, for example, *Annual Report, FY 1982*, US Department of Defense (1981), pages 55, 57.
4. US Joint Chiefs of Staff, *US Military Posture, FY 1979* (1978), page 28.
5. *The Fiscal Year 1969–'73 Defense Program and the 1969 Defense Budget*, US Department of Defense (1968), page 57.
6. P. H. Nitze, Congressional Record, 20 July 1979, page S10077.
7. *Los Angeles Times*, 16 January 1982.

FEIVESON/VON HIPPEL *continued*

submarines could be replaced by quieter models equipped with all the latest antisubmarine countermeasures. Only the capabilities of the submarine-launched missiles would be frozen. The one capability of these missiles that is relevant to antisubmarine warfare is range, and this is already great enough so that about half the equivalent megatonnage on both US and Soviet submarine-launched ballistic missiles can reach the capital city of the other from home waters. Indeed, when asked recently about the superiority of the range of the Trident II missile over that of the Trident I, Rear Admiral William A. Williams III, director of the US Navy's strategic and theater nuclear warfare division, stated that[14]

> we are not advocating the D-5 [Trident II] because of its greater range. The C-4 [Trident I] has a very comfortable range.

Another problem that would arise in the long run, argues the Administration, is that US strategic systems will begin to wear out earlier than their Soviet counterparts because they were generally deployed years earlier. In fact, most US systems are not creaking with age. The Minuteman III and Poseidon missiles have all been deployed since 1970, and all Trident I missiles since 1980. The submarines are older, but as we pointed out above, they can be replaced. Only the bombers—both US and Soviet—would have a real problem with aging, and then only if a freeze lasted for decades. But, of course, the freeze is not designed to last forever. It would serve best as a transition period between the arms race and genuine arms reduction.

Finally, we should emphasize once more that a freeze would slow the erosion of the stability of the nuclear balance by dramatically slowing the counterforce race, if not entirely stopping it Contrary to the impression given by Administration spokesmen, significant parts of the US strategic program that a freeze would stop, including the MX and Trident II programs, are aimed primarily at threatening Soviet nuclear missiles—not at reducing the vulnerability or our own.

The Reagan Administration asserts that a freeze would weaken chances for deep reductions in strategic weapons, a goal of the on-going "START" (Strategic Arms Reduction Talks) negotiations. It contends that only the threat of a vigorous US buildup of strategic weapons will adequately motivate the Soviets to make a deal. The Administration's declared passion for reductions, however, is belied by its refusal in START to offer to stop the deployment of any of the planned US counterforce weapons. Only quantitative limits

have been proposed: 850 ballistic missiles carrying a total of 5000 warheads not more than 2500 of which could be carried by ICBMs.

Within these quantitative limits the Administration proposes to pursue the technological arms race without constraint. The START proposal would allow the United States to deploy the MX, Trident II, Pershing II and cruise missiles while the older and less-threatening Poseidon, Trident I and Minuteman missiles are retired to bring the total warhead and missile count down to the proposed limits. The counterforce race could, therefore, continue unconstrained under a START agreement and could indeed become more dangerous as the number of targets for a first strike were reduced faster than the number of warheads that could be directed against them. This would compound the historic mistake that the United States and Soviet Union made when they decided against seeking to include a ban on multiple independently targetable reentry vehicles in the SALT-I agreement.[22]

Role of scientists

Scientists and engineers are not only the designers of nuclear weapons and their delivery systems, they also play key roles in developing "scenarios" of how these weapons might be rationally used. Gerard Smith, chief US negotiator in the SALT-I talks, recalls that[22]

> I sensed that civilian scientists and engineers in the office of the Secretary of Defense were more influential with Secretary [Melvin] Laird than professional military officers. These men would never have to be users of nuclear weapons. They were not members of military services with experience in fighting wars but a kind of elite which knew or gave the impression of knowing the new secrets of the nuclear-missile age.

McGeorge Bundy has also testified[21] to the unwillingness of national leaders to challenge these "nuclear gamesmen":

> Presidents and Politburos may know in their hearts that the only thing they want from strategic weapons is never to have to use them; in their public postures they have felt it necessary to claim more. They may not themselves be persuaded by the refined calculations of the nuclear gamesmen— but they do not find it prudent to expose them for the irrelevance they are. The public in both countries has been allowed by its leaders to believe that somewhere in ever growing strength there is safety and that it still means some-

thing to be "ahead." The politics of internal decision making has not been squared with the reality of international stalemate.

Independent scientists have, therefore, on occasion played a key role in challenging the rationalizations that have been proposed for continuing the arms race.

Perhaps the most recent debate in which independent scientists have been involved on a large scale in critiquing official US policy in the arms race was the national debate that occurred in the period 1967–1970 over proposals by the Johnson and Nixon Administrations to deploy a nationwide antiballistic-missile system. As with the current debate, the ABM debate began with a citizens' uprising—developing first around the suburban sites originally proposed for the nuclear-armed antimissile missiles. However, the opposition continued to broaden even after the Nixon Administration shifted the ABM sites away from populated areas—in part because critical scientists used the forums created by the initial uprising to focus public and congressional attention on the ABM's implications for the arms race.[23] Hans Bethe, for example, opened his 4 March 1969 talk at a teach-in at MIT as follows:[24]

> I believe that most of the audience here is against the ABM, and I believe that I am here to tell you why.

The ABM debate in the US educated both the US public and the Soviet leadership about the difficulties of missile defense, difficulties the military leadership on both sides had refused to admit. As a result, while the political leadership on neither side was strong enough to impose unilateral abstention from an ABM race on their militaries, together they were able to prevent the nuclear-arms race from expanding into this new dimension.

Today, rising public concern has created an opportunity for scientists to explain to the public the even greater dangers and futility of the counterforce race. In addition, scientists can help to define specific bilateral freeze agreements that are adequately verifiable and that can be used as a starting point for nuclear-arms reductions.

References

1. *Fiscal Year 1981 Arms Control Impact Statements*, US Senate Committe on Foreign Relations and House Committee on Foreign Affairs (1980), pages 46, 50, 56, 338.

2. R. Forsberg, *Call to Halt the Nuclear Arms Race*, American Friends Service Committee, Clergy and Laity Concerned, Fellowship of Reconciliation, In-

stitute for Defense and Disarmament Studies. Available from Nuclear Weapons Freeze Campaign, National Clearinghouse, St. Louis (1980).

3. *New York Times*, 18 May 1982, page A1.

4. *The Military Balance*, 1981–1982, International Institute for Strategic Studies, London (1981), page 104.

5. *Annual Report, Fiscal Year 1980*, US Department of Defense (January 1979), page 117.

6. *Department of Defense Authorization for Appropriations for Fiscal Year 1982: Part 7, Strategic and Theater Nuclear Forces, Civil Defense*, US Senate Armed Services Committee, (February–March 1981), pages 3802, 3880, 3924.

7. R. DeLauer, Astronautics and Aeronautics, May 1982, page 39.

8. *The New York Times*, 29 May 1982, page A1.

9. J. D. Steinbruner, Foreign Policy, Winter 1981–82, page 16.

10. S. Glasstone, P. J. Dolan, eds., *The Effects of Nuclear Weapons*, 3rd ed., US Departments of Defense and Energy, (1977), pages 111, 115.

11. See, for example, *Analyses of Effects of Limited Nuclear War*, Senate Committee on Foreign Relations, Subcommittee on Arms Control, International Organizations and Security Agreements, Committee Print (1975), page 51; and S. D. Drell, F. von Hippel, Scientific American, November 1976, page 27.

12. *MX Missile Basing*, US Office of Technology Assessment (1981), page 106; and *The Effects of Nuclear War*, US Office of Technology Assessment (1979), page 91.

13. A. A. Tinajero, *US/USSR Strategic Offensive Weapons: Projected Inventories Based on Carter Policies*, US Congressional Research Service, Report No. 81–238F (1981).

14. *Hearings on Strategic Force Modernization Programs*, US Senate Armed Services Committee, Subcommittee on Strategic and Theatre Nuclear Forces, October–November 1981, pages 43, 168, 179, 187, 203, 254, 405.

15. B. Bennet, J. Foster in *Cruise Missiles: Technology, Strategy, Politics*, R. K. Betts, ed., The Brookings Institution, Washington, D.C. (1981), page 152.

16. *World Armaments and Disarmament*, Stockholm International Peace Research Institute (1980), pages XLII, XLIII, 182.

17. P. H. Nitze in the Congressional Record, 20 July 1979, page S10077.

18. D. F. Ustinov in Pravda, 12 July 1982; quoted in *The New York Times*, 13 July 1982, page A3.

19. H. York, *Race to Oblivion*, Simon and Schuster, New York (1970), page 232.

20. See, for example, R. Perle, Assistant Secretary of Defense for International Security Policy, *The New York Times*, 7 September 1982, page A23.

21. McG. Bundy, Foreign Affairs, October 1969, page 1.

22. G. L. Smith, *Double Talk: The Story of the First Strategic Arms Limitation Talks*, Doubleday, Garden City, N.Y. (1980).

23. See, for example, J. Primack, F. von Hippel, *Advice and Dissent, Scientists in the Political Arena*, Basic Books, New York (1974), New American Library, New York (1975).

24. J. Allen, ed., *March 4, Scientists and Society*, MIT, Cambridge, Mass., (1970), page 142. □

What price security?

A National Academy panel evaluates trade-offs between dangers to national security that arise from technology transfers and threats to the openness of scientific communication that are caused by too much secrecy.

Dale Corson

PHYSICS TODAY/FEBRUARY 1983

"There is an overlap between technological information and national security which inevitably produces tension. This tension results from the scientist's desire for unconstrained research and publication on the one hand, and the Federal government's need to protect certain information from potential foreign adversaries who might use that information against this nation. Both are powerful forces. Thus, it should not be a surprise that finding a workable and just balance between them is quite difficult." So said Admiral Bobby R. Inman, then Deputy Director of the Central Intelligence Agency, in a speech at the 7 January 1982 meeting of the American Association for the Advancement of Science.

Dale Corson, a physicist and former president of Cornell University, led the National Academy panel.

Inman's speech has since sparked widespread discussions aimed at delineating the differing needs of these two forces and suggesting ways to balance them. In fact, the tension about which Inman spoke, and the dilemma it poses, were the focus for a study recently completed under my chairmanship, entitled "Scientific Communication and National Security" (PHYSICS TODAY, November, page 69). The study, conducted under the auspices of the National Academies of Science and Engineering, considered the interests of both national security and scientific communication; attention focused on the control mechanisms now being used to restrict the flow of information and on the application of these controls; the committee also recommended specific improvements to the system.

The underlying conflict between the drive for security and the drive to open communication is not a new issue. Recently, however, concerns about national security as well as concerns about the free flow of information among scientists have increased. Why?

Recent events increase concerns

Although administrative concern over the technology-transfer problem increased during the last Administration, it has escalated sharply in the current one. This new sense of alarm has emerged, to some degree at least, from a change in perceptions. The US intelligence community, in fact, has identified four trends as significant.

▶ The US lead in at least some areas of military technology has diminished. The intelligence community sees this diminishing lead as a result of Soviet absorption of Western technology.

▶ Military systems are depending more and more on such high technol-

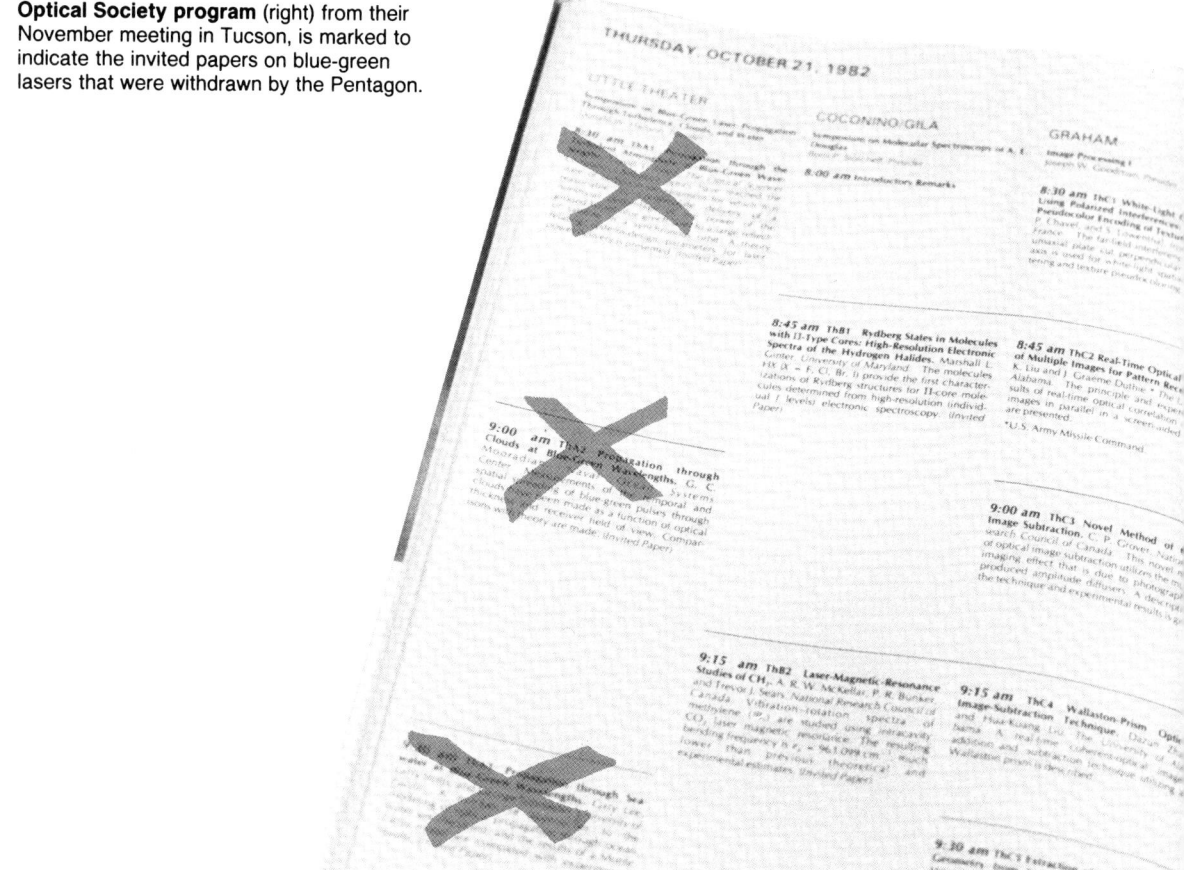

Optical Society program (right) from their November meeting in Tucson, is marked to indicate the invited papers on blue-green lasers that were withdrawn by the Pentagon.

ogies as state-of-the-art microelectronics, lasers and so forth.

▶ A steadily increasing share of these technologies has both military and nonmilitary applications; there is substantial difficulty in controlling leaks in non-military systems.

▶ Recent American foreign policy has multiplied the number of routes for leakage. Significant expansion of East/West trade in the 1970s, for example, has resulted in a variety of agreements that further encourage the transfer of technology.

Adding further to the alarm is a sense that the Soviet Union is making a concerted effort to acquire scientific and technical information. This view was expressed strongly by Lawrence J. Brady, Assistant Secretary of Commerce, in a speech before the intelligence community last March. He said:

Operating out of embassies, consulates, and so-called 'business delegations,' KGB operatives have blanketed the developed capitalist countries with a network that operates like a gigantic vacuum cleaner, sucking up formulas, patents, blueprints and know-how with frightening precision. We believe these operations rank higher in priority even than the collection of military intelligence... This network seeks to exploit the "soft underbelly"—the individuals who, out of idealism or greed, fall victim to intelligence schemes;

our traditions of an open press and unrestricted access to knowledge; and finally, the desire of academia to jealously preserve its prerogatives as a community of scholars unencumbered by government regulation. Certainly, these freedoms provide the underpinning of the American way of life. It is time, however, to ask what price we must pay if we are unable to protect our secrets?

The question of what price the Administration is willing to pay to keep information out of the hands of adversaries, particularly the Soviet Union, is perhaps the central concern of the scientific community. And now this concern has been heightened, primarily because of recent events and what they imply regarding further restrictions on scientific communication.

Notable among these events have been efforts to elicit the cooperation of universities in restricting the movements of visiting Soviet scientists. In addition, there have been repeated instances in which the Pentagon or the Department of State has sought to prevent scheduled papers from being presented at scientific conferences. One such incident that recently received wide publicity took place at the Society of Photo-optical Instrumentation Engineers' conference in San Diego in August: The Pentagon had nearly 150 papers withdrawn several days before the meeting. It now appears that many of these papers will, after all, receive clearance and be included in the published proceedings from this meeting. Similar incidents in which scheduled papers have been withdrawn from scientific meetings have taken place before and apparently will continue to take place, as the Optical Society of America discovered in November when several papers were withdrawn from its meeting in Tucson. These events stem, in part, from a confusion over how to apply the Federal regulations to the scientific and academic community.

Panel studies key issues

Our panel of 19 people included a former Under Secretary of Defense, a former Under Secretary of Energy, a former Director of the National Science Foundation, a former Presidential Science Advisor, four former members of the President's Science Advisory Committee, five members or former members of the National Science Board, six current or former university presidents, one former Director of the National Security Agency, four execu-

tives of high-technology industry, several present or former members of the Defense Science Board and two lawyers.

Our charge included four tasks:

▶ An examination of national-security issues and scientific communication interests within the context of certain fields of science and technology

▶ A review of the controls used in restricting scientific communication as well as identification of the issues arising from the use of the controls

▶ A rigorous evaluation of the critical issues concerning the application of controls, and

▶ The development of ways to make the system operate more effectively.

Although the panel's mission was to investigate the effects of restrictions on scientific communication in general, it found in reaching its recommendations that the university requires separate consideration within the context of the US research community. Restrictions on open communication have categorically different implications for universities than they do for industrial, governmental and other realms of the community; there are two main reasons for this distinction:

▶ Universities integrate research and education; thus, any adverse effects on research will also adversely effect the quality of education for the next generation of scientists and engineers.

▶ Unlike other research institutions, universities have never established broad controls on access to information to ensure that sensitive information be protected. Such restrictions, therefore, would present an unfamiliar and unwelcome challenge to the university.

Because the potential national security concerns are most likely to arise in work that is funded by the government, the panel's conclusions concentrate on government-supported research.

While much of our report applies to basic industrial research just as much as it applies to university research, there are important questions bearing on industry that we have not addressed at all. For example, how does one treat the problem of communication with a multinational company that has laboratories abroad and foreign subsidiaries? For many, this may be the most important question of all; I regret I cannot help, for this question requires study by a new group constituted in a different way.

Due to both the current level of concern and the panel's limited time and resources, study focused on technology transfer to the USSR from the US.

To study these issues, the panel had

to begin by learning exactly what the nature of the technology-leakage problem was. We realized early on that we would have to operate on a classified basis; consequently we arranged for security clearance for all panel members at the secret level. In addition, six of our members, who held security clearanes at the highest level, arranged for intelligence briefings and discussions at the very highest security levels and reported back to the full panel at the secret level. They also produced a Secret report which is on file in the National Academy of Sciences. In addition, they produced an unclassified report, which is included in our panel report as an appendix, and which gives a clear picture of the technology-leakage problem.

The panel is unanimous in its conclusions and recommendations.

Major suggestions and conclusions

The evidence from all sources suggests that indeed there is a substantial and serious technology-transfer problem. There is a continuing flow of products, processes and ideas from the US and its allies to the Soviet Union, through both overt and covert means. Although much of this unwanted transfer has mattered little to US security, either because the US did not enjoy a monopoly on a particular technology or because the technology in question had little or no military significance, a substantial portion of the transfer has been damaging to national security (See the table for some evidence presented by the Central Intelligence Agency). These damaging transfers have taken place through the legal as well as illegal sale of products, through transfers via third countries and through a highly organized espionage operation.

Although a good deal of information has been transfered through open scientific communication, the panel concludes that, in comparison with other channels of technology transfer, open scientific communication involving the research community does not threaten our near-term military position. Given both this conclusion and our concern for finding an approach that will maintain the vitality of our universities and their roles in education and research, while at the same time protecting the security of our advanced technology, how should we proceed?

The panel believes that scientific research and technological development are best nurtured in an environment where such efforts are dispersed but interdependent. Openness and a free flow of information are essential aspects of such an environment. The technological leadership that the US enjoys is based in no small part on a

scientific foundation whose vitality in turn depends on effective communication among scientists, and between scientists and engineers; the short-term security achieved by restricting the flow of information is purchased at a price.

After weighing the alternatives, the panel concludes that the best way to ensure long-term national security lies in a strategy of "security by accomplishment," and that an essential ingredient of technological accomplishment is open and free scientific communication. Such a policy involves risk, because new scientific findings will inevitably be conveyed to US adversaries. Nonetheless, the panel believes the risk is acceptable because American industrial and military institutions are able to develop new technology swiftly enough to give the US a continuing advantage over its military adversaries.

Against this general background, the panel comes to three specific conclusions:

▶ The vast majority of university research programs, whether basic or applied, should be subject to no limitations on access or communications.

▶ Where specific information has direct military relevance and must perforce be kept secret, it should be classified strictly and guarded careful-

ly. The decision to accept or reject classified research projects, or to establish off-campus classified facilities, is a matter to be decided by individual universities.

▶ There are a few gray areas of research that are sensitive from a security standpoint, but where classification is not appropriate. These areas are at the ill-defined boundary between applications and basic research and are characteristic of fields where the time from discovery to application is short. (At present, a portion of the field of microelectronics is the most visible of these technologies.)

While it is impossible to specify these gray areas with precision, there are some broad criteria that help to define the few areas in question. The panel recommends that *no* restrictions of any kind that limit access or communication should be applied to any area of university research, basic or applied, unless it involves technology meeting all of the following four criteria:

▶ The technology is developing rapidly and the time from basic science to application is short; and

▶ The technology has identifiable direct military applications, or is dual-use, and involves process- or production-related techniques; and

▶ Transfer of the technology would give the USSR a significant near-term

Acquisitions from the West affecting Soviet military technology

Key technology area	Notable success
Computers	Purchases and acquisitions of complete systems designs, concepts, hardware and software, including a wide variety of Western general purpose computers and minicomputers, for military applications.
Microelectronics	Complete industrial processes and semiconductor manufacturing equipment capable of meeting all Soviet military requirements, if acquisitions were combined.
Signal Processing	Acquisitions of processing equipment and know-how.
Manufacturing	Acquisitions of automated and precision manufacturing equipment for electronics, materials, and optical and future laser weapons technology; acquisition of information on manufacturing technology related to weapons, ammunition, and aircraft parts including turbine blades, computers, and electronic components; acquisition of machine tools for cutting large gears for ship propulsion systems.
Communications	Acquisition of low-power, low-noise, high-sensitivity receivers.
Lasers	Acquisition of optical, pulsed power source, and other laser-related components, including special optical mirrors and mirror technology suitable for future laser weapons.
Guidance and Navigation	Acquisitions of marine and other navigation receivers, advanced inertial-guidance components, including miniature and laser gyros; acquisitions of missile guidance subsystems; acquisitions of precision machinery for ball-bearing production for missile and other applications; acquisition of missile test-range instrumentation systems and documentation and precision cinetheodolites for collecting data critical to postflight ballistic-missile analysis.
Structural Materials	Purchases and acquisitions of Western titanium alloys, welding equipment, and furnaces for producing titanium plate of large size applicable to submarine construction.
Propulsion	Missile technology; some ground-propulsion technology (diesels, turbines, and rotaries); purchases and acquisitions of advanced jet-engine fabrication technology and jet-engine design information.
Acoustical Sensors	Acquisitions of underwater navigation and direction-finding equipment.
Electro-optical Sensors	Acquisition of information on satellite technology, laser range finders, and underwater low-light-level television cameras and systems for remote operation.
Radars	Acquisitions and exploitations of air defense radars and antenna designs for missile systems.

Table adapted from a Central Intelligence Agency report entitled "Soviet Acquisition of Western Technology," April 1982.

military advantage; and

▶ Either the US is the only source of information about the technology, or other friendly nations that could also be the source have control systems at least as secure as ours.

The panel recommends that in the limited number of instances in which all of the above criteria are met, but where classification is unwarranted, the values of open science can be preserved and the needs of government can be met by written agreements or contracts no more restrictive than the following:

▶ Prohibition of direct participation in government-supported research projects by nationals of designated foreign countries but with no attempt to limit physical access to university space or facilities or to limit enrollment in any classroom course or study. The danger to national security lies in the immersion of a suspect visitor in a research program over an extended period, not in casual observation of equipment or research data.

▶ Submission of stipulated manuscripts simultaneously to the publisher and to the Federal agency contract officer, with the contract officer having 60 days to seek modifications in the manuscript if he so wishes.

The review period is not intended to give the government the power to order changes. The right and freedom to publish remain with the university as they do with all unclassified research. The government nonetheless is a powerful negotiator in these discussions; it has the ultimate power to classify the research or to cancel the contract.

Knottier problems

The panel recognized the difficulty of limiting the access of foreign visitors on campuses to sensitive information, particularly when universities typically have people who are not working on federally-funded projects but who have free access to the laboratories and all that goes on within the university.

Let me simplify the problem by suggesting what might happen in a specific case. Visitors come to universities with restictions on their visas. Such restrictions may include travel restrictions, restrictions on what they can work on, and currently there might also be restrictions on what they can see. The contract officer occasionally checks up on the visitor and he also asks the university to report on what these particular visitors are up to. Certainly, according to our recommendations, the university would be alerted to the problem and notified that the visitors should not be supported with project funds over an extended period of time.

In the case of the similar research laboratory next door, performing non-

government-funded research, we suggested that it would not be inappropriate for the university to respond affirmatively to requests from government agencies for information about possible attempts by the visitors to gain support to work with the nongovernment-funded project over an extended period. We reasoned that if the researchers did obtain that type of support, in doing so they would be presumably violating the terms of their visas. Thus we think it's appropriate for the university to respond affirmatively if asked, when those visa restrictions are being violated. Such requests, however, should not require surveillance or monitoring of foreign nationals by the universities.

It is important for the welfare of the country that universities' educational and research programs remain vital. The procedures recommended by the panel for dealing with the gray areas of research are intended to protect university interests, and at the same time to be responsive to the government's requirements.

The panel believes that the provisions of Export Administration Regulations and International Traffic in Arms Regulations should not be invoked to deal with these gray areas in government-funded university research. Rather, the appropriate procedure should be incorporated in research contracts or other written agreements in those rare cases where some measure of control is required. Furthermore, the panel believes that universities and industrial research laboratories should be treated in exactly the same way insofar as EAR and ITAR are concerned.

Writing the contract ahead of time poses two problems. The first is that one never knows what is going to happen; perhaps something will come up that was not anticipated in the contract. The second is that Federal contracting officers may act overzealously in protecting themselves by writing in restrictions that are unnecessary. Both are real concerns. To address the first problem—not knowing what's going to come up—we'd like to have the rules clearly understood ahead of time, insofar as they can be, so that everybody knows what the rules are and can play by the same rules. When cases come up where it is necessary to elaborate, we believe that constructive discussion can take place and problems can usually be resolved if there exists an atmosphere of good communication.

As an example of such a resolution, I can cite the situation that began several years ago in the field of cryptography. There were several instances; one in particular occurred in about 1978. A young researcher at the University of

Wisconsin in Milwaukee applied for a patent on a cryptographic invention he had made. He didn't hear from the Patent Office for a long time. Eventually he received a post card as the only response to the application—a post card saying that his research program had been classified Secret and that he was not to talk to anybody about it. This action was authorized under the Invention Secrecy Act.

Admiral Inman played a major role in resolving that issue and reducing a tense situation to one that is now handled on a voluntary basis. The American Council on Education also played a lead role by convening a study group on the cryptography problem, in which the mathematicians participated. I also participated in the very first discussion of that problem at the American Council on Education, where I first met Inman. As a result of these discussions, people working in cryptography now submit their papers to the National Security Agency for comment; simultaneously they submit their papers to the publisher. Some 50 papers have been submitted under this voluntary arrangement. I think changes or suggested changes have been proposed by NSA in a couple of cases, but I have not heard of any great dissatisfaction. I also believe that there are some people working in the field who have declined to cooperate and are going ahead on their own. We spoke both with the National Security people and with people from universities with researchers in the field, and all of them expressed satisfaction with the current system. This is an example of what can happen when people get together and talk about the problem.

The panel believes, however, that one cannot extend this particular system to other research. Cryptography is a very narrow field in which everybody working in it knows everybody else working in it, and the focus of the research is limited and generally well-defined. This is not true for most other fields of research.

The second problem—the overzealous contract officer writing in unnecessary restrictions—is harder to deal with. I suspect that this problem is part of what happened at the San Diego SPIE Conference in August. In that instance, however, it wasn't the contract officer who was overzealous, but rather it was somebody in the Pentagon; I don't know how to protect against Pentagon intervention.

The Defense Department supports a significant amount of first-rate basic research, their so-called 6.1 research. Traditionally, research supported by 6.1 funding is unclassified, unrestricted, and free for publication. I suspect that now there is a move to restrict 6.1 supported research in various ways,

and there are many contract officers who are writing individual contracts for this research. Consider, for example, a situation in which somebody in the 6.1 office in the Defense Advanced Research Project Agency decides to support a certain program but he personally doesn't write the contract. Somebody at Wright-Patterson Air Development Center writes the contract. The person who writes the contract is eager not to get in any trouble, so he writes restrictions in. I don't know how to deal with that problem, except by starting at the highest level, setting major policy issues and establishing educational programs for contract officers. I am glad that the Office of Science and Technology Policy is now interested in this kind of problem.

Although these are major problems, and we recognize them, the panel felt that if we could write the agreements ahead of time, so that everybody knew the rules, we would have gained something.

The panel has studied the control system now in effect, and the report has some substantial discussion of the system and its problems. The panel's suggestions apply equally to industrial and university research. The current system is undergoing rapid change. Because the perceived nature of the technology leak problem has shifted only recently, government control mechanisms themselves are still being adjusted to meet the new perceptions.

In a fundamental sense, government is still in the early stages of the learning process as it reorients existing laws, policies and programs—designed for other purposes—to achieve a new objective, the dimensions of which are not yet fully determined. The adjustment is particularly difficult because the current effort to understand and control unwanted technology transfer is unavoidably fractionated within the Federal establishment. Four intelligence agencies—the FBI, the CIA, the Defense Intelligence Agency and the National Security Agency—share the job of gathering intelligence on the nature, extent and significance of unwanted transfers.

Major regulatory authority is also split among three separate offices: the Department of Commerce's EAR administrators, the Department of State's ITAR administrators, and the Department of State's Visa Processing Office. These offices depend heavily on outside units in the defense and intelligence communities for advice as they reach their judgments.

Similarly diffuse is the government's authority for classifying information and for monitoring results from the research and development that it funds. Regulatory enforcement shows similar diversity and includes yet another agency, the Department of Treasury's Customs Service. The panel discovered, not surprisingly, that few people inside or outside the government truly understand the government's technology-transfer control effort.

The panel believes that there is much room for improvement in targeting the government's efforts to prevent unwanted technology transfer. Priorities must be set and communicated. The panel believes that the government should concentrate on the most feasible forms of control and should avoid regulations that impose compliance burdens without significantly affecting leakage. The government should concentrate its resources more systematically on those technologies that are of greatest relevance to near-term Soviet military strength.

Finally, the panel addressed problems of inadequate staffing in agencies that deal with control measures, as well as problems of inadequate communication between the research community and the Federal agencies. The panel also identified areas where the research community might help the government assess the nature of the technology-transfer problem more reliably.

In assessing the current policies and procedures, we heard the word "confusion" from just about everybody we spoke to about both the ITAR and EAR.

Let me give you an example of the complexity of the system. In the Export Administration Act of 1979, an act which has been revised regularly and is the underlying legislation for EAR, it was specified that the Commodity Control List should be based on something called the Militarily Critical Technologies List. The Commodity Control List is the basis for licensing exports and the Militarily Critical Technologies List is now undergoing its second revision. The third version of this list is going to be issued some time in the immediate future. The second version was a 700-page book, all of which is classified Secret. If one wants to take this to its logical end, it means that the people who are going to be subject to heavy fines through the implementation of these regulations will not be able know what it is that the violation is based on. The regulations are administered somewhat more intelligently than this sounds, but nonetheless individual parts of the Commodity Control List are classified individually. For example, some are Confidential, some are Secret and some are Unclassified. Regardless of classification, all are subject to export restrictions determined by EAR. Among the unclassified technologies are such things as high-vacuum technology, or manufacturing techniques for the mass production of ultra-high frequency generators, and techniques for making certain kinds of magnets which industrial people are making every day of the week.

The list has been developed by dedicated people who have taken a military system apart piece by piece to see what went into it; those people have taken their work seriously and they've done an excellent job of finding what underlies every military system that exists.

Due to the comprehensiveness of this list and its classification, however, there seems to be no way to start from that list and arrive at a straightforward and clear definition of what it is that the regulations are going to apply to. Thus one of our recommendations is to streamline the MCTL. Our general suggestion was to build high walls around narrow areas that are clearly defined, with priorities established in words that everybody can understand. I don't have any great hope, however, that tomorrow's mail will bring such a list to my desk. □

Signing SALT I.
Richard Nixon and Leonid Brezhnev are shown signing the Strategic Arms Limitation Talks agreement, which includes the ABM treaty. The ceremony took place in Moscow on 26 May 1972, after two and one-half years of negotiations. (United Press International photograph.)

Arms-limitation strategies

Treaties, protocols, international conventions and unilateral restraint, while not true solutions, can buy time and move us in the right direction.

PHYSICS TODAY/MARCH 1983

Herbert F. York

In 1934, when the Nazi terror was just beginning, and artificial radioactivity and the neutron had just been discovered, Leo Szilard invented the idea of a neutron chain reaction. He believed there must be some atomic nucleus that would, upon absorbing one neutron, emit two neutrons and energy, and that these two neutrons would lead to four, and so on. He did not know what nuclear process would be involved or what element would support that process. Nevertheless, he quickly went on to sketch two distinct devices based

Herbert F. York, former director of research and engineering at the Department of Defense, is professor of physics at the University of California, San Diego. He was a member of the General Advisory Committee on Arms Control and Disarmament, 1962–69, and US Ambassador to the Comprehensive Test Ban Talks in Geneva, 1979–81.

on such a neutron chain reaction.

One of these devices would release energy in a steady and controlled fashion, resembling in principle today's nuclear reactors; the other would release energy suddenly and in an uncontrolled fashion, similar in principle to today's nuclear bombs. Szilard's intuition told him that the former would be a boon to mankind, and that the latter would constitute a most serious threat. His views concerning this threat were strongly influenced by the rise of Nazi terror and his projection of how the political situation would develop. (Szilard had only recently arrived in England as one of the very first of what was to become a flood of refugees.) His solution to the dilemma inherent in all this was classic: He took out a public patent on the reactor-like device and he managed, with some difficulty, to obtain a secret patent on the bomb-like device.

Thus, one might say secrecy was the

first nuclear-arms limitation strategy. Indeed, governments still consider secrecy useful, although the approach to arms control they now favor most is the treaty. But because treaties to date have fallen far short of eliminating nuclear weapons or removing nuclear weapons from the realm of national sovereignty, they are, at best, only partial measures against the threat that Szilard foresaw. Nevertheless, these partial measures may buy time, so it is important that physicists who wish to participate in research and education concerning the prevention of nuclear war know about the measures that have been taken, and about those being proposed or negotiated. My goal in this article is to present some of this information, emphasizing international agreements that bear on arms control. Some readers will probably be surprised to find that there are quite a few relevant treaties in force or in various stages of development.[1]

Ever since Szilard's secret patent, governments have used secrecy to attempt to separate the development and promotion of nuclear power from the development and acquisition of nuclear weapons. In particular, they have used secrecy to prevent the spread of nuclear weapons to other powers and, more recently, even to subnational groups. Among governments, secrecy is not only a highly regarded means to this end, it is the only universally agreed-upon means. Obviously, secrecy does not "work" in the absolute sense—weapons have already spread to other countries and it seems certain that they will continue do so. However, secrecy does inhibit the whole process by sharply limiting the total number of people—scientists and inventors—who can engage in the kind of personal interaction that speeds the development of this or any technology, and it cuts off the interaction between groups of such people in different countries. Indeed, Edward Teller's well-known advocacy of reduction in the secrecy barrier is based in part on his belief that with less secrecy qualitative improvements in weaponry would come at a faster pace, and that would be a net benefit to the United States. In the early years, the American and Soviet programs did, in fact, unavoidably trade information by providing each other with radioactive fallout to analyze; but since the atmospheric test ban of 1963, that channel of communication has been closed.

Peril anticipated

In December of 1938, Otto Hahn and Fritz Strassman in Germany discovered that neutrons induced in uranium what was soon named the fission process. Other experiments and theoretical elucidations quickly followed, and it was immediately and widely recognized that here was the process that would make the neutron chain reaction possible in both its controlled and violent forms.

Programs to explore and, if possible, exploit these new discoveries started in at least six countries. However, only the American program was to succeed. Only in America was it possible, given the violent circumstances then prevailing most everywhere else, to marshall the intellectual and material resources necessary to produce an atomic bomb.

But, even before the Manhattan Project achieved its goal, many among its leadership were able to find some time to contemplate the future they were striving to bring about. They again took it as almost axiomatic that nuclear power would be a boon to mankind, and that nuclear research in general had an exciting and fruitful

future before it. Many of them also believed that no matter what short-term benefits their work might bring, in the long run the advent of nuclear weapons, and the nuclear arms race that they feared would build up around them, constituted a most serious peril for the future of man and his civilization. President Harry Truman, and many other knowledgeable statesmen soon came to share similar views.

Immediately after the war, as one of its responses to these growing concerns, the government established a special committee to explore means and generate proposals for coping with this great problem. David Lilienthal chaired the committee, which reported to the highest levels of government. Members included Robert Oppenheimer, Chester Barnard, Charles Thomas and Harry Winne. Because of his knowledge and intellect, Oppenheimer was the committee's central figure, just as he was in other similar instances at the time.

The radical solution

The Lilienthal committee considered the general situation to be unprecedented both in its seriousness and its substance and they proposed what they held to be an appropriately novel and radical solution. They proposed to create an international agency that would "conduct all intrinsically dangerous operations in the nuclear field, with individual nations and their citizens free also to conduct, under license and a minimum of inspection, all non-dangerous, or safe, operations." They also proposed to eliminate nuclear weapons—eventually—saying that nothing less would suffice. In sum, they said the rules and customs that derive from the concept of national sovereignty must not be allowed to prevail in the realm of nuclear fission and all the processes and activities that derive from it.

It was not to be. The American plan, which was presented to the United Nations in 1946 by Bernard Baruch, and whose central substance consisted of the proposals of the Lilienthal Committee, was promptly rejected by the Soviets. Also, it seems unlikely that the American body politic would have accepted it in the end, even though it had the full blessing of the President himself.

Historians and analysts have often questioned the sincerity of Oppenheimer and Lilienthal, Truman and Dean Acheson, and Baruch. Surely, it has been said, they must have known the plan was too radical to succeed. In fact, the authors of the plan anticipated just such a reaction. In the report itself they wrote:

The program we propose will undoubtedly arouse skepticism when it is first considered. It did among us, but thought and discussion have converted us.

It may seem too idealistic. It seems time we endeavor to bring some of our expressed ideals into being.

It may seem too radical, too advanced, too much beyond human experience. All these terms apply with peculiar fitness to the atomic bomb.

In considering the plan, as inevitable doubts arise as to its acceptability, one should ask oneself, "What are the alternatives?" We have, and we find no tolerable answer.

I believe Oppenheimer and the others were right: There are no alternative solutions to the problems created by the advent of nuclear weapons. The other approaches discussed in the remainder of this article are all palliatives, means for moderating or delaying the nuclear holocaust, but inadequate for preventing it altogether. Of course, in a situation as dreadful as the one in which we find ourselves, it is very definitely worthwhile to pursue and promote palliative and partial measures in the hope that they will both buy time and eventually lead to a real solution. But while we are pursuing such measures, we should bear in mind their true nature.

During the first decade following World War II, no progress toward arms control was possible, but in the mid-fifties the outlook improved. A number of factors contributed importantly, including the achievement of the atomic bomb by the Soviets, a development that made it possible for them to deal with the West at least nominally from a position of equality. In addition, Stalin died and was replaced by leaders more open to dealing with the outside world. At about the same time, the international dialog, which had never totally ceased, changed from one with a strong rhetorical emphasis on total measures—"general and complete disarmament"—to a more serious search for partial measures that could be taken up individually and in a serious manner more likely to lead to success in each case.

Since then, governments have proposed and attempted many such partial measures, and some have been achieved. These measures have come in several quite different formats and have involved a wide variety of substantive issues. The formats have included bilateral treaties such as SALT, multilateral treaties such as the Nonproliferation Treaty, United Nations Conventions such as the Convention on

Leo Szilard at an American Physical Society meeting in Washington, DC, 27 April 1935. Through his use of a special patent, Szilard was the first to use secrecy as a nuclear-arms limitation strategy.

Biological Warfare, carefully matched unilateral actions such as the test moratorium of 1958–61, simple unilateral actions such as the 20 years of US restraint in the development of antisatellite weapons, and the establishment of special international regimes such as the International Atomic Energy Agency. The substance of these measures has varied even more widely, and has included a variety of limitations on nuclear testing, restrictions on providing assistance to others, prohibitions on the misuse of assistance from others, the denuclearization of specific regions such as Latin America and outer space, numerical ceilings on specific deployments, the legitimization of certain unilateral intelligence-gathering techniques and the establishment of some cooperative means of verification.

Treaties now in force

The arms-limitation format most favored by governments, statesmen and diplomats is the treaty. The principal advantage of a treaty over, say, paired unilateral actions, is that is spells out in detail all the limitations and restrictions that are to be undertaken, the means to be used for verifying compliance, the duration, and the conditions under which it may be terminated. Only by providing such details is it possible to avoid the misunderstandings (which often seem deliberate) that so commonly characterize relations between states with strongly differing political systems. Moreover, treaties between the US and the USSR are developed step by step and in detail by the superpowers themselves. As a result, all of their provisions are carefully tailored to be in the mutual interest of both parties, thus ensuring a higher probability of compliance than in the case of UN conventions and other very broadly international arrangements.

Conversely, the principal disadvantage of the treaty approach is that negotiation and ratification often take a very long time, during which external events can take place that cause the process to abort. One recent case in point is SALT II, which was slowed and delayed by a series of such events— Deng Xiao-Ping's visit to Washington, the controversy over the brigade of Soviet troops in Cuba, the Teheran embassy capture—and finally aborted by the invasion of Afghanistan. Another case is that of the Comprehensive Test Ban Treaty negotiations during the Carter Administration. Internal opponents of the treaty were able to search out and use bureaucratic maneuvers to slow the negotiating process until it too was finally aborted by the same events that killed SALT II.

Let us take a quick look at some of the arms-control treaties now in force.

▶ The Antarctic Treaty of 1959 in essence demilitarizes Antarctia. It was signed by all the parties having territorial claims or other special interests in the region. A special feature is its provision for on-site inspections of the various research bases maintained in the region to assure that they are not conducting banned activities. The United States has frequently exercised its rights under this provision.

▶ The Limited Test Ban Treaty of 1963 prohibits nuclear testing in the atmosphere, outer space and under water. The original intent was to ban nuclear tests altogether, but in the end the US view was that no adequate means were available for verifying a ban on underground tests so these could not be included in the treaty. Numerically speaking, most of the public support for this treaty came from those whose principal stated concern was the health problem posed by radioactive fallout, and their concern was, of course, satisfied by the limited test ban.

▶ The Outer-Space Treaty of 1967 and the Seabeds Treaty of 1972 ban the placement of "weapons of mass destruction" in those two locations.

▶ The Nonproliferation Treaty of 1968, which went into force in 1970, is designed principally to bar the nuclear-weapons states from helping other states to get or to build nuclear weapons, and to bar the non-nuclear weapons states from seeking to obtain or build such weapons. The treaty contains several additional provisions designed specifically to make it more palatable to the nuclear have-nots. One such provision calls for the nuclear-weapons states to "pursue negotiations in good faith on effective measures relative to cessation of the nuclear arms race." Others call for the nuclear-weapons states to help the non-nuclear-weapons states acquire the benefits of nuclear energy, including benefits deriving from the so-called "peaceful uses" of nuclear explosives. The treaty calls for a review of the nonproliferation regime at five-year intervals. In the past two such reviews, there have been widespread charges that the superpowers have not lived up to their obligations under all of the additional provisions.

▶ The Latin American Nuclear Free Zone Treaty of 1968 is designed to eliminate all nuclear weapons from Latin America. It has been ratified by most but not all of the states of the region. It also includes protocols placing certain consistent requirements on states outside the region when those states own Latin American territories or possess nuclear weapons. This treaty has served as a model for as yet unfulfilled proposals for "nuclear-free zones" in several other parts of the world.

▶ The Antiballistic Missile Treaty was ratified in 1972. Together with its later amendments, this treaty defines an ABM system and places numerical and geographical limits on its deployment. It contains provisions designed to limit certain characteristics of ABM subsystems and to inhibit, but not prevent, the development of new types of ABMs. In addition, it contains provisions legitimizing and protecting so-called "national technical means" of verification and restricts practices that would confound such systems. The treaty is of indefinite duration, but it is to be reviewed every five years. It also calls for the establishment of a "Standing

Consultative Commission," which would meet in private and consider complaints and questions raised by either side. This commission has met as required and has reported that all questions raised have been resolved satisfactorily.

▶ The Executive Agreement Covering Certain Offensive Systems, also ratified in 1972, places limits on the numbers and sizes of specific offensive systems, including ICBMs and sea-launched ballistic missiles. This agreement was in essence a "freeze," because the limits were set at the numbers of systems already deployed or in the process of being deployed. It restated the provisions dealing with national technical means of verification. It was to run for five years, at which time it was to be replaced by a treaty that would be broader and more far reaching in scope. The United States and the Soviet Union have not achieved this, but through a continuing series of understandings they have kept the executive agreement in force.

Uncompleted treaties

The SALT II Treaty was negotiated and signed during the Carter Administration. It was called for in the executive agreement of SALT I, and was based in large part on certain general "guidelines" presented in the Vladivostok agreement worked out by Presidents Ford and Brezhnev in 1974. It sets numerical limits on all the major types of strategic offensive systems, and it sets sublimits that restrain the total number of warheads by limiting the "MIRVing" of delivery vehicles. The treaty contains a data base and numerous working definitions, all of which are designed to make it easier to

achieve further agreements, which the treaty tacitly assumes will follow, and which would eventually lead to a real reduction in numbers. The SALT II treaty was finally signed and submitted to the Senate in late 1979 after several long delays. But before it could be ratified, the events mentioned above intervened and put the necessary two-thirds majority in the Senate beyond reach.

The Threshold Treaty of 1974 and the Peaceful Nuclear Explosions Treaty of 1976 were worked out and signed in the Nixon and Ford Administrations respectively. The first was designed to prohibit tests of nuclear weapons exceeding 150 kilotons in yield, and the latter was designed to allow multiple nuclear explosions conducted for peaceful purposes to exceed this limit in the aggregate but not individually. Two very important features of these treaties are that they provide for an exchange of geophysical data concerning the test site, and that they call for on-site observations and measurements by personnel of the other principal in certain special circumstances. This was the first instance in which the Soviets ratified a treaty allowing such intrusive procedures to take place on their home territory, but these treaties have not been ratified by the United States, so these special provisions are not in force. Both states, however, have said they would for now comply with the 150-kt limit, and both appear to be doing so.

Treaties attempted or proposed. A Comprehensive Test Ban, which would prohibit all nuclear weapons tests everywhere, has been a stated goal of both the US and the USSR since 1958. The pursuit of this goal has led to

several partial measures—the Limited Test Ban and Nonproliferation treaties, and the signed-but-not-ratified Threshold and Peaceful Nuclear Explosions treaties—but the goal itself remains elusive. Negotiations during the Carter Administration did achieve several advances of fundamental importance in the area of verification. The US and USSR reached agreement in principle on a system of voluntary on-site inspections. These would be based on a series of challenges and required responses and the deployment of national seismic stations (the "black boxes" of earlier years), which would be placed on each other's territory and which would be constructed and operated so as to provide a continuous, unadulterated, stream of data in near real time. In addition, the Soviets agreed to forego at least temporarily their program to develop and use so-called "economic nuclear explosions," analogous to our "peaceful nuclear explosions." After some important early progress, this negotiation encountered a steadily and seriously deteriorating political climate (Teheran, Kabul, the 1980 US elections), and was eventually stalled by the bureaucratic maneuvers of its opponents. As a result, the details underlying these agreements-in-principle were never fully elaborated. When President Reagan took office in 1981, the negotiations were cancelled.

The anti-satellite, or ASAT, negotiations of 1978–79 were intended to forestall the construction of anti-satellite weapons systems and the eventual preparation to conduct warfare of various novel kinds in space. Three negotiating sessions took place during the early Carter years, but it eventually

Negotiators in Geneva. Four of the members of the author's delegation to the Comprehensive Test Ban talks appear in this photo. From left to right, they are Gerald W. Johnson, former special assistant to the Secretary of Defense for atomic energy, Warren Heckrotte of Lawrence Livermore Laboratory, York, and Edward Giller, a retired Air Force major general and former head of the military applications division of the Atomic Energy Commission. All four are physicists. (United States International Communication Agency photograph.)

became evident they were leading no-where and they were abandoned even before the international scene serious-ly deteriorated. In my view, this hap-pened largely because neither side was able to develop a clear and generally internally acceptable view of its objec-tives.

During the last quarter century, several proposals to limit certain nu-clear-weapons activities have been put forward but never seriously negotiated. These included proposals to cut off production of the special materials needed to construct nuclear weapons, to limit seriously launches of large rockets so as to inhibit greatly their further development, and to nip in the bud the MIRV development program. In addition, there have been proposals for many other "nuclear-free zones," as well as proposals for making certain parts of the ocean safe "havens" for nuclear submarines as a means of supporting the so-called mutually as-sured destruction, or "MAD," doctrine.

Conventions and the like. United Na-tions conventions have the advantage of being universal, or nearly so. And, to the extent that they work and exercise a positive influence, they tend to rein-force other peacekeeping and peace-supporting activities of the UN. The corresponding disadvantage, as com-pared to bilateral treaties between states of roughly equal size and power (or multilateral treaties involving two distinct sides or two dominant parties), is that they are not taken quite so seriously. Thus, while the record of compliance with arms-control treaties is generally quite good, the record in the case of these more broadly based arrangements is quite different, as evidenced by the many apparent viola-tions of the Chemical Protocol and the Biological Convention, and the cavalier way in which many states, the US and USSR included, fail to live up fully to the purposes and regulations of the International Atomic Energy Agency. The reason for this difference, I believe, lies in the fact that the superpowers have much less control over the details of universal conventions, and they fre-quently end up facing a "take it or leave it" situation. They often choose to "take it," but then are less likely comply with the detailed provisions than when they are in full control of the details. I believe Soviet behavior in connection with the Chemical and Bio-logical Conventions is a case in point. Both are broadly multilateral conven-tions produced by an international organization, not by direct principal-to-principal negotiations, and the latter in fact contains no provisions for verifica-tion.

The 1925 Geneva Protocol on Chemi-cal Weapons and the 1977 Biological Weapons Convention were sponsored by the League of Nations and the United Nations, respectively. The for-mer seeks to prohibit the use of poison-ous gases and "bacteriological meth-ods" of warfare, and the latter seeks to prohibit the development, production and stockpiling of bacteriological and toxic weapons as well. The Geneva Protocol went into effect in 1928, but the United States did not ratify it until 1975, when it also ratified the Biologi-cal Weapons Convention. The parties to the latter also agreed to destroy such stocks as then existed within nine months after it went into force.

The International Atomic Energy Agency, which was created under the aegis of the UN, grew out of President Eisenhower's "Atoms for Peace" speech of 1953. Its twin purposes were to promote the spread of nuclear ener-gy throughout the world and at the same time prevent the spread of nu-clear weapons technology along with it. While the system of inspections and other procedures cannot prevent the spread of nuclear weapons—as those knowledgeable of but hostile to the Agency like to repeat—this system can and does greatly inhibit the process. Similar but more limited arrange-ments with the same objective have also been carried out in other contexts. For instance, the European Atomic Energy Community has similar pur-poses within Europe.

Unilateral actions

The defense programs of all coun-tries involve a continuous stream of unilateral actions, many of which have the effect of moderating the nuclear arms race. But nearly all of the moder-ating actions are the result of fiscal, not political, restraints. Occasionally, however, a unilateral restraint is im-posed for the sole or principal purpose of moderating the military confronta-tion or avoiding some particular exa-cerbation of it.

A particularly interesting and im-portant example was the 20-year period of restraint imposed by US presidents on the development of anti-satellite weapons. The United States, almost from the very beginning of the space age, has used its military space assets—reconnaissance satellites, for exam-ple—for a variety of very important national-security purposes. From Ei-senhower onward, our Presidents have concluded that we would derive a net benefit from a situation in which no state had any anti-satellite capability. The US, therefore, was willing to forego a program to develop such weapons, in the hope that the Soviets would follow suit. As a result, proposals arising in the US Air Force and aerospace indus-try to develop a general-purpose anti-satellite weapon were continually re-jected.

In the meantime, and despite US restraint, the Soviets launched their first anti-satellite experiment in 1968, and they have continued since then to conduct further tests in an on-again off-again fashion. Finally, in 1977 Presi-dent Carter decided to make a three-pronged response to those Soviet actions: first, to initiate the develop-ment of a US anti-satellite weapon;

Entry into force of the Limited Nuclear Test Ban Treaty, 10 October 1963. Photograph shows the Washington signing of the Protocol of Deposit of the Instruments of Ratification of the treaty. The treaty itself was signed about two months earlier. Seated: British Ambassador David Ormsby Gore, US Secretary of State Dean Rusk and Soviet Ambassador Anatoliy F. Dobrynin. Aides from each of the three countries stand behind their respective representatives. Instruments of ratification were also deposited in Moscow and London. (Department of State photograph.)

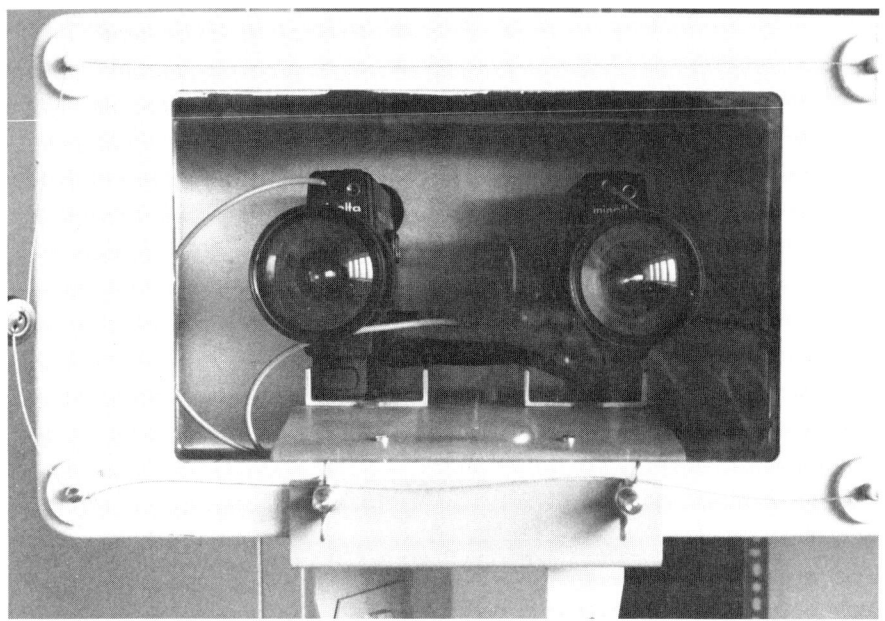

Surveillance system, sealed in a tamper-proof enclosure. The International Atomic Energy Agency monitors about 850 nuclear installations in the non-nuclear-weapons states to guard against the diversion of fissionable material. In 1981, about 160 surveillance systems took about 8 million pictures. (Photograph courtesy of IAEA.)

second, to explore means for defending satellites against attack; and third, to begin negotiations with the Soviets to forestall all such developments. Beginning at Helsinki in 1978, three negotiating sessions took place, and, as I mentioned above, they got nowhere. Now both the US and the Soviets have anti-satellite programs underway.

Another important unilateral action took place in October 1958 when, after a succession of slowly converging unilateral statements by Eisenhower and Khrushchev, the US and the Soviet Union both suspended nuclear tests. This moratorium, as it was called, was in essence based on a matched set of unilateral statements in which each party pledged not to test if the other party did not test. The purpose was to create an atmosphere suitable for working out a formal treaty on the subject, and Eisenhower estimated that one year should suffice. The negotiations did not work out, however, and after fourteen months, on 29 December 1959, Eisenhower denounced the moratorium, saying the US was no longer bound by it, but would not begin testing without giving notice. Three days later, the Soviets denounced it also, but said they would not test unless the "Western powers" did so first. When France tested its first nuclear weapon three months later, Khrushchev took formal and public note of it. Hence, as of that time there was no longer any *de jure* basis for a nuclear moratorium, but testing did not in fact resume until 18 months later, when an extensive Soviet series put a final end to the then purely *de facto* moratorium.

Although there were no external legal restraints—formal or informal—on such testing by any party at that time, US official opinion has always held that we were in an important sense entrapped by the Russians. In the words of one current high-level White House official, "they used the moratorium to gain a full lap on us" in the cycle of nuclear testing and development. Ever since then, unverifiable morato-

ria, whether bilateral or not, have found little support in US government circles.

In recent years, the Chinese and the Soviets have made unilateral pledges to the effect that they would not be the first to use nuclear weapons, and many have urged that the United States also make such a "no first use" pledge. United States policy has always been to reserve the right to initiate the use of nuclear weapons, particularly in situations where the conventional balance may be very heavily weighted against us, and so we have always declined to join in such a pledge, whether on a unilateral or a multilateral basis. Another basic argument has been that such a pledge, if unaccompanied by any further, more concrete actions that might really make first use more difficult or less desirable, would be too easily reversed or disregarded.

More recently, a number of Europeans and Americans have proposed that we unilaterally and as a "first step" either eliminate all so-called "battle-field nuclear weapons" (such as nuclear artillery) or at least remove them from close proximity to the inter-German border. They argue that the actual use of such weapons for resisting an armed attack would not be to our net advantage, that we can successfully develop and deploy conventional means for resisting such an attack, and that removing such battlefield "nukes" would substantially reduce the probability of nuclear war in Europe. They argue further that the longer-range nuclear systems, which would presu-

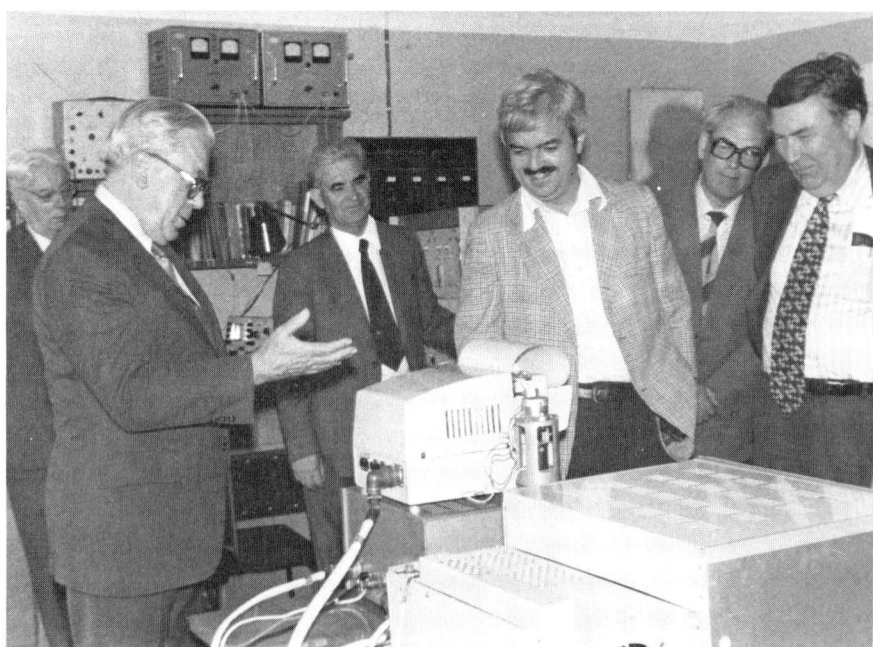

Moscow lab tour, 1979. Photo shows three Americans and three Soviets in a Soviet geophysical laboratory involved in the design of equipment capable of monitoring a test ban. In the foreground are Americans Gerald W. Johnson, John Marcum, a physicist employed by the White House as part of the National Security Council staff, and York. At the far left is the academician M. A. Sadovsky, director of the Soviet Institute of Geophysics.

Inspectors from the International Atomic Energy Agency at work in the Austrian nuclear power station at Zwentendorf. (Photograph courtesy of IAEA.)

production and deployment of all nuclear weapons systems. Only through the treaty-writing process discussed above would it be possible to elaborate and negotiate the details necessary to achieve and sustain such a goal. A number of treaties encompassing partial freezes have, in fact, already been fully or partially negotiated—SALT I, SALT II, the Comprehensive Test Ban—so there are some precedents for such an action. However, some of the experts who commonly advocate nuclear restraints in general have argued that the freeze proposals include too many elements that are too widely disparate to serve as the basis for a unitary bilateral or multilateral negotiation.

It is clear that in its first purpose, the freeze movement is succeeding. In November 1982, in eight of the nine states where a freeze resolution appeared on the ballot, the electorate favored it. I believe these and similar actions will stimulate some real progress by governments in moderating the nuclear confrontation and achieving formal agreements further limiting the development and deployment of nuclear weapons, especially in Europe. However, it is not at all clear to me at this time precisely what final and formal form these new limitations will take. Nor is it clear whether any such progress can happen soon, or whether it must await the arrival of an administration in Washington (and perhaps in Moscow) more closely attuned to the public's concerns in this vital matter.

mably still remain available after such a "first-step," would be fully adequate for continuing the state of nuclear deterrence that has persisted the last 35 years.

The freeze movement

The people behind the current wave of "nuclear freeze" proposals have two distinct purposes. One is to provide a simple, easily understandable focus for public opinion; the other is to outline the substance for an international agreement to stop the nuclear arms race in its tracks.

In their first role, the proposals have taken the form of resolutions placed before the Congress and before the voters in many state and even local elections. In their second role, the goal is usually said to be a "bilateral and verifiable" freeze in the development,

Reference

1. *Arms Control and Disarmament Agreements*, 1982 edition, United States Arms Control and Disarmament Agency, Washington, D.C. (1982). □

SECTION 4

THE DEFENSE

This section primarily covers President Reagan's Strategic Defense Initiative. Reagan directly appealed to scientists for a technical solution, the SDI, in his March 1983 speech:

> Wouldn't it be better to save lives than to avenge them? ... What if free people could live secure in the knowledge that their security did not rest upon the threat of instant retaliation to deter a Soviet Attack, that we could intercept and destroy strategic ballistic missiles before they reached our soil or that of our allies? ... I call upon the scientific community, those who gave us nuclear weapons, to turn their great talents now to the cause of mankind and world peace, to give us the means of rendering these nuclear weapons impotent and obsolete.

The articles in this section span the decade 1980–89. They are listed chronologically to give a flavor for the evolution of strategic defense. In the first article, Bertram Schwarzschild describes the rail-gun technologies. These were initially being developed for a variety of reasons: to transport mass into space from moon colonies, to initiate fusion, and to create new types of weapons. The SDI Organization expanded on the weapons concepts with space-based hypervelocity railguns for projecting pellets at 25 km/sec at incoming reentry vehicles. The issues of power needs in orbit and countermeasures were soon raised by the opponents of SDI's rail-gun program. Shortly after President Reagan's speech, Barbara Levi examined the status of the various directed energy weapons, and where they were headed. SDIO considers the free-electron laser as one of the more promising technologies. Since FELs are too heavy to put into orbit, they would have to be used with space-based mirrors; they might also be used from the ground to shoot down satellites. Phillip Sprangle and Timothy Coffey describe the basic physics of sending beams of electrons through magnetic wigglers or undulators to produce laser beams.

In a debate, Gerold Yonas, as the former chief scientist of SDIO presents the case for SDI and describes the technical progress SDIO has made. Wolfgang Panofsky responds that further research within treaty limits is a good idea, but that the SDI program is too uncertain, too large, and too politicized, and that it could create grave dangers to world security. William Sweet describes SDIO's shifts in November 1985 away from space-based chemical lasers and land-based excimer lasers. Four years later, Yonas now comments that:

> The difficulty with US strategic defense programs has always been one of problem definition: not will it work, but what is it? The real barrier to deployment has been the lack of a broad, non-partisan consensus for a consistent strategy of offense, defense, and arms control to deal with the problem. We just can't seem to figure out what to protect, how well, and against what threat. The arms control advocates argue that a successful defense would convince the Soviets that we are considering a

first strike, and they would then be tempted to preempt before or after our deployment. While the arguments continue, the technology has improved but the opposition continues against any defense even with limited value and cost and in spite of its potential as a bargaining chip. The Soviets argue, as do the opposition in the US, that SDI will not work, that the cost will bankrupt the American people, that it will work too well, that it will kill arms control, and all of the above. The opposition to Strategic Defense is as experienced, capable and determined as ever, and the advocates are still at a real disadvantage; nothing has changed.

Over the past 15 years the APS has produced a number of technical studies, mostly on energy related issues. Because directed energy weapons systems rely so much on the basic mechanisms of physics, it was only natural that the APS would examine the science and technology of the directed-energy weapons. The DEWS report concluded that: "We estimate that even in the best of circumstances, a decade or more of intensive research would be required to provide the technical knowledge needed for an informed decision about the potential effectiveness and survivability of directed energy weapon systems." The executive summary of the report contains many technical conclusions; it is included in this volume. Irwin Goodwin describes the wide press coverage given to the APS report, with lead editorials in such diverse papers as the New York Times and the Wall Street Journal, the former, counseling a quick trade of the SDI bargaining chip, while the latter called it a "naysayer report [that] had nothing to do with the near-term prospects [for SDI]" since it didn't cover kinetic-kill vehicles. The findings of the APS DEWS report were debated in PHYSICS TODAY and they are presented in this section, Nicolaas Bloembergen and Kumar Petal defending the report, and Gregory Canavan attacking it. Bleombergen recently has said that:

> The attention of the reader is called to the complete report which has been published in the *Reviews of Modern Physics* **59**, S1-S202 (July 1987). This report has become a generally accepted document in the scientific literature. The only published criticism with some basis, together with rebuttal by the authors of the report, has appeared in PHYSICS TODAY (and in this book).

Canavan, the author of the criticism, writes:

> Since the debate, John Hammond, then Director of SDIO's Directed Energy Office, and I prepared a report on the 'Constellation Sizing for Modest Directed Energy Platforms' (LA-11238-MS), which shows that smaller platforms and scalings can be used with little penalty in performance, cost, or survivability. While further development is required, an order of magnitude scaling and a decade of research need not be. Thus, lasers could be available when needed to supplement ki-

114

netic energy interceptors, as argued in 'Directed Energy Concepts for Strategic Defense' (LA-UR-1658, May 1987), the basis for the debate.

The article by Irwin Goodwin describes attempts by SDIO to attack the APS study in public and within the bureaucracy. Levi describes more recent progress on FELs, and the choice being made between the FELs driven by induction linacs and those driven by radio-frequency linacs. During the last years of the Reagan administration, the SDI's emphasis moved from directed-energy weapons to kinetic-kill vehicles. Goodwin describes how the Defense Science Board was skeptical about the early deploy-

ment of "smart rocks," or kinetic-kill vehicles. A higher authority in the DOD, the Defense Acquisition Board, was less skeptical of smart rocks in 1987, urging the Milestone 1 development of Phase 1 of SDI, but they became less enthusiastic in 1988, as the concept of "brilliant pebbles," smaller KKVs was introduced. Lastly, in a pair of articles on ASATs, Sweet lays out some of the issues on ASAT treaties, the relationship between SDI weapons as ASATs, and the plans to test ASATs. Since a few, soft satellites are much easier to destroy than thousands of missiles, or ten-thousand, hard, reentry vehicles (RV), it is likely that some of the SDI technologies could evolve into ASAT weapons.

CONTENTS

Electromagnetic guns and launchers

PHYSICS TODAY/DECEMBER 1980

While we accelerate elementary particles and ions by the most advanced electromagnetic means, our standard techniques for propelling macroscopic objects—from birdshot to interplanetary vehicles—are not very different from those in use since the introduction of gunpowder from China. But propulsion by chemical combustion suffers from severe limitations that are keenly felt by people interested in space travel, inertial-confinement fusion, and even such prosaic pursuits as artillery.

The Germans made an abortive attempt to use a "rail gun," a kind of linear dc motor, as an antiaircraft launcher during World War II. But for the next three decades very little was done about electromagnetic acceleration schemes for macroscopic projectiles. Now we are seeing a surge of interest and activity in this field, attested to by a DOD-sponsored conference on electromagnetic guns and launchers, held last month in San Diego.

Although such "guns" do have military applications, they are also of particular interest to solid-state physicists interested in the behavior of materials at extreme pressures, and to those thinking about initiating thermonuclear fusion with beams of "macroparticles." This latter was the subject of the DOE-sponsored Impact Fusion Workshop at Los Alamos last year.

Using rail guns a few meters long, groups at the Australian National University (Canberra), Los Alamos (in collaboration with Livermore), and Westinghouse have in the past few years succeeded in accelerating projectiles weighing a few grams to speeds approaching 10 km/sec—the escape velocity from the Earth. Attaining such a speed in so short a distance involves a steady acceleration on the order of a million g.

A Princeton–MIT collaboration is currently building a "mass driver," a $2\frac{1}{2}$-meter-long travelling-magnetic-wave accelerator, intended to accelerate a 1-kilogram vehicle to about 110 meters/sec (250 miles/hr). Although this device is thought of as a prototype launcher for interplanetary transport, the same ac linear-synchronous-motor scheme is being considered for the ignition of thermonuclear fusion—by accelerating small superconducting projectiles to speeds in excess of 100 km/sec.

Conventional artillery is limited by the chemical energy of the explosive (about 1 electron volt per atom) to muzzle velocities of one or two km/sec. One can't push the projectile in the gun barrel to speeds higher than those of the molecular combustion products. Rockets, though they are also driven by chemical combustion, do not suffer this velocity limitation, because the combustion takes place in the projectile itself. But one pays for this by having to carry aloft a great mass of propellant. The useful payload put into the Earth orbit turns out to be less than one percent of the launched weight, with a correspondingly horrific cost. Henry Kolm, head of the mass-driver group at MIT, looks forward to reducing the cost of putting payloads into Earth orbit to less than a dollar per pound—three orders of magnitude less than today's cost, and fifty times less than the early promises of the space-shuttle program.

The rail gun is conceptually the simplest electromagnetic launcher currently under consideration—and the one that has thus far achieved the highest velocities. It consists of two conducting rails mounted in a gun barrel. A pulsed dc current is sent down one rail and comes back along the other. In the original rail-gun design, the conducting bridge between the rails

is a sliding metal conductor, analogous to the armature of a dc motor. The armature is propelled forward by the Lorentz force of the magnetic field generated by the current in the rails acting on the current in the armature. The propelling force is thus proportional to the square of the current. The problem is to achieve sufficiently high and steady currents—and to keep the armature from disintegrating.

Several important steps toward the solution of these problems were taken between 1968 and 1977 by Richard Marshall and his colleagues at the Australian National University.[1] The Canberra group had access to the world's largest homopolar generator, a 500-megajoule dc storage generator built by Sir Mark Oliphant with parts cannibalized from a synchrocylotron that was never completed. The rotors of this generator are giant Faraday disks that act as flywheels to store energy accumulated from an external source over several minutes. When the rotor is up to speed, the generator can dump its stored energy in about a second—as a half-megampere dc current pulse. Marshall's student John Barber found that one could compress this output into a few milliseconds by interposing an inductive storage between the homopolar generator and the rail gun. It is important that the current remain roughly constant over

Livermore–Los Alamos rail gun at the Los Alamos firing site. Two 8-foot-long parallel copper strips (foreground) constitute the magnetic-flux-compression generator. Sheet explosive driving the upper strip down generates a megawatt current pulse in the 6-foot-long small-bore rail gun (background). The $\frac{1}{2}$-inch plastic cube is launched at speeds up to 5.5 km/sec.

the several milliseconds during which the armature is accelerated in the gun barrel—to maximize the final muzzle velocity and to prevent the device from blowing apart. Earlier attempts to energize rail guns by discharging capacitor banks had failed because the current was dumped too precipitously.

The original idea of using a sliding metal armature to complete the circuit between the rails and push a projectile turned out to have difficulties in practice. The mass of the accelerating armature limited the speeds that could be achieved, but if one tried to reduce the mass of accelerating metal, the armature would melt under the high currents to which it was subjected. The Canberra group demonstrated that one could dispense with the metal armature, replacing it by a plasma discharge arc between the rails. A nonconducting (plastic) projectile is pushed ahead of the plasma by the Lorentz force on the discharge current.

With a rail gun 5 meters long, the Canberra group was able in this way to accelerate half-inch Lexan cubes to speeds up to 6 km/sec in a square-bore rail gun. The gun barrel serves to keep the projectile from bursting under the acceleration stress of half a million g, and to keep the rails from flying apart despite the strong magnetic repulsion produced by the 300-kA current. Marshall is now at the University of Texas (Austin), working on rail guns with distributed energy storage. If one feeds the current into the rail gun only at the breech end, as was done with the Canberra homopolar generator, one suffers a lot of resistive energy loss in the rails. He is therefore building a system that will feed energy into the gun from capacitors and inductors arrayed all along the length of the barrel. The primary application he has in mind for such rail guns is inertial-confinement fusion.

Westinghouse and Los Alamos. Experimental rail guns are currently in operation at Westinghouse and Los Alamos. The small and large-bore guns at Los Alamos were built by a Livermore–Los Alamos collaboration headed by Max Fowler and Dennis Peterson (Los Alamos), and Ronald Hawke (Livermore). The current in these guns is generated by explosively driven magnetic-flux compression rather than by a homopolar generator. This flux-compression generator is an inductor consisting of two parallel conductors, one of which is lined with a sheet of explosive. After the inductor is energized from a capacitor bank, the explosive is detonated, propelling the one conductor toward the other, thus compressing the magnetic flux and inducing a long-

er and more powerful current pulse than one could get directly from the capacitor bank.

The small-bore Livermore–Los Alamos gun is 2 meters long and has a $\frac{1}{2}$-inch square bore. With a plasma-arc armature and a peak current of 800 kA, it has succeeded in launching a 3-gram plastic projectile at 5.5 km/sec. At a still higher current (1.2 MA), Hawke told us, the group believes it has recently achieved a projectile velocity of 10 km/sec. This would be the world's speed record for electromagnetic macroparticle accelerators. But Hawke points out that they were unable to verify this speed because the projectile disintegrated under the acceleration stress (5×10^6 g) as it departed the muzzle. In order to have the projectile survive such a stress (15 times its elastic limit), he told us, they will have to arrange things so that acceleration ceases well before the projectile leaves the protective confines of the barrel. But the group was pleased to find that the plasma-arc armature continued to function well at four times the highest currents employed in the Canberra gun.

The Westinghouse group, led by Ian McNab, Dan Deiss and John Mole, has a 2-meter-long gun driven by a capacitor bank, with an intervening inductor to lengthen the discharge current pulse. With plasma-arc and metallic armatures they have accelerated half-inch plastic projectiles to 1 km/sec, and heavier metallic projectiles to a few hundred m/sec. The group is now building a 15-megajoule homopolar generator for a rail gun that is expected to accelerate 3/4-pound projectiles to 3 km/sec by next year. At such intermediate speeds McNab believes that metal armatures will prove to work better than plasma arcs. The rail-gun work at Los Alamos, Livermore and Westinghouse is funded by the DOD, which is interested in electromagnetic accelerators as an alternative to conventional launchers and guns in applications where high velocity is important.

The mass-driver was suggested in 1974 by Gerard O'Neill (PHYSICS TODAY, September 1974, page 32), as a device for launching raw materials from the Moon. Because the mass driver concept is quite similar to the Magneplane superconductively levitated train developed by Kolm and his collaborators at MIT (PHYSICS TODAY, July 1977, page 34), O'Neill and Kolm have undertaken a collaborative effort to build a prototype mass driver, a traveling-magnetic-wave dipole accelerator. A persistent superconducting current loop on the projectile is propelled by magnetic-dipole interaction with a closely spaced row of synchronously activated current

loops on the launcher. Each accelerating loop is activated as the projectile approaches. The projectile thus rides a magnetic wave—much like a surfer, Kolm told us. A similar linear dipole accelerator had been suggested in 1929 by the rocket pioneer Hermann Oberth—with ferromagnets rather than superconducting coils.

In O'Neill's conception, payloads are launched from a train of reusable "buckets" (each with superconducting coils) that circulate through the system. Though the original idea was to launch cargoes of aluminum-rich soil from the airless Moon, Kolm's group has recently calculated, rather surprisingly, that one could launch a vehicle as light as a ton ("in the shape of a telephone pole") from the Earth with only a 3% ablation loss as it traverses the atmosphere in about a second.

Mass drivers for space launchers are envisioned as being several kilometers long. Three years ago, Kolm and his students built the first prototype—Mass Driver I. The 2-meter-long device accelerated a half-kilogram bucket to about 100 miles/hr. Mass Driver II, now being constructed by the Princeton–MIT collaboration, may eventually accelerate kilogram vehicles to about the speed of sound.[2]

Impact fusion. Fifteen years ago Friedwardt Winterberg (University of Nevada, Reno) suggested using a traveling-magnetic-wave accelerator to propel small superconducting projectiles to speeds high enough to ignite thermonuclear fusion in deuterium–tritium targets. The use of macroparticle accelerators for "impact fusion" has recently been looked at in some detail by a number of investigators. To deposit the requisite megajoule of impact energy in a one-cm^3 target volume in about 10 nanoseconds requires, it is generally agreed, impact velocities of between 150 and 200 km/sec—an order of magnitude faster than anything that will be achieved in the immediate future.

In a critical review of the various impact-fusion options,[3] based on the 1979 Impact Fusion Workshop, Fred Ribe (University of Washington) and Alfred Peaslee (Los Alamos) conclude that segmented rail guns with distributed energy stores appear to offer the greatest promise. Richard Muller (Berkeley), Richard Garwin (IBM) and Burton Richter (SLAC) had proposed to the workshop a kilometer-long segmented rail gun firing 0.05-gm projectiles at 200 km/sec. Hawke presented a similar impact-fusion gun design at the Workshop.

Ribe and Peaslee believe that rail guns are well suited to deliver 10–100

MJ to a fusion target with a relatively simple and inexpensive technology. Unlike light-ion and electron beams, macroparticles are very easy to focus on a target pellet, and the accelerating apparatus is easily shielded from the thermonuclear explosions in a reactor. Hawke believes one may be able to ignite fusion targets with a pair of rail guns firing from opposite sides—each only 30 meters long.

Garwin, somewhat cynically, told us that the main virtue of impact fusion is that it will teach us faster and cheaper than any other technology that inertial-confinement fusion won't work—at least in an economic sense. The basic problem, he believes, is that all such schemes require the concentration of large amounts of energy into 10-nanosecond pulses. Winterberg is skeptical that rail guns can achieve velocities high enough to ignite conventional fusion targets, because friction-generated radiation losses increase as v^8. Magnetic-wave accelerators, with superconductively levitated projectiles, suffer no such friction losses, but they provide significantly less acceleration. He points out, however, that with magnetized target designs one might achieve ignition at impact velocities less than 50 km/sec. Magnetized targets have been suggested by Ribe and his Seattle colleague George Vlases, and independently by Shyke Goldstein and Derek Tidman of Jaycor (Alexandria, Va.). In such fusion targets, a 10-megagauss pulsed magnetic field would thermally insulate a plasma from the walls of its confining cavity.   —BMS

References

1. S. C. Rashleigh, R. A. Marshall, J. Appl. Phys. **49**, 2540 (1978).
2. H. Kolm, P. Mongeau, F. Williams, IEEE Trans. Mag. **16**, 719 (1980).
3. Proceedings of the Impact Fusion Workshop, Los Alamos report #8000-C (1979).

Directed energy weapons— where are they headed

PHYSICS TODAY/AUGUST 1983

A futuristic look at the nation's defensive weapons systems sometimes includes visions of space-based lasers or particle beams, able to direct their energy precisely and devastatingly upon any target. The Defense Department has several technology programs that deal with these directed-energy weapons, although practical implementation appears to be at least a decade away. President Reagan may now wish to place more stress on such programs. In a speech to the nation on 23 March, the President proposed a "comprehensive and intensive effort to define the long-term R&D program" on defensive weapons against ballistic missiles. He hoped to ensure peace by rendering such nuclear weapons launchers "impotent and inoperative." One ingredient of such a program might be a land- or space-based system of lasers and particle beams.

Responding to the President's initiative, officials within the White House and within the Departments that sponsor directed-energy research have begun to evaluate the various technology programs. They intend to identify the most promising ones, to decide the appropriate level of emphasis for each and to develop a program plan that might be incorporated into the budget proposal for fiscal year 1985. Their review is being conducted away from the public eye.

The President's speech has also rekindled the controversy over a defense based on directed-energy systems and has sparked curiosity about the status of work on such weapons. What missions might such weapons systems fulfill? What hurdles currently block the way to practical realization of these systems? What arguments are being made for and against directed-energy weapons systems as the cornerstone of the US defense against nuclear attack? We set out to survey the programs in progress and to listen to the ongoing debate.

Possible defensive missions. The role most often envisioned for directed-energy weapons is to defend the US against enemy nuclear missiles. Both lasers and particle beams show promise

Defense of ships is one possible role for charged-particle beam weapons. Because of problems with beam stability, they may have ranges of only a few kilometers in the atmosphere.

for such a role because the beams arrive at their target at or nearly at the speed of light and because they deliver a destructive quantity of energy in a very short time. The first characteristic means that the weapons can be aimed along the line of sight (although at long ranges, the beam must lead the target by a slight amount). Evasive actions by the target are not possible. The second enables the weapon to engage many targets very quickly, provided, of course, that the power supply is adequate for numerous refires. Laser beams can be directed at targets attacking from any direction, but particle beams are restricted to a narrower angular range.

A typical weapons system (as discussed in the laser- and particle-beam fact sheets published by the Department of Defense),[1,2] whether based on land or ship, in air or space, consists of the beam source, a beam-control sub-

system to direct the beam and to keep it pointed on its target, and a fire-control subsystem to identify each target, discriminate it from decoys, determine whether a given shot was successful and to redirect the beam accordingly. To exploit fully the unique advantages of a directed-energy weapon, these subsystems have stringent requirements placed upon them. First, the targeting must be precise because these weapons can damage an object only by making direct contact with them. The distances over which these shots are fired in space make this targeting task especially difficult. The control systems must keep the beam focused on one spot of the target long enough to do the requisite damage. Furthermore, the fire-control system must be very fast, especially if one envisions defense against a massive attack.

The two types of missions might be classified roughly according to whether the weapon is based within the atmosphere or outside it. Endoatmospheric missions can include the point defense of a large ship such as an aircraft carrier, the hard-site defense of intercontinental ballistic missiles, or airborne strikes with laser beams against satellites. The range required in the first two applications is typically only a few kilometers. Exoatmospheric missions could involve defense against ballistic missiles or enemy satellites. For such space missions a number of directed-energy weapons stations is required so that at least one remains in sight of every potential ballistic missile trajectory. The higher the orbit, the fewer the stations needed but the longer the strike range. Some estimates of the number of stations required were made as part of a study undertaken by physicists at MIT several years ago.[3] For a synchronous orbit, at an altitude of about 40 000 km, only a handful would be required. For a low Earth orbit, the range might be reduced to about 1000 km but the number of weapons stations might be increased to around 150 to 200.

Another type of mission might involve a combination of endo- and exoatmospheric basing. According to testimony[4] given by Edward Teller before a subcommittee of the Senate Armed Services Committee, defense planners have proposed "pop-up" defense systems that are sited on Earth but launched into space in times of crisis. Such systems would be less vulnerable than permanently orbiting space stations but, as some critics say, might not necessarily be launched in sufficient time to engage an enemy target.

Laser systems encounter several sources of degradation when shot near the Earth. Some of these were examined in a study of laser weapons done[5] at MIT in 1980. Molecules and particulates within the atmosphere absorb and scatter the radiation, limiting the weapon's range in clear weather and virtually neutralizing it on foggy or rainy days. Even in good weather, the beam would diverge through an effect known as "thermal blooming." In this effect, the beam spreads out at the edges because the heated air in the beam channel has a lower index of refraction. Turbulence in the air further bends and defocuses the laser beam. Although current studies are exploring the problems of beam propagation in the atmosphere, the existence of such problems makes the space-based applications of lasers more attractive. For space applications, the challenge is to design an efficient and powerful laser and an optically precise, suitably large pointing mirror.

The missions envisioned for particle beams vary with the type of particle. Charged-particle beams cannot remain focused over the distances required in space because the mutual repulsion of the charged particles is greater than the focusing effect of the magnetic fields generated by the beam. Charged-particle beams are further limited in space because of the deflection by Earth's magnetic field and because of the space charge that the accelerator develops outside the atmosphere.[3] Thus weapons systems based on streams of particles such as electrons can only be used within the atmosphere and over very short ranges. In the atmosphere, the beam becomes electrically neutralized by ionization of the ambient atmosphere; thus the magnetic field generated by the beam helps to focus it. One might be able to ameliorate the atmospheric attenuation of the beam by pulsing it: Each successive pulse might further evacuate a channel in which the next pulse could travel. Still, problems such as beam stability have created considerable doubt about whether particle-beam weapons can ever be used for ranges beyond a few kilometers in the atmosphere. The study of beam propagation in the atmosphere is one of the major items on the agenda of the particle-beams program.

Neutral beams might be used for weapons systems based in space. They would not be susceptible to the spreading caused by repulsive forces nor to the bending by the Earth's magnetic field. The major technical problems are to produce a neutral beam of low initial divergence and high current density, and to design a fire-control system that accounts for the inability to determine the error in directing the beam.[3]

The challenges posed by these various missions are being addressed in the technology programs sponsored by the Defense Advanced Research Projects Agency, the Air Force, the Army and the Navy. These organizations are exploring numerous laser concepts and have built several particle-beam accelerators as well as systems for beam and fire control. This spectrum of programs is closely coordinated by the Assistant for Directed Energy Weapons in the Office of the Under Secretary of Defense for Research and Engineering. The level of funding in FY 1983 is $375 million for the high-energy laser program and $47 million for the particle-beams program. The Department of Defense request for FY 1984 is $469 million and $50 million for the two programs, respectively.[1,2]

The high-energy laser program. DARPA has undertaken a three-pronged approach to study the practicality of a laser weapons system in space. The first component of its program, Alpha, is aimed at the fabrication and testing of a high-powered chemical laser. The second component, the large optics demonstration experiment, is directed at the problem of constructing the high-quality mirror. The third component, Talon Gold, is a planned spaceborne pointing and tracking experiment. The Air Force has an Airborne Laser Laboratory, which consists of a carbon dioxide laser mounted into a NKC-135 cargo aircraft. The Air Force has used this Laboratory to engage air-to-air missiles in flight. The Navy has also been conducting tests of laser systems, using chemical hydrogen or deuterium fluoride lasers. In a series of tests at San Juan Capistrano in 1978, a laser of this type together with a pointer–tracker successfully intercepted and destroyed some TOW antitank missiles. The Navy is planning another series of tests, called Sea Lite, in the mid 1980s at the High Energy Laser National Test Range now being developed at the White Sands Missile Range in New Mexico. From such programs, the Air Force and Navy should learn more about the damage to targets from lasers as well as about the behavior of specific system designs. The Army is interested in a laser as a possible weapon in the forward-area battlefield.

The key ingredient in all these programs is, of course, the laser. We spoke about some of the laser research with Douglas Tanimoto (Maxwell Labs), who was formerly head of the directed-energy weapons program at DARPA. In addition to developing high-power chemical lasers (the quantitative power goal is classified), the advanced research is directed to short wavelengths, to limit the restrictions imposed by diffraction spreading of the beam. In space, a laser beam reflected from a mirror has an angle of divergence that varies directly with the ratio of its wavelength to the diameter of the mirror. To keep the size of the mirror as small as possible, the laser wavelength should be as short as possible. Another advantage of the shorter wavelengths is that the radiation couples better to metal targets. The other side of the coin is that the specifications on the optical quality of the reflecting surface become more stringent for shorter wavelengths.

The first high-power laser was the carbon-dioxide laser, which radiates at 10.6 microns. In this gas-dynamic laser, the population inversion results from the rapid expansion of a gas. In a chemical laser, the molecules are formed in the excited state and held there long enough for the lasing to occur. One common type of chemical laser is the hydrogen fluoride laser, which operates at a wavelength around 2.7 microns. Using deuterium instead of hydrogen changes the wavelength to about 3.8 microns, making it less susceptible to atmospheric absorption.

Sandia's radial line accelerator (RADLAC) and Livermore's Advanced Test Accelerator are exploring charged-particle beam technology. In this photo of RADLAC, a generator charges an intermediate storage capacitor (left) to energize the injector. Accelerating voltage is applied synchronously by radial pulse lines (right rear). The accelerating gradient is 5 MeV/meter.

The Navy and the Army appear interested in both HF and DF lasers. A general review[6] of laser and particle-beam weapons in *Aviation Week and Space Technology* pointed out that because the toxic gases pose a potential health hazard for the Army's land-based missions, the Army has developed a chemical pump to operate the laser in a closed cycle. Another type of laser is the electric-discharge laser in which an electron beam imparts its energy to excite the gas molecules to vibrational states.

Two more advanced concepts are also being studied within the directed-energy weapons program. They are the excited dimer (excimer) laser and the free-electron laser. In the excimer laser, electrons bombard a noble gas such as xenon or krypton and strip the outer electrons. The resulting ions, in their excited state, react chemically to form such compounds as xenon fluoride or krypton fluoride, which constitute the inverted population for lasing. The wavelength of excimer lasers is in the submicron region.

The free-electron laser is more like a traveling-wave tube than a laser. It consists of a beam of relativistic electrons sent into a wiggler magnet, which produces a static transverse magnetic field whose value varies with distance in a periodic fashion. The motion of the electron in the periodic field produces electromagnetic radiation whose wavelength equals that of the wiggler divided by a factor proportional to the square of the kinetic energy of the

electrons. Therefore, if one uses high-voltage accelerators to produce the beam, not only can the wavelength be short but it can be tunable. When there is an initial field of the appropriate frequency, the free-electron laser acts as an amplifier.

We spoke with Charlie Brau (Los Alamos National Lab) about the experiments with free-electron lasers there. About a year ago they operated such a device and claim to have demonstrated an efficiency of 3.5% for conversion of energy from the electron beam to the laser radiation. With an energy-recovery scheme they have devised, they hope to reach an efficiency of 20%. Many other labs are also working on free-electron lasers.

The newest laser on the horizon is the x-ray laser. Lawrence Livermore Laboratory may have tested an x-ray laser that was pumped by a small nuclear explosion (PHYSICS TODAY, June, page 61). Apparently the idea is to have one such explosion power numerous x-ray lasers in space, all pointed at different targets.

Particle-beam weapons. Particle beams can transfer a larger fraction of their energy to the target in a given amount of time than laser beams, which are subject to reflections and other interactions with the surface. Particle beams deposit their damaging energy further beneath the surface than do lasers and would thus be more effective for damaging internal components but less efficient for heating the outer surface of a target. The techno-

logy for particle-beam weapons lags behind that of laser beams by a considerable extent. While the US has extensive experience in accelerator design, the requirements for a military application are considerably different.[2] The particle-beam weapon facility must deliver a high current at a rapid pulse rate.

One of the major components of the program in charged-particle beams is the Advanced Test Accelerator, completed in Fall 1982 at Livermore (PHYSICS TODAY, February 1982, page 20). This machine is an induction linear accelerator that is designed to produce 50-MeV electrons at a current of 10 kiloamps. Its main purpose is to test the propagation of a pulsed charged-particle beam in the atmosphere. A second charged-particle accelerator, called RADLAC, is being operated by the Air Force Weapons Laboratory and Sandia National Laboratory; it uses pulsed power-transmission lines to apply voltages to the linear accelerating gaps. This design lends itself to very high accelerating gradients—between 3 and 5 MeV per meter—but cannot be pulsed rapidly. Like ATA, it is being used for studies of atmospheric propagation.

The neutral-beam program, called the White Horse program, is centered at Los Alamos. Neutral beams of hydrogen are produced by accelerating negatively charged hydrogen ions and stripping them of their charge at the end. The heart of the Los Alamos program at this time is a radiofrequency quadrupole preaccelerator that forms micropulses at 420 Hz acceptable for input into a drift-tube linac (not yet on line). Harald Dogliani, program director of White Horse, described the design of the rf quadrupole to us by comparing the vanes of the quadrupole to bread knives: Each "blade" is longitudinally parallel to the accelerator axis and set at right angles to the others; the serrated edges point into the axis and modulate the electric field, which accelerates and focuses the beam. The modulator produces both beam bunching and acceleration in a relatively compact space. The quadrupole is 2.9 m long and can accelerate a beam up to 2 MeV. The Los Alamos team based their designs for both the ion source and the rf quadrupole on concepts that were first described in the open Russian literature. Pierre Grand (Brookhaven) told us that V. V. Vladimirskij and I. M. Kapchinskij (Institute for Theoretical and Experimental Physics, Moscow) hold a Soviet patent for the rf quadrupole.

Dogliani said the purpose of the program is to determine the military feasibility of particle-beam accelerators; the potential applications are for ballistic-missile defense or for antisatellite weapons. The White Horse group is striving to obtain a high-intensity beam with a low effective temperature. The effective temperature is a measure of the divergence of the beam.

Controversy over the weapons. Does the idea of "bullets" fired by laser or particle-beam "guns" make any sense? Critics have often made assessments of the weapons and their missions and have concluded that they are not practical. For example, Richard Garwin (IBM) has estimated that the ballistic-missile defense by lasers based in space would require an increase in laser brightness of at least a factor of 10^6 over what had been demonstrated in any ground-based system.[7] For an antisatellite beam weapon in synchronous orbit he calculated that the energy required would be some 80 tons of fuel (with concomitant transportation demand of at least 8000 tons of rocket fuel). In response, defenders of the directed-energy weapons program state that such calculations of the numerical requirements of the mission are too often based on numbers that err by orders of magnitude on the conservative side. Tanimoto noted that, for instance, estimates of the required laser brightness have ignored the great strides that have been made in pointing stabilities. He mentioned that the Soviet Union has a considerable effort in directed-energy weapons technology; he feels that the US also needs to move ahead with the technology to understand better its limits and capabilities.

Beyond the question of whether an individual directed-energy weapon can now or in the future destroy its assigned target lies the larger question of whether an entire defense system based on such weapons can be effective against the full range of attacking missiles. To many critics, these capabilities are separated by a large gap. Wolfgang Panofsky (SLAC) commented to us that a defense system costs too much if the other side can defeat it for less (and it might simply fuel the arms race in the process). The offense has available a wide variety of relatively inexpensive countermeasures to any directed-energy defense. For example, the offense might coat all satellites or launchers with highly reflective material to minimize absorption of laser energy, use decoys to confuse the fire-control systems, or jam the communications links to the space-based weapons. The ultimate countermeasure would be the destruction of the space-based weapon itself with space mines. Panofsky said that in the past the cost of defending a country against nuclear attack has always been much larger than the price of countermeasures to nullify that defense; he knows of no studies indicating that a directed-energy defense would be any different.

A final concern of critics of the directed-energy weapons is that they may represent the introduction of weapons into space. Garwin has pointed out the valuable and stabilizing roles served by our present communications and surveillance satellites. He and others would not like to see those satellites threatened by introduction of destructive machines into space. Garwin and Carl Sagan (Cornell) prepared a "Petition For a Ban on Space Weaponry," which was signed by more than 40 scientists and strategists. They presented the petition to Congress in May. At the same time, Garwin, together with Kurt Gottfried (Cornell), presented a draft treaty prepared under the auspices of the Union of Concerned Scientists that calls for a ban on the testing and use of weapons in space.   —BGL

References

1. Fact Sheet, DOD High Energy Laser Program, February 1983.
2. Fact Sheet, DOD Particle Beam (PB) Technology Program, February 1983.
3. G. Bekefi, B. T. Feld, J. Parmentola, K. Tsipis, Nature 284, 219 (1980).
4. E. Teller, in testimony before subcommittee on Strategic and Theater Nuclear Forces of the Committee on Armed Services, US Senate, 2 May 1983.
5. M. Callahan, K. Tsipis, "High Energy Laser Weapons: A Technical Assessment," Program for Science and Technology for International Security Report No. 6, MIT, November 1980.
6. "Technical Survey: Particle Beam, Laser Weapons," Aviation Week and Space Technology, 28 July 1980, page 32; 4 August 1980, page 44.
7. R. L. Garwin, Bull. At. Sci., May 1981, page 48.

New sources of high-power coherent radiation

**Free-electron lasers and cyclotron-resonance masers
show considerable promise for producing previously unattainable levels of power
at wavelengths ranging from millimeters to the ultraviolet.**

PHYSICS TODAY/MARCH 1984

Phillip Sprangle and Timothy Coffey

Recent progress in novel techniques for generating high-power coherent radiation promises to make available sources with a variety of new and exciting applications. Interestingly, the new techniques have more in common with those used in the earliest sources of coherent radiation—the various microwave generators—than with those used in the more recent optical lasers. Development of new sources based on these techniques is proceeding rapidly at research centers around the world, because the new sources have a great potential for extending the currently available range of wavelengths and levels of power, while maintaining high operating efficiencies. The areas of application that stand to benefit include spectroscopy, advanced accelerators, short-wavelength radar, and plasma heating in fusion reactors.

Conventional sources of coherent radiation, such as the magnetron, the klystron and the traveling-wave devices, have limited power output and efficiency at short wavelengths. To circumvent these limits, researchers have proposed many new concepts and mechanisms, as well as variations on the conventional approaches. Two types of sources that were first demonstrated around 1960 are currently the focus of much attention: free-electron lasers and cyclotron-resonance masers, both of which are powered by relativistic electron beams.

We begin this article with a brief description of the physical mechanism of these and other novel sources of radiation. Then we look at some of the present and future areas of application and give an overview of the relevant experimental programs. Free-electron lasers offer operating efficiencies of

over 20% at millimeter wavelengths, and one can extend their operation to ultraviolet wavelengths while maintaining relatively broad tunability. Figure 1 shows free-electron-laser radiation at visible wavelengths. Cyclotron-resonance masers offer high efficiency and high power at centimeter to millimeter wavelengths.

Physical mechanisms

The terms free-electron laser and cyclotron-resonance maser refer to mechanisms. Each denotes a wide class of coherent sources that operate over a wide range of wavelengths, the words maser and laser having long lost their original limitations to microwaves and light. Although amplification by stimulated emission of radiation is fundamentally quantum mechanical, classical models are sufficient to understand both free-electron lasers and cyclotron-resonance masers.

Electrons radiate when they are accelerated. When an electromagnetic field of proper polarization and phase is imposed on a beam of electrons, the electrons will accelerate in such a way as to radiate coherently. The condition for coherence is that the radiation from the electrons reinforce the original imposed electromagnetic field. The electrons must be moving initially because they radiate at the expense of their kinetic energy. We call a coherent source an amplifier if the imposed field is from an external source, and an oscillator if the imposed field is generated internally by spontaneous radiation from individual electrons.

The free-electron laser, such as the one shown in figure 2, consists of an electron beam, an external "pump field" and the imposed radiation field.[1-4] The pump field, typically a static periodic magnetic field, can be any field that causes the moving electrons to oscillate transversely. Although the basic mechanism of emis-

sion does not rely on relativistic effects, the electrons must be highly relativistic to produce short-wavelength radiation.

The radiation wavelength in the free-electron laser, unlike in most conventional sources, is not fixed by the size of the structure. Furthermore, the lasing medium, being a pump field, cannot break down. Hence, in principle, large structures can generate short wavelengths at high power levels.

We will consider only the common pump field: a static periodic magnetic "wiggler" with its primary field component transverse to the electron and radiation beams, as shown in figure 3. Injected monoenergetic electrons stream through the periodic magnetic field and wiggle, or oscillate transversely, in the same direction as the radiation electric field, and thus can lose energy to the radiation field or gain energy from it. At the point of injection, the electrons are randomly phased and radiate incoherently, generating spontaneous bremsstrahlung radiation. However, the so-called "ponderomotive" wave produced by the beating of the wiggler and radiation fields bunches the electrons within the interaction region and generates coherent radiation in the process, as we will see. The ponderomotive, or trapping wave, originates from the $\mathbf{v} \times \mathbf{B}$ force on the electrons and causes them to bunch in the axial direction. The longitudinal ponderomotive wave, which excites a density wave, acts like the slow-traveling electromagnetic wave in conventional traveling-wave sources. The ponderomotive wave bunches the electrons by decelerating some and accelerating others. If the axial velocity v_0 of the electrons is slightly greater than the velocity of the ponderomotive wave, the average energy of the electrons decreases and the radiation field is enhanced. An excessive spread in the velocity of the electrons can greatly

Phillip Sprangle is head of the plasma theory branch, and Timothy Coffey is director of research, at the Naval Research Laboratory, in Washington, D.C.

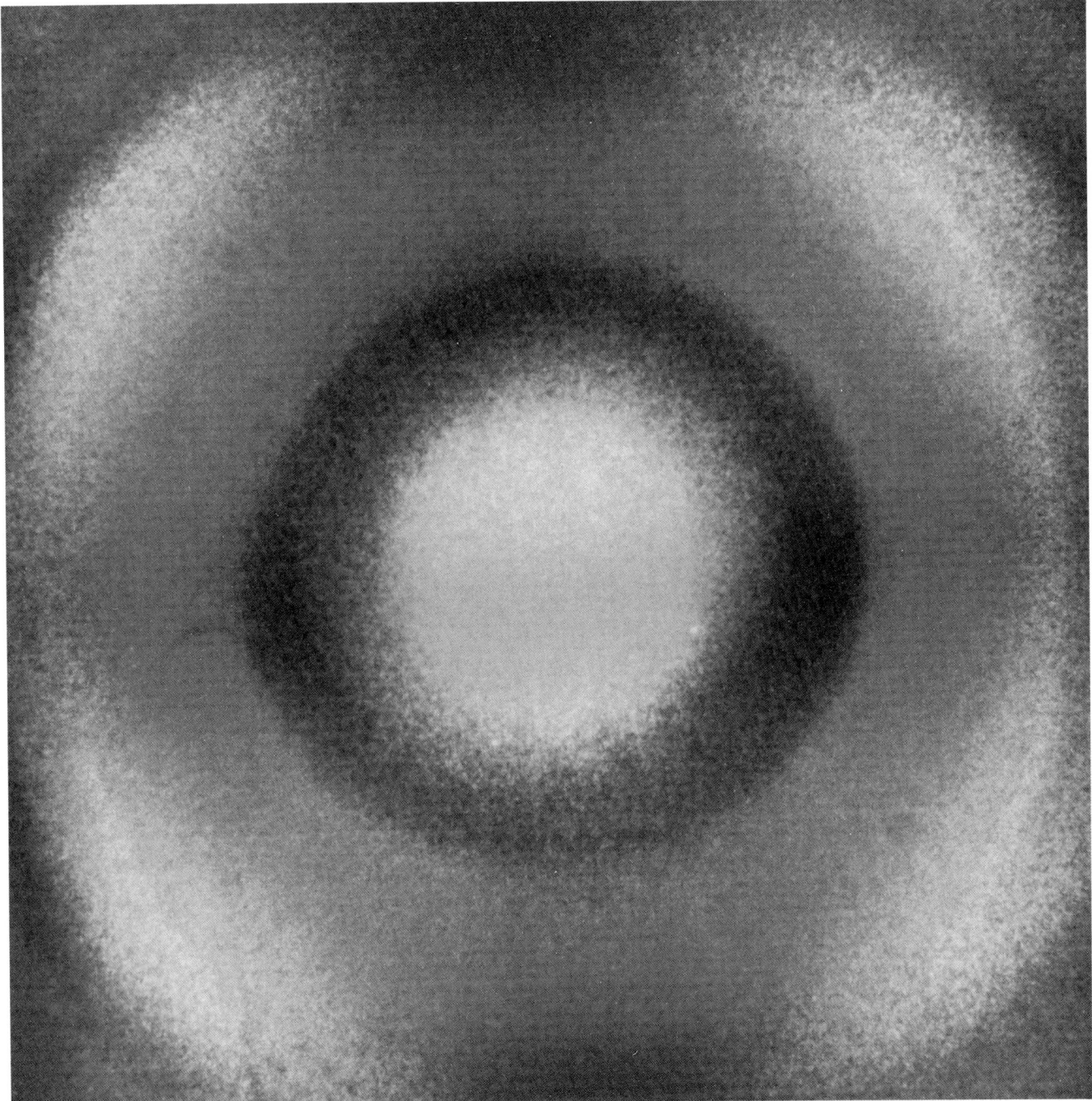

Far-field visible spontaneous radiation pattern from a free-electron laser. The laser used a wiggler magnet and was powered by electrons from the ACO storage ring at Orsay, France. The emission wavelength varies as the point of observation shifts off the beam axis. Figure 1

reduce the bunching and the extraction of energy, especially at shorter wavelengths.

The $\mathbf{v} \times \mathbf{B}$ force that gives rise to the ponderomotive wave involves the electron wiggle velocity v_w, which is typically much less than the axial velocity v_0, and $\mathbf{B}_R$, the strength of the radiation magnetic field. The radiation frequency and wavenumber are related by the vacuum relation $\omega = ck$. The phase velocity v_{ph} of the ponderomotive wave is $\omega/(k + k_w)$, which is about equal to the electrons' axial velocity v_0; here k_w is the wiggler wavenumber. For the bunched streaming electrons, sychronism requires that the radiation

frequency satisfy the relation

$$\omega = v_0 k_w / (1 - v_0/c)$$

For highly relativistic electrons the radiation wavelength λ is approximately $\lambda_w / 2\gamma_0^2$, where γ_0 is the relativistic gamma factor $(1 - v_0^2/c^2)^{-1/2}$ and λ_w is the wiggler wavelength shown in figure 3. The radiation wavelength is much smaller than the wiggler wavelength, and can be varied by changing the energy of the electron beam.

One alternative configuration features an additional magnetic field, oriented longitudinally.[5] As we will see, this hybrid-field arrangement is one of the basic features of the cyclo-

tron-resonance maser.

The nature of the electron-beam source and the way the electrons interact divide free-electron lasers into categories.[1-4] Free-electron lasers based on low-current beam sources such as rf linacs, microtrons or storage rings usually operate in the "Compton regime," in which primarily single particles interact, so that we can neglect collective or space-charge effects. In this regime the radiation gain—the growth rate along the interaction length—typically is low, so that practical radiation sources would be oscillators, which do not require high gain. Without enhancement, operating effi-

Free-electron-laser experiment employing a radio-frequency linear accelerator and a tapered wiggler. Here we see equipment that measures the energy spread in the electron beam (right), the tapered wiggler (center) and a time-resolved electron spectrometer (left). (Photograph from Mathematical Sciences Northwest Inc. and the Boeing Aerospace Company.)   Figure 2

ciencies are generally less than 1%. With high-energy electron beams of high quality, or small spread in energy, free-electron lasers in the Compton regime can operate at optical or ultraviolet wavelengths.

Other free-electron lasers are based on intense relativistic electron beams from sources such as induction linear accelerators or pulsed transmission-line accelerators, which Hans Fleischmann has described[6] in these pages. These lasers operate in the "Raman regime," in which collective effects influence the radiation growth rate and the interaction efficiency. Nevertheless, the operating wavelength λ remains at about $\lambda_w/2\gamma_0^2$, as in the Compton regime. Pulse-line-generated beams from plasma-induced field-emission diodes typically have relatively flat voltage and current pulses that last a few tens of nanoseconds. The electron beam's low quality and low energy—typically in the MeV range—limit the free-electron laser to millimeter wavelengths. But beam currents in the kiloamperes allow the laser to operate as an amplifier.

A third operating regime, known as the high-gain Compton regime, has features of both the Compton and Raman regimes. Here the wiggler field is so strong that the ponderomotive force on the electrons is dominant over the space-charge forces, and the radiation gain is large.

Cyclotron-resonance masers are far more developed than free-electron lasers and are among the most efficient devices for generating coherent high-power radiation at centimeter and mil-

limeter wavelengths.[7] The mechanism was proposed independently by Richard Q. Twiss, Jurgen Schneider, Andrei Gapanov and Richard Pantell in the late 1950s. Their early theoretical studies demonstrated that relativistic effects associated with monoenergetic electrons gyrating in a magnetic field could result in stimulated cyclotron emission rather than absorption. The first clearly defined experimental confirmation of the mechanism was reported by Jay Hirschfield and Jonathon Wachtel in 1964. Devices based on this mechanism, whether oscillators or amplifiers, are referred to as gyrotrons.

Scientists in the Soviet Union developed the gyrotron concept into a practical source of radiation during the 1960s and 1970s, primarily at Gorkii State University.[8] In the 1970s there were also major advances in the United States at the Naval Research Laboratory as well as at MIT, Yale University, Varian and Hughes. The demonstrated efficiencies and power levels in the millimeter regime are impressive. The Gorkii group, for example, as early as 1975 developed a 22-kW cw oscillator that produced 2-mm radiation with a 22% efficiency.

At the heart of the cyclotron-resonance maser is a beam of nearly monoenergetic electrons streaming along and gyrating about an external magnetic field $B_0\mathbf{e}_z$, as figures 4 and 5 indicate. An imposed electromagnetic field of the form $\mathbf{E} = E_0\cos(\omega t)\mathbf{e}_y$, closely approximates the field of the transverse-electric mode of a cavity or waveguide when ω is near one of the cutoff

frequencies of the structure. This allows the electrons to radiate coherently. The electrons behave as individual oscillators in the cavity, gyrating about the magnetic field B_0 with a rotation frequency $\Omega_R = \Omega_0/\gamma$, where Ω_0 is the nonrelativistic cyclotron frequency eB_0/m_0c and γ is the usual relativistic mass factor calculated from the transverse electron rotation velocity $v_\perp$.

To understand the process of amplification, we will for simplicity consider only eight electrons, initially distributed uniformly and rotating in clockwise circular orbits as shown in figure 5a. With the radiation field polarized initially as shown, and with the electron orbiting frequency Ω_R slightly lower than the radiation frequency ω, those particles in the upper half plane ($x > 0$) will move closer to resonance with the radiation field and, therefore, lose energy to that field and increase their rotation frequency. Those in the lower half plane will move farther from resonance and hence gain energy and decrease their rotation frequency. The overall result is known as "phase bunching"; viewed after an integral number of wave periods $2\pi/\omega$, most electrons will be in the upper half plane, losing energy and amplifying the field. This mechanism requires only that the radiation frequency slightly exceed the rotation frequency, that the rotation frequency is energy dependent, and that all the electrons have roughly the same transverse rotation velocity $v_\perp$. The cavity length is chosen such that the electrons exit the cavity when their average energy is a minimum.

High-efficiency operation requires a large ratio of transverse to longitudinal electron velocity $v_\perp/v_z$. This ratio is typically between 1 and 3, and demonstrated efficiencies are as high as 60%.

Potential applications

Ultimately, the importance of new sources of coherent radiation will be determined more by their utility than by their novelty. Free-electron lasers and cyclotron-resonance masers have exciting potential applications as sources for spectroscopy, accelerators, radar and plasma heating.

Spectroscopy. A National Academy of Sciences study concludes[9] that the free-electron laser is a promising source for spectroscopy at far-infrared wavelengths greater than 25 microns, and at ultraviolet wavelengths less than 200 nm. This laser's coherence, narrow bandwidth, tunability and stable high power would be especially important in condensed-matter physics and surface chemistry, and in the spectroscopy of atoms, molecules and ions. The short time duration and thus high peak power available from some free-electron-lasers would allow

Table 1. Free-electron-laser experiments

Using rf linacs and microtrons

Laboratory	Class	Wavelength (microns)	Beam energy (MeV)	Peak current (A)
Stanford U.	Amplifier	10.6	24	0.1
Stanford U.	Oscillator	3.3	43	1.3
Los Alamos	Amplifier	10.6	20	10
Los Alamos	Oscillator	10.6	20	30–60
Mathematical Sciences Northwest/Boeing	Amplifier	10.6	20	5
TRW	Amplifier	10.6	25	10
TRW/Stanford U.	Oscillator	1.6	66	0.5–2.5
NRL	Oscillator	16.0	35	5
Bell Labs*	Amplifier	100–400	10–20	5
Frascati*	Amplifier	16	20	0.6

Using pulse-line-generated beams**

Laboratory	Peak power (MW)	Wavelength (mm)	Beam energy (MeV)	Beam current (kA)
NRL	1	0.4	2	30
NRL	35	4	1.35	1.5
NRL/Columbia U.	1	0.4	1.2	25
Columbia U.	8	1.5	0.86	5
Columbia U.	1	0.6	0.9	10
MIT	1.5	3	1	5
Ecole Polytechnique	2	2	1	2

Using electrostatic and induction linacs

Laboratory	Accelerator	Wavelength (mm)	Beam energy (MeV)	Peak current (A)
UCSB	Electrostatic accelerator	0.1–1	6	2
UCSB	Electrostatic accelerator	0.36	3	2
NRL	Induction linac	8	0.7	200
LLNL	Induction linac	3–8	4	400

Using storage-ring beams

Laboratory	Storage ring	Wavelength (microns)	Beam energy (MeV)	Beam current (A)	Gain per pass (%)
Orsay	ACO	0.5	240	2 (peak) 0.03 (average)	0.07 (measured)
Frascati	ADONE	0.5	600	10 (peak) 0.1 (average)	0.02 (measured)
Novosibirsk	VEPP-3	6	340	20 (peak)	0.4 (measured)
Brookhaven	VUV	0.35	500	108 (peak) 1.0 (average)	2 (calculated)
Stanford U.	ARRL (planned)	0.5	1000	200 (peak) 1.0 (average)	—

All entries in tables 1 and 2 represent typical values.
*Microtron beam source.
**Typical pulse times are tens of nanoseconds.

important new applications of radiation ranging in wavelength from 25 to 1000 microns. Pulses as short as tens of picoseconds could probe the dynamics of charge carriers in semiconductors and the dynamics of phonons, plasmons and superconducting gaps. High-power tunable picosecond pulses at wavelengths under 200 nm would substantially strengthen studies of fast chemical kinetics, photochemistry and vibrational relaxation processes that involve more than one photon.

Accelerators. As noted[10] at a recent workshop on laser acceleration of particles, affordable high-power sources of centimeter waves could lead to shorter high-energy accelerators. Conventional rf accelerators use microwave klystrons that generate about 25 MW of peak power. Recent developments indicate that free-electron lasers and cyclotron-resonance masers could generate gigawatts. These higher powers would mean fewer power tubes and possibly lower total cost. Researchers must resolve practical and scientific questions, however, before they can demonstrate that these new sources would deliver this power with acceptable efficiency and stability.

It may be possible to accelerate particles by reversing the dynamics of the free-electron laser.[10] The electric field of an intense laser beam, such as from a CO_2 laser, together with a wiggler could produce a large-amplitude ponderomotive wave to trap and accelerate electrons. One could energize the trapped electrons by increasing the wiggler field's period or amplitude or both. Accelerating gradients could exceed 100 MeV/m, but laser-beam diffraction would limit the acceleration length, so that electrons would gain at most a few GeV in a single stage. A major question is how to refocus the laser beam for multistage acceleration.

Radar. Most radar operates at microwave frequencies primarily because centimeter-wave power tubes and components are available and atmospheric losses are low. The free-electron laser and cyclotron-resonance maser promise millimeter-wave radar. Atmospheric absorption, although generally higher at millimeter wavelengths, has minima at 35, 94, 220 and 325 GHz. Relative to conventional microwave radar, millimeter-wavelength radar would have narrow beamwidths, large bandwidths and small antennas. Narrow beams permit tracking at low angles of elevation. Large bandwidths enhance resistance to interfering signals—or to electronic countermeasures in the case of military radar—and permit high-range resolution. Millimeter waves are less affected by fog, clouds, rain or smoke than are optical or infrared waves.

A number of issues concerning millimeter-wave radar remain to be resolved. The typical cyclotron-resonance maser uses a high magnetic field and requires superconducting magnets. Free-electron lasers, even at millimeter wavelengths, are now too large for most radar applications, and their high operating voltages are also a problem. The lack of millimeter-wave components has been another practical problem, but these components are developing rapidly.

Fusion power. The problems of plasma heating still prevent practical magnetic-confinement-fusion power reactors. Practical high-power sources at millimeter wavelengths could solve some of these problems.

Recent experiments on the Oak Ridge ISXB Tokamak, using a 35-GHz cyclotron-resonance maser developed at the Naval Research Laboratory, demonstrated large absorption through electron-cyclotron resonance.[7] The ab-

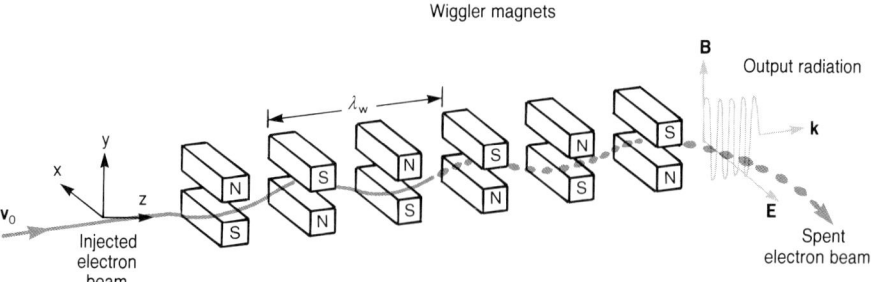

Basic components of a free-electron laser. The pump field is produced by a periodic arrangement of magnets—a "wiggler"—in which the electrons undergo transverse oscillations. The small transverse component of velocity, known as the wiggle velocity, is in the direction of the radiation electric field **E**, and can cause the electrons to lose energy and amplify the radiation field. The interaction between the electrons and the radiation field occurs over the entire length defined by the wiggler magnets.   Figure 3

sorption heated the electrons significantly, but because of the low plasma density, the ions, as expected, were not measurably heated. These results imply that high-power cyclotron-resonance masers can heat fusion-reactor plasmas at the required high densities and long confinement times. Free-electron lasers are expected to be less efficient than cyclotron-resonance masers in producing millimeter waves and thus less suitable for plasma heating.

The success of high-power sources of coherent-radiation in any of the potential applications discussed above would be an important development. However, as with any new technology, we can expect the unexpected, implying that we have yet to identify the most important applications.

Enhancing efficiency

In free-electron lasers operating in the Compton regime, a radiation gain per pass of only 0.1 and an intrinsic efficiency of only 1% are typical. In this regime the intrinsic efficiency is given by the reciprocal of twice the number of wiggles in the interaction

length. In the Raman regime, high gains are possible, with efficiencies as high as 15%.

One can increase operating efficiencies substantially either by converting the electron kinetic energy to radiation energy with greater efficiency or by recovering a portion of the electron kinetic energy after the electrons interact with the trapping wave. In principle, one can dramatically improve the efficiency with which the electrons transfer their energy to the wave, increasing it from about 1% to 20% in the Compton regime. One approach is to decrease gradually the trapping wave's phase velocity v_{ph}, which is $\omega/(k + k_w)$ or approximately $c(1 - \lambda/\lambda_w)$. By spatially decreasing the wavelength λ_w of the wiggler field, one decreases the phase velocity v_{ph}. In this approach, the electrons remain trapped and lose a large fraction of their kinetic energy to the radiation field.

Instead of decreasing the phase velocity v_{ph}—or in addition to doing so—one can apply a longitudinal accelerating force to the trapped electrons to enhance the efficiency. An external

Table 2. US experimental cyclotron-resonance masers

Oscillators

Laboratory	Frequency (GHz)	Power (kW)	Efficiency (%)	Pulse duration
Varian	28	340	45	Continuous
Varian	60	120	38	Continuous
NRL	35	340	54	1 μsec
MIT	140	180	30	1 μsec
Hughes	60	240	30	100 msec

Amplifiers

Laboratory	Frequency (GHz)	Power (kW)	Efficiency (%)	Pulse duration (microsec)	Bandwidth (%)
NRL	35	10	8	1.5	2–13
Varian	28	65	9	1000	1
Varian	5	120	26	50	6
Yale U.	6	20	10	1	11

uniform axial electric field can provide this accelerating force. A more practical approach is to decrease the spatial amplitude of the wiggler field. In either case, the resulting phase shift of the trapped electrons is such that they perform work on the trapping wave, enhancing the radiation. Norman Kroll of the University of California at San Diego and Marshall Rosenbluth of the University of Texas at Austin have found that instabilities can arise in the trapped-particle mode of operation, resulting in sideband radiation.[4]

In cyclotron-resonance masers, one approach to achieving higher than intrinsic efficiency is to contour the radiation field in the cavity by varying the radius of the cavity wall, as shown in the left half of figure 6. Electrons entering the cavity start phase bunching where the field amplitude is small. The electrons give up relatively little energy until they enter the high-field region. There they radiate more efficiently because they are highly bunched and closer to resonance.

In an alternative approach to efficiency enhancement, experimenters contour the external longitudinal magnetic field axially, as shown on the right side of figure 6. They make the magnetic field near the input smaller than normally required, allowing the electrons to bunch in phase without losing or gaining much energy. Again, as the phase bunching increases, the electrons go into a higher magnetic field and move closer to resonance. In addition, one can recover as much as 90% of the longitudinal energy of the spent electrons.

Experiments

Pioneering in free-electron-laser experiments in the Compton regime is a Stanford University group using the Superconducting Linear Accelerator.[11,12] One such experiment, headed

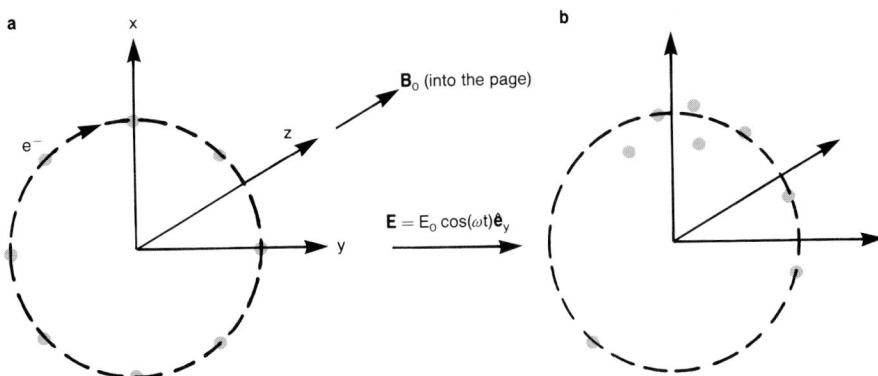

Phase distribution of gyrating electrons. This simplified representation shows the initial phase distribution (at **a** in figure 4) and the distribution after an integral number of wave periods (at **b** in figure 4).   Figure 5

by John Madey,[12] is sketched in figure 7 and listed second in table 1. The 43-MeV electron beam macropulses used in this experiment were 1.5 msec in duration and consisted of 1-mm-long micropulses spaced 25.4 m apart. Spontaneous incoherent radiation from the electrons built up into intense coherent radiation at 3.3 microns, because the gain of the free-electron-laser process peaks near that wavelength. Madey's group carefully separated the optical resonator mirrors so that the round-trip bounce time of the radiation pulses just matched the time between electron micropulses. In the presence of the electron micropulse, the radiation pulse velocity is slightly less than the velocity of light in a vacuum, an effect known as "laser lethargy."[4] In the Stanford experiment, the measured peak output power through a mirror of 1.5% transmittance was 6 kW; hence, the peak radiation power within the resonator was 400 kW. The measured linewidth $\Delta\lambda/\lambda$ of the saturated radiation was about 0.006. The 6% measured gain per pass was in fair agreement with the theoretical value of

about 10%.

At Los Alamos, Charles Brau and his coworkers are developing a highly efficient free-electron-laser oscillator source.[4,12,13] (See PHYSICS TODAY, August 1983, page 17.) This free-electron laser will employ an rf linac accelerator and radiate at 10.6 microns. The experimenters will vary the wiggler wavelength and amplitude spatially and recover part of the energy of the spent electron beam. They anticipate a 20% overall efficiency and an average output power of 100 kW.

At Mathematical Sciences Northwest and Boeing Aircraft, a group led by Jack Slater is developing[12,13] an optical free-electron-laser oscillator that will radiate at 0.5 microns. It employs a radio frequency linac beam with a peak current of 100 A.

Experimenters at the Naval Research Laboratory, in a recent free-electron-laser experiment[13] using an intense relativistic electron beam from a pulse-line generator, produced 35 MW of 4 mm radiation with 2.5% efficiency. (See table 1.) The energy spread of the injected electron beam was uniquely low. Experimental programs at Columbia University, MIT and Ecole Polytechnique are also employing high-current beams generated by pulse lines, as the list in table 1 indicates.[4]

At the University of California, Santa Barbara, Luis Elias and Gerald Ramian are conducting experiments[4,12] with a 6-MeV Van de Graaff accelerator to evaluate a dc energy-recovery scheme. (See table 1.) Their free-electron laser is designed to operate at 200 microns and achieve an output power of 12 kW.

Two free-electron-laser experiments powered by induction linacs are underway in the US, as table 1 indicates. Both operate in the high-gain regime. The Naval Research Laboratory's induction linac experiments, headed by Chris Kapetanakos and John Pasour, feature a uniquely long pulse duration

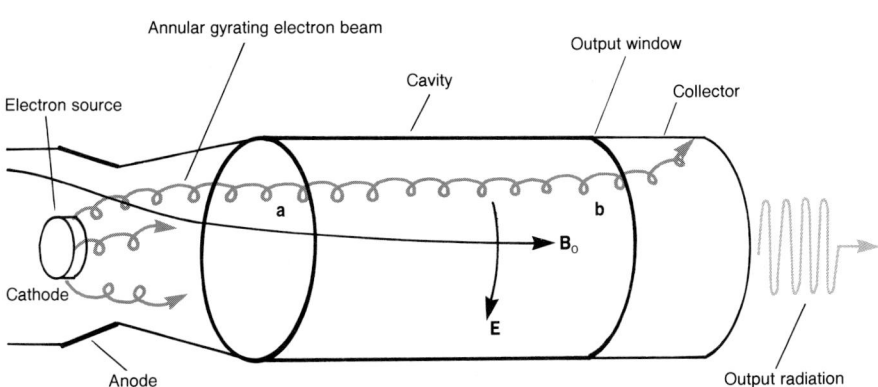

Cyclotron-resonance-maser oscillator, in schematic view. The electron source is a magnetron injection gun. The cathode emits an annular beam that gyrates about an applied magnetic field B_0 as it propagates through a cavity. The cavity operates in a transverse-electric mode near its cutoff frequency. The spent electron beam is collected, and radiation is emitted through an output window. Figure 5 compares the uniform electron phase distribution at **a** with the bunching at **b**.   Figure 4

of about 2 microsec, enabling the free-electron laser to operate as an oscillator. At present, this free-electron laser operates as a superradiant amplifier generating 4.2 MW at a wavelength of 8 mm and an efficiency of 3%. At Lawrence Livermore National Laboratory, experiments led by Donald Prosnitz and Andy Sessler use the laboratory's 5 MeV Experimental Test Accelerator. Because of this accelerator's short beam pulse of 30 nsec, the laser will operate as an amplifier.

A number of free-electron-laser experiments use electron storage rings (See table 1), with the wiggler being in one of the straight sections. One such storage-ring experiment, headed by Claudio Pellegrini[4] at Brookhaven National Laboratory, will operate at 500 MeV and a peak current of 108 A. The radiation gain should be a few percent at a wavelength of 0.35 microns.

Future direction of research. Charles Roberson, of the Office of Naval Research, and his coworkers suggest[4] the possibility of powering free-electron lasers with intense cyclic electron beams generated by racetrack induction accelerators or modified betatrons. Such sources, however, are still in a proof-of-principle stage of development.

Because wiggler wavelengths are typically at least a few centimeters, optical free-electron lasers require electron beams with energies of at least 50 MeV. Beam energies could be lower with use of a high-frequency electromagnetic pump field, such as an intense laser beam or the output of another free-electron-laser. With a CO_2 laser pump and a 1-MeV electron beam, a free-electron laser could in principle radiate at optical frequencies.

Another interesting possibility for avoiding high beam energies is a two-stage free-electron laser using a single electron beam. The radiation produced in the first stage, which would employ a wiggler field, would become the pump field for the second stage. However, in this scheme, and in the scheme using lasers to generate an electromagnetic pump field, the gain per pass is low, and because beams with extremely low energy spreads are necessary, the trapping efficiency is low.

The electron beam from the Advanced Test Accelerator at Livermore is expected to produce 500-GW pulses of electrons, which could, in principle, generate tunable multigigawatt pulses of radiation at near-optical frequencies.

Maser experiments. Experimenters commonly use magnetron injection guns to produce electron beams for cyclotron-resonance masers (see figure 4). These thermionic sources can generate several amperes and electron energies as high as 100 keV.

To generate millimeter-wavelength

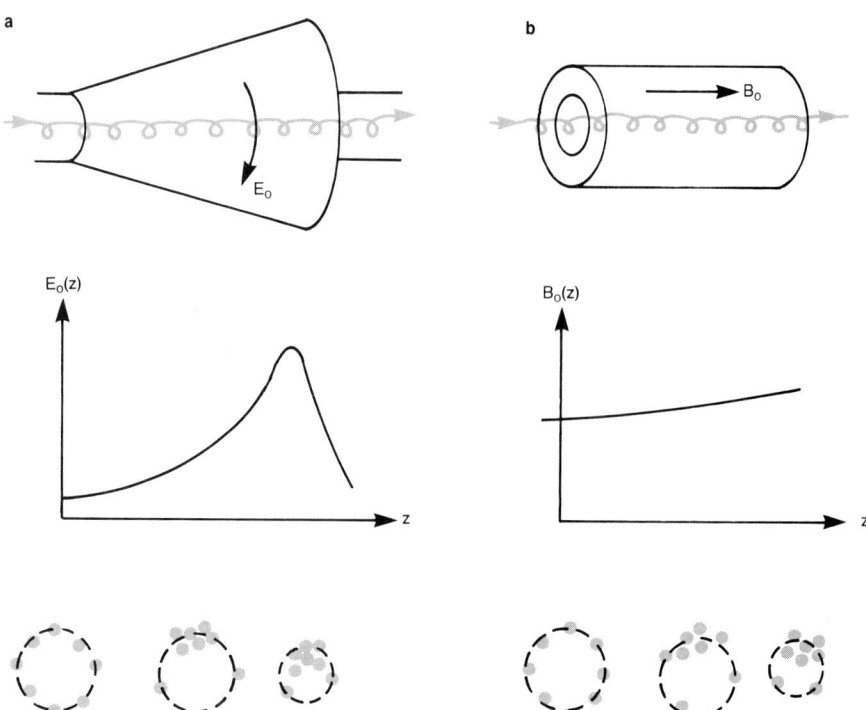

Efficiency enhancement methods used in cyclotron-resonance-maser oscillators. These are the two most common enhancement techniques. In **a**, the radius of the cavity wall increases longitudinally so that the field in the cavity varies as shown. In **b**, the radius of the cavity wall is constant, but the longitudinal external magnetic field varies. The diagrams at the bottom show the electron phase distributions at various points within the cavities.   Figure 6

radiation with cyclotron-resonance masers, experimenters usually use superconducting sources for the magnetic field. At the fundamental cyclotron harmonic, one needs a 34-kG magnetic field to generate radiation at 94-GHz, or 3 mm. It is possible to overcome the need for superconducting magnets in the generation of millimeter waves by operating at higher cyclotron harmonics, because the required magnetic field is reduced by a factor approximately equal to the harmonic number. The efficiency at the second harmonic remains high and with some designs can be higher than at the fundamental frequency. Generally, however, the efficiency falls sharply beyond the second harmonic.

An example of the state of the art in high-power devices comes from Soviet scientists at Gorkii State University,[14] who have generated 1.25 MW of 45-GHz (6.7 mm) radiation with a pulse duration of 1 to 5 msec and have produced 1.1 MW of 100-GHz (3.0 mm) radiation with a pulse duration of 100 microsec. Both of their oscillators operated at the fundamental cyclotron harmonic with efficiencies of 34%. Another impressive accomplishment of the Gorkii group[15] is a 120-kW cyclotron-resonance maser operating at 375 GHz (0.8 mm) with pulse durations of 0.1 msec. Recently, in the US, Richard Temkin and his coworkers at MIT achieved impressive power levels of over 180 kW at 140 GHz.

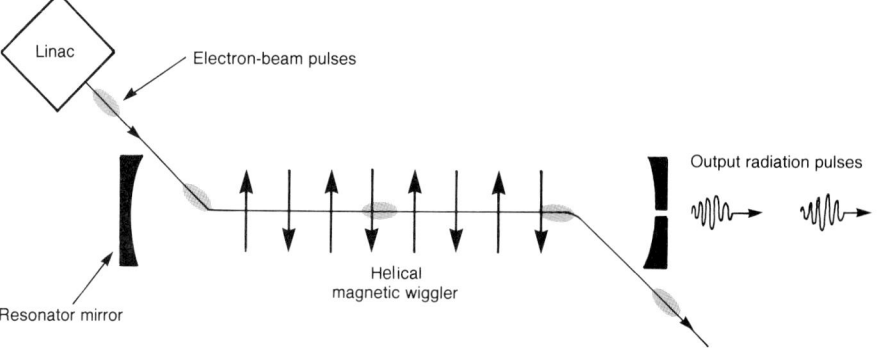

Oscillator using a pulsed electron beam. This is a schematic diagram of a typical free-electron-laser oscillator. The source is a radio frequency linear accelerator. The spacing of electron pulses and the length of the resonator are such that the reflecting pulses of radiation are synchronized with the incoming beam pulses. The mirror on the right is partially transmitting, so that a small fraction of the radiation pulses transiting the resonator can escape. The wiggler magnets are 3.3 cm apart over a total distance of 5.3 m. The magnetic field is 2.3 kG.   Figure 7

To increase the cavity volume and thereby avoid excessive thermal loading and competition among modes, high-power millimeter-wave cyclotron-resonance masers will have to operate in highly "overmoded" cavities, that is, in cavities that support many frequency modes at the same time. Recent results[16] show that one can stabilize highly overmoded cyclotron-resonance masers by adding a small prebunching cavity in front of the large energy-extraction cavity.

Prospects for practical cyclotron-resonance amplifiers are great. Researchers at the Naval Research Laboratory, for example, have achieved[7,17] impressive gains of 18 to 56 dB over large useful bandwidths, this at a frequency of 35 GHz (8.5 mm) and a typical power of 10 kW. Table 2 highlights experimental results in the United States with cyclotron-resonance oscillators and amplifiers.

Replacing the cavity with an open resonator,[18] allows one to operate at submillimeter wavelengths, to select modes and to handle extremely large powers. Preliminary experiments[19] at Yale University on this approach are encouraging.

Other novel sources. Many groups are actively pursuing other concepts for producing high-power radiation. One concept is the nonisochronic reflecting electron system. Here a high-current beam forms a virtual cathode; the emitted electrons oscillate between the actual and virtual cathodes, bunch in phase and generate radiation copiously. Our own experiments[20] on this source, which is compact, tunable and simple, have produced over 100 MW of 3-cm radiation.

Coherent Cherenkov radiation is a less novel but interesting source of millimeter waves. Some experiments produce this radiation by directing a relativistic electron beam along a dielectric surface. In one such experiment[4] at Dartmouth College, John Walsh, Kevin Felch and their coworkers achieved efficiencies of 10% and power levels of 100 kW at a wavelength of 4 mm.

One novel relativistic magnetron can generate unprecedented levels of coherent radiation at centimeter wavelengths. In experiments[21] headed by George Bekefi at MIT, a relativistic magnetron driven by an electron beam from a pulse line produced microwaves at a power level of 10 GW.

The history of science shows that applications, technology and theoretical concepts usually evolve together. In the case of high-power sources of coherent radiation, we are now witnessing rapid evolution in all these areas. Our ultimate understanding and the most important applications probably remain to be seen.

★ ★ ★

We gratefully acknowledge assistance from Cha-Mei Tang in preparing this article.

References

1. N. M. Kroll, W. A. McMullin, Phys. Rev. A **17**, 300 (1978).
2. A. A. Kolomenskii, A. N. Lebedev, Sov. J. Quantum Electron. **8**, 879 (1978).
3. P. Sprangle, R. A. Smith, V. L. Granatstein in *Infrared and Millimeter Waves*, Vol. 1, K. J. Button, ed., Academic, New York (1979).
4. *Free-Electron Generators of Coherent Radiation*, Physics of Quantum Electronics series, S. F. Jacobs, H. S. Pilloff, M. Sargent III, M. O. Scully, R. Spitzer, eds., Addison-Wesley, Reading, Mass. (1980), volumes 7, 8 and 9.
5. R. Davidson, W. McMullin, Phys. Fluids **26**, 840 (1983).
6. H. Fleischmann, PHYSICS TODAY, May 1975, page 35.
7. V. L. Granatstein, M. E. Read, L. R. Barnett in *Infrared and Millimeter Waves*, vol. 5, K. J. Button, ed., Academic, New York (1982).
8. IEEE Trans. Microwave Theory Tech. (special issue) **MTT-25**, No. 6 (1977).
9. *The Free Electron Laser*, the report of the free-electron-laser subcommittee of the Solid State Sciences Committee, National Academy of Sciences, National Academy Press, Washington, D.C. (1982).
10. P. J. Channell, ed., *Laser Acceleration of Particles*, AIP Conf. Proc. No. 91, Am. Inst. Phys., New York (1982).
11. L. R. Elias, W. M. Fairbanks, J. M. J. Madey, H. A. Schwettman, T. I. Smith, Phys. Rev. Lett. **36**, 717 (1976).
12. Bendor Free Electron Laser Conf., Journal de Physique **44**, C1 (1983).
13. IEEE J. Quant. Electron. **QE-19**, a special issue on free-electron lasers (1983).
14. A. A. Andronov, V. A. Flyagin, A. V. Gaponov, A. L. Goldenberg, M. I. Petelin, V. G. Usov, V. K. Yulpatov, Infrared Physics **18**, 385 (December 1978).
15. V. A. Flyagin, A. G. Luchinin, G. S. Nusinovich, Int. J. Infrared and Millimeter Waves **3**, 765 (1982).
16. Y. Carmel, K. R. Chu, M. Read, A. K. Ganguly, D. Dialetis, R. Seeley, J. S. Levine, V. L. Granatstein, Phys. Rev. Lett. **50**, 112 (1983).
17. L. R. Barnett, Y. Y. Lau, K. R. Chu, V. L. Granatstein, IEEE Trans. Electron Devices **ED-28**, 872 (1981).
18. P. Sprangle, J. Vomvoridis, W. Manheimer, Phys. Rev. **A23**, 3127 (1981).
19. N. A. Ebrahim, Z. Liang, J. L. Hirschfield, Phys. Rev. Lett. **49**, 1556 (1982).
20. R. A. Mahaffey, P. Sprangle, J. Golden, C. A. Kapetanakos, Phys. Rev. Lett. **39**, 843 (1977).
21. G. Bekefi, T. J. Orzechowski, Phys. Rev. Lett. **37**, 379 (1976). □

It is still not much more than a gleam in President Reagan's eye, but the possibility of intercepting and destroying offensive ballistic missiles in space has stirred a contentious debate. Physicists are expected to do much of the research for the Strategic Defense Initiative, more popularly known as "Star Wars." They have also been contributing to the controversy on whether the exotic concept makes scientific and technological sense. This article, by Gerold Yonas, chief scientist of SDI, and the following article on page 34, by Wolfgang K. H. Panofsky, director emeritus of SLAC, are part of our continuing coverage of SDI.

Strategic Defense Initiative: The politics and science of weapons in space

A new research program is under way to study how lasers, particle beams and homing projectiles could destroy ballistic missiles to protect populations against a massive first strike.

Gerold Yonas

PHYSICS TODAY/JUNE 1985

Where there is no vision, the people perish.
Proverbs XXIX, 18

On 23 March 1983, President Reagan announced on national television that he was "launching an effort which holds the promise of changing the course of human history"—a scientific research program to determine whether a defense against ballistic missiles could be feasible. The President characterized the program, now known as the Strategic Defense Initiative, as a "formidable technical task" that might not be attainable before the end of this century. "Would it not be better to save lives than to avenge them?" he asked that night. "What if free people could live secure in the knowledge that their security did not rest upon the threat of instant retaliation to deter a Soviet attack, that we

Gerold Yonas, who had directed pulsed energy sciences at Sandia National Labs since 1972, became chief scientist of the Strategic Defense Initiative Organization last August.

could intercept and destroy strategic ballistic missiles before they reach our soil or that of our allies?" At the end of his talk, he called on the scientific community that invented nuclear weapons "to turn their great talents now to the cause of mankind and world peace, to give us the means" to devise ways of "rendering nuclear weapons impotent and obsolete."

The President's proposal took many people by surprise. It isn't every day that a president challenges the prevailing wisdom within the political and defense establishments about nuclear strategy. His ultimate goal, often expressed, is that science and technology will enable the US and its allies to protect themselves against ballistic missiles and thereby liberate the world from the specter of nuclear war. President Reagan has speculated that the research effort for SDI "could pave the way for arms control measures to eliminate the weapons themselves." Indeed, he recently stated: "We seek to render obsolete the balance of terror or mutual assured destruction, as it is

called, and replace it with a system incapable of initiating armed conflict or causing mass destruction, and yet effective in preventing war."

While I cannot profess to know President Reagan's precise motivations for advancing SDI, I can cite several factors that were significant. One involves nuclear arms control, which began as an effort to limit offensive ballistic-missile systems. It was conventional wisdom in the late 1960s that we should forgo ballistic-missile defense to remove the incentive to continue the arms race. In fact, in June 1967, President Johnson tried to persuade Soviet Premier Aleksei Kosygin that deploying ABM systems would imperil the quest for what defense experts call "arms-race stability" by encouraging each side to produce more offensive weapons to ensure retaliatory capability. The existing ABM agreement, which is part of the 1972 SALT I accords, prohibits the development, testing and deployment of ballistic-missile defenses that are based in space or in the atmosphere, on the sea or in

Multilayered space defense, in an artist's conception, indicates the successive events and objects that could be involved in an ICBM attack. In the boost phase, the missile's exhaust flame would be detected by infrared sensors, and lasers, particle beams or other space weapons would be aimed to destroy the ballistic missiles— eliminating the need to deal with warheads and decoys later. In the post-boost phase, the "bus" would be intercepted before it releases the warheads and decoys. During the missile's midcourse, outside the atmosphere, defense weapons must distinguish between warheads, penetration aids, decoys and debris in a "threat cloud" of reentry vehicles. As the cloud reenters the atmosphere, friction slows the light decoys and debris, leaving warheads easier to spot, but these need to be destroyed many miles up, far from cities.

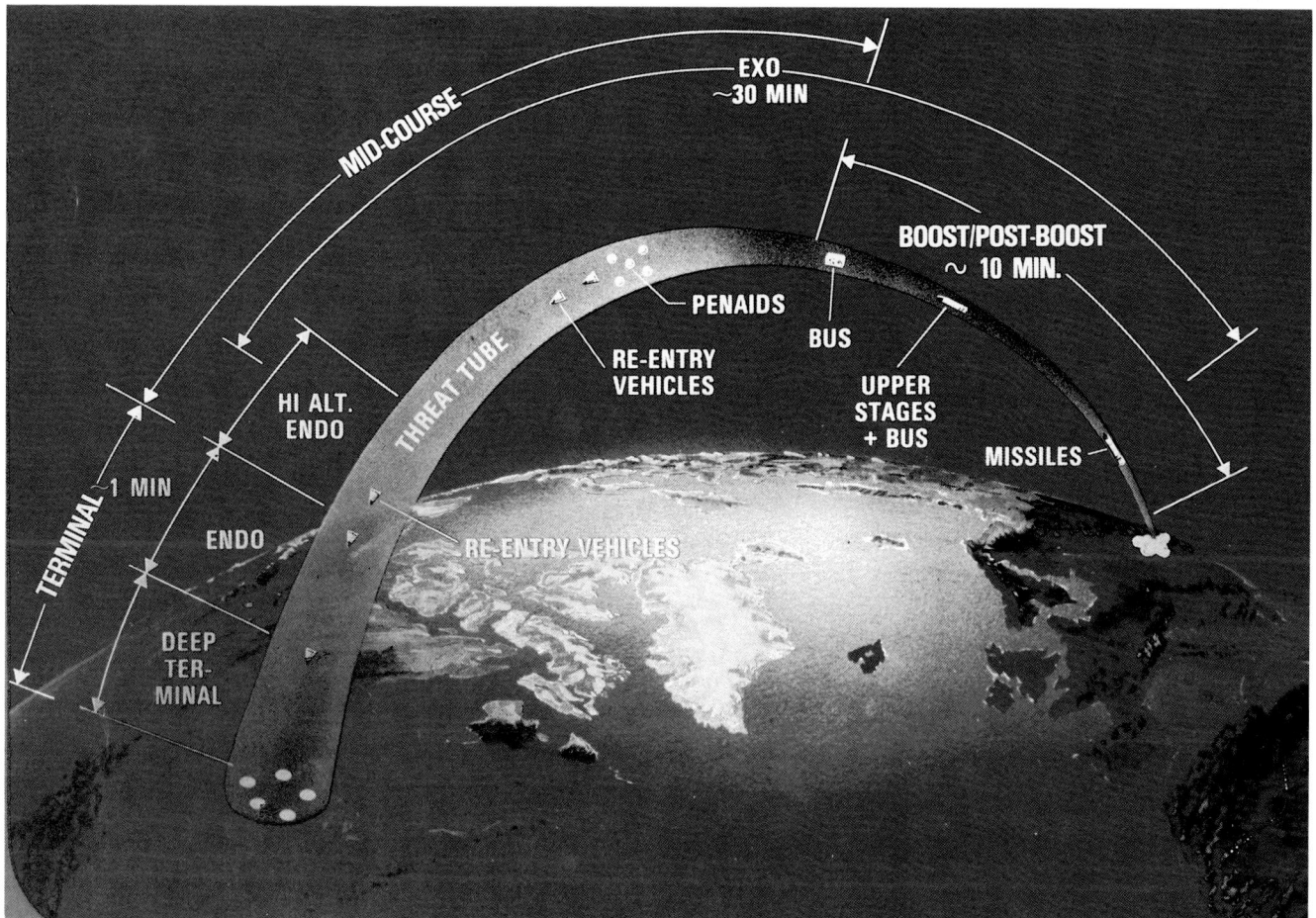

movable positions on land; but it permits research on these mobile concepts. The US interprets this provision as allowing research short of field testing prototype components or systems that are mobile in space, in the air, at sea or on land. The treaty even allows deployment of 100 long-range interceptor missiles, launchers and radars at one fixed site on the Earth's surface.

For our part, the expectation that the ABM Treaty would lead to strategic arms limitations has been shattered by the consistent buildup in Soviet strategic forces, predominantly directed toward ICBMs that are capable of destroying hardened targets as well as

toward submarine-launched ballistic missiles, or SLBMs. In numerical terms, the Soviet Union now has close to 3000 long-range ballistic missiles, compared with about 1800 in 1970, and in the same 15 years its nuclear warheads have quadrupled to more than 10 000. These weapons, our intelligence experts tell us, are capable of rapid response and great accuracy, making our Minuteman missiles and other military assets vulnerable to a first strike. In addition, the protracted search for a survivable basing mode for the MX has frustrated many who work in the strategic weapons arena.

Even before President Reagan's SDI

speech, there were those who advocated a revolutionary change in our approach to the nation's security. Many voices were calling for a stepped-up research program. In the late 1970s, Senator Malcolm Wallop, a Wyoming Republican, proposed a crash program to develop space-based chemical lasers as a missile defense. His optimism about laser defenses, undoubtedly known to the President, was expressed right after the SDI speech in an open letter to the Senate (8 July 1983). Another voice was that of retired Army Lieutenant General Daniel O. Graham, a former director of the Defense Intelligence Agency, whose High Fron-

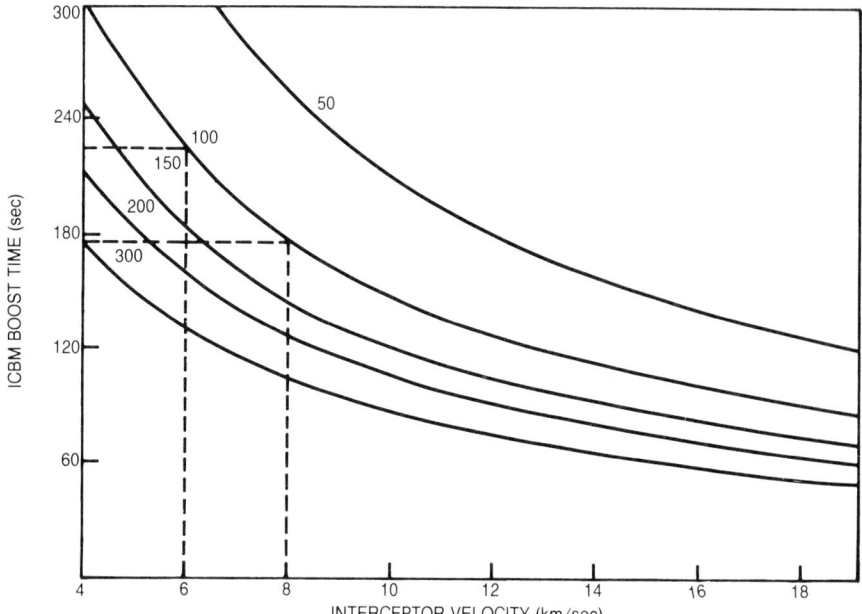

Boost-phase defense, with space-based kinetic-energy interceptors, requires a large satellite force with high closing velocities (on the order of 10 km/sec) to engage attacking ballistic missiles. The size of the satellite force (in numbers of satellites) is indicated for each of the curves. Shorter boost-phase intercept times require higher projectile velocities.

tier organization actively advocates a defense plan using hundreds of orbiting battle stations that would carry sensors and rockets to destroy missile warheads in space.[1] Then there was Edward Teller, who has advised Reagan on scientific matters for more than two decades. Teller urged the President to increase support of classified research on nuclear-driven directed-energy weapons at Department of Energy weapons laboratories.

It also became increasingly obvious that the Soviet Union is creating ground-based defensive systems to augment its long-range offensive weapons. It is continuing tests of an operational antisatellite system, improving a ballistic-missile defense capability around Moscow and completing a large phased-array radar facility at Krasnoyarsk in Siberia, far from its borders and facing inward, as if to direct ABM weapons. It is this radar that President Reagan has criticized as a violation of the ABM Treaty. The Soviet Union also is making large investments in advanced long-term R&D for ballistic-missile defense, such as ground-based lasers.[2] In addition, the USSR is developing an antitactical ballistic missile that may have capabilities against strategic missiles as well. Taken together, the Soviet ABM and ABM-related activities indicate the USSR may be preparing a nationwide defense.

By contrast to the Soviet activities, it appeared to many of us in weapons research, the US was pursuing an inadequately supported program in ballistic-missile defense. Worse, the program seemed to lack definition, strategy and goals. It could be characterized, we said among ourselves, by fitful starts and stops, changes in direction, disputes over funding and turf

battles between government departments and military services. Despite rather significant expenditures, the full potential of research on ballistic-missile defense was not being realized. We were aware that research and development had advanced considerably in the previous 10 years, but the nation was not acquiring the maximum returns on its investment in ballistic-missile defenses.

Research on the back burner

Such research and technology has been largely a back-burner issue in the US since the 1972 ABM Treaty, though interest was revived briefly during the MX debate over a "survivable" basing scheme for that 10-warhead missile. During the Carter Administration, the Defense Department considered plans to develop the Army's low-altitude defense system, which was intended to protect MX shelters. Such a system, designed to be used with mobile launchers and radar facilities, could not be developed or deployed without amending the treaty.

A variety of directed-energy research has been carried out by the Defense Advanced Research Projects Agency. Among these was the DARPA triad that consisted of the "Alpha" program to build a hydrogen-fluoride chemical laser, the "Lamp" program to demonstrate that a large, light-weight mirror for space applications can be used to direct a laser at a target, and the "Talon Gold" program that was expected to use the space shuttle to demonstrate an acquisition, tracking and precision pointing capability against objects in space. Other programs involved excimer lasers and free-electron lasers, as well as charged- and neutral-particle beams. The Air Force, for its

part, has explored the use of an airborne laser against conventional weaponry such as air-to-air missiles.

In the midst of these advanced technology efforts, the Army ballistic-missile division at Huntsville, Alabama, made a startling advance. Just one year ago this month, the Army showed that an ICBM can be stopped by a new device operating in space. In a test called the Homing Overlay Experiment, a ballistic missile with a simulated Soviet warhead was launched from Vandenberg Air Force Base on the California coast in a ballistic trajectory over the Pacific toward Kwajalein Island. Radars detected the missile, and computers calculated its course, so that 20 minutes after the launching, a Minuteman I first stage—fitted with a homing interceptor—was fired from Meck Island, near Kwajalein, to destroy the dummy warhead. Using a longwave infrared sensor while closing in on the target at 7.5 km/sec, the homing overlay vehicle tracked the approaching missile across hundreds of miles and relayed maneuvering instructions to an onboard computer. The homing vehicle positioned itself for a head-on collision with the warhead and unfurled a steel-ribbed umbrella-like device (see cover) that hit the dummy, destroying it.

Though this test was a brilliant success, it does not prove that a full ballistic missile defense system is possible. What if there had been many warheads, accompanied by decoys? The test exemplifies the state of today's technology. It is an example of what scientists and technical people have been able to achieve when challenged by what appears to be impossible tasks. The homing device is clearly a sweet technology to its inventors and build-

Homing Overlay Experiment conducted high above Kwajalein Atoll range in the central Pacific. A missile fired from Vandenberg Air Force Base in California was detected and destroyed by a homing projectile (see cover) mounted atop a Minuteman first stage launched from Meck Island.

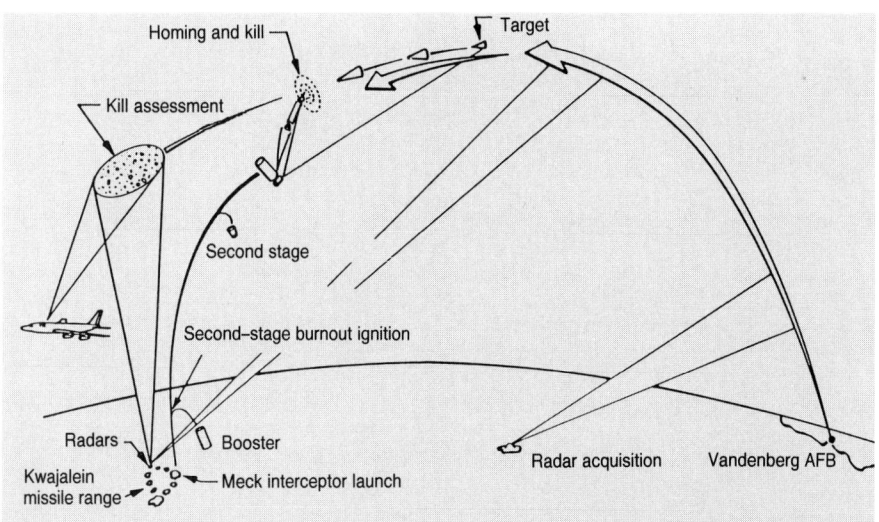

ers, though it will scarcely meet the requirements of a complete and reliable missile defense system.

Even though we have every reason to be optimistic about the potential of such nonnuclear interceptors, we should not be complacent. The Soviet Union has spent as much on strategic defense as on strategic offense over the past 12 years, and, in consequence, its defense capability is awesome. It includes 7000 air-surveillance radars, some 10 000 surface-to-air missile launchers, 12 000 interceptor aircraft and about 100 ground-level launchers located around Moscow in the world's only operational ABM system. Moreover, the DOD's latest edition of *Soviet Military Power* indicates that the Soviet strategic defense capability consists of a multilayered system, "compensating for shortcomings in individual systems and for the likelihood that neither offensive strikes nor any one layer of defense will stop all attacking weapons." The DOD report states that since the 1960s the Soviet Union has performed research on ground-based and space-based particle beams, chemical and gas-dynamic lasers and radio-frequency signals that could damage or destroy components of missiles, satellites and reentry vehicles. At a proving ground near Sary Shagan, the DOD report asserts, a laser is undergoing tests for antisatellite defense in the 1990s.

Our own SDI program, as directed by the President, will be conducted in ways that are fully consistent with all US treaty obligations, including the 1972 ABM Treaty. The purpose of the program is to explore a range of different concepts and technologies that hold potential for ballistic missile defense. The program is designed to answer

many fundamental scientific and engineering questions before the promise of a BMD system can be fully evaluated.

Decision in the early 1990s

No single concept or technology has been identified as best or most appropriate. The SDI research program was initiated by President Reagan with the knowledge that his term in the White House would be over before a "go or no-go" decision is made, most likely in the early 1990s, on developing and deploying a full ballistic-missile defense.

Accordingly, in April 1983, when the President signed National Security Study Directive 6-83 calling for studies on policy implications and on science and technology, he was well aware that SDI would mark a departure from the nation's approach to nuclear defense. A policy study team, headed by Fred S. Hoffman of Pan Heuristics, a California think tank, concluded[3] first of all that "US national security requires vigorous development of technical opportunities for advanced ballistic missile defense." Technologies for BMD are advancing rapidly, "together with those for precise, effective and discriminate nuclear and nonnuclear offensive systems," says the Hoffman report. The common assumption that the US alone is capable of putting up a realistic ballistic-missile defense, the report continues, "is completely unjustified. The Soviets give every appearance of preparing for such a deployment whenever they believe they will derive strategic advantage by doing so." The policy panel also anticipated some of the criticism of SDI. Unless the world is made aware of Soviet actions in ballistic-missile defense, the panel states, "the US will probably be blamed for initiating 'another round in the arms

Critics assert that any ballistic-missile defense system would have to be perfect. They say this knowing that no technological system of such scope and complexity could operate with quartz-clock precision. Simple logic argues that some percentage of warheads, although potentially a small number, would get through.

A reliable and effective missile defense is certain to prevent plans for a first strike by drastically increasing both the uncertainty of success and the ultimate cost of such an operation. In addition, other delivery systems do not present the same unstable characteristics of fast-reaction, high-precision ballistic missiles. Effective ballistic-missile defense, says the Hoffman panel, would "cast doubt on the feasibility of the entire attack plan and so contribute to deterrence, . . ." as well as to provide opportunities for offensive arms control.

The defensive technology study team, led by James C. Fletcher, a former administrator of NASA, produced its still-classified report[4] from a wide variety of views that were almost certain to come from 50 experts drawn from industry, government, universities and national laboratories. Because the team consisted of people still engaged or with previous experience in ballistic-missile defenses, it was not surprising they held strong convictions, including a great deal of skepticism, on the subject. A myriad of countermeasures were suggested to defeat any defense system. The toughest issue was the cost–exchange ratio: weighing the cost of a ballistic-missile defense system against the cost of producing ballistic missiles and delivering them on targets or the cost of countermeasures to penetrate the defense.

Directed-energy R&D facility at the Sary Shagan proving ground in the Soviet Union is shown in this drawing taken from the Defense Department's latest edition of *Soviet Military Power*. The illustration depicts the firing of a laser beam that the report says "could be used in an antisatellite role today and possibly a BMD role in the future."ᴸ

President Reagan's call for a new national effort to develop ballistic missile defense did not suggest any single approach to technology. Nor did the Fletcher panel. As a member of this panel, I know we agreed at the start that there was no perfect defense against a determined adversary, and it is not likely there ever will be. Nor did the President direct us to design a crash program, as some critics claim. Our understanding was, rather, that we were to evaluate the existing research programs, recommend speeding up or slowing down the present efforts and propose innovative options that might yield entirely new and effective technologies for ballistic-missile defenses.

The Fletcher panel was reasonably clear about the problem. Every missile defense system begins with a scenario of a nuclear attack: It takes roughly 30 minutes flight time from a silo or launch pad in Siberia to an explosion above a Minuteman silo in North Dakota, say, or any other potential target. During this period, a Soviet warhead will go through four stages: boost, post-boost, midcourse and terminal (or reentry). A defense system would try to destroy these missiles at any of these stages. A resolute opponent, of course, would attempt to defeat a defense system with countermeasures, which might range from attacking the defense system itself to massive proliferation of existing weapons.

Exploring technological limits

The approach the Fletcher panel adopted emphasized high-payoff, long-range capabilities rather than systems that might be more easily attained but might be, in turn, more easily countered. We developed a five-year research program that would thoroughly explore the limits of technology for both offensive and defensive systems to winnow out the least promising defense options. Only in this way could sufficient knowledge be acquired for an informed decision in the early 1990s on whether to proceed with engineered prototypes and full-scale development.

An effective defense against a massive missile attack would require multiple tiers of defense, countering the offense at each phase so that the system would allow only a small fraction of the warheads to strike. Boost phase is the ideal time to hit a missile. Here the advantage to the defense is that the bright flame from the booster motors can be easily detected by infrared sensors. The flame provides a reference point for aiming defensive weapons at a booster, so that the independently targeted warheads and decoys can be destroyed before they are released. The challenge is that boost phase lasts about 150–300 seconds for today's ICBMs and 150–200 seconds for submarine-launched ballistic missiles. The defense must be able to see the blastoff and fire quickly from afar. In addition, the offense may expand its missile force to overwhelm the defense or redesign or rebuild its force with radically different features to reduce the duration of the boost phase or harden the missiles against damage.

During the post-boost phase, lasting some 10–300 seconds, the defense still has an advantage if it can intercept the vehicle or "bus" before it releases all of the warheads and decoys. The challenge is that the bus's maneuvering engines, being relatively small, are more difficult to detect than the plume of the huge booster. Furthermore, an aggressor could shorten the maneuvering stage of the bus, or shroud the deployment of warheads and decoys.

In midcourse, the advantage to the defense is the relatively long period of roughly 10–15 minutes for ICBMs and 7–10 minutes for SLBMs when the warheads and decoys can be located and destroyed by space-based and ground-based interceptors far away from such targets as cities or industrial sites. The challenge is that warheads and effective decoys would look more or less alike to sensors. The sheer numbers of warheads and decoys—in the worst case, tens of thousands of warheads and hundreds of thousands of decoys—could swamp the sensors and interceptors. The offense can try to complicate the job of the defense by deploying not only clever decoys but warheads inside of decoys and so on. The defense must either discriminate warheads from decoys or make interceptions so cheap and easy that it can attack practically all of them.

On reentry, the decoys would be separated from the heavier reentry vehicles or burn up, thereby solving the discrimination problem. However, there is scarcely time, possibly only 30–40 seconds, to carry out the intercep-

Neutral-particle beam accelerator, part of the White Horse program centered at Los Alamos National Laboratory, produces neutral beams of hydrogen by stripping the negative charge of hydrogen ions. The heart of this accelerator is a radiofrequency quadrupole that preaccelerates a beam to 2 MeV. The rf quadrupole is an invention by Soviet physicists.

tion of any warheads the ballistic-missile defense system has missed in the previous phases before they damage or destroy their targets. Moreover, warheads can be "salvage fuzed" to detonate if struck in the upper atmosphere, stressing the already difficult job of the terminal sensors.

The criteria for a reliable ballistic-missile defense are survivability, lethality and cost effectiveness. The defensive system must be able to knock down offensive weapons but avoid being knocked out itself. The system has to be flexible and responsive enough to deal with the probability of a changing threat. The Soviet Union could expand and improve its missile force to meet our ballistic-missile defense system. To provide an incentive for the Soviets to reduce their offensive weapons, we must show that our defensive technologies enjoy a clear advantage in a cost-exchange analysis. Put more clearly, we have to impose an unacceptably high cost on the Soviets to respond to our missile defense system with greater investments in their own ballistic missiles and countermeasures.

Shorter-range tactical ballistic missiles, which pose a threat to our allies, as well as missiles launched from submarines off our coast, present similar challenges of boost and reentry. Such missiles have a much shorter midcourse phase, though, making midcourse interception more difficult, but also making it harder for the aggressor to deploy decoys and, concomitantly, making it easier for the defender to discriminate the decoys from the war-

head.

Decisive issues of cost

A ballistic-missile defense would require substantial expenditures, of course, and so would a Soviet offensive force to deal with that US defensive system. So the issues of cost effectiveness and marginal-cost exchange are extremely important, but it is at this time admittedly difficult to come to grips with such matters because no single system construct, architecture or set of specific technologies has been determined. The relevant technologies for ballistic-missile defense are in various stages of maturity. Some concepts are in basic research, while others are close to engineering validation. We see, for instance, that kinetic-energy interceptors, as evidenced by the homing experiment, are more mature than directed-energy systems. Still, directed energy, even in a relatively early period, could be used in an interactive mode to assist in midcourse discrimination. Finally, as the various technologies evolve, we envision the perfection of very bright beam weapons that would be able to respond to even the most advanced ballistic missiles.

The SDI program is now grouped into four major elements:
▶ Surveillance, acquisition, tracking and kill assessment.
▶ Interception and destruction
▶ Battle management
▶ Survivability, lethality and key technologies

Surveillance, acquisition, tracking and kill assessment. Acquisition and dis-

crimination during boost phase will be greatly aided by the hot flame from the booster's engines; detecting colder warheads and decoys against the background of space and discriminating between them during post-boost and mid-course phases may be possible by using a variety of sensors (optical, infrared and radar) in different locations (based in space, in the atmosphere and on the ground) and integrating the data picked up by the sensors. Perhaps we may be able to distinguish decoys from warheads, if we can observe decoys at their "birth," the instant they are deployed, and afterwards keep track of the objects that are warheads to be intercepted and those that are debris or decoys to be ignored—the so-called "birth-to-death" tracking. All this will depend on advances in sensor technology and the computer software to process the data generated, as well as on studies of the characteristic signatures of warheads and decoys. This element includes some of the most as well as least mature technologies in the program and accounts for 40% of the current SDI budget—or $546 million in the fiscal 1985 appropriation. The Defense Department seeks $1.4 billion for this part of SDI in fiscal 1986.

Interception and destruction. The ability of any defense to respond effectively to a ballistic-missile attack is largely dependent on the feasibility and reliability of boost-phase and post-boost interception. Two types of concepts are being considered: directed-energy, or speed-of-light weapons, and kinetic-energy weapons. Directed-energy wea-

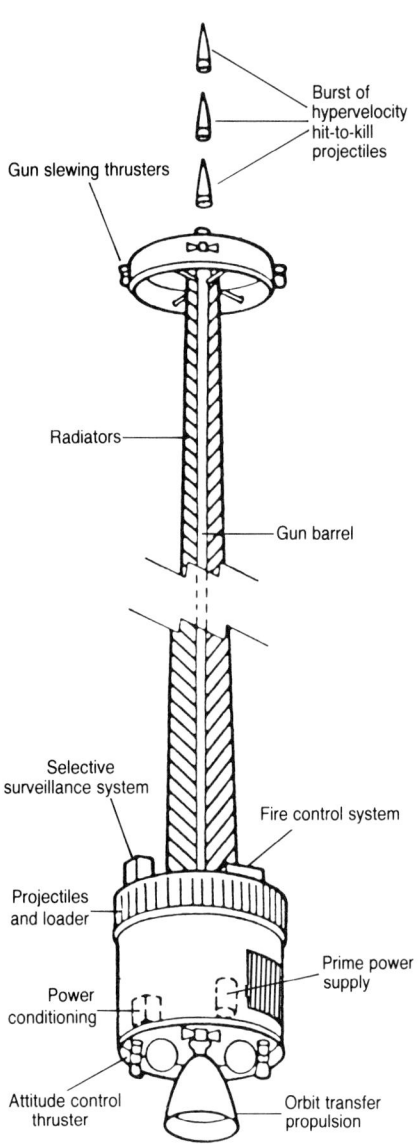

Gun slewing thrusters

Burst of
hypervelocity
hit-to-kill
projectiles

Radiators

Gun barrel

Selective
surveillance system

Fire control system

Projectiles
and loader

Power
conditioning

Prime power
supply

Attitude control
thruster

Orbit transfer
propulsion

Hypervelocity railgun, the popular name for an electromagnetic launcher. In concept, it is simplicity itself, using the familiar Lorentz force (created by the interaction of an electric current with a magnetic field) to accelerate projectiles to velocities as great as 25 km/sec.

pons could include space-based and ground-based lasers, space-based particle beams and x-ray lasers powered by nuclear explosives. The current goal is to construct laboratory-scale beam generators by the late 1980s or early 1990s and to demonstrate the feasibility of scaling them up to weapons-level performance by the early 1990s.

The critical parameter for beam weapons is beam brightness. It determines the weapon's range and how long the beam needs to dwell on each target. The necessary beam brightness (measured in watts per steradian) is simply the target's lethal irradiance times the square of the distance to the target. Brightness, therefore, determines the number of weapons needed in orbit to cover Soviet missile launching sites. Beams of intense brightness may become available from such short-wavelength lasers as the electrically powered excimer and free-electron lasers, but both are in relatively early stages of development. The free-elec-

tron laser may be extremely efficient and therefore a candidate for eventual deployment in space. A much simpler technological approach involves more mature chemical lasers, utilizing the reaction of gases such as hydrogen and fluorine, and the possible invention of entirely new short-wavelength lasers, based on different chemical reactions and pumping mechanisms.

It is useful to consider a specific example to understand the nature of the problem with lasers. For instance, let's say we want to deliver a burst of energy of 10 000 joules/cm^2 to burn through a missile skin in 1 sec at a distance of 3000 km. That's a reasonable level of lethality and an acceptable range in terms of the size of the constellation of platforms to deal with a massive attack. The parameters demand a brightness of 10^{21} W/sr. Now, the brightness B of a laser system is a function of the wavelength λ, the aperture of the output mirror or beam director D, along with optics precision and the actual jitter of the pointing system. Thus:

$$B < PD^2/\lambda^2$$

where P is the laser power. So with a 5-MW infrared laser using output optics 4 m in diameter, brightness would be less than 10^{19} W/sr. This would not be lethal to a hard target requiring 10 000 joules/cm^2 in 1 sec. The solution to this problem is to shorten the wavelength, going from 3 microns to, say, a fraction of a micron, because brightness increases as the inverse of the square of the wavelength. But shifting to a shorter wavelength is not enough without also providing precision optics as well as the requisite pointing and tracking. In addition, if the size of the threat and the rate of booster launchings demand even faster target destruction, the brightness would have to be increased further, using phased arrays of small optical elements.

High-power excimer lasers that operate at short wavelengths and in a pulsed mode could destroy a missile by focusing on it for a fraction of a second, but their efficiency is so low and

generating apparatus so bulky (even though the optics could be a reasonable size) it is unlikely that they and their fuel supply could be lifted into space in cost-effective ways. The laser stations would have to be at ground level, perhaps at high elevations above the densest layers of the atmosphere. There would need to be enough ground stations to allow for cloud cover and, in addition, the laser optics would need to correct for atmospheric perturbations. At an altitude of about 10 km the atmosphere is so turbulent that the beam of an excimer laser would be diffracted, for the same reason that the light from stars appears to twinkle. A beam propagated from the ground straight through the atmosphere would have little intensity at a distance of several hundred kilometers. Scientists are investigating ways to correct for this by determining the deformation of the wave front and using deformable optics to make the correction. We are in the initial stages of investigating such phenomena at low power.

Because the free-electron laser could operate potentially at short wavelengths with high efficiency, it could be based in space (as well as on the ground, of course). In the past year, development of such a laser has been encouraging (PHYSICS TODAY, April, page 17).

At the High Energy Laser Test Facility at White Sands in New Mexico, a 1.5-m-diameter beam director is operating alongside the 3.8-micron MIRACL infrared laser. We are beginning to study the issues of generating high-quality beams and transporting the energy through large optical systems. We are also investigating laser effectiveness against various techniques for hardening targets. Such lethality tests are crucial to our evaluation of possible Soviet countermeasures that could conceivably make certain laser approaches impractical.

Neutral-particle beams would draw on the well-established technology of high-energy physics accelerators. The approach being followed is to accelerate negative ions and then strip off the extra electron after the beam is accelerated and magnetically aimed in the desired direction. Such beam weapons would have the advantage of being difficult to shield against because of the penetrating power of atoms accelerated to nearly the speed of light. The sophisticated electronics of a missile or warhead might be damaged or disrupted by even low beam currents, and high currents could melt the warhead structure or detonate its high explosive. We are also considering the use of such beams for discriminating warheads from decoys. The critical issue, in fact, may not be whether or not the necessary brightness can be reached. Instead, the problems that dominate this approach are likely to involve accelerating the particles with compact accelerators and understanding the potential adverse effects of different countermeasures.

Once the beam has developed a large transverse velocity, it is impossible to control precisely over long distances. So it is important to devise methods of maintaining low beam divergence in the initial part of the accelerator and the electron stripping section. Beams of electrically neutral particles, given low enough transverse velocity, could propagate over thousands of kilometers and destroy ICBMs or their warheads. Incidentally, significant advances in developing neutral-beam accelerators come from Soviet innovations—the radiofrequency quadrupole (PHYSICS TODAY, August 1983, page 17) and the negative-ion source. These enable physicists to produce a beam of high quality and low divergence with compact accelerators. Even so, neutral-particle beams face potential problems. The problems include providing the necessary power, making sure of kills and avoiding possible countermeasures. Neutral-particle beams, it so happens, are rapidly ionized and then deflected by the Earth's magnetic field once they enter the thin upper layers of the atmosphere at 100 km. So the development of fast-burning boosters would defy any boost-phase neutral-particle-beam intercepts below that

altitude, but would still allow the bus to be attacked while deploying decoys or reentry vehicles.

Directed-energy weapons powered by nuclear explosive are being studied. Such concepts as the x-ray laser are being examined at DOE laboratories to determine their feasibility. At the same time, we need to answer the critical questions of how and where such weapons could be based, as well as the implications of such weapons if the Soviet Union were to use them to attack US strategic defenses.

The entire DOD directed-energy program accounts for about 20% of SDI, or $376 million, this fiscal year. The Defense budget for this element next year calls for $966 million, while the Energy Department seeks $280 million for 1986.

Kinetic-energy weapons are the other devices under examination for interception in all phases of a ballistic-missile trajectory. Such weapons include electromagnetic launchers, also known as hypervelocity guns, as well as more conventional chemical-fueled rockets, both of which would propel homing projectiles toward attacking boosters at speeds up to 10 km/sec. Electromagnetic launchers are potentially capable of producing much greater velocities. When based on orbiting platforms, such devices could serve in boost-phase and mid-course interceptions, and in defending space-based equipment from direct attack. The goal would be to design weapons that were not only effective but also cheap, so that the defense could afford to engage all warheads and many of the decoys if sensor technologies could not discriminate among them reliably.

There is little question of the lethality of an 8-kg projectile, say, hitting a target at 10 km/sec (the approximate force of 100 kg of high explosive). But the development of a low-mass guidance and control system for such a homing projectile, capable of surviving initial acceleration, and of low-cost chemical rockets or a space-based power supply to deliver hundreds of megawatts of electricity for electro-

magnetic guns may be among the most difficult jobs before the SDI research program. In this year's budget, kinetic-energy weapons received $256 million from Congress; for fiscal 1986, the program requests $860 million.

Battle management. Integrating multiple layers of defensive weapons into a highly reliable and fault-tolerant system will require high-performance computers, sophisticated software and adaptable communications networks far beyond existing capabilities. The system will need to track tens of thousands of objects, from launch to destruction, and allocate defense weapons most efficiently. The defense system will require a network of space-based computers capable of performing millions of operations per second, surviving virtually maintenance-free for years in space and adapting to flaws and failures within the network. Equally exacting will be the requirements for software. Programs must be written and tested exhaustively to make sure they are free of errors. In fact, the job of producing the necessary software may have to be done by advanced computers, which would mean developing computer-controlled programming and debugging. Battle management now takes $99 million of the SDI budget; for fiscal 1986, we have proposed $243 million.

Survivability, lethality and key technologies. The strength of any defense system rests squarely on its ability to withstand a direct assault and continue to function effectively even if damaged or degraded. Components of any ballistic-missile defense may face many threats, including direct-ascent antisatellite weapons, lasers in space or on land, space mines, particle-beam weapons and the effects of nuclear explosions. The tactics of survivability are familiar—hardening, active self-defense, concealment, proliferation and maneuvering out of harm's way. Applying those tactics, however, especially in cost-effective ways, poses a difficult challenge and may be pivotal to the outcome of the entire ballistic-missile defense endeavor.

A beam director, such as this 1.5-m diameter facility, operates alongside the the 3.8-micron MIRACL infrared laser at the White Sands Proving Ground in New Mexico. It is used to focus the beam onto distant targets.

By no means does this short list of critical challenges exhaust the range of issues that must be dealt with in the SDI research and technology program. The program will need to pursue a whole set of support projects, including finding generic means for hardening boosters and investigating the lethal effects of weapons against those means (especially because we are not likely to get specific engineering design data on Soviet boosters); developing multi-megawatt electric-power supplies (both nuclear and nonnuclear) for operating weapons, sensors, computers and other BMD components in space; providing launchers capable of lifting 100 tons or more into orbit (the space shuttle's current capability is no more than 30 tons); as well as other unforeseeable needs of a ballistic-missile defense. The survivability, lethality and key technologies program of SDI operates on a budget of $112 million in the fiscal 1985 appropriation; the request for this element in fiscal 1986 is $258 million.

Four phases of SDI

The first phase of SDI, from now until the early 1990s, will be devoted to research. After this period, informed decisions could be made on whether or not to begin engineering development of specific weapons and components. The second phase would center on systems development, when prototypes of actual defense components could be designed, built and tested, assuming the political and strategic decisions are made to proceed with ballistic-missile defense. The third phase is known as the transition, when presumably both the US and USSR would begin to deploy their defenses in sequential, incremental ways. In the transition phase, the scenario runs, both superpowers would make significant reductions in offensive missile forces. The final phase would be reached when ballistic-missile defenses are fully in place and offensive missiles are at a negotiated low point.

By now it is quite clear that we have a long way to go before achieving a thoroughly reliable defense against ballistic missiles. As the research evolves, however, new discoveries are likely to be made that may radically change the emphasis of the defense program. It is therefore vital to stimulate, encourage and investigate fully any innovative approaches as early as possible in the program. Good ideas are just as likely to come from younger as well as older members of the scientific and technical communities, from established as well as unexpected places, from laboratories and classrooms where the challenges have not yet been contemplated. To stimulate the widest involvement of technical talent within government, industry and universities, the departments of Defense and Energy will provide specially earmarked funds to encourage the investigation of wholly new and different concepts that suggest better defense capabilities. And because such discoveries might help in developing offensive capabilities as well, some of the approaches we now think about for defense may be discarded or modified along the way. We need to eliminate all impractical or unfeasible defensive approaches as quickly as possible to focus our talent and resources on those worth doing. It is essential to approach the entire SDI program with a healthy skepticism, as well as with the sense of discovery and excitement we associate with venturing into a new frontier.

Many of the underlying issues and technologies discussed here are far from being fully resolved. They are being examined and debated vigorously every day. SDI has prompted widespread attention to the dilemma of survival in a nuclear age, and some members of our scientific and technical communities are already mobilizing. The SDI effort will take time and patience. Most important, if it is to succeed, it must receive wide support among the American people. The hope of a more secure and stable future motivates many of us to strive to solve the problems of ballistic missile defense.

References

1. D. O. Graham, *The Non-Nuclear Defense of Cities: The High-Frontier Space-Based Defense Against ICBM Attack*, Abt Books, Cambridge, Mass. (1983).

2. *Soviet Military Power 1985*, Department of Defense (1985).

3. F. S. Hoffman, Study Director, *Ballistic Missile Defenses and US National Security*, Summary report for Future Security Strategy Study, Department of Defense (Oct. 1983).

4. For nonclassified accounts of the study see J. C. Fletcher, testimony before Subcommittee on R&D, Armed Services Committee, House of Representatives, 1 March 1984, and J. C. Fletcher, "The Technologies for Ballistic Missile Defense," *Issues in Science and Technology* **1**, No. 1 (Fall 1984), p. 15.   □

The Strategic Defense Initiative: Perception vs reality

ABM defense technology deserves further research within treaty limits, but the "Star Wars" program is too large, too political, raises false hopes and poses grave dangers to national and world security.

Wolfgang K. H. Panofsky

PHYSICS TODAY/JUNE 1985

We have witnessed over the last year or so an enormous growth in the political impact, if not the technical impact, of the President's Strategic Defense Initiative, popularly called "Star Wars." Some argue that without this initiative the Soviets would not have "returned to the bargaining table" to resume talks on strategic and intermediate-range weapons. Others argue[1] that *with* SDI the United States cannot possibly reach an arms-control agreement with the Soviet Union. The true believers assert that SDI points the way to a future free of nuclear weapons, while opponents claim that it is a sure-fire prescription for a major escalation in the arms race, the militarization of space and a collapse of the current, albeit limited, arms-control regime. (See Gerold Yonas's article on page 24.)

One of the purposes of this piece is to point out that political perceptions as to what SDI is all about are running wildly ahead of the technical realities of strategic missile defense. Overblown expectations alone are usually not too serious a matter; after all, such expectations are characteristic of many ambitious initiatives. The reason expectations are a grave issue in the case of SDI stems from the very nature of the nuclear-arms buildup. The arms race has been and continues to be fueled by

just such overblown expectations of the military utility of nuclear weapons. We know that if only a small fraction of the nuclear weapons stored worldwide—50 000 or so—were used in war for whatever reason, then the future of civilization as we know it would be gravely in doubt. Yet the so-called rational decision-making processes of the world's nations, in particular the Soviet Union and the United States, have let this insane number accrue. One of the main reasons for this history is that both sides have exaggerated the leverage associated with large numbers of nuclear weapons, letting their perceptions go well beyond what is justified by the military and technological utility of the weapons. In consequence, decisions within the councils of the two superpowers frequently do not reflect proper military questions such as "Is this weapon needed to *counter* a weapon just fielded or about to be fielded by the adversary?" but rather deal with the political perceptual issue of *matching* or *exceeding* what the other side has.

The current MX debate is a case in point. Opponents and proponents alike tend to agree that MX missiles based in existing Minuteman silos are vulnerable, expensive and add little to the deterrent capacity of existing US strategic forces. Yet the argument that we must "match Soviet modernization," demonstrate our resolve, and support the Geneva arms-control negotiations by "arming in order to disarm" persuaded Congress to approve the addi-

tion of 210 more warheads to the nearly 10 000 nuclear weapons that the United States can now deliver against Soviet territory from air, land or sea. Thus again the political perception of nuclear weapons prevailed over technical reality. Indeed, in general, the political question of real or imagined nuclear superiority, equivalence, "gaps" and so on has come to take priority over the question of the actual military–technical usefulness of weapons systems.

With SDI, it is not so much the real technological potential as it is the political, "hyped" perception of that potential that is so threatening. In the SDI program itself, there is actually "less there than meets the eye." Yet, thanks to the many diverse if not outright conflicting statements of US national leaders, the aggressive salesmanship of the SDI management, the ambiguous foreign response and the generally uncritical media analyses, the Strategic Defense Initiative has sown the seeds of an extraordinarily costly and dangerous further military expansion. Mark Twain said, "There is something fascinating about Science. One gets such wholesale returns of conjecture out of such trifling investment of fact." I could add to this, "The Strategic Defense Initiative is truly impressive; one can base so much political and strategic posturing on such limited technical and military potential."

In this article I discuss the background, the current status and the

Wolfgang Panofsky is director emeritus and professor at the Stanford Linear Accelerator Center. He was president of The American Physical Society in 1974.

Spartan missile. The Army developed this long-range interceptor in the late 1960s for the Sentinel antiballistic missile system. (US Army photograph.)

goals of the Strategic Defense Initiative. I argue that the present program of SDI research and salesmanship may threaten to fuel the arms race, violate the ABM Treaty and upset strategic stability. I find no technical or stategic justification for anything beyond a limited program of studies and experiments, designed not to promote a project but to examine the possibility of defense and insure against technological surprise.

Seeking a technological fix

The history preceding President Reagan's talk of 23 March 1983 is characterized by lack of consultation commensurate with the importance of the topic. The President, like his predecessors, became extremely uncomfortable with the current balance of terror. He was rightfully repelled by the possibility of having to make a decision that could cause the death of perhaps hundreds of millions of individuals if nuclear or even nonnuclear attack against the United States or its European allies occurred—or even worse—if he were assured it was about to occur. The President's discomfort was reinforced by some members of the military leadership, and a narrow segment of the scientific community dangled the possibility of a "technological fix" of this abhorrent situation before his eyes. While past Presidents responded to these very same discomforts by initiating formal interdepartmental or independent studies, President Reagan reacted in an instinctive "there must

be a better way" mode, and the famous "Star Wars" speech was the result. There were no studies preceding that speech even remotely indicating that nuclear weapons could be made "impotent and obsolete" through defense, let alone indicating a stable path of how to get there from here. Neither the technical defense community nor the allies were consulted.

In comparison let me quote a statement by President Nixon on 14 March 1969 reflecting his reaction to the same dilemma:

Although every instinct motivates me to provide the American people with complete protection against a major nuclear attack, it is not now within our power to do so. The heaviest defense system we considered, one designed to protect our

major cities, still could not prevent a catastrophic level of US fatalities from a deliberate all-out Soviet attack. And it might look to an opponent like the prelude to an offensive strategy threatening the Soviet deterrent.

Is there any new emergent technology that justifies the difference between the two presidental statements? I conclude there is not.

Work on defense against delivery of nuclear weapons, be it by aircraft, by missiles, or even by "suitcase," has proceeded almost from the day when nuclear weapons entered the world's arsenals. All the studies had to face one overwhelming fact: The advent of nuclear energy has multiplied by about a million the amount of explosive power that can be packed into a single

weapon of given weight and size. This completely upsets the traditional calculus that compares the profitability of adopting offensive measures with that of adopting defensive measures. Historical arguments no longer hold. Indeed the Battle of Britain was "won" by that courageous island's radar-guided fighter planes and antiaircraft *defenses.* However, "winning" meant that perhaps 10% of the incoming aircraft of the Luftwaffe were shot down in each raid. In the prenuclear age this yielded intolerable losses for Hitler's forces, because after ten sorties only one-third of the original attacking craft would remain. Yet, had the Nazi airplanes carried nuclear weapons, there is no question that even if Britain had a defense capable of stopping 90% of the attacks, it would have been totally destroyed as a viable civilization by those few penetrating German aircraft.

As long as countries continue to stockpile nuclear weapons in the numbers we are seeing today, we are condemned to live in an "offense dominated" balance. This is a situation where all-out war is inhibited by the fact that either side can absorb a first strike and retaliate with a blow strong enough to destroy the initiator as a functioning society. Under this situation the populations of the United States and the Soviet Union are in fact hostages for the maintenance of peace, or at least for the absence of all-out hostility between those two powers.

This situation is repugnant to many, including Presidents, who must visual-ize their own role in it, but it cannot be abolished by exhorting that "there must be a better way." There is no new technology on the horizon that realistically makes the President's speech anything but an expression of hope. Possibly the most dramatic advances in military technology are those associated with rapid and reasonably error-free data processing and transmission. However, this advance affects defensive as well as offensive measures and countermeasures. "Directed energy weapons" are an old concept as such. Particle-beam devices, laser weapons and means to direct and focus the radiation from nuclear explosives have been studied for decades.

One genuinely new technological development is the laser pumped by nuclear explosives. Such a laser can in principle concentrate a certain fraction of a nuclear bomb's energy into a narrow cone of x rays or other electromagnetic radiation. Making the cone narrower would extend the "reach" of the radiation in a vacuum, but would require more precise and delicate target detection and guidance systems. To use J. Robert Oppenheimer's phrase, this possibility is "technologically sweet" and involves some truly challenging problems in physics. However, in the context of the Strategic Defense Initiative, this new technology offers limited opportunities—in fact, the Administration's recent emphasis that SDI is to lead to non-nuclear defenses has pushed it even further into the background.

The President expressed in his speech the hope of escaping from the bondage of Mutual Assured Destruction. Yet MAD is a condition brought about by the numbers and powers of the world's stockpiled nuclear weapons; it cannot be abolished by decree of political authority.

What is new is that the Commander-in-Chief has made a policy decision that a maximum effort shall be manned to convert the present condition of mutual deterrence based on offense to one where the protection is based on defense. Yet policy in itself cannot compel or generate technology. On the contrary, policy should not be established in disregard of or in conflict with technical realities, or in anticipation of highly uncertain future technical results.

The current status

The President has directed the establishment of a program whose goal is to convert the "vision" expressed in his 23 March 1983 speech into reality. If this Presidential decision were implemented literally it could unleash a research and development effort of almost unlimited scope and cost. While nothing as yet in the SDI research and technology program violates the ABM Treaty of 1972, the President's declared intent contradicts the expressed principles of that treaty, according to which

Each Party undertakes not to deploy ABM systems for a defense of the territory of its country and not to provide a base for such a defense, and not to deploy ABM systems for the defense of an indi-

vidual region except as provided for in Article III of the Treaty.

The President has, however, formally directed that SDI not violate the treaty during his term in office.

It is well known that the Soviets are currently spending considerably larger sums than the United States on overall strategic defensive activities, in particular, air defense and civil defense. Yet this Soviet effort hardly threatens any meaningful attrition of a US retaliatory nuclear strike. Both the Soviet Union and the United States have been pursuing research and technology programs in ballistic missile defense for decades. Such efforts include "conventional" ABM programs designed to improve terminal intercept systems that use radar-guided, nuclear-tipped interceptor missiles, as well as more esoteric programs based on various forms of directed energy and nonnuclear interceptors. The Soviet program is at least as large as that of the United States, but on balance no more advanced.

In outline the current ABM activities of the two sides involve the following:
▶ Soviet upgrading of the existing ABM deployment around Moscow, limited by treaty to 100 interceptors
▶ Various construction programs by both sides, including the installation of phased-array radars (see the figures on pages 38 and 39)
▶ Ongoing ABM research and technology programs on both sides, with a large expansion, in the form of SDI, proposed by the United States
▶ The oratory on the US side touting the virtues of SDI.

The United States has unilaterally abandoned its ABM deployment at the one site permitted under Article III of the 1972 treaty as amended in the 1974 Protocol, while the Soviets are continuing to improve the corresponding Moscow deployment. Some critics have accused the Soviets of having failed to reciprocate such "self restraint" of the United States. This charge is unfair. The US abolished the Grand Forks, North Dakota, missile-silo defense site after spending about $7 billion and finding we were getting a negligible amount of defense for our money. (See the photograph on page 37.) This was fiscal prudence and not military self restraint. One can argue with merit that the Soviets are wasting many rubles on their Moscow ABM defense, which US missiles can obviously penetrate. Note that starting in the 1960s the principal consequence of the Moscow defenses was that the United States targeted more missiles on Moscow!

There have also been highly orchestrated charges that the Soviets are violating or have violated the ABM Treaty. The Soviets have made counterclaims. The most substantial charge against the Soviet Union concerns the purpose of the yet uncompleted phased-array radar located in Krasnoyarsk, Siberia. (See the drawing on page 39.) The Soviets claim that this radar, which probably will not radiate observable signals for several more years, is a legal space-tracking radar. This explanation is highly improbable.

There is little question that the Soviets are much more inclined to risk "gray area" violations of existing arms control treaties than is the US. Yet no one can claim with merit that any of the alleged Soviet, or as far as that goes, the alleged US violations of the ABM Treaty are of sufficient military significance to affect the strategic situation.

Nothing in the US Constitution dilutes the responsibility of a president to comply with existing treaties in force. Most treaties, and the ABM Treaty is no exception, contain abrogation provisions, if one party finds its supreme national interests challenged. Alternately, a treaty can be renegotiated by mutual consent of the parties. This is not as simple a matter as is often implied. If we wish to modify a treaty over Soviet opposition, we would have to pay a stiff price in other areas; abrogation has, of course, many highly negative aspects.

Notwithstanding the debate within the US about whether SDI should be a bargaining chip at Geneva or an immutable cornerstone of US strategic policy, there is a real question as to what specific measures to inhibit SDI the Soviet Union will introduce in the Geneva talks. Specific negotiations cannot simply deal with the goal of reaffirming a valid treaty against whose erosion both parties have thus far publicly protested. The negotiating parties cannot buy again the same horse they bought in 1972. Unless the Soviets are more forthcoming about their own research and technology programs, it will be difficult for them to

Antenna of the Pave Paws radar at Beale Air Force Base, 40 miles northeast of Sacramento, California. This phased-array warning system, designed to detect sea-launched ballistic missiles, has been operating since 1980. Another Pave Paws system is in operation on the East Coast, at Otis AFB near Cape Cod, Massachusetts. Two are planned for the South, in Georgia and Texas. (US Air Force photograph.)

propose specific measures in Geneva for inhibiting US SDI programs beyond the limits already set by the ABM Treaty.

The Geneva negotiations or other Soviet–US meetings could lead to further erosion of the ABM Treaty. It is not inconceivable that the US will push to give SDI latitude beyond present treaty constraints, and the Soviets, notwithstanding their present public posture, may wish to legitimize some of their "gray area" ABM activities.

In contrast, the Geneva negotiations could, and I hope will, strengthen rather than weaken the ABM Treaty. This could be done by tightening some of the ABM Treaty's language to leave fewer ambiguities and openings regarding space testing and the development and deployment of defenses against tactical ballistic missiles. Probably the most constructive measure would be to restrict ABM activities to laboratory-type research; to go further than that would probably require non-negotiable verification measures.

Directly connected to the question of strengthening or weakening the ABM Treaty is, of course, the issue of prohibiting weapons in space. The Outer Space Treaty of 1967 prohibits weapons of mass destruction, including nuclear weapons, in space. (Herbert York sur-

veys the arms-control treaties in his PHYSICS TODAY article, March 1983, page 24.) The Soviets have proposed various forms of a draft treaty inhibiting military use of space. Note that there is a profound difference between military use of space and weapons in space. There are many military uses of space, in particular, satellite reconnaissance—including verification of arms-control agreements—and military command, control and communications, that are to some extent stabilizing and that technically have broad overlap with commercial uses of space.

The technologies of ABM and ASAT partially overlap. It is technically easier to dedicate any directed-energy weapon to the offensive antisatellite mission than it is to incorporate it into a complex, defensive ABM system. Prohibiting further ASAT tests would prevent governments from using the ASAT label as a vehicle for bypassing the ABM Treaty's strictures on testing. Banning weapons in space is thus associated with the separate issue of whether impeding ASAT development and deployments is in the interest of the United States and global security. I conclude that it is. The US depends deeply on its military and civilian space activities, and surveillance satellites have made this a more open world. Yet

a prohibition of all military uses of space would definitely be against US interests, would raise exceedingly difficult verification issues and might well be generally destablizing. In contrast, prohibiting just *weapons* in space would help to protect our existing and future highly vulnerable space systems. Others have discussed these issues in detail.[2]

To summarize, it is unclear what specific measures the Soviets had in mind when they insisted that discussions of limits on strategic defense had to be part of the Geneva talks on strategic and intermediate-range missiles. The best one can expect is a deepened commitment to the existing ABM Treaty and further measures limiting ambiguous activities.

The introduction of SDI into the Geneva negotiations is therefore in some respects "spooky." The President has initiated a major new effort whose expressed purpose would invalidate the very foundation of the ABM Treaty. Yet, when discussing alleged Soviet violations on the fringes of the treaty, the US proclaims opposition to *any* action weakening the ABM Treaty. The actual technical activities now carried out by either side and specifically projected by the US side for the next few years are in general com-

Soviet phased-array radar under construction in Krasnoyarsk, Siberia. The US claims that this radar will violate the ABM Treaty. (Defense Department drawing.)

pliance with the existing ABM Treaty. Although US oratory is probably the most aggressive and provocative component of the situation, it is hardly a subject for discussion across the table in Geneva.

SDI's divergent goals

It was only after the President's speech that a group headed by former NASA administrator James C. Fletcher began a major technical study (see the bibliography on page 41). However, Fletcher's panel was charged with designing an optimum research program to implement the President's decision, not with reviewing the wisdom or even the feasibility of redirecting US nuclear strategy into a defense-dominated pattern. Even though the President's major policy address was not preceded by the usual internal staff analyses, it obliged the US Government to get in step with the Commander-in-Chief. However, because the motives of the different sections of government implementing SDI are extremely varied, government statements show gross inconsistencies about SDI's goals and missions. Let me list some of these divergent avowed purposes of SDI.

▶ *To increase US bargaining strength in Geneva.* The "bargaining strength" argument for acquiring a new weapons system is the last refuge for an otherwise bad military concept. Again, we should learn from history. In 1969 the

technically poor "Safeguard" ABM system passed by one vote in the Senate, with the "bargaining leverage" argument possibly being decisive. The US later abandoned the resulting Grand Forks installation. With over 10 000 deliverable US nuclear weapons facing the Soviet Union, one must resist the argument that just one more US move will change Soviet policy.

▶ *To begin research and development aimed at establishing the feasibility of totally defending the nation against ballistic missiles, with deployment starting not much sooner than the end of the century; offensive forces are to become "obsolete."* This is in essence the President's "vision" as echoed by the Secretary of Defense and other high officials. Yet at the technical working level this goal has been largely pushed into the background. The reason is clear: Total protection implies an almost leakproof defense against all means of delivery of nuclear weapons, and most members of the military technical community consider this to be totally unrealistic, particularly if the Soviet Union makes countermoves. No one has identified an even plausible scenario through which we can make a safe transition from the current offense-dominated nuclear balance to the proposed defense-dominated stability.

▶ *To protect the US against "limited" missile attacks due to terrorist acts and accidental or unauthorized launches.*

Historically, this was the mission of the "Sentinel" system proposed in 1967 by then Secretary of Defense Robert McNamara (see the photograph on page 35), and it may possibly be a mission of the now-deployed Moscow ABM system as limited by the ABM Treaty. The problem with a "light" area-wide defense is that it will always be a source of concern to an opponent that it might readily thicken into a heavy defense. Moreover, the goal of generating a light defense is a clear mismatch to the singular emphasis and special organizational structure of SDI. Finally, third-country terrorists, should they acquire some nuclear weapons, will most likely be not capable of delivering them with ballistic missiles.

▶ *To enhance deterrence by defending the landbased missile force.* If the fixed US landbased missiles—Minuteman and maybe MX—were defended with ABMs, they would better survive an attack by high-accuracy Soviet reentry vehicles. Technically, this is a much simpler task than defending the entire nation, because the enemy need be denied entry into only a very small region near the silos to be defended. It would clearly be extremely inefficient if the principal benefit of SDI, with its multilayer systems protecting the entire United States and even Europe, were a "spinoff," namely, the protection of one thousand or so highly localized areas in the United. States. It makes no sense to deploy a defense system much more expensive than the value of the assets defended and more costly than the increased cost the enemy would have to bear to defeat the system. At any rate, defending missile silos does not in any way meet the President's concern about lifting the balance of terror.

▶ *To enhance deterrence through partial defense by "complicating the task of the attacker."* Here the ABM system is added to offensive forces. Without abandoning total defense as a long-range goal, most supporters of SDI argue that the existence of even a leaky defense would make it more costly and less predictable for an opponent to attack the United States, and would therefore help deter nuclear war. This mission is currently the dominant justification for the SDI program. Yet it falls far short of promising to fulfill the President's vision. It offers no hope of relieving the balance of terror and is apt to aggravate the problem. Taken at face value, this mission is an escalation, adding another layer to offensive strategic systems. The validity of this mission depends critically on the performance of the systems actually considered and, in particular, on the Soviet response to any deployed systems. While a partial defense does not protect population and industry from a massive strike, it might be perceived by the opponent as blunting a second strike by the remnants of the strategic forces surviving a first-strike attack.

Paralleling the diversity in the avowed purposes of SDI is the diversity in estimates as to when the initiative can actually lead to the deployment of some defenses. Most discussion focuses on deciding in the 1990s whether or not to deploy a nationwide ABM system. However, there are a few strong voices advocating earlier deployment of some limited defenses, thus creating a more imminent confrontation with the ABM Treaty.

As the bibliography on page 41 indicates, scientists both sympathetic and not sympathetic to SDI have given extensive technical descriptions of the program.

There has been too much talk about "technological breakthroughs" and the need for "demonstrations" of new technologies to convince the SDI skeptics of the error of their ways. Such demonstrations are just what we do not need. The problem of understanding the merit of defenses is detailed and complex in the extreme. Total protection of the

Homing Overlay Experiment, 10 June 1984 (GMT). A few minutes after this experimental interceptor took off from Kwajalein missile range in the Marshall Islands, it destroyed its target, a reentry vehicle from an ICBM launched from Vandenberg AFB in California. (US Army photograph.)

nation requires deployment of nearly impenetrable systems—a systems and operational problem of vast proportion. Within the more limited goal of "enhanced deterrence" we need an objective and realistic evaluation of SDI's marginal advantage of defense vs offense. What we do not need is the hype that goes with the demonstration of a single intercept of a cooperative target by some new contraption. The Army's **Homing Overlay Experiment was an example of this kind. (See the photograph above.) Such demonstrations tend to emphasize the easy rather than the complex and difficult problems.** This is the surest way to waste money, to mislead Congress and the public, to steer the program in the wrong direction and to produce an unnecessary confrontation with the ABM Treaty. At best, a premature demonstration hypes the impractical; to quote Samuel Johnson,[3] it is "... like a dog walking on his hind legs. It is not done well but you are surprised to find it done at all."

In view of the foregoing there is a certain "house of cards" atmosphere around the current SDI program. There is a solid core of valuable technical activity, most of which is not new

and could be pursued and improved within existing policies.

The issues

Along with the ambiguities of the SDI mission there is a wide spectrum of opinion on the nature of the issues themselves. Naturally there is a large range of opinion regarding the technical potential of an ABM research and technology program. Most of the criticism of SDI focuses not as much on the technical content of the research program as on the implications of that program should it lead to demonstrations, prototypes or deployment. If the momentum of SDI were to drive the program beyond the technology phase, the results could likely be adverse to US national security. However, if it were not for the raised expectations built up by the current SDI promotion, the research program itself could weigh the merit of strategic defense against its costs and dangers.

The issue should not be whether governments can now replace mutual assured destruction in a predictable manner by protection based on defense—political authority does not have the technical power to do this, and would not have the power even if

Annotated technical bibliography

● *Aviation Week and Space Technology* (17, 24, 31 October 1984). Summarizes findings of the Fletcher panel, including a description of candidate technologies for boost-phase, mid-course and terminal intercept. Discusses sensor and battle management developments and overall system concepts.

● H. Brown, *The Strategic Defense Initiative: Defensive Systems and the Strategic Debate*, California Seminar on International Security and Foreign Policy, Santa Monica, Calif. (1985). Clarifies the main issues of debate, including change of doctrine following deployment of population defense, defense of retaliatory forces, limiting damage through limited defense, the relationship of defense and the ABM Treaty, and future arms priorities for research and development.

● A. B. Carter, *Directed Energy Missile Defense in Space*, Office of Technology Assessment, Washington, D.C. (April 1984). Primarily a technical description and analysis of directed-energy technologies as applied to boost-phase intercept. Discusses associated technological issues such as sensing, aiming, command and control. The paper also assesses various countermeasures available to the offense.

● A. B. Carter, D. N. Schwartz, eds., *Ballistic Missile Defense*, Brookings Institution, Washington, D.C. (1984). Deals with the technologies of nuclear and non-nuclear ballistic-missile defense systems for a variety of threats. Considers the political and strategic ramifications. This book also presents the personal opinions of several authorities with a spectrum of views on defense issues.

● *Defense Against Ballistic Missiles: An Assessment of Technologies and Policy Implications*, Department of Defense, Washington, D. C. (6 March 1984). Describes the major technical elements of the SDI program and the components of a defensive system. Gives rationale for SDI in terms of protection against Soviet breakout from the ABM Treaty, eventual defense dominance of the strategic relationship between the US and the USSR, and the effect upon arms control.

● S. D. Drell, P. J. Farley, D. Holloway, *The Reagan Strategic Defense Initiative: A Technical, Political, and Arms Control Assessment*, Stanford Center for International Security and Arms Control, Stanford, Calif. (July 1984). Detailed technical analysis of candidate technologies for boost-phase intercept, including possible countermeasures. Discusses major problems in discrimination and battle management for mid-course and terminal intercept. Report also deals with present and possible future Soviet responses, and the implications for arms control in general and the ABM Treaty in particular.

● S. D. Drell, W. K. H. Panofsky, Issues in Science and Technology 1, 45 (Fall 1984). A critical discussion of ABM technologies and their expected performance and implications.

● *The Fallacy of Star Wars*, Union of Concerned Scientists, Cambridge, Mass. (1983). Discusses various candidate technologies for intercept, discusses systems implications of using these technologies, and discusses some possible countermeasures available to the offense. A condensed version of this report appears in *Scientific American*, October 1984, p. 39.

● *Final Report of President's Commission on Strategic Forces*, US Government Printing Office, Washington, D. C. (21 March 1984). Advocates research on defense technologies within the constraints of the ABM Treaty so that the US can understand the technological limitations and guard against Soviet breakout.

● J. C. Fletcher, chairman, *Report of Defensive Technologies Study Team* (classified except for volumes on Battle Management, Communications and Data Processing), Department of Defense, Washington, D.C. (1983). Study reviews all relevant areas of technology for defense against ballistic missiles, as-

sesses their current status and recommends research programs leading to technology demonstrations by the early 1990s.

● J. C. Fletcher, Issues in Science and Technology 1, 15 (Fall 1984). Describes the technologies for ballistic missile defense, identifying the most promising approaches and system components, as well as critical technologies that must be demonstrated before an effective defense would be feasible.

● D. O. Graham, *High Frontier: A New National Strategy*, High Frontier, Inc, Washington, D.C. (1982). A schematic description of a three-tiered, non-nuclear defense system for protecting hard sites. Also discusses the strategic rationale for such a defense.

● F. S. Hoffman, study director, *Ballistic Missile Defenses and US National Security: Summary Report*, prepared for the Future Security Strategic Study, Department of Defense, Washington, D.C. (October 1983). Primarily assesses the strategic and political implications of defensive systems of various technical capabilities.

● *The New York Times* (2–7 March 1984). A series of six articles summarizing the genesis, technologies and vulnerability of a Star Wars system. Considers Soviet response options and the political and strategic consequences.

● *Report to the Congress on the Stategic Defense Initiative*, Department of Defense, Washington, D.C. (1985). This report is the result of various Congressional acts that require the Strategic Defense Initiative Office to submit a detailed program description and budgets. Of particular interest is appendix B, which deals with the DOD position on the contentious issue of whether or not the planned demonstration projects under the SDI program are in compliance with the ABM Treaty. The DOD position hinges on the interpretation that the components demonstrated are not actual subsystems of workable ABM systems.

Congressional testimony:

● J. A. Abrahamson, director of the Department of Defense SDI office, Senate Appropriations Committee, 15 May 1984. General description of program elements, phases and funding for SDI.

● H. M. Agnew, vice-chairman of the Defensive Technologies Study Team, 8 March 1984, Senate Armed Services Committee. General rationale for the SDI program.

● R. D. DeLauer, Under Secretary of Defense for Reserch and Engineering, House Armed Services Committee, 1 March 1984. Summarizes the technical findings of the Fletcher study and makes recommendations for organizational structure and funding.

● S. D. Drell, Senate Foreign Relations Committee, 25 April 1984. Discusses technical shortcomings of candidate technologies, barriers to achieving feasibility, hazards to the strategic relationship with the Soviet Union, and implications for arms control.

● J. C. Fletcher, Senate Armed Services Committee, 8 March 1984. General overview of the work of the Defensive Technologies Study Team. Gives the study's conclusions and rationale for the SDI program.

● G. C. Smith, former US ambassador to SALT I, Senate Armed Services Committee, 24 April 1984. General discussion of the potential deleterious political and strategic consequences of SDI and its implications for arms control and the strategic relationship between the United States and the Soviet Union.

● G. Yonas, chairman of the directed-energy-weapon panel of the Defensive Technologies Study Team and chief scientist of the Department of Defense SDI office, Senate Armed Services Committee, 8 March 1984. General description of major elements and requirements of the directed-energy-weapon portion of the SDI program. Emphasis is on long-range development against a responsive threat.

nuclear weapons were drastically reduced worldwide. The real current questions are these:

▶ On what scale and within what framework should the United States conduct strategic defense research?

▶ Should strategic defense research be limited more strictly than it is by the currently binding constraints of the ABM Treaty? Alternately, should SDI go "full speed ahead" in disregard of the treaty?

▶ When should or when can the US reasonably make policy decisions based on the expected results of such research?

The principal issue is not whether strategic defense research should be done at all. The scale and method of carrying out defense research work should be governed by considerations of resource constraints, available talent, expected dangers and expected utility—considerations similar to those that affect the level of support by the Federal government of other areas of science and technology.

The media have dedicated much space to the debate over whether or not "it" could work. Yet, as we have seen, it is unclear what mission "it" is to accomplish. In such debates those critics who are pessimistic that SDI is a promising undertaking expose themselves to charges of being "naysayers"; one can cite many historical instances where highly reputable scientists said that certain achievements are impossible, only to be shown incorrect by later developments. Yet here the issue is not whether or when scientists can accomplish some ill-defined "it." The issue is that the current program—and particularly its accompanying oratory—is beset by gross inconsistencies and poses great dangers to national and world security and to strategic stability.

Economics and performance do matter. Paul Nitze, the President's senior adviser on arms control, brought this into perspective succinctly in a speech to the Philadelphia World Affairs Council in February. Nitze reiterated that a ballistic missile defense, to be useful, would have to pass two stringent tests:

▶ It must be based invulnerably. This means that it should be no easier for the attacker to destroy the defense bases than for the defense to destroy vehicles of the attacker.

▶ The marginal economic cost of a defense must be less than what it would cost the opponent to either increase its offensive forces or upgrade its ability to penetrate defenses, or both, so as to leave the defended country just as vulnerable as it was without a defense.

No past ABM system considered has come even close to passing these criteria, and I am extremely doubtful that honest analysis will project that a future system will be luckier. The reasons are clear on physical grounds: Space-based defenses are more vulnerable than ballistic missiles in any part of their orbits, yet almost all "Star Wars" technologies involve space-based components. Even land-based terminal defenses involve the problem that the eyes of the system—its radars—are generally harder to protect than the interceptors that they guide. A full ABM system is untestable and requires invulnerable and exceedingly precise sensing and data processing on a truly unprecedented scale in a hostile environment.

I know of no responsible study, either within or outside government, that indicates that the competition between the marginal cost of ballistic missile defense and that of offensive countermeasures even remotely favors the defense.

Any objective evaluation of the above "Nitze criteria" would require that the government, separate from SDI management, establish and maintain a "Blue Team" and a "Red Team," with the Blue Team charged with conceptual design of defensive systems and the Red Team charged with defeating those systems at minimal cost. The Fletcher report made a start in this direction. The government must consider the whole spectrum of defense countermeasures, countercountermeasures, and so on. In the absence of such a critical process, the present SDI promotion tends toward unbridled salesmanship without meaningful checks and balances.

The Commander-in-Chief decided on an intensive ABM research program before defining the missions for the product of that research. Without premature demonstration, without confronting the ABM Treaty and without changing current strategies, how large should ABM research be and what should it contain? If this were a rational world, there would be no reason why strategic defense research would require any significant increase above the $1.4 billion per year that the Defense Department has been spending on such programs for some time. These programs are yielding valuable developments, some of which involve challenging technology. Any program can gain in effectiveness by better coordinated management and by the elimination of clearly unpromising technologies. For instance, the program would lose little if it did not include expensive systems based on long-wavelength infrared lasers, which require vulnerable basing of extremely massive and power-hungry delivery devices in space. I maintain that one cannot justify to the American people and Congress any significant expansion of the strategic defense program until

there is good reason to believe that the prospects of passing the "Nitze criteria" are better than they have ever been in the past.

Ballistic missile defense research should continue to protect against technological surprise, that is, to keep our eyes focused on emerging technologies so that if the Soviets do begin to reveal some new system, we can understand and counter it. We do need a limited program so that the Soviets cannot hope to gain significant advantage by clandestine preparations for a "breakout" from the ABM regime while our technology base has withered.[4] Due to the enormity of the destructive power of the offensive forces, "breakout" activities by either side could only have a significant impact on the strategic balance over a period of a decade or decades. The specter of suddenly waking up with defenses in operation on the other side is pure fantasy. Finally, we must continue research to maintain a realistic and up-to-date assessment of the marginal costs of offense and defense.

What is frightening at this time is the blatant salesmanship, which does not focus on SDI's military merits, but which appeals to economic self interest. The SDI management in its presentations throughout the nation has attempted to make it clear that Defense Department "high technology" money will flow very heavily into SDI channels and that US scientists and high-technology contractors should get into the act early. Predictably, this in turn has generated strong lobbying pressures on Congress.

Foreign reaction to SDI

The matter is even more insidious in respect to persuading America's allies to join the SDI effort. On strategic grounds, Europe's reaction to the President's speech on 23 March 1983 has been generally cool if not outright negative.[5] There are good reasons for this cautious attitude, and the merit of these reasons depends on the course SDI takes. If SDI expands but does not lead to predictable consequences, then the strategic–tactical situation in Europe would not be changed as such: Resources would be diverted from the higher-priority but less glamorous needs facing the allies. The President has emphasized that his proposed umbrella would unfold over Europe as well as the US. Yet, as is predictable, if the Soviets would match such an effort on their own—in particular, if the Soviets were to accept the President's offer of sharing defense technology—then Europe's security problems would grow. If, as a result, the ABM Treaty were abrogated and the Soviets and the US were to deploy increasingly large anti-ballistic missile defenses, then the first

Nike Zeus at White Sands, New Mexico, 25 years ago. This missile was designed to be an exo-atmospheric antimissile interceptor. (US Army photograph.)

consequence would be that the relatively small European independent missile forces, such as the French *force de frappe* and the British missiles, could not penetrate. In other words, European independent nuclear deterrence would be the first to lose credibility.

If we make the more extreme assumption of an effective *total* defense of the United States, Europe and the Soviet Union, then Europe faces the ultimate threat of "decoupling"—a situation in which a war, either nuclear or non-nuclear, could be fought on European soil without risk to the home countries of the superpowers. This universal umbrella—although technically unrealistic in any foreseeable future—remains the avowed goal of SDI in the President's pronouncements.

Notwithstanding occasional newspaper headlines of the "Allies endorse Star Wars" type, no European government leader has endorsed SDI beyond research that conforms to the ABM Treaty.[6] However, the US promoters have identified SDI as heralding the wave of the future for military technology and have made it clear that unless Europe joins the SDI effort now, "the technology train will have left the station." This argument, its lack of substance notwithstanding, bears a great deal of weight. European leaders are acutely aware of a widening gap in military technology between the US and Western Europe, and this frankly commercial SDI appeal has fallen on some receptive ears.

Paradoxically, the Defense Department's heightened barriers against certain high-technology exports to Western Europe have exacerbated Euro-

pean problems. Thus we are facing the spectacle of Defense Department spokesmen raising their voices against sharing military technology with the allies, while at the same time promoting that very sharing in connection with SDI. Negotiations now in progress are attempting to reconcile these conflicting values.

Soviet response to SDI has been strongly negative, even though the Soviets have traditionally devoted a considerably larger part of their strategic military effort to defense than has the US. In fact, in the late 1960s it was US diplomatic and informal initiatives that persuaded the Soviets that countrywide ballistic missile defenses would be escalatory and destabilizing. The unfavorable economic competition between ABMs and offensive countermeasures then pertaining, together with the obvious technical difficulties of ABM, persuaded both the Soviets and the US that the 1972 ABM Treaty enhanced the national security of both parties as well as world security. It is indeed paradoxical that now, starting with the President's speech, it is again the US taking the initiative, but this time in precisely the opposite direction. Yet there is no persuasive evidence whatever that new technical or economic factors have reversed or are likely to reverse the earlier rationale.

On its face, the Soviet reaction coincides with that of many domestic critics of SDI: They express skepticism that ABM can ever offer total protection, yet they feel threatened by a partial system reinforcing the potential of US offensive systems. Soviet responses accuse SDI of leading inevitably to the weaponization of space and signaling

the beginning of a new high-technology arms race.

Paradoxically, the Soviet response may have reinforced the high-technology promotion that SDI proponents have made in Europe. The Soviets appear concerned that SDI signals a new jump in the technology of military systems. It is this factor, paced by the lag in Soviet computer achievements, that has lent some credence to the statement that SDI has been a factor—how large a factor is difficult to ascertain—causing the Soviets to return to the Geneva negotiations. Although the Soviet reaction may be understandably motivated by concern that SDI signals the beginning of a new high-technology arms race, their response has given US technology much more bargaining leverage than is readily justifiable on its merits.

Outlook

The President's speech of 23 March 1983 was motivated by easily understandable and even idealistic goals, but was ill prepared as to the means to attain those goals and to deal with the consequences of the announcement itself. There has evolved from that starting point a new organizational structure dedicated to defense against the delivery of nuclear weapons by ballistic missile, but not by cruise missile or aircraft. (See the figure on page 36.) There is highly organized political promotion of the SDI program, both nationally and internationally. There is a growing public debate among "experts" differing in their projections of the likelihood of "success" of the program without any accepted definition of what the goals of the program are. In fact there is no genuinely new program on the technical level and no prospect for a change in the basic offense–defense relationship in the nuclear age from what it has been for decades.

There exists at this time no technical basis that justifies expanding research and technology programs in ballistic-missile defense beyond a program of limited experimentation and of studies examining the promise of defense in an objective rather than promotional manner. I object strongly to the politicization and "high technology" salesmanship of SDI.

The nuclear-arms race has been fueled by permitting nuclear weapons to become symbols of political strength, resolve and bargaining power, thereby deflecting attention from the great dangers and lack of usefulness of nuclear weapons in real military situations. It is this triumph of perception over technical reality that has denied the American and Soviet decision makers from answering the fateful question, "When is enough enough?" We must not permit this to happen again in connection with SDI. Otherwise we are going to embark on another upward spiral in the arms race without any rational understanding as to means or purpose.

In spite of the lack of a solid technical basis for SDI, a situation of enormous danger has evolved that includes a threat to strategic stability, a threat to the viability of the ABM Treaty which has served the security needs of the Alliance well, a threat of new major military expenditures, a threat of deploying a whole new generation of offensive and defensive weapons systems, and a threat of distorting technical manpower and financial priorities both within the military and within the entire national budget. Scientists worldwide have a responsibility to remind their fellow citizens and national leaders that nuclear weapons are real and lethal. If we permit overblown expectations of protection from nuclear weapons to conceal the stark reality that even a small fraction of the world's nuclear weapons can end civilization, then the world will move closer to disaster.

References

1. McG. Bundy, G. F. Kennan, R. S. McMamara, G. Smith, Foreign Affairs **63**, 264 (Winter 1984–85).

2. See R. L. Garwin, K. Gottfried, D. L. Hasner, Sci. Am., June 1984, p. 45.

3. See J. Boswell, *The Life of Samuel Johnson*, quotation from 31 July 1763, Oxford U. P., Oxford (1982).

4. For an outline of a sensible nonconfrontational ABM research program, see S. D. Drell, T. Johnson, co-chairmen, *Strategic Missile Defense: Necessities, Prospects and Dangers in the Near Term*, Stanford Ctr. for Intl. Security and Arms Control, Stanford, Calif., April 1985.

5. For an outline of allied response to SDI, see P. E. Gallis, M. M. Lowenthal, M. S. Smith, *The Strategic Defense Initiative and United States Alliance Strategy*, Congressional Research Service, Library of Congress, Washington, D.C., 1 February 1985.

6. For a thoughtful European response to the issue of working within the ABM Treaty, see the 15 March 1985 speech of British Foreign and Commonwealth Secretary Geoffrey Howe, British Embassy Information Department, Washington, D.C.   □

SDI attempts to zap APS directed-energy weapons report

PHYSICS TODAY/JUNE 1987

One of the most striking chapters of The American Physical Society's report on directed-energy weapons appeared even before the document was formally issued. The chapter never appeared in print. It revealed the way the Strategic Defense Initiative office tried to influence the press accounts of the report, which bears the title *Science and Technology of Directed Energy Weapons*. (See PHYSICS TODAY, May, page S1, for summary of report.)

As soon as the SDI organization learned that copies of the 424-page report were being handed out to news reporters for background reading before the official release the next morning, the agency distributed a statement of its own. Doing this, SDI apparently reasoned, would allow reporters to have an instant response to the findings of the report in time for the APS news conference at 10 am on 23 April. In fact, SDI had been tipped that both *The New York Times* and *The Washington Post* intended to publish accounts of the report before its release. Accordingly, the Pentagon's decision to put out its own message as early as possible seemed right in the circumstance. The surprising thing was that in the more extensive press coverage that followed (see the news story on page 55) the SDI response was virtually ignored.

Acclaim. This turned out to be a serious deficiency in the coverage of the APS report because the SDI statement and the answers of panel members at the morning news conference were revealing. In its news handout, SDI acclaims the nine technical chapters as "an objective independent appraisal of various technologies." Among the reasons for praising the report, the organization says, is that by performing the study APS "has responded to the President's challenge to scientists and engineers . . . to join together to seek defensive solutions to the ballistic missile threat. The report offers a challenge to the APS membership to help us seek innovative solutions to the technical issues we must resolve to develop effective directed-energy weapons."

"JUST WHEN WE WERE READY FOR LAUNCHING"

HERBLOCK/THE WASHINGTON POST

The cheering stops abruptly, however, when the SDI announcement boos the report's summary and major conclusions as "subjective and unduly pessimistic." What's more, says the SDI office, the report bears "the additional problem of being a snapshot in time" that shows it already out of date upon publication. According to SDI, "we have made significant progress" in free-electron lasers and neutral particle beams in the seven months since the physicists completed their study last summer. Those advances, SDI asserts,

are "several orders of magnitude" better than the performance reported by the study group. The obvious implication is that directed-energy weapons technologies are developing faster than the APS panel found, thus approaching the levels of feasibility required to transmogrify President Reagan's dream of a ballistic missile defense into the reality popularly known as "Star Wars."

SDI's statement claims that in the intervening months it has scaled the first induction free-electron laser from

Press conference for DEW report attracted some 75 journalists to the American Physical Society meeting in Washington on 23 April. Cochairmen of the study were C. Kumar N. Patel of Bell Labs (left inset) and Nicolaas Bloembergen of Harvard (right).

0.8 cm to 10.6 microns. According to members of the APS panel, the figures are a spurious comparison of power levels and wavelengths, which are unrelated. SDI also asserts that the brightness of the electron beam injector for the free-electron laser at Los Alamos has improved by two orders of magnitude and that the injector, designed to reach 5 MeV, has achieved full beam current without significant emittance growth. The injector now operates at 1 MeV. Further, the SDI statement boasts that a new cw ion source for neutral beam machines, developed at Culham Laboratory in the United Kingdom, now meets the goal for beam quality.

Irony. It is ironic that the Pentagon considers the report outdated, because the publication was held up for some seven months while drafts underwent a circuitous process of security clearance through the SDI office, various other units in the Pentagon, the departments of Energy and State and even the White House Office of Science and Technology Policy. Not every agency considered it politically prudent to release the report, especially during budget-cutting season on Capitol Hill, but only two government offices actually opposed its unclassified publication. In the end the view of SDI's director, Lieutenant General James A. Abrahamson, prevailed. He insisted all along that the report be approved for dissemination, though he acknowledged that some cuts and changes had to be made for security reasons.

In a statement separate from the report, APS's review team, headed by George Pake of Xerox, observed that

there had been "small but significant deletions"—notably in references to possible vulnerabilities and potential countermeasures that might be used against laser light and particle beam technologies of a Star Wars system. The review panel termed the deletions a "minuscule" fraction of the study group's original report. When a reporter asked at the 23 April press briefing whether the government's review had dragged on too long, the cochairman of the study group, C. Kumar N. Patel of Bell Labs, replied that it could have been "compressed" and that many of the Pentagon's complaints were "onesies and twosies corrections," but that none of the changes affected the group's essential conclusions.

The draft report was submitted to security reviews because the study group, whose members all hold clearances, had access to classified information provided in detailed briefings and visits to restricted laboratories. If it had refused classified information about SDI work, the panel would have invited criticism that it lacked essential knowledge or understanding of directed-energy technologies.

Of course the 15-member panel already possessed wide familiarity with the research. Its cochairmen were Patel, who invented the carbon dioxide laser, a possible directed-energy weapon, and Nicolaas Bloembergen of Harvard University, who won a Nobel Prize for his work on nonlinear optics and laser interactions. The other members also are preeminent in directed-energy technologies, working at government or industrial laboratories and at leading universities.

Two A's. Their report had been eagerly expected for months by arms control experts, members of Congress, the Pentagon and the press. When it finally appeared it was greeted with the expected partisanship—though there were some unexpected exceptions. Louis Marquet, SDI's deputy director for technology, for instance, characterized the report as "very responsible" at a news conference called at the Pentagon and underscored the study group's conclusion about the need to achieve a thorough understanding of the physics underlying directed-energy weapons. That a panel of nonpartisan outsiders was able to explore heretofore inaccessible territory and draw detailed charts for future work is really remarkable, observed Marquet. "I think it's probably unique in government annals for an open society to review a classified program." He concluded that "both of us gave each other A's. I think they gave us an A from the standpoint of understanding the technical aspects of the program. . . . There was nothing in their report that says we're completely out of our minds, that some things are beyond the laws of physics."

Gerold Yonas, who was SDI's first chief scientist until he joined Titan Corporation last summer, calls the report "an important contribution, providing understanding and perspective to the debate over SDI. It is the most useful balance sheet for seeing where we are and where we ought to go with directed-energy defenses and, as such, goes well beyond the bumper stickers and newspaper slogans that seem to entertain and do so much to misinform the public."

It was the need to inform public opinion that led Richard DeLauer in 1983, when Under Secretary of Defense, to welcome the APS study and offer DOD's cooperation. DeLauer's enthusiasm was followed a year and a half later by letters of encouragement from Abrahamson and George A. Keyworth II, then the President's science adviser. All three asked the study group to limit its examination of the SDI program to directed-energy weapons (or DEWs, as they are known in defense circles). DeLauer wrote that the panel should concentrate on "the areas of the society's recognized expertise, namely physical analysis, and to refrain from clouding the study with policy evaluations which could detract from its technical credibility." The panel did precisely that.

ETA and ATA. Since President Reagan revealed his grandiose vision of perfect defenses and impotent missiles, laser light and particle beams have occupied a central place in both popular conceptions of Star Wars and SDI's own strategy. SDI's annual report to Congress, made public just two days before the APS report was released, says, for instance, that DEWs "are critical to providing a wide selection of defense options" and "the key to defeating the more serious threats that might be deployed in response to first-generation US defenses." The SDI report speaks of "significant accomplishments" for DEWs, consisting mainly of early stages of design, construction or tests of free-electron lasers (see the news story on page 17), such as the induction linac Experimental Test Accelerator at Livermore and the Paladin experiment using the Advanced Test Accelerator at Livermore; hydrogen fluoride chemical lasers; laser mirrors; and other optical components.

Still, the latest "breakthroughs" in DEWs appeared in SDI's response to the APS report. When a reporter attending the APS news conference asked about the SDI claim of improving the brightness (by two orders of magnitude) of an rf injector for free-electron lasers at Los Alamos, a panel member, Andrew Sessler of Lawrence Berkeley Laboratory, replied that this referred only to a component, a photocathode source, and while it represented a significant improvement, "it is still a long way from a cathode to an accelerated beam and a longer way from an accelerated beam to a free-electron laser." As for the induction linac approach, the report points out, brightness still needs to be improved by a factor of 4. What's more, both rf and induction techniques suffer from sideband instabilities and harmonic generation, and not much is known about the

---

APS Council speaks out on SDI

A day after The American Physical Society issued a 424-page report on the directed-energy weapons that are being considered as part of the Strategic Defense Initiative, the APS Council, the elected governing body of the society's more than 38 000 members, released the following statement. The statement, adopted after considerable discussion at the council meeting on 24 April, is more far-reaching than the conclusions in the report, Science and Technology of Directed Energy Weapons. *It reads:*

A major study of the science and technology of directed-energy weapons, conducted by a study group of The American Physical Society, found that:
▶ The development of an effective ballistic missile defense utilizing directed-energy weapons would require performamnce levels that vastly exceed current capabilities.
▶ There is insufficient information to decide whether the required performance levels can be achieved.
▶ A decade or more of intensive research would be required to provide the technical knowledge needed for an informed decision about the potential effectiveness and survivability of directed-energy weapons systems.
▶ The important issues of system integration and effectiveness depend critically on information that does not now exist.

The Council of the APS believes that it has a public responsibility to express concerns about SDI that go beyond the issues of directed-energy weapons covered in the study.
▶ Even a very small percentage of nuclear weapons penetrating a defensive system would cause human suffering and death far beyond that ever seen on this planet.
▶ It is likely to be decades, if ever, before an effective, reliable and survivable defensive system could be deployed.
▶ Development of prototypes or deployment of SDI components in a state of technological uncertainty risks enormous waste of financial and human resources.

In view of the large gap between current technology and the advanced levels required for an effective missile defense, the SDI program should not be a controlling factor in US security planning and the process of arms control.

It is the judgment of the Council of the APS that there should be no early commitment to the deployment of SDI components.

---

relative importance of such phenomena in oscillators and amplifiers. Accordingly, SDI's developments, said Sessler, are "a small step, an important step, but still a small step."

Nor was another APS conclusion invalidated by SDI's claim of higher beam currents for an ion source for neutral particle beam accelerators, said Bruce Miller of Sandia National Laboratories. The conclusion that such accelerators need to be scaled up by at least two orders of magnitude in voltage and duty cycle was still correct. Another member of the study group, Richard Zare of Stanford University, observed that it is tricky to extrapolate from existing performance levels to higher ones, which SDI officials often did in describing to Congress the case for directed-energy systems.

Reactions. Within hours of its release, the report was praised by Star Wars adversaries in the Senate, such as Wisconsin's William Proxmire and Louisiana's J. Bennett Johnston. Senate advocates of SDI, including Indiana's Dan Quale and Wyoming's Malcolm Wallop, indicated they doubted the study group's conclusion that at least a decade would be needed before a realistic decision could be made about the feasibility of DEWs. A group calling itself the Science and Engineering Committee for a Secure World, under the chairmanship of Frederick Seitz,

onetime president of the National Academy of Sciences, Rockefeller University and APS, complained that the APS report is simply irrelevant, because it does not deal with kinetic-energy weapons. KEWs, the group notes, might be developed and deployed by 1993 or 1994, as a first phase in the evolution of a missile shield. Indeed, since the publication of a report on KEWs by the George C. Marshall Institute (see PHYSICS TODAY, January, page 47), which Seitz also heads, Defense Secretary Caspar Weinberger has urged the President and Congress to adopt a phased SDI program for starters.

The APS panel once discussed expanding the study to include kinetic-energy technologies, but decided to stick to the issues designated by DeLauer, Keyworth and Abrahamson. To have included such matters as kinetic kill weapons, as well as battle management, say, and computer software, would have meant adding more panel members with expertise in those matters and taking more time to conduct the study.

Even limited to DEWs the APS report is certain to have a significant impact on the fierce political debate currently being waged on Capitol Hill over how soon to deploy a DEW defense system and how much to fund SDI in the next few years. The report prompted Representative Henry J. Hyde, Republican of Illinois, to send a "Dear colleague" letter on 6 May under a two-line title:

Falsehood flies on falcon's wings
While truth shuffles along in wooden
    shoes!

The letter declares that "the report is egregiously flawed in some very important respects" and claims it is a "rather misleading piece of work." To support his assertion, Hyde attached an "SDI watch" column from the 22 May issue of *National Review*, based, according to Hyde, on a paper written by Gregory Canavan of Los Alamos. The article states that the study group's estimates for the electric power required to maintain DEW-equipped satellites in orbit are too high by a factor of 30. In fact, the difference between the report's estimate and Canavan's estimate is about a factor of 2—a reasonable disagreement because no detailed study of scaling SDI architecture has been done by anyone.

Protests. Other objections were raised on 19 May by Seitz, accompanied by Lowell Wood of Lawrence Livermore, during a combined briefing for a scattering of Republican members of the House of Representatives and the press. Seitz, who is chairman of SDI's scientific advisory committee in addition to his other affiliations, protested that the APS report "is not worthy of serious consideration" because it contains "numerous errors, inconsistencies and unrealistic assumptions"—"always in the direction of making . . . SDI seem farther from achievement of its objectives than it actually is."

Wood's qualms were of a different order. Though Wood based his technical comments on a 60-page paper written by Canavan, his discussion of the APS report often sounded like political polemic. He charged that some members of the study group and the review committee either opposed SDI or "expressed deep reservations in private" about the program. The APS report, Wood argued, is therefore not

an objective examination of DEWs. He then attacked the APS Council's statement (see box, page 45) on early deployment, saying the group was passing judgment "on the basis of no technical studies or reviews done by it or under its auspices."

In a telephone interview, Canavan disavowed such remarks. "My paper is strictly a technical review of the APS report," he observed, "and it says very clearly that some of the calculations are wrong. The numerical errors have nothing to do with the personalities behind the report. There was no cabal. It's not even an issue of scaling or architecture. The fault is not in the panel's politics but in its math."

'Boo-boo.' One of the glaring errors Canavan caught is the inconsistency between the first conclusion and the full text. The conclusion says chemical lasers have attained power levels exceeding 200 kW with acceptable beam quality—a factor of 100 below the minimum requirements of an SDI system. In an early draft, it turns out, the APS study group had used the correct number for chemical lasers, only to have this altered by the SDI office during a classification review. Later in the review cycle, however, SDI officials relented and allowed the panel to write that chemical lasers had attained more than 1 MW. The Pentagon's about-face took place because in the meantime *Aviation Week* had printed the correct figure. The APS panel changed chapter 3 but failed to change conclusion 1 accordingly. "It was an editorial boo-boo," says an APS panelist. If the conclusion had been changed, it would have read, "greater than 1 MW and thus a factor of 20." Even so, notes Patel, the APS conclusion is correct in stating that chemical laser output will encounter problems in scaling to higher power levels, probably by two orders of magnitude.

In citing errors relating to laser lethality effects on ablative shielding used to harden ballistic missiles in their boost phase, Canavan argues that the study panel was in error by a factor

of 160 in power performance. The trouble with Canavan's recalculations, Patel asserts, is that he specified excimer lasers as the kill mechanism where the report speaks of x-ray impulses. "This is a case where the report doesn't say explicitly x-ray fluence," says Patel. "We should have made sure the section was headed 'Structural damage from impulse loading by x rays.' Canavan caught us out here, but if he had read the report in sequence he would have known we were speaking of x rays."

Many of Canavan's other quarrels with the APS report—particularly those involving the length of time that a particle beam would take to penetrate a missile, the power needs of neutral particle beams and the number and wattage of orbiting nuclear reactors for peacetime "housekeeping" of an SDI system—depend largely on differing assumptions rather than on faulty physics. "My intention was not to discredit the report," says Canavan. "It was always to make it more accurate and complete. I think of myself as a reviewer."

Hearings. In Washington there are those who believe that Canavan's cavils as well as arguments against the APS report by the conservative Heritage Foundation will lead inevitably to hearings about DEWs on Capitol Hill. Indeed, editorials criticizing the report in the *National Review* and *The Wall Street Journal* are likely to hasten the hearings, especially while both houses of Congress are determined to cut SDI's budget request for fiscal 1988 by $2 billion or more.

In a letter to the *Journal*, Val Fitch of Princeton University, currently president of APS, writes that the committee members "performed a monumental service to the country. We of The American Physical Society believe that the report is as objective as humanly possible, that it will make an important contribution to the formulation of policy and that it will substantially raise the level of discussion."

—IRWIN GOODWIN

DEBATE ON APS DIRECTED-ENERGY WEAPONS STUDY

Is it unduly pessimistic to conclude that it would take a decade or more of intensive research just to determine the feasibility of shielding the US with a system of high-intensity laser and energetic particle beam weapons?

PHYSICS TODAY/NOVEMBER 1987

Gregory H. Canavan, Nicolaas Bloembergen and C. Kumar N. Patel

In April, The American Physical Society issued a 424-page report saying that "even in the best of circumstances, a decade or more of intensive research would be required to provide the technical knowledge needed for an informed decision about the potential effectiveness and survivability of directed-energy weapon systems." (See PHYSICS TODAY, May 1987, page S1.) In May, Los Alamos National Laboratory published a 70-page paper by Gregory Canavan that is optimistic about a directed-energy weapons system and critical of the APS study. (See PHYSICS TODAY, June, page 43.) In this debate Canavan criticizes the APS study on ten technical issues, and Nicolaas Bloembergen and Kumar Patel, the cochairmen of the APS study group, respond point by point.

PHYSICS TODAY wrote brief introductions to each of the ten issues after receiving the manuscripts from both sides in the debate. These introductions appear in italics.

The debaters make frequent reference to the following documents:
▷ *The 424-page APS report. This report, which was published with some modifications as a supplement to the July issue of Reviews of Modern Physics, is reference 1 in the debate.*
▷ *Canavan's 70-page response to the APS study, and a shorter response to the APS study by Lowell Wood (Lawrence Livermore National Laboratory) and Canavan. These two documents, titled Directed Energy Concepts for*

Strategic Defense *(or Concepts for short) and Analysis of the APS Report, respectively, are reference 2 in the debate.*
▷ *The APS study group's response to the critiques by Wood and Canavan. This is reference 3 in the debate.*

Opening statements

Canavan: The document *Directed Energy Concepts for Strategic Defense*[2] provides a context for evaluating the role of directed-energy weapons in strategic defense. *Concepts* was written as an independent and objective evaluation of the role of directed energy in strategic defense. *Concepts* identifies a number of inconsistencies in The American Physical Society's report.[1] Ten of those inconsistencies were developed into issues by the APS study group and transmitted to Congress in June.[3] Congressman Curt Weldon of Pennsylvania and 50 of his colleagues requested that I respond to those issues, and my paragraphs below are substantially those sent to Congress. APS selected the ten issues and publicized them in the process of criticizing my paper.

Bloembergen and Patel: We appreciate the opportunity offered us by PHYSICS TODAY to respond to the material presented here by Gregory Canavan. Our complete response[3] to Wood and Canavan's initial critique[2] of the APS study[1] is available in preprint form from APS.

Chemical laser power

Chemical lasers emit continuous beams of infrared radiation, which is generated by the mixing of steady streams of reactive gases such as hydrogen and fluorine. Is the APS report correct when it says that ground-based chemical lasers would have to produce powers of 20 MW if their beams, after redirection by orbiting mirrors, are to structurally disable missiles in the boost phase?

Canavan: The executive summary at the beginning of the APS report states that chemical lasers require power

Gregory Canavan is assistant leader of the physics division at Los Alamos National Laboratory. His work on this article was performed under the auspices of the US Department of Energy. **Nicolaas Bloembergen** is Gerhard Gade University Professor at Harvard University. **Kumar Patel** is executive director of the division of research, materials science, engineering and academic affairs at AT&T Bell Laboratories.

Gregory Canavan

Nicolaas Bloembergen

Kumar Patel

levels "increased further by at least two orders of magnitude," but the body of the report acknowledges "measured power in excess of 1 MW" (section 3.2.2), so *Concepts* observes that the increase is "one order of magnitude rather than two" (page 16). The APS response acknowledges that "the APS report erroneously states its first conclusion of its executive summary. . . . Wood and Canavan are correct." The Strategic Defense Initiative Organization has stated that it commented on classification only, and that content and editing were APS's responsibility alone.

Bloembergen and Patel: The currently demonstrated power levels of the MIRACL chemical laser as stated by Canavan are correct. However, as reference 3 explains, the demonstrated power levels for the hydrogen fluoride–deuterium fluoride chemical lasers, and therefore the needed improvements, were inconsistently stated in the preprint version of the APS report because the Strategic Defense Initiative Organization's classification requirements changed in the seven-month period during which the classification review took place. The clerical error has been corrected in the *Reviews of Modern Physics* version of the APS report. Furthermore, Wood and Canavan's many inferences in reference 2 that this error was deliberate are incorrect. The needed improvement in the HF–DF laser output is a factor of 20. More importantly, as the APS report makes amply clear in section 3.2.2, the laser that achieved the best performance (the laser referred to both in the APS report and in Wood and Canavan's response[2]) cannot be scaled to significantly higher powers. Thus, the true measure of scaling must start from a different chemical laser design, and no such design has yet been demonstrated.

Chemical laser scaling

Orbiting chemical lasers, too, could potentially destroy missiles in the boost phase. How close are today's lasers to having the required power, and how many laser satellites would be needed?

Canavan: The APS response only changes the *Reviews of Modern Physics* article to say that power must be "increased by at least an order of magnitude" (page 1), but *Concepts* showed that such an increase is not necessary for useful applications (pages 9–10). The *Analysis of the APS Report*[2] notes that the APS report's "current" power levels are "within a factor of 3 of the beam power needed" (page E2006). The analysis in the APS report supports the factor of 3 (appendix B). *Concepts* shows on page 9 that in early, low constellations the number N of satellites would scale as about $1/B^{1/2}$ with laser brightness B. If a satellite's mass is proportional to its brightness, the constellation's mass and cost would then scale as NB, or $B^{1/2}$, and smaller lasers would be *preferred*. The exact solution[4] changes the details, but not the general observation that smaller lasers can be useful as well as less sensitive to the APS's "short ranges [and] rapid retargeting."[3]

Bloembergen and Patel: Reference 2 states that current chemical laser power levels on the ground are within a factor of 3 of those needed to oppose current threats. Conclusion 1 in the APS report is for power levels needed in space to meet realistic future threats, as the body of the report makes abundantly clear. Even against the current threat, Wood and Canavan's analysis includes a number of assumptions and judgments favorable to the defense's satellites, such as short ranges, rapid retargeting, high beam quality, low losses and very effective lethality. None of these is close to being demonstrated at scale. Canavan specifically argues that large numbers of low-power, low-altitude satellites will suffice, but this design would maximize the constellation's survivability problems, as we mention below.

Satellite constellations

What restrictions on the defense arise from the relationships among the ranges of orbiting weapons, their altitudes in space and the locations of their intended targets?

Canavan: The "standard reference"[5] on satellite constellation scaling cited by the APS response is derived from earlier work,[6] as the response acknowledges. The APS study group derives its analysis from that reference, but makes unreliable predictions of sensitivities. A useful discussion of constellation scaling and sensitivity requires the exact solution, but even the APS report's approximate analysis indicates that modest lasers are effective. The APS response questions the 1000-km range used in an illustration in reference 6, but that is about the properly averaged range for nominal threats and lasers, as confirmed by standard analyses and the analysis in the APS report. A satellite over the launch area is responsible for the missiles in a zone of radius R equal to $2R_{\rm E}/(zN)^{1/2}$, where $R_{\rm E}$ is the Earth's radius and N is the number of satellites; the constellation concentration[5] z is about 3. For 100 satellites, the zone radius is 740 km, which is reduced to 520 km by averaging over the range to targets.[6] For a typical constellation altitude of 500 km, that gives an average range of 720 km, which is less than 1000 km, even neglecting the shortening of the range as the boosters climb. In the more exact calculation the interpretation is more subtle, but the results are similar.[4] Other departures from ideality reduce the spacing, so the APS's "range of 2000 km, the value most commonly mentioned in presentations to the study group," either applies to lesser threats, for which smaller numbers of lasers would suffice, or is inconsistent with the APS study.

Bloembergen and Patel: It is straightforward to lower the requirements on laser power by decreasing the operating range and increasing the number of platforms. Mirror platforms, however, cannot be positioned in low Earth orbit, say at 300-km altitude, because their survivability would be a serious issue. The minimum plausible constellation altitude, as section B.1.1 of our report correctly states,[1] is about 500 km, corresponding to ranges of about 1000 km. A "typical" constellation altitude, from the majority of the presentations made to our study group, is about 1000 km, with ranges corresponding to about 2000 km. Our previous response[3] stands.

ICBM shielding

Attacking missiles could be shielded to protect them from directed-energy weapons, but at what cost in payload?

Canavan: *Concepts* states that "the uniform hardening of all stages would require that material be added in proportion to their *areas* rather than their *masses*" (page 30). The APS response states that "Canavan's computation is correct within the bounds of that assumption," but states that the APS report gives "a general methodology . . . keeping in mind the fundamental purpose of the group's work as a tutorial" and questioning whether "apportioning shielding by stage area is self-evidently the option that would be chosen by the Soviets." *Concepts* notes that under "the [APS] report's prescription the first stage would be harder than the upper stages, which would leave the upper stages, the ones most susceptible to attack, relatively unhardened" (page 30), the error of which is self-evident.

The literature contains idealized discussions of shielding only the first stage, the second, the bus or combinations. The first corresponds to the case treated in the APS report; the second cannot be justified operationally either.[7] Lasers can deliver lethal energies to the cloud tops, so leaving the first stage unhardened would gratu-

itously reduce the defense's requirements by an order of magnitude. Uniform hardening requires about twice the payload penalties given in the APS report—5 to 10 of the 10 reentry vehicles carried. Thus "hardened missiles should probably be regarded as carrying only 30–50% of the current number of reentry vehicles" (*Concepts*, page 33). Because the APS report's calculation is irrelevant to the cases of interest, is misleading by a factor of two or more on payload penalties, and is neither simpler nor more general than that in *Concepts*, there is no reason why the APS report's tutorial should not be replaced by the admittedly correct calculation in *Concepts*.

Bloembergen and Patel: As stated previously,[3] the weight penalty of shielding was illustrated for tutorial purposes in our report by two limiting cases, which require the least amount of algebra. Canavan's computation is correct for uniform hardening if the total shielding mass is fixed at 6 metric tons. Because our discussion of required laser powers throughout[1,3] has been based on a nominal hardening with 1 g/cm^2 of ablative material, the logical boundary condition is that no segment should have less than this amount. Using the model rocket geometry of Canavan, one finds that the total shielding mass for uniform shielding is now only 3 tons. Missiles hardened in this manner could carry 60–70% of the current number of reentry vehicles, rather than the 30–50% Canavan calculates on the basis of 2 g/cm^2 of uniform hardening. Note that our example[1] with shielding proportional to the mass of each stage and a total shielding mass of 6 tons still leaves about 1 g/cm^2 on the upper stage and bus and about 5 tons on the first stage. This example, from section 2.3.2 of the APS report, illustrates how much less the payload penalty is due to shielding on the first stage compared with Canavan's estimate.

Even the opposite example of no shielding at all on the first stage cannot be dismissed by Canavan's simple statement that this "would gratuitously reduce the defense's requirements by an order of magnitude." If the burnout altitude of the first stage is below 40 km, and if the defense's lasers are repetitively pulsed excimer or free-electron lasers, then according to the analysis of stimulated Raman scattering on the downlink (section 5.4.9.3 of the APS report), it may be impossible to achieve the focal spot sizes indicated by Wood and Canavan. In conclusion, a nominal uniform hardening of 1 g/cm^2 is achievable with much smaller reentry vehicle penalties than Wood and Canavan state.[2] The tutorial examples of reference 1 fairly illustrate these penalties for any distribution of shielding between the first and second stages.

Spinning boosters

It is more difficult to destroy a missile by laser beam heating if it is rotating, but what are the necessary rotation rates, and are they practical?

Canavan: The APS report states that "rotations of missiles at angular rates of the order of 1 rps have been studied and shown to extract little or no penalty to the offense" (section 2.3.2). *Concepts* calculates that "missiles would have to rotate at least once per second to have any impact" (page 34), so the upper limit of the rotation rates studied is the minimum required to have any impact. The APS response does not disagree; it only states that "whether it would be practical to accomplish booster rotation by retrofit depends on detailed design features"— the point made by *Concepts*.

Bloembergen and Patel: Wood and Canavan's initial critique is now missing from Canavan's argument. Their objections based on 10-rps booster rotation have no bearing on the 1-rps rotation discussed in the APS report; 1-rps rotation would increase the required kill threshold power for cw lasers, as stated in the APS report. Also,

Canavan is incorrect in stating that the initial APS response agrees with his initial statement that 1-rps rotation is impractical.

Fast-burn boosters

A missile that burns its fuel very rapidly can separate from its weapons bus earlier and at a lower altitude. What impact would this technology have on the defense?

Canavan: *Concepts* does not state, as the APS response suggests, that the APS report "inconsistently rules out neutral particle beams . . . as boost-phase kill mechanisms" (reference 3, page 5). *Concepts* observes only that "the early Soviet and Union of Concerned Scientists reports on SDI erroneously concluded that neutral particle beams could not propagate below 200–300 km" (footnote, page 25), an observation no one challenges. Buses are not intrinsically harder than missiles—in fact, they may be softer because they must open up to deploy reentry vehicles. The APS response's main point, on which there is strong *agreement*, is that "destroying [the bus] would indeed still provide the leverage of destroying many reentry vehicles at once" (page 6)—a key point missed by Soviet, UCS and Office of Technology Assessment reports, which assumed that buses became invulnerable at burnout. The APS report's discussion of the impact of drag (section 2.3.4 and figure 2.9) corrected the early arguments that missiles could deploy weapons and decoys too low for the defenses to reach. There is no significant disagreement—or impact on the defense.

Bloembergen and Patel: Wood and Canavan's long discussion of fast-burn boosters is now seen to be a non-issue for Canavan. The original version of APS report is correct as written.

Free-electron laser power and efficiency

A free-electron laser produces coherent radiation by sending an intense, energetic beam of electrons through an undulating magnetic field. Is the APS report correct when it says that a ground-based FEL would have to produce an average power of at least 1 gigawatt to be effective?

Canavan: Our *Analysis of the APS Report*[2] states (on page E2006) and *Concepts* derives (on page 19) that a 4-MW visible free-electron laser with a 4-meter mirror, or "4–4" platform, should have "performance roughly equivalent to a 20–10 chemical laser at 2.7 microns." For space basing, the cost and mass of the FEL platform are minimized by operation at the shortest wavelength compatible with simple optics, which the APS report indicates is in the visible (sections 5.2.2 and 5.2.3). *Concepts* used 0.5 microns (0.4 on page 18 is a typo), a wavelength at which FEL operation compatible with efficient scaling to 4 MW has been demonstrated. Power scales as wavelength, so that at 0.5 microns the power is reduced to about $(0.5/2.7) \times 20$ MW, or 3.7 MW. If a visible FEL were required to have the same power and mirror area as a 20–10 chemical laser, its brightness would be greater by a factor of $(2.7/0.5)^2$, or 29, and thus would be 7×10^{21} W/sr. A single laser of that brightness would be oversized to handle nominal engagements, by the APS report's own analysis. For meaningful comparisons brightness should be held constant, so the APS response's observation that a 4-MW FEL at 40% efficiency would require "only 10 MW of delivered power, rather than the 1 GW stated by the APS report," only reflects that the APS report's 1-GW power is unsupportable (*Concepts*, page 19).

Thermal kill requirements for free-electron lasers are similar to those for chemical lasers, so the APS report overestimates the power per FEL by a factor of $(20 \text{ MW})/(4 \text{ MW})$, or 5, and incorrectly assumes that a single FEL would engage all missiles, for a total error of a factor of 50. (*Note added in proof:* This point was debated in a House

Armed Services Committee hearing on 15 September 1987. At that hearing SDIO concurred with the position that there is no basis for APS's assumption that one laser would have to engage all missiles.) *Concepts* uses the 40% efficiency of recirculating FELs (APS report, chapter 3, references 79 and 80), but halving it would only increase the input power to (4 MW)/0.2, or 20 MW—a factor of 50 below the APS's 1 GW. The APS's ground-based FELs propagate better at wavelengths around 1 micron (APS report, section 5.4.1 and figure 5.15).

Bloembergen and Patel: The statements that Canavan quotes from reference 2 refer to space-based laser platform configurations. The 1-GW power mentioned in conclusion 3 in the executive summary of the APS report holds for thermal kill by one ground-based laser. The laser beam would be steered by a relay mirror in a geostationary orbit to a constellation of fighting mirrors, as discussed in chapter 5 of the APS report. Alternatively, one could consider architectures with several relay mirrors in high-altitude orbits lower than geostationary orbits. The much longer optical path required for ground-based laser systems invalidates Canavan's scaling arguments.

The architecture that Canavan alludes to on page 19 of his response[2] to the APS study involves many smaller ground-based FELs and is ill defined because the requirement for relay mirrors is completely ignored. Reference 2 puts the number of FELs at the number of fighting mirrors: 50–100. This may actually be more FELs than are necessary, because of the mirrors' absentee ratio—the ratio of space-based platforms over the launch area at the time of attack to the total number of platforms deployed in space. A number of relay mirrors at least equal to the number of lasers in action appears to be indispensable. But one must recall that five to seven lasers must be built for every one in the battle because of cloud cover restrictions. This and a number of other factors, such as dwell times, low retargeting rates and mirror filling errors, have led the Strategic Defense Initiative Organization's own systems architects to eliminate the option of many small ground-based lasers, and no detailed version was ever presented to the APS study group, for excellent reasons.

Returning to the architecture based on one ground-based free-electron laser, such a laser must indeed produce the same total power as the 50 ground-based lasers assumed by Canavan. In addition to the losses in atmospheric propagation, one must also allow for the fact that the long optical path and limitations on the size of the relay mirrors make it unlikely that one could achieve the small spot sizes on target that would follow from the scaling argument of reference 2. This adds another factor of at least 5 to the required ground-based power. Contrary to Canavan's claim, the power figure of 1 GW is correct for a ground-based FEL system of appropriate size and realistic efficiency using thermal kill mechanisms.[3] The alleged error of a factor of 100 simply does not exist.

Neutral particle beam power

A beam of fast-moving atoms emitted by a space-based accelerator could penetrate deep into a missile or warhead, but how powerful would such a beam need to be?

Canavan: The APS response indicates that the 1-GW power estimated by the APS report for "beam particle accelerators" is actually for lasers, but power is only an issue for space-based FELs. The power of a properly scaled FEL is 4 MW, so at the APS report's 20% efficiency, the power required is (4 MW)/0.2, or 20 MW, 2% of the 1 GW assumed by the APS response. The 50-MW FEL posited in the APS response is too powerful by a factor of 12.5. If its mirrors were held at 10 m, its brightness would be too large by a factor of $12.5 \times (10/4)^2$, or 78. However, it is unimportant that in this irrelevant configuration the power is 80 times too large; in a properly scaled platform, the power is 10–20 MW.

The APS response suggests that the correct neutral particle beam power is intermediate between that in the APS report and that in *Concepts*, but the APS report uses 100 J/g for "massive upset" (section 4.3), which actually corresponds to the energy for melting weapon components or detonating high explosives (section 6.4.2). Depositions of a few millijoules per gram can upset hardened circuits; neutral particle beam dose rates are high and have been difficult to harden against. According to SDIO a practical value for upset is about 1 J/g, and for the destruction of electronics, which is what the APS report presumably meant, the requirement is about 10 J/g.

The APS report deals with tradeoffs within a single platform, so its statement that "a few hundred MeV probably is a reasonable compromise" (section 4.3) is without basis; 100 MeV could be useful for early applications. Even the beam energy of 250 MeV given in the APS response only increases the beam power to 0.1 amp $\times$ 250 MeV, or 25 MW, and the input power to about 80 MW— over an order of magnitude below the 1 GW of the APS report and response. The APS response cites current losses in the stripping cell that neutralizes the beam, but on the time scale the APS group assumes (section 4.3.4), there is no fundamental reason for such losses. The ranges given in the APS response are inappropriate; where brightness is concerned, neutral particle beams scale much like other directed-energy weapons. *Concepts* (page 25) uses the APS report's analysis to calculate that the electronics kill time is about 0.1 sec. The configuration in *Concepts* is not a "system minimum"; reasonable variations do not change *Concepts'* statement that there would be enough time to engage all missiles.[8]

Bloembergen and Patel: Canavan's estimates are based on system parameters "useful for early applications." The APS response clearly states the assumptions on which we derive a power requirement of 200 MW for space-based neutral particle beams. The range of powers would be from 100 MW to 1000 MW depending on the range and the retargeting rate. Conclusion 21 and the paragraph that follows it in the APS report were reworded for publication in *Reviews of Modern Physics* to avoid confusion.

Excimer laser energy and power

An electric discharge in a mixture of rare gas atoms and halogen molecules can cause the formation of electronically excited, ionically bound molecules such as krypton fluoride. When these so-called excimers make a radiative transition to unbound, ground state atoms, they emit at short wavelengths. Ground-based excimer lasers could emit intense pulses of radiation, but what energy and repetition rate must these pulses have to be of strategic military use?

Canavan: The APS report provides only one calculation of the "structural damage from impulse loading" at issue (section 6.3.5). This calculation was presented to the APS study group as a calculation of the impulse required for *excimer lasers*.[9] The APS report states (section 6.3.1) that "coupling coefficients for short-wavelength lasers have recently been measured. . . . The parameters agree with theoretical calculations for one-dimensional spot sizes." Experiments have therefore confirmed the excimer laser coupling and scaling used. The APS report uses the confirmed 5-kJ/cm^2 fluence coupling (equation 6.35) to predict a 10-cm spot radius for a one-dimensional interaction (equation 6.38), so the equations and parameters in the *body* of the APS report give a lethal energy of (5

kJ/cm^2) π (10 cm)^2, or 1.57 MJ, which with the APS's factor-of-4 transmission loss corresponds to 6 MJ, as *Concepts* states. However, the executive summary in the APS report states that "ground-based excimer lasers for strategic defense applications must produce at least *100 MJ* of energy in single pulse."

The APS report's analysis prescribes 6 MJ per kill, so destroying 1000 missiles in 100 seconds, or 10 missiles/sec, with the 10 mirrors that would typically be in range would require a power of (6 MJ/kill)$\times$(1 kill/sec), or 6 MW for each laser. Hence the total power would be 60 MW rather than the 1 GW given in conclusion 2 in the executive summary of the APS report, which assumes that a kill requires 100 MJ and that one laser must engage all targets. Pulsed lasers need not dwell; they need only hit the missile somewhere, so their control requirements are less stringent. APS responds by saying that "the SDIO has by its actions disagreed with Canavan's assertion," in that it has relegated excimer lasers "to being a 'backup' technology." But all directed-energy weapons play a backup role to kinetic-energy weapons (*Concepts*, page 3). APS argues that the SDIO action justifies the APS report's assessment, which appears to imply that further SDIO reductions on the basis of errors in the APS report would justify those errors. Physics determines what is right; budgets, only what is popular.

Bloembergen and Patel: Canavan quotes equation 6.38 of the APS report, which predicts a minimum spot size of 10 cm for a valid one-dimensional interaction. This implies that laboratory tests need at least this spot size. It is incorrect to equate this with the actual spot size on a target in space during a strategic defense engagement. A ground-based excimer laser system faces problems similar to those of a ground-based free-electron laser system: propagation through the atmosphere and long optical paths involving relay mirrors. In light of the problems, we estimate that achievable minimum spot sizes on target lie between 30 cm and 1 meter. This would require a minimum of about 25 MJ on target because the spot area is at least 10–20 times larger than Canavan assumes. As with a free-electron laser, each ground-based excimer laser requires one or more relay mirrors as well as the constellation of fighting mirrors. An architecture with an individual laser engaging all targets is justified. There are no errors of a factor of 160, as reference 2 alleges and Canavan reiterates.

Space-based power

Would orbiting strategic defense platforms require nuclear reactors to supply "housekeeping" power?

Canavan: Conclusion 20 in the executive summary of the APS report states that "housekeeping power [would] necessitate nuclear-reactor-driven power plants [producing] 100–700 kW of continuous power," but the *Analysis of the APS Report*[2] states that the requirements could be met by "a few thousands to tens of thousands of watts for housekeeping—a requirement readily met by solar panels and storage batteries, *without use of nuclear reactors*" (page E2006). The APS response says instead that it estimated the power "to be in the 100-kW range . . . about a factor of two higher than the estimates presented to us by SDIO officials," so there is no longer any disagreement. SDIO's 50 kW is a few "tens of thousands of watts," which could be satisfied without nuclear reactors.

Bloembergen and Patel: No detailed response is necessary except to say that a few tens of kilowatts of electrical power *necessitates* nuclear power reactors for two reasons. First is survivability: The large area needed for solar cells would make a satellite very vulnerable to actions of the offense. Second is reliability: The long expected stay in orbit could reduce the availability of power because of the radiation damage that occurs over 10-year time scales, as discussed in the National Research Council report *Electric Power from Orbit: A Critique of a Satellite Power System* (National Academy Press, Washington, DC, 1981).

Closing statements

Canavan: The points in the APS response are not difficult to address individually, and taken together they demonstrate that while the APS report may have good physics, its analysis and scaling projections are sufficiently flawed that they should be redone or removed. Even in a tutorial one does not wish to use calculations that are off by factors of 2^n, if n ranges[10] up to 20. The APS response had no argument with the discussion in *Concepts* (pages 37–55) of the role of directed-energy weapons in midcourse, which is probably more important than the boost phase. The ability of directed-energy weapons to interactively discriminate decoys from real weapons (*Concepts*, pages 37–50; APS report, sections 7.7–7.8) and intercept the latter (*Concepts*, pages 54–55) is pivotal. These issues could be usefully discussed at the level of this debate, as could survivability, which the APS report outlines (chapter 9) and *Concepts* bounds (pages 37–38). But they need a proper forum, which has been a problem (*Concepts*, pages 55–69). I thank APS and PHYSICS TODAY for providing one for this exchange.

Bloembergen and Patel: There is no disagreement on the basic physics and the equations to be used in discussing directed-energy weapons. All of this is made available to all physicists, scientists and engineers by publication of the APS report in *Reviews of Modern Physics*. Surprisingly, large disagreements can nevertheless arise, based on different assumptions about values of parameters in the equations. The most salient disagreement still existing relates to the minimum spot size that ground-based laser systems can achieve at the target. Here Canavan's analysis in reference 2 needs a much more thorough consideration of relay mirrors and atmospheric optics along lines similar to those used in chapter 5 of the APS report.

The official response of the Strategic Defense Initiative Organization to our report acknowledges our accurate description of the state-of-the-art directed-energy weapons technology at the time of the study's completion. SDIO's response goes on to say that the wording of many of our conclusions is "unduly pessimistic"; Canavan pushes estimates of achievable parameter values further toward more optimistic numbers. There are, undoubtedly, many other physicists, scientists and engineers, including some working actively on SDI-related projects, who believe instead that our conclusions are unduly optimistic. We believe our conclusions are realistic.

The APS study group, comprising 16 individuals with widely diverging political views, heard numerous presentations from many dedicated scientists working on SDI-related problems. They probably had an equally wide range of personal convictions. It is remarkable that a unanimous publication emerged from this effort. This was possible only because the group eliminated all politically motivated statements and narrowed the choice of scientific parameters to a reasonable range. We arrived at the final version of our report, published in *Reviews of Modern Physics*, only after careful consideration of the criticisms in reference 2. We reject Canavan's statement that the "analysis and scaling projections [in the APS report] are sufficiently flawed that they should be redone or removed."

We reject in stronger terms the additional, gratuitous statement, "Even in a tutorial one does not wish to use calculations that are off by factors of 2^n, if n ranges up to 20." Here Canavan endorses and lends scientific credibility to a letter that appeared in *The Wall Street Journal*.[10] We had until now declined public comment on this letter, as it is political rather than scientific in character. However, an allegation in PHYSICS TODAY of an error of 2^{20} cannot be ignored.

The starting point for the *Journal* letter was the statement in section 5.2.1 of the original version of the APS report that the total system cost of the 2.4-meter primary mirror for the Hubble Space Telescope is about 1.2 billion 1984 dollars. In this statement we made reference to an article in *Science*.[11] It should be clear from this reference that the cost refers to the entire system, and this was stated explicitly in the APS report. Russell Seitz, the author of the *Journal* letter quoted by Canavan, chose to read this as the mirror cost even after explicit discussions with us in which we pointed out the phrase "system cost." The fabrication cost of the 2.4-meter mirror alone is $5 million. To avoid misunderstanding, the *Reviews of Modern Physics* version of the APS report spells out explicitly both the mirror cost and the system cost, and Seitz was informed of this before publication of his cost estimates in *The Wall Street Journal*. The APS study made no cost estimates whatsoever. Canavan's allegation that our report contains an error of 2^n has no basis in fact.

Canavan's ten points do not necessitate any changes in the APS report as published in *Reviews of Modern Physics*. As noted above, the *RMP* version differs from the April 1987 report only in the correction of a clerical error regarding chemical laser power and minor rewording in a few places to avoid ambiguity or misinterpretation.

References

1. American Physical Society Study Group (N. Bloembergen, C. K. Patel, cochairmen), *Report to The American Physical Society of the Study Group on Science and Technology of Directed Energy Weapons*, APS, New York (April 1987); Rev. Mod. Phys. **59**(3), part II (July 1987).

2. G. H. Canavan, *Directed Energy Concepts for Strategic Defense*, report no. LA-UR 87-1658, Los Alamos National Laboratory, Los Alamos, N. M. (May 1987). L. Wood, G. Canavan, *Analysis of the APS Report*, Congressional Record, 20 May 1987, p. E2005.

3. American Physical Society Study Group (N. Bloembergen, C. K. Patel, cochairmen), *APS Directed Energy Study Group Responses to Critiques by Wood and Canavan*, APS, New York (18 June 1987).

4. G. Canavan, A. Petschek, *Satellite Allocation for Boost Phase Missile Intercept*, report no. LA-10926-MS, Los Alamos National Laboratory, Los Alamos, N. M. (April 1987), and NTIS document no. DE 87007719, National Technical Information Service, Springfield, Va. (1987); also submitted to Nature.

5. R. Garwin, Nature **315**, 286 (1985).

6. G. Canavan, H. Flicker, O. Judd, K. Taggart, *Comments on the OTA Paper on Directed Energy Missile Defense in Space*, report no. LA-UR 85-3572, Los Alamos National Laboratory, Los Alamos, N. M. (6 May 1984), p. 6 and appendix A.

7. Letters to PHYSICS TODAY, July 1986, pp. 13–15, 90–96. R. Jastrow, *How to Make Nuclear Weapons Obsolete*, Little, Brown, Boston (1985).

8. G. Canavan, F. Seitz, *Comments on Directed Energy Concepts for Strategic Defense*, report no. LA-UR 2150, Los Alamos National Laboratory, Los Alamos, N. M. (June 1987); to appear in Nature.

9. G. H. Canavan, August 1985 briefing to APS study group. G. H. Canavan, letter to N. Bloembergen and C. K. Patel, Los Alamos letter P/AC:2O3, 12 August 1987.

10. R. Seitz, The Wall Street Journal, 15 July 1987, p. 29.

11. G. Field, D. Spergel, Science **231**, 1387 (1986). ■

Free electron lasers take small steps toward distant goal

PHYSICS TODAY/JUNE 1987

Until recently, no free-electron laser was able to convert more than 5% of the energy from an electron beam into microwave radiation. Last fall, however, a team from Lawrence Livermore National Laboratory and the University of California at Berkeley published a (previously classified) report that they had operated a free-electron laser with an unprecedented 40% efficiency at a wavelength of 8.6 mm. The peak power exceeded 1 gigawatt. Meanwhile, researchers at Los Alamos National Laboratory who operate a free-electron laser of different design at a shorter wavelength (10 microns) have reported a modest increase in efficiency from 1.3% to 2% with peak power output of 20 MW, as well as development of several promising enhancements of their laser operation. In yet another experiment, a Stanford–TRW team has operated an FEL in the visible region of the spectrum (525 nm) with a peak power as high as 170 kW.

Each of these steps moves FELs a bit closer to some of their anticipated applications in medicine, electronics and basic research. Still, they remain orders of magnitude away from the the most publicized of applications—defense against ballistic missiles. The APS Study Group on Directed Energy Weapons (see PHYSICS TODAY, May, page S1) estimated that FELs for strategic defense applications might need to reach average power outputs in the GW region at wavelengths of about 1 micron. To explore whether FELs can scale to the required performance levels, the Army is building a ground-based FEL facility at White Sands Missile Range, New Mexico. In the next year, they will choose which of two competing FEL concepts will be the basis for the large-scale experiment: an FEL whose electron beam is accelerated by an induction linac, such as the facility operated at Livermore, or one based on a radiofrequency linac, like that at Los Alamos.

In any free-electron laser, an accelerator generates an energetic electron beam, which then travels down a wiggler or undulator, a region subjected to a spatially periodic, transverse magnetic field. (In an undulator the radiation from successive bends is coherent, whereas in a wiggler it is not. Nevertheless, "wiggler" has become a common term to describe either situation.) As the relativistic electrons oscillate in this wiggler, they can radiate coherently with an imposed (or spontaneously emitted) radiation field provided the radiation field maintains the right phase with respect to the electrons. The resonant condition is essentially that the optical wave travel one optical wavelength further than an electron during the time that the electron travels one wiggler period. If λ_s is the wavelength of the radiation and λ_w is the "wavelength" of the wiggler, then

$$\lambda_s = \frac{\lambda_w}{2\gamma^2}\left(1 + \frac{a_w{}^2}{2}\right)$$

Here γ is the electron's Lorentz factor and a_w is a term that varies directly with the product of the wavelength and magnetic field of the wiggler. Note that the wavelength of the signal output varies as the inverse square of the electron energy. The two main types of FELs differ in the design of the linear accelerator that provides the electron beam. From the different beam characteristics stem important differences between the two laser types. An induction linac typically produces widely spaced beam pulses with high current in each pulse. Because the interval between pulses is so long, an induction FEL must operate as an amplifier, increasing the power of an applied radiation field in one pass. The relatively high currents of the induction linacs at Livermore have enabled FELs there to operate at powers in the GW region, but the moderate energy of the linac on which Livermore did its first FEL work restricted their operation to wavelengths of about 1 cm.

By contrast, the rf linacs that have been available for FEL work to date have higher energies and lower currents than induction linacs. Thus the FELs they power typically operate now at shorter wavelengths but lower power levels. The electron beam consists of packets of thousands of micropulses, with each of the micropulses lasting perhaps tens of psec. Existing rf FELs are operated as oscillators, building up the spontaneously emitted radiation by successive reflections off mirrors in optical cavities. The output approaches that of a continuous wave laser.

The induction and rf technologies being considered for the ground-based FEL face distinctly different problems as they scale up. To tackle these problems, the Army has directed Livermore Lab to have its industrial partner, TRW, develop a proposal for an induction FEL at the White Sands facility. This month the Army will select a firm to collaborate with Los Alamos in preparing an rf FEL proposal.

Colonel James McNulty, director of the GBFEL program, said that the Army will monitor progress by technical milestones between now and early spring 1988, at which time the industrial firms must submit specific design proposals. The Army will make its final selection by 1 June 1988. It plans to have the facility operating in the early 1990s.

McNulty could not comment on some of the specific design criteria such as power, but he did say that the wavelength would have to be in the range from 0.8 to 1.6 microns. (The wavelength is dictated in part by requirements of atmospheric transmission.) The Army requires that the device be tunable over a small region near the selected wavelength, even while lasing. And the selected FEL technology would have to be scalable to a weapons class system.

Induction FELs. The Livermore–Berkeley FEL collaboration has been using

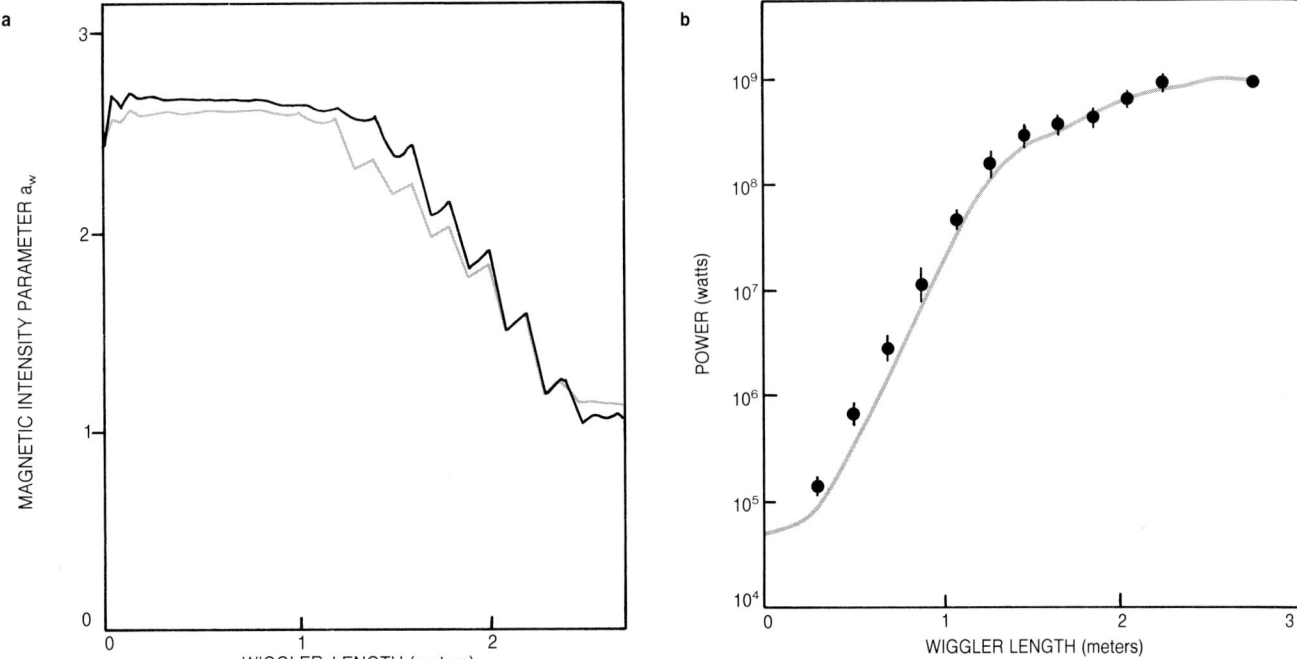

Wiggler field profile (**a**) and power output (**b**). Decreasing the magnetic field strength of the wiggler in their free-electron laser enabled a Livermore–Berkeley team[1] to reach efficiencies higher than 40%. The magnetic field intensity, measured by parameter a_w, is constant for about 1 meter, then decreases to maintain resonance as the electrons lose energy. As a result, power grows along the wiggler length. Red curves are numerical evaluations. Experimental results are shown as black curve and data points.

the 5-MeV Experimental Test Accelerator at Livermore to generate microwave radiation with efficiencies of about 5%. (See PHYSICS TODAY, April 1985, page 17.) Last fall, they reported[1] achieving much higher efficiencies by tapering their wiggler—a technique pioneered by the Stanford–TRW group and followed in other FELs as well. In a tapered wiggler, either the strength or the period of the magnetic field is decreased along the wiggler length to maintain the resonance condition given above as the electrons lose energy. In the Livermore experiment, the ETA operated at 3.5 MeV to match the 34.6-GHz resonance in the operating range of the wiggler. With a uniform magnetic field in the 3-m-long wiggler, the researchers found that the output signal saturated at 180 MW after a length of 1.3 m. Once they tapered the field strength in the last 1.7 m of the wiggler down to 45% of its initial value, they got a peak output power above 1.0 GW, corresponding to a 40% efficiency in extracting energy from the electrons. The graphs above show the wiggler field profile and the amplified signal output as a function of wiggler length in their experiment.

The challenge will be to maintain a high efficiency as the group moves toward shorter wavelengths. Meeting it will require an accelerator that provides a beam with high brightness, that is, one with both high current and low transverse spread of electron veloc-

ities, even at the higher energies needed to yield shorter wavelengths of radiation. The Livermore–TRW team is researching improvements in several features of an induction accelerator. One such feature is the electron injector. Existing field-emission cathodes produce beams of sufficient brightness but do not operate at the required duty factor of 10^{-4}. The Livermore group has tested a dispenser cathode that has provided almost the required brightness at 1 kiloamps and which, they feel, should work well at higher currents.

The Livermore–TRW team is also exploring an accelerator design with a higher gradient and a pulse repetition rate much higher than the current 1 Hz. To produce adequate average power for an FEL for weapons applications, the induction linac would have to operate at rates of thousands of pulses per second. For this reason, Livermore has designed a nonlinear magnetic pulse compressor. This device relies on a very rapid change of permeability in ferromagnetic materials to produce a sudden pulse of current.

Another requirement is to maintain the horizontal focus of the electrons as they are focused vertically by the wiggler field. The external quadrupole magnets used in the 3-m wiggler will not preserve velocities over longer wigglers, so the Livermore experimenters are shaping the wiggler magnets to provide the required focusing.

One of the most severe problems for

the induction FEL, however, will be to prevent divergence of the radiation down the length of a wiggler that may have to be as long as 100 m to give the required amplification. Many hopes rest on a theoretical prediction that optical guiding occurs in the wiggler: The electron beam may act as an optical fiber along which the radiation is guided without diverging outward. So far this crucial prediction has not been experimentally verified. Livermore is now conducting a series of tests called the Paladin experiments to measure the divergence of the laser beam as the length of a tapered wiggler is increased in stages from 5 m to 25 m by adding more 5-m wiggler sections. Donald Prosnitz (Livermore) told us the Livermore researchers are already seeing some gain in the amplifier with a 5-m wiggler. For this experiment, they are using the 50-MeV Advanced Test Accelerator at Livermore and operating at a wavelength of 10.6 microns with a CO_2 laser as the input.

Radiofrequency FELs. A team at Los Alamos led by Roger Warren has recently made several improvements in their FEL to help overcome some barriers to scaling to higher power levels.[2] By modifying their accelerator, they increased the beam current to 300 A and raised the FEL efficiency from 1.3% to 2% for operation with a tapered wiggler at a wavelength of 10 microns. The corresponding intracavity peak power level and output were

2000 and 40 MW, respectively. At efficiencies of just a few percent, it pays to recover the unspent energy in the electron beam. So Los Alamos installed an energy recovery loop that circles the electron beam back to the linac, where deceleration of the electrons pumps up to 70% of their energy back into the rf cavities. Charles Brau of Los Alamos told us the design goal had been 50% recovery. This loop functioned stably even while the FEL was lasing.

Last fall Los Alamos operated a new type of injector based on photoemission technology, which uses a laser to generate very short pulses of electrons that give a high-quality beam. Their current injector produces a long pulse which must then be compressed, with subsequent degradation of the beam quality. The new injector has produced a peak current of 400 A for an 80-psec pulse with an energy of 1 MeV. (The average current over a 10-μsec burst of pulses was 3 A.) The brightness at a peak current of 120 A exceeded the specifications for the White Sands demonstration. Brau emphasized that the injector may be of potential interest for many accelerator applications.

Perhaps the major limitation on rf FELs is the capability and survivability of the optical resonator at high power levels. The high-reflectivity mirrors at either end of the cavity are readily damaged by intense radiation. Such damage has so far prevented the Los Alamos team from operating at the design point of the wiggler. The group plans to try copper mirrors with increased cavity length. In a separate experiment with an argon ion gain medium, the Los Alamos team has shown that they can fabricate and align a 64-m-long optical resonator with 4° grazing incidence mirrors.

Another obstacle to high-power rf FEL operation has been the low power available from existing accelerators. The Boeing Aerospace Company has just finished constructing a 120-MeV rf accelerator designed to operate with a short-wavelength (0.5 micron) FEL. They hope to produce average currents as high as 100 A and increase power output by generating more closely spaced packets of micropulses. Spectra Technology and Rockwell's Rocketdyne division have provided the 5-m wiggler and the optical resonator, respectively. Boeing project director Donald Shoff-

stall told us the company hoped to get the device to lase by late May and would gradually work toward high-power operation. Boeing is competing for selection as the contractor to work with Los Alamos to develop an rf FEL design for the White Sands facility.

More FEL work. While the efforts just described focus on defense applications, other FEL work is directed more toward such applications as medicine or materials science, where average power levels of a few watts suffice. A Stanford–TRW collaboration has recently reported operation of an FEL at a wavelength of 525 nm. The Stanford FEL uses the electron beam from Stanford's superconducting linear accelerator, whose energy is 60 MeV. To get wavelengths in the visible, the Stanford–TRW team needed an electron beam energy above 100 MeV. Thus they extracted the beam emerging from the linac at 60 MeV and piped it back for a second pass through the linac to boost its energy to 115 MeV. Only one other FEL—at Orsay, France—has operated in the visible before and it did not achieve the power obtained in the experiment at Stanford, where the peak power output was 170 kW. The collaborators on this experiment are Todd Smith and Alan Schwettman of Stanford and George Neil, John Edighoffer and Steven Fornaca of TRW.

Smith said they are now refurbishing the laser and hope to push for ultraviolet wavelengths when it is running again in October. The group has also developed an energy recovery scheme in which they take the electron beam from the wiggler exit and feed it back into the linac 180° out of phase with the field. As the electrons are decelerated they give their energy back to the radiofrequency field for acceleration of a fresh bunch of electrons.

The White Sands GBFEL program will explore whether ground-based FELs can direct high levels of radiant power to mirrors in space, which would in turn direct that energy toward enemy targets. The program must evaluate not only the potential for high-power FEL operation but also the performance of beam control systems and problems of atmospheric transmission.

The Army has asked a team of experts called the Technical Advisory Group to recommend which of the two FEL concepts—induction or rf—should

be chosen. The group is headed by Lieutenant Colonel Tom Johnson of the US Military Academy (West Point). Because they must consider not only which laser is most feasible but which will give the beam with the most promise for atmospheric transmission, the 14 members and 5 consultants on the advisory group include not only experts on accelerators, FEL physics and optics but also experts on atmospheric propagation. Five of these experts, including Johnson, participated in or reviewed the APS DEW study. Johnson said his group would be looking at FEL technologies now in even greater detail.

The APS DEW study group stated in its report that "FELs suitable for strategic defense applications, operating near 1 micron, require validation of several physical concepts." Among those they cited were the demonstration of optical guiding and transverse sextupole focusing for the induction FEL, and control of sideband instabilities and harmonic oscillations for the rf-type FELs. Andrew Sessler (University of California at Berkeley), a member of the DEW panel, feels it is unlikely that enough technical information will be developed for a proper choice to be made between the two approaches by June 1988.

While recognizing that the Army might have to make a decision before all the questions are completely answered, McNulty nevertheless feels it is necessary to take some risks and set challenging goals to move the FEL time schedule ahead. Asked which type of FEL now held the edge, McNulty said that he felt they were tied at this point. He cited optical guiding as the critical issue for induction FELs and optical resonators as the challenge for rf FELs.

—BARBARA GOSS LEVI

References

1. T. J. Orzechowski, B. R. Anderson, J. C. Clark, W. M. Fawley, A. C. Paul, D. Prosnitz, E. T. Scharlemann, S. M. Yarema, D. B. Hopkins, A. M. Sessler, J. S. Wurtele, Phys. Rev. Lett. **57**, 2172 (1986).

2. D. W. Feldman, R. W. Warren, B. E. Carlsten, W. E. Stein, A. H. Lumpkin, S. C. Bender, G. Spalek, J. M. Watson, L. M. Young, J. S. Fraser, J. C. Goldstein, H. Takeda, T. S. Wang, K. C. D. Chan, B. D. McVey, B. E. Newman, R. A. Lohsen, R. B. Feldman, R. K. Cooper, W. J. Johnson, C. A. Brau, submitted to IEEE J. Quant. Electron. (1987).

Modified SDI stresses rockets and free-electron lasers

PHYSICS TODAY/JANUARY 1986

As 1985 drew to a close, the debate about the President's Strategic Defense Initiative continued to spread and intensify in the physics community. Petitions opposing Star Wars picked up more signatures in physics departments around the country, but some physicists and computer scientists entered the fray on the other side, issuing public statements that strongly supported the SDI research program (see page 79). Meanwhile, significant changes in the SDI program were announced that seem to indicate that Star Wars will be debated on different terms in the coming years than it was in 1983–85.

In a nutshell, some of the technologies that were much debated in the first phase of the SDI program—high-power space-based chemical and excimer lasers, in particular—are being put on the back burner. Instead, priority is to be given to rapid development of a system of heat-seeking or command-guidance rocket interceptors suitable for deployment in space and on the ground, and to ground-based free-electron lasers.

To experts who had been following the program closely, the changes in the program did not come as a great surprise. In fact, Lieutenant General James A. Abrahamson, the director of the Strategic Defense Initiative Organization, had given pretty clear advance notice that a streamlining of the program was to be expected.

In a speech delivered at a space-technology conference in Colorado Springs on 20 November—the day President Reagan was conducting final discussions with Soviet leader Gorbachev in Geneva—Abrahamson said that he expected to receive instructions after the summit to move ahead "much more quickly and effectively" with research into the design of a space-based defense system against nuclear missiles. While Abrahamson's speech was not prominently reported in many newspapers, it was striking because it expressed confidence in the President's intentions at a time when other Administration officials were betraying considerable nervousness about what Reagan might agree to with Gorbachev.

On 26 November, two days before

A rocket-powered interceptor, the second stage of an air-launched missile that was tested against a satellite this fall, is shown in the top photo. Similar but more advanced devices are now considered the technology closest to being ready for use as a component in a potential US missile-defense system. In the photo, one can see the rocket engines surrounding the homing device, the infrared sensor at center left and the laser gyroscope at center right. The bottom photograph is of the Advanced Test Accelerator at Lawrence Livermore National Laboratory, an 80-meter-long induction linac that will provide the high-intensity electron beam for a high-powered free-electron laser. ATA is an enlargement of the prototype Experimental Test Accelerator, which has produced impressive results in the past year.

Thanksgiving, it was announced that Abrahamson would hold a press conference following the daily DOD briefing at the Pentagon. This was not billed as an event where significant policy changes would be announced, and to the extent Abrahamson's remarks were reported, they tended to appear on the inside pages of the final preholiday newspapers.

Abrahamson briefing. Abrahamson opened the briefing by remarking that he was there "not to talk about some of the policy issues on the program nor to talk about Geneva." What he intended to do, he said, was to give "an overall impression of some of the very steady progress that we are making ... and some of the neat things that are coming out."

Emphasizing that SDIO remained committed to the principles of a multilayered defense system and boost-phase interception, Abrahamson proceeded to enumerate in a free and broad-ranging style various developments in the program and problems that are to receive urgent attention. He made, for example, the following points:

▶ The development of effective ground-based terminal and midcourse defenses will depend on better techniques to discriminate between decoys and real warheads.

▶ Some but not thousands of terminal defenses will be needed around certain areas or targets considered particularly important.

▶ There has been "incredible" progress with free-electron lasers at Lawrence Livermore Laboratory (see PHYSICS TODAY, April 1985, page 17), and while such lasers could not be based in space, test results have indicated that they could deliver enough energy from the ground to destroy boosters and that the problem of atmospheric distortion could be overcome.

▶ Test results from the antiballistic-missile "homing overlay experiment" in June 1984 and a recent test of an antisatellite weapon have indicated the effectiveness of heat-seeking interceptor rockets.

▶ A ground-based heat-seeking interceptor rocket is being developed for terminal defense, and under the ABM Treaty "we could go all the way even to deployment for 100 of these if it made sense," though it would be better to wait until there is a basis for determining the viability of a full multilayered system.

▶ Emphasis needs to be placed on development of radiation-immune data collectors and processors containing thousands, tens of thousands or even millions of elements.

▶ Highly decentralized signal-processing systems are being considered particularly thoroughly.

▶ In addition to the 300-kW reactor

being developed at Hanford for use as a space-based power source, other innovative ways of storing and converting very large quantities of energy will be needed.

▶ Cheaper methods of making very large mirrors will be needed if systems based on ground-based lasers are to be viable.

Amid all the detail, it was easy to miss Abrahamson's central message, which he brought up about a third of the way into his extensive opening remarks. In the context of a discussion about how it is useful to have different systems in place to perform the same functions, Abrahamson pointed to a slide and said: "For example, [we see here] a ground-based laser ... going up and bouncing off a mirror in space and going forward to what we call a fighting mirror and then going down to destroy a missile in the boost phase. The advantage of that kind of a system is clearly that it strikes with the speed of light. On the other hand, you can't deliver as much energy. So therefore on the right-hand side you see something called an SBKKV, a space-based kinetic kill vehicle. Those are simple rockets in space."

Such rockets could be command-guided or they could rely on the infrared heat-seeking systems used in missile-defense and antisatellite experiments in 1984 and 1985 (PHYSICS TODAY, November 1984, page 99, and November 1985, page 96). In the September ASAT test, Abrahamson reminded the press, the Air Force "hit a satellite with something weighing a little less than 50 pounds. Our job now is to get it down to less than 10 pounds. We can use a rocket to get it up or use a railgun."

Asked following his formal presentation why it had been decided to decelerate or drop space-based lasers and railguns, Abrahamson said: "We didn't

ABRAHAMSON

get the money that we needed, either in fiscal 1985 or 1986. And we had a choice. One choice is to try to take this broad range of technology and just slow it all down evenly. ... The other one is to try to take the knowledge that we have developed in our experiments so far, in the research so far, in the architectural studies so far, and begin to make decisions. So we are making decisions. ... So the way we're doing this is that we're not cutting out all of that research, we're just throttling way back to what I'm calling a backup technology."

The emphasis of the program, Abrahamson reiterated, would be on ground-based free-electron lasers and the "space-based kinetic kill vehicle." The space-based rocket "appears right now to be the simplest way to proceed," Abrahamson said, "and we see some very real cost reductions. ... But in the future, we haven't cut out railgun activity, not at all. We can see that that one is more than just a backup technology, but we haven't proceeded with it at the rate at which I think there is real opportunity to proceed."

Import of decisions. Abrahamson's announcement was not a major surprise in light of many developments. The Fletcher study team on defensive systems concluded two years ago that chemical lasers should be emphasized only if there were no fiscal constraints on the SDI program (PHYSICS TODAY, December 1983, page 43). SDI contractors were rumored to be uneager to make large R&D efforts in areas where they saw little or no chance for large equipment orders before the next century. An influential SDI panel on computer software recommended that a space-based ABM system consist of a very large number of coordinated but semiautonomous components. In keeping with that recommendation, Gerold Yonas, chief scientist of SDI, was known to be increasingly in favor of "pods"—space platforms carrying a small number of interceptors, which could be deployed by the hundreds or thousands.

Yonas denies that there has been a "sea change" in SDI and he says that some of what are normally thought of as Star Wars technologies will continue to be emphasized. He predicts, for example, that a neutral-beam weapon will be tested in space by the end of the decade, and he says that the SDI organization is trying to get added funds for nuclear-pumped x-ray-laser research, despite reports that recent test results have been of dubious validity. Yonas says the money is needed precisely in order to subject these results to closer scrutiny.

Some of the reporters present at Abrahamson's press conference came away with the impression that he was only *threatening* to downgrade many of the Star Wars projects as a budgetary ploy to squeeze more funds out of Congress for 1986 or to get more written into the 1987 budget. But Abrahamson did not say he was threatening to cut back certain programs; he said he *was* cutting them back.

Among opponents of SDI, it was tempting to view Abrahamson's announcements as a setback for the program as a whole, in that some of the more highly trumpeted technologies were downgraded. But it would be rash to characterize the announcements as a victory for SDI "doves" over SDI "hawks." Richard Garwin of IBM notes that the kind of SDI program now taking shape is very much the kind favored by retired general Daniel O. Graham, who is regarded as one of the most hawkish Star Wars proponents.

Graham, like the President himself, has often expressed irritation about the term "Star Wars," and it may indeed be fair to ask whether a system that relies mainly on more or less conventional rocket interceptors deserves to be associated with the far-out technologies of the George Lucas films. Garwin argues that "Star Wars" is still an appropriate label because, after all, the interceptors would be based in space, where they would be vulnerable to space mines and ASATs, so that war indeed would occur in space if such a system ever were used.

Regardless of how Abrahamson's announcements are interpreted or characterized, it seems apparent that the terms of the Star Wars debate have been substantially transformed. Early last year, Sidney Drell of SLAC observes, a Stanford workshop on Star Wars put considerable stress on its opposition to chemical lasers, which were very controversial at the time. Now, Drell points out, such concerns are moot.

Drell, together with many other physicists ranging from Edward Teller to Garwin, has expressed skepticism about any ABM system that relies heavily on vulnerable space-based components. But he thinks ground-based free-electron lasers would make an excellent ASAT system, an observation he makes reluctantly because he opposes the development of ASATs capable of striking targets in high orbits.

It would be quite an irony if the weapon type that was generally thought of as the nation's future ASAT—infrared homing vehicles based on Earth—instead turned out to be based in space as the nation's principal ABM system, and the technology that was generally thought of as the basis for Star Wars—lasers based in space—turned out to be based on Earth as the the nation's principal ASAT.

—WILLIAM SWEET

DOD Science Board finds SDI Phase I reasonable but 'sketchy'

PHYSICS TODAY/SEPTEMBER 1987

The latest contribution to the long-running debate over the feasibility of an operational defense against Soviet ballistic missiles is a report prepared by a special task force of the Defense Science Board, the Pentagon's senior scientific advisory group. The report, completed by the panel in late June and immediately stamped "secret," though still called a draft, leaked to the press in dribs and drabs until Representative James B. Olin, a Virginia Democrat, inserted the full text into the *Congressional Record* on 14 July. It not only endorses the main conclusion of *Science and Technology of Directed Energy Weapons*, issued last April by a study group of The American Physical Society (see PHYSICS TODAY, May, page S1), but in about 2000 words it goes well beyond that 422-page report to state that the Strategic Defense Initiative is far too unconventional, uncertain and undirected for the government to consider elaborate early demonstrations, let alone deployment by the mid-1990s, of any SDI system.

The task force, under the chairmanship of Robert R. Everett, a former president of Mitre Corp, examined SDI's achievements and shortcomings as seen by some experts close to or within military circles. Its report card gives SDI fairly low marks. At one point in the report, written as a memorandum to the under secretary of Defense for acquisition, Robert P. Godwin, the panel asserts that the concept of an SDI system using space-based and ground-based kinetic kill weapons that could be launched in 1994 is "quite sketchy" and "takes the form more of a list of components than of a consistent design."

The panel reached its judgment after eight sessions in which by SDI officials, contractors and scientists presented classified briefings over a period of three months. Although the panel was told that technology for the kinetic vehicles, which are designed to ram enemy missiles and warheads, is either in hand or well along, it finds that "much remains to be done before a confident decision can be made to proceed with the implementation of an initial phase." The task force goes on to say that technology for survivable rocket-powered kinetic kill vehicles positioned on platforms in space is "still uncertain." Indeed, that space-based interceptors (the term the Pentagon now uses instead of "space-based kinetic kill vehicles") are vulnerable to attack from antisatellite weapons and ground-based lasers at virtually any time is "particularly disturbing," the report states.

Questions. The panel argues that precise targeting of an ICBM booster amid the fire and smoke of the launch plume cannot be achieved with certainty right now. In its report, the APS team observed that even in some later phase, when lasers and particle beams might be used, hitting a spot perhaps half a meter across and several meters above the top of the plume of a booster rising at several kilometers per second would be difficult if not impossible. Tracking an ICBM's plume, said the APS document, will require reliable space-based conventional or optical radar, incorporating a feedback loop to determine if the target is struck and to make corrections automatically if it is not. The APS report suggested that fast-burn boosters would be an effective countermeasure to all directed-energy weapons in an early phase. The Pentagon panel admits that much more needs to be known about various US and Soviet boosters before the problem is solved.

Other serious questions for a Phase I system involve passive infrared sensors to discriminate warheads in space from even the most primitive decoys and debris. Technology for fabricating large infrared focal planes is not at hand, states the Pentagon task force. Accordingly, "there is a major need to create an adequate data base of the phenomenology involved in SDI," the group points out. "There is very little available information on how objects look in space or how rockets look in boost phase. Components and system design are proceeding on the basis of assumptions and calculations which may or may not prove reliable."

The panel was formed to assist Godwin in a formal review of a plan to deploy a limited "Star Wars" Phase I, which has been urged by Defense Secretary Caspar W. Weinberger. The scheme has the backing of Lieutenant General James A. Abrahamson, SDI's director. But among the Defense Department's Joint Chiefs of Staff and its senior political appointees, as well as within the White House and Congress, many contradictory voices are heard about the idea. Godwin appointed the panel with the approval of the Defense Science Board and Weinberger.

Besides Everett, a computer engineer who once worked at MIT's Lincoln Laboratories, the group consists of General Russell E. Dougherty, retired commander of the Air Force's Strategic Air Command; Harry Gray, retired chair-

EVERETT

man of United Technologies; Harry Haynes, retired chairman of Chevron; Ralph Lee, retired executive vice president of Hewlett–Packard; Walter Morrow Jr, director of Lincoln Labs; General Samuel C. Phillips, former head of the Air Force Systems Command and a former vice president of TRW Inc; and William J. Perry, a top Pentagon scientist in the Carter Administration and now an official at H&Q Technology Partners in Menlo Park, California. Morrow was a member of the APS team that conducted the study of directed-energy weapons.

Tactic. Approval by this task force would be necessary if early deployment is to gain the support of the Defense Science Board and the Defense Acquisition Board. This is considered important to Weinberger and Abrahamson as a means of legitimizing the controversial Phase I system in the Pentagon and Congress and raising SDI's entire stature among US allies and the public. It also is seen as a tactic for speeding up the program so that SDI is fully converted from theology to technology when the next Administration arrives in January 1989. Even SDI diehards in Congress believe the blessing of the Joint Chiefs is vital before the Pentagon can be given the go-ahead to "bend metal" and actually produce components for Phase I.

Whenever Weinberger and Abrahamson talk about Phase I, they describe the "architecture" outlined in a report by the George C. Marshall Institute (PHYSICS TODAY, January, page 47). This calls for a three-layered antimissile defense using some 11 000 space-based interceptors consisting of small rockets and electromagnetic rail guns, which could hurl projectiles at Soviet ICBMs in the boost and post-boost stages; another 10 000 exoatmospheric reentry-vehicle interceptors set off from the ground against warheads in midcourse; and 3000 high-endoatmospheric defense interceptors to strike warheads that make it through the first two layers. The entire system would need to be supported by additional satellites for communication, surveillance and battle management. Another vital element in the system is supercomputers to feed data to the command and control components.

Abrahamson has told Congress that the cost of such a system would be $40 billion to $60 billion. Phase I would be designed to protect a limited number of military installations, not cities, although it could later be supplemented by more elaborate—and presumably more effective—antimissile defense systems. But the Congressional Research Service reported as recently as 1 August that simply launching Phase I might run as high as $32 billion, not counting the cost of R&D and manufacturing the system, and that deployment of additional phases later on could put the bill up to $1 trillion. The CRS projections vary widely because the launch cost could drop from the current $3000–5000 per pound to something like $400 per pound for low-Earth orbit and from $18 000 to $3000 for geosynchronous orbit if an advanced heavy-lift system is developed.

Evolution. The Pentagon task force, like the APS panel, supported continued research on SDI. Asked to review the prospects of proposed space-based interceptors, the Everett panel evaluated such matters as systems design, cost estimates, development schedules and "milestone decisions." Its report cautions that defensive systems are never built to an immutable architecture. "Enemy reactions, new technology and changing requirements all lead to continual evolution," it says. "The plan to build SDI in phases is therefore reasonable and customary."

Before Weinberger, Abrahamson and other Phase I cheerleaders can find comfort in that statement, however, the Everett panel goes on to warn that none of the current cost estimates are reliable, "even assuming that the current Phase I concept holds. By the time the necessary system and underlying technology work is complete, the design may change considerably and costs change as well. There are also sizable uncertainties in such matters as learning curves for space hardware produced in modest quantities, launch costs and production costs for ir focal planes and hardened high-speed data processing." As for scheduling deployment in 1994 or thereabouts, the panel observes, Congressional support is so uncertain that anything said now is not likely to hold up.

One section addressing milestone decisions that have to be made by the Defense Acquisition Board was deleted from some versions of the panel report. That section was deleted, according to one of the panel's members, because there was no way of evaluating the gaps in either the current design or the key technologies to enable the Joint Chiefs to be sure that the system would meet their requirements. However, an earlier draft of this section appears in the version that Congressman Olin placed in the *Record*.

Defiance. The report provides additional ammunition to members of Congress who would like to zap or cap the SDI budget. Though a Defense appropriations bill is unlikely to be passed before fiscal 1988 begins on 1 October, the House and Senate have both indicated where they stand on SDI. The House voted to reduce President Reagan's request for $5.9 billion to $3.1 billion, which is more than $600 million below SDI's current account. In the Senate, the Armed Services Committee recommended $4.5 billion. Senate Democrats are holding the entire military budget hostage to SDI—in open defiance of Reagan and, surprisingly, public opinion. Polls have shown that between 60% and 82% (depending on the way questions are asked) of the US public favors developing the President's vision of a missile shield.

The Pentagon, meanwhile, operating on the strategy that the best defense is an offense, released the first formal description of Phase I, presumably now given the official title of Strategic Defense System—1. Like the Everett report, SDS-1, contained in a document 2 inches thick, was issued for use by the Defense Science Board and Defense Acquisition Board. It calls for at least 13 major tests of six different systems, including a space-based interceptor rocket, sensor satellites and a communications network, over the next five years. The experiments would provide the first glimpse of technologies needed for a low-tech Star Wars, and all are designed to comply with the "narrow" or traditional interpretation of the 1972 US–Soviet Antiballistic Missile Treaty.

None of the tests involve the exotic laser or particle beams that the public mind usually associates with Star Wars. More than half of the proposed space experiments for SDS-1 would consist of attempts to intercept missiles in flight using infrared guidance to direct a small rocket, sometimes called a "smart rock." Two tests would involve launching state-of-the-art satellites to detect and track missiles in their boost and post-boost phases.

The SDS document covers the environmental effects of testing on 15 DOD sites, including the Kwajalein Atoll in the Pacific.

—IRWIN GOODWIN

Anti-satellite treaties

Model agreements prepared by scientists have helped make anti-satellite weaponry the focus of efforts to revive US–Soviet talks on arms control

William Sweet   PHYSICS TODAY/NOVEMBER 1984

One of the most surprising developments in arms control during the past two years was the emergence of anti-satellite weaponry as the main focus of efforts to revive serious negotiations. Anti-satellite weapons (ASATs) can be based on a variety of technologies, including ground-based interceptor systems, space mines and lasers; and their evaluation presents formidable difficulties. The story of how the ASAT issue came to be the central focus of arms-control efforts, despite its technical complexity, is an unusual tale in which physicists have played prominent roles.

At the beginning of 1983, all eyes were fixed on the impasse in the parallel Geneva talks over intermediate-range and strategic missiles, the meteoric growth of the European peace movement and the ascendant freeze movement in the United States. The issue of anti-satellite weaponry was generally regarded as not particularly interesting, important or promising. While the Russians had proposed an ASAT treaty in 1981, the Reagan administration seemed unshakable in its conviction that an ASAT agreement would be unverifiable.

In mid-1984, the Reagan administration rather abruptly adopted a more flexible attitude, signaling Soviet leaders that it might, after all, be willing to enter into talks on an ASAT ban. After the Soviet government extended an invitation to meet in Vienna for talks in September, complicated maneuvering followed in which each side seemed intent on keeping the possibility of negotiations open without making a commitment to actually begin them before the US election.

The emergence of the ASAT issue as the focus of efforts to reopen US–Soviet arms-control talks can be attributed, in part, to pressure from scientists specializing in arms control and to public-interest science organizations, notably the Union of Concerned Scientists and the Federation of American Scientists. Starting last year, in an effort to rebut the Administration's claim that an ASAT treaty could not be verified,

specialists connected with UCS and FAS drafted model ASAT treaties, in which they spelled out in legalistic detail how such an agreement could be written. The scientists got their case heard in Congress, which in turn restricted the Adminstration's freedom to forge ahead with ASAT testing. At every stage, Congressional Fellows sponsored by The American Physical Society helped legislators grapple with technical aspects of the ASAT question.

Needless to say, UCS and FAS do not speak for all scientists, and not all the physicists in APS favor promotion of an ASAT treaty. Some physicists with very considerable experience in weapons and arms-control issues consider negotiation of an ASAT treaty unimportant, unnecessary, improbable or even impossible.

The skeptics include physicists Harold M. Agnew, former director of the Los Alamos Scientific Laboratory, and Herbert York, director of defense research and engineering under President Eisenhower and President Kennedy. Agnew, taking note of the fact that any long-range missile can be targeted against an object in space, wrote in *The Washington Post* on 28 August that "as long as any nation has an intercontinental missile, it will possess an ASAT capability—either with a nuclear or a conventional 'kill' capability." Agnew, who currently is president of GA Technologies (formerly General Atomic), concluded that "today's flurry of proposed treaties barring ASATs is really an unwarranted effort." In an interview, Agnew said that "the whole world is militarized, and to put such emphasis on one minor part is silly . . . I don't know whether it is pride of authorship or what, but we're so desirous of making treaties that we'll sign anything, and the Russians will just sit and wait."

York, current director of the Institute on Global Conflict and Cooperation at the University of California, San Diego, is inclined to think that negotiation of an ASAT treaty may be impossible because satellites are be-

coming too important militarily to be left alone in the event of war. Given the fact that satellites are used or soon will be used not only for photoreconnaissance, but also for the coordination of conventional and strategic sea and land forces, York believes that "trying to eliminate ASATs might be like trying to eliminate submarines." York's opinion is especially noteworthy because he does not generally subscribe to the philosophy that negotiations must always bow to the march of technical innovation. In his book on the decision to build the H-bomb (*The Advisors: Oppenheimer, Teller and the Superbomb*), York argued that the United States could have afforded to delay development of the hydrogen bomb, pending the outcome of exploratory talks with the Soviet Union.

If, as York and Agnew observe, space is becoming highly militarized and both superpowers already possess overpowering ASAT capabilities in the form of reprogrammable intercontinental missiles, what indeed is the point of promoting an ASAT treaty? In a nutshell, treaty advocates argue that a wasteful and possibly destabilizing race in new ASAT technologies would be worth preventing or inhibiting, even if countries retain some residual ASAT capabilities. Beyond that, treaty proponents believe that an agreement barring ASATs also would make it much more difficult to develop "Star Wars" defense systems based on directed-energy technologies (see PHYSICS TODAY, August 1983, page 17). As the treaty proponents see it, an ASAT agreement would complement and fortify the 1972 treaty limiting anti-ballistic missile systems—not to mention the Outer Space Treaty of 1967, the SALT I agreement on strategic arms and the (unratified) SALT II treaty, all of which guarantee the integrity of satellites used for arms-control verification. "The existence of any ASAT weapons poses a real and symbolic threat to these guarantees," Donald M. Kerr, current director of Los Alamos, observed in a paper on the ASAT–ABM link that he presented to a symposium

A US Air Force F-15 fighter carrying an anti-satellite missile under its belly. The missile is designed to carry a miniature homing interceptor vehicle into space.

at the Stockholm International Peace Research Institute in September 1983.

Anti-satellite weapon types

Existing ASAT technologies consist of ground-launched intercept vehicles developed by the United States and the Soviet Union, starting in the 1960s. From 1963 to 1967, the United States deployed an operational ASAT system based on the Nike–Zeus ABM missile system at Kwajalein Atoll in the Pacific, and from 1964 to 1975, it had a thrust-augmented Thor ASAT system on Johnston Island, though the Thor system was on standby status from 1970 on. Both systems were ditched in part because they relied on high-yield nuclear weapons that would damage US communications if they ever were used against Soviet satellites.

The Soviet Union began to test an ASAT interceptor on an augmented SS-9 missile in 1968, using an active radar to home in on the target after two orbits with a conventional warhead. According to John Pike, associate director of the Federation of American Scientists, the two-orbit interceptor was successful in five of seven tests between 1968 and 1971. In 1976, the Russians began testing an active radar interceptor designed to attack its target after one orbit, and it apparently failed in two of four tests, Pike believes. In 1976, they also started to test a more advanced heat-seeking ASAT, but it failed in at least five of six tests. Renewed testing of the two-orbit interceptor in 1976 produced two successes and one failure, Pike says.

The Soviet direct-ascent anti-satellite system is relatively slow and cumbersome, and it has been targeted only against satellites in low orbits—the satellites that are militarily used by the United States mainly for photo- and radar-reconnaissance and electronic intelligence. A Soviet interceptor has not been tested against targets in high-altitude geosynchronous orbits, where the US has stationed most of its satellites for early warning of nuclear attack and for military communications. (The USSR, in contrast, still has most of its early warning satellites in elliptical orbits with perigees close to the Earth.)

The renewal of Soviet ASAT testing in 1976 prompted President Ford to authorize a new US program, involving the development of a missile that would be launched from an F-15 fighter and would destroy its target by direct impact. Like the Soviet interceptor, the US "air-launched miniature vehicle" can reach only low-orbit satellites, but it would be faster and more versatile than the Soviet system. After President Carter took office in 1977, the development program was continued but the Administration also proposed to the Soviets talks on limiting ASATs.

Carter and his top military officials were ambivalent about the importance of the Soviet program, and talks were delayed until 1978, only to be left in suspension after the Soviet invasion of Afghanistan in late 1979. Physicist Harold Brown, Carter's defense secretary, said he found the Soviet program "somewhat troublesome" but that he hoped it would be possible to "damp down" the race in ASAT weaponry, if not completely "stop it."

UCS and FAS model treaties

The first legislative move to stimulate ASAT negotiations occurred in spring 1982, when Democratic Representative Joe Moakley and some 50 cosponsors offered a resolution calling for immediate talks. The resolution never came to a vote, but in early 1983, Congressional aides formed a space working group and began to meet regularly with lobbyists and outside experts in Moakley's office to exchange information and plan strategy. By this time, the Union of Concerned Scientists and the Federation of American Scientists were preparing model ASAT treaties. The UCS version, a highly polished document published with explanatory materials as a booklet last year, was written by a committee chaired by Kurt Gottfried of Cornell. The committee included physicists Richard Garwin (IBM), Hans A. Bethe (Cornell) and Henry W. Kendall (MIT), astronomer Carl Sagan (Cornell) and chemist Franklin Long (Cornell). The FAS treaty, which has circulated much less formally, was written by the Federation's space weapons specialist, John Pike.

Both model treaties are formal documents that contain the standard types of provisions found in all arms-control agreements: preambles stating general intentions, specific prohibitions, methods for resolving disputes over suspect activities, procedures for ratification, and a withdrawal clause that can be activated when extraordinary events jeopardize a country's supreme interests. The main difference between the two documents is that Pike's version contains detailed prohibitions specified for a variety of potential ASAT systems, including bans on deployment of ground-based directed-energy systems and "any system that has been tested in a prohibited mode," while the UCS version confines itself to a general ban on tests of weapons against objects in space and a specific ban on deployment of ASATs in space.

Despite their different approaches and the greater caution of the UCS scientists on the question of deployment, the authors of the FAS and UCS model treaties seem to be in agreement on the main points affecting verification, namely that:

▶ Significant expansion of the current Soviet system to threaten more US satellites could be detected because the SS-9 launcher must be modified to carry an ASAT, and because the very large modified missile is readily recognized from space with its support facilities.

▶ Further testing of the US miniature homing vehicle would be readily detected by the Soviets, but a ban on deployments of the very small vehicle would be very hard to verify once it were fully tested.

▶ Deployment of space mines disguised as satellites might escape detection, as long as the number of mines were small, but deployment of mines near a significant proportion of the adversary's satellites would not escape detection.

▶ Any directed-energy weapon in space would be detectable and highly vulnerable to attack, including attack by nuclear explosives detonated at a large distance away.

▶ Testing of a ground-based directed-energy system also could be detected (though possibly not without special precautions) because the system would have to be quite large and have distinctive characteristics to overcome disruptive atmospheric effects, and because there is only a small number of suitable locations where such a system could be built.

The last point, concerning ground-based laser systems, is the most controversial of the verification issues connected with ASATS. Allegations about Soviet work on directed-energy weapons have given rise to heated arguments in recent years, but hard evidence is extremely scarce. The 1984 edition of the Pentagon's booklet on "Soviet Military Power" is surprisingly cautious on the subject. "The Soviets could test a prototype laser and anti-satellite weapon as soon as the later

1980s," the Pentagon says, and a particle-beam weapon "designed to destroy the satellites could be tested in space in the mid-1990s."

Congressional action

The Soviets proposed an ASAT treaty to the US in August 1981, but this draft was "seriously flawed" and "indeed...did not preclude the deployment of either the US or Soviet ASATs," as James Treglio observed in a paper on the issue. Treglio, an APS Fellow, served on the staff of Senator Paul E. Tsongas from fall 1982 through summer 1983, when he was succeeded by another APS Fellow, Aviva Brecher. (Tsongas, now retired, was a Democratic Senator from Massachusetts and was actively interested in science and high technology issues as well as arms control.)

In May 1983, Treglio arranged with other Senate aides for UCS to present its case for an ASAT treaty to the Senate Foreign Relations Committee. The Committee also heard testimony from Kenneth Adelman, who had just been confirmed as director of the US Arms Control and Disarmament Agency after a bruising Senate battle. Adelman said that an ASAT treaty posed "daunting problems" of verification and that "we should not rush into negotiations on these subjects unless we are ready with verifiable proposals that will enhance national security."

According to Treglio, members of the Arms Control Subcommittee and their staff came away from the hearing "with a feeling that something had to be done to slow the US program" because otherwise it would "reach a point where a low-orbit ban would no longer be verifiable from the Soviet point of view." Taking the initiative, Tsongas introduced an amendment to the defense authorization bill barring tests of an ASAT against a target in space unless the President (1) sought ASAT negotiations with the Soviet Union, or (2) certified that testing was necessary to prevent irreparable harm to US national security.

To their surprise, the backers of the Tsongas amendment found that Senator Henry Jackson (now deceased) was willing to go along with the amendment, provided the wording was modified to specify testing of an "inert or explosive anti-satellite warhead," so as to leave open the possibility of laser ASAT tests. In negotiations with Senator John W. Warner (R-Va.), chairman of the Subcommittee on Strategic and Theater Nuclear Forces, backers of the amendment agreed that the President would not be required actually to enter negotiations, but merely to express his willingness to do so.

With these two changes, the Tsongas amendment passed the Senate by a 91–

0 vote in July 1983. Even now, backers of the amendment are somewhat baffled about how they won without opposing votes, but the general consensus is that they owed a good deal to the element of surprise. Apparently, White House aides did not grasp initially how hard it would be to claim that ASAT tests against targets in space were vital to US national security. Following the Senate vote, the White House reversed its position and lobbied, with success, to get the House to defeat the amendment. Despite that, Congress retained the amendment in the defense authorization conference.

In August 1983, one month after the Senate adopted the Tsongas amendment, the Soviet Union submitted a new draft ASAT treaty to the United Nations. While it contained provisions considered unverifiable or unnegotiable, such as a clause banning military uses of the US space shuttle, the new draft represented a considerable improvement over the earlier one in the eyes of arms control specialists. The Reagan administration, however, continued to insist that an ASAT treaty could not be negotiated.

White House position

In a report to Congress, submitted on 31 March 1984 in response to a mandate contained in the 1984 defense appropriation bill, the White House said that "no arrangements beyond those already governing military activities in outer space have been found to date that are judged to be in the overall interest of the United States and its Allies. The factors that impede the identification of effective ASAT arms control measures include significant

difficulties of verification, diverse sources of threats to US and Allied satellites, and threats posed by Soviet targeting and reconnaissance satellites that undermine conventional and nuclear deterrence." The report said that "in present circumstances, a US capability to destroy satellites clearly responds to the need to deter such Soviet attacks on US satellites in a crisis or conflict."

Specific problems highlighted in the White House report to Congress included the following:
▶ "The satellites which serve US and Allied security are few in number. Cheating on anti-satellite limitations, even on small scale, could pose a disproportionate risk to the United States."
▶ "The Soviet interceptor is relatively small and is launched by a type of space booster that the Soviets use for other space launch missions.... The USSR could maintain a covert supply of interceptors."
▶ "Tests of a ground-based laser ASAT weapon could be concealed."
▶ "Breakout potential could exist even if the Soviets, upon agreeing to a ban on ASAT systems, were to destroy all of their existing systems. The Soviets could retain the capability to redeploy quickly a system in which they would have confidence. If prior to the ban the United States had not tested its [maneuverable vehicle] ASAT system, the Soviets alone would possess such proven technology."

The White House report made few concessions to backers of an ASAT treaty. The report did say that the US could adopt a defensive strategy by "procuring sufficient satellite and boos-

In a US Army test in June 1984, an interceptor launched at Kwajalein missile range destroyed a reentry vehicle from an intercontinental ballistic missile. On the photo, the first stage of the rocket carrying the interceptor is seen piercing the clouds at roughly 10 000 feet and rising to a level of 150 000 feet, where it detaches from the second stage and starts falling back to Earth, leaving a dotted track. The interceptor relied on an infrared guidance system similar to the one used in the US miniature homing anti-satellite weapon.

ter spares," but it mentioned this possibility without enthusiasm, noting that it would run "counter to current US trends of developing space systems of greater sophistication and longer expected useful mission."

In January 1983, two months before the White House issued its report, the Office of Technology Assessment held a workshop on ASAT issues at the request of Senator Larry Pressler (R-S.D.), chairman of the Senate Subcommittee on Arms Control. Participants in the workshop represented a wide range of viewpoints and institutions, and when a summary of their proceedings was published last May, the consensus among the experts differed in many respects from the conclusions of the White House. The participants agreed that "no arms-control agreement can eliminate all anti-satellite capability," but they thought that a "ban on testing ASAT weapons would greatly increase the difficulty of developing a high-confidence, high-quality dedicated ASAT system." They disagreed "regarding how much significance can be attributed to residual or covert ASAT capability," but "the idea that the United States needs an ASAT weapon in order to deter enemy ASAT attacks was not strongly supported."

Recent legislative moves

In 1984 Congressional action, the Senate passed what most observers regarded as a watered-down version of the Tsongas amendment permitting ASAT tests against targets in space provided the President certified he was endeavoring to "negotiate in good faith the strictest possible limitations." Meanwhile, the House passed a tighter limit barring ASAT tests against space targets for a year, provided the Soviet Union did the same. The House victory was attributable to sustained work by treaty backers such as Representatives Moakley, George E. Brown Jr (D-Cal.) efforts by Representative Lawrence Coughlin (R-Penn.) to win support for negotiations among Republicans; and support from key "arms control moderates" in the House, notably Representatives Les Aspin (D-Wis.), Norman D. Dicks (D-Wash.), and Albert Gore Jr (D-Tenn.). Brown sponsored the 1983 ASAT amendment in the House, and Coughlin joined him as cosponsor in 1984.

The compromise Senate amendment was engineered largely by Senator Sam Nunn, a Georgia Democrat who ordinarily acts as a leader of hard-line arms-control critics. Days after the Senate vote on the Nunn compromise, President Reagan unexpectedly said at a press conference that "we don't have a flat no" on the possibility of negotiating ASAT verification measures and that

"we haven't slammed the door on that [ASAT talks] at all." Two weeks later, the Soviet Union formally invited the United States to open ASAT talks in Vienna, only to pull back when the Reagan administration promptly accepted.

The Soviets accused the Reagan administration, which wanted to link ASAT talks to a resumption of negotiations on intermediate-range and strategic missiles, of trying to set preconditions. Reagan aides made similar charges about Soviet demands that the talks embrace all space weapons, that the United States join in a moratorium on ASAT tests and that it agree in advance of talks that a treaty would be concluded. Thus began the parrying that may, or may not, lead to serious negotiations in 1985.

If negotiations are resumed, some kind of provisional agreement on ASATs is very likely to be the starting point, but it is questionable how important the ASAT issue will be overall. Solomon J. Buchsbaum, Executive Vice-President of AT&T Bell Labs, who is head of the White House Science Council, thinks that "arms-control negotiations are very important," but he says that "the ASAT issue would not be at the top of my list of priorities," though it is an "issue that must be taken into account at the table."

Reviewing the record of political maneuvering on ASATs, the technical complexity of the issue stands out as an important explanatory factor in its own right. In 1983, when the issue first came up, most people in Congress knew little about it, and the few who were able to learn fast secured an advantage quickly. Tsongas, who together with Pressler was the key initiator, gives a lot of credit to APS Fellows Treglio and Brecher for educating his staff. "Without their technical expertise we would never have been competitive on that issue," Tsongas says. "Whatever difference we made . . . is almost exclusively attributable to their efforts."

When the key action shifted to the House, the level of education members had attained again played an important role. According to Sybil Francis, an aide to Representative George Brown, and Jim McGovern, who works for Moakley, members of the House responded strongly to the idea that space should not become the next arena for a new arms race, but they did not generally understand the technical links between the ASAT and ABM technologies.

Treaty prospects

In the next round of legislative maneuvering over ASATs, if there is one, members of Congress may have to confront the ABM–ASAT links head-on for the first time. Specialists are in-

creasingly convinced, as Garwin observes, that you "can't have an ASAT ban without a Star Wars ban." In Garwin's view, if either side were to begin deployment of a Star Wars ABM system, the obvious and immediate reaction of the other side would be to surround it with space mines—relatively cheap conventional or nuclear explosives that could be detonated by remote control at the first hint that a nuclear war might break out. Because of the transcendent importance of the ABM issue, Garwin considers it clear that neither side would agree to give up ASAT systems such as space mines as long as the Star Wars issue were open. On the other hand, Garwin does not consider the ASAT–ABM connection so close that ratification of an ASAT treaty would lead ineluctably to termination of all funding for research on directed energy weapons.

Ultimately, of course, whether or not a treaty is concluded will depend on US and Soviet political leaders and the domestic constituencies they are beholden to. It bears noting though, that nuclear arms-control issues have become increasingly sensitive in both the Western alliance system and the Soviet bloc.

The militarization of space is an issue of growing concern to scientists in many countries besides the US and the Soviet Union. Earlier this year, a group of German scientists prepared yet another model ASAT treaty, which goes further than the UCS and FAS drafts in that it seeks to bar all military uses of space except for reconnaissance, verification and early warning. At the annual Pugwash meeting, which took place in Sweden this year, physicist H. P. Dürr of the Max Planck Institute circulated the German model treaty. Meanwhile, in early July, the treaty was given prominent attention at an international scientific conference on the militarization of space that was held at the University of Göttingen. Participants in the Göttingen conference included physicists Daniele Amati and Jack Steinberger of CERN. Victor Weisskopf of MIT, and Nina Byers of UCLA. Representative George Brown also attended and reports that he was "amazed to find such a high level of interest in weapons issues," not just among the conference participants, but among Germans from all walks of life.

Pike, commenting on the German draft ASAT treaty, says he is "sympathetic" to the goal of trying to prevent the militarization of space on a broad front, but he considers it impolitic to set such an ambitious objective. If one takes on all kinds of space weapons simultaneously, Pike points out, the military constituencies "you're up against are orders of magnitude greater."

Report to the APS of the Study Group on
Science and Technology of Directed Energy Weapons

PHYSICS TODAY/MAY 1987

Executive Summary and Major Conclusions

The American Physical Society (APS) convened this Study Group to evaluate the status of the science and technology of directed energy weapons (DEW). The evaluation focuses on a variety of lasers and energetic particle beam technologies for their potential applications to the defense against a ballistic missile attack. This action by the APS was motivated by the divergence of views within the scientific community in the wake of President Reagan's speech on March 23, 1983, in which he called on the U.S. scientific community to develop a system that " . . . could intercept and destroy strategic ballistic missiles before they reach our soil. . . ." Directed energy weapons were expected to play a crucial role in the ballistic missile defense (BMD).

The APS charged the Study Group to produce an unclassified report, which would provide the membership of the Society, other scientists and engineers, as well as a wider interested audience, with basic technological information about DEW. It is hoped that this report, detailing the current state of the art and the future potential of DEW for strategic defense purposes, will serve as a technical reference point for better-informed public discussions on issues relating to the Strategic Defense Initiative.

The study concentrated on the physical basis of high intensity lasers and energetic particle beams as well as beam control and propagation. Further, the issues of target acquisition, discrimination, beam–material interactions, lethality, power sources, and survivability were studied.

The technology of kinetic energy weapons (KEW) is not explicitly reviewed, but the role of space-based KEW in support of DEW systems is considered in the report where appropriate. Further, many important issues concerning command, control, communication, and intelligence (C³I), computing hardware, software creation and reliability for battle management, and overall system complexity have been identified but not discussed in detail. Other issues, which were recognized but not addressed, include manpower requirements, costs and cost-effectiveness, arms control and strategic stability, and international and domestic policy implications.

DEW technology is considered in BMD applications both for midcourse discrimination between decoys and reentry vehicles, and for kill in the boost phase and the post-boost phase of ICBMs. Such consideration has become serious because of numerous technological advances during the past decade in DEW technologies. Although the achievement of an effective defense of the entire nation may require a substantial boost phase intercept component, other strategic defense scenarios, including discrimination for hard point defense purposes, would place less demanding requirements on DEW systems. The Study Group deemed it important to describe the current state of the art in DEW technology, and to evaluate it with respect to substantial boost phase intercept and midcourse discrimination roles.

Although substantial progress has been made in many technologies of DEW over the last two decades, the Study Group finds significant gaps in the scientific and engineering understanding of many issues associated with the development of these technologies. Successful resolution of these issues is critical for the extrapolation to performance levels that would be required in an effective ballistic missile defense system. At present, there is insufficient information to decide whether the required extrapolations can or cannot be achieved. Most crucial elements required for a DEW system need improvements of several orders of magnitude. Because the elements are inter-related, the improvements must be achieved in a mutually consistent manner. We estimate that even in the best of circumstances, a decade or more of intensive research would be required to provide the technical knowledge needed for an informed decision about the potential effectiveness and survivability of direct-

DEW study group

Nicolaas Bloembergen, *cochair*
    Harvard University
C. K. N. Patel, *cochair*
    AT&T Bell Laboratories
Petras Avizonis
    Air Force Weapons Laboratory
Robert Clem
    Sandia National Laboratories
Abraham Hertzberg
    University of Washington
Thomas Johnson
    US Military Academy
Thomas Marshall
    Columbia University
Bruce Miller
    Sandia National Laboratories
Walter Morrow
    MIT Lincoln Laboratories
Edwin Salpeter
    Cornell University
Andrew Sessler
    Lawrence Berkeley Laboratory
Jeremiah Sullivan
    University of Illinois
James Wyant
    University of Arizona
Amnon Yariv
    California Institute of Technology
Richard Zare
    Stanford University

Charles Hebel, *executive officer*
    Xerox Corporation
Alex Glass, *principal consultant*
    KMS Fusion Inc

Scope of the DEW study

The following was included in the APS proposal for grants from the Carnegie Corporation and the John D. and Catherine MacArthur Foundation.

The study will review and evaluate the scientific and technological foundations of directed-energy weapons based on high-intensity laser beams of several types (the characteristics and effects of x-ray lasers will be included, but no attempt will be made to evaluate pumping by small nuclear explosions, an inherently classified subject) or on high-intensity charged and neutral particle beams. Selected aspects of projectile (pellet) approaches also will be examined for comparison. The study will emphasize technologies suitable against targets at long range, in other words thousands of kilometers.

The work will address two main areas:
▶ The physical basis of high-intensity laser and particle beams and the interactions that can affect beam control and target vulnerability. Representative topics include high-intensity beam production, beam instabilities and spreading mechanisms, long-range beam propagation and focusing, and interaction of beams with gaseous and solid matter. The study will emphasize current understanding of the underlying science, identifying physical questions that remain unclarified and limitations on performance that arise from intrinsic or fundamental physical constraints.
▶ The technological feasibility of full-scale components and subsystems for space- or ground-based applications of directed-energy weapons. Representative topics include energy sources and beam generators; beam aiming and control subsystems; target acquisition, command and control subsystems; vulnerability and countermeasures; and system integration. The feasibility evaluation will emphasize prospects for the technology as it may evolve over the next decade in order to improve delivered beam intensity and deal with the limitations or vulnerability of key components. Where possible, the study will identify R&D requirements and alternatives and also point out the implications inherent in the various approaches.

Assessment of component and subsystem feasibility requires a system context. For this purpose, the system performance of the components will be estimated against representative targets, taking into account the requirements and constraints inherent in a strategic defense mission. (Pertinent issues include the scale of directed-energy systems required for coverage against ballistic missile attacks of various types, and the possible effects on surveillance and communications satellites. The study also should recognize possible technological constraints that existing treaties may impose on field testing of directed-energy systems.) No attempt will be made to evaluate any particular existing or prospective system in detail. Instead, the study will emphasize key features of components and subsystems that govern system practicality, including order-of-magnitude cost estimates.

Beam director for the Mid-Infrared Advanced Chemical Laser, a medium-power chemical laser now used in experimental programs at the White Sands Missile Range.

ed energy weapon systems. **In addition, the important issues of overall system integration and effectiveness depend critically upon information that, to our knowledge, does not yet exist.**

The following observations elaborate on the above finding.

We estimate that all existing candidates for directed energy weapons (DEWs) require two or more orders of magnitude (powers of 10) improvements in power output and beam quality before they may be seriously considered for application in ballistic missile defense systems. In addition, many supporting technologies such as space power, beam control and delivery, sensing, tracking, and discrimination need similar improvements over current performance levels before DEWs could be considered for use against ballistic missiles.

Directed energy weapon candidates are currently in varied states of development. Among the many possibilities, infrared chemical lasers have been under study for the longest period and several high power laboratory models have been built. However, because of their long wavelengths and other tech-

nical features, these lasers are perceived to be less attractive candidates for BMD weapons even though they are closest to the required performance levels in a relative sense. Free electron lasers and excimer lasers are currently perceived as more attractive candidates for BMD missions; but few high power laboratory models have been operated, and the scaling required to reach relevant power levels is estimated to be greater than that for chemical lasers. Nuclear-explosion-pumped X-ray lasers, although the subject of much public discussion, are currently under study at the research level. In our opinion their BMD potential is uncertain.[1] Charged and neutral particle beam devices build on an existing base of accelerator technology but require considerable extrapolations beyond current performance levels.

Supporting technologies are also in varied states of development. In many areas, research is progressing at a rapid pace, for example schemes for rapid steering of optical beams, and active systems for tracking to microradian class or better.[2] Other critical technologies, such as the techniques for interactive discrimination, are being conceived and addressed. The same cau-

High-power chemical laser battle station in a low Earth orbit is portrayed in this artist's conception. Their comparatively compact power sources make chemical lasers favorable for space basing. Their long wavelengths necessitate using large optics over very long ranges. Atmospheric absorption makes these lasers unsuitable for ground basing.

tion described above for DEWs applies here; namely, proposed supporting technologies need to be systematically studied before their performance at parameter levels appropriate to BMD applications can be realistically evaluated.

Like any defensive system an effective DEW defensive system must be able to handle an evolving and unpredictable missile threat. In addition to retrofit and redesign of the missiles themselves, decoys and other effective penetration aids can be developed by the offense over the long times required to develop and deploy ballistic missile defenses. In contrast to the technical problems faced in developing DEWs capable of boost phase kill for defense systems, the options available to the offense, including attacks on DEW platforms, may be less difficult and costly to develop and may require fewer orders-of-magnitude performance improvements.

A successful BMD system must survive, but survival of high value space-based assets is problematic. Ground-based assets of DEW systems are also subject to threats. Architectures which address the responsive threat are still in their infancy. As an overall

BMD system employing directed energy weapons becomes more complex, the currently unresolved issues of computability, testability, and predictability become increasingly critical.

For directed energy weapons to have an important role as a kill mechanism in a strategic defense system, designed to defend the entire nation against a ballistic missile attack, the following requirements need to be met:

I. For operations in the boost phase:

A. Sufficient power/energy from the directed energy weapons to kill the ballistic missile in the boost phase, or to kill the post-boost vehicle during the deployment phase.

B. Sufficient beam quality, pointing accuracy, and agility (retargetability) to deliver lethal powers or energies to targets within the available engagement time provided by the system.

C. For lasers, optical systems for transmitting beams from sources to targets.

D. Accurate detection, location of the booster in its plume, and precision tracking from launch detection until kill is accom-

plished.

E. Reliable kill verification.

II. For operations during the midcourse:

A. Reliable means of discrimination between reentry vehicles and decoys unless all objects can be destroyed.

B. Accurate detection, tracking of a very large number of objects in the midcourse flight, and kill verification.

C. Rapid retargeting and sufficient delivered power/energy from the DEW to destroy the reentry vehicles.

III. For terminal phase:

We do not expect DEW to play an important role in the terminal phase of the trajectory of ballistic missiles.

IV. For space-based operation:

A. Nuclear reactors or other means to supply adequate electrical power for housekeeping functions.

B. Adequate burst power for operation of DEW during engagements.

C. Space qualified reliability of all components and subsystems on the platform notwithstanding long periods of dormancy.

How leading candidate DEW sources work

Chemical lasers (see the figures on pages S4 and S5) operate by using an exothermic chemical reaction to pump the molecular vibration level, which lases. Typically, separate oxidizer and fuel streams are rapidly mixed inside an optical cavity. For instance, in the HF laser a combuster is used to produce atomic fluorine, F, which then reacts with molecular hydrogen, H_2, to produce the upper laser level, vibrationally excited HF.

To make an **excimer laser** (see the figures on page S7), one uses an electron beam or a gas discharge to produce excitations or ionizations from electron impact in a gas. These excitations or ionizations cause a chain of chemical reactions that results in an excited molecule whose ground state is either repulsive or very weakly bound. Thus, in an excimer species every upper-state molecule formed is, effectively, a population inversion. One family of excimers, rare-gas monohalide molecules (such as KrF and XeCl), produces radiation in the soft uv (249 nm for KrF, 308 nm for XeCl) at relatively high efficiencies (2–7% overall), and is of principal interest for weapons applications. Individual laser cavities use unstable resonators to extract high power without optical damage.

A **free electron laser** (see the figures on pages S10 and S11) operates by sending an intense, energetic electron beam through an undulating magnetic field to produce coherent radiation. The "extraction efficiency" of an FEL is the transfer of beam power to electromagnetic wave power. The high-energy electron beams of FELs can be produced either by accelerating electrons in rf cavities (as is done in most conventional particle accelerators) or by passing them through linear acceleration stages in which the beam acts, in essence, as the secondary to a series of primary transformer windings. The accelerators are known as rf linacs and induction linacs, respectively, and the same nomenclature applies to the FELs themselves. Because they have relatively lower acceleration gradients, induction accelerators must be very long (hundreds of meters for electrons with energies of hundreds of MeV), too long to make them practical as laser oscillators; rf FELs can be operated either as oscillators or as amplifiers, but high-energy oscillators raise questions about damage to optical components.

A **neutral-particle beam generator** (see the figures on pages S12 and S13) consists of a negative hydrogen ion (H⁻) source, an rf quadrupole that provides initial acceleration and transverse focusing of the ion beam, a drift-tube linac that provides the major fraction of acceleration, a beam expander that takes the small transverse beam to a large radius (and hence very small angular divergence) and a stripper that removes one electron (but not two electrons) so as to produce neutral hydrogen, H⁰. The beam must be directed and its direction sensed before the final stripping.

V. For system survivability:

A. DEW must be able to operate in a hostile environment during a conflict.

B. DEW must be integrated in an overall system that includes a survivable command, control, communication, and intelligence (C^3I) system.

We have examined most of these issues in some detail, except for items III, IV.C, and V.B. The following major conclusions are based on detailed considerations in the main body of the report indicated by relevant section numbers in parentheses.

1. **We estimate that chemical laser out powers at acceptable beam quality need to be increased by at least one order of magnitude for HF/DF lasers for use as an effective kill weapon in the boost phase. Similarly for atomic iodine lasers, at least five orders of magnitude improvement is necessary.**

The HF/DF cw chemical lasers have been stated to yield power levels exceeding 200 kilowatt with acceptable beam quality.[3] Based on these data, we estimate that even the least demanding strategic defense applications require power levels to be increased further by

at least two orders of magnitude while retaining beam quality. However, the laser geometry which achieved the above demonstration will have scaling problems to higher power levels; thus, the combination of power scaling and adequate beam quality remains an open issue. A chemically pumped atomic iodine laser at 1.3 μm has been developed, although at this point only 5 kW of continuous wave power has been demonstrated. Because of atmospheric absorption, the HF laser ($\lambda = 2.8\ \mu$m) would have to deployed on space platforms, while the DF laser ($\lambda = 3.8\ \mu$m) and the atomic iodine laser ($\lambda = 1.3\ \mu$m) could also operate on the ground. When based in space, chemical lasers face a special set of problems arising from vibrations and the exhaust of the burnt fuel (Section 3.2).

2. **We estimate that the pulse energy from excimer lasers for strategic defense applications needs improvement by at least four orders of magnitude over that currently achieved. Many advances are needed to achieve the required repetitive pulsing of these lasers at full scale.**

The pulsed excimer lasers have demonstrated single pulse energies of about 10 kJ in 1 μsec pulses from a single

module[4] (Section 3.3). This laser currently uses krypton fluoride ($\lambda = 249$ nm); the other principal contender excimer species is xenon chloride ($\lambda = 308$ nm). From our estimates, assuming an overall propagation loss factor of four (relay mirror losses, Rayleigh scattering losses, and atmospheric losses), ground-based excimer lasers for strategic defense applications must produce at least 100 MJ of energy in a single pulse or pulse train with a total duration between several and several hundred microseconds (Section 6.3). To kill multiple targets a firing rate of ten per second would be desirable. For thermal kill 1 GW of average power would be required (Section 6.2). The gap of four orders of magnitude might be bridged by first combining lasers into modules at the hundreds of kilowatt level, then combining many modules optically. To produce high optical quality beams from the modules, the output from low optical quality amplifier apertures may be combined using stimulated Raman scattering or other means (Section 3.3). We estimate that the techniques for Raman beam combination must be scaled up by two orders of magnitude or more in combined laser power and efficiency from that which has been demonstrated in the laboratory. The technology for phase locking a large number of modules is not yet demonstrated (Section 5.4).

3. **Free electron lasers suitable for strategic defense applications, operating near 1 μm, require validation of several physical concepts.**

The free electron laser (FEL) is one of the newest laser technologies to be demonstrated. Peak powers of approximately 1 MW have been produced at a wavelength of 1 μm; peak powers of approximately 1 GW have been produced at a wavelength of 8 mm, demonstrating high gain and high efficiency at that wavelength.[5] Scaling to short wavelengths at high powers is a more difficult technical problem than simply increasing average power. Obtaining high efficiency, high power free electron laser operation at 1 μm requires experimental verification of physical concepts which thus far are only theoretically developed, e.g., optical guiding and transverse sextupole focusing for the amplifier configuration, and sideband and harmonic control for the oscillator configuration.[6] We estimate that for strategic defense applications, a ground-based free electron laser should produce an average power level of at least 1 GW at 1 μm wavelength, corresponding to peak powers of 0.1–1.0 TW (Sections 3.4 and 6.3).

4. **Nuclear-explosion-pumped X-**

ray lasers require validation of many of the physical concepts before their application to strategic defense can be evaluated.[7]

A sub-committee of the Study Group reviewed the progress in X-ray lasers. A nuclear-explosion-pumped X-ray laser has been demonstrated. This is a research program where a lot of physics and engineering issues are still being examined. What has not been proven is whether it will be possible to make a militarily useful X-ray laser[7] (Section 3.5). Atmospheric interaction limits the use of nuclear-explosion-pumped X-ray lasers to altitudes greater than about 80 km (Section 5.10). The high energy-to-weight ratio of the nuclear explosives makes it possible for these devices to be considered for "pop-up" deployment (Section 9.3).

5. **We estimate that neutral particle beam (NPB) accelerators operating at the necessary current levels ($\geqslant$100 mA) must be scaled up by two orders of magnitude in voltage and duty cycle with no increase in normalized beam emittance. The required pointing accuracy and retargeting rate remain to be achieved. These devices must be based in space to avoid beam loss via atmospheric interactions.**

Structural kills with NPB devices require an equivalent charge of about 1 coulomb (e.g., 100 mA for 10 seconds) delivered at a few hundred MeV, with a beam divergence of 0.75–1.5 microradian (as discussed and calculated in Sections 4.3 and 6.4). Disruption of electronic function because of radiation dose could occur at significantly lower beam parameters, although this kill mechanism is system dependent, and kill assessment may be more difficult (Chapter 4).

Existing radio frequency (rf) ion accelerators have achieved particle kinetic energies of several hundred MeV, but at beam current levels two orders of magnitude below the required levels (Section 4.3). New negative ion sources have achieved the necessary peak currents and low beam emittances, but such sources have not been reported to operate continuously. Additional is-

sues are emittance growth of the high current beams in the low energy accelerator sections, and the development of large bore magnetic optics. Power requirements and weight are also significant issues (Chapter 8).

Ionization of the neutral beam atoms via atmospheric collision (and subsequent ion deflection in earth's magnetic field) establishes a minimum operating altitude of about 120 km for beam kinetic energies of a few hundred MeV (Section 4.1).

In order to take advantage of the absentee ratio of a NPB device platform constellation designed for booster kill, NPB devices have been suggested for use in an interactive midcourse discrimination mode (identifying massive reentry vehicles in a postulated threat cloud which includes lightweight decoys). In this case the beam power requirements will not change significantly, but the target dwell times may be reduced by a factor of 10–1000, and retargeting rates of > 10 sec^{-1} may be necessary. Hence, device issues which will require new ideas and further exploration for this mission are development of rapid retargeting mechanisms using magnetic beam steering and fast accurate methods for beam direction sensing (Section 7.7).

6. **Energetic electron beams require propagation in laser-created plasma channels in order to avoid beam deflection in the earth's magnetic field; this restricts the operational altitude at the low end by beam instability and at the high end by ion density starvation. We estimate that booster kill applications require a scale-up in accelerator voltage by at least one order of magnitude, in pulse duration by at least two orders of magnitude, and in average powers by at least three orders of magnitude. Active discrimination applications require scale-up in pulse duration by at least two orders of magnitude, and in average power by at least two orders of magnitude. The lasers needed for the creation of plasma channels require develop-**

contents of the APS report on directed-energy weapons

Cavity of a large KrF laser under construction at Los Alamos National Laboratory. The large oval magnet coils provide uniform deposition of electron beam energy in the lasing gas. The laser oscillates in the left–right direction in this photo.

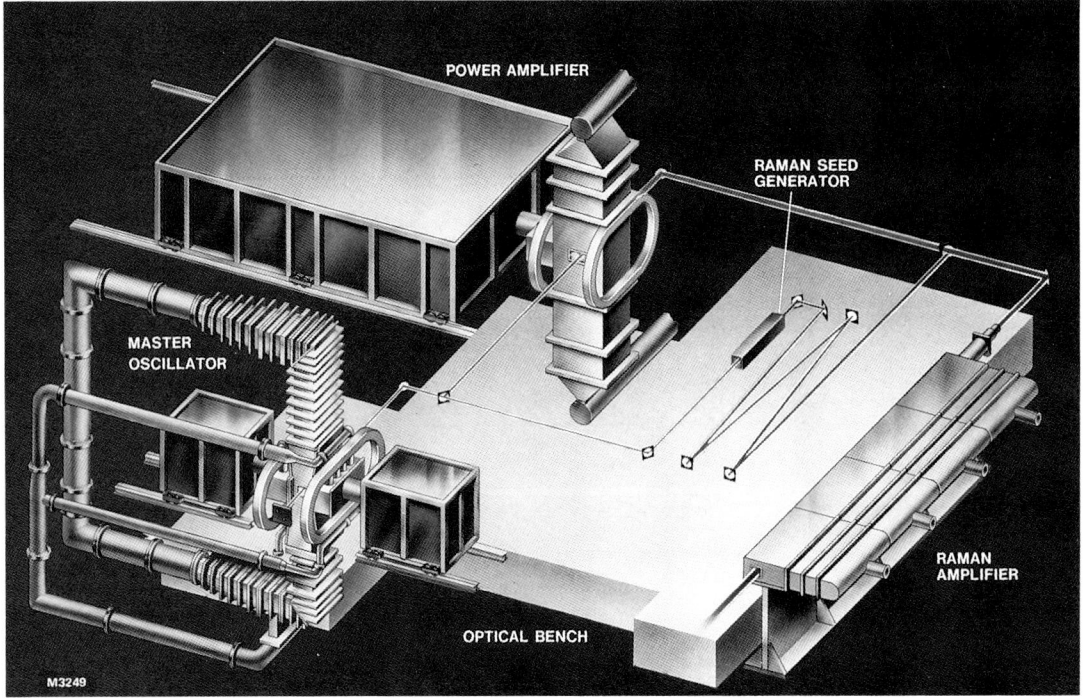

Artist's conception of an excimer laser facility designed to provide good beam quality by Raman cleanup, in which the output from the excimer laser pumps a Raman-active gas (hydrogen); then a seed beam with high optical quality extracts the power from the medium, shifting the wavelength of the original beam slightly and retaining the optical quality of the seed beam.

Wiggler section of a free electron laser device now operating at Los Alamos National Laboratory. An electron beam from an accelerator is guided into one end of the wiggler; as the beam propagates through the vacuum channel in the wiggler, it is subjected to magnetic fields of varying intensity perpendicular to its path. The electrons oscillate and radiate in phase as the magnetic fields accelerate them.

ment. We estimate that propagation distances must be increased by several orders of magnitude.

Propagation through a laser-created plasma channel is necessary to prevent beam space-charge blow-up and beam bending in the earth's magnetic field. This implies both a lower and an upper altitude operational limitations. The lower bound arises from beam stability considerations, while the upper bound results from ion density starvation. This mechanism for beam guiding has been successfully demonstrated in the laboratory, but over distances of 95 meters[8] (Section 4.2). For optimum beam currents of a few kiloamperes, delivering lethal pulses to distances in excess of 1000 kilometers will require beam kinetic energies of several hundred MeV. Useful ranges for some suggested interactive discrimination applications could be as small as a few hundred kilometers, in which case the particle energy requirement would decrease by an order of magnitude (Section 7.7). Existing linear induction accelerators have demonstrated the necessary peak power capability (tens of MeV at peak currents of tens of kiloamperes and pulse repetition rates of a few hertz), although not for required pulse lengths of microseconds (Section 4.2). Although several approaches have been suggested, the laser technologies required for creating the plasma channel have not been demonstrated. Because of the limited engagement space, rapid retargeting (~ 0.1 sec) and high repetition rates ($\geqslant 10$ Hz) are essential.

7. **Phase correction techniques are required for obtaining near diffraction limited performance of most types of laser weapon devices. Further, phase control techniques are required for coherently combining outputs from different modules in a multiple laser system into a single diffraction limited beam. These techniques, demonstrated at low powers, must be scaled up by many orders of magnitude in power.**

High power laser systems are likely to require active control and correction of the optical phase of the output beam to reach the nearly diffraction limited performance desired for strategic defense applications. Several techniques are available for these purposes. These include correction of slowly varying phase errors with low spatial frequencies through use of adaptive optics and self-correction of phase errors using nonlinear phase conjugation techniques, such as stimulated Brillouin scattering, or four-wave mixing; and combining beams from multiple apertures by phase locking of multiple laser modules, or through stimulated Raman scattering. Each of the laser technologies under development may use different types of phase corrections. All of these approaches for phase correction have been demonstrated on a laboratory scale, but extensions to high power systems and large apertures remain to be demonstrated (Section 5.4).

8. **Dynamic phasing of arrays of telescopes requires extensive development in order to obtain large effective aperture optical systems. As calculations indicate (Section 5.4.5), the number of phase correcting elements must be increased by at least two orders of magnitude over currently demonstrated values.**

Optical laser systems will require large effective optical apertures in order to achieve the necessary beam intensity on target. Such radiating apertures have to provide near diffraction limited beams which can be rapidly retargeted. The state of the art for ground-based monolithic telescope primaries for astronomical applications is about 8 meters.[9] Torque requirements for rapid steering of large telescopes limit such telescopes to approximately 8 meters aperture; the larger "effective aperture" primaries have to be synthesized by dynamically phasing a number of smaller telescopes. Such phasing of a number of telescopes has been accomplished[10] by dynamically controlling the wavefront "piston," tilt, and focus of the laser beams feeding each telescope of the array. This adds complexity to the system but allows beam pointing in terms of target tracking

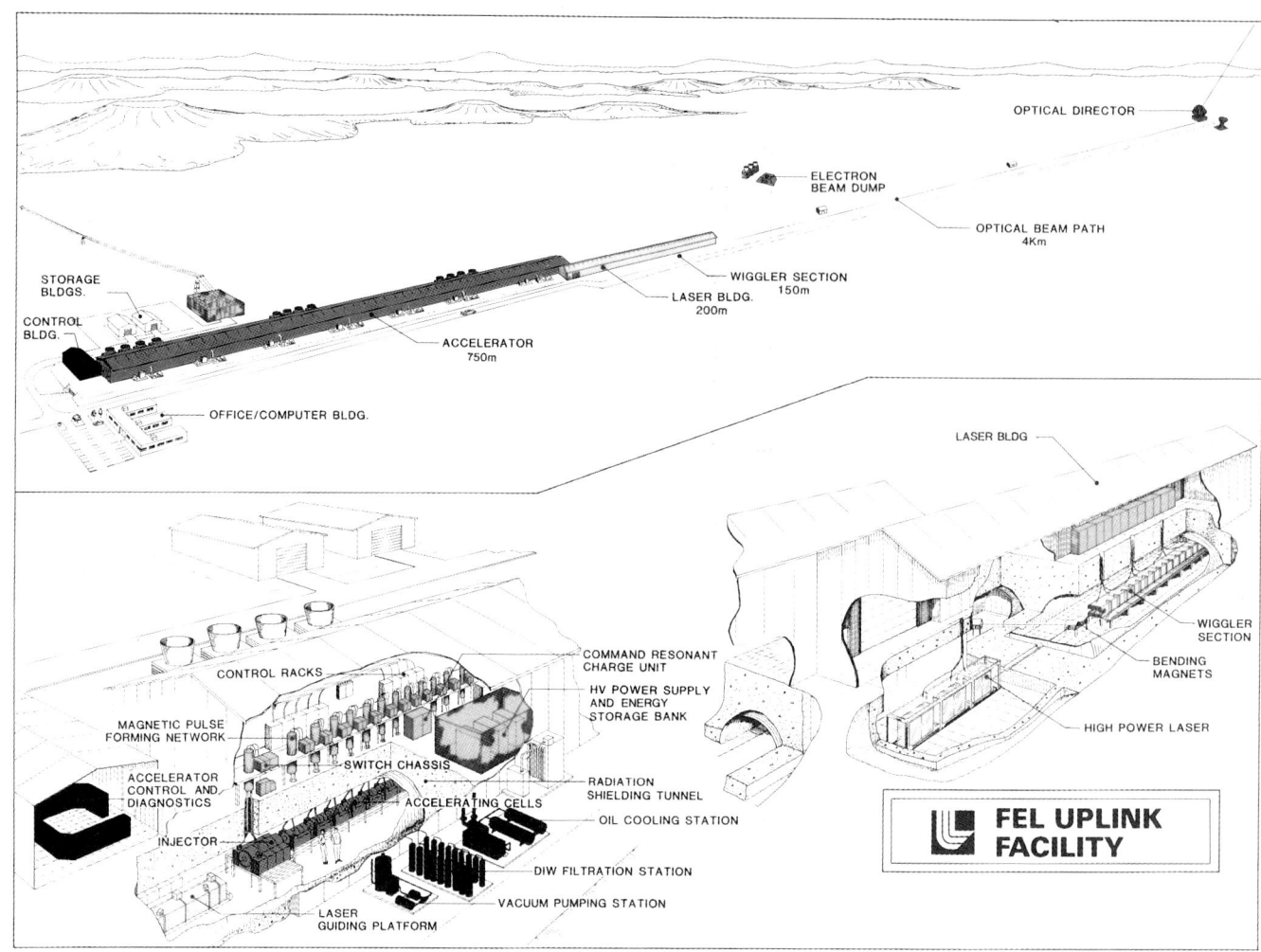

Schematic of the Lawrence Livermore National Laboratory proposal for one of the two
FEL approaches that are being considered for construction at White Sands Missile Range.
This FEL uses an induction accelerator; the Los Alamos FEL shown on the facing page
uses an rf accelerator.

without requiring slewing of telescopes (Section 5.2).

The phase front of the outgoing wave is monitored in such phasing schemes, and corrections are applied via electrically driven actuators. Components for control of about several hundred such actuators are commercially available. For the large apertures contemplated for BMD applications the number of actuators needed lies between ten thousand and one hundred thousand, a substantial extrapolation. The technology of phase-controlling an array of primary mirrors is in an early stage of development. Scaling of such arrays to high power has not been accomplished (Section 5.4).

An alternative approach is to use telescopes where the primaries are made out of single large flexible membranes which are appropriately distorted by many actuators. The concept has been demonstrated only for small flexible primaries at low powers. Extensions to large mirrors at higher powers remains to be shown (Section 5.4).

9. The optical coatings of large primary mirrors are particularly vulnerable in space-based optical subsystems.

The large primary mirror, which directs the laser beam towards the target, is particularly vulnerable to radiation from other lasers (from any direction) (Section 5.6). Based on discussions with commercial vendors, we find that the cw power loading threshold for reflective coatings is about 100 $kW\,cm^{-2}$. For laser pulses of a few microseconds or less, the damage threshold will be about 8 $J\,cm^{-2}$ of absorbed energies, corresponding to peak powers of 10 $MW\,cm^{-2}$. These damage thresholds are for operation at a nominal laser wavelength of 3 μm (Section 6.2). If attacked by lasers at other wavelengths in the visible, near ultraviolet (UV), or X-ray region, the damage threshold may be significantly lower. Further, there is a possibility of damage to the high reflectivity coatings from energetic particles in the ambient background, i.e., MeV protons

and electrons, during long term residence of the high reflectivity mirrors in space.

10. Small secondary mirrors in the optical trains of high power lasers will need very low absorptivity coatings and will have to be cooled.

The requisite power levels for ballistic missile defense lethality will necessitate cooling of the small mirrors in the optical train of high power lasers to prevent damage. A beam power of 1 GW on a mirror of 100 cm^2 area implies an incident power of $10^7\,W\,cm^{-2}$. High reflectivity coatings with less than 10^{-4} absorptivity are needed. Such mirrors have been demonstrated, and lead to an absorbed power of 1 $kW\,cm^{-2}$. Cooled silicon or silicon carbide mirrors show promise for raising this threshold (Section 5.5).

11. Ground-based laser systems for BMD applications need geographical multiplicity to deal with adverse weather conditions.

Whitehorse Test Stand, an accelerator for production of neutral-particle beams at Los Alamos National Laboratory. The accelerator produces beams of negative ions, then strips the extra electrons in a gas cell, leaving a neutral beam.

For each ground-based laser system which must be available in battle, a number of geographically separated laser sites are needed to provide availability of at least one site in the system when the others are obscured by adverse climatic conditions. These locations must be separated by distances greater than the coherence length scale for weather patterns. Based on weather statistics, a multiplicity of five independent ground-based lasers could provide a 99.7 percent availability. By going to 7 climatically isolated locations in the continental U.S. availability of 99.97 percent is possible. At each of these sites, local cloud cover conditions require further multiplicity of the large ground telescopes, separated by few km (Section 5.4).

12. **Ground-based laser systems require techniques for correcting atmospheric propagation aberrations. We estimate that these techniques must be extended by at least two orders of magnitude in resolution (number of actuators) than presently demonstrated. Phase correction techniques must be demonstrated at high powers.**

Ground-based laser systems will require either linear or nonlinear adaptive optics of a very sophisticated nature in order to precompensate the laser beam for atmospheric aberrations caused by atmospheric turbulence and by thermal blooming induced by the laser beam itself. A retroreflector or a low power laser located at an appropriate point-ahead position in front of a space-based relay mirror would provide a reference source for transmission through the atmosphere to the ground telescope, where the wavefront would be analyzed for acquired aberrations due to the atmosphere. This analysis would be used to actuate adaptive optics of high resolution ($\geqslant 10{,}000$ actuators per aperture) at high bandwidths (≈ 1.0 kHz). This technique requires an extensive computational capability. Such atmospheric compensation experiments have been successfully demonstrated at low powers (no thermal blooming in the atmosphere) and at average atmospheric viewing conditions for Mt. Haleakala, Maui (moderate turbulence), with a small number of actuators (< 100). At high power levels, the turbulence may be high enough to cause a beam intensity redistribution which could be uncorrectable (Sections 5.2 and 5.4).

The incorporation of phase correction schemes in pulsed induction linac FEL amplifier is particularly stressing because the atmospheric compensation must be carried at high power levels. Atmospheric compensation techniques are needed for point-ahead angles which are large and for targets which may be non-cooperative.

13. **Uplink in a ground-based laser system faces transmission losses in the atmosphere.**

The uplink of high power output from a ground-based laser system faces natural atmospheric losses such as Rayleigh scattering, which stress the short wavelength systems, and atmospheric absorption losses, primarily from water vapor, which stress the longer wavelength systems. The optimum wavelength region is 0.4–1.0 μm. Even in this region, nonlinear effects such as stimulated Raman scattering and thermal blooming force the use of large final transmitting optics on ground (Section 5.4).

14. **Nonlinear scattering processes in the atmosphere impose a lower limit on the altitude at which targets can be attacked with a laser beam from space.**

Power delivery downward through the atmosphere to rising targets may be limited by stimulated Raman scattering and thermal blooming by ozone absorption. These phenomena limit the minimum attack altitude to 80 km for very short pulses, or require a longer pulse length (1–10 ms), because the laser beam must be focused to a small, ~ 1 m^2, spot size on the target. At the required high laser intensities, nonlinear effects may throw the optical power out of the focused beam before reaching the target (Section 5.4).

15. **Detection and acquisition of ICBM launches will pose stringent requirements for high detection probability and low false alarm rates.**

The achievement of boost phase kill probabilities of 90% implies booster detection and acquisition probabilities of better than 90%. In addition, successful operation of a midcourse system depends importantly on being given

Deployment of a neutral-particle beam battle station in space is envisioned in this artist's conception. Neutral-particle beams cannot propagate in the atmosphere; they would be stripped and the resulting ions would be deflected by Earth's magnetic field. Unlike lasers, particle beams deposit their energy inside their targets, rather than on the surface. Low-intensity neutral-particle beams could be used to discriminate between RVs and decoys or to damage a missile's firing mechanism. High-intensity beams might, for example, cause structural failure of the RV or effectively disarm its nuclear weapon.

good booster trajectory information. Of even greater importance, low false alarm rates are required so that a BMD system is not activated in peacetime because of the false alarms (Section 7.2).

16. For boost phase, infrared tracking of missile plumes will have to be supplemented by other means to support submicroradian aiming requirements of DEWs.

Tracking of missiles by detecting the intense short wavelength infrared (SWIR) radiation from booster plumes is a technology which has been pursued for some time. The plume brightness greatly exceeds that of the missile, and the position of the missile within the plume depends in a complex manner on altitude, missile type, rocket motor, fuel characteristics, etc. and is susceptible to variation by the offense in a manner which cannot be predicted by the defense. Other passive means of accurately locating and tracking missiles in boost phase are in early stages of study (Section 7.5).

Active means of tracking may be required. Of the likely candidates, microwave radars are the most developed although electronic countermeasures for them are also well developed. Optical radars may be more promising, if the illuminating beam can be rapidly retargeted, and if an imaging capability can be achieved (either range–Doppler or angle–angle systems would be sufficient). If rapid retargeting cannot be developed and if power–aperture requirements for microwave radars be-

come too severe hundreds to thousands of space platforms will be needed (Section 7.6).

17. For post-boost and midcourse, precision tracking will require active sensor systems.

Observation of PBVs [post-boost vehicles] and RVs [reentry vehicles] (at 300 K) will require detection of weak thermal signatures since these signatures vary as T^4. Similar signatures are associated with objects in midcourse. Thermal detectors in the long wavelength infrared (LWIR) can be used only above the earth's limb against a cold sky background. Low noise LWIR detector assemblies having the appropriate resolution, i.e., large element arrays, are being developed. Because of the long wavelengths involved (8–12 μm), submicroradian tracking accuracy is not feasible in LWIR without using telescopes with apertures in excess of ten meters (Section 7.2). Thus, thermal detectors will have to be supplemented by some active means such as microwave or optical radars. A large number of space-based platforms will be required. These might be the same platforms that are performing similar duties in the boost phase (Section 7.3).

18. For midcourse, when the RVs are interspersed with penetration aids, interactive discrimination may be required. At present the application of DEW technologies to this task is in the conceptual and early experimental stage.

Missiles which survive the boost

phase can deploy large numbers of decoys and other penetration aids. Since LWIR and radar signatures depend largely on surface phenomena, there are many options available to the offense desiring to confuse or saturate the defense (use of balloons for example). Directed energy technologies may offer the possibility of "mass" discrimination by interactive, perturbing means, e.g., detection of particle-beam-induced secondary emissions or velocity changes caused by laser-ablation-induced impulse. DEW platforms absent from the boost phase intercept theater might be useful in this function. Such interaction discrimination is in a conceptual and early experimental stage, and would require large numbers of additional sensor/detector platforms, plus the ability to function in nuclear-disturbed backgrounds (Section 7.7).

19. The development of an effective boost phase defense is highly desirable, perhaps essential for limiting the number of objects with which the midcourse and terminal defense elements must cope.

Given the present number of Soviet boosters and their capability, the offense can deploy half a million or more threat objects (reentry vehicles and decoys). Boost phase attrition is required if midcourse discrimination systems can deal with only a limited number of threat objects. Even an 80% effective boost phase defense would leave 100,000 or more objects entering the midcourse phase. If further in-

Adaptive optics, or "rubber mirror," component. To propagate efficiently through the turbulence of the atmosphere, laser beams must have their phase fronts adjusted. Adaptive optics achieves this compensation by moving the mirror segments individually to produce a phase front that exactly adapts to the turbulence-induced phase changes.

creases in the offensive threat or degraded performance of the boost phase tier overloads the tracking and discrimination capabilities of later tiers, then the overall performance of the defensive system would degrade catastrophically rather than linearly when saturation is approached. The tracking and discrimination of tens to hundreds of thousands of objects during the midcourse phase poses formidable challenges to sensors and battle management computers. If discrimination requires birth-to-death tracking of all threat objects, these problems become even more demanding (Section 2.3).

20. **Housekeeping power requirements for operational maintenance of many space platforms for strategic defense applications necessitate nuclear reactor driven power plants on each of these platforms.**

The power requirements for "housekeeping," i.e., the requirements for a space platform to control attitude, to cool mirrors, to receive and transmit information, to operate radars, etc. are estimated to be in the range of 100–700 kW of continous power. This would require a nuclear reactor driven power plant for each platform, necessitating perhaps a hundred or more of these nuclear reactors in space. These foregoing needs require solving many challenging engineering problems not yet explored. Cooling of large space-based power plants is a very difficult task (Chapter 8).

21. **During engagements prime power requirements for electrically driven space-based DEW present significant technical obstacles.**

The prime power required for electrically driven DEW, e.g., particle beam accelerators, is estimated to be 1 GW. This power could be provided by large chemical or nuclear rocket engines and generators, deployed at considerable distances or otherwise decoupled from the DEW platforms in order to avoid mechanical disturbances and effects of exhaust gases. This may require complex power transfer systems comprising cables, microwave systems, etc. Correspondingly, chemical fuel consumption would be more than five tons per minute of operation per platform (Section 8.3).

22. **Survivability is an essential feature of any BMD system employing space-based assets; such survivability is highly questionable at present. Evaluation of these issues requires a systems approach that includes hardening, active defense, and operational tactics. During the deployment phase, the space-based assets are especially vulnerable.**

The space platforms carry sensors, optical mirrors, or radar dishes, many of which have considerably lower damage thresholds than do the hardened boosters, post-boost buses, and RVs. While sensors and optical mirrors on satellite platforms may be shielded during long periods of inactivity, they would be exposed when put on alert prior to an impending ICBM attack. Such an attack could be preceded by an attack on these platforms by space-based and ground-based DEW, space-based kinetic energy weapons (KEW), space mines, or direct ascent nuclear

and non-nuclear anti-satellite (ASAT) weapons of the offense. Moreover, the system must be developed by a process of accumulation of space assets while the system is less capable of defending itself (Sections 9.3 and 9.4).

The ground-based laser systems for strategic defense applications require a substantial number of space-based optical elements and space-based sensors. The space-based optical elements include telescopes with large primary mirrors, the size and numbers of which will depend on the basing modes for the relay and the fighting mirrors. These space-based elements entail the same vulnerability as any other space-based components (Section 9.3).

23. **Survivability of ground-based facilities also raises serious issues. The relatively small number of large facilities associated with ground-based laser sites makes these facilities high value targets.**

The ground-based laser BMD facilities must be successfully protected from direct attack from many threats (e.g., cruise missiles, sabotage, etc.), in addition to ballistic missiles. Thus, any strategic defense system depending on ground-based lasers, or on other ground-based facilities which cannot be extensively proliferated, must be effective in defending against more threats than just ballistic missiles (Section 9.3).

24. **Directed energy weapons with capabilities below those needed for many ballistic missile defense applications can threaten space-based assets of a defensive system.**

If a DEW falls short of ballistic missile defense requirements, it may still be a credible threat to space-based assets. Space-based platforms move in known orbits and can therefore be targeted over much longer time spans than ballistic missile boosters, post-boost buses or reentry vehicles. The defense platforms may have key components that are more vulnerable than the boosters and the reentry vehicles. Furthermore, space-based platforms in low earth orbits can be attacked from shorter ranges than those required for boost phase intercepts (Sections 9.3 and 9.6).

25. **X-ray lasers driven by nuclear**

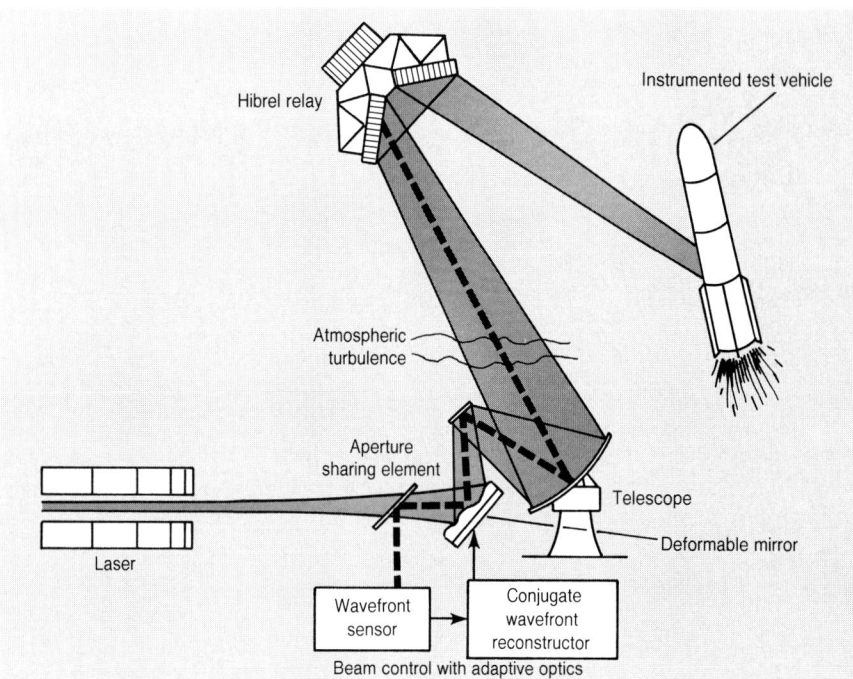

Beam control with adaptive optics

Schematic drawing showing how adaptive optics (pictured in figure on facing page) could be integrated into a ground-based laser system. A downlink signal from the relay mirror, passing through the turbulence, tells the wavefront sensor how to cancel the atmospheric effects. A computer reconstructs the wavefront that will perform the compensation and drives the actuators that deform the rubber mirror segments by the precise amounts required.

explosions would constitute a special threat to space-based sensors, electronics, and optics.

The high energy-to-weight ratio of nuclear explosive devices driving the directed energy beam weapons permits their use as "pop-up" devices. For this reason the X-ray laser, if successfully developed, would constitute a particularly serious threat against space-based assets of a BMD (Sections 3.5 and 9.3).

26. Since a long time will be required to develop and deploy an effective ballistic missile defense, it follows that a considerable time will be available for responses by the offense. Any defense will have to be designed to handle a variety of responses since a specific threat cannot be predicted accurately in advance of deployment.

A thorough understanding of practical responses, such as attacks on the defensive assets, hardening of offensive systems, and rapid deployment of large number of decoys, must be established before conclusions about the technical feasibility and cost-effectiveness of a defensive system can be made. A DEW system designed for today's threats is likely to be inadequate for the threat that it will face when deployed (Section 2.3 and Chapter 9).

References

1. "X-ray Lasers for Missile Defense," Defense Science and Engineering, November 1986, pp. 17–19.
2. U.S. Congress, Office of Technology Assessment, *Ballistic Missile Defense Technologies,* OTA-ISC-254 (Washington, D.C.: U.S. Government Printing Office, September 1985).
3. IEEE Spectrum, September 1985 [special issue on "SDI: The Grand Experiment"], and references cited therein.
4. See J. A. Mangano, J. H. Jacobs, Appl. Phys. Lett. **28**, 724 (1976).
5. T. J. Orzchowski *et al.*, Phys. Rev. Lett. **57**, 2172–2174 (1986).
6. See N. M. Kroll *et al.*, IEEE J. Quantum Electron. **QE-17**, 1496 (1981).
7. E. Walbridge, "Angle Constraint for Nuclear Pumped X-Ray Laser Weapons," Nature **310**, 180–182 (1984), and reference cited therein; George Miller (Associate Director, Lawrence Livermore National Laboratory), quoted in "Experts Cast Doubt on X-Ray Lasers," Science **230**, 647 (1985).
8. G. J. Caporaso, F. Rainer, W. E. Martin, D. S. Prono, and A. G. Cole, "Laser Guiding of Electron Beams in Advanced Test Acceleration," Phys. Rev. Lett. **57**, 1591–1594 (1986).
9. C. H. Townes, private communication.
10. See L. Marquet, in *Lasers '85: Proc. Int. Conf. on Lasers '85,* C. P. Wang, ed. (McLean, Va.: STS Press, 1986), pp. 247–252; and J. S. Fender, "Synthetic Aperture Systems," SPIE Proc. **440** (1983).

NAVY AND ARMY INITIATE ASAT PROGRAMS, WITH TEST BAND LIFTED

PHYSICS TODAY/APRIL 1989

Because of the affinities between anti-satellite and missile defense technologies, forces opposing Star Wars have made it their business since the mid-1980s to block testing of ASAT systems. Specifically, from 1985 to 1988 a Congressional coalition barred testing in space of the Air Force's fighter-launched rocket interceptor (see PHYSICS TODAY, November 1984, page 99). Late last year, when the Air Force withdrew its request for further funding for the F-15 ASAT program, it seemed to be a significant victory for the coalition.

But apparently the cancellation of the F-15 program led to a certain complacency among ASAT opponents on the Capitiol Hill, causing them to overplay their hand. Representative George Brown, the Democrat from California who has been a chief sponsor of legislation barring ASAT tests, tried at the end of last year to get the House to adopt a permanent ban on testing of the F-15 ASAT. This provided ASAT proponents an occasion to reopen the testing issue and swing votes; the result is that the test ban has been lifted. This shift in the balance of power to the forces favoring Star Wars represents a political

sea change, and it clears the way for the Pentagon to start testing sensitive missile-defense elements as ASAT systems.

The Army has asked for $100 million in the 1990–91 budget to spend on free-electron laser ASAT work; the Air force wants to spend $62.8 million on excimer and oxygen–iodine ASAT lasers; and the Navy—causing eyebrows to rise—would like to have $218.9 million to spend on rocket-launched ASAT interceptors based on the Exoatmospheric Reentry-Vehicle Interceptor Subsystem, or ERIS, developed by Lockheed Corp for the Strategic Defense Initiative.

"A lot of people might be wondering what interest the Navy, of all services, has in ASATs," observed David C. Morrison, national security correspondent for the *National Journal*, one of the two leading weeklies covering government in Washington. One explanation, Morrison wrote, is that the USSR has been putting its naval reconnaissance satellites into higher orbits, beyond the reach of the Air Force's F-15-launched ASAT. Another explanation, a writer pointed out in a recent issue of the *Bulletin of the Atomic Scientists*, is that the Navy

is especially well placed to attack Soviet reconnaissance satellites, many of which are in elliptical polar orbits that bring them to their lowest points when they pass over the South Pacific. The broadest explanation, given to this magazine five years ago by Herbert York, has to do with the Navy's increasing reliance on satellites for the command and control of naval operations. Because of the military's ever-growing dependence on satellites for reconnaissance, command and control, York thought that trying to permanently ban ASATs might be "like trying to eliminate submarines."

The new ASAT programs requested by the services would not have been affected by the F-15 test ban, and the only ASAT test that is likely to occur in 1989 will involve using the Mid-Infrared Advanced Chemical Laser, or MIRACL at White Sands against a target in space. ASAT testing of ERIS—the system the Reagan administration wanted to deploy as the first tier of a missile defense system—is expected to begin next year.

—WILLIAM SWEET

SECTION 5

NUCLEAR PROLIFERATION

Not surprisingly, the five nuclear weapons states, as defined in the Nonproliferation Treaty, are the "big five" winners from World War II, not the losers, Japan and Germany, and not the less-developed countries. This situation changed dramatically when India, a less-developed country by many standards, exploded a nuclear weapon in May 1974. This section examines several technical aspects of nuclear nonproliferation. In the first article, the APS Nuclear Fuel Cycle study group examines waste disposal, and the reprocessing of spent fuel. Ernest Moniz and Thomas Neff examine the broader aspects of proliferation, with Neff recently commenting that:

> A lot has changed: Despite a slowing of actual proliferation, incipient proliferation has moved ahead considerably in that the mix of capability needed is more widely available. The danger is that broader world or regional instability would precipitate a 'phase transition' to weapons development. The other major change is geopolitical, other nations, not the US, have the most responsibility, but Japan and Europe are not accustomed to such a leadership role. Plutonium technology is mostly in Japan and Europe, and it is from there it will spread if the other nations involved do not integrate the nonproliferation concerns into their business and foreign policy matrix.

Craig Waff describes how the Carter Administration explored other nuclear fuel cycles that might be more resistant to proliferation. Michael Jacobs then describes the International Nuclear Fuel Cycle Evaluation, a 46-nation study, which concluded that there is definitely a connection between nuclear power and proliferation, and that the thorium fuel cycles are at a disadvantage for several reasons. Gloria Lubkin describes a study which concluded that the enrichment of isotopes by lasers is approximately similar to centrifuges in terms of potential prolifera-

tion. Paul McCloskey describes the "born secret" aspects of the Atomic Energy Act of 1954, which were invoked after the *Progressive Magazine* published the approximate designs of the H bomb. Barbara Levi describes the Israeli destruction of the Iraqi research reactor, a response by a state widely assumed to have nuclear weapons to activities by a neighbor that might lead to further proliferation of nuclear weapons.

Fred Donath and Robert Pohl then debate the merits of disposal of radioactive wastes in salt, and other nuclear-waste issues. Irwin Goodwin describes the challenge by intervenors to prevent DOE and university reactors from using highly-enriched uranium (93% enrichment) which could be made into nuclear weapons. The intervenors requested that the reactors operate with fuels that were more densely packed with uranium atoms, but with reduced enrichments of about 20%. Because of these challenges, UCLA ultimately closed its reactor. Goodwin describes the symptoms of potential proliferation in a number of nations, and the response of governments and supplier groups to that threat. In 1976 and 1977, Presidents Ford and Carter decided not to reprocess spent fuel, and in 1977 President Carter canceled the US Clinch River Breeder Reactor.

Since that time, the focus of the debate on these issues has moved to Europe and Japan, as described by William Sweet. The April 1986 accident at the unit 4 reactor at Chernobyl terrified much of eastern Europe and Scandinavia. Levi describes the accident sequence, the distribution of released radioisotopes, and the estimates of the number of additional cancer mortalities. Because of the similarities with the Chernobyl reactor and other reasons, DOE shut down its plutonium and tritium production reactors (see Section 1). The timing of these events could be affected by a strategic arms reduction treaty or by the introduction of other approaches of making tritium. Lastly, Sweet describes Japan's growing nuclear program, with its emphasis on plutonium, breeders, and safeguards.

CONTENTS

The nuclear fuel cycle: an appraisal

PHYSICS TODAY/OCTOBER 1977

The APS Study Group
finds existing technology and
straightforward extensions sufficient
for managing nuclear wastes,
but unresolved economic, institutional
and political questions cloud the
commercial use of plutonium.

The Study Group on
Nuclear Fuel Cycles and Waste Management

A fuel cask is unloaded and moved to a pool for underwater storage at the General Electric fuel-reprocessing plant at Morris, Illinois. According to the APS study, present technology is well developed for handling and transport—but insufficient for long-term isolation of the spent fuel.

On 25 April at the Washington meeting of The American Physical Society, the principal conclusions and recommendations of a year-long study of nuclear fuel cycles and waste management were released. The study group consisted of a dozen physicists, chemists, engineers and geologists; it was chaired by L. Charles Hebel of the Xerox Corporation and reported through the APS Panel on Public Affairs and a review committee consisting of Hans Frauenfelder (chairman), Wolfgang K. H. Panofsky, Theodore L. Cairns and M. Gene Simmons. The study was financed by the National Science Foundation.

The current public debate on nuclear power has focussed on two fuel-cycle issues: safe management of nuclear waste and commercial recycling of the plutonium produced as a by-product of normal light-water-reactor operation. The nuclear-power industry has planned towards commercial plutonium recycle for many years; the Energy Research and Development Administration also is developing plutonium-fuelled fast breeder reactors to extend the Nation's energy resources. Public concern over plutonium centers on its potential contribution to proliferation of nuclear weapons and on adequate safeguards against plutonium theft by subnational groups. The Nuclear Regulatory Commission was evaluating possible approval of plutonium recycle, but recently commercial use of plutonium has been deferred indefinitely by Presidential order.

The issues in nuclear waste management arise either from the spent fuel itself or from waste created by fuel reprocessing and refabrication. Such waste contains fission products and long-lived transuranic elements created by the nuclear reaction; both must be isolated until the radioactive isotopes have decayed to insignificant levels. For fission products this requires less than a thousand years; however, for some of the transuranic elements hundreds of thousands of years may be required for adequate isolation from the biosphere. The long time scales have raised institutional as well as technical questions about the viability of nuclear waste management.

The study was undertaken as an independent evaluation of technical issues in the use of fissionable materials in nuclear fuel cycles, together with their principal economic, environmental, health and safety implications. The focus of attention was on the "back end" of the fuel cycle—spent-fuel storage, reprocessing, recycling and waste disposal with additional examination of uranium resource considerations. The study emphasized the light-water reactor cycle, including evaluation of selected technical measures proposed as safeguards to prevent misuse of fuel-cycle materials. It also devoted attention to fuel cycles for some advanced reactors that have potential resource or safeguards significance—especially fast-breeder, advanced converter and so-called "denatured" fuel cycles. Much of the work centered on the principal alternatives for storage and disposal of radioactive wastes—in particular, high-level and transuranic wastes, and tailings from uranium mills. Occupational exposures in fuel-cycle facilities and public-health effects from fuel-cycle effluents were studied. The group also examined the research and development programs sponsored by government agencies, along with associated relationships among agencies and between government and private industry.

The effort centered on normal fuel-cycle operation and did not examine reactor accidents; thus it is complementary to the study of light-water reactor safety carried out for The American Physical Society in 1974–75. Overall, the study group looked for features and differences among various fuel-cycle assumptions that could influence choice of waste management alternatives, improve use of nuclear fuel resources or influence effective implementation of safeguards. The study group did not address the complex political and institutional considerations necessary for complete evaluation of fuel cycles on a national or international scale. Neither did they examine nuclear versus non-nuclear energy alternatives. The group felt that, as members of the technical community, they could make their most valuable contribution by independent evaluation of the present technical foundation of nuclear fuel cycles as well as prospects for future improvement.

●

PHYSICS TODAY here summarizes the principal conclusions and recommendations of the report, beginning with its overall conclusions. Discussions taken from the text of the report on some important issues are also included. The report is being published in its entirety as a supplement to Volume 49 of the Reviews of Modern Physics.

For all LWR fuel-cycle options, safe and reliable management of nuclear waste and control of radioactive effluents can be accomplished with technologies that either exist or involve straightforward extension of existing capabilities. However, technical choices, including those for geologic waste disposal, require further delineation of regulatory policies. For normal operation of all fuel-cycle options studied, potential radiation exposures from either wastes or effluents do not appear to limit deployment of nuclear power.

The decision to reprocess nuclear fuel does not depend significantly on waste-management considerations; instead it depends on resource and economic incentives and on international and domestic safeguards constraints. The technology for LWR reprocessing–recycling options is well advanced. However, the present recycle economics are uncertain; they depend on technical, regulatory and policy choices. In the future, as the availability of uranium ore becomes an important constraint, advanced reactors—including breeders—could provide more effective options than do present LWR's; such options require reprocessing.

Safeguarding the fuel cycle raises important unresolved institutional and political issues. Pending their resolution, recoverable storage of spent fuel rods ("stowaway") offers a fuel-cycle alternative that requires minimal safeguards and preserves energy resources; safe interim stowaway measures exist, and geologic stowaway could be safely continued indefinitely. If permitted by resolution of safeguards issues, reliable industrial-scale operation could be attained for reprocessing and refabrication for recycle in LWR's or for use in advanced reactors, with safe isolation of high-level and transuranic waste in geologic repositories.

Management of nuclear wastes

Federal regulations and standards are not yet complete concerning required solidification, processing, transport and subsequent storage or isolation of high-level and transuranic wastes. Encapsulating the spent fuel can create acceptable waste form. The technology exists for recoverable storage of the spent fuel with minimal deterioration to preserve the associated resources, but full-scale implementation is required. Where reprocessing is employed, solidification of liquid high-level waste is currently a required step. Technology is well developed to immobilize such waste in borosilicate glass cast in stainless-steel canisters for handling and transport. Present understanding is not sufficient to rely upon this technique, by itself, as a principal long-term barrier to release.

Effective long-term isolation for spent fuel, or for high-level or transuranic waste can be achieved by "geologic emplace-

L. Charles Hebel, the chairman of the Study Group on Nuclear Fuel Cycles and Waste Management, is with the Xerox Palo Alto Research Center. The other members of the study group are: Eldon L. Christensen, Los Alamos; Fred A. Donath, University of Illinois; Warren E. Falconer, Bell Labs; Leon J. Lidofsky, Columbia University; Ernest J. Moniz, MIT; Thomas H. Moss, staff, US House of Representatives; Robert L. Pigford, University of Delaware; Thomas H. Pigford and Gene I. Rochlin, University of California, Berkeley; Robert H. Silsbee, Cornell University, and McDonald E. Wrenn, New York University Medical Center.

ment," that is, deep underground burial. A waste repository can be developed at a site chosen by appropriate criteria to ensure a low probability of erosion, volcanism, meteorite impact and other natural events breaching the repository. The possibility of inadvertent human intrusion also can be made remote and limited in its consequences.

Hydrogeologic transport is the most important mechanism for potential transfer of radionuclides from a geologic repository to the biosphere. We conclude that many waste-repository sites with satisfactory hydrogeology can be identified in the continental US in a variety of geological formations. Bedded salt, proposed for the first repository in current ERDA plans, can be a satisfactory medium for a repository, but certain other rock types, notably granite and possibly shale, could offer even greater long-term advantages. Irrespective of the time scale adopted for reprocessing, two geologic test facilities in different media should be completed. As a possibly superior disposal option for the future, rock-melting concepts are attractive, and techniques associated with superdeep drilling could provide an effective means of waste emplacement.

The dominant fuel-cycle gaseous effluents that affect long-term public exposure are C^{14}, Kr^{85} and, to a lesser extent, tritium and I^{129}. Even when the number of facilities becomes large, the increment to background dose rate is small, and it can be reduced even further by suitably engineered controls. More attention should be given to the control of collective occupational dose, which exceeds collective public dose. We find no evidence to justify major reductions in plutonium inhalation concentration limits, proposed with the "hot-particle" hypothesis. For long-term waste management, the hazard associated with radium is more significant than that for plutonium. In addition, for regional population exposure, radionuclides in uranium mill tailings are potentially at least as important as the actinide chain elements in high-level waste; the relative accessibility of mill tailings contrasts with the isolation proposed for other actinide-containing wastes.

Reprocessing, refabrication and recycle

An essentially complete technology is at hand for industrial-scale chemical reprocessing of present LWR fuel. The chemistry and engineering are well understood and have been tested in several plants. Mixed-oxide fuel fabrication also has been demonstrated at the level of a pilot-scale batch process. Subsequent reprocessing of mixed-oxide fuel requires (and is receiving) further work. Reliable operation of an industrial-scale reprocessing–refabrication system has not yet been attained and represents an important step early in industrial deployment.

Operation of the industrial-scale reprocessing–refabrication system would yield a much broader understanding of the operation features and a firmer design base for future facilities.

Recycle would provide significant reduction in ore requirements, but resource considerations alone for LWR fueling provide little urgency to begin industrial-scale reprocessing within the next decade. Currently, various uncertainties preclude an unequivocal assessment of the possible economic benefits of recycle options. Sensitivity analysis of the fuel-cycle cost–benefits indicates that the dominant uncertainty is the cost of reprocessing and the second most important uncertainty is the future price of uranium ore. The reference case defined in chapter 4 of the complete report indicates a net benefit approximately 9% of the fuel-cycle cost of 1% of the net cost of

---

Waste isolation and hydrogeologic modeling

After studying alternatives for the long-term isolation of high-level and transuranic wastes, the APS Study Group concludes that for the immediate future the only acceptable choice is isolation of the waste in a deep continental geologic formation. The time scale is determined by the decay (over a few hundred thousand years) of the actinides in the waste. The group concentrated its efforts on identifying and evaluating those factors that might compromise the integrity of such a repository and the isolation of radioactive waste emplaced in it—in particular the criteria for appropriate site selection (depth and location, geologic medium, groundwater hydrology) that would be necessary and sufficient to define the basic geological integrity of a repository for the long term. The group concluded that current knowledge and technology are adequate to design and locate a suitable waste repository of the conventional mined type such that it would not be breached instantaneously as a consequence of either surface nuclear explosions or meteorite impact and would not be compromised by erosion; moreover, utilization of appropriate criteria would make it highly unlikely that the repository would be compromised by inadvertent human intrusion or by future tectonic or volcanic activity.

The APS group therefore concentrated its attention on the most important transport mechanism for radioactive waste emplaced in a geologic repository—ground water. They recognized that a favorable ground-water regime could provide isolation as stable and effective as that provided by the physical integrity of the geologic unit in which the waste is emplaced, in that radionuclides would be transported so slowly that they would not reach the biosphere during the desired period of isolation. The group examined in some detail those factors that control radionuclide transport by ground water as well as the effects of possible conditions that might change an initially favorable ground-water regime into an unfavorable one.

The movement of ground water implies the existence of a three-dimensional potential field that provides the driving force. The configuration of this field depends upon the potentials acting within the region (for example, gravitational, chemical or temperature potentials), the geometry of the region in which the potential field exists, the existing boundary conditions, and the nature and distribution of properties that control flow within the region. A ground-water basin is a three-dimensional region, the lateral and bottom boundaries of which are no-flux boundaries; water moves into and out of the region only through the upper boundary, which is the water table. Geologic observations and theoretical analysis permit a reasonably accurate definition of the geometry of a ground-water basin and of the water-table configuration. Detailed observations can fix more precisely the initial geometry and boundary conditions. Thus, the principal unknowns in the ground-water flow system are the particular potentials that may act in addition to gravity, and the detailed distribution of relevant properties (the geologic configuration) within the region.

To illustrate the problem of waste isolation and to show the essential characteristics of a predictive methodology, the group undertook computer modeling of an isothermal ground-water system for which gravity is the most important driving force. For purposes of illustration, two ground-water basin models were used. The first, a favorable system, illustrates ground-water flow and contaminant distribution in a layered sequence of homogeneous and isotropic geologic units of different permeabilities; the second, a potentially unfavorable system, illustrates the effect on the ground-water flow and contaminant distribution associated with a major high-permeability zone that passes vertically through the layered model. The second model was used to explore possible effects that disruptive events, such as faulting, might have on the hydrogeologic flow regime and transport of contaminant. Excerpts from the results are illustrated in figures 1 and 2.

The permeabilities of the horizontal layers in both models were identical. The hydraulic conductivity in the middle layer is 10^{-7} cm/sec, typical of shale, siltstone or argillaceous limestone. The layers immediately above and below the middle layer have an hydraulic conductivity an order of magnitude higher; the lowest layer, two orders of magnitude higher, and the uppermost, three orders of magnitude higher (10^{-4} cm/sec). The vertical high-permeability zone has the same hydraulic conductivity as the lowest layer, and it provides a conduit between the

nuclear electric energy; other reasonable cases lead to an increased benefit or even a loss. None of these cases include the cost of safeguards.

Another important reprocessing consideration is provided by the fast-breeder program; ERDA and the nuclear industry planned to start up the reactors with plutonium from LWR reprocessing. Breeder commercialization in 1993 would require industrial-scale reprocessing of

LWR fuel by the late 1980's. If breeder commercialization were delayed, or if start-up is based on other resources, present estimates of uranium-ore resources and projections of LWR growth indicate that industrial-scale reprocessing would not be necessary for resource extension until near the end of this century. LWR reprocessing experience provides a useful background for the reprocessing required for advanced fuel cycles, but

breeder fuel and thorium fuel reprocessing and refabrication need further development before industrial-scale operations can be undertaken.

Fuel-cycle safeguards

Technical measures for safeguards, by themselves, do not provide adequate protection against theft or diversion of fissile material, but such measures can play an important role in complementing

upper and the lowest higher-permeability layers.

The figures illustrate the contaminant spread after 200 000 years (a), after 400 000 years (b) and after 800 000 years (c). The contaminant source location is indicated by ⊕, and the intermediate contoured area includes between 2 and 10% of particles in the system. In figure 1 the outermost contour encloses all particles in the system; 99.8 percent of the contaminant particles are restricted to a zone 3800 meters lateral by 630 meters vertical even after 800 000 years. Figure 2 illustrates the effect, with increasing periods of time, on contaminant distribution when a vertical zone of high permeability lies to the discharge side of the contaminant source location; the altered velocity field results in an appreciably more extensive distribution of contaminant for the same period, although no particles have yet reached the boundary for the assumed conditions.

The assigned values for the hydraulic parameters and the general geologic configuration depicted by the layered model are considered sufficiently ordinary that the APS Study Group anticipates no difficulty in locating several sites with suitable hydrogeology in the immediate future. The group cautions, however, that these models have not incorporated the effects of other possible potentials, which could have significant influence in certain site-specific situations. Inasmuch as the modeling deliberately did not incorporate the effects of cation exchange and related concentration attenuation factors, the results are conservative. Contaminant distribution can be reduced significantly by such factors.

The study group considers it unreasonable to expect a model to yield a precise description of contaminant distributions in a real subsurface situation, but believes a predictive methodology provides the only way for simultaneous evaluation of all significant transport and attentuation processes. The range of consequences of possible future changes in the geology can be explored and bounded. By selecting realistic and conservative parameter values, the minimum and maximum extent of subsurface contamination can be obtained from deterministic model trials. *Conservative* modeling of the type undertaken by the group would be

appropriate for establishing a site-suitability criterion. Incorporation of the effects of various concentration-attenuation mechanisms and potentials other than gravity would provide more *realistic* modeling, which could provide a quantitative basis for comparing the *relative desirability* of acceptable repository sites.

The APS group strongly recommends that such modeling, combined with the associ-

ated geologic exploration and measurements needed to validate it, be an essential part of repository site selection and risk assessment. An adequate data base does not yet exist to permit completion of the recommended analysis of ground-water flow and mass transport, but the study group foresees no difficulty in obtaining the required data for a specific site and completing the analysis within a few years.

After they're used—what then? These fuel pellets are leaving the sintering press at the Hanford Engineering Development Laboratory, where they were formed from a powder of 80% thorium dioxide and 20% uranium dioxide. After fueling nuclear reactors, the spent pellets contain fission products and long-lived transuranic elements, creating a storage and management problem.

Taking a close look. Shirley Mayhan, a specialist in microstructural analysis at Westinghouse Hanford, photographs thorium with a scanning electron microscope for detailed examination and characterization. Together with x-ray fluorescence, this method assays the fuel rapidly during automatic pilot-plant production for breeder reactors or advanced converters. Fast breeders achieve greater economy in the use of resources than advanced converters; however, the commercialization of fast breeders with plutonium is presently undergoing a worldwide reexamination.

and reinforcing the conventional physical security measures that are necessary. For this purpose real-time accountability systems are attractive and merit further development. Coprecipitation of plutonium and uranium at the reprocessing plant is technically feasible, albeit more costly, and would have the important safeguards advantage that plutonium never would appear separately in the entire fuel cycle. Effectiveness of either domestic or international safeguards depends as much on political agreements and choice of institutions as it does on physical security and technical safeguards measures. We regard long-term and short-term considerations as equally important; design for long-term effectiveness of safeguards appears to be the more difficult task.

We have considered isotopically denatured fuel cycles involving uranium and thorium cycles with denatured reactors located at "national sites" and with plutonium-burning reactors, reprocessing and enrichment restricted to "international sites." Such cycles might contribute to safeguards by diminishing the threat of sub-national theft and by impeding the use of fuel-cycle facilities and materials for weapons production. Even so, nuclear-weapons capability can be attained independently of the spread of fuel-cycle facilities. Denatured uranium cycles, with low-enriched uranium in national reactors, provide near-term alternatives. Plutonium is inevitably present in the spent fuel; denatured thorium–uranium cycles reduce such plutonium production by about a factor of seven and may have long-term safeguards benefits. However, thorium-cycle costs are uncertain; in addition, potential safeguards problems are introduced by the use of highly enriched uranium make-up fuel and denatured U^{233}–Th fuel, which can be enriched to weapons grade with comparatively little effort. The nature of the necessary institutional and political safeguards arrangements are not substantially different from those required for denatured uranium fuel cycles.

Advanced fuel cycles

In the long term, significant resource extension could be obtained from several advanced reactors utilizing mixed-oxide fuels with uranium and thorium cycles. Fast breeders are the most resource efficient of all options. Breeders can be started with Pu, U^{233} and U^{235}; plutonium start-up is much more economic than U^{235} start-up; U^{233} is intermediate but supplies are not available. Advanced heavy-water reactors appear to be the next most resource efficient. Such options require development and would become competitive as uranium ore costs rise.

Shorter-term advantages are possible through improved LWR's, especially with thorium. Significantly improved uranium resource utilization can be achieved thereby in LWR's with higher conversion ratios in the 0.7 to 0.8 range. However, versions with even higher conversion ratios, such as the light-water breeder reactor, have no short-term advantage—primarily because of the long time required to achieve a net resource saving.

Major recommendations

In the area of **waste and effluent management,** our recommendations can be summarized thus:

1. High priority should be given to timely completion of Federal regulations and standards, including the required data base, concerning appropriate waste forms, transport, and subsequent storage or isolation of high-level and transuranic wastes and spent fuel elements.

2. Waste-management programs should complete and implement the technology for recoverable interim storage, either surface or geologic, as well as provide for long-term geologic isolation for spent reactor fuel.

3. Whatever time scale is adopted regarding a decision on reprocessing, two test facilities for geologic isolation should be developed in different media. The program should include hydrogeologic measurement and modeling, as well as geological exploration. Upon completion of two satisfactory test facilities, procedures should be initiated for licensing of the more favorable of the two. When the needs for storage dictate, that facility can be expanded into a full repository.

4. Criteria for selection of a waste-repository site should include specifications of appropriate hydrogeologic parameters that must be satisfied by present-day hydrogeology and by the projected bounds of the future hydrogeology of any specific site.

5. The waste-solidification program should complete the implementation of present technology for the treatment of high-level wastes and continue to develop advanced technologies. Transuranic waste deserves comparable attention, especially with regard to compaction and the form of the waste most suitable for geologic disposal. The goals of such programs, especially regarding immobilization, should be defined more clearly and quantitatively.

6. We urge active research on rock-melting concepts as a possible future alternative for waste disposal. The technology should be evaluated for emplacement of waste canisters using superdeep drilling.

7. We support re-examination of the criteria and practices for management of uranium mill tailings, to make their treatment consistent with that for other actinide wastes.

8. The major sources of collective occupational dose in the entire fuel cycle

---

Uranium resource utilization and future fission-power options

The study group analyzed several fast-breeder fuel-cycle alternatives and finds fast breeders to be the most resource efficient of all the long-term nuclear-fission alternatives. The group did not examine the reactor-safety issues of the liquid-metal fast breeder, which would make up a complex study in itself. The main-line US program has emphasized a liquid-sodium-cooled breeder fueled with plutonium and uranium, which produces new Pu^{239} in a blanket of depleted U^{238} by neutrons leaked from the plutonium–uranium mixed-oxide fuel pins in the reactor core. The US program has planned to obtain the plutonium required for fast-breeder start-up by reprocessing spent fuel from light-water reactors.

The study group examined other ways to fuel the fast breeder with plutonium. U^{235} enriched to 20% can also be used for start-up, but the economic penalties relative to plutonium start-up would be large. The group estimates that the fuel-cycle cost of start-up for first-generation fast breeders using enriched uranium initially is 2.5–3.3 times that of using plutonium, because larger quantities of fissile material are required with U^{235} and because the breeding gain is substantially lower than for plutonium. As an alternative, fast breeders may also be started with U^{233} and operated on a stable cycle by breeding this isotope from thorium. Such an alternative is intermediate in efficiency, but supplies of U^{233} do not exist in Nature; the U^{233}–Th fast-breeder cycle could be initiated if thorium fueling of thermal reactors were to precede the introduction of fast breeders. In any event, to realize the benefits of fast breeders the reactor-safety, fuel-cycle safeguards and plutonium utilization issues must be resolved.

Advanced converter reactors would not be necessary for resource conservation in the US if the fast breeder were to be commercialized in this century. Therefore, the study group recommends that development of advanced converters should proceed with the near-term goal of identifying well-defined options, which can be selected for more intensive development, should there be a significant deferral in commercialization of the fast breeder and/or a significant shortage in uranium resources.

Among the non-breeder alternatives, the study group finds that the advanced heavy-water reactor or converter appears to be the most resource efficient. The group considered several modes for a CANDU-type reactor (a pressure-tube heavy-water reactor already commercialized in Canada) and concluded that very significant resource extension could be obtained from a thorium cycle with recycle of the fissile material left in the spent fuel. Versions can even be designed that would operate as true converters producing as much fissile material as they consume. If the fast breeder is significantly delayed or cancelled, some version of the CANDU reactor may become important to conserve uranium resources. However, before commercialization in the US, this cycle would require additional redesign and development of CANDU fuel for higher burn-up, development and deployment of industrial-scale thoria fuel reprocessing and refabricating technology, and evaluation of acceptability under US licensing criteria. The group recommends work on all three of these aspects.

Another advanced reactor considered by the study group is the high-temperature gas reactor, which is a graphite-moderated design with highly enriched uranium operating on a thorium cycle. This option is intermediate in uranium resource efficiency between LWR's and CANDU, and advanced designs could approach advanced heavy-water reactors in resource efficiency. However, this reactor is no longer commercialized in the US and its graphite-based fuel requires different reprocessing and refabrication technology than oxide fuels. For this reason it would require considerable further development.

Improved light-water reactors can also provide near-term alternatives to use uranium ore more efficiently, although not as efficiently as do heavy-water designs. The study group found that "spectral shift" operation of an LWR, using some heavy water mixed with the ordinary light-water moderator, could provide at least a factor-of-two improvement in lifetime ore commitment relative to present LWR's, if thorium cycles are introduced. Moreover, such improved operation with conversion ratios in the 0.7 to 0.85 range (compared with the present 0.6) could be implemented with fairly minor modification to existing LWR designs. The study group recommends consideration of improved LWR's for the near term, while evaluating long-term alternatives such as fast breeders and advanced heavy-water converters.

All such cycles require reprocessing to derive significant resource extension. The study group notes that reprocessing technology for LWR fuel is quite advanced and provides useful background for reprocessing required for advanced, more resource-efficient cycles; but breeder-fuel and thorium-fuel reprocessing require further development before advanced cycles can be commercialized. The study group recommends that development of improved reactors and advanced-converter alternatives include the development, at the pilot-plant scale, of associated reprocessing and refabrication operations.

should be identified, and design effort should be directed to dose minimization. Regulatory guides should be developed for *collective* occupational dose in future facilities.

9. Technology should be completed for control and sequestering of C^{14}, Kr^{85} and I^{129}, and eventually H^3; steps should be taken toward development of international agreements for the degree and timing of control.

Our major recommendations for **reprocessing, refabrication and recycling** are:

1. If reprocessing is to be a major component of the US nuclear industry in the near future, appropriate existing reprocessing facilities should be completed and operated to gain experience with integrated technology on industrial scale; further, the corresponding refabrication facility should be built and operated with a similar goal. We emphasize that resolution of the issues involved in the GESMO decision and in international

fuel-cycle safeguards strongly influence the timing of such operation. [GESMO is the generic environment statement on mixed-oxide fuel being drafted by the Nuclear Regulatory Commission.]

2. The technology and engineering of the dissolution and separations steps needed to reprocess fuels for an advanced fuel cycle should be carried forward to a state of readiness sufficient for future engineering scale-up.

Recommendations for **fuel-cycle safeguards** include:

1. We urge evaluation of safeguards fuel cycles using low-enrichment uranium fuel, especially the economic, institutional and technical arrangements associated with co-located reprocessing, refabrication and plutonium-fueled reactors.

2. Real-time nondestructive assay–accountability systems should be developed further for possible use for control of special nuclear materials in industrial-scale reprocessing and refabrication

plants. Design criteria should be set, and costs involved in meeting these criteria should be evaluated.

3. The economic impact on the fuel cycle of coprecipitating to a fixed plutonium–uranium ratio should be evaluated.

Finally, our main recommendations concerning **advanced fuel cycles** are:

1. We urge evaluation of advanced heavy-water thermal reactors, and also of improved light-water reactors with higher conversion ratio, as candidates for development in addition to the already existing program to develop fast-breeder reactor technology.

2. The development programs for future reactors should include their associated reprocessing–refabrication technology and should emphasize providing options, so that future commercialization can be chosen from several nuclear-fission power alternatives to fit circumstances now too uncertain to assess precisely. ◻

Administration pushes alternative nuclear fuel cycles

PHYSICS TODAY/FEBRUARY 1979

Early in his presidency, Jimmy Carter made it clear that his Administration intended to discourage nuclear-weapons proliferation by attempting to slow the spread of commercial nuclear technologies that would make directly usable weapons-grade materials more readily available. One such technology is the reprocessing of spent fuel for recycling of plutonium. President Ford in October 1976 had already announced a decision to defer reprocessing of spent reactor fuel. In April 1977 Carter announced deferral of development of the Clinch River plutonium breeder reactor and later asked the Nuclear Regulatory Commission not to license the commercial reprocessing facility at Barnwell, South Carolina.

The breeder has long been regarded as an attractive energy option by many countries, especially those that do not have large supplies of uranium. Current fission power plants (mostly light-water reactors) principally use U^{235}; a plutonium breeder reactor, in contrast, would derive most of its energy by the transmutation of U^{238} into fissile plutonium following neutron capture and the subsequent fissioning of the plutonium. The breeder, however, has come in for increasing criticism on economic risk and proliferation grounds.

A major spokesman for Carter Administration nonproliferation policy over the past two years has been Joseph S. Nye, who has just returned to Harvard's Center for International Affairs after serving as deputy to the Undersecretary of State for Security Assistance, Science and Technology for two years. Nye points out that the government does not oppose breeder R&D programs at home or abroad, but does advocate that commercialization of this system be delayed until proliferation-resistant technological and institutional arrangements can be developed.

To encourage such a development, the Carter administration has convinced 52 countries and four international organizations to participate with the US in a two-year cooperative International Nuclear Fuel Cycle Evaluation. INFCE is composed of eight working groups looking at different parts of the fuel cycle: uranium resources, enrichment, supply assurances, reprocessing, breeders, spent fuels, radioactive wastes, and alternative

systems and research reactors. At a plenary meeting at Vienna in late November, INFCE delegates agreed to issue a final report on their activities in February 1980.

Apart from the INFCE activity, a group of representatives from the US government and American industry, universities and research institutions met for two weeks last August, to explore "ways of developing nuclear energy for civilian purposes while limiting its possible use for nuclear explosives." The Non-Proliferation Summer Study Group discussed this subject at the Aspen Institute for Humanistic Studies in Colorado, with the support of the US Department of Energy and the US Arms Control and Disarmament Agency. The Aspen conference was not an official part of INFCE—it was limited only to Americans—but the participants did comment on draft papers prepared for submission to INFCE and on materials prepared by DOE's Non-Proliferation Alternative System Assessment Program.

The principal findings of the study group are discussed in *Exploring Nuclear Futures: Report of the NASAP/INFCE Summer Study on Alternative Nuclear Systems,* prepared by Henry S. Rowen (Graduate School of Business, Stanford University), the conference chairman, and Fred Hoffman (RAND Corporation) and Herbert Kouts (Brookhaven), the chairmen, respectively, of the institutional and technical panels of the group.

The assessment of alternative fuel cycles carried out by the Aspen group was based

on information generated mostly by the Non-Proliferation Alternative System Assessment Program. The group noted that NASAP analysis of light-water reactors has concentrated on pressurized water reactors; a General Electric analysis of boiling-water reactors is expected to yield comparable results in most cases. Observing that the minimization of specific uranium consumption on a "once-through" fuel cycle (one not involving plutonium and uranium recycle) has not in the past been a high-priority reactor-design goal, the group examined NASAP-sponsored studies of methods for reducing consumption. In the near and intermediate term, the only options they view as attractive are ones involving the current fuel cycle, such as the redesign of fuel for longer burnup and improved fuel management. These methods, NASAP analysis indicates, might improve fuel efficiency about 15% in the next 10–15 years and possibly another 10% by about the year 2000. The study group sees longer-term value in methods leading to quite different core concepts and further reduction of the U^{235} assay of enrichment plant tails, and they encourage research in these areas.

The Aspen group believes that a heavy-water reactor with slightly enriched fuel—natural uranium currently fuels Canadian heavy-water reactors—would be superior to the light-water reactors in uranium efficiency and could be roughly competitive with it in terms of power costs. A heavy-water reactor with a denatured uranium–thorium recycle—a

Discussion at the nonproliferation summer study between Spurgeon Keeny (left), deputy director, ACDA, and Albert Carnesale (Harvard), chief US delegate to technical committee of INFCE.

cycle in which thorium and uranium are produced through chemical reprocessing of the spent fuel—would yield improvements in fuel efficiency, but not at competitive power costs. The group foresees several difficulties connected with heavy-water reactors. These reactors have continuous on-line refueling, and each individual fuel element is about two feet in length. In contrast, the light-water reactors are refueled about once a year and their fuel elements are about 12 feet in length. Because of the larger number of elements and the continuous refueling, it is more difficult to safeguard the heavy-water reactor spent fuel. The group also estimated that 20–30 years would be needed to build a substantial role for these reactors in industry.

NASAP calculations have indicated that high-temperature gas-cooled reactors would also have a better uranium efficiency than light-water reactors. The study group cautioned, however, that high-temperature gas-cooled reactors would, like the heavy-water reactors or breeders, require a large program for commercialization.

The Aspen group viewed the spectral-shift control reactor (a version of the pressurized-water reactor in which changes in the percentage of heavy water in the coolant adjusts the energy of the neutrons in the reactor core) as having a marginal advantage over light-water reactors in terms of uranium-utilization efficiency at a somewhat higher electrical power cost.

The study group said that current estimates of the capital cost of a breeder, which has the advantage of being able to burn most of its uranium, are generally between 1.25 and 1.75 times that of a reference light-water reactor. Frank von Hippel, who participated in the study group, told us that the projected capital-cost disadvantage is much larger now than when it was originally proposed.

US nuclear strategy. The Aspen group believes that technologies that increase uranium-utilization efficiency can be economically advantageous, especially if uranium prices rise. Improved light-water reactors could then compete with the breeder, possibly for a long period of time, in an era of rising uranium prices. The group urges the highest funding priority be given to modifying the fuel cycle in existing light-water reactors. Saul Strauch, deputy director of NASAP, told PHYSICS TODAY that DOE believes industry can make these improvements by the late 1980's or 1990's.

The report of the study group also proposes an international strategy for the US. Non-proprietary parts of the American program, especially fuel-stretching measures in light-water reactors, should be offered to other countries as a way of reducing pressure on them to introduce recycling of plutonium. An international program for heavy-water reactors, according to the group, should have the US concentrate on improved fuel (slightly enriched uranium and Th^{232}–U^{233}) and Canada on overall reactor design. They also view the participation of Germany, France and Japan as essential in an international high-temperature gas-cooled reactor program. The study group observes that US participation in an internationalized breeder program will improve US prospects for influencing the technical features of breeders built in other countries to minimize proliferation.

The report notes that there was disagreement among the study participants as to which new system—liquid-metal fast breeder reactors or advanced converter reactors (devices, such as heavy-water reactors and high-temperature gas-cooled reactors, where the ratio of new fissile atoms produced to those destroyed approaches one)—are closer to practical reality in the US and elsewhere. Those favoring the first system point to major development programs in the US, France, USSR, Japan, the United Kingdom and Germany. Other study-group participants, according to the report, note that the introduction of breeders will necessitate the parallel development of a fuel recycle industry. This will make such a reactor "more difficult, most costly, and more risky (both in terms of proliferation and economics)" to introduce than advanced converters, which could be operated initially on the once-through cycle.

Von Hippel and two of his colleagues at Princeton University's Center for Environmental Studies, Harold A. Feiveson and Robert H. Williams, favor use of the advanced converter. In a report, *An Evolutionary Strategy for Nuclear Power* (*Alternatives to the Breeder*), issued last September, they recommend continued reliance (where practical) on once-through fuel cycles, with a shift to more uranium-efficient advanced converters after the year 2000. They also advocate the maintenance of an option to shift this type of reactor to an isotopically denatured uranium–thorium recycle mode if the uranium supply becomes limited. Feiveson, von Hippel and Williams point out that if advanced converters and the denatured uranium–thorium recycle both were introduced shortly after the year 2000, it would take 50–100 more years before the greater uranium efficiency of the breeder would be significant.—CBW

Study finds no proliferation— proof nuclear fuel cycle

PHYSICS TODAY/JULY 1980

The International Nuclear Fuel Cycle Evaluation, which concluded its two-year study earlier this year, and a US study along the same lines, seem to have ended once and for all the myth that nuclear weapons and nuclear energy have nothing to do with one another. INFCE was set up at the instigation of the Carter Administration in 1977 to explore ways we can minimize proliferation of nuclear weapons resulting from the development of nuclear energy, and some consider it a major defeat for US international nuclear policy.

Carter hoped that INFCE would convince the international nuclear community to restrict severely any activity or technology that would involve the production of weapons-usable plutonium, such as the breeder reactor or reactor fuel reprocessing, and come up instead with a proliferation–proof technical alernative. That did not happen. But the 66-nation group did reach broad agreement that there are proliferation risks associated with nuclear power as well as measures that can be taken to make such risks more tolerable and manageable.

The study was divided into eight working groups, each with responsibility for a specified aspect of the fuel cycle. In all, 519 experts from 46 countries and five international organizations participated in the working groups. They produced more than 20 000 pages of documents over the two years of the study. A technical coordinating committee was also set up, composed of the 22 co-chairmen of the working groups, to coordinate the study from the technical point of view.

The INFCE reports do not contain policy recommendations, because INFCE was intended as a nonpolitical technical and analytic study rather than as a negotiation. They do define the problems, however, and provide suggestions that include technical, legal and institutional measures. INFCE also analyzed the special needs of developing countries.

Among the major conclusions of INFCE:
▶ There is no hope of developing a thorium fuel cycle for nuclear reactors

Clinch River Breeder Reactor vessel has been completed but the Carter administration wants to abandon the project. The recently completed INFCE report said that breeder reactors in routine operation are less environmentally hazardous than light–water reactors.

that could be available on a commercial basis until well after the year 2000.
▶ Sensitive nuclear facilities, such as reprocessing plants and enrichment plants, that provide direct access to weapons-usable materials, should be limited to a few countries.
▶ Reprocessing is not, as some had contended, necessary for the permanent disposal of nuclear wastes, and the reprocessing of fuel for use in non-breeder reactors is at best marginal economically.
▶ Reprocessing looks particularly unattractive economically for developing countries with small nuclear power programs.
▶ Breeder reactors, in routine operation are less environmentally hazardous than are light water reactors and are most appropriate in industrialized nations with large nuclear power grids.
▶ Effective international safeguards are an essential feature of the nuclear

power industry and should receive substantial attention.

INFCE examined the economics of what it considered to be the three major fuel cycle options: fast breeder cycles, thorium–uranium fuel cycles and other advanced reactor systems and concluded that "no one fuel cycle can be said to have an economic advantage in all cases." There are conceivable circumstances under which each of the three cycles considered may have an economic advantage over the others.

A parallel study to INFCE was performed by the Department of Energy and completed about the same time as the international review. The DOE's Non-proliferation Alternative Systems Assessment Program, which focussed primarily on evaluating the proliferation resistance of alternative nuclear systems, but also considered other factors that influence national decisions on the acquisition or expansion of nuclear power programs. It drew many conclusions similar to those of INFCE, but went a step further and put forth recommendations for the US nuclear research, development and demonstration programs. Both NASAP and INFCE set a high priority on improvement of light water reactors. Both groups also agreed that, although there is no technical "fix" to the problem of proliferation, the relative risks of proliferation resulting from fuel cycles can be reduced by applying certain technical and institutional measures, and recommended continued research in areas such as co-conversion and co-processing.

Even on the issue of the breeder reactor, the US and the international groups were in close agreement, arguing for further breeder research. But they disagreed as to when the breeder might first become a necessity. INFCE assumed a fairly high rate of growth in nuclear generating capacity. By the year 2000, INFCE concluded, the non-communist world could be generating as much as 1200 GW from nuclear reactors, compared with today's capacity of about 125 GW. Based on INFCE's long-term uranium supply assessment, they concluded that the uranium industry would be able to achieve annual supply levels adequate only for the requirements up to the year 2000, assuming the highest growth projections. NASAP estimated that uranium supply will satisfy demand in the US at least until 2010 without resorting to the breeder. During that two-year interval, forecasts of future world energy demand dropped sharply, and many countries revised downward their plans for installing nuclear power.

One interesting approach to breeder deployment has been put forth by IBM physicist Richard Garwin. He says that we can greatly reduce the hazard of proliferation associated with the breeder reactor by delaying its deployment and designing reactors that produce less weapons-usable plutonium. Nuclear power experts have traditionally encouraged early deployment of the breeder and development of high breeding ratios (the ratio of isotopes consumed to isotopes produced) in order to generate enough plutonium to fuel succeeding generations of breed-

ers. But Garwin's calculations show that by fueling not only the first generation but each deployed liquid-metal fast breeder reactor initially with U^{235} rather than with reprocessed plutonium, we can sustain a reactor for more than 2000 years of operation at 1000 MW(e), even if there were no excess plutonium production in the mature LMFBR (that is, a breeder ratio of one). According to Garwin, his proposal to start the reactor operating with enriched uranium would allow a large, rapid deployment of breeders when they are economically desirable without the necessity of premature commercial breeder operation or plutonium separation. An available uranium resource of 3.5 million short tons of U_3O_8 would fuel 1000 LMFBR's for more than 2000 years even if their breeding performance were far worse than has already been demonstrated, he says.

Garwin's approach to breeder deployment has several nonproliferation ad-

vantages: We would not need to separate plutonium now to fuel future breeders. Nor would we need to deploy first-generation breeders now, when they are not economically competitive with either LWR or fossil plants. Finally, opting for lower breeding ratios can eliminate the production of excess Pu that could be diverted for weapons uses.

The emphasis of breeder research should therefore shift from trying to raise the breeder ratio to modifying the design of the LMFBR to make it cheaper, safer and reduce the uranium investment required to fuel a new LMFBR, Garwin says.

Garwin's proposal will have a hard time winning acceptance. Both the INFCE and the NASAP studies recognized that breeder deployment decisions will be made for reasons of energy security as well as for economic considerations, and that different countries will reach different conclusions as to the timing and need for the breeder.   —MEJ

Laser enrichment process called proliferation resistant

PHYSICS TODAY/JULY 1979

Although many technically advanced nations are attempting to develop laser isotope separation of uranium, detailed information has always been difficult to obtain. Much of the work is classified because of its potential weapons applications; still other work is considered "company confidential" by the industrial firms involved.

Last year, in a highly unusual move, Exxon Nuclear Company and Avco Corporation convened a group of 12 experts in science, foreign policy and arms control to evaluate a laser isotope separation process being developed by their wholly owned subsidiary—Jersey Nuclear–Avco Isotopes Inc (JNAI). This Laser Enrichment Review Panel, headed by T. Keith Glennan (and including Peter Auer, Hans Bethe, Harvey Brooks, Richard Garwin and Gerald Tape), recently concluded that the JNAI process, were it to be developed commercially in the US by JNAI, "would be consistent with US nonproliferation objectives."

Since 1971, JNAI has spent $50 million in developing their uranium enrichment technique, which is based on an invention of Richard H. Levy and G. Sargent Janes (US Patent 3 772 519, granted 13 November 1973). Some discussion of the atomic vapor technique has appeared in the open literature,[1] but in September 1977 the Department of Energy retroactively classified the JNAI project. The project uses selective excitation and ionization of atoms in multiple steps.

Selective excitation and ionization of rubidium atoms was reported[2] by V. S. Letokhov and his collaborators at the Institute for Spectroscopy in Moscow in 1971. Letokhov points out (PHYSICS TODAY, May 1977, page 23) that an Avco Everett Laboratory group did similar experiments with uranium atoms later in 1971 but did not report[1] their results until 1975.

Because JNAI is nearing the stage where it has to decide whether to invest $50 million more on an Experimental Test Facility for integrated testing of prototype components, the firm undoubtedly wanted indications from the US government that JNAI would not be prevented from developing its enrichment process commercially. No definitive decisions are available at this writing.

The uranium-enrichment processes now in use leave significant amounts of

U^{235} in the process waste stream. U^{235} from these depleted tails stands a good chance of being recovered economically from some method of laser isotope separation.

Both Los Alamos and Livermore have large laser isotope separation programs, but results on uranium are rarely discussed. Livermore has been doing experiments with atomic vapor, and in 1974 (PHYSICS TODAY, September 1974, page 17), reported using a technique similar to that of Levy and Janes. Los Alamos has been emphasizing the molecular approach and in 1976 reported making slightly enriched uranium with UF_6; however experimental detail was withheld.

In addition to the atomic process being developed by JNAI, Exxon is said to be working on a laser isotope separation process involving molecules at its research and engineering center in Linden, N.J.

Janes, who is vice president for isotope research at Avco Everett Research Laboratory, recently told us, "Unfortunately, perhaps because of classification and proprietary requirements, the open literature tends to be unrepresentative of

the real situation wherein a significant fraction of the problems involve tough engineering requirements rather than clever physics. This is particularly true of the atomic processes and has led to the publication of a number of papers suggesting schemes which concentrate on solving the wrong problems."

The JNAI approach takes advantage of the fact that the absorption lines of U^{235} and U^{238} atoms have very small shifts in some transitions in the visible range (roughly one-fourth of a wave number or 3×10^{-5} eV), which are, however, larger than the bandwidths of individual transitions for each isotope. By choosing the right transitions, a collection of lasers can be tuned to make the shifted absorption wavelengths accessible to selective excitation and ionization. JNAI uses four different wavelength dye lasers. [Has JNAI found another solution to the four-color problem?]

Primarily because of its high Z, uranium has one of the most complex optical spectra of any element. Janes, Harold K. Forsen and Levy note[3] that in uranium over 900 levels and 9000 transitions have

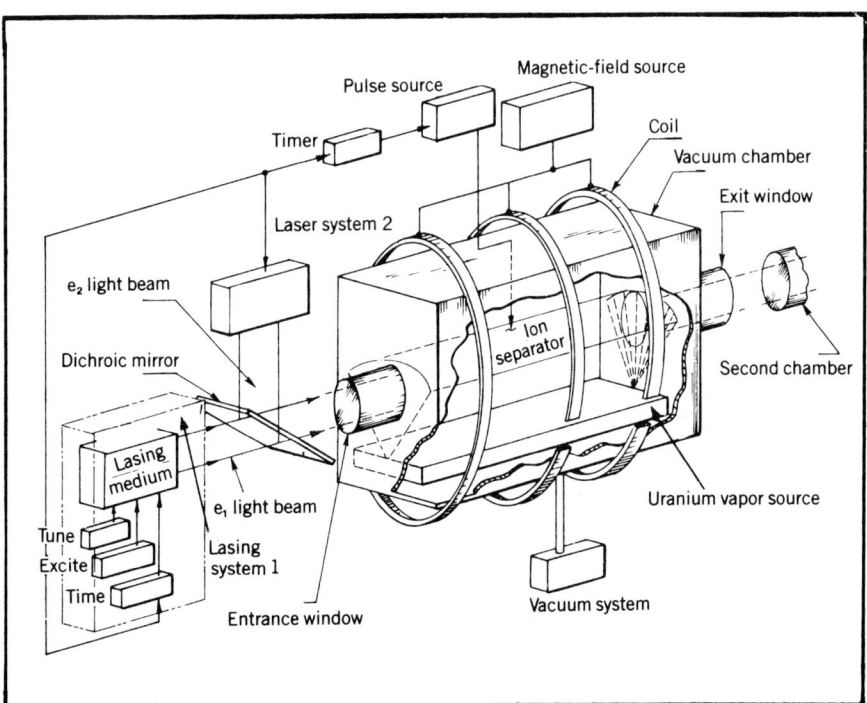

Atomic isotope separation technique used by JNAI. The laser system sends pulses into the vacuum chamber, which contains a uranium vapor source and an electromagnetic type of ion separator to remove U^{235} ions from the neutral U^{238} background vapor. (Figure from a JNAI patent.)

been identified; perhaps as many as 300 000 visible lines are present. As the Glennan report notes, identifying the specific transitions useful for isotope separation is time consuming and exacting. So, the report goes on, the frequencies used by JNAI are classified. A further difficulty is to produce and maintain precisely tuned light to less than one part in 10^5 and still cover the entire U^{235} absorption spectrum for a selected transition.

As shown in the figure on page 17, taken from one of the 35 existing JNAI patents, the laser system sends carefully tuned and timed pulses into the vacuum chamber, which contains both a uranium-vapor source and an electromagnetic or plasma type of ion separator to remove U^{235} ions from the neutral U^{238} background vapor. Because collisions limit the vapor density, one needs a long path length for a reasonable fraction of laser light to be absorbed. So in practice the process would have several such modules.

The module is surrounded by a 100–200 gauss magnetic field parallel to the laser beams; the magnetic field is needed for both the vapor source and the electromagnetic ion extraction process. A preferred approach is to use four lasers, in which three are for excitation and one for ionization. Three-step processes allow the use of lasers in the red–orange portion of the spectrum, where dyes are more efficient. Such a three-step, four-color process is shown in the figure at the top of this page.

The vapor source is a water-cooled crucible plus a high-energy electron beam that is focused by the magnetic field along a narrow line on the surface of molten uranium. The electron beam heats the uranium to 3000 K, producing a vapor that is then allowed to expand radially to speeds comparable to that of sound. After the vapor enters the ion extraction structure, it is illuminated by the laser beams. Once the U^{235} atoms are selectively ionized, an electrical pulse is applied to the ion deflector plates. The resulting electric field produces electron currents within the vapor which, together with the magnetic field, deflect ions out of the neutral stream onto the product collection surfaces. Provided the density is low enough that neutral-ion collisions can be neglected, the neutral vapor will continue to flow through the ion-extraction structure and collect on the tails-collection surface.

For a high U^{235} ionization probability, the laser energy needed for each laser step is fairly high—tens of millijoules per cm². Thus, the lasers are pulsed—10 000 pulses per second. An average power for the laser system of several kilowatts is required. Assuming a 0.2% efficiency, megawatts of input power would be needed.

In the JNAI experiments, single-stage product enrichments of 6% were pro-

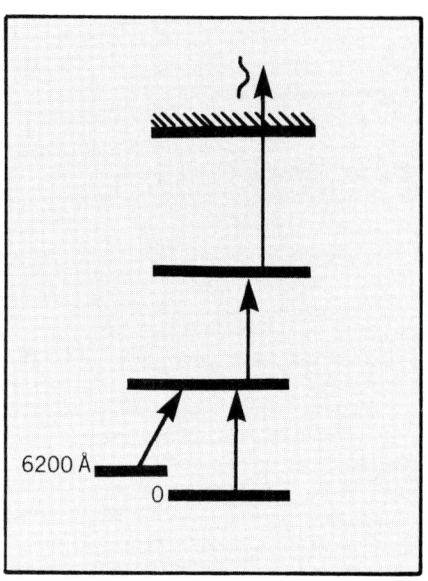

A preferred approach for atomic separation uses four lasers—three for excitation and one for ionization. Three-step processes allow the use of lasers in the red–orange part of the spectrum, where dyes are more efficient than in other portions of the spectrum.

duced[3]—the limit imposed by scattering considerations.

The JNAI program is aimed at producing low enriched uranium (2–3% U^{235}) for use as a light-water reactor fuel. It would operate in a single stage. To make highly enriched uranium would probably require cascading.

If JNAI decides to go ahead with its Experimental Test Facility, according to Harold K. Forsen, vice-president for laser enrichment at Exxon Nuclear Co, it would be aimed at, among other topics: the engineering demonstration of laser systems control; large-scale uranium handling; component and systems lifetime studies; long-path light propagation, and development of the necessary data to support a commercial plant license application.

Risks of nuclear proliferation exist in any enrichment process, the Glennan report notes. To use the JNAI process to make highly enriched uranium would first require substantial development and then modifications of an existing facility. If a JNAI separation facility is kept under safeguards, it would be simple to detect conversion. On the other hand, the report notes, a centrifuge plant can be converted without major modification—in less time and with far less uncertainty.

The characteristics of the JNAI process are such that one could detect clandestine plant construction or operation "through appropriate national intelligence measures which include monitoring the export of critical components and electronic intelligence for detection of plant electromagnetic emissions," according to the report.

Laser isotope separation is being developed in several countries. So even if the JNAI program is cancelled or even all US development, it is unlikely to stop foreign nations from continuing their efforts, the report says.

If the JNAI process turns out to be as economical as anticipated, the report says, it can contribute to US nonproliferation objects in these ways:

▶ It would allow the recovery of additional U^{235} from the growing stockpile of diffusion and centrifuge plant tails. This one-time addition to the U^{235} supply is equivalent to 60 000 tons of natural uranium, an amount sufficient to supply the operation of ten (1000 MWe) power reactors for their expected lifetimes. Laser isotope separation could reduce requirements for natural uranium by 20%, approximately the same benefit as from plutonium recycling. The process would allow nations with light-water reactors to send their tails to the US for enrichment.

▶ The JNAI process might be cheaper than enrichment processes not involving lasers. Any reduction in cost would affect the relative economic attractiveness of reprocessing and plutonium recycle in light-water reactors because these operations would have to compete with the lowered price of fresh enriched fuel.

▶ By reducing natural uranium requirements, the JNAI process would tend to stabilize yellow-cake prices.

The report notes that "the JNAI process is anything but 'garage technology.' The vaporization of metal by electron beam, the laser system, the optical system, and the extraction of ions by electric or magnetic fields are all high-technology operations which only a country with sophisticated scientific and technological capabilities could successfully achieve. Conversion of UF_6 into metal, the removal of the pyrophoric deposit from collector and tails plates, chemical processing of the metal and tails, and several other parts of the materials handling, might be accomplished by a country with medium technological capabilities but not by a subnational group . . ."

"Perhaps the best indication of the technical difficulty of the process is that after more than seven years of research and development, the JNAI process is just nearing the stage of integrated testing of prototypical components. On JNAI's own schedule, its first demonstration plant is at least a decade away from operation."   —GBL

References

1. G. S. Janes, I. Itzkan, C. T. Pike, R. H. Levy, L. Levin, IEEE J. Quant. Electr. **QE-11**, 101D (1975).
2. R. V. Ambartzumian, V. P. Kalinin, V. S. Letokhov, JETP Lett. **13**, 217 (1971).
3. G. S. Janes, H. K. Forsen, R. H. Levy, A.I.Ch.E. Symposium Series 73, no. 169, page 62 (1976).

"Born secret" disclosure law

Paul N. McCloskey, Jr.   PHYSICS TODAY/JULY 1980

For the first time in my experience, honorable and distinguished scientists are falling under the shadow of possible dishonor at the hands of that very agency of government which, perhaps more than any other, bears responsibility for the peace of the world and relies on scientists to meet that responsibility.

I speak of the Department of Energy, and its responsibilities for nuclear fission and fusion. The honorable and distinguished scientists I speak of are men like Edward Teller, Ted Taylor, George Rathjens, Ray Kidder, Hugh DeWitt, Ted Postol, George Stanford, Gerold Marsh and Alexander DeVolpi.

The problem is that of the "born secret" concept contained in the Atomic Energy Act of 1954. It has three elements: (1) the classification procedures and policies of the Department of Energy, (2) the ambiguity of the present law as it is being interpreted by the Energy and Justice Departments, and (3) the increasing public dispersion of scientific data bearing on construction and use of weapons that can destroy mankind.

These factors have led to a situation whereby the Government is now depending on the threat of criminal prosecution to cow scientists, both government and private, into restraint in the communication of ideas, while conceding that if criminal prosecution were attempted, it would probably fail.

At stake is the ability of scientists, both inside and outside government, to communicate with each other and thus advance scientific knowledge. Balanced against this goal, which has historically been considered as a highly laudable one, looms our growing uneasiness, if not conviction, that uncontrolled advancement of science in the fields of nuclear weaponry, biological warfare and perhaps other areas such as genetics can destroy the world.

Clearly, a *balance* is required between advancing science and protecting the public against a *too-easy* creation or possession by terrorists of hydrogen bombs, nerve gas and so on.

Congressman Paul N. McCloskey, Jr. is a Republican representing the twelfth Congressional District in California.

The balance struck in the 1954 Atomic Energy Act, however, seems no longer adequate, but if the law is to be changed by legislative action, the changes should be fully considered and advanced by scientists as well as lawyers and politicians.

I can assure you that *without* serious participation and action by scientists neither the Department of Energy nor the Congress will have the courage, ability and will to attack the problem; and it will continue to cause concern and sleepless nights for able people in your profession.

The situation created by *The Progressive Magazine* case decision last year is this:

Whenever an individual, public employee or private citizen, generates a new concept of nuclear weaponry, that concept is "born secret" and, under the law, *must* be classified as soon as it is placed on paper.

But only the *government* can classify.

And, I might add, only the government apparently can declassify under current law.

The relevant section of the Atomic Energy Act is Section 2014 (11) (y), which defines restricted data—that which is prohibited from publication—as:

"*all* data concerning (1) design, manufacture, or utilization of atomic weapons; (2) the production of specific nuclear material; or (3) the use of special nuclear material in the production of energy, but shall *not* include data declassified or removed from the Restricted Data category pursuant to section 2162 of this title." [*Emphasis added.*]

That last phase is crucial. Restricted data does not include data that the government had declassified, that is, allowed to be published in the public domain.

A great deal has been published in the public domain in recent years, most of it by distinguished government scientists—men like Teller, Taylor and Rathjens.

From that mass of published information and following questioning of several government scientists, an astute but untrained newspaper reporter, Howard Morland, prepared an article for publication in *The Progressive Magazine*. Morland contends that he had described nothing more than information which had lawfully been "removed from the 'Restricted Data' category" through information released into the public domain. The Court felt otherwise, finding that the article contained "*concepts*" not previously published.

The Court's language focused on a stark reality:

"Faced with a stark choice between upholding the right to continued life and the right to freedom of the press, most jurists would have no difficulty in opting for the chance to continue to breathe and function as they work to achieve perfect freedom of expression."

In effect, the court found that the threat to life represented by unrestricted publication of H-bomb justified prior restraint of that publication.

During the court proceedings, a constituent of mine, Chuck Hansen, an amateur nuclear-weapons enthusiast of Mountain View, California, called my attention to his belief that the Department of Energy was managing its classification program improperly. It was Hansen's contention that DOE had permitted leading government scientists in years past to publish the basic concepts of nuclear weaponry, but was now applying a different standard to private citizens such as Howard Morland.

Hansen felt that any "concepts" presented by Morland were easily deducible from already published data.

Hansen had sent an 18-page letter to three Members of Congress and several newspapers containing his views. When the government learned of the letter, the Justice Department sought to prohibit the *Daily Californian* from publishing it. However, the *Madison Press Connection*, on 16 September 1979, published Hansen's letter in its entirety, rendering moot the *Progressive* controversy and causing the Justice Department to drop its lawsuit.

To my knowledge, Hansen had done nothing more than any private citizen might have done in reading public journals and studying data available to the public.

At least seven other people during the past several years have sent communications relating to nuclear-weaponry concepts which they believed, in good faith, to contain information that had been properly declassified, only to have it promptly classified by DOE.

For example, in May 1978 Dmitri Rotow, a Harvard student, found in the Los Alamos Public Library a document called "Final Design Status of the TX-7" and proceeded to write a paper on its content. Subsequently his paper was classified by DOE. A year later, Rotow returned to the Library to research the public availability of information on nuclear weapons for the American Civil Liberties Union. He then discovered document number UCRL 4725 and made copies of it. A librarian observing his work discovered that the document had been wrongly declassified.

Postol, Marsh, Stanford and DeVolpi at the Argonne National Laboratory have also maintained that DOE is manipulating the classification procedures for political purposes.

On 25 April 1979, during the *Progressive* case, they wrote a letter to Senator John Glenn concerning alleged misuse of DOE's classification procedures. DOE then classified *their* letter. Significantly, DOE did not *remove* such classification until one day before our Government Operations Subcommittee held public hearings on the question last month.

Two scientists at the Lawrence Livermore National Laboratory, Hugh DeWitt and Ray Kidder, sent a memo to the Regents of the University of California concerning the declassification of the Inertial Confinement Fusion Program. DOE immediately classified

their memo.

Note now that *nine* reputable individuals, six government employees and three private citizens, have therefore been able to generate communications in *good faith*, believing them to contain already published information which the government subsequently chose to classify.

The six government employees all believe that DOE has used its classification procedures for political purposes rather than in checking bona-fide attempts to protect nuclear secrecy.

DOE contends it *must* classify all previously unpublished new concepts *under the law*.

In any event, it seems clear that the cat is out of the bag.

If, indeed, a Harvard student, a newspaper reporter and an untrained nuclear amateur can generate articles gathered from publicly available sources, which the government feels must be kept secret because those articles threaten the peace of the world, *the law is no longer adequate to protect the national security*.

Yet DOE contends that no *change* in the law is necessary. To date both Armed Services Committee chairman Mel Price and Ranking Republican Member Bob Wilson agree that no change is necessary.

It is of course possible that the cure would be worse than the malady. Most of the lawyers and scientists who have written on the subject argue that no law can be written to prevent the *private* citizen, at least, from publishing new concepts in this area, even though what that private citizen conceives and publishes may be used to permit some nation or terrorist group to build and detonate their own hydrogen bombs.

None of the nine individuals who have published information subsequently classified by DOE have been prosecuted.

It certainly must be considered an anomaly that the government can *prevent* publication of information, yet not prosecute the person who wishes to publish it.

In view of this anomaly, I prepared and introduced H.R. 6024 which would make it a crime for a private citizen, in the field of nuclear weaponry, to publish his concepts without first checking which government agency would have jurisdiction. I did this primarily to prompt the focusing of debate on the issue.

DOE responded that if every citizen who had a concept of nuclear weaponry was required to submit his plan to the government before publication, DOE could conceivably spend all of its time evaluating the concepts of the private citizens, with none left over to solve the

nation's energy problems. DOE announced that it has now adopted a policy of "no comment" on such information, conceding that any comment seeking to prevent publication merely gives validity to the concepts involved, and that there is no real way to prevent publication anyway, save by waving the *threat* of criminal prosecution.

In a letter to an Oakland newspaper dated 17 October 1979, DOE's Assistant Secretary for Defense Programs, Duane Sewell, said:

"It is the Department of Energy's policy not to comment on the accuracy (or similarly, the classification) of published information concerning nuclear weapon designs . . ."

We thus appear to be back on dead center. Any citizen who conceives a nuclear concept in the future is free to print it without concern as long as he believes he is using whatever the government has allowed to be published in the past. The government is, however, entitled to try, under the *Progressive* rule, that the proposed publication would include new concepts but, as in the *Progressive* case, the resulting furor will probably result in their being published anyway. Duane Sewell's position indicates they will merely make "no comment" in the future.

The prevailing congressional sentiment is expressed in a letter from Bob Wilson:

"Notwithstanding the adequacy of existing laws, to prevent or deter the spread of Restricted Data, there has been a reluctance by the FBI and the Department of Justice to agressively investigate and prosecute alleged violations. Without enforcement no amendment to the law would make any difference."

Congressman Wilson's view is shared by top DOE officials who claim the FBI and Department of Justice have been remiss in meeting their responsibilities under the 1954 Act. DOE apparently feels that Morland, Hansen and Rotow, at least, could not have derived their new concepts had not someone at DOE "leaked" information. Both the Department of Justice and the FBI, at the congressional hearings last month, however, indicated that they held little hope of successfully prosecuting private citizens under the 1954 Act, since to be guilty of any offense the private citizen has to be proven guilty beyond a reasonable doubt, in the unanimous view of twelve fellow citizens, that he published his ideas "with intent to injure the United States . . .or with reason to believe such data will be utilized to injure the United States." (42 U.S.C. §2274).

As a former prosecutor and defense attorney, I agree completely.

The chilling effect on *government* scientists is far more deadly, however. A *government* scientist can be prosecuted if he "knowingly communicates, or whoever conspires to communicate or to receive, any Restricted Data, to any person not authorized to receive Restricted Data...upon conviction thereof, be punishable by a fine not more than $2,500." (42 U.S.C. §2277).

Under these circumstances, we have the unusual circumstance of DOE threatening its own scientists with the black cloud of dishonor and criminal prosecution, not because the law is clear, but because it is *not* clear.

This is, in effect, the rule of men, not of law. Respect for the law has never been more necessary, but that respect has traditionally been based on the premise that the law is clear and unambiguous and everyone is presumed to understand it. The law should not be a tool whereby an embarassed bureaucrat can threaten an honorable employee in order to cover his own political hindquarters.

As a legislator, I am offended both by DOE's position and by Justice's position, since it would seem to me that their head-in-the-sand attitude will encourage great scientists to remain in private life rather than serve the government. It also seems to me they owe a duty to come forward with recommendations for a change in the law. I have little confidence that they will, however.

What is needed is a full debate on the scientific community as to a new law setting out a new balance between the communication of scientific concepts and the protection against every American being able to add a nuclear weapon to his handgun collection. What that balance should be is clearly beyond the will or desire of the lawyers of DOE, the Justice Department and Congress, at least thus far.

Whatever that balance should be, however, should be clearly set forth *in the law*. As one legislator, I should like to help draft the new law in a way which scientists find to be a fair compromise between the two basic principles.

I hope scientists will therefore understand my plea for help. I know of no group more affected, no group who have more to lose, and no group who can more capably further this debate towards an ultimate resolution, by legislative action, if possible, but if not, by a conscious and informed judgment that the issue is one, like religious beliefs, which is not susceptible of legislative solution.

* * *

This guest comment is based on an invited paper presented at the 1980 Spring Meeting of The American Physical Society in Washington, D.C.

Iraqi reactor damaged in Israeli bombing raid

PHYSICS TODAY/AUGUST 1981

The Israeli destruction of an Iraqi nuclear research reactor on 7 June has challenged the world once again to reconcile the peaceful and destructive potentials of the atom. The recent drama in the Middle East has stimulated many physicists to think beyond the particular actors involved to some more general, technical questions: What opportunities does a reactor such as Iraq's provide for a nation to acquire significant amounts of nuclear weapons material? How can such attempts be detected or deterred? What types of research are appropriate to a reactor of that size? What radiation hazards might result if the reactor were bombed after it was fueled and operating?

The physicists who have tried to answer such questions, either privately or publicly, have reached differing assessments that result both from incomplete information and from the necessity for applying some individual judgment.

The target of the Israeli bombs was a research reactor being built by the French at the Iraqi Nuclear Research Center in Tuwaitha, 14 km from Baghdad. The large reactor, named Tamuz 1, was patterned after the 70 MW (thermal) Osiris reactor in France. (See figure.) The reactor is cooled and moderated by light water in a swimming-pool configuration. The fuel elements consist of 93% enriched uranium in an aluminum–uranium alloy sandwiched between plates of aluminum. A grid array of seven rows of eight slots each supports either fuel or experimental elements. From 25 to 32 of the slots must be filled with fuel elements for the reactor to go critical.

The French have reported that the power level of the reactor is 40 MW, although *Science* cites a small-reactor expert in the Department of Energy who maintains that the reactor could run at 70 MW with no major mechanical changes. According to Herbert Kouts (Brookhaven), the actual power will depend on the cooling-tower efficiency (which may be higher in the arid climate of Iraq) and on the margin of conservatism in coolant temperatures. A second reactor of only ½MW

(thermal) had been built alongside the larger one. The Tamuz reactor was not Iraq's first: Since 1967 that nation has owned and operated a 2-MW pool-type reactor supplied by the Soviet Union and fueled by 10%, 36% and 80% enriched uranium.

Opportunities for diversion. A facility like the Tamuz reactor provides its operator with access to strategic quantities of nuclear fuel both at the front and at the back end of the reactor operation. Up front, Iraq was to receive from France the highly enriched uranium fuel. No public consensus

has been reached concerning the number of nuclear weapons, if any, that can be made from the amount of enriched uranium in one core loading (12 to 15 kg) but Ted Taylor (independent consultant to Princeton) has told us that it is sufficient for at least one fairly sophisticated bomb. (A crude bomb would require more material.)

Weapons-grade material could potentially emerge from the back end of the reactor operation if the Iraqis placed natural uranium around the core. Absorption of core neutrons by the uranium-238 in such a "blanket"

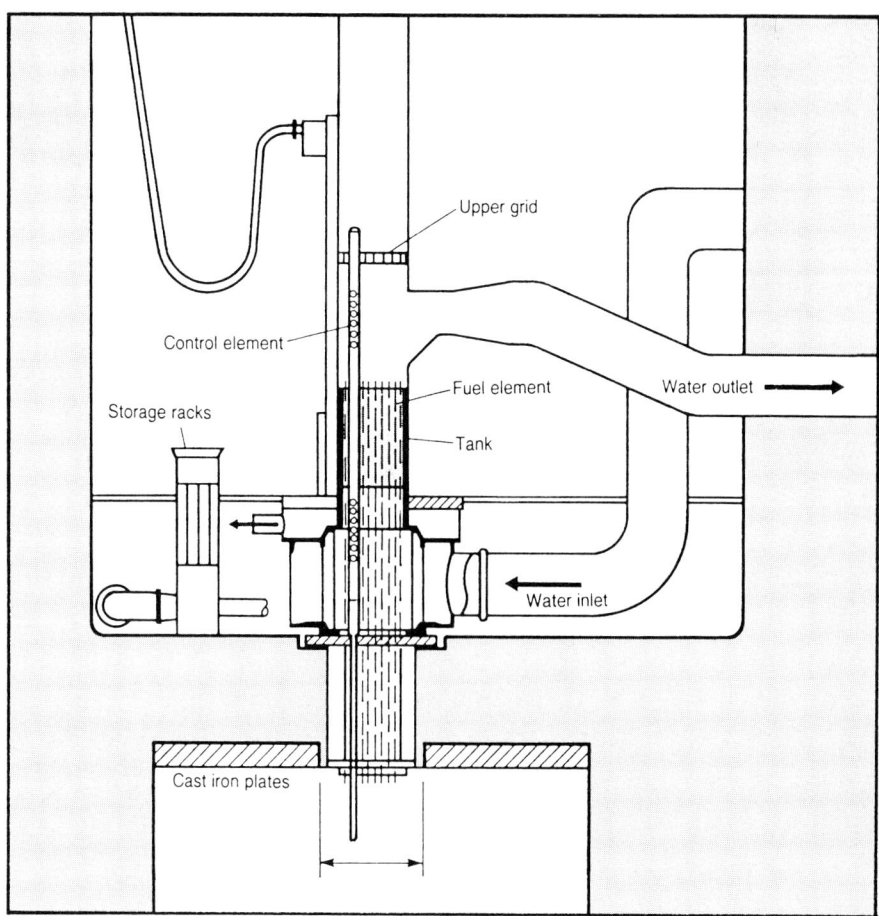

Simplified vertical section of France's Osiris reactor is similar to the research reactor that was recently destroyed in Iraq. A grid of 7 rows of 8 slots each supports both experimental elements and fuel elements (shaded) in the reactor tank. The core is cooled by circulating water, as shown.

could convert it to plutonium. A reactor of the size of the Tamuz facility could produce plutonium sufficient for about one bomb per year. Iraq had recently acquired 100 tons of natural uranium oxide from Portugal and possibly bought more elsewhere.

Whether the Iraqis could have, if desired, clandestinely diverted some of this material to military purposes depends on the success of the international controls being applied to it. As a party to the nuclear Non-Proliferation Treaty, Iraq had promised not to develop nuclear weapons and had agreed to accept safeguards on any strategic quantities of nuclear material in its possession. The document specifying the safeguards applied to a given facility is confidential. The deputy director general for safeguards at the International Atomic Energy Agency, Hans Gruenn, has stated however, that the Iraqi reactor was to be visited by IAEA inspectors about every two weeks. (While the Tamuz facility was under construction, the inspectors came only about three times a year.) The possibility of surprise visits is somewhat limited by the host country through its issuance of visas. Inspectors would observe directly that all fuel elements were present in holding vaults, the reactor pool and spent fuel storage.

The French were keeping an additional watch on the Iraqis, as they revealed after the Israeli attack: Iraq had consented in a secret agreement to allow French technicians to remain on site through 1989 and to irradiate the fresh fuel for 20 days in the small Tamuz 2 reactor after it arrived from France. The irradiation was intended to deter diversion attempts. The French did not, however, succeed in getting the Iraqis to agree to accept a fuel of lower enrichment (but higher cost) called Caramel. In this experimental French fuel and other fuel being developed in the US the density of the uranium is increased so that its enrichment can be correspondingly lowered to achieve the same neutron flux.

Taylor expressed to us his concern that nations can divert the highly enriched uranium fuel from such safeguarded reactors as Iraq's. He pointed out that the uranium in one core directly represents more effectively useful weapons material than the plutonium that can be produced by one core. He said that a reactor that size may require refueling up to five or six times a year, which may preclude the timely presence of an IAEA inspector. The nation operating the facility may also wish to stockpile several core loadings on site, thus giving it access to material for several weapons even if diversion is detected and all further fuel shipments

cut off. Herbert Goldstein (Columbia) told us that France made an agreement in 1975 with Iraq to provide up to 80 kg of the enriched fuel at one time. The radiation from the fuel may complicate but not necessarily prevent the handling of the fuel.

Other physicists such as Kouts and Goldstein are more concerned with the possible production of plutonium in a reactor such as Tamuz 1, although they disagree about the ease of detecting such attempts. (The two spoke on 24 June to a press briefing in New York sponsored by the American Professors for Peace in the Middle East.) Kouts noted that even if the natural uranium itself were well hidden within the reactor pool, its presence might still be evidenced by the cooling system it requires. (Kouts estimated that the blanket in a 40-MW reactor might generate 10 MW of heat.) In addition the core might have an unusual size in these circumstances: Because natural uranium absorbs neutrons, its presence would necessitate more fuel elements. The appearance of the natural uranium elements would most likely be different from those of the fuel elements.

Goldstein felt that a nation might both conceal and cool natural uranium fuel by inserting it into some experimental locations within the grid plate. If these blanket elements were within the existing forced cooling loop, they may not require additional cooling. Goldstein felt that the natural uranium elements could easily be removed before the arrival of IAEA inspectors.

Still others feel it would be extremely difficult for Iraq to undertake any type of diversion as long as the French remained at the reactor.

If a nation such as Iraq did acquire the requisite amount of fuel, few doubt that they could have constructed a nuclear device with it, probably within three years. For making a plutonium bomb, the effort would require a hot lab and a fuel fabrication lab, which the Iraqis had probably already received from Italy, and other special equipment that would not be difficult to obtain. Less time and equipment would be required for making a uranium bomb, Taylor said.

Research capabilities. The Osiris and Tamuz reactors are often called materials testing reactors because of their most common application. The high neutron flux can be used to test materials and their behavior under irradiation, with the output of such tests being of primary concern for the design of advanced nuclear reactors. Other possible applications of this type of reactor could be the production of radioisotopes or the acquisition of experience with reactor operation, but both of these functions could be equally well served

by a reactor of lower power. Although some question the need of such a large reactor by a country like Iraq, others maintain that the desire of a developing nation to advance technologically may be sufficient motivation for it to purchase a large reactor.

One other interesting question was posed by the Israeli raid in Iraq. Prime Minister Begin defended the timing of the attack by predicting that a bomb dropped on a fueled reactor would spread lethal radiation doses all the way to Baghdad. Since then others have disputed Begin's claims. For example, Kouts has concluded that the radiation danger would be negligible outside a radius of perhaps 1000 feet around the plant. His estimate is based on the much lower fission-product inventory (down by a factor of 100) in this research reactor compared to a commercial reactor; on his reasoning that the bombs may not breach the integrity of the reactor pool (due to the bulk of heavy equipment above it); and on the expectation that even if the fuel elements were scattered they would not melt if their integrity is destroyed. (The radiative heat loss from any of the individual plates that comprise a fuel element is sufficient to prevent melting, Kouts estimates.)

A very different estimate of the radiation hazard predicts that some radiation doses would reach Baghdad while lethal doses might prevail in the vicinity of the reactor. That estimate, according to Goldstein, is based on IAEA release factors for fuel rods from commercial reactors: 60% for noble gases and 25% for volatiles such as iodine. The analysis must also make some assumption about meteorological conditions.

Similar release factors were assumed in a radiation-release calculation done as part of a safety analysis for Brookhaven's similarly sized, high-flux neutron reactor. That study assumed that the core was melted but that the containment was not breached so that 95% of the iodine is filtered. The result was that the radiation levels at distances equivalent to that of Baghdad would be barely detectable, William Higinbotham (Brookhaven) told us. The release factor for iodine that has been considered standard in such calculations may be subject to great uncertainty: Measurements at Three Mile Island indicate that in the accident there, the amount of iodine released was many orders of magnitude below that expected.

Fortunately, these estimates of the radiation hazard were not put to the test. One can hope that the other speculation about the proliferation potential of the research reactors will remain just that—speculation. —BGL

Debate on radioactiv

Do we know enough to dispose safely of waste that will remain radioactive for tens of thousands of years, or will any disposal program inevitably end up as an albatross around the neck of future generations?

No technical barriers

PHYSICS TODAY/DECEMBER 1982

Fred A. Donath

The "energy crisis" of the 1970s brought with it an unparalleled awareness of this country's energy needs for the future. It became clear that an acceptable standard of living would require the use of all energy alternatives and significant expansion of specific ones. For any reasonable projection this translates into considerable dependence on nuclear energy to meet the energy demands of the next two decades while new technologies (such as solar) and improvements in existing technologies are brought to levels that can meet future demand.

With this realization, increasing concern has developed over the growing volume of radioactive waste produced by nuclear power. Indeed, critics have advocated no further expansion of nuclear energy until a satisfactory means of waste disposal has been demonstrated. This essay discusses several of the important questions that people want answered, such as how we can ensure that radioactive waste will be safely isolated from the environment for hundreds of thousands of years.

Methods of waste disposal

Because spent fuel from a nuclear reactor contains a significant amount of unused fissile uranium and plutonium in addition to the unusable "waste" products, it can also be regarded as a potential energy resource. Three options exist for handling the spent fuel. The first is to dispose of it permanently; this option is commonly referred to as the "throwaway cycle." The second is to store the spent fuel temporarily, pending a decision on whether or not reprocessing of spent fuel will be allowed. The third is to reprocess the

spent fuel and recover the unused fissile uranium and plutonium for use in other nuclear reactors. Should we choose to develop breeder reactors to help meet energy demands at the turn of the century, it would certainly seem prudent to remain flexible and view reprocessing as a viable option.

Many intriguing disposal techniques have been proposed[1] and seriously considered, but most have been discarded as impractical, certainly before the year 2000. Ejecting waste into outer space would be enormously expensive; even the small chance of a launch mishap makes this option unacceptable. Development of the technology for transmutation, which consists of extracting the transuranics from the remainder of the waste through reprocessing and then "burning" them in commercial or breeder reactors, is not anticipated before the year 2000, if then. Burying nuclear waste in the Antarctic ice sheets raises questions about ice sheet stability. Not only might water within and beneath the ice sheets transport the waste to the biosphere, but the ice sheets themselves undergo rapid surges roughly every 10 000 years, and this could even be triggered prematurely by waste-generated heat. Subseabed isolation, whereby waste is emplaced in thick sediments or rock underlying the ocean, is a future possibility, but not until more is learned about possible thermal currents and any sediment or rock behavior that could cause the waste to move back into the biosphere. Another technique would be to drill superdeep holes (as deep as 20 000 feet) and dispose of highly concentrated liquid or solid waste. Heat from the waste would initially melt the surrounding rock; later the rock would resolidify, sealing in the waste which would become an integral part of the rock structure. However, drilling such wide holes to accommodate canisters still poses prac-

tical problems, and retrieving the waste would be virtually impossible.

Even if the technology were demonstrated for these disposal techniques, they would still be unacceptable because present government policy rules that waste must be recoverable from any storage site for the first few decades.[2] Not only must we be able to retrieve the waste in the event of leakage, but also to recover valuable unused plutonium in spent fuel, if that is the waste form. Although a few of these and related techniques show promise as future permanent disposal methods, government policy has eliminated all but one technique: burying the waste in excavated cavities in deep geologic formations such as salt beds, granite, or basalt.

Underground cavities

Construction plans for an underground repository are impressive. A network of tunnels and storage rooms would be excavated 2000 feet underground and connected to the surface by access and ventilation shafts (see figure 2). If the spent fuel is reprocessed, the waste would be contained in solid form, packaged in corrosion-resistant canisters, and then placed ten meters apart in holes dug into the floors of the facility. The annual waste from 400 commercial nuclear plants (the number of plants we would have in the US if all our electricity were derived from nuclear[3]) could be stored in an area no larger than half a square kilometer.[1] In fact, thirty tons of spent fuel from a 1000-megawatt reactor operating for one year would, after reprocessing, be reduced to two cubic meters, an amount that would fit easily under a dining table. The principal reason that the waste cannot be stored so compactly, but rather would need to be stored with space between canisters, is to prevent unacceptable heat buildup.

continued on page 208

Fred A. Donath is president of a consulting firm, CGS, in Urbana, Illinois.

into glass blocks, which would provide a barrier against dissolution (even by hot brine). In 1973, however, J. E. Mendel and I. M. Warner at the Battelle Pacific Northwest Laboratories, discovered that the dissolutioning of glass was strongly temperature dependent.[4] At temperatures above 100 °C the leach rate of glass in water increased very rapidly. These measurements have since been extended to higher temperatures, and also to hot brine, and it has been shown that under these conditions glass provides essentially no barrier against dissolution. These findings have influenced many of the current research activities. Materials scientists and geochemists are working to encapsulate the waste in more stable (that is, less leachable) waste forms, such as crystalline ceramics ("Super Calcine") or synthetic rocks ("SYNROC"), but none has yet progressed beyond the laboratory stage, and the debate continues about their relative merits.

Also, according to the plans of the AEC in 1971 the solidified wastes were to be sealed in steel canisters that were only expected to provide protection during transportation and emplacement in the Lyons disposal site. This view has also changed. Recently the NRC has proposed that waste packages should instead by designed "so that there is reasonable assurance that radionuclides will be contained for at least the first 1000 years after decommissioning."[5] I doubt whether any of the waste forms considered can be shown to satisfy this criterion alone, and disposal methods will consequently call for further engineered barriers, such as canisters, overpacks and buffers, to prevent a chemical attack of the waste by groundwater for at least the first 1000 years.

While rock salt had been considered since 1957 to be the prime geological disposal medium, attention is now also directed toward other rocks, including granite, basalt, gneis, clay, tuff, and anhydrite, so as to avoid brine, which has been recognized as the "universal solvent."

While this brief historical review cannot be anything but a thumbnail sketch of the evolution of the present thinking, it does demonstrate that the method of disposal that appeared adequate to the planners of the Lyons repository less than ten years ago—glass blocks in steel canisters buried in an abandoned salt mine—indeed has numerous serious shortcomings when judged by our present knowledge. With that experience in mind, it would be imprudent to predict what the waste-disposal technique may look like when it is finally accepted by the scientists and the public, and also when this will be. During the past three years the Federal government's assumed opening date for a high-level waste depository has been moved back fully nine years, from 1988 to 1997, while in the Federal Republic of Germany the decision about the suitability of a single salt dome (at Gorleben) is expected to require twelve more years of investigation.[6]

Heat dissipation problems

As a specific example of the geotechnical problems encountered with the disposal of high-level waste, let us consider the decay heat given off by the waste, which will be trapped for very long times in the rock surrounding the waste.

Because of this temperature rise, the heated rock will expand. This in turn will lift the rock above the waste, and can cause the rock to fracture. If water penetrates through the cracks in the fractured rock, it can disolve the waste and carry it back ot the biosphere. It is difficult to predict the likelihood of these events, or to determine the maximum levels of heat the repository can handle safely. To illustrative, we mention an intriguing point that has been raised by P. R. Dawson and J. R. Tillerson from the Sandia Laboratories[7] and which may be important for a depository in rock salt: Because of its thermal expansion, the heated region will become lighter. Like hot air rising because it is less dense than cold air, the entire heated zone will be pushed upward. Rock salt is also plastic, that is, it flows and can deform to fill voids, which is usually considered to be an advantage since salt formations are therefore self-sealing. In this particular instance, however, the higher plasticity of salt, and the fact that it becomes more plastic as it is heated, are definitely drawbacks. Dawson and Tillerson have estimated that with the kind of waste loading presently considered for waste depositories, the entire depository would rise approximately ten meters in 600 years. However, the surface would be lifted much less, and consequently some sideways motion would also take place. Such movements could rearrange the waste, making estimates of future safety of the depository extremely difficult. In re-

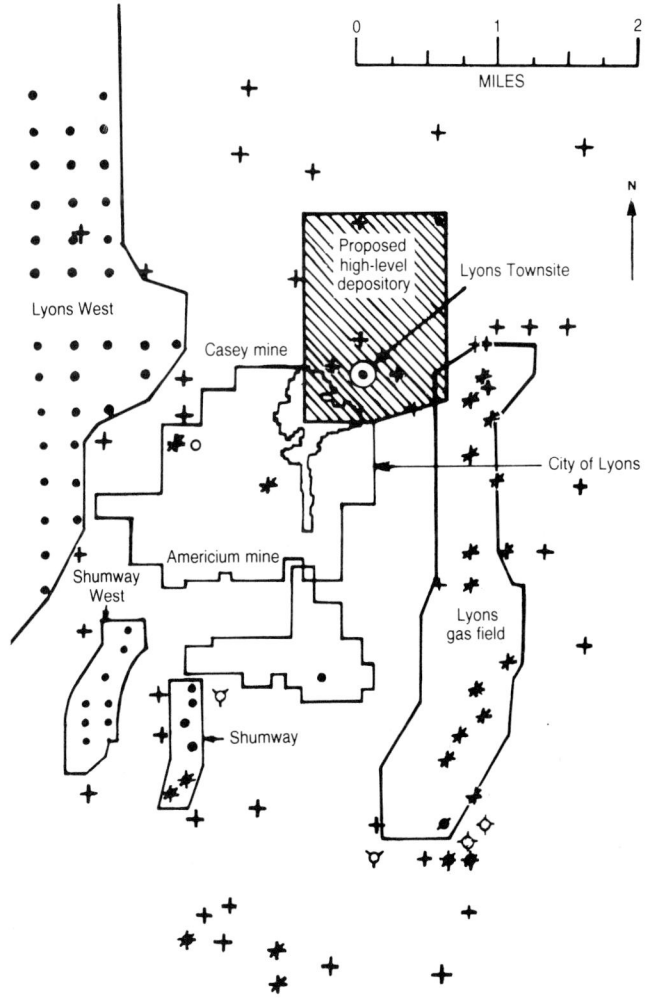

0   1   2

MILES

N

Proposed high-level depository

Lyons Townsite

Lyons West

Casey mine

City of Lyons

Americium mine

Shumway West

Lyons gas field

Shumway

Abortive attempt to store nuclear waste. The proposed federal repository for waste at Lyons, Kansas, proved a major source of embarassment to the AEC when it was discovered that numerous forgotten oil and gas drillings riddled the area. Not only would holes compromise the security of the radioactive wastes but their existence suggests that, in the future, mining companies looking for mineral deposits might disturb site.   Figure 3

⛉ Brine disposal well
● Oil well
✹ Plugged well
⬦ Dry hole

viewing this work Wendell Weart, director of the nuclear-waste disposal efforts at Sandia Laboratories, declared in 1978 that "validation of these calculations will require several years of precision measurements on an experimental area."[8] Any answers derived from such studies are likely to be specific to the experimental site, and results obtained on a small scale in an experimental area may not be meaningful on the much larger scale of the entire depository. For example, the rising of the heated rock salt depends critically on the plasticity of the rock. It is not clear, however, how to relate the plasticity of the rock salt obtained in small-scale experiments to the effective plasticity of a large and nonhomogeneous geologic formation.

The questions mentioned here are only a few examples of the many unknowns that have been pointed out by geoscientists, and which have to be answered before we can decide what constitutes a safe thermal loading, that is, how far the waste needs to be spread, or how long it must be aged until it may be buried. In other words, we cannot talk about conservatively loading a depository until we know precisely what "conservative" means. Geology as a predictive rather than a descriptive science is a new endeavor, and much painstaking work still needs to be done.

In the foreword to the Geological Survey Circular 779,[9] W. A. Radlinski, acting director of the Geological Survey, summarized his judgment as follows: "The many weaknesses in geologic knowledge noted in this report warrant a conservative approach to the development of geologic repositories in any medium. Increased participation in this problem by Earth scientists of various disciplines appears necessary before final decisions are made to use repositories."

Uranium mill tailings

It is generally accepted that the spent nuclear fuel and high-level waste are very toxic and have to be isolated permanently from the biosphere. Far less well recognized is the fact that some low-level wastes, which are generated in vast quantities in the nuclear fuel cycle, also present considerable hazards to future generations. As an example we will discuss uranium mill tailings.[10]

Uranium is a relatively rare element—the ore that is currently exploited contains only 0.1–0.2% by weight of uranium—and hence several hundred thousand tons of ore must be mined to provide the fuel for one large reactor to operate for one year. In the milling process the ore is ground and the uranium is extracted chemically. The residues, called tailings, which contain all nonuranium isotopes of the decay series, are discarded in a tailings pond; the pond is eventually allowed to dry to form a pile. In the tailings the long-lived thorium isotope Th-230 (its half-life is 80 000 years) is the source of the radium isotope Ra-226, the radon isotope Rn-222, and so on. Radon is a chemically inert gas. Some of it escapes from the finely ground mill tailings into the air, where it can be carried over long distances before it decays into its radioactive daughters, polonium, lead and bismuth.

Radon gas and radium can both cause cancer, by inhalation and ingestion, respectively. Removed from their geologic confinement in the ore body, the unprotected mill tailings pose a considerable environmental hazard. On top of a tailings pile that contains an average concentration of thorium and its daughters, a person will be exposed to a whole-body gamma-ray dose rate of 1.34 millirems per hour. This corresponds to 12 rems per year, or more than twice the maximum permissible dose for occupational exposure (5 rems per years).[10] If a building were constructed on a tailings pile with a basement dug into it, lung dose rates of 30 rems per year would be expected from the inhalation of the radon that diffuses through the concrete, and from its daughters.[11] People living in this house for approximately ten years would be twice as likely to die from lung cancer as they otherwise would, according to the Environmental Protection Agency. Their annual exposure would be twice the current limits for uranium miners. (These limits are now being attacked by the National Institute for Occupational Safety and Health as being too high!)

While living on top of an unprotected mill-tailings pile would represent an extreme case of radon exposure, even the radon carried by the wind would increase the exposure for people living in its neighborhood. According to an investigation performed for the Department of Energy, the radon-induced lung-cancer rate within one mile from an existing tailings pile is expected[12] to result in an increase of occurrences of lung cancer from all other sources by 14%. Since every twentieth American today dies from lung cancer (according to the American Cancer Society), a 14% increase of the probability of dying from lung cancer in the vicinity of the piles appears to be significant.

One might argue that these numbers demonstrate that distance apparently provides adequate protection. Unfortunately, however, mill tailings often occur as sand and hence are attractive as construction material for houses, roads, or as admixture to tight clay soils. Thus, people unaware of the

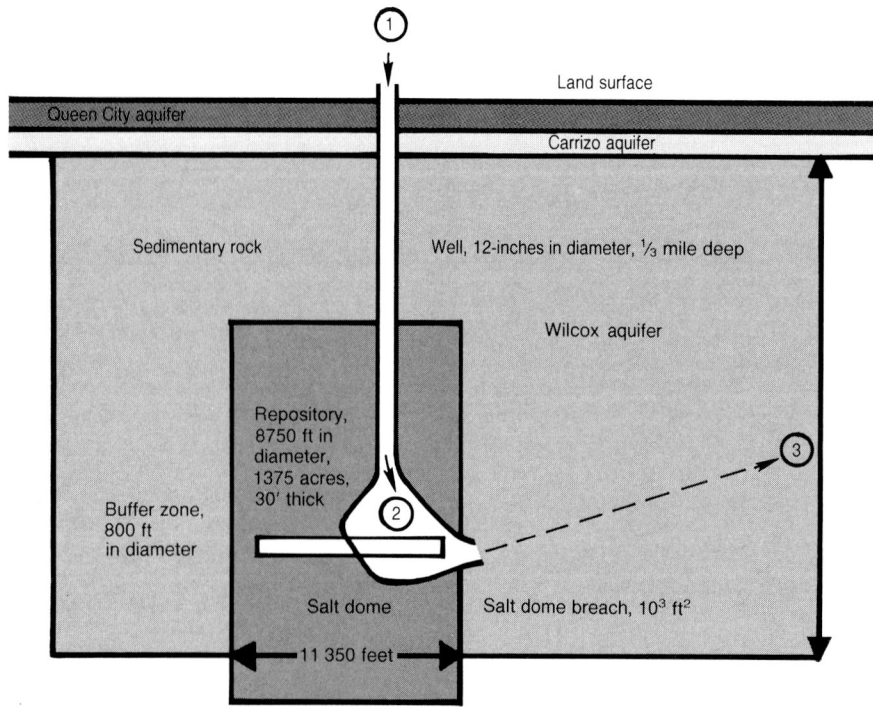

(not to scale)

Hainesville salt dome, one of 26 salt domes in the Northeast Texas salt dome basin within 200 miles of four large cities (populations exceeding 100 000). Worst case of human intrusion assumed that solution mining would destroy the repository, resulting in the following scenario: Water is blown into the salt dome through a hole drilled by an unsuspecting mining company 1000 years from now; the salt disolves, breaching the depository; the brine finds its way to a nearby aquifer, which would allow radioactive wastes to contaminate drinking water systems or reach the surface.[16]   Figure 4

toxicity of the tailings may actually seek them out. Numerous cases of such misuse have become known already in Grand Junction, Colorado; Salt Lake City, Utah; Edgemont, South Dakota; Port Hope, Ontario; and in the Cane Valley area of southeastern Utah and northeastern Arizona. In Grand Junction, tailings sand was used extensively for over fifteen years throughout the city as construction material, until in 1966 the Colorado Department of Health found excessive radiation levels in some buildings. Current estimates are that 800 individual structures require clean-up, at a cost to the public of $16 960 000.[13] So far such remedial action has been taken on 289 individual structures, 14 schools, and 22 business or church locations. This rather substantial contamination was produced by a relatively small amount of mill tailings: Only about 50 000 tons of tailings were used as foundation material containing an estimated 40 curies of radium. The uranium extracted from these tailings would have been enough to operate one 1000 megawatt electric nuclear power plant for only about six months. Fifty thousand tons is only 0.03% of the present inventory of all mill tailings in the US (estimated to be 150 million metric tons). It is also interesting to note that, per unit energy produced, the amount of mill tailings equals that of the ash produced in a coal-burning power plant, assuming 0.1% uranium in the ore, and 10% ash in the coal. The often-made claim that the nuclear fuel cycle produces only small amounts of waste is therefore incorrect.

It has been estimated that by the year 2000 the mill tailings will reach 1.5 billion metric tons, ten times more than today, enough to bury the entire District of Columbia under fifteen feet of tailings.[10] How should one deal with such massive amounts of toxic waste? It appears obvious that the criterion for proper waste disposal must be a protection of future generations equal to that required for themselves by those who produce it. It follows that we have to safeguard against incidents like the one in Grand Junction where people simply did not know that the sand they used was toxic. The present practice of dumping the tailings and leaving them unprotected is therefore unacceptable.

The NRC has proposed, over the vigorous objection of the mining industry which fears the additional expense, that future mill tailings should preferably be buried below the ground surface, but above the groundwater table. They should be covered with soil such that the atmospheric release rate of the radon would be reduced to 2 pCi/m²sec, which is twice the average radon release rate of soil. In cases where burial below grade would not be feasible,

Uses for a salt dome

Compressed-air energy storage (CAES)
Solution mining
Dry salt mining
Hydrocarbon storage
Strategic petroleum reserve (SPR)
Hazardous-waste storage
Natural gas storage
Oil and gas development
Nuclear and non-nuclear explosions
High-energy neutrino astronomy
Radioactive-waste isolation

abovegrade disposal and a soil cover might also be permissible.[14]

While these measures would be improvements over the present practice, there are several reasons why they would still be inadequate. First, no soil or rock cover can be relied on to last hundreds, or even thousands of years. The mill tailings, however, will lose only one-half of their toxicity in 80 000 years (the halflife of the parent isotope thorium-230). For further protection the NRC has suggested that there be continued surveillance of the disposal sites; this is even more likely to fail on the time scales involved. Second, burial below grade will reduce the distance between tailings and groundwater. Radium leached from the tailings would be a serious contaminant for the latter. In the extreme case, the groundwater table could rise into the mill tailings pile (perhaps as a result of a change in climate), and people would withdraw drinking water from it. Resulting exposure rates for individuals have apparently not yet been estimated; if we consider that the 80 curies of radium contained in the mill tailings produced in providing the fuel for operating a large reactor for one year could contaminate eight billion cubic meters of water (the average annual amount of water carried by a good-size stream) to the maximum permissible concentration for the general population (called MPC_w), we expect the dose rates in this scenario to be very high. To reduce the risks of groundwater pollution, the NRC has recommended the use of liners made of clay or plastic. Their lifetimes are at least as doubtful as those of the covers.

If we consider the risks posed for practically all future generations by a billion tons of tailings in hundreds of piles of mill tailings, distributed over many of our western states, only two disposal techniques seem adequate. One would be to bury the tailings mixed with a binder, like cement or asphalt, in deep, dry mines that would approximate the original situation of the ore body to the best possible degree. This technique might be suitable, although the availability of adequate space may be a problem, and the longevity of the binder would also be

doubtful. Alternatively, one could chemically extract the thorium and radium in the milling process, and dispose of them in a mined geologic high-level waste repository. The extraction technique has been demonstrated on the laboratory scale but is receiving very little attention at present.[10] The NRC has discarded both techniques as too expensive, but has reached this conclusion without including the costs of perpetual surveillance and remedial action required for its preferred disposal modes.[15]

It can only be hoped that the public insistence on a solution to the mill tailings disposal problem will be successful before the sheer enormity of the amount of waste precludes any technical solution whatsoever.

Human intervention

My fourth and last example deals with the problem of the unpredictability of future human actions, which probably presents the most serious threat to the long-term isolation of high-level waste.

Let us inspect again the mining and drilling record of the bedded salt formation at Lyons, shown earlier in figure 3. At and near the site of the proposed high-level waste repository the salt beds are literally riddled with underground salt mines and with bore holes, shafts, and a variety of wells, giving the area the appearance of a piece of Swiss cheese. Apparently this is what must be expected if an area containing natural resources is inhabited by a technological society. It is obvious that these human intrusions will have a great impact on the long-term behavior of the salt formation in that area: Water can enter the unplugged or inadequately plugged holes, can dissolve more salt, and can cause subsidence of the overlying formations. More water can enter through the new pathways, can dissolve more salt, and finally some of the brine can return to the surface of the Earth via some aquifers. The variety of scenarios is almost limitless. It was largely the recognition of the risks posed by these man-made pathways that led to the cancellation of the Lyons project, because a reliable, long-term seal of all existing holes could not be guaranteed.[3]

Researchers at the Battelle Pacific Northwest Laboratories have recently studied a worst-case scenario for the disposal of spent nuclear fuel in a salt dome,[16] shown schematically in figure 4. It was assumed that either 100 or 1000 years after the waste had been buried, unsuspecting mining companies would start solution mining of rock salt, and that the dissolved waste would be incorporated into the salt used by 15 million people for culinary purposes.

This scenario would lead to a total body dose to the population over a fifty-year time span of 1.6×10^{11} and 1.3×10^9 person-rems if intrusion were to occur after 100 or 1000 years, respectively. This exposure would cause 29 million (!) or 230 000 cancer fatalities (based on the dose–health-effect conversion rate of 180 cancer fatalities per million person-rems, as suggested in the BEIR report.[17] Although this calculation was performed using some rather extreme assumptions, it nevertheless demonstrated that the integrity of a depository must be guaranteed for periods longer than 1000 years.

It is often argued that after 1000 years the nuclear wastes are no more toxic than the uranium ore from which the fuel was produced; consequently there would be little reason to worry about the toxicity of the nuclear waste beyond that time. The fallacy of this argument is that the unmined uranium ore is well locked up in its natural geologic setting, where it has been for millions of years, while the nuclear waste in a repository is in a highly artificial form and in an unnatural setting. The same is to be said about the leftovers from the uranium mining, the mill tailings, as we saw earlier. Furthermore, the repository will contain the wastes resulting from a hundred reactors operating for several decades, in a single place covering an area of a few square miles; the ore from which the corresponding fuel was made, however, came from many different ore bodies scattered over hundreds of thousands of square miles.

While in principle it is possible to seal all holes that exist at the time the waste depository is filled, the real problem will be to prevent future generations from drilling new ones after the location of the depository has been forgotten (see table). If companies drilling for materials were to hit the depository or even a waste canister, they would soon rediscover the depository, which could then be resealed. More serious would be a near-miss not leading to the discovery of the depository, such as the establishment of a new solution mine nearby, its operation and final abandonment, which might lead to serious consequences for the future integrity of the waste repository. Based on historic longevity of past civilizations, and considering the great instability of present human institutions, I doubt that any specific event, in particular one that is as unglamorous as waste disposal, will be remembered for more than one hundred, or at best a few hundred years. Thereafter we must suppose that our enterprising descendants would resume their drilling activities. It has recently been estimated that out

of the 150 salt domes in the US Gulf Coast region which would be at the proper depth to be potentially suitable for nuclear-waste disposal, 95, or almost two-thirds, have undergone some form of industrial development in the roughly 100 years of industrial activity in that area.[18] Based on these facts, a human intrusion into the general area of a nuclear-waste depository within a few hundred years must be considered a near certainty, if this depository is located in a geologic medium containing salt or any resources that humans may want to make use of (although the high water solubility of rock salt certainly compounds the problem).

In light of these considerations, the following recommendations,[19] made recently by the National Academy of Sciences, make a great deal of sense: "no area with a present or past record of resource extraction, other than for bulk materials won by surface quarrying, should be considered as a geological site for radioactive wastes. This restriction rests on one or more of three possible considerations: (a) present or predictable future importance as a potential source of needed raw materials; (b) disturbance of the natural hydrologic regime in consequence of present or past underground development and exploration, such as tunneling, hydraulic fracturing, etc., resulting in greater uncertainty as to the paths and volumes of fluid flow; and (c) potential attractiveness to future developers and explorers for natural resources who may be drawn to the area by evidence of past activities of resource extraction."

If we further consider the high water solubility of salt and the corrosiveness of the resulting brine, we cannot help but conclude that, quite apart from any geologic considerations, salt formations must be judged unsuitable for the disposal of nuclear wastes (or, in fact, any highly toxic wastes). Based on the same argument, formations of the abundant silicate rocks like granite, gneis, tuff or basalt should be more acceptable, in particular if they occur at great depth, making them far less attractive for mining than the same material close to the surface.

The proposed rule-making on geologic disposal of high-level radioactive waste published recently by the NRC contains the following sentence: "The human intrusion issue is a difficult one that is far from having been resolved."[5] I cannot see how it can be resolved at all, except by relying far more on man-made barriers of isolation rather than on the geologic medium. Work on the man-made barriers is still in its early stages, as we saw earlier. As far as the proper choice of the geologic medium is concerned, it appears that rock salt,

apparently still a favorite in the US and the Federal Republic of Germany, is likely to put the most stringent requirements on waste packaging.

Our legacy to future generations

The need to provide effectively permanent isolation from the biosphere is not unique to the radioactive wastes; in fact, right now the legacy we are leaving behind in the form of chemical wastes probably represents a far more serious threat to future generations. However, if we consider the use of nuclear energy as a major source of our future energy needs, we must find adequate solutions now, rather than delaying the efforts until it is too late, as appears to be the case with many of our chemical wastes.

In discussing adequate protection from the nuclear waste, the proper yardstick, in my opinion, is not how many people will be killed by it on a statistical basis (sometimes even expressed as the number of cancer fatalities per megawatt-year of electrical energy produced). Rather, the point is whether we want to impose on future generations the need to live permanently with radiation monitors, something we do not have to do right now—apart from some unfortunate exceptions. In my opinion, we should make every effort to avoid subjecting our descendants to this additional concern. This would require finding permanently safe disposal methods for all forms of radioactive wastes, since any disposal of long-lived radioactive species in shallow landfills, or through ocean dumping, as is currently practiced for the many forms of low-level waste arising in the nuclear fuel cycle, would be unacceptable. Properly isolating these myriads of different wastes from the biosphere might pose equal, or even greater difficulties for the nuclear industry than the disposal of the high-level waste. The disposal of the mill tailings is only one example of the large category of low-level wastes.

The solution of the technical problems involving the disposal of nuclear wastes will require a great deal of additional research, in particular where geologic processes occurring over long time periods have to be considered. Many of these questions cannot be answered through laboratory experiments and field studies alone, but also require models for extrapolation into the distant future. The validity of these models must always remain doubtful to some degree. By far the greatest challenge for the technical community, however, will be to convince a distrustful public that remaining uncertainties constitute an acceptable risk. The technologists themselves are responsible for this lack of trust, because for decades they have failed even to recognize that the permanent isolation of nuclear wastes is a major technical problem, and have consequently been unable to deliver on their promises that these wastes would be managed safely. The Lyons debacle or the mismanagement of uranium mill tailings are only two examples. Other incidents that have contributed to the erosion of the public confidence are, to name a few, the well-known leaks from the high-level waste tanks at the Hanford reservation, the leaching from the poorly engineered low-level waste burial grounds in West Valley, New York, and Maxey Flats, Kentucky, and the catastrophic dispersal of nuclear waste in the Urals, which contaminated hundreds of square miles.[20]

We must stop belittling the technical problems of nuclear-waste disposal. The only way of regaining public confidence will be through candid discussions of the problems, and through painstaking and critical research and development efforts. In addition a most meticulous clean-up of "hot spots" of radioactive waste which exist already in many parts of the country should receive the highest priority in order to give convincing evidence that the nuclear community is doing something to solve the waste problem right now. This path will be long and expensive, but it will be the only one by which the nuclear community can hope to restore its credibility, and thus hope to assure its own survival.

References

1. A. Wolman, A. E. Gorman, Waste Materials in the United States Atomic Energy Program. WASH-8, US, AEC, 1950.

2. US AEC Authorizing Legislation, Fiscal Year 1972, Hearing before the Joint Committee on Atomic Energy, part 3, March 16–17, 1971.

3. D. S. Metlay, in *Essays on Issues Relevant to the Regulation of Radioactive Waste Management*, W. P. Bishop, ed., NUREG 0412, US Nuclear Regulatory Commission (1978).

4. J. E. Mendel and I. M. Warner, in Battelle Pacific Northwest Laboratories Quarterly Progress Report BNW-1761, A. M. Platt, ed., June 1973, page 4.

5. US Nuclear Regulatory Commission, Technical Criteria for Regulating Geologic Disposal of High-Level Radioactive Waste, Federal Register **45**, no. 94 (13 May 1980) page 32400.

6. M. Knapic, Nucl. Fuel, 8 December 1980, page 12; Nucl. Fuel, 13 April 1981, page 13.

7. P. R. Dawson and J. R. Tillerson, in *Proc. Int. Conf. on Evaluation and Prediction of Subsidence*, Pensacola, Fla., 15–20 January, 1977 (report no. CONF 780136-1).

8. W. D. Weart, in *Radioactive Waste in Geologic Storage*, S. Fried, ed., ACS Symposium Series 100, American Chemical Society, Washington (1978), pages 14, 32.

9. J. D. Bredehoeft et al., Geologic Disposal of High-Level Radioactive Wastes—Earth-Science Perspectives, Geological Survey Circular 779, US Geologic Survey (1978).

10. E. Landa, Isolation of Uranium Mill Tailings and Their Component Radionuclides from the Biosphere—Some Earth Science Perspectives, Geological Survey Circular no. 814, US Geologic Survey (1980).

11. J. A. Adams, V. C. Rogers, A Classification System for Radioactive Waste Disposal—What Goes Where? NUREG-0456, US Nuclear Regulatory Commission (1978) page 140.

12. Engineering Assessment of Inactive Mill Tailings, Mexican Hat Site, Utah, Ford Bacon and Davis Utah, Inc., Phase II, Title I. (1977).

13. Progress Report on the Grand Junction Uranium Mill Tailings Remedial Action Program, DOE/EV-0033, US Department of Energy (1979).

14. US Nuclear Regulatory Commission, "Uranium Mill Licensing Requirements," Final Rules, Federal Register **45**, no. 194 (3 October, 1980), page 65521.

15. Cited in Final Generic Environmental Impact Statement on Uranium Milling, NUREG-0706, US Nuclear Regulatory Commission (1980).

16. Reference Site Intital Assessment for a Salt Dome Repository, Report PNL-2955, Battelle Pacific Northwest Laboratory (1979).

17. *The Effects on Population of Exposure to Low Levels of Ionizing Radiation*, report of the Advisory Committee on the Biological Effects of Ionizing Radiation, National Academy of Sciences, Washington (1972).

18. K. S. Johnson, S. Gonzales, Salt Deposits in the United States and Regional Geologic Characteristics Important for Storage of Radioactive Waste, Y/OWI/SUB-7414/1, Office of Waste Isolation (1978), page 174.

19. Geological Criteria for Repositories for High-Level Radioactive Wastes, National Academy of Sciences Committee on Radioactive Waste Management, Washington (1978).

20. Z. A. Medvedev, *Nuclear Disaster in the Urals*, Norton, New York (1979); J. R. Trabalka, L. D. Eyman, S. I. Auerbach, Science **209**, 345 (1980).

* * *

Nuclear power and nuclear-weapons proliferation

The danger that fissile isotopes may be diverted from nuclear power production to the construction of nuclear weapons would be aggravated by a switch to the plutonium breeder—but future uranium supplies are uncertain.

Ernest J. Moniz and Thomas L. Neff

PHYSICS TODAY/APRIL 1978

For decades, nuclear power has been considered a major component in the energy supply plans of some countries and an important option for the future in others. Like other energy sources, especially oil, nuclear power has become linked to national security and economic health in many countries; the magnitude of fuel reserves and the assurance of supply have become issues of intense international concern. However, nuclear power raises an additional issue: its potential for contributing to the acquisition of nuclear weapons by nations or even by terrorist groups. The goals of adequate energy supply and nuclear-weapons nonproliferation are therefore potentially in conflict.

Proliferation risks differ with the forms of nuclear technology, and the political and technical opportunities for international control vary for different fuel cycles. Nuclear power reactors in the United States and in most other countries now operate on uranium enriched to about 3% in the fissile isotope U^{235} (see the Box on page 51 for a brief description of fissile materials and fuel-cycle technology), and such fuel can not be used in a nuclear weapon without further isotopic enrichment. Commercial enrichment, which demands very advanced technology, is still restricted to a few supplier countries and has not contributed directly to weapons proliferation.

On the other hand, the vigorous international pursuit of advanced fuel-cycle technologies has stimulated serious con-

cerns about increased proliferation risks. One source of concern would be widespread deployment of *isotopic-enrichment* technologies, such as gas centrifuges. Another arises from the long anticipated worldwide shift to *plutonium* fuels, first in thermal reactors of the type now operating and eventually in fast breeder reactors. Plutonium is bred from U^{238} during reactor operation and, if chemically separated from the spent fuel, can be used to extend naturally available nuclear fuels. Pilot plants built to gain experience for future commercial activities could be immediate sources of weapons material, as the basic technology of plutonium separation is, unlike enrichment, known and widely published.

These concerns have led to extensive reexamination of the technical, economic and political assumptions underlying both national and international nuclear policies. In the United States several extensive studies of nuclear-power issues[1-3] have been completed within the last year, while an International Nuclear Fuel Cycle Evaluation involving more than 40 nations will take place over the next two years.

There is yet little agreement on the balance between energy benefits and proliferation risks associated with different fuel-cycle choices. This lack of agreement stems from differing views of proliferation risks and from wide variations in the energy supply and the security problems of different nations. The resolution of these differences will be largely a political process, and we do not pretend to offer solutions in this article. However, we will seek to clarify the basic technical and political issues, and to set forth the connections between various fuel cycles and their possible proliferation risks.

Fissile materials that can, in principle, be used to make nuclear weapons are

present in all nuclear fuel cycles, in quantities large compared to the amount needed for making explosives. The ease with which such material can be recovered for weapons use, however, varies greatly with fuel cycle. Uranium and plutonium fuels differ in this regard: The former can be isotopically denatured, the latter can not. The thermally fissile isotopes U^{233} and U^{235}, when diluted to isotopic content less than 15–20% with U^{238}, can not be used in a nuclear weapon, because sufficiently rapid supercritical assembly becomes impractical. This rapid increase of critical mass with decreasing U^{235} isotopic fraction is indicated in the figure on page 44.

Uranium versus plutonium

For plutonium, on the other hand, the critical mass of any isotopic composition is quite small, as this figure also shows, so weapons material can be obtained by chemical rather than isotopic separation. Plutonium with thermally fissile Pu^{239} content greater than 93% is generally defined as "weapons-grade," but reactor-grade plutonium, with a fissile content of about 60–70%, has a critical mass only a factor of two greater. This does not mean that an efficient, high-yield explosive is manufactured easily with reactor-grade plutonium. In particular, Pu^{240} creates a substantial neutron background because of spontaneous fission, and premature initiation of the chain reaction is apt to occur. However, an explosion will still result—and the yield from a nuclear "fizzle" can be extremely large compared with that from conventional explosives.[3]

The plutonium-bearing spent fuel from the uranium cycle, intensely radioactive from fission-product activity, requires remote-handling facilities for any subsequent processing. If the plutonium is

Ernest J. Moniz, an associate professor of physics at MIT, was a member of the American Physical Society Study Group on Nuclear Fuel Cycles and Waste Management; Thomas L. Neff, a policy analyst in the MIT Energy Laboratory, was senior staff member for the Ford Foundation's Nuclear Energy Policy Study.

recycled, the fission products would be removed during reprocessing, and this would eliminate the greatest technical barrier to diversion of plutonium from the fuel cycle to weapons use. Enrichment technology is tightly held by a few technologically advanced nations, and developmental research programs or direct technology transfer would be needed to diffuse this technology. On the other hand, many nations are now able to construct pilot-scale fuel-reprocessing plants, which provide plutonium suitable for weapons. India demonstrated this capability, and the capacity to construct a nuclear explosive clandestinely, in setting off a nuclear explosion in 1974, shown in the photos on pages 46 and 47.

Given the possibility of increased security risks arising from the adoption of plutonium fuel cycles, it is important to review their potential benefits critically, as these will determine the extent and rate of national commitments to plutonium use. The most obvious benefit of recycling is resource extension; the lifetime uranium requirements of light-water reactors are reduced by about 32% if the uranium and plutonium in the spent fuel are recycled. For an expanding reactor system this reduces to perhaps 20–25% in the US over the next few decades.[2] With current uranium prices, recycling would increase or decrease the cost of electricity by at most a few percent.

Enough uranium?

The importance of resource extension depends on the availability of uranium. A country's concern about its future supply centers on several factors:
▶ the magnitude of the domestic and foreign resource base,
▶ the ease with which these resources may be produced and
▶ secure access to an equitable international system for allocating uranium.
All of these are uncertain. Projections of

Reprocessing plant in limbo. The completed separation and UF$_6$ facility awaits licensing to begin operations, but the Nuclear Regulatory Commission has terminated the study on which this decision depends. The operators of the Barnwell, S.C. plant, Allied-General Nuclear Services, are seeking to turn the plant over to the US government. Nuclear-fuel reprocessing is likely to aggravate the risk of weapons proliferation.

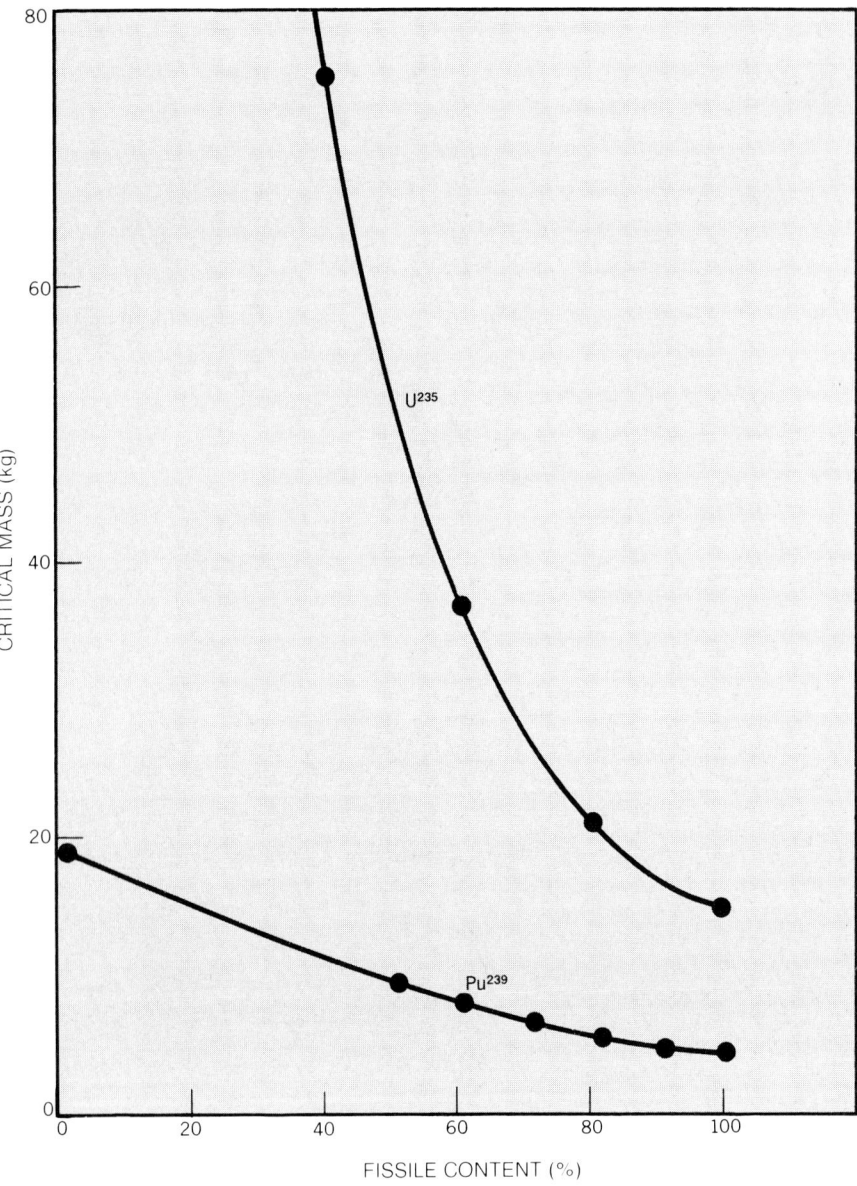

The critical masses of uranium and plutonium as functions of fissile content. The two metals are in the form of spheres enclosed in thick neutron reflectors of natural uranium. The rapid increase in its critical mass makes isotopically dilute uranium unusable as a bomb. This is not so for plutonium, making it a greater proliferation hazard. Data from Theodore Taylor, Ann. Rev. Nucl. Sci. **25,** 407 (1975), and derived from a personal communication by Robert Selden.

low-cost uranium resources, summarized in table 1, are based upon sketchy data. Uranium appears in many geological environments, and exploration has been limited by periodically unhealthy conditions in the industry. Estimates of uranium availability at costs higher than the $30 a pound used in the table are even more uncertain, but some projections of $100-a-pound uranium range as high as 13 million tons.

There is considerable controversy about the meaning of these projections for long-term planning. If we accept such projections as upper bounds on producible uranium, use of this rapidly depleting resource would be increasingly unattractive. On the other hand, resource economists view the uranium industry as immature and contend that, as with other minerals, new reserves and lower-grade

deposits will be produced on demand.[1] It is not yet clear which of these perspectives will prove correct.

For some decisions the present information is adequate. For example, a conservative projection of low-cost uranium available[4] in the US (about 1.9 million tons of U_3O_8) shows that there is at least enough uranium to supply the lifetime requirements (of 30 years at 65% capacity) of light-water reactors with approximately 350 GWe of capacity; the current projection for installed capacity in the year 2000 is 380 GWe. Consequently, resource considerations alone do not appear to require plutonium recycling during the next two decades.[1,2] However, recycling may be more attractive in countries without indigenous uranium and with a perceived insecurity of access to external supplies.

Uncertainties in long-term uranium supplies make it difficult to plan long-term research and development programs for new nuclear technologies. For example, the plutonium breeder would use resources at least 30 times as efficiently as current reactors. However, there would be economic penalties if uranium prices had not risen enough to overcome the additional capital cost. The potential social penalties that argue against early commitments will be discussed below. The rate at which uranium resources are discovered and consumed also has a bearing on opportunities for introducing advanced converter reactors or alternative breeders.

The problem of nuclear wastes is another element in the debate over early reprocessing and recycling. Because plutonium is a major source of radioactivity in wastes after about 500 years, some have concluded that reprocessing and subsequent "burning" of the plutonium is an important waste-management step. However, recent studies[1,2] reject this view. Only a small reduction of long-term actinide activity can be accomplished in this way, and the efficacy of long-term geologic isolation of the wastes is not demonstrably dependent on waste form.

Thus far we have focussed on the uranium light-water cycle with low enrichment and recycling of the plutonium extracted from the spent fuel. This path, usually seen as leading eventually to the plutonium breeder, has received by far the greatest research and development effort throughout the world. However, alternative fuel cycles might be preferable for reducing proliferation hazards. More efficient "once-through" cycles would diminish the pressure for moving ahead to plutonium breeders. Such cycles, especially those based upon thorium, might be more amenable to international measures for reducing access to weapons material. We shall discuss some of these more specifically in dealing with strategies for control of fissile material.

In table 2 we list several thermal-reactor options and their associated lifetime uranium commitments. Note that large resource extensions are possible; for example, modification of reactors for "spectral-shift operation" with thorium fueling can reduce ore requirements by more than a factor of two. In spectral-shift operation, the cooling system of the reactor is modified to include both light and heavy water, with the ratio of D_2O to H_2O decreasing through a fuel-burn cycle. However, except for the natural-uranium heavy-water reactor, CANDU, now operated in Canada, these alternatives are not yet ready for widespread commercial use.

Proliferation risks

Proliferation risks resulting from expansion and evolution of nuclear-power systems should be evaluated in the context of alternative routes towards weapons capability. Some have argued that there are no essential connections between the development of nuclear power and of weapons, because countries deciding to acquire nuclear weapons could achieve these capabilities through facilities dedicated to that purposes. For example, a uranium-enrichment program, free of commercial demands, could use comparatively simple means to support a very small weapons program. More likely would be construction of a plutonium production reactor fueled by natural uranium and an associated reprocessing plant; the design and operating characteristics for both are openly available. These dedicated routes would yield weapons material of higher quality than would be obtained by diversion of commercially produced plutonium, though several years would be required to construct and use dedicated facilities.

However, the widespread use of enrichment facilities, highly enriched uranium or plutonium fuels may alter drastically the political context of weapons-acquisition decisions in several ways:

▸ Countries deciding to develop weapons could use such fuel-cycle steps as a cover for the most time-consuming and detectible phase of a weapons program—the acquisition of fissile material.

▸ The scale of a weapons program fed by fissile material from a commercial fuel cycle could be much greater than that deriving from dedicated facilities.

▸ Many countries could move very close, in time and technical capability, to weapons without having to make and sustain the kind of political decisions required for dedicated programs.

What would be necessary to make a transition from non-weapons to weapons status rapidly—within a few days or weeks—would be preparatory steps involving weapons design and ancillary technical development, all of which are allowed under current agreements and may be conducted relatively easily in secret. This "latent" proliferation can adversely affect international relationships: For example, the possibility that a nation has made such preparation could pressure potential adversaries to do the same.

The risks associated with terrorists or other "subnational" theft are also greatly magnified by plutonium recycling. Stringent security measures are needed from the time plutonium is separated from fission products at the reprocessing plant until it is inserted back into the reactor as part of fresh fuel. There is considerable disagreement about the ability of a subnational group to construct a reliable, effective nuclear explosive. Nevertheless, there is consensus that a small competent group, given enough time, would have a reasonable probability of obtaining a significant yield.[3] This risk is clearly sufficient to warrant extensive

safeguards for the control of plutonium in the fuel cycle.

Strategies for limiting proliferation risks must take into account not only the nature of the risks involved but also the technical and political opportunities for dealing with them presented by different fuel cycles. In the next decade, fuel-cycle choices are limited to natural uranium, uranium of low enrichment and recycled plutonium. The problem of theft by a subnational group and diversion by a nation are quite different, and deserve separate consideration.

Safeguards against theft

Safeguards against subnational theft include physical security and technical measures intended to deter theft, to increase the chance of detection if it occurs, and to make weapons fabrication more difficult and time-consuming. With plutonium recycling the most vulnerable points for covert diversion are at the reprocessing and mixed-oxide-fabrication plants, while the transportation links between these plants and the reactor might be targets for overt theft.

Physical security measures might include armed guards, massive transportation casks and special communications during transportation and surveillance, tightly controlled access to process streams and storage areas at reprocessing and mixed-oxide-fuel fabrication plants. These measures are qualitatively similar to those used in the protection of other valuable or dangerous commodities.

Technical measures aimed at complementing and reinforcing physical security include isotopic accountability schemes and fuel-form modification, specifically suited to the control of nuclear materials.

▸ In the accountability approaches neutron and high-resolution gamma-ray measurements monitor accurately the

flow of fissile materials through the facilities. Under development are systems in which the measuring devices are coupled to a central computer for real-time analysis. Such accounting is complicated by the great variety of physical and chemical forms in which plutonium appears in process and waste streams. Fortunately, accurate accountability is achieved most easily at the same fuel cycle points at which the fissile material is most accessible for diversion, and automated accountability systems may therefore be important in maintaining security over the lifetime of a fuel-cycle facility. Such systems are not yet available.

▸ Fuel-form modification could involve pre-irradiating, "spiking" with radioactive isotopes (to make theft and subsequent handling more dangerous), incorporation of intense neutron sources (complicating weapons design), or dilution of plutonium with uranium (to force chemical processing of the material).

Although none of these measures can prevent misuse of fissile material, they can significantly complicate matters, and so gain time for the recovery forces after a theft. Because these measures must be consistent with the safe normal operation of the fuel cycle, the simple dilution approach is particularly attractive. Dilution can be be accomplished by mixing after processing or preferably, by adjusting the reprocessing chemistry so that plutonium and uranium are processed together and therefore never completely separated.

These measures for plutonium fuel cycles do entail additional fuel-cycle costs. For example, "coprocessing" results in the need to handle substantially larger quantities of plutonium-bearing materials and somewhat complicates mixed-oxide-fuel fabrication. Political and social costs are also involved, including the impact of security measures on the civil

---

Table 1. Estimated uranium resources recoverable at $30/lb
(1000 short tons U_3O_8)

	Reserves	Reasonably assured[b]	Estimated additional[c]
United States[a]	680	1090	1600
Canada		218	510
Africa		500	160
Europe		520	140
Australia		430	100
Asia		60	30
South America		40	60
		3538	2600

a. Estimated reserves and resources available at less than $30/lb forward cost. Figures, from reference 4, do not include byproduct uranium from phosphate production.

b. For the US, the figure is for probable resources; elsewhere, the figures are for U_3O_8 at less than $30/lb from *Uranium Resources, Production and Demand*, OECD–NEA/IAEA (1975), updated for Canada by R. Wright (Uranium Industry Seminar, DOE–GJO–108 (1977).

c. For the US, the figure is for possible and speculative resources; elsewhere, figures are from the OECD–NEA/IAEA report.

rights of workers in the nuclear industry.

The economic costs involved in implementing physical security are certain to be small compared to the overall cost of nuclear-generated electricity.[5] However, the cost/benefit calculation is made difficult by the unquantifiable nature of the threat and by the difficulty of agreeing on what constitutes an "acceptable" level of risk. In the US, this determination awaits the establishment of safeguards performance criteria by the Nuclear Regulatory Commission.

Strategies for international control

Safeguards play a somewhat different role in inhibiting national proliferation than in restraining subnational theft. The physical security and other measures relied upon in meeting the threat within a country are largely ineffective in dealing with national risks, because governments not only have control of nuclear facilities and materials but also have considerable resources for overcoming physical barriers to access to fissile material. Consequently, safeguards against national proliferation involve very important political components and operate primarily through the threat of detection and subsequent international response. Three elements are essential: an appreciable chance of detection, suitable international response mechanisms and time to respond before completion or use of weapons.

International safeguards are now based on a combination of bilateral constraints imposed by suppliers, and an international safeguards system administered by the International Atomic Energy Agency, a United Nations affiliate agency. The IAEA safeguards regime implements the surveillance function agreed to by the more than 100 nations that have ratified the Non-Proliferation Treaty. Bilateral constraints include restrictions on transfers—and subsequent retransfers— of nuclear technologies and materials. Efforts by supplier governments to achieve greater uniformity in these restrictions, particularly in the export of sensitive facilities such as reprocessing plants, resulted in the formulation of common supplier guidelines late in 1977. These partially extend IAEA safeguards to other countries.

Safeguards are far from universally applied. As an intrusion into national sovereignty, IAEA inspection is sometimes resisted. Even with treaty-signatory countries, IAEA inspectors do not have access to the cascade area of enrichment plants.[3] A number of countries have refused to sign the Non-Proliferation Treaty, citing the inequality it institutionalizes between weapons and nonweapons states and the sacrifice of national sovereignty involved.

The technology and procedures of safeguards are also imperfect: Inspections have at times been infrequent and lacking in accuracy at the levels desirable; they are subject to human failures, do not involve real-time monitoring of nuclear activities and vary considerably between countries. Furthermore, the international process by which detection of diversion or misuse could be verified and brought to the timely attention of the international community is uncertain in its efficacy. Despite these problems, the present safeguards regime has been successful in the sense that misuse of fissile materials in the once-through uranium fuel cycle has not occurred. On the other hand, the opportunities for misuse have been rather limited.

Diffusion of technology and technological change present major challenges to the continued success of the safeguards system. Increased technical sophistication worldwide will magnify the importance of international monitoring of all nuclear facilities. The greatest challenge comes from the spread of enrichment and plutonium-separation technologies. While there is little experience with enrichment plants in non-weapons states, technologies requiring a large number of stages—gaseous diffusion and aerodynamic nozzle—appear to offer the best opportunities for safeguards, especially if direct inspection of the cascades is allowed. Technologies with fewer stages—gas centrifuges and eventually lasers—present greater problems, because high enrichment can be achieved more easily.

Restrictions on the export of these high-technology devices could effectively close off this route to weapons for many years in all but a few advanced nations. However, there is disagreement between some consumer and supplier states as to whether such a policy is consistent with the obligations of suppliers under the Non-Proliferation Treaty: The consumer states emphasize the supplier pledge to provide technology, while some suppliers say that to provide sensitive technologies is inconsistent with their overall Treaty responsibilities.

Plutonium fuel cycles also magnify the problems of detectibility and response time, because of the large amounts of potential weapons material involved and its relatively quick accessibility. Considerably more intrusive, and hence politically sensitive, international controls would be needed. Resident inspectors and automated internationally monitored real-time accountability systems have been proposed but these proposals have serious inherent limitations in preventing national diversion. If utilization of plutonium fuels makes it possible for countries to build bombs quickly, a primary source of international leverage on the national

India tests a bomb. The sequence shows the landscape before the explosion, the mound rising seconds after, and an aerial view of the crater. The 1974 underground test emphasized worldwide concerns that the spread of nuclear power technology would hasten the entry of additional nations to the nuclear club. Photographs courtesy of the Consulate General of India, New York.

proliferation problem would be undermined. The ultimate ability of safeguards to deal with these problems is still very uncertain.

Internationalization of fuel-cycle activities has been widely discussed as potentially fruitful in curbing latent proliferation. The basic idea is that all enrichment, reprocessing and fuel fabrication take place in internationally or regionally operated fuel-cycle centers so as to reduce the opportunities for any nation to divert fissile material and to enhance assurance of fuel-cycle services. Although there are formidable political realities to be confronted in establishing such centers, there are also incentives, such as fuel assurance, which may help to overcome these barriers.

Alternative fuel cycles

Alternative fuel cycles may provide a way to avoid some of the proliferation problems associated with use of plutonium fuels. A primary nonproliferation requirement for such cycles is that they minimize the presence of separated, or easily separable, weapons-usable material. There are a number of fuel cycles that satisfy this requirement, including the presently used cycle. Within the next two decades it would be possible to use once-through fuel cycles optimized for uranium and thorium utilization. These include light-water reactors modified for spectral-shift operation and thorium fueling, as well as modifications of the CANDU reactor. Beyond the next decade it may be possible to deploy once-through high-temperature gas-cooled reactors operating on moderately enriched uranium (20% or less) and thorium. Use of such cycles would help stretch fuel resources without appreciably increasing proliferation risks and buy time for possible development of proliferation-resis-

tant longer-term recycle fuel cycles, for developing uranium resources and for international institutional changes allowing safer use of advanced cycles.

If nuclear power is to contribute to energy production in the long term, it will be necessary to turn eventually to reprocessing and recycling for high conversion or breeding rates. The U^{233}–thorium cycle offers an alternative to the uranium–plutonium cycle, with both substantial resource extension and enhanced opportunity for international safeguards control.[6] National reactors would operate on denatured fuel, containing about 15% fissile uranium. The spent fuel would contain some plutonium and so reprocessing would be under international control. Plutonium production is substantially reduced and, if the plutonium is "burned" at an international fuel cycle center, the ratio of nationally to internationally generated nuclear power would be about ten in a light-water cycle.[2,6] An additional nonproliferation advantage is that the produced U^{233} is accompanied by U^{232}, which leads to the emission of energetic gamma rays in its decay and thus to greater difficulty in handling spent fuel. This same feature of course increases occupational hazards in reprocessing and recycling.

This concept of denaturing might be carried to the extreme of eliminating entirely transfers of fissile material once a reactor is in operation.[3] An example of such a reactor is the *molten-salt breeder*, a technology which received considerable support in earlier US programs before being eliminated in favor of plutonium breeders. If the molten-salt reactor can be engineered to have a breeding ratio of unity, then once it is loaded with fissile material, its annual fuel makeup would consist only of thorium. Fission products would be removed by a small on-line re-

processing plant which would not have the capability for removing fissile material. Fissile material, after its initial loading, would not be accessible without shutting down the reactor. While molten-salt reactors would have lower breeding gain, they make more efficient use of fissile resources during a period of growth, because they have a smaller initial fissile inventory (about 2500 kg versus 6000–7500 kg for a comparable liquid-metal breeder).[2] A plutonium breeder would have to operate for more than twenty years to breed enough fissile material to overcome this initial inventory disadvantage. *Gaseous-core reactors*, with similar advantages, have also been proposed; these alternatives to the liquid-metal fast breeder all require, and all deserve, further technical development.

The relative nonproliferation advantages of these alternative cycles require further study; it is not difficult to foresee possible technical and political problems. For example, the relatively large isotopic mass difference between U^{233} and U^{238}, used in some denatured fuel cycles, may make separation of weapons material possible with even crude centrifuges (though this would still involve a certain national dedication and time). However, the greatest difficulty in reducing proliferation hazards through the use of alternate fuel cycles may be in achieving the necessary international cooperation.

Except for once-through cycles, all advanced cycles appear to require some form of international organization. For example, the denatured U^{233}–thorium cycle would require international fuel centers (or continued dependence of non-weapons states on supplier countries already possessing weapons); the molten-salt cycle would require a source of initial fuel and perhaps international su-

pervision of initial fueling and subsequent reactor operation. However, it is possible that the institutional problems associated with alternative cycles have solutions that are more easily achieved than those required by plutonium fuels.

Choices and prospects

Basic technological choices in nuclear power have a strong bearing on prolifer-

ation risks. In the past, choices have been made on the basis of economic and technical factors, with a conviction that technical and political measures would be found that would allow new technologies to be accommodated safely. The recognition that proliferation concerns should enter explicitly into technology choice represents a new phase in the 30-year effort to establish distinctions between

peaceful and non-peaceful uses of the atom.

Bringing nonproliferation concerns into the process of technology choice is made difficult by the narrowing of technical options that has already occurred in response to uncertainties in uranium supply and projections of high nuclear growth rates. The result in most supplier countries has been a decision to pursue early commercialization of plutonium recycling and plutonium breeders. In developing and less-developed countries, the primary locus of many proliferation concerns, energy problems are severe, but the ability of advanced fuel cycles to alleviate these problems will be much lower or much delayed, resulting in less real need to commit to such cycles. Nevertheless, the expectation that plutonium breeders would soon provide significant energy benefits has led some of these countries to attach high value to acquisition of pilot reprocessing plants and plutonium stockpiles. Because of the long lead times associated with nuclear power development, proliferation problems can precede, by decades, the actual utilization of the technology.

There is as yet no worldwide agreement on whether it is possible or desirable to reshape the nature and pace of evolution of nuclear power technology.

In the US, research and development efforts are shifting away from early commercial demonstration of the plutonium breeder towards broadened consideration of fuel cycle options. The Carter Administration has deferred plutonium recovery from spent light-water reactor fuel, citing the negative effects a US commitment would have on decisions being made in other countries and on US efforts to restrain the pace of plutonium commitments, and pointing to threefold reductions in the nuclear growth projections that motivated early commitments to plutonium.

The long-term international impact of these program shifts is unclear. Several advanced countries have indicated their intentions to proceed with reprocessing and plutonium-breeder development. This has suggested to some that the US is pursuing an isolationist course, which will deny it the opportunity to help in devising technical and political solutions to the proliferation problems posed by what is seen as an inevitable use of plutonium fuels.

In contrast, the proponents of the Administration strategy argue that the real source of US leverage is through considerations other than commercial rivalry, such as fuel assurance or security arrangements. They also observe that proliferation risks arise in the great majority of countries without plutonium commitments, countries where US restraint may at least prevent accelerating commitments before plutonium fuel cycles have been shown necessary or con-

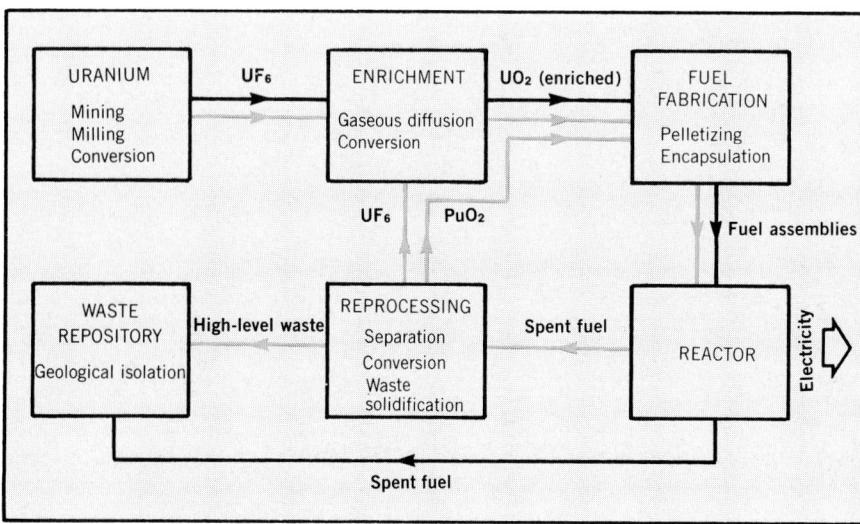

Table 2. Lifetime uranium commitments for several thermal-reactor options

Options	Uranium commitment (short tons)
Light-water reactors	
U, no recycling	6410
U, with U recycling	5280
U and Pu recycling	4340
U + Th, U recycling	3650
U + Th, spectral shift	<3000
Heavy-water reactors	
Natural U, no recycling	5263
U (1% U^{235}), no recycling*	3800
Natural U, Pu recycling	2861
Pu–Th, U recycling	2210
High-temperature gas-cooled reactors	
U^{235}–Th, U recycling	2970
Pu–Th, U and Pu recycling	4990

The data are for a 1000-MW(e) reactor operating at 80% capacity factor for 30 years. Enrichment is at 0.2% tails assay for those cycles utilizing enriched uranium. From APS report, reference 2. The capacity factors achieved so far have been considerably lower than 80%; for average capacity factors in the 60% to 75% range, the uranium commitments can be approximated by scaling down the results shown above.
* This result has been obtained by adjusting for different capacity factor the result of Y. I. Chang *et al* (Argonne) and assumes a fuel burnup of 16 MWD(th)/kg.

trollable by international institutions.

Those arguing for a broader domestic research and development program also point out that the US, with extensive uranium deposits and research capabilities, is in a unique position to develop nuclear power alternatives. If worldwide uranium supplies prove much greater than the conservative assumptions used in planning nuclear-power programs abroad, the US may have advanced other reactor concepts to the point where they can be made available at a net advantage in economics as well as nonproliferation. The possibility that new technologies may become available is also seen as possibly restraining near-term commitments to plutonium, especially in less developed countries.

How this discussion will be resolved, and what the consequences will be for nuclear power and for nonproliferation, are yet uncertain. In the longer term—beyond the year 2000—there are reasons for pessimism: Increasing technological sophistication and technology transfers, especially isotope-separation techniques, may undermine technological approaches to avoiding proliferation. The relative importance of political and institutional approaches will thereby be increased. In the nearer term, it is possible to be more optimistic about opportunities to deal with proliferation risks and perhaps avoid largely those associated with nuclear power. Over the next decade, nonproliferation goals would be served by deferrals of plutonium commitments, improved worldwide fuel assurance and examination of alternative fuel cycles from a nonproliferation perspective.

Goals for the eighties

It is vital to future nonproliferation efforts that the time available prior to widespread plutonium utilization be used well. We can list three primary technical goals for the next decade:

▶ a comprehensive worldwide assessment of uranium and thorium resources, using advanced exploratory techniques and supported by governments at levels commensurate with the long-term value of this information;

▶ immediate efforts to improve efficiency of resource utilization in once-through reactor fuel cycles, and

▶ advancement of various breeders (including the plutonium breeder) and advanced-converter reactors to the point where the merits of each, including nonproliferation advantages, can be assessed on a common ground by governments and industry.

It is too early now, and the basis for choices too narrow, to make decisions that would put major limitations on our future abilities for dealing with long-term energy supply and proliferation problems. For at least a few years, the costs of developing better information are not high compared to the potential benefits.

References

1. The Nuclear Energy Policy Study, *Nuclear Power: Issues and Choices*, Ballinger, Cambridge (1977).

2. APS Study Group on Nuclear Fuel Cycles and Waste Management, Rev. Mod. Phys. **50,** part 2, S1 (1978).

3. Office of Technology Assessment, *Nuclear Proliferation and Safeguards*, Praeger, New York (1977).

4. D. L. Hetland, in *Proceedings* of the Uranium Industry Seminar, Department of Energy report DOE–GJO–108, Washington, D.C. (1977).

5. M. Willrich, T. B. Taylor, *Nuclear Theft: Risks and Safeguards,* Ballinger, Cambridge (1974).

6. H. A. Feivson, T. B. Taylor, Bulletin of the Atomic Scientists, Dec. 1976, page 14.   □

Reactor fuel cycles

The source of energy in the nuclear fuel cycle is the neturon-induced fission of uranium or plutonium in a nuclear power reactor, each fission releasing about 200 MeV of energy and several neutrons. The useful fissile isotopes are U^{235}, U^{233} and Pu^{239}, although only U^{235} is available in Nature (approximately 0.7% of natural uranium is U^{235}, the rest being U^{238}). The other fissile isotopes can be bred by neutron capture on fertile isotopes:

$$U^{238} \xrightarrow{(n,\gamma)} U^{239} \xrightarrow[23.5 \text{ m}]{\beta^-} Np^{239} \xrightarrow[2.4 \text{ d}]{\beta^-} Pu^{239}$$

$$Th^{232} \xrightarrow{(n,\gamma)} Th^{233} \xrightarrow[22 \text{ m}]{\beta^-} Pa^{233} \xrightarrow[27 \text{ d}]{\beta^-} U^{233}$$

These reactions offer a considerable resource extension because the fertile isoptes U^{238} and Th^{232} are fairly common in Nature and because a sufficiently large number of neutrons are given off in fission to breed new fuel as well as to sustain the chain reaction. In a breeder reactor more fissile material is produced than is consumed.

The average fission-neutron energy is about 1 MeV, but the cross section for fission is orders of magnitude larger at lower energy. Reactors therefore currently operate on a thermal-neutron spectrum, meaning that the fission neutrons are moderated by collision with a light element. This is either hydrogen (in light-water reactors) or deuterium (in heavy-water reactors).

The number of neutrons emitted per neutron capture by thermally fissile isotopes varies from 2.1 to 2.3 with thermal neutrons, but is as high as 3.0 for Pu^{239} with fast (1-MeV) neutrons. The U^{233}–Th cycle offers the greatest potential conversion ratio for thermal reactors, while U^{235}–Pu cycle in a fast-neutron spectrum offers the greatest potential resource efficiency. This is one reason why most research and development has emphasized development of the plutonium-fueled fast-breeder reactor. However, that reactors can be designed to operate on virtually any combination of fissile, fertile and moderating materials.

Although the power reactor is clearly at the heart of the fuel cycle, a considerable number of supporting facilities are needed. A flow sheet for the light-water-reactor fuel cycle, with and without plutonium recycling, is shown in the figure opposite; the other fuel cycles are not qualitatively different.

The uranium ore is mined, milled, converted to gaseous UF_6 and fed to an isotopic enrichment facility. Here the isotopic fraction of thermally fissile U^{235} is raised from 0.7% to around 3% by gaseous diffusion through porous barriers. This technology is very capital- and energy-intensive and future enrichment capacity will likely rely on the gas-centrifuge process. The enriched gas is converted to solid UO_2, which is fed to the fuel-fabrication plant.

When spent fuel is discharged from a reactor, it is intensely radioactive because of fission products. A light-water reactor with capacity 1000 MW (electric) discharges about thirty tons of spent fuel per year, containing about 250 kg of plutonium (about 70% fissile) and a comparable amount of U^{235}. With or without recycling of this fissile material, it is envisioned that these fission products will be sent to a Federal nuclear waste repository for long-term geological isolation from the biosphere. A pilot-plant repository is scheduled for operation in 1985. The spent fuel is now being stored in cooling ponds.

With plutonium recycling, the spent fuel would be sent to a reprocessing facility. There, the plutonium and uranium would be separated chemically from the fission products and, if desired, from each other. The plutonium would then be converted into solid PuO_2 and sent to a mixed-oxide-fuel fabrication plant for combination with enriched uranium and incorporation into fuel assemblies. The uranium would be converted into UF_6 and recycled. The stream of high-level waste would contain the radioactive fission products (plus residual amounts of plutonium and other actinides). After cooling, these would be incorporated into a solid matrix (for example, of borosilicate glass) and transported to the waste repository. The recycled plutonium and uranium would improve resource utilization by reducing uranium requirements.

We stress that this fuel cycle is just one of many options, albeit the one that has received the most research and development. The flow sheet for any of these cycles is generically the same as the above, except that some employ thorium as a fertile material and that some do not require uranium enrichment.

Different thermal reactor options may lead to different breeder choices; for example, a U^{233}–Th cycle might supply initial fueling for molten-salt breeders (thermal breeders operating on the thorium cycle), in the way the U–Pu light-water cycle was expected to fuel liquid-metal plutonium breeders.

DOE and universities oppose reducing U²³⁵ in campus reactors

PHYSICS TODAY/DECEMBER 1984

Three months before the Olympics began in Los Angeles last summer, UCLA announced that its 100-kW Argonaut research reactor would be shut down during the games to avoid any acts of vandalism or terrorism directed against it. The reason for UCLA's concern: The reactor core contains 3.56 kg of uranium enriched to 93% of uranium-235, which is the enrichment level for weapons. In August, UCLA officials decided to close down the reactor permanently, not because it was a target for terrorists but because of worries about meeting the requirements of the Nuclear Regulatory Commission.

The decision was based on the implications of a proposed rule change the NRC published in the Federal Register on 6 July, requiring those research reactors at university campuses and corporate centers that use highly enriched fuel to convert to low-enriched fuel containing less than 20% uranium-235. High costs and long delays involved in converting the fuel and relicensing the reactor are too much to handle, UCLA administrators figured, especially with the nuclear science department already having trouble recruiting graduate students. On campus since 1960, the reactor was shut down once before for changes. What's more, UCLA has experienced four years of hearings and paperwork for a license renewal, over objections from anti-nuclear groups and environmentalists, at a cost to the university of $500 000 in legal fees.

Costly conversion. So far, none of the other research, testing and training reactors with highly enriched uranium have followed UCLA's action. Throughout the US, high-enriched uranium is used to fuel 69 nonpower research reactors, with the Department of Energy operating 36, which account for about 90% of weapons-grade fuel. DOE estimated in 1978, during Congressional hearings on the Nuclear Nonproliferation Act, that it would take two years and some $10 million to convert all research reactors from high-enriched to low-enriched uranium. Since then, DOE has spent about $24 million on its Reduced Enrichment Research and Test Reactor Program and reckons another $24 mil-

UCLA reactor for research and training is shut down rather than converting it to use low-enriched uranium fuel.

lion more will be needed to develop new fuel elements with low-enriched uranium in aluminide, oxide and silicide plates or zirconium-hydride rods. The new elements are under development in France, West Germany and the US, mainly at Argonne National Laboratory. The 2000-kW Ford Reactor at the University of Michigan has already been converted to a uranium-aluminide fuel with 20% enrichment as part of a DOE-funded demonstration of what can be done for campus reactors. Still, the cost of converting all the high-enriched plate-type university reactors and four of 25 rod-type university reactors with enrichments of 70% is variously estimated by the National Association of State Universities and Land Grant Colleges as totaling $8 million to $35 million.

At hearings before two House energy subcommittees on 25 September, A. Francis DiMeglio, director of the University of Rhode Island reactor and chairman of a national organization of research reactors, said the estimated cost of conversion did not include expenditures for relicensing; but, even so, "while something like $76 million may not seem a large amount of money to members of Congress in terms of nonpower reactors, it is sufficient to keep dozens of them running many years."

NRC and State Department officials claim the new rule is necessary for two reasons: to reduce the risk that weapons-grade uranium could be stolen from storage or in shipment at university facilities and turned into a nuclear explosive by some foreign government or terrorist group, and to set an example to foreign countries to convert their small nonpower reactors to low-enriched uranium fuel. The end result in both cases would be to reduce the worldwide threat of nuclear weapons proliferation.

Currently there are 137 research and test reactors using high-enriched uranium in 34 countries in the Western bloc. In the US, high-enriched U²³⁵ is used to fuel 36 DOE research reactors, three at other government agencies, 25 at universities and five at commercial firms. Only the NRC-licensed reactors at universities, companies and the National Bureau of Standards would be affected by the new NRC rule because reactors located at national laboratories and government facilities are not licensed by the agency and presumably have stricter security precautions. The NRC Advisory Committee on Reactor Safety has suggested that the conversion to low-enriched uranium be made "in a gradual and orderly manner as funds and fuels become available" and that operators have the option of converting or retaining high-enriched fuel and meeting security requirements. While two university reactors, at MIT and the University of Missouri, would not be exempt from the rule, they may be granted a delay in implementation because the design of their fuel cores and heat-transfer systems does not permit conversion with fuel elements now available.

The US is responsible in part for the situation. Under the Atomic Energy Act of 1954 the government encouraged universities to build research reactors to train scientists, engineers and technicians. In the early 1960s the US exported low-enriched uranium for research reactors abroad as part of the "Atoms for Peace" program. As greater power levels and neutron fluxes were needed to perform new experiments, the quantity of U^{235} was stepped up, either by using fuels of high uranium density or by increasing uranium enrichment. The Nuclear Proliferation Act of 1978 was intended to reduce high-enrichment inventories both at home and abroad as a principal way of decreasing the risk of weapons-grade uranium from falling in the hands of agitators and adversaries.

At the 25 September hearings before the House Committee on Science and Technology, arguments were presented against the NRC rule. "I believe the evidence is compelling," said Edwin L. Zebroski of the Electric Power Research Institute, "that a number of research reactors are likely to cease operation and cease to be available for training, research, testing or the production of radioisotopes for research and medicine" if the rule is imposed. "Our society—and with it our regulatory, legislative and judicial establishment—is in considerable disarray on how to define or specify 'how safe is safe enough'... The proposed rule on conversion is an apparent attempt to reach an objective the NRC uses in the phrase 'preferably zero risk.'"

No known threat. Safety and security measures, including increased physical barriers to access, alarms, guards and television monitoring systems, have been upgraded in the past few years, said Zebroski. Indeed, NRC staff testified that there has never been an attempted theft of fuel at a research reactor facility in the nearly three decades that they have been running and that no known threat exists.

For DOE, James S. Kane, deputy director of energy research, observed that if as much unirradiated fuel as possible were to be stored at secure DOE facilities, "theft or diversion ceases to be an issue. We have never considered the diversion of irradiated fuel to be an issue. The difficulties in surreptitiously handling a highly radioactive element, plus the technical and financial resources needed to separate the uranium in sufficient amount for a nuclear device, make such threats, if any, extremely remote. Our main disagreement with the proposed rules centers on the treatment of reactors that have essentially a lifetime supply of fuel. It is worth noting that the US government is not seeking the conversion to low-enriched uranium of foreign reactors with lifetime cores. With the exception of the four 1-mW TRIGA reactors at Texas A&M, Washington State, Oregon State and the University of Wisconsin, which use a different type of fuel with 70% enrichment, our research reactors are low powered with in-core inventories of less than 5 kg of U^{235}. The total amount of high-enriched uranium currently in storage at all of these reactors is less than 15 kg." So, to gather enough U^{235} for a workable weapon would require the theft of all the fuel in storage either simultaneously or in rapid succession. "That is extremely improbable," said Kane.

'Negligible benefit.' Kane went on to observe that DOE holds university reactors to be "very important" in training nuclear physicists and chemists, as well as reactor engineers and technicians. "If conversion of all US university reactors is mandated," he said, "some are almost certain to be shut down. For negligible benefit, we will have lost a valuable nuclear training and research capability."

Of those reactors that comply with the low-enrichment order, argued Robert S. Carter, chief of the Reactor Radiation Division at the National Bureau of Standards, their beam intensities would be decreased by 10%–15% and undesirable radiation would increase by a similar amount. Reducing the enrichment from 93% to 20% implies, whatever the fuel type, adding 10 kg or more of U^{238} to the reactor core. While studies of core physics show this is feasible, there is marginal loss of fast neutrons and substantial loss of thermal neutron flux, according to Carter. Accordingly, performance of some campus and commerical reactors would decrease.

For his part, DiMeglio told members of Congress that the expense of converting university reactors, presumably to be funded by DOE, "will have no discernible education benefit." DOE now provides about $200 000 per year for the use of university reactors. The $8 million to $15 million to implement the new rule could be spent more productively to upgrade obsolescent university laboratories, he observed, "which would have a major effect on improving the quality of science and engineering in this country."   —IG

New reports warn of mounting nuclear proliferation threat

PHYSICS TODAY/JANUARY 1985

Of all the issues that divide the US and Soviet Union, nuclear nonproliferation is not among them. It is in the mutual interests of both superpowers to limit the spread of nuclear arms. "We and the Soviets are of a common mind on that," says Richard T. Kennedy, a US ambassador-at-large who advises Secretary George P. Shultz on nonproliferation policy. On 28–30 November, Kennedy met in Moscow with Andronik M. Petrosyants, chairman of the USSR atomic energy committee, in the fourth round of a series of wide-ranging talks that began in Washington in 1982 to strengthen the Nuclear Nonproliferation Treaty.

The importance of the meeting was heightened for two reasons: It enabled the Reagan administration to disclose what took place at a secret meeting last summer in Luxembourg with representatives of 11 industrial nations that are actual or potential exporters of nuclear material and technology. The US had tried to convince the supplier countries to refrain from major new commitments or to require full safeguards for any nuclear deals. By several accounts, the US proposal was defeated. The results of the meeting take on increased significance with the release of two reports in November cautioning that international safeguards and political persuasion are unlikely to prevent a few nations from gatecrashing the nuclear "club" by the end of the decade—thereby joining the US, Soviet Union, Britain, France and China as full-fledged nuclear weapons states. The US–USSR meeting in Moscow, furthermore, is a prelude to the Nonproliferation Treaty Review Conference of 125 nations that signed the treaty, taking place in Geneva next August to review the progress and problems of nuclear arms control.

The second reason is related in part to the first. The nonproliferation treaty requires the US, Soviet Union, and other club members to take steps to end the nuclear arms race "at an early date." The Kennedy–Petrosyants exchange, in the event, became a curtain-raiser for the resumption of arms-control talks between Shultz and Foreign Minister Andrei A. Gromyko in Geneva on 7–8 January.

Nonproliferation ethic. In a speech to the United Nations Association of the US on 1 November, Shultz characterized the nations represented at the Luxembourg meeting as the world's main nuclear suppliers—besides the US, of course, Australia, Belgium, Britain, Canada, France, Italy, Japan, the Netherlands, Sweden, Switzerland, and West Germany. In addition to these countries, says one of the new reports, *Nuclear Proliferation Today*, by Leonard S. Spector, senior associate at the Carnegie Endowment for International Peace, the suppliers include Argentina, China, and South Africa. "As we pursue our dialog with emerging suppliers," Shultz said in his speech, "we will work to assure that they, too, come to understand and adopt the nonproliferation ethic that traditional suppliers have developed over the past quarter century."

When the US held a monopoly on nuclear technology in the early years of the atomic era, such authorities on the subject as Dean Acheson, David Lilienthal and Bernard Baruch considered the proliferation of nuclear power plants and supporting fuel-cycle facilities to be unthinkable. They advocated international ownership and oversight of nuclear facilities with weapons potential. The official US position changed as nuclear power turned commercial in the 1950s. In 1953 President Eisenhower launched the Atoms for Peace program. Forseeing the risks involved, Eisenhower proposed creating the International Atomic Energy Agency to provide a system of safeguards against diverting nuclear material to bombs while, at the same time, to promote peaceful uses of nuclear energy. Organized in 1957, the agency now inspects nuclear installations in 50 nations and accounts for the amount of enriched uranium and plutonium to make sure that the material is not diverted to make nuclear weapons.

The Nth-country problem that worried Western physicists and other scientists in the 1960s became a reality in 1964 when China exploded its first nuclear bomb. Soon afterward, largely through US diplomacy, efforts began to negotiate an international nonproliferation treaty—a prodigious matter as it turned out. In 1968 the Nuclear Nonproliferation Treaty was signed; it was placed in force in 1970. In adhering to it, non-nuclear nations agree not to acquire "nukes" in their arsenals and to accept IAEA inspections of their nuclear facilities.

'Greek tragedy.' Even so, there were few illusions about keeping the nuclear club exclusive. At the time President Kennedy signed the Limited Test Ban Treaty with the Soviet Union in 1963, defense policy experts such as Albert Wohlstetter of the University of Chicago agreed that by the mid 1980s as many as 15 or 20 nations would possess nuclear weapons. It was accepted wisdom, recalled Shultz in his address to the UN Association, "that the spread of nuclear weapons was inexorable, advancing like a Greek tragedy to some preordained disaster."

Accordingly, once China entered the club, it seemed predestined that India would follow. India consistently opposed IAEA as an invention of the superpowers to prevent developing

Will the nuclear club be enlarged?

Here is a list of eight countries that "took important steps" toward nuclear-arms capability in the past year, according to a report by Leonard S. Spector of the Carnegie Endowment for International Peace. Details are based on the report.

▶ Argentina, with the most advanced nuclear program in Latin America, disclosed after five years of secrecy it had a gaseous-diffusion enrichment plant capable of producing weapons-grade U^{235} soon. The inauguration in December 1983 of Raul Alfonsin, marked the end of the military junta and raised hopes of change in the nation's nuclear program. Still, Argentina refuses to sign the Nonproliferation Treaty or to ratify the Treaty of Tlateloco, which would make all Latin America a nuclear-free zone.

▶ Brazil, though considerably behind Argentina in nuclear programs, possesses the scientific and industrial underpinnings necessary to design and fabricate nuclear weapons. Currently, Brazil lacks a source of plutonium or highly enriched uranium, but during the early 1980s it began to accelerate a series of nuclear programs without safeguards and outside the ambit of its projects with West Germany. Among these is a plant at the Instituto des Pesquisas Energeticas e Nucleares (IPEN) at the University of Sao Paulo to develop a centrifuge enrichment capability for weapons-grade material by 1990.

▶ India exploded a nuclear "device" in 1974 with Pu^{239} produced at unsafeguarded facilities at the CIRUS reactor and a reprocessing plant at the Bhabha Atomic Research Center (formerly called Trombay). A new reprocessing plant at Tarapur has a maximum output of 135 to 150 kg of plutonium per year, three times larger than the Trombay plant. It can process uranium oxide fuel from all of India's nuclear power plants, which now consist of Tarapur I and II, Rajasthan I and II and the unsafeguarded Madras I. India is not believed to have deliverable nuclear arms, but has the talent and is expanding its facilities.

▶ Iraq may have aimed to develop nuclear weapons by the mid-1990s, but its program has been essentially dormant since Israel destroyed the French-supplied Osirak reactor in June 1981. France has agreed in principle to rebuild the reactor under stringent specifications calling for low-enriched fuel and French technicians to operate it as a regional research center. Still, no significant reconstruction work has started. Meanwhile, since the Israeli raid, Iraq has adopted a clandestine strategy to obtain weapons-grade plutonium through black-market sources in Europe

▶ Israel has the technology, materials and talent to produce nuclear weapons quickly—though US intelligence officials have claimed for the past decade that it probably already has an arsenal of 10 to 20 Hiroshima-sized bombs. A recent report by Georgetown University Center for Strategic and International Studies concludes that Israel will produce enough nuclear material in its Dimona reactor to make 100 warheads by the end of the century. Other possible sources of bomb-grade material, according to the CIA, are France, credited with providing 14 kg of plutonium in 1967—enough for one or two weapons—and a clandestine cache of about 100 kg of highly enriched uranium supplied by the US Atomic Energy Commission to the Nuclear Materials and Equipment Corporation in Apollo, Pennsylvania, for processing and subsequently "lost." Despite investigations, the uranium has not been located.

▶ Libya has attempted to buy nuclear technology from China, Pakistan, Argentina, the Soviet Union, Belgium and elsewhere, according to reports. In 1974, Libya sought assistance for a nuclear program from General Atomics Corporation, but the State Department and Congress opposed this. In 1977, Libya signed an accord with the Soviet Union for a 440–MW nuclear power reactor, but construction is hung up in protracted negotiations between the countries. Libya is considered to be decades away from having the technology to build a bomb.

▶ Pakistan may be able to produce weapons-grade uranium at its enrichment facility near Kahuta, "thereby surmounting the final obstacle on its 12-year quest for nuclear arms," writes Spector. Its plan to build a 900-MW nuclear power plant at Chashma has met a total boycott by all nuclear supplier nations—an important milestone in US efforts to establish a common front in curbing the spread of nuclear arms. In the recent past China has reportedly provided Pakistan with material and information for nuclear weapons.

▶ South Africa has the capability to produce highly enriched uranium at its secret enrichment plant at Valindaba, which it claims uses a process that is "an invention of our own." The plant can yield enough for two or three nuclear weapons per year. Therefore, South Africa may have an many as 15 to 25 Hiroshima-sized bombs. If reports that it imported substantial quantities of low-enriched uranium from China in 1981 are true, South Africa could rapidly enrich this material to weapons grade, substantially increasing its stockpile of nuclear explosive material. Under a 1957 agreement, the US has trained South Africans in nuclear science and engineering and supplied two small research reactors in the 1960s. In August 1977 the Soviet Union notified the US that one of its surveillance satellites had spotted what appeared to be a nuclear test site in South Africa's Kalahari Desert. After US experts confirmed the finding, the two superpowers, in a rare display of unity, warned Pretoria against exploding a nuclear device. In 1979 a US satellite detected a distinctive two-pulse flash over the South Atlantic near a South African naval exercise. South Africa denied it had detonated a nuclear device, and no corroborating evidence of fallout has been obtained. Rumors persist that South Africa and Israel cooperate in nuclear weapons development. —IG

countries from obtaining nuclear energy. Homi Bhabha, the first head of India's Atomic Energy Commision, spoke of the "inalienable rights of states to produce and hold the fissionable material required for their peaceful power program." On 18 May 1974, India detonated a nuclear device of about 15 kilotons (in the range of the Hiroshima bomb) at the Pokharan Range in the western Rajasthan Desert. India characterized its test as a peaceful explosion intended for studying the cratering and cracking effects on rocks—a purpose that it claimed would not violate its assurances to Canada and the US that plutonium-239 produced at its 40–MW CIRUS reactor would not be used for weapons.

London Group. India's test came shortly after the US Central Intelligence Agency circulated reports that Israel made a few nuclear weapons from Pu^{239} separated in its reprocessing facility from spent fuel in a 24-MW reactor at Dimona in the northern Negev Desert. With reactor hardware, nuclear fuel and heavy water supplied by many countries, including the US, Canada, West Germany and France, both India and Israel, possessing ample talent and technology of their own, seemed at the threshold of the nuclear club by 1974—and beyond control by the IAEA, because neither had signed the Nonproliferation Treaty. To make matters worse, fuel-cycle facilities for producing fissile uranium-233 had been supplied to Brazil by West Germany and to Pakistan and South Korea by France. Alarmed by the situation in 1974, the US and 14 other nations, including Canada, Japan, Western European countries and the Soviet Union, Czechoslovakia, East Germany and Poland, convened in London to tighten the policies and procedures for exporting nuclear supplies, components and technology. By 1976, the London Suppliers Group, as the nations came to be called, agreed on a "trigger list" of technologies that could be transferred to other countries only if the customers agreed to IAEA controls over imports.

Another group, known as the Zangger Committee, after its chairman, Claude M. Zangger, a Swiss physicist and government official, consists of 21 nations that have established similar lists of nuclear-weapons technologies. At the urging of the US State Department, the Zangger Committee recently enlarged its lists to include components of ultra-high-speed gas centrifuges.

Despite the international apparatus to curtail the traffic in nuclear material and knowhow, the number of nations considered technologically capable and politically motivated continues to increase. Spector names eight emerging nuclear-weapons countries—India, Pakistan, Israel, Libya, Iraq, Argentina, Brazil and South Africa (see box). A report sent to members of Congress on 24 September by Warren H. Donnelly of the Congressional Research Service lists 23 nonnuclear-weapons states with the potential, based on the capacity of their nuclear industry, to produce plutonium or highly enriched uranium. Donnelly also examines political pressures to acquire nuclear weapons and countervailing restraints that could reduce incentives to join the club.

Donnelly's report states that a reasonably industralized state with sufficient determination and financial resources, and without interference by other nations. could produce some nuclear weapons within perhaps three to five years. Among the countries with the greatest nuclear capability—meaning the installed nuclear facilities and scientific and technical talent—Donnelly lists Israel, India, Italy, Argentina, West Germany, Japan and South Africa. Those countries with the greatest pressure to joint the club, according to Donnelly, are Israel, Pakistan, Cuba, India, Iran, Iraq and South Africa. "Just as many non-nuclear weapons states may feel pressure pushing them toward nuclear weapons," writes Donnelly, "so too these states may be subject to ... constraints." Thus, a nation that believes itself threatened by a neighbor's developing nuclear capacity may be tempted to follow Israel's example when it destroyed Iraq's Osirak reactor in an air raid on 7 June 1981. Iran had failed to demolish Osirak on 30 September 1980, eight days after the Iran–Iraq War started. Without technicians and technology of its own, Iraq relied upon France for the construction and operation of Osirak. It bought a fabrication facility from Italy and uranium ore or "yellowcake" from Brazil and Portugal.

The countries most likely to join the club next, Donnelly concludes, are Argentina, Israel, India, Pakistan and South Africa, in that order. Though Japan, West Germany, Belgium, Italy and the Netherlands possess nuclear capability, writes Donnelly, these countries are constrained by culture and politics from developing nuclear arms.

Scientific imperative. In his recent book, *Weapons and Hope*, Freeman Dyson argues that in the case of each member nation of the club, scientists rather than generals took the initiative in getting nuclear-weapons programs started. "In each case of which we have knowledge, scientists were motivated to build weapons by feelings of professional pride as well as of patriotic duty. The construction of a bomb was a technical challenge which aroused their fiercest competitive instincts. . . . The nuclear arms race from 1940 to 1960 was powerfully reinforced by the professional ambitions of scientists who saw nuclear weapons technology as a grand arena for the exercise of their talents. The walls of official secrecy which surrounded these exercises made professional rivalries more intense and gave a false glamour to the new technology," Dyson writes.

"But now no more," claims Dyson. "Since 1964, anybody who wanted to know the general principles of fission weapon design could find them explained by Robert Serber in the declassified *Los Alamos Primer*. Since 1979, anybody who wanted to know how hydrogen bombs work could refer to Howard Morland's article in the November 1979 issue of *The Progressive*. Nuclear weapons design has been stripped of its mysteries, and there is no longer any scientific glory attached to it. . . . From now on there will be no more first-rate scientists driving the nuclear arms race with their rivalries. Even in scientifically backward countries, young people of talent now know that nuclear weapons have ceased to be a scientific challenge."

Political factors. Dyson's idea of scientific imperative appears to gain support from an observation by J. Robert Oppenheimer on the US decision to develop the hydrogen bomb. Oppenheimer was quoted as saying the design was "technically so sweet" that development of the bomb was inevitable. By contrast, a recent analysis of the factors determining whether countries acquire or forgo nuclear weapons, presented in *The Dynamics of Nuclear Proliferation* by Stephen M. Meyer of MIT, contradicts the primacy of the scientific or technological imperatives. Meyer argues that political and military factors dominate—that the will is more important than the way in the decision to join the club.

Secretary of State Shultz appears to agree with Meyer's theory. In seeking to prevent new entrants to the club, he told the UN Association last November, "we have employed a range of political, economic and security measures." These, along with IAEA safeguards and efforts by the London Suppliers Group, have not decreased the risk of proliferation, however. According to Spector, the Nuclear Nonproliferation Act signed by President Carter in 1978 has been violated by the US itself. While the act is intended to make suppliers more sensitive to curtailing nuclear exports, it did not prevent two American brokers, Edlow International and swuco Inc, from arranging the sale of uranium fuel processed in Belgium and France to South Africa through a consortium of Swiss utilities. During 1982 hearings on exports to South Africa before a committee of the House of Representatives, it was revealed that the State Department had been advised of the deal before it was completed but had taken no steps to discourage Edlow and swuco from proceeding.

According to Shultz, the Reagan administration's approach is to consult and cooperate with other nations to thwart the destabilizing effects of proliferation. In keeping with this policy, the US has been discussing with Japan and the European Atomic Energy Community (EURATOM) long-term arrangements on reprocessing nuclear fuels and using plutonium. Although President Reagan initialed an agreement on nuclear energy with China last April, formal cooperation has foundered, because China has yet to provide assurances it will not help other nations build nuclear arms. Washington is concerned about China's nuclear exports to Pakistan and South Africa.

Such actions and the black-market traffic in nuclear material and technology, warn State Department officials and Ambassador Kennedy, cause the specter of more nuclear weapons states to haunt the world. "Pakistan and India may be poised on the brink of a major arms race," writes Spector in his report for the Carnegie Endowment. Moreover, Argentina, with the most advanced nuclear program in Latin America, could produce a deliverable weapon in "several years"—a situation that causes Brazil to accelerate its program.   —IG

Breeders and reprocessing are challenged in Germany, UK

PHYSICS TODAY/OCTOBER 1986

On the weekend of 30–31 July 1977 anti-nuclear activists from all over Europe gathered at Creys–Malville, a small town in the upper Rhône valley where ground was being broken for the Superphénix fast breeder reactor. In a violent confrontation with French police, several demonstrators reportedly lost limbs and one, a German, died. In an attempt to brand the demonstration as the work of outside agitators and play on anti-German sentiment, a local politician declared it was the second time in a generation that the town had been occupied by Germans.

The demonstration was in fact more characteristic of confrontations that had been taking place over nuclear energy in Germany than it was of the situation in France. Construction of the Superphénix proceeded smoothly in the following years, and the reactor currently is approaching full power, raising questions about how the high-grade plutonium produced in its blanket will be used (see PHYSICS TODAY, September, page 53). Even so, the Superphénix probably will stand isolated for some decades, a feat of engineering with some prospect of being the technological basis of our electricity supply in the middle of the next century, but of dubious current value.

When construction began on the Superphénix ten years ago, its builders thought that it would be the prototype for a series of breeders that would be deployed throughout the European community by the end of the century. But the case for using plutonium has lost force, partly because world uranium prices are currently very low and are expected to remain low for many years, and partly because the technologies that employ plutonium are proving to be more expensive than their promoters had hoped they would be.

The situation is a disappointment for those who staked their careers on the expectation that breeder reactors would be the most important source of our electricity supply by the end of the century. But it is a blessing for those who have the job of devising methods to assure that the world would have timely warning if plutonium were sto-

AP–WIDE WORLD

Demonstrators at the Wackersdorf reprocessing plant in Bavaria, West Germany, march around the fence at the construction site on 31 March. Demonstrations at Wackersdorf were an almost weekly occurrence in the spring and summer and sometimes turned violent, resulting in police action and arrests. Individuals coming to protest from nearby Austria have been denied entry at the border.

len or "diverted" for use in nuclear weapons. As Director General Hans Blix of the International Atomic Energy Agency put it last year, "Fortunately there is still time to tackle these problems before the safeguards system has to meet them in practice."

A second full-scale breeder reactor is in theory to be built in Germany, but the French apparently are having second thoughts about that understanding and to the annoyance of the German government would prefer to put it in France. It is a strange argument considering the circumstances: The French are getting sticky about where to build a reactor they admit could not be built to any sound economic purpose for at least 20 years, while the German government is insistent on having this hypothetical reactor built in Germany, even though this very same government currently is unable to commission a 300-MW breeder that an international group has just finished building

in Kalkar, a small town near the Dutch border.

The politics of the SNR-300 capture much of West Germany's whole political scene. Kalkar is located in the state of Nordrhein–Westfalen, the Federal Republic's largest and most highly industrial region, dominated by aging coal and steel industries and labor unions, which have never been fond of nuclear energy. The premier of Nordrhein–Westfalen, Johannes Rau of the Social Democratic Party, happens to be the SPD's candidate to run against Chancellor Helmut Kohl in the national election next January.

Since the Soviet reactor accident at Chernobyl last April, intra-party discussions in the SPD have been largely taken up with the question of whether it would be possible and advisable to terminate reliance on nuclear energy altogether. A special committee headed by Volker Hauff, a former minister of research and technology, was established by the Socialists to evaluate the issue, and in late summer the party adopted the position that nuclear power plants should be phased out within ten years.

The SPD is firmly against breeder reactors and highly critical of reprocessing and recycling spent nuclear fuels. Rau personally opposes commissioning the SNR-300. The conventional wisdom is that the SNR-300 certainly will not run before next year and may well never run.

The rationale for reprocessing is increasingly controversial in West Germany now that the future of the breeder program is in question. Until recently, it was almost universally agreed among the European nuclear authorities that reprocessing was an essential element of spent-fuel disposal, and the European publics were largely sold on that position. By the late 1970s, however, experts on nuclear-waste disposal were increasingly agreed that reprocessing of spent fuel elements from light-water reactors is not essential to their disposal and might indeed complicate the disposal problem more than it would help. The only really compelling reason to reprocess, then, would be to prepare the ground for breeders, which of course do not serve their intended purpose unless the plutonium bred in their blankets is extracted. But if breeders are not built, it becomes debatable whether reprocessing should be done at all.

The German government's original plan in the mid-1970s was to build an integrated waste-management complex in Gorleben, a site near the East German border in the state of Niedersachsen. The complex was to include both a repository for the permanent disposal of high-level waste in a salt dome and a reprocessing plant with the capacity to handle 1500 tons of spent fuel per year. Because of very hefty opposition from environmentalists, plans for the national reprocessing plant (but not the waste repository) had to be scrapped, and instead a smaller reprocessing facility with an annual capacity of 350 metric tons is being built at Wackersdorf, a site in Bavaria not far from the Czechoslovakian and Austrian borders. Since Chernobyl, this site too has become the scene of frequent demonstrations.

If, contrary to most current expectations, the Social Democratic Party were to win the election in January 1987 and take power in coalition with the Greens or with the support of the Greens, there is little doubt that breeders and reprocessing would be rejected in their entirety. Regardless of the results next January, the political situation will be dramatically different from the situation that prevailed in the Federal

The reprocessing complex at Barnwell, South Carolina, was built and ready to operate when President Jimmy Carter announced in April 1977 that he would not permit it to open, saying he wanted to set an example to the world. Two previous commercial reprocessing ventures in the United States, one in upstate New York and one in Illinois, were economic and technical failures.

Republic ten years previously, when President Jimmy Carter took office in the United States determined to dissuade US allies from reprocessing and recycling nuclear fuels. At that time, the Greens did not exist, plutonium was not a household word and the established German parties were united in regarding Carter's initiatives as an obnoxious intrusion into their domestic affairs.

Carter's plutonium policies first emerged in a campaign speech delivered to the United Nations Association in New York in spring 1976. Carter said he would seek "a voluntary moratorium on the national purchase or sale of enrichment or reprocessing plants" and he stated that "surely this whole matter of plutonium recycling should be reexamined on an international basis."

In an otherwise somewhat lackluster campaign, President Gerald Ford felt sufficiently threatened by Carter's position on plutonium and by pressure from Capitol Hill, where restrictive nuclear export legislation was being prepared, to announce, very shortly before the election, that he would defer plans to open a newly built commercial reprocessing plant in Barnwell, South Carolina.

In April 1977, barely two months after taking office, Carter announced that he would cancel all plans for commercial reprocessing and block construction of a proposed demonstration breeder reactor at Clinch River in Tennessee. It was his intention, he said, to set an example that he hoped

other countries would follow.

The following year supporters of tough anti-proliferation measures in Congress enacted the Nuclear Nonproliferation Act, a complicated bill that required, among other things, US approval for reprocessing of spent nuclear fuels that originated in the United States. The bill was widely denounced by conservatives in the United States as interventionist and unilateralist, a complaint more commonly made by foreign-policy liberals about conservative policies.

As a candidate in 1980, Ronald Reagan voiced doubts about whether nuclear nonproliferation was really "any of our business" in the United States, implying that Carter's plutonium policies would be modified or even reversed when he took office. The surprising thing, in hindsight, is the extent to which the Reagan Administration has ended up retaining and even building on Carter's policies, notes Spurgeon Keeny. Keeny, president of the Arms Control Association in Washington, was the chairman of an influential Ford Foundation group that came out against breeders, reprocessing and recycling in early 1977.

The Reagan Administration made it known at the outset that it was going to stop making a fuss over every single request it got from a friendly or allied country for a transfer of nuclear material for reprocesssing. Instead it would negotiate blanket long-term agreements with governments that presided over sufficiently large economies and sufficiently sophisticated nuclear programs to warrant such transfers, provided the country's nonproliferation credentials were in good order. But the Reagan Administration did not seek repeal of the Nuclear Nonproliferation Act, and it did not put its full weight into an effort to revive commercial reprocessing and the Clinch River breeder project. An Administration plan to extract plutonium from spent commercial reactor fuel for use in the US weapons program was blocked by an amendment introduced by Senator Gary Hart.

In the first important instance involving a transfer of nuclear materials—a shipment of plutonium from a reprocessing plant in France to Japan—the Reagan Administration ended up taking an extremely tough stand on security measures for the shipment: When the shipment finally left France last year after nearly two years of negotiations, it went under military escort, with elaborate procedures for monitoring its position by satellite at every stage.

The other major decision on a plutonium transfer by the Reagan Administration involved a shipment last year from France to Switzerland of plutonium from Swiss spent fuel reprocessed in France. The Administration stalled on the request, apparently because of annoyance about assistance a Swiss company gave to Argentina in building a secret uranium-enrichment facility and to South Africa in building a heavy-water production plant

Last year Switzerland and France started to recycle plutonium from spent reactor fuels, and Germany and Japan have recycled on a modest scale for several years. Sweden and the United States, on the other hand, have decided against recycling.

Commercial reprocessing, ever since the United States got out of the business, has been virtually the exclusive preserve of United Reprocessors, an international partnership in which the major facilities are the French plant at Cap la Hague operated by COGEMA (Compagnie Générale des Matières Nucléaires) and the British Nuclear Fuels Ltd plant at Sellafield (previously called Windscale), UK. In the late 1970s and early 1980s, BNFL and COGEMA negotiated reprocessing contracts with foreign utilities in Japan, West Germany, Belgium, the Nether-lands, Italy, Sweden and Switzerland. For countries such as Germany and Sweden, where legislation had been enacted in the mid-1970s saying that no further reactors could be commissioned until there were concrete steps to solve the problem of nuclear-waste disposal, the United Reprocessors contracts provided apparent evidence that such steps indeed were being taken.

To handle the extra business, CO-GEMA started to build a third reprocessing plant at Cap la Hague, UP3, designed to handle 800 metric tons of spent fuel a year, and Britain started to build THORP, a thermal oxide reprocessing plant designed to handle 1200 metric tons of fuel annually, at Sellafield. Altogether, COGEMA and BNFL have contracted to treat about 10 000 tons of spent fuel from abroad.

BNFL's contracts are cost-plus, a good deal for the government-owned company that liberates it from financial risk—but not from technical, environmental and political risk. The Windscale–Sellafield plants have a long history of accidents and unintended radiation releases and still give rise to "constant disclosures of leaks or news of leaks that were worse than originally reported," according to John Surrey, a member of the science-policy unit at the University of Sussex. Walter Patterson, a leading writer on nuclear energy based in the UK, says that most people he knows are convinced that the name Windscale was changed to Sellafield because Windscale had come to have such a negative public image.

Partly but not mainly because of Chernobyl, which has had less impact in the UK than in Germany, plans for THORP are increasingly controversial. The plant is scheduled to start operations in four or five years, and spent fuel from British advanced gas-cooled reactors and foreign light-water reactors already is accumulating at Sellafield.

A recent report prepared by an all-party parliamentary committee recommends, however, that the whole project should be reconsidered. While the report concedes that reprocessing is good business for BNFL, it raises questions about whether it is good for the general public. The report finds no overriding technical reason for reprocessing and says that recent research indicates that solidified oxide fuels may be more stable and more leach resistant than vitrified waste. The parliamentary report says that "it is not safe to justify THORP on the basis of an experimental fast reactor program which may or may not become commercially successful sometime in the next century." The report takes note of another recent parliamentary report, in which it is said that Britain already has spent £2.4 billion on breeder R&D in the first 25–30 years of the program, that Britain seems to be about halfway through a 60-year development program and that a cumulative total of £5.7 billion would likely be spent before a commercially successful breeder is built.

The Thatcher government has rejected the all-party committee's recommendations, saying that abandonment of THORP would waste nearly $1 billion already spent and that reprocessing is a necessary preparation for the 21st century.

A second reprocessing plant, designed for extraction of breeder-blanket plutonium, is slated for construction at Dounreay, the breeder-research facility in northern Scotland. British sources indicate that this proposed plant is part of a web of agreements with France, Germany, Italy and Belgium in which the deal, roughly, was that France was to get the first breeder, Germany the second and the UK the reprocessing facilities. Conversations about the proposed reprocessing plant tend to resonate with the Franco-German argument about the second breeder.

When one asks French engineers where the fuel from the Superphénix is to be reprocessed, the usual answer is that a new facility probably will be built at Marcoule, France's plutonium-production complex on the lower Rhône north of Avignon. Nobody mentions the supposed agreement with the UK. British officials, on the other hand, are talking about a facility that would be built to treat spent fuel from three breeders initially, and ultimately six or seven. They seem oblivious to the news that a second breeder may not be built anytime soon and unaware that the French do not appear to be thinking of having the Superphénix fuel treated in Britain.

—WILLIAM SWEET

Soviets assess cause of Chernobyl accident

PHYSICS TODAY/DECEMBER 1986

Some—but by no means all—of the Chernobyl story can now be told. Soviet scientists have assembled enough information from interviews, plant records and computer simulations to describe the chronology of events leading to a severe accident that destroyed a nuclear reactor at Chernobyl on 26 April and spread radiation over much of Europe. A Soviet delegation of 28 specialists headed by V. A. Legasov (Kurchatov Institute) presented a very frank report on the accident's causes and consequences to an experts' meeting sponsored by the International Atomic Energy Agency in Vienna on 25–29 August. Their report places heavy blame on the reactor staff for committing numerous violations of operating rules, one of which left the reactor operating at a power level where the consequences of further violations were greatly exacerbated.

The material presented at the meeting has since been summarized and evaluated by IAEA's International Nuclear Safety Advisory Group, a team of 14 reactor specialists from as many

nations, headed by A. P. Vuorinen of Finland. The INSAG report stated that the information exchange at the meeting exceeded expectations. Overall, INSAG felt participants recognized the need for international cooperation on nuclear safety and radiological protection. (Herbert Kouts, a member of INSAG, discusses Chernobyl in his editorial on page 136.)

Sequence of operator errors. As had been rumored, the accident developed when the reactor staff was conducting a test at unit 4 of the Chernobyl nuclear power station (PHYSICS TODAY, July, page 17). The staff wanted to evaluate the turbogenerator's ability to power some electrical equipment even when the turbine is coasting down after its steam supply has been cut off. The equipment normally powered by the turbine in Chernobyl-type reactors includes some of the pumps from the fast-acting emergency core-cooling system. A previous test had indicated that the voltage from the generator fell off long before the mechanical energy of the turbine rotor was spent. The test of 25

April was to determine whether a special magnetic-field regulator for the generator could help capture that mechanical energy more effectively. In retrospect, the Soviet review team judged that the test plans specified safety regulations in only a cursory way, and they noted that the officer in charge was an electrical engineer, not a reactor specialist.

At 1:00 am on 25 April the staff began the test by reducing power to half its nominal peak value of 3200 thermal MW. (See the time line below.) At half power, they switched off one of two turbogenerators driven by the reactor. Soon after, they shut off the emergency core-cooling system in accordance with the test plan so that it would not interfere with the experiment. Operating procedures forbid disabling this system, and the Soviet report identifies this as one of the six most dangerous violations of operating rules during the test.

After a nine-hour delay in further power reduction caused by a continued demand for power, the staff began to

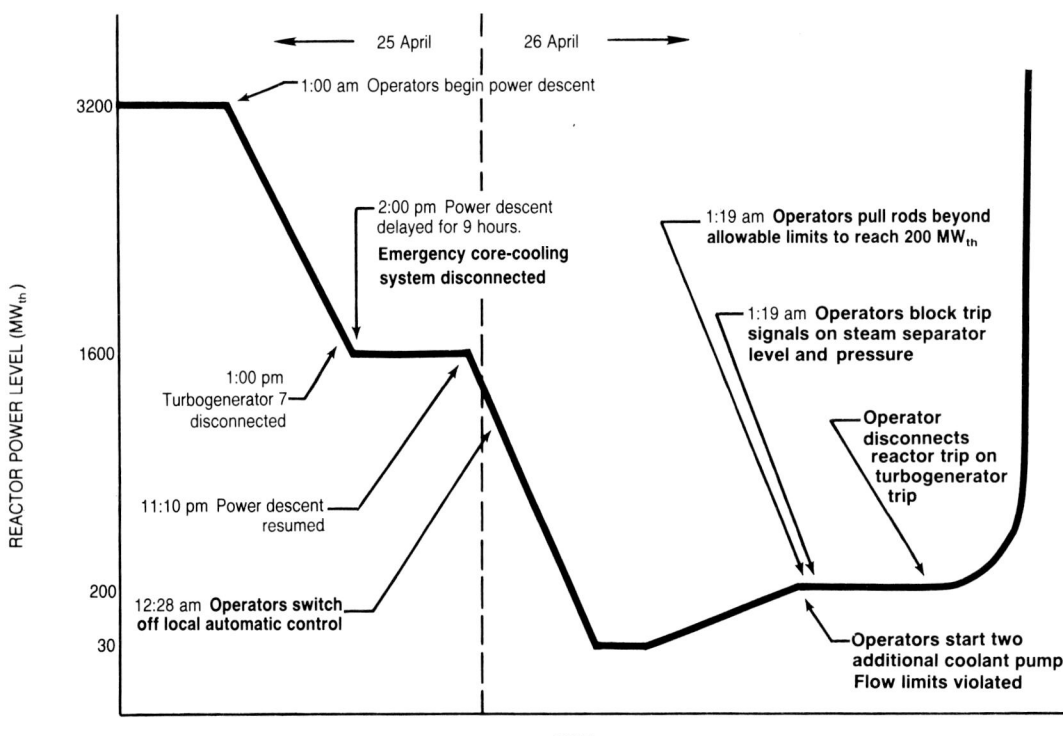

Time line portrays key events leading to the accident at the Chernobyl unit 4 power plant. The six actions identified by the Soviets as violations of operating procedures are indicated in boldface.

TIME ⟶

reduce power again, toward the 700–1000 MW_{th} at which the test was to be conducted. When the power level becomes this low, the computers regulating the reactor are supposed to receive their input from detectors located in the reflector region around the perimeter of the core rather than from those located within the core itself, because the core detectors are less sensitive at low power. As one operator made this switch, he neglected to enter a required instruction for the computer to hold the power at its current level. As a direct result of this second error, the power plunged to 30 MW_{th} before operators acted to reverse it.

Plant operators withdrew control rods to increase the power but could only bring the reactor back up to 200 MW_{th}. Further power increases were hampered by an excess of xenon-135, a fission-product daughter that absorbs neutrons and hence poisons the reaction. In steady-state operation, the xenon concentration is at equilibrium, but as the power level decreases, less xenon is removed by neutron absorption and subsequent decay, and its concentration temporarily increases. Just to boost the power back to 200 MW_{th} required a third violation of safety rules: the withdrawal of too many manual regulating rods. The regulations specify that the reactor core must always contain at least 30 such rods in effective positions to compensate for excess positive reactivity.

Dangerous operation. At this point the reactor was operating at a power far below the minimum of 700 MW_{th} specified by operating procedures. In this unstable region, the net power coefficient is positive; that is, any increase in power is likely to cause a further increase in power. The dominant term in the net power coefficient at low power is the void coefficient of reactivity: If the amount of void, or steam, increases, fewer neutrons are absorbed by water so that the flux, and hence the reactivity, increases.

Nevertheless, the staff proceeded with the test. They hooked an additional two pumps to the six then connected to the primary cooling loop to allow safe cooling after the trial rundown of the turbogenerator, which was powering four of the pumps. Although the test procedures called for this action, it increased the risk of pump cavitation, and the Soviets cite it as a fourth safety violation. The combination of excess flow rate and low power created difficult conditions, and the operators had to make many manual adjustments in an effort to keep the right steam pressure and water level in the steam drum. Because these parameters were fluctuating considerably and the operators wanted to avoid

shutting down the reactor, they decided to disconnect the emergency protection signals relating to the steam pressure and water level. That action was safety violation number five.

About the same time—1:22 am, 26 April—a computer printout indicated that the available excess reactivity had fallen to just six to eight rods, a level requiring immediate shutdown. The operators nevertheless persisted with the test and made their last fatal error by blocking the trip signal on the turbogenerator. They did not want the reactor to shut down when they stopped the turbine, just in case they needed to conduct the test a second time. Such concern for completion of the test led to five of the six operator errors. Ed Purvis of DOE said that "although the Soviets characterize these as operator errors, they might more accurately be described as management errors in that operators appeared to be following instructions given them by those in charge."

When the test began at 1:23 am, the power began to rise slowly. The positive power coefficient accelerated this rise while the available excess reactivity was insufficient to control it. Operators pressed the "scram" button, which triggers the insertion of absorbing rods to stop the reaction.

Soviet data indicate that the power started its precipitous rise at this point, and that some factor in addition to the increased void was now adding reactivity. A team of scientists from various DOE laboratories feels that the extra reactivity could have come from what they term a "positive scram." Below the absorbing material in each scram rod is a 5-m section of graphite, which moderates the reaction when the scram rods are withdrawn. As the many scram rods were inserted simultaneously, the graphite sections may have increased the reactivity in the bottom of the core just enough to make it go critical from prompt neutrons alone. The Soviets have calculated that the power may have soared to 100 times its nominal value of 3200 MW_{th} within four seconds.

The huge release of energy (the Soviets estimate over 300 cal/g) from this power excursion essentially fragmented the fuel into minute (millimeter diameter or less), hot particles. The fuel cladding failed and allowed these particles to contact the coolant in the channels. Rapid steam generation together with expansion of the volatile fission products from the fuel raised the pressure enough to destroy the cooling channels. Steam erupting from these channels caused the vault containing the reactor core to fail catastrophically, lifting and tilting the cover plate above the core and opening the

enclosure to air. These events led to a worsening of reactor-power transients, and a second explosion was reportedly heard two to three seconds later. No one yet knows whether there was in fact a second explosion and, if so, what caused it. The Soviet report said that "witnesses observed these reactions in the form of a fireworks display of glowing particles and fragments escaping from the units."

Handling the aftermath. Flaming pieces of core and hot graphite landed on several of the roofs of the reactor housing, starting fires in 30 places (see the photo on page 20). The fires were out by 5 am on 26 April. For about ten more days, however, reactions within the graphite continued to produce dark smoke. Starting on 28 April, military helicopters dropped about 5000 tons of boron, dolomite, sand, clay and lead on the smoldering core to absorb and filter the aerosol particles. On 4–5 May the Soviets pumped nitrogen under pressure into the space beneath the reactor vault to cool the fuel. Radioactive releases were continuing all the while. By the end of June the Soviets had constructed a flat heat exchanger in a concrete slab under the reactor.

Work has begun on a structure to entomb the damaged reactor and separate it from the adjoining but undamaged unit 3. The structure will be ventilated by an open system with air filters. The goal is to reduce the radiation level below 5 millirads per hour at the roof and 1 mR/h at the walls. (Normal background levels are about 0.015 mR/h.) The government has removed contaminated soil and laid concrete around the plant site in an effort to reduce the radiation to rates acceptable for resumed operation of units 1 and 2, and possibly unit 3. Both units 1 and 2 had been restarted by the end of October, staffed by operators from other Soviet reactors, who will be rotated to this duty for two-week shifts. The Soviet government does not plan to modify the projected growth in its nuclear-power program.

Clearly the accident might have been averted if the reactor had had certain additional safety features. As a short-term measure, the USSR is implementing several modifications in all its graphite-moderated reactors. The number of control rods required to be partly inserted into the core will be increased to 70–80 and shutdown protection will be installed to prevent operation below 700 MW_{th}. In the longer term, the Soviets may increase the fuel enrichment level from 2% to 2.4% while adding more absorbers. The combination would reduce the problem of the positive void coefficient. Some observers feel that the impact of these design changes on overall safety

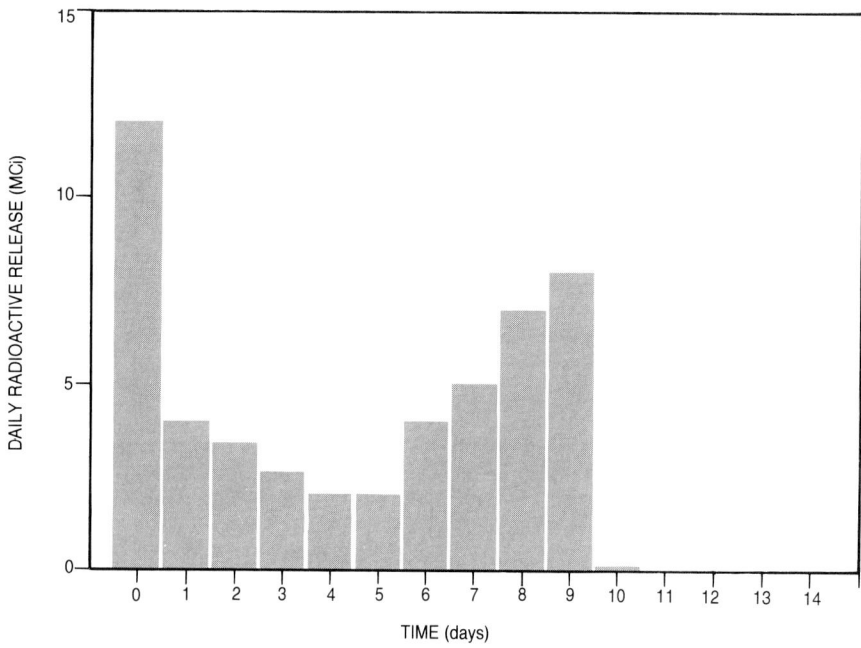

Daily release of radioactive nuclides after the Chernobyl accident shows first a drop and then rise. Not included are 45 and 5 MCi, respectively, of krypton and xenon, volatile gases presumed to have been released immediately.

is not clear, and worry that while these fixes might mitigate some of the circumstances that led to the accident, they could possibly aggravate others.

Radionuclide releases. The radioactive material in the Chernobyl core would have had an activity of 1000 megacuries if measured on 6 May. Because the accident exposed nearly all the fuel, the Soviets have estimated that virtually 100% of the noble-gas fission products xenon-133 and krypton-85 escaped from the reactor, amounting to 45 and 5 MCi, respectively. To estimate the releases of radionuclides other than noble gases the Soviets surveyed the ground contamination within their borders with aerial gamma photography. They concluded that the amount of radioactive fallout in the European territory of the USSR was between 30 and 50 MCi, or about 3.5% of the total activity accumulated within the reactor core. Of this, iodine-131 and cesium-137 accounted for about 7.3 and 1 MCi, corresponding to releases of 20% and 13%, respectively, of their core inventories. To put these numbers in perspective, the Three Mile Island accident released 15 Ci, millions of times less. The cesium from all atmospheric nuclear-weapons tests is estimated to be on the order of 30 MCi.

Both the composition and timing of the radionuclide release were considerably different from what reactor experts often expect from a catastrophic accident in a US light-water reactor. Postulated disaster scenarios frequently involve a core melt and rupture of the containment structure, so that the fission products are distilled from the

molten core, releasing higher fractions of the more volatile isotopes. At Chernobyl, however, much of the early release consisted of fine fuel fragments explosively expelled from the reactor. Thus the radioisotopic composition in the total release corresponded roughly to that of the irradiated fuel, but was enriched in volatile isotopes of iodine, tellurium, cesium and inert gases. About 3–5% of the core inventory of relatively refractory elements such as strontium, plutonium and ruthenium escaped from the reactor—much more than would be expected in a light-water-reactor core melt. Surprisingly, although ruthenium is less volatile than strontium, it appeared in appreciable quantities in filter samples throughout Europe, while strontium did not.

Only about one-fourth of the total release of non-noble gases—about 12 MCi—left the reactor during the first day, according to Soviet data (see the figure above). The release rate declined in the next few days, only to rise again to about 8 MCi on 5 May and fall sharply the next day. The high release on 26 April was no doubt driven by the explosion and intense heat, and the decrease probably reflects the measures taken to extinguish the fire and filter the releases. The subsequent rise has not been fully explained, although many suspect that the blanket of boron and lead insulated the core and caused its temperature to rise, boiling off more radionuclides.

Long-term health effects. The Soviets have made a very preliminary estimate of the long-term health consequences of the accident, based on environmental

monitoring supplemented by predictive modeling. They have directly measured the uptake of I^{131} in the thyroids of about 100 000 persons and the whole-body doses from Cs^{137} in about 10 000 people. The largest contribution to the collective long-term dose is expected to come from cesium, whose halflife is 30 years. People will be exposed over many years both to external radiation from cesium on the ground and to internal radiation from cesium ingested in food. The Soviets estimate that over the next 50 years the external radiation will expose the 75 million people in the western USSR to a collective dose of about 29 million person-rems. If one assumes a no-threshold linear dose–response relationship and takes the commonly used risk coefficient of roughly one cancer death for every 10 000 person-rems, this collective dose might result in about 3000 cancer deaths above the 9.5 million cancer deaths expected from other causes in the same population.

The 135 000 people evacuated from a zone 30 km in radius around the reactor had a collective dose from external radiation of 1.6 million person-rems. The average dose of 12 rems per person is surprisingly low for such close proximity to the accident. That may reflect the evacuation procedures, meteorological conditions and the special circumstances of the Chernobyl accident, which pumped material high into the atmosphere.

The Soviets had originally evaluated the dose commitment—that is, the dose expected from future exposure—from ingestion of cesium to be 210 million person-rems, a factor of ten higher than the external dose. In arriving at this estimate, the Soviets followed the procedures formulated by the United Nations Scientific Committee on the Effects of Atomic Radiation. However, other participants in the Vienna meeting felt this estimate was overly pessimistic. There is considerable uncertainty about how much cesium has fallen on different soil types, how cesium much will pass into the crops and how much contaminated food will be eaten. Analysis of atmospheric nuclear-weapons tests suggests[1] that the cesium dose from the food chain is somewhat less than that from external radiation. Marvin Goldman (University of California, Davis) told us that the Soviets recognized that their methodology would predict Cs^{137} body burdens, and hence dose levels, ten times higher than those they were measuring. Thus they lowered their estimate of the dose commitment from ingestion of cesium.

The Soviets' estimates of total radionuclide releases and of dose commitment apply to their own country only. To get a complete picture, one must

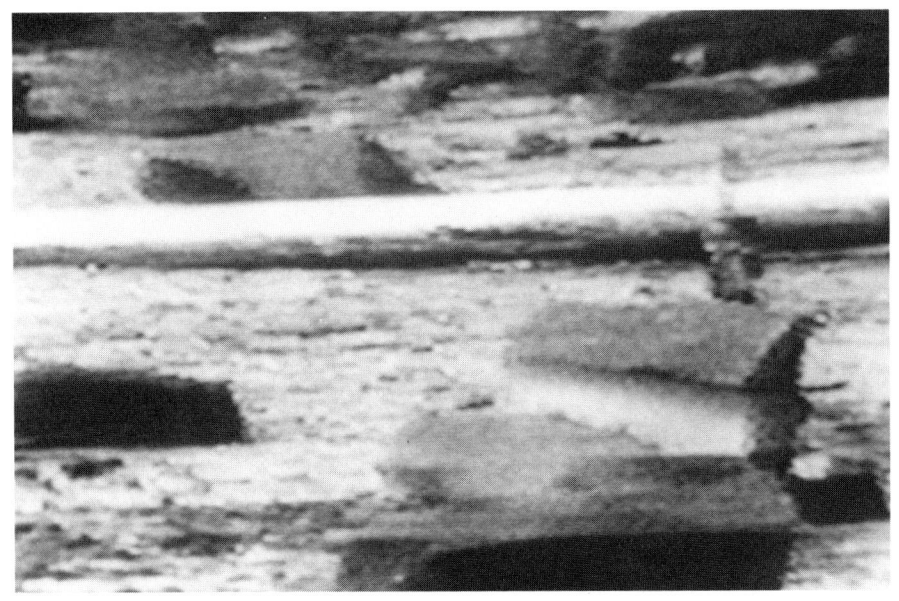

Debris thrown up by force of the Chernobyl explosion litters roof of adjacent building. A large chunk at right, in front of a length of 6″ pipe, appears to be half of a graphite block, with indentation left by cylindrical cooling channel. (Photo from Soviet tv, provided by Ed Purvis.)

match their numbers with those being assembled outside the USSR. Much work remains to be done before all the pieces are in place. The few simulations that have been done are as preliminary as the Soviet estimates.

Helen ApSimon and her colleagues at Imperial College, London, presented some initial findings to a World Health Organization conference last June in the Netherlands. They simulated the transport of 21 MCi of iodine and 1.5 MCi of cesium from the reactor across Europe, accounting for both winds and rainfall in the days after the accident. They estimate that the 500 million Europeans outside the USSR might have a collective dose from both ground-deposited and ingested cesium of 24 million person-rems.

Other simulations have been run by a collaboration in the Netherlands (the National Institute of Public Health and Environmental Hygiene and the Royal Netherlands Meteorological Institute) and by a team at Lawrence Livermore National Lab.[2] In these models the source terms have been adjusted to agree with surface-air concentrations measured at many places in eastern and western Europe as well as some points in the Middle East and Asia. Marvin Dickerson of the Livermore Lab says there is a need to work with the Soviets to arrive at a better estimate of the overall radionuclide release.

These simulations can only predict depositions averaged over very large regions. Chris Hohenemser (Clark University) points out, however, that Europeans have measured many "hot spots," such as places in southern Germany where rainfall produced dose levels up to 40 times those in places with no precipitation.

Studies under way. The implications of the Chernobyl accident are just beginning to unfold. Groups in many countries have already issued assessments of the radiological impact. Oth-

er institutions have assessments now in progress. UNSCEAR, with the cooperation of IAEA and the World Health Organization, will try to determine the overall radiological consequences. The US Department of Energy has established a task group under William Bair (Battelle Pacific Northwest Lab) in the Office of Health and Environmental Research. One of the group's four committees, under the leadership of Goldman, was scheduled to report at the end of October on its estimates of the radiological consequences. Another committee in the task group is assessing the opportunities presented by Chernobyl to validate existing models. A third committee is studying what the US might learn from the Soviet experience to improve emergency response preparedness.

Brian Sheron of the Nuclear Regulatory Commission told us that an interagency group in the US was preparing a factual report on the accident, from which each agency is to draw implications for its particular area of responsibility. Sheron is participating in a committee under the aegis of the Organization for Economic Cooperation and Development that is evaluating the implications of the accident for OECD member countries.

By the end of this year DOE was scheduled to receive six separate consultants' reports about the safety of the N-Reactor, a plutonium production reactor in Hanford, Washington. In late July, the National Research Council asked a panel of experts under Richard Meserve (Covington and Burling, Washington, DC) to report in nine months on the safety of plutonium production reactors and another nine months on US research reactors.

—BARBARA GOSS LEVI

References

1. UN Scientific Committee on the Effects of Atomic Radiation, *Sources and Effects of Ionizing Radiation*, report to the General Assembly (1977).
2. P. Gudiksen, R. Lange, Nature, to be published.

JAPAN'S NUCLEAR PROGRAM STRESSES BREEDERS, PLUTONIUM AND SAFEGUARDS

PHYSICS TODAY/JANUARY 1988

Japan's Atomic Energy Commission has prepared a new long-term program for nuclear energy that anticipates a doubling of installed generating capacity between now and the year 2000, implying that nuclear power plants will produce about 40% of the nation's electricity at the end of the century.

The "Long-Term Program for Development and Utilization of Nuclear Energy" was submitted to the Prime Minister last June by the head of Japan's Atomic Energy Commission, Y. Mitsubayashi, who also serves in the Cabinet as Science Minister. The program, which is binding on the Prime Minister under legislation governing Japan's AEC, recently became available in English translation.

The program puts great emphasis at the outset on improving safety, by building on what it claims is already an excellent record by international standards, and on helping to strengthen barriers to the proliferation of nuclear weapons or the misuse of nuclear materials. Even as Japan is "required to play a still more important role in promoting the development and utilization of nuclear energy based on a philosophy of harmonization and mutual understanding with foreign partners," the program says, "it becomes particularly important for Japan to ... take the initiative in promoting and demonstrating [in] practice [the] peaceful use [of nuclear energy] compatible with nonproliferation."

The program calls upon the Japanese government to become a member of the Convention on the Physical Protection of Nuclear Material, an agreement that was negotiated under the auspices of the International Atomic Energy Agency with strong support from the United States, and it calls for the "establishment of a domestic system to cope with it." The convention was opened for signature in the early 1980s and took force on 8 February 1987, after 21 countries—the required minimum number—had signed and ratified it. Twenty-five other countries and EUR-ATOM have signed but not yet acceded to the convention.

Japan's pilot reprocessing plant, located at Tokai-Mura and owned by the Power Reactor and Nuclear Development Corporation, opened over opposition from the Carter Administration.

The prominence Japan's long-term program gives to the protection of nuclear materials represents a marked change in attitude for a country that initially adopted a very skeptical attitude toward safeguards and joined the Nuclear Non-Proliferation Treaty only after obtaining extensive assurances on the subject of safeguards. Leslie Thorne, a high IAEA official who was responsible for safeguards in Asia for about 15 years, considers Japan to be one of the agency's "big success stories." He attributes Japan's more forthcoming attitude toward the safeguards system partly to the IAEA's growing technical professionalism and partly to the Soviet Union's more positive attitude toward the agency, but also to Japan's own growing technical self-confidence. A country that once worried that safeguards might cripple its competitiveness vis-à-vis the nuclear-weapons states now "is probably the world leader in advanced nuclear technology," Thorne believes.

Japan's 33 light-water reactors provided nearly 25% of the country's electricity in 1986, and they operated that year at 76% of their rated capacity, one of the world's best records. The long-term program calls for

nuclear power plants to generate 40% of Japan's electricity by 2000 and 60% by 2030.

Plans for plutonium

At a time when the Europeans are revising plans for breeder reactors in the wake of an accident at France's Superphénix (see box at right), Japan remains firmly committed to the development of breeders and to the commercial use of plutonium as fuel for both breeders and light-water reactors, as well as to vigorous research on fusion (see next story). "The commercial deployment of fast breeder reactors will be determined by the market mechanism, and it is difficult to forecast their epoch at the present time," the program says. "But ... basic policy will be to take a positive attitude with the objective of establishing the plutonium utilization system by means of fast breeder reactors...."

The long-term program anticipates that the 280-MW Monju breeder reactor will go critical about 1992 and that construction will start in the late 1990s on a demonstration reactor comparable to the Superphénix. The plan aims to establish an energy system involving use of plutonium in fast breeder reactors between 2020 and 2030.

In November, Japan and the United States signed a controversial long-term agreement concerning the transport of Japanese nuclear fuels to and from Europe for reprocessing, a practice that has been a bone of contention between the two countries since the late 1970s. In recent years, Japan has contracted with reprocessors in France and Great Britain to treat more than 5500 metric tons of spent fuel from Japanese light-water reactors. "An estimated 25 000 kg of plutonium will be separated [from the fuel] over the next 15–20 years," according to a report by the US Arms Control and Disarmament Agency.

Critics of plutonium recycling have argued for many years that plutonium would be most vulnerable to theft by terrorists or "diversion" by wayward governments following its separation from the highly radioactive components of spent nuclear fuel and during its transport to fuel fabrication, storage or power plants. This consideration, combined with the darkening outlook for breeder economics, prompted Britain's Royal Commission on Environmental Pollution and the Ford Foundation's Nuclear Energy Policy Group to issue reports in 1976 that were highly critical of reprocessing and recycling. The same year, a group at Princeton University that included physicists Frank von Hippel, Theodore B. Taylor and Robert H. Williams and political scientist Harold A. Feiveson published a series of influential articles in the December issue of the *Bulletin of the Atomic Scientists*, collectively entitled "On the Brink of the Plutonium Economy," in which they recommended the evaluation of alternative technologies.

The following spring, months after President Carter took office, he unilaterally terminated commercial reprocessing in the United States and curtailed breeder development work, saying that he wanted to set the world an example. During the following decade, some countries such as Sweden and Canada chose to follow the US example, and everywhere plans to introduce recycling operations were severely delayed because of adverse economic and technological developments (see PHYSICS TODAY, September 1986, page 53, and October 1986, page 115). On the other hand, countries such as France, Great Britain, West Germany, Belgium, Italy, Switzerland and—not least—Japan stuck with plans to reuse plutonium in mixed-oxide fuels for light-water or breeder reactors. According to research done last year by Feiveson and

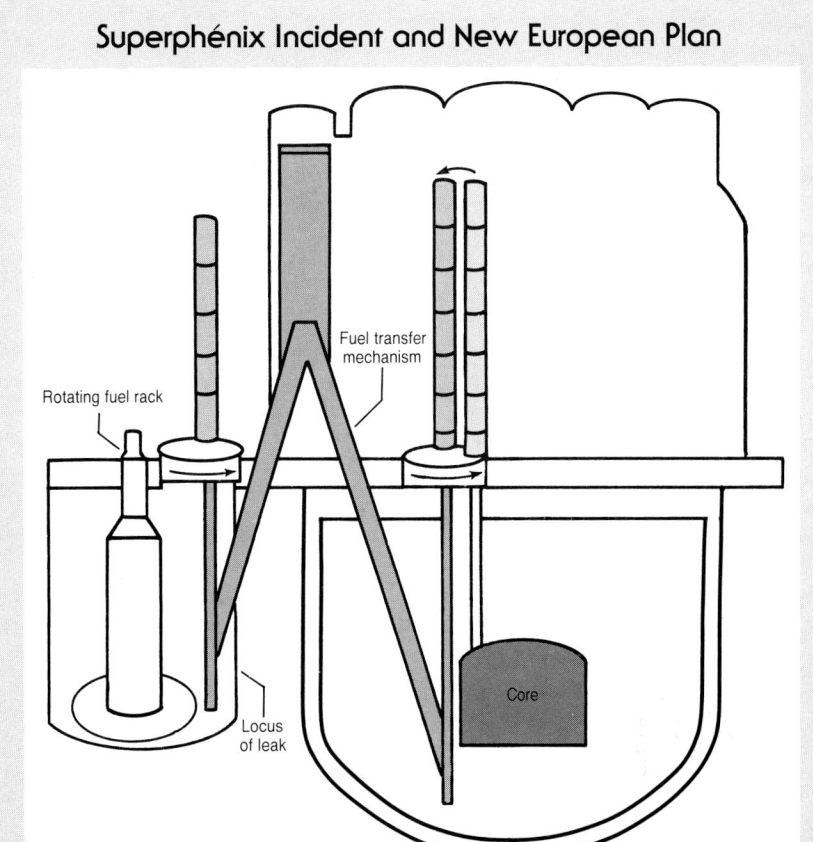

Superphénix Incident and New European Plan

A leak of liquid sodium, which explodes on contact with water and burns on contact with air, was detected and reported last May at France's 1200-MW Superphénix, the world's first full-scale commercial breeder reactor. At the beginning of September the source of the leak was identified as a hole near the base of a tank adjacent to the reactor in which spent fuel elements are stored. The hole is near the point where spent fuel enters the tank via an elaborate fuel transfer mechanism (yellow-shaded areas). Spent fuel rods are lifted from the core through a plug, rotated to the compass-shaped transfer device, inserted into the storage tank and finally removed through another rotating plug.

From the point of view of safety, the leak is not nearly as serious as it would be if it had taken place in the much more sensitive reactor vessel or in the secondary circuits, which are equipped with devices for fast dumping of sodium in the event of trouble. But repairing the leak is greatly complicated by the fact that the storage tank can be accessed only via the plugs at the very top. The entire storage tank may have to be replaced, at a cost of around 400 million francs, or it may be possible to do without the tank and store spent fuel assemblies in the containment structure itself.

The accident has drawn renewed and closer attention in France and other European countries to the failure of the Superphénix to live up to early hopes for its economic performance. Because the Superphénix produces electricity at about twice the cost achieved by French light-water reactors, Electricité de France, the customer for all French nuclear power plants, is reevaluating its long-term plans for breeders. Meanwhile, the nuclear authorities in the United Kingdom, West Germany and France have delayed plans to build a demonstration breeder in each country and instead for now will build just one successor to Superphénix.

On 1 December, representatives of the European Fast Reactor Utilities Group met in London and decided to launch a two-year program to develop a single conceptual design for the next European breeder reactor. In a follow-up meeting that took place a week later in Lyon, representatives of European R&D organizations, utilities and design companies met to hammer out details for the collaborative design program.

David Albright, a physicist with the Federation of American Scientists, if these countries continue with current plans, by the end of the century more than 300 000 kg of plutonium—enough to build about 30 000 small atomic bombs—will have been separated from spent fuel and placed into commerce.

In early November, the same week the United States and Japan signed the agreement granting Japan blanket approval to reprocess and recycle US-origin nuclear fuels for the next 30 years, the US Department of Defense delivered a report to the House Foreign Affairs Committee in which it warned that "opportunities for terrorist acts, including attempts to steal civil plutonium, will increase substantially as a result of the increased commercial use of plutonium." The report estimated that by the late 1990s as many as 300 shipments of plutonium annually will leave reprocessing plants in Europe for destinations in Europe and Japan.

The Nuclear Control Institute, a watchdog organization in Washington headed by Paul L. Leventhal, a veteran Congressional staffer who has concerned himself with proliferation for many years, claims that under the US–Japan agreement there will be flights of cargo planes carrying about 230 kg of plutonium to Japan from Europe via Alaska every two weeks. The institute has raised questions about the adequacy of current plutonium casks to survive realistic flight tests. In September, the State of Alaska filed suit in a Federal district court to block the flights, arguing that there is no way to guarantee safe air transport.

Toichi Sakata, who is responsible for nuclear affairs at the Japanese embassy in Washington, has the impression that the first transport flights will not take place for a couple of years in any case.

Fuel-cycle facilities

However Japan's near-term plans for transport of plutonium are affected by action in US courts or Congress, the fate of its program to establish a plutonium-based economy in the 2020s may ultimately depend more on its own development efforts and on its stated commitment to keep building a credible safeguards system. The long-term program seems in fact to adopt a somewhat standoffish attitude toward the international transport of fuels. It says that "reprocessing of spent fuels should be undertaken in Japan, in principle," and that "new commitments [for] overseas reprocessing should be treated with careful consideration."

Japan brought its first small commercial reprocessing plant into operation at Tokai-Mura in the late 1970s (over objections from the Carter Administration), but the plant only recently has begun to operate close to its design capacity of 700 kg of heavy metal (uranium and plutonium) per day. The program calls for Japan's private sector to construct a reprocessing plant with a capacity of 800 tons at Rokkasho-Mura, in Aomori

Prefecture. The plant is to start operating by the mid-1990s. Meanwhile, at the same site, the country's first large-scale commercial enrichment plant is to be constructed. Also a private sector venture, the plant is to have a capacity of 1500 tons separative work units and is to start partial operations around 1991 or 1992. The program calls for a total Japanese capacity of 3000 tons SWU by 2000.

Among the major fuel-cycle technologies, only fuel fabrication has been commercialized so far in Japan. A major plant is located near Osaka, and Japan has the capacity to produce fuel rods for all 35 of its nuclear power plants currently operating. In addition, a test facility coupled to the Tokai reprocessing plant produces fuel for the experimental test-bed breeder Joyo (100 MW thermal), and for a prototype advanced thermal reactor (165 MWe) that runs on mixed- oxide fuel. A larger plant at Tokai-Mura will produce fuel for the Monju breeder and for a 600-MW advanced thermal reactor.

Under an agreement concluded recently by the US Department of Energy and Japan's Power Reactor and Nuclear Development Corporation, each of the two parties contributes $5 million annually to a joint research program on the breeder fuel cycle. The program involves the transfer to Japan of breeder-fuel technology developed at Oak Ridge National Laboratory. William D. Burch, director of the fuel recycle division at Oak Ridge, has argued in an Oak Ridge publication that the collaboration "will allow the United States to maintain a core of expertise" in reprocessing and advanced reactor fuel technologies. A similar US–Japan collaboration covers fuels for high-temperature gas-cooled reactors.

Safeguards

Bulk fuel-cycle facilities such as reprocessing and enrichment plants have always been considered the most difficult to safeguard. The conventional wisdom in recent years has been that nuclear materials can be accounted for at such facilities with an accuracy of no more than $\pm 1\%$—implying that if ordinary methods of material accountancy were used in a sizable fuel-cycle facility, enough material to build a bomb could disappear from the plant without international inspectors being able to draw any conclusions. Accordingly, innovative techniques, procedures and arrangements are called for.

Thorne, currently director for safeguards standardization, training and administrative support at the IAEA, concedes that "it's a bit of a business

getting everything you want" at a facility such as the Tokai-Mura reprocessing plant, even given Japan's improved posture toward safeguards. But Thorne believes that significant strides have been made. IAEA inspectors are at the plant, where they have their own offices, 24 hours a day. Certain points are under continuous surveillance by cameras, and increasingly sophisticated devices are promising to reduce the margin of error in material accountancy. For example, high-performance neutron time correlation counters developed by Howard Menlove and James Swansen at Los Alamos National Laboratory can measure plutonium in any facility processing nuclear materials. This type of equipment is being interfaced with the robotic systems in a new automated fuel fabrication facility in Japan.

Similar instruments and arrangements will be employed when Japan completes its first large centrifuge enrichment plant, Thorne says. Under agreements negotiated between 1980 and 1983 by the six major countries that do enrichment, a common instrumentation approach has been developed involving, among other things, a "device for measuring enriched uranium that was invented in Australia, tested in the United States, refined in Japan, and first applied in the Netherlands." The six countries have agreed that IAEA staff will have the authority to make snap inspections on one hour's notice at all enrichment plants.

At a time when the Soviet Union and United States have just concluded an agreement on short- and medium-range missiles amid much fanfare about provisions for on-site inspection of rocket production plants, it is noteworthy that on-site inspection of sensitive facilities already has been pioneered in countries like Japan under the auspices of the IAEA. But the IAEA's system is only as strong as its weakest link, of course, and facilities such as enrichment plants can only be subjected to snap inspections if additional countries building such facilities agree to accept safeguards in the first place.

A noteworthy aspect of Japan's long-term program for nuclear energy, in this context, is its endorsement of "full-scope safeguards"—the principle that nuclear equipment, fuel or services will be exported to a non-nuclear-weapons state only if that country agrees to accept IAEA safeguards on all its nuclear facilities, regardless of whether it is a party to the Non-Proliferation Treaty or not. "Effective measures will be taken for full-scope safeguards . . . to be accepted by the non-member countries of the NPT," the program says. More generally, the program says that nuclear cooperation with other countries will depend on each country's nonproliferation credibility, that Japan will pay close attention to sensitive items contained on lists prepared by the so-called London supplier group, that "cooperation in fields related to the utilization of plutonium will be considered with utmost care" and that "handling of sensitive nuclear technology will be studied hereafter also with utmost care."

When the principle of full-scope safeguards first was incorporated by the US Congress into the Nuclear Non-Proliferation Act of 1978 as the result of efforts by individuals like Leventhal, it was widely dismissed as unilateralist and unrealistic. Its growing acceptance by supplier countries such as Japan is one measure of the safeguards system's increased strength.

—WILLIAM SWEET

SECTION 6 _____

HISTORY OF NUCLEAR WEAPONS

This section contains a mix of personal reminiscences, historical analysis, and modern issues. The articles are listed chronologically, beginning with the time of Albert Einstein in Germany, and ending with the recent efforts of Andrei Sakharov and Roald Sagdeev in the Soviet Chamber of Deputies. In December 1989, Andrei Sakharov died at the age of 68. Sakharov won the 1975 Nobel Peace Prize, having spoken out forcefully for human rights and political freedom, and against nuclear terror. The APS Forum on Physics and Society awarded Sakharov the Szilard Award in 1983. On a happier note, Roald Sagdeev and Susan Eisenhower were married in Moscow during the spring of 1990.

CONTENTS

Einstein and Germany

The native German physicist, unlike many of his colleagues, had an early antipathy to German nationalism, so that for him, Hitlerism was a confirmation of an earlier intuition.

Fritz Stern   PHYSICS TODAY/FEBRUARY 1986

There was nothing simple about Albert Einstein, ever. His apparent simplicity concealed an impenetrable complexity. Even the links to his native Germany were prematurely ambiguous. At a time when most Germans thought their country a hospitable home, a perfect training ground for their talents, Einstein was repelled: In 1894, as a 15-year-old, he left Germany and became a Swiss citizen. Twenty years later, a few weeks before the outbreak of the Great War, he returned to Germany and remained for 18 years of troubled renown, years in which he appreciated what was congenial and opposed what was antipathetic in Germany. Long before Hitler's rise, he felt unease.

Einstein's fame, his capacity for homelessness, and the degradation of his country made him a citizen of the world, seemingly detached from Germany. But I believe that his early encounters with Germany and his consequent hostility to its official culture shaped his public stance. The German experience haunted Einstein to the very end, as it haunted so many of his generation later. It was the text of his political–moral education, the back-

ground against which he came to mold his unorthodox views and play his controversial public role.

In Einstein's time, Germany was the promise and later the nemesis of the world, a country that had decisive bearing on world politics and where, for a moment that seemed a lifetime, the moral drama of our era was enacted. At certain critical moments, Einstein's responses differed radically even from those of his closest colleagues. Documenting this diversity will complicate our understanding of Germany, and this is desirable because Germany's past has often been treated with didactic simplicity. Einstein and Germany: They illuminate each other.

A rebel from the start

Einstein grew up in southern Germany. We know little of his early life. He was no child prodigy; rather, his reticence in speaking for the first three years, his difficulty with learning foreign languages and his mistakes in computation have been a source of endless comfort to the similarly afflicted and their parents—though affinity in failure may not suffice for later success. He went through a brief but

intense religious phase, the end of which, he said, left him suspicious of all authorities. His parents, secularized Jews, had little to do with his intellectual development; an uncle fed his mathematical curiosity. His father was an amiable failure, mildly inept at all the businesses he started. In 1894 his parents went to Italy to start yet another business, leaving the 15-year-old Albert behind in a well-known Munich *Gymnasium*. The authoritarian atmosphere and the mindless teaching appalled him. There was more than a hint of arrogance about the young Einstein, and hence it does not strain one's credulity that a teacher exclaimed, "Your mere presence spoils the respect of the class for me." He was a rebel from the start.

Encouraged by his teachers' hostility, he decided to quit school and leave Germany. His unsuccessful career facilitated his later fame in Germany: Erik Erikson has rightly referred to "the German habit of gilding school failure with the suspicion of hidden genius." It is often said that Einstein left school because he objected to its militarism. I find this unpersuasive: Bavarian militarism? I would suppose

KEYSTONE

Addressing a meeting. Einstein is at the podium delivering the principal address at a meeting sponsored by the Academic Assistance Council, a group formed to aid refugees. Seated at the right are Ernest Rutherford, Austin Chamberlain and the Bishop of Exeter. The October 1933 meeting was held in Royal Albert Hall, London.

that there might have been a stifling Catholicism; an insolent, thoughtless authoritarianism; a repulsive tone—any of which would have sufficed to discourage a youth like Einstein.

I suspect Einstein left Germany so precipitously to escape serving in the German army; by obtaining Swiss citizenship in time, he could do so without incurring the charge of desertion. His first adult decision, then, was to escape the clutches of compulsion—the image of Einstein as a recruit in a field-gray uniform does boggle the mind. He left Germany without regrets. His encounters with that country had not been happy.

There followed the obscure and difficult years in Switzerland, the failures, the marginal existence, the Zurich Polytechnic and, finally, the security of the patent office in Berne. From there in 1905 emerged the four papers destined to revolutionize modern physics and cosmology. They were published in the *Annalen der Physik*, and Max

Planck was the first man to recognize the genius of the unknown author. The international scientific community took note as well, and Einstein finally received his first academic appointments. In 1913, while he was a professor at the Zurich Polytechnic, two German scientists appeared, Walter Nernst and Planck, to offer him an unprecedented position: salaried membership in the Prussian Academy of Sciences and a professorship at the university, but without the obligation to teach. When Nernst and Planck left, Einstein turned to his assistant, Otto Stern, and said, "The two of them were like men looking for a rare postage stamp." The remark was perhaps an early instance of that self-depreciatory humor, that modesty of genius, that was to characterize Einstein.

Einstein began his new German life in April 1914. Berlin was the world's preeminent center of the natural sciences, and Planck, Fritz Haber and a dazzling array of talent rejoiced at

having this young genius at the head of their circle. Three months later the war shattered the idyllic community. Einstein had returned to Germany in time to see the country seized by the exaltation of August 1914, when almost all Germans were caught up in an orgy of nationalism, gripped by a joyful feeling that a common danger had at last united and ennobled the people.

The intoxication passed; the business of killing was too grim to sustain the

Fritz Stern is Seth Low Professor of History at Columbia University, where from 1980 to 1983 he served as provost. He is the author of *Gold and Iron: Bismarck, Bleichröder, and the Building of the German Empire* (Knopf, New York, 1977) and other books on European history. He is on the editorial committee of the Collected Papers of Einstein and is currently working on a book on Einstein and the German public.

unbridled enthusiasm of August 1914. The elite rallied to the nation, as it did elsewhere too. In the fall of 1914, 93 of Germany's best-known scientists and artists, including Planck, Haber and Richard Willstätter, signed a manifesto that was meant to repudiate Allied charges of German atrocities, but that by tone and perhaps unconscious intent argued Germany's complete innocence and blamed all misfortunes and wrong-doing on Germany's enemies. The manifesto of the 93 has often been seen as a warrant for aggression, a declaration of unrestrained chauvinism. I suspect it was as well the outcry of people to whom the outside world mattered and who intuitively sensed that the Allies would come to cast Germans as pariahs again. Some of the 93 probably hoped for continued respect across the trenches—and signed a document that had the opposite effect. It was not the last time that Germans confirmed the sentiments they set out to deny.

With but few exceptions, intellectuals everywhere joined in this chorus of hatred and in the cry for blood. So did the guardians of morality and the servants of God, the priests who sanctified the killing as an act of mythical purification. In time, some of the 93 turned moderate, or perhaps remained the patriots they had been, but others passed them on the right in the nation's wild leap to pan-German madness.

Politicized by the war

Einstein was alone and disbelieving. The war that was to politicize everyone as the cause of universal grief politicized him as well. Before 1914 Einstein had never concerned himself with politics; his very departure from Germany had been a youthful withdrawal from the claims of the state. Now, for the first time, he ventured forth from his study, convinced of the insanity of the war, shocked by the ease with which people had broken ties of international friendship and mutual respect. A pacifist asked him to sign a countermanifesto addressed to Europeans, demanding an immediate, just peace, a peace without annexations. It was the very first appeal he ever signed. It was published only in 1917 and then only

abroad, for want of sufficient signatures. Somewhat later he joined a tiny group of like-minded democrats and pacifists. In November 1915 the Berlin Goethebund asked for his opinion about the war, and he sent a message with this rather special ending: "But why many words when I can say everything in one sentence, and moreover in a sentence that is particularly fitting for me as a Jew: Honor your Master Jesus Christ not in words and hymns, but above all through your deeds."

Intermittently Einstein forsook his work—his central passion—to bear witness in an unpopular cause he took to be right. He had been a pacifist and a European of the first hour, never touched by the frenzy that ravaged nearly all. Convinced of Germany's special responsibility for the outbreak and the continuation of the war, he hoped for the nation's defeat.

To understand Einstein's isolation, one must look at the responses of his friends and colleagues to the war. Haber, for example, became the very antithesis of Einstein. Einstein's senior by 11 years, Haber was a chemist of genius, a born organizer and, in wartime, an ardent patriot. Without Haber's process for fixing nitrogen from the air, discovered just before the war, Germany would have run out of explosives and fertilizers in the first six months of the war. During the war he came to direct Germany's scientific effort; in 1915 he experimented with poison gas and supervised the introduction of the new weapon at the western front. To enable him to operate within a military machine that had no understanding of the need for a scientist, Haber received the effective rank of captain. He relished his new role; marshaling all one's talents and energies in a cause one believes in and under the shadow of danger is a heady experience. Einstein, the lonely pacifist who had come to feel his solidarity with Jews, and Haber, the restless organizer of wartime science and a convert from Judaism—the contrast is obvious. For all their antithetical responses, Haber and Einstein remained exceptionally close and, on Haber's side, loving friends. Haber's

life was a kind of foil to Einstein's, and it encompassed the triumphs and the tragedy of German Jewry. I shall return to him because his relations with Einstein were so important—and because he happened to have been my godfather and paternal friend of my parents.

Einstein had been horrified at the beginning of the war, but I doubt that even he could have imagined the full measure of disaster: the senseless killing and maiming of millions, the starving of children, the mortgaging of Europe's future, the rupture of a civilization that appeared ever more fragile. For what? Why? Einstein blamed it on an epidemic of madness and greed that had suddenly overwhelmed Europe—and Germany most especially. The old German dream of greatness had turned into a nightmare of blind and brutal greed. During the later phases of the war, Einstein was again totally absorbed in his work, but whiffs of hysteria would reach him— and always from the German side. I doubt that he knew of the excesses on the other side.

Einstein had been right about the war. At its end many felt as he had at the beginning. The war was a great radicalizing experience, pushing most people to the left and some to a new, frantic right. If there had been no war, Bolshevism and fascism would not have afflicted Europe. The war discredited the old order and the old rulers; antagonism to capitalism, imperialism and militarism appeared everywhere. Lenin's Bolsheviks offered themselves as the receivers of a bankrupt system; Bolshevism was a speculation in Europe's downfall. Liberal Europeans pinned their hopes on Woodrow Wilson, but those hopes faded in the vengeful spirit of Versailles. The logic of events had brought many Europeans to share Einstein's radical–liberal, faintly socialist, thoroughly internationalist views.

A public figure

For a short time Einstein had hopes for Germany. Defeat had brought the collapse of the old and the rise of a new, democratic regime, as he had expected. In November 1918, at the height of the

Fritz Haber and Einstein. The two men first met in 1911 and remained friends until Haber's death in 1934.

Über

Relativitätsprinzip, Äther, Gravitation

Von

P. Lenard

in Heidelberg

Neue, vermehrte Ausgabe

Ladenpreis
für Deutschland und Deutsch-Österreich 5 Mk.;
für das Ausland nur nach der Valutaordnung
des deutschen Buchhandels

Verlag von S. Hirzel in Leipzig 1920

Anti-relativity pamphlet. This is the title page of Philipp Lenard's 1920 pamphlet "On the relativity principle, ether, gravitation," in which he argues that relativity theory is false. This eminent physics professor attacked Einstein not only as a publicity-seeking theorist but also as a Jew.

German Revolution, he cautioned radical students who had just deposed the university rector: "All true democrats must stand guard lest the old class tyranny of the right be replaced by a new class tyranny of the left." He warned[1] against force, which "breeds only bitterness, hatred and reaction," and he condemned the dictatorship of the proletariat in the first of his occasional bitter denunciations of the Soviet Union as the enemy of freedom. However, at other times and in different contexts, he would sign appeals of what we have come to call "front organizations."

We now come to a fateful coincidence in the rise of the public Einstein. In March 1919 a British expedition headed by Arthur Stanley Eddington had observed the solar eclipse. In November it was announced that the results confirmed the predictions of the general theory of relativity. In London the president of the Royal Society, Nobel laureate J. J. Thomson, hailed Einstein's work, now confirmed, as "one of the greatest—perhaps *the* greatest of achievements in the history of human thought." Almost overnight Einstein became a celebrated hero—the scientific genius, untainted by war, of dubious nationality, who had revolutionized man's conception of the universe, redefined the fundamentals of time and space, and done so in a fashion so recondite that only a handful of scientists could grasp the new, mysterious truth.

The new hero appeared, as if by divine design, at the very moment when the old heroes had been buried in the rubble of the war. Soldiers, monarchs, statesmen, priests, captains of industry—all had failed. The old superior class had been found inferior; *Disenchantment* was, appropriately, the title of one of the finest books written about the war. "Before 1914," Noel Annan has asserted,[2] "intellectuals counted for little." After the war, and in a sense in the wake of Einstein, they counted for more. Einstein now became a force, or at least a celebrity, in the world.

After 1919 Einstein appeared more and more often as a public figure. His views were continually solicited, and

Einstein and Zionism

Chaim Weizmann (bearded) and Einstein at a reception at City Hall, New York, April 1921.

During World War I, Einstein became a champion of Zionism, the effort to found a Jewish homeland in Palestine as a secular haven for the persecuted and as a means of moral regeneration. By the early 1920s he had become a public advocate of Zionism—to the surprise and likely dismay of many of his colleagues. Assimilated Jews must have found this reminder of Jewish apartness painful; internationalists must have boggled at the implied argument for a new national community. But Einstein had come to feel a sense of solidarity with other Jews, especially with Jewish victims of discrimination, and he appeared to believe in the existence of an ineradicable antagonism between gentiles and Jews, especially between Germans and Jews—with the fault by no means all on one side. Hence his view that Jews needed a spiritual home and a possible haven. He specifically cited the discrimination that talented Jews from Eastern Europe and from Germany suffered at German universities.

In 1921 Chaim Weizmann persuaded Einstein to join him on a trip to the United States to raise money for the projected Hebrew University in Jerusalem (see the photograph at right). For Weizmann, Einstein's support was critical; for Einstein, the visit to Jerusalem in 1923 was a deeply moving experience. Still, there were conflicts. Einstein railed against the mediocrity of the American head of the university; he saw him as a creature of the crass American–Jewish plutocrats for whom Einstein had contempt even as he helped to lighten their financial burden. He quarreled publicly with Weizmann over the policies of the Hebrew University and repeatedly threatened to withdraw his sponsorship. He urged a Jewish presence in Palestine that would promote, not injure, Arab interests. In 1929, at a time of major attacks on Jewish settlements, he again pleaded with Weizmann for Jewish–Arab cooperation and warned against a "nationalism à la prussienne," by which he meant a policy of toughness and a reliance on force:

> If we do not find the path to honest cooperation and honest negotiations with the Arabs, then we have learned nothing from our 2000 years of suffering and we deserve the fate that will befall us. Above all, we should be careful not to rely too heavily on the English. For if we don't get to a real cooperation with the leading Arabs,

then the English will drop us, if not officially, then *de facto*. And they will lament our debacle with traditional, pious glances toward heaven, with assurances of their innocence, and without lifting a finger for us. [Letter to Chaim Weizmann, 29 November 1929, in the Weizmann Archives, Yad Chaim Weizmann, Rehovot, Israel]

Weizmann replied instantly, at the height of the violence in Palestine, with a four-page handwritten letter. He expounded his views, which were somewhere between those of Zionist extremists and those of the irenic Einstein—who, in the meantime, had criticized the Jewish stance publicly. Weizmann pointed to the recalcitrance of the Arab leaders, their fanaticism, their inability to understand anything but firmness. He pleaded with Einstein to cease his injurious attacks on the Zionists. Of course they would negotiate in time, Weizmann insisted, but "we do not want to negotiate with the murderers at the open grave of the Hebron and Safed victims." Einstein remained skeptical. Weizmann, desperate to retain his support, had written to Felix Warburg a year earlier, "There is really no length to which I would not go to bring back to our work the wonderful and lovable personality—perhaps the greatest genius the Jews have produced in recent

centuries and withal so fine and noble a character."

At the time of the greatest need for a Jewish home in Palestine, immediately after Hitler's seizure of power, Einstein formally broke with the Hebrew University and with Weizmann. The correspondence between the two men suggests all the intractable issues about Jewish–Arab relations, all the differences between the safe outsider and the practical statesman. In April 1938 Einstein resigned his position on the Governing Body of the Hebrew University and again warned against a "narrow nationalism." Once again Weizmann explained that at the moment when five million Jews faced, as he put it, "a war of extermination," they needed the support of the intellectual elite of Jewry, and not, by implication, public criticism.

Einstein was not an easy ally. To some he must have appeared as a man of conscience and of unshakable principle; to others, as an uncompromising fanatic impervious to practical exigencies. As Robert Oppenheimer put it in his memorial lecture on Einstein (*New York Review of Books*, 17 March 1966, page 4): "He was almost wholly without sophistication and wholly without worldliness. . . . There was always with him a wonderful purity at once childlike and profoundly stubborn."

---

he obliged with his ideas about life, education, politics and culture. He had a special kinship with other dissenters from the Great War. Like Bertrand Russell, Romain Rolland and John Dewey, he became what the French call *un homme de bonne volonté*. His views—rational, progressive, liberal, in favor of international cooperation, condemnatory of the evils of militarism, nationalism, tyranny and exploitation—described as well a cast of mind

characteristic of the Weimar intelligentsia.

The intellectuals of Weimar—and this needs to be said at a time when Weimar is often portrayed as some sort of Paradise Lost—were a shallow lot in their moralizing politics. Their views often were utopian and simplistic, pious and fiercely polemical by turns. They were cynical because, as Herbert Marcuse once said to me about himself, they knew how beautiful the world

could be. They lived in a world peopled by George Grosz caricatures and three-penny indictments of bourgeois falsehood. It is perhaps too simple to say that they lived off the bankruptcy of the old order, but they did rather revel in the crudity of their opponents. It is not good for the mind to have dumb, discredited enemies. The real strength of Weimar lay in clusters of talent: in Heidelberg around Max and later Alfred Weber; at Göttingen in mathemat-

BROWN BROTHERS

Einstein in a motorcade on the occasion of his arrival in New York City, 1921.

ics; the Bauhaus and the Berlin circles.

Einstein stood above these progressive intellectuals, in consonance with them, but usually more complicated and less predictable and always more independent than they. But he too was a theorist without a touch of practical experience. Einstein offered his prescriptions the more readily because he had been so overwhelmingly right when the multitudes had been wrong. By 1919 he had not only overthrown the scientific canons of centuries; he had also defied conventional wisdom and mass hysteria in wartime. His views were often deceptively simple; they were not so naive as has often been alleged nor quite so profound as admirers think. There is no reason to think that a scientific genius will have special insights into other realms. He had reflected on some issues and felt strongly about others; as for the rest, his views showed clearly that genius is divisible and can be compartmentalized.

Einstein's views and prescriptions were unassailably, conventionally well intended, but they often lacked a certain *gravitas*, a certain reality—in part, I think, because he approached the problems of the world distantly, unhistorically, not overly impressed by the nature or intractability of the obstacles to ideal solutions. He was not a political thinker; he was a philosopher, moralist, prophet, and the travails of the world would prompt him to propose or support social remedies. Sometimes those remedies would be blueprints of utopia addressed to people who had lost their footing in a swamp and were sinking fast.

Much later, in fact at a moment when Einstein had attacked the Nazi government, Max von Laue questioned whether the scientist should deal with political issues. Einstein rejected such considerations:

. . . you see especially in the circumstances of Germany where

such self-restraint leads. It means leaving leadership to the blind and the irresponsible without resistance. Where would we be if Giordano Bruno, Spinoza, Voltaire and Humboldt had thought and acted this way?

Laue pointed out in a letter to Einstein that Einstein's examples were not exact natural scientists and that physics was so remote as not to prepare its practitioners for politics in the same way that law or history did. On that letter Einstein simply scribbled, "Don't answer."

Like so many thinkers of the 1920s Einstein underestimated the force of the irrational, of what the Germans call the demonic, in public affairs. That is what left them so ill prepared for an understanding of fascism. In their innocence they thought that men were bribed to be fascists, that fascism was but frightened capitalism; in its essence it was something much more sinister and elemental.

A democratic rebuke to authority

What gave Einstein's views exceptional resonance was the magic of his person and his incomparable achievement. He was taken by many as a sage and a saint. In fact, as I have said before, he was an unfathomably complex person. In the complexity of nature he found simplicity; in the complexity of his own nature, the principle of simplicity ranked high. Indeed, it was his simplicity, his otherworldliness, that impressed people. His clothes were simple, his tastes were simple, his appearance was meticulously simple. His modesty was celebrated—and genuine, as was his unselfishness. He was a lonely man, indifferent to honors, homeless by his own admission, solicitous of humanity and diffident about his relations with those closest to him. At times he appeared like a latter-day St. Francis of Assisi: a solitary saint, innocently sailing, those

melancholy eyes gazing distractedly into the distance. At other times he played with the press, finding himself in the company of the famous and the powerful despite himself.

In some ways, I believe, he came to invest in his own fame, perhaps unconsciously to groom himself for his new public role. He lectured in distant lands, "a traveler in relativity." In 1921, after his first visit to the United States, he said:[3]

The cult of individuals is always, in my view, unjustified. . . . It strikes me as unfair, and even in bad taste, to select a few [individuals] for boundless admiration, attributing superhuman powers of mind and character to them. This has been my fate, and the contrast between the popular estimate of my powers and achievements and the reality is simply grotesque.

This admiration would have been unbearable except that "it is a welcome symptom in an age which is commonly denounced as materialistic that it makes heroes of men whose goals lie wholly in the intellectual and moral sphere. . . . My experience teaches me that this idealistic outlook is particularly prevalent in America." He knew that he had become a hero—and was endlessly surprised by it. In 1929 he described himself as a "saint of the Jews." He played many roles by turns, each, I think, completely genuinely; he was a simple man of complex roles.

In the simplicity and goodness that were his, I detect, perhaps wrongly, a distant echo of his encounters with German life. Could one imagine a greater contrast between the Germans surrounding him—people so formal in their bearing, so attentive to appearance, so solicitous of titles, honors and externals—and himself? Did the insolence of office, the arrogance of the uniform, push him into ever greater idiosyncratic informality? Was not his appearance a democratic rebuke to

League of Nations commission. Einstein is seated fourth from the right at this 1927 meeting of the International Commission on Intellectual Co-operation. Hendrik A. Lorentz is at the far right.

authority?

In the immediate postwar era, Einstein was friendly to the governments of Weimar and appalled by the vindictiveness of the Allies, who seemed to have caught what he had thought was a German disease. In all his public stands he had what Gerald Holton has called a "vulnerability to pity," and in the early 1920s he had a fleeting moment of pity for Germany. He refused to leave it in its time of trial. For years he was an uncertain member of the International Commission on Intellectual Co-operation of the League of Nations, intermittently resigning whenever he thought the commission too pro-French, too *Allied.* He hoped to restore an international community, Germans included. In the end he asked Haber to take his place. Successive German governments regarded him as a national asset, perhaps their sole asset in a morally and materially empty treasury. They saw in his travels and in his fame the promise of some reflected glory. But his own hopes gradually faded. He had warned Walter Rathenau against assuming the foreign ministry; Jews should not play so prominent a role, he felt. When right-wing assassins killed Rathenau and were widely hailed in Germany as true patriots, Einstein had reason to fear for his own life. The inborn servility of the Germans, he thought, had survived the successive shocks of 1918.

Immediately after the war and at the beginning of his popular fame, Einstein embraced several causes. Having embraced them, he would often embarrass and repudiate them as well. He was the antithesis of an organization man. Unstintingly he would help individuals and chosen causes, but I doubt that he

listened to them. He remained a detached theorist who thought the rational order of the world wantonly violated, but at times his commandments contained visionary practicality. A pacifist during the war, he now became Germany's most prominent champion of organized pacifism. He hated militarism—as blindly as its defenders loved it. He condemned[4] "the worst outgrowth of herd life, the military system. . . . I feel only contempt for those who take pleasure marching in rank and file to the strains of a band. . . . Heroism on command, senseless violence and all the loathsome nonsense that goes by the name of patriotism—how passionately I despise them!" This, surely, is exemplary of the spirit of the 1920s, formed by the experience of the first war and soaked in the we–they antithesis that precludes understanding. It precluded the understanding that had led William James to plead for a moral equivalent of war, for something practical that would make peaceful use of the old martial virtues. Einstein insisted that "the advance of modern science has made the delivery of mankind from the menace of war . . . a matter of life and death for civilization as we know it." But Einstein did not grapple with the psychological issues, with people's desire for danger and comradeship. In his exchange with Freud about the nature of war he acknowledged[5] that "the normal objective of my thought affords no insight into the dark places of human feeling and will." For Einstein war was a disease, a disorder planted by men of greed, to be abolished by men of good will through the creation of international sovereignty or through a revolutionary pacifism, that is, through the refusal of men to bear arms in peace or

war. He called for resistance to war, but in 1933, almost immediately after Hitler's assumption of power, he renounced pacifism altogether—to the fury of his doctrinaire followers. In fact he urged the Western powers to prepare themselves against another German onslaught.

Humane collegiality

Unlike many academics, Einstein took education with the utmost seriousness—and academics with magnificent irreverence. He had great faith in the possibilities of primary and secondary education; at one point he said that if the League of Nations improved primary education, it would have fulfilled its mission. His ironic contemplation of universities found expression in private letters. He once complimented[6] his close friend Max Wertheimer, the Gestalt psychologist: "I really believe there are very few who have been so little harmed by learning as yourself." In 1924 he wrote:[7] "In truth, the university is generally a machine of poor efficacy and still irreplaceable and not in any essential way improvable. Here the community must take the point of view that the biblical God took toward Sodom and Gomorrah. For the sake of very few, the great effort must be made—and it is worth it!"

Einstein's success—the enormous acclaim, especially abroad at a time when most German scientists were still banished from international meetings—caused much ill will at home. His opinions enraged the superpatriots. Some physicists condemned the fanfare surrounding the dubious theory of relativity; one fellow Nobel laureate, Philipp Lenard, attacked it as "a Jewish fraud." For anti-Semites, Einstein became a favorite and obvious target.

ULLSTEIN

Meeting at Harnack House, Berlin, 1931. Left to right: Max Planck, British Prime Minister Ramsay MacDonald, Reichsminister Gottfried Treviranus (from behind), Einstein, Privy Counselor Hermann Schmitz (?) of I. G. Farben, Vice-Chancellor Hermann Dietrich and (partially obscured) Foreign Minister Julius Curtius (?).

The waves of hatred spilled from the streets into the lecture halls, and Einstein's occasional and sometimes ill-considered deprecations made things worse.

It would be hard to imagine three causes less pleasing to the bulk of the German professoriat than liberal internationalism, pacifism and Zionism (see the box on page 44).

Germany frightened Einstein again. His hopes for the Weimar Republic had dimmed. As early as 1922 his life was threatened. He traveled even more than before, but still he refused handsome offers from Leiden and Zurich, the universities with which he had the closest ties. He stayed in Germany despite his misgivings; he stayed because Berlin in the 1920s was the golden center of physics; he stayed because proximity to Planck, Laue, Haber and others was a unique professional gift, because, as he wrote Laue in 1928, "I see at every occasion how fortunate I can call myself for having you and Planck as my colleagues." In 1934 he wrote Laue that "the small circle of men that earlier was bound together harmoniously was really unique and in its human decency something I scarcely encountered again." In 1947 he wrote[8] Planck's widow that his time with Planck "will remain among the happiest memories for the rest of my life."

The unpublished correspondence among these men suggests even more than a professional tie. The letters bespeak a degree of humane collegiality, a shared pleasure in work, as well as a delicacy of sentiment, a candid avowal of affection, that in turn allowed for confessions of anguish and self-doubt, of melancholy as well as high spirits. They spoke of joys and

torment, in close or distant friendship, in an enviable style. The letters also breathe a kind of innocence, as if science was their insulated realm, nature the great, enticing mystery and one's labors of understanding exclusively an intellectual pursuit, remote from social consequences. Such clusters of collaboration and of friendship have always existed, I suppose, and they have made life better and infinitely richer. Germany may have had a special knack for breeding them.

Einstein's Germany included gentiles and Jews working together in extraordinary harmony. Still, one can state categorically that none of the Jewish scientists escaped the ambiguity, the intermittent hostility, that being a Jew entailed in imperial and Weimar Germany. Neither fame nor achievement, neither the Nobel Prize nor baptism, offered immunity. Passions were fiercer in Weimar, that cauldron of resentments. Official barriers against Jews had been lowered, but new fears and hatreds came to supplement old prejudices.

Three incidents may illustrate the uncertain temper of the time. In 1921 Haber begged Einstein not to go to America with Chaim Weizmann on the ground that Germans would take amiss Einstein's traveling in Allied countries with Allied nationals at the very time when the Allies were once again tightening the screws on Germany. To persuade Einstein, Haber warned that German anti-Semites would capitalize on Einstein's seeming desertion and that innocent Jewish students would be made to suffer; anti-Semitism, rampant as it was, did not need to be goaded. Einstein's warning to Rathenau originated in a similar apprehension. Or take another incident. In 1920 a well-

known physicist opposed the university appointment of the later Nobel laureate Otto Stern: "I have high regard for Stern, but he has such a corrosive Jewish intellect."

Or consider this last example. In 1915 the king of Bavaria, confirming the Nobel laureate Willstätter's appointment to a professorship at Munich, admonished his minister, "This is the last time I will let you have a Jew." Ten years later, discussing with his colleagues a new academic appointment, Willstätter proposed a candidate. A murmur arose: "another Jew." Willstätter walked out, resigned his post and never entered the university again, the unanimous pleas of his students notwithstanding. For the next 14 years he had daily, hour-long telephone calls with his assistant so that she could conduct the experiments in a laboratory that he would no longer enter. A man of conscience and of courage, someone who did not blink at the reality of anti-Semitism. But his stand in 1924 was his undoing a decade later. A devoted German, but no longer a civil servant, he assumed that the Nazis would leave untouched a private scholar. He believed that some Jews had contributed to this new storm. He could not comprehend the radical newness of the phenomenon. In February 1938 he wrote my mother urging her not to leave Germany without the most careful reflection. He himself refused exile until the aftermath of the *Kristallnacht* forced him into it.

I cite Willstätter's example in particular precisely because of its contradictory nature: Awareness of anti-Semitism could cloud one's perception of Nazism. If anti-Semitism had always existed, then perhaps Nazism was but an intensification of it. It is not uncommon these days to hear summary judgments about German Jewry, about their putative self-surrender, their cravenness or their opportunism. These judgments often have a polemical edge and they are likely to do violence to the past and to the future: The myth of yesterday's self-surrender could feed the delusion of tomorrow's intransigence. If our aim is to understand a past culture, we must note that German–Jewish scientists thought Germany their only and their best home, despite the anti-Semitism that crawled all around them. They may have loved not wisely but too well, and yet their sentiments are perhaps not so much an indictment of themselves as a tribute to the appeals of Germany. We owe that past no less than what we owe any past: a sense of its integrity.

The denouement

Let me hasten to the denouement. In 1932 Einstein left Germany provisionally, with the intention of returning to Berlin for one semester each year. Hitler's accession to power the next year changed all that. Einstein immediately denounced the new regime, and in response the Prussian Academy expelled him. His books were burned, his property seized. The first Nazi decrees on the purification of the universities would have allowed some Jews to maintain their positions. Einstein's non-Aryan friends spurned such sufferance and resigned. German physics was decimated, and a few remaining masters battled to defend some shreds of decency, some measure of autonomy. Laue once wrote Einstein that in teaching the theory of relativity he had sarcastically added that it had of course been translated from the

Hebrew. Even such jokes—to say nothing of Laue's eulogies of Jewish colleagues—aroused Nazi wrath. The Nazis proscribed the very mention of Einstein, even in scientific discussions. They wished him to be a nonperson.

For most, exile was hard; the habits of a lifetime are not easily shaken. For others, as the physicist Max Born put it,[9] "a disaster turned out to be a blessing. For there is nothing more wholesome and refreshing for a man than to be uprooted and replanted in completely different surroundings." Resiliency was a function of age and temperament. For Haber exile was a crushing blow and led to a final irony in his relations with Einstein. In mid-1933 he wrote to Einstein that as soon as his health would allow it, he would go to Palestine, but in the meantime he begged Einstein to patch up the public quarrel Einstein had had with Chaim Weizmann (see box on page 44). Einstein replied[10] at length: "pleased...that your former love for the blond beast has cooled off a bit. Who would have thought that my dear Haber would appear before me as defender of the Jewish, yes even the Palestinian, cause. The old fox [Weizmann] did not pick a bad defender." He then lashed out against Weizmann and concluded:

I hope you won't return to Germany. It's no bargain to work for an intellectual group that consists of men who lie on their bellies in front of common criminals and even sympathize to a degree with these criminals. They could not disappoint me, for I never had any respect or sympathy for them— aside from a few fine personalities (Planck 60% noble, and Laue 100%). I want nothing so much for you as a truly humane atmosphere in which you could regain your happy spirits (France or England). For me the most beautiful thing is to be in contact with a few fine Jews—a few millennia of civilized past do mean something after all.

The German patriot Haber died a few months later in Basel, en route to Palestine. And Einstein found a refuge at the Institute for Advanced Study at Princeton under conditions not dissimilar from what the Prussian Academy had offered him 20 years earlier. For as Erwin Panofsky has said[11] of the institute, it "owes its reputation to the fact that its members do their research work openly and their teaching surreptitiously, whereas the opposite is true of so many other institutions of learning."

Einstein's public life continued to be

Henry A. Wallace, Einstein, journalist Frank Kingdom and singer Paul Robeson during the 1948 Presidential campaign. Wallace was the Progressive party candidate.

BETTMANN ARCHIVE

Rudolf W. Ladenburg and Einstein, on the occasion of Ladenburg's retirement from Princeton University.

dominated by his fear of Germany. He warned the West against a new German onslaught. He abandoned the pacifism he had so fervently espoused and in 1939 signed the famous letter to President Franklin D. Roosevelt urging the Administration to prepare the United States because Germany might develop nuclear fission for military purposes. In the winter of 1945, when Germany was desolate in defeat and when the Morgenthau spirit, if not the plan, had a considerable grip on American thinking, a fellow laureate and old friend, James Franck, asked Einstein to sign a manifesto of exiles that would appeal to the United States not to starve the German people. Einstein vowed that he would publicly attack such a plea. Franck pleaded with him that to give up all hope for a moral position in politics would be tantamount to a Nazi victory after all. But Einstein, who had signed so many appeals that he himself once said he was not a hero in no-saying, scathingly rejected Franck's plea. For him, genocide was Germany at its most demonic; after Auschwitz he could muster no magnanimity. Even the righteous could not redeem the "country of mass murderers," as he called Germany. He rebuffed Laue's plea to help a young German physicist. He knew that Planck, who lost one son in the first war, had now lost another, whom the Nazis murdered because of his participation in the plot against Hitler. The serene Einstein, always the champion of the rights of the individual against the collectivity, now proclaimed the principle of collective guilt. At that moment, of course, the world shared Einstein's horror at German inhumanity. But in him the violence of senti-

ment, the total absence of that vulnerability to pity, puzzles, for it shows how desperately deep and all-consuming his antipathy to Germany had become.

Even Einstein's postwar laments about America, his horror at McCarthyism, were shaped by his image of Germany. America, he believed, was somehow following the path of Germany. The world of politics he saw through German eyes—always.

But his deepest feelings also retained something of a German cast, echoed some very German themes. When Rudolf Ladenburg, a physicist and fellow exile, died in 1952, Einstein spoke[12] at the graveside:

Brief is this existence, as a fleeting visit in a strange house. The path to be pursued is poorly lit by a flickering consciousness, the center of which is the limiting and separating "I."

The limitation to the I is for the likes of our nature unthinkable considering both our naked existence and our deeper feeling for life. The I leads us to the Thou and to the We—a step which alone makes us what we are. And yet the bridge which leads from the I to the Thou is subtle and uncertain, as is life's entire adventure.

When a group of individuals becomes a We, a harmonious whole, then the highest is reached that humans as creatures can reach.

★   ★   ★

This essay is a revised version of a lecture I gave at the Einstein Centennial Symposium in Jerusalem in 1979. Papers from the symposium appear in Albert Einstein, Historical and Cultural Perspectives, *G. Holton,*

Y. Elkana, eds., Princeton U.P., Princeton, New Jersey (1982).

I benefited from conversations with Marshall Clagett, Felix Gilbert, Gerald Holton, Martin Klein, I. I. Rabi and Malvin Ruderman. It was in long and frequent talks with Otto Stern that I first sensed how extraordinary those early days in Zurich must have been.

References

● In writing this essay I found *Einstein on Peace* and *Ideas and Opinions*, references 1 and 3 below, particularly pertinent. I was also fortunate enough to be allowed to use the Albert Einstein Archives when they were still at the Institute for Advanced Study in Princeton, New Jersey, a treasure made still more valuable by the always helpful advice and recollections of Helen Dukas, who was in charge of them. I also read the unpublished correspondence of James Franck and Albert Einstein, deposited at the University of Chicago Library. In addition to the books cited in the preceding notes, I found the following particularly useful: A. D. Beyerchen, *Scientists under Hitler: Politics and the Physics Community in the Third Reich*, Yale U.P., New Haven (1981); G. Holton, *The Scientific Imagination: Case Studies*, Cambridge U.P., New York (1978); R. Willstätter, *Aus Meinem Leben*, Verlag Chemie, Weinhein (1949); and H. Zuckerman, *Scientific Elites: Nobel Laureates in the United States*, Free Press, New York (1977).

1. O. Nathan, H. Norden, eds., *Einstein on Peace*, Schocken, New York (1960), p. 25.

2. N. Annan, Daedalus, Fall 1978, p. 83.

3. A. Einstein, *Ideas and Opinions*, new trans. and rev. by S. Bargmann, Crown, New York (1954), p. 4.

4. O. Nathan, H. Norden, eds., *Einstein on Peace*, Schocken, New York (1960), p. 111.

5. A. Einstein, S. Freud, *Why War?*, International Institute of Intellectual Cooperation, League of Nations, Paris (1933), p. 12.

6. A. Einstein, letter to Max Wertheimer in the Einstein Archives, Boston Univ., Boston, Mass.

7. A. Einstein, letter to Julius Schwalbe, 18 July 1924, in the Einstein Archives.

8. Letters in the Einstein Archives.

9. M. Born, *My Life and My Views*, Scribner's, New York (1968), p. 38.

10. Letter in the Einstein Archives.

11. E. Panofsky, *Meaning in the Visual Arts*, U. of Chicago P., Chicago (1983), p. 322.

12. Copy in the Einstein Archives.   □

PHYSICS AND SOVIET– WESTERN RELATIONS IN THE 1920s AND 1930s

After seven years of isolation brought about by world war
and the Russian Revolution, Soviet physicists rejoined
the international scientific community in the early 1920s,
only to have these contacts restricted again under Stalin.

Paul R. Josephson

PHYSICS TODAY/SEPTEMBER 1988

The recent intensification of Soviet interest in participating in international scientific activities is not merely a result of new policies introduced under Mikhail Gorbachev. Soviet scholars have sought out contacts with Western scholars since the first years of the Russian Revolution, although under Joseph Stalin and during times of heightened political tensions with the West these contacts have been reduced. On the eve of the revolution, the Russian empire had no more than a hundred physicists—including professors, docents and laboratory assistants with the equivalent of graduate degrees, but not including primary- or secondary-school teachers—and very few well-equipped laboratories. While creating conditions propitious to the long-term growth of physics as a dispcipline, the revolution led to short-term disruptions of research. Making matters worse, World War I cut physicists off from their customary contacts with Western scholars and laboratories, and the 1918–20 civil war between the Reds (the Bolsheviks and their allies) and the Whites (the monarchists and their sympathizers) atomized the domestic physics community.

This article examines how Soviet physicists reestablished scientific relations with the West during the 1920s after almost eight years of isolation, and how in the 1930s, under Stalin, Soviet scholars once again lost those vital

scientific ties they had worked so hard to resurrect. The story involves many great physicists, including:

▷ Lev Davidovich Landau, the greatest Russian theorist of the 20th century. Landau obtained first-rate results in virtually every area of theoretical physics, ranging from quantum electrodynamics to solid-state physics. He was the author with Evgenii Mikhailovich Lifshitz of an internationally known series of texts on theoretical physics.

▷ Petr L. Kapitsa, who worked under Ernest Rutherford at the Cavendish Laboratory in Cambridge, England, and after returning to Russia worked on high magnetic fields, superconductivity, superfluidity and other areas of low-temperature physics at the Institute of Physical Problems, which was created for him in 1937. Kapitsa designed liquefiers for helium and other gases, and created the Soviet liquid oxygen industry.

▷ Sergei I. Vavilov, head of the department of physical optics at the Institute of Physics and Biophysics in the 1920s, then scientific director of the State Optical Institute and later president of the Soviet Academy of Sciences. It was under Vavilov's guidance that Pavel A. Cherenkov did his work on radiation from electrons moving quickly through matter.

▷ Nikolai N. Semenov, deputy director of the Leningrad Physico–Technical Institute, then director of the Institute of Chemical Physics and vice president of the Soviet Academy of Sciences, and one of the founders of chemical physics.

▷ Abram Feodorovich Ioffe (pictured on page 57), founder

Paul Josephson teaches science policy and the history of science at Sarah Lawrence College, in Bronxville, New York.

Talking physics on the Volga in 1928 during the sixth congress of the Russian Association of Physicists are (from left to right) Paul Dirac from England, Iakov I. Frenkel' of the Soviet Union and Alfred Lande from Germany. The general makeup and program of the congress reflected the youthful vitality of Soviet physics and the growing international respect accorded Soviet physicists. Much of the congress was held during a steamship trip down the Volga, in an atmosphere that promoted free discussion of controversial questions of contemporary physics.

of the Leningrad Physico–Technical Institute, which became the leading Soviet physics institute between the two world wars, and organizer of the Russian Association of Physicists. Ioffe, the dean of Soviet experimental physics for many years, worked on the physics of semiconductors and the electrical, mechanical and other properties of crystals.

▷ Iakov I. Frenkel' (shown in the photo above), a leading theorist of the early Soviet period whose work covered the electronic theory of solids, quantum mechanics, magnetism, the physics of liquids and nuclear physics.

War, revolution and isolation

World War I and the revolution isolated Russia from other countries, created shortfalls of materials, supplies and reagents, disrupted publication and interrupted regular government funding for scientists. The Czar had abdicated in February 1917, and physicists met with little success in getting Alexander Kerensky's provisional government to turn attention to these difficulties. The Bolshevik coup in October initially made matters worse. Throughout the winter of 1917–18, as research and living conditions grew ever more precarious, physicists, who had sought to avoid political involvement, opened formal talks with the new government. By the spring of 1918 attendance at periodic meetings of the physics section of the Russian Physico–Chemical Society had dropped to 25, less than half the normal figure. With sources of support dwindling and cold and hunger taking hold of Leningrad, physicists approached Anatoli V. Lunacharskii, Commissar of Enlightenment, for financial support. They needed funding for all areas of the physics enterprise: publications, congresses and the several newly founded physics institutes. By the end of March, the physicist Orest D. Khvol'son had secured emergency publication subvention from Lunacharskii, but this went little beyond short-term needs. Leading Soviet scholars, and particularly those in the Leningrad physics community, therefore began to work through a newly founded organization, the Russian Association of Physicists, to lobby the government for the financial resources necessary to reopen domestic and international scientific contacts.

Publication of major Soviet physics journals, 1917–23

	Number of journals published	Total number of issues
1917	1	3
1918	2	7
1919–20	3	12
1921	1	2
1922	3	9
1923	2	7

Standing at the forefront of this effort was Ioffe, who enlisted the help of Dmitrii S. Rozhdestvenskii, founder of the State Optical Institute, Khvol'son and Aleksei N. Krylov from the Soviet Academy of Sciences. In 1919 they and other Leningrad and Moscow physicists had organized the Russian Association of Physicists to attack the problems that frustrated development of the nascent discipline. The group was set up as an adjunct to the physics section of the Czarist-era Russian Physico–Chemical Society, a body of some 200 members, and soon took its place. The association pressed the Bolshevik government to support research and publication. There was a critical need for materials and apparatus from abroad because Russia's factories were as yet unable to produce them.

At its first congress, held on the 50th anniversary of the Mendeleev periodic chart in February 1919, in what a report in the *Journal of the Russian Physico–Chemical Society* called a "melancholy Leningrad of iced-over buildings without food or heat or the promise of reimbursement of expenses," the Russian Association of Physicists established a committee of four—Ioffe, Krylov, V. A. Anri and Petr P. Lazarev, a Moscow physicist who founded the Institute of Physics and Biophysics—to coordinate the receipt of foreign literature, instruments and equipment, and to resurrect foreign research travel. They called for government support for new institutes, reestablishment of ties between physicists in Russia and abroad, and resumption of publication. Physicists at the congress also discussed such diverse topics as the Bohr model of the atom, optics, spectrography, relativity and quantum theory.

Members of the association encountered significant difficulties in following through on their professional mandate. During the 1921–22 academic year, the position of the association deteriorated. It lost government funding and had to turn to its 17 institutional members for support. At national congresses in Moscow in 1920, in Kiev in 1921 and in Nizhnyi Novgorod in 1922, the association discussed the continuing scientific and organizational challenges to the discipline, but it failed until the fourth congress in Leningrad in 1924 to achieve the national flavor and strength it had desired from the start.

The major Soviet physics journals of the time were the following:
▷ *Zhurnal Russkogo Fiziko–Khimicheskogo Obshchestva*, or *Journal of the Russian Physico–Chemical Society*; in 1931 the physics section of this journal became *Zhurnal Eksperimental'noi i Teoreticheskoi Fiziki*, or *Journal of Experimental and Theoretical Physics*.
▷ *Vestnik Rentgenologii i Radiologii*, or *Herald of Roentgenology and Radiology*, which appeared for three issues before ceasing publication.
▷ *Uspekhi Fizicheskikh Nauk*, or *Successes of the Physical Sciences*.
▷ *Trudy Gosudarstvennogo Opticheskogo Instituta*, or *Works of the State Optical Institute*.
▷ Academy of Sciences publications such as *Doklady*, or *Reports*, and *Izvestiia*, or *Proceedings*.

Between 1917 and 1923 these physics journals were published irregularly, as the table at the top of this page indicates. The receipt of 50 tons of paper from the United States in 1921 helped reestablish Academy of Sciences publications, although problems persisted until 1924.

Reestablishing ties with the West

Although Russian physicists made progress in the domestic organization of physics research and publication after the revolution, their international isolation continued to be a pressing concern until the mid-1920s. With the end of the civil war in 1920, scientists turned to the government for approval of foreign travel for research and for funds to purchase instruments, books and journals. In petitioning the Commissariat of Enlightenment about "the necessity of foreign travel for scholars for scientific purposes," physicists emphasized the international character of science, which "demanded uninterrupted interaction" of scholars of all countries.

Ioffe assumed a prominent role in this process. He had studied with Wilhelm Röntgen for three years in Munich, where he used x rays to investigate the mechanical strength of quartz crystals. In 1906 he returned to St. Petersburg. There he developed a close friendship with Paul Ehrenfest, who taught at Petersburg University from 1907 to 1912 before taking over Hendrik A. Lorentz's chair in physics at the University of Leiden. Ioffe wrote Ehrenfest in the summer of 1920, asking him to help Soviet scholars catch up on recent developments in the West by sending important recent books to Ioffe through an address in Finland. Ehrenfest did what he could, but could not single-handedly reestablish scientific relations. Soviet physicists needed above all else firsthand contact with Western achievements.

In February 1920 physicists in Leningrad had requested approval from the Main Scientific Administration of the Commissariat of Enlightenment to send the radiologist Mikhail I. Nemenov and Ioffe abroad. The commissariat approved the request, but two problems remained. First, it was necessary to establish diplomatic relations with a Western country to get visas. The USSR established diplomatic relations with Germany in 1922 and with the rest of Europe only after 1924. Second, because the Bolshevik government had canceled all Czarist debt obligations, only hard currency would suffice for foreign purchases. Ioffe was abroad from February through August of 1921 and again from April through September of 1922, and corresponded with his wife in Leningrad during these periods. The letters are extant and reveal the successes Soviet physicists encountered in establishing foreign contacts in the face of the many problems.

Ioffe in Western Europe

On his first trip, Ioffe spent two weeks in Estonia waiting for visas to Holland, Sweden and Germany. He received only the last one. By the end of March he had made it to Berlin and ordered 391 journal subscriptions, for three- to five-year periods, and 350 books, but he was still waiting for Kapitsa to arrive with the letters of credit for these purchases. A large number of the books and journals ordered never arrived; journal subscriptions received at the Leningrad Physico–Technical Institute library in fact dropped from 54 in 1922 to 42 in 1925 and 36 in 1926. While in Berlin, Ioffe gave lectures at several physics colloquia, to which Max Planck, Walter Nernst and others reacted favorably. To Ioffe's delight, Ehrenfest arrived in Berlin in April to renew their friendship, discuss issues in physics and plan how to normalize Soviet scientific exchanges with the West. These personal successes only somewhat buoyed Ioffe's spirits in light of the continued difficulties in Moscow in getting the Commissariat of Enlightenment to release the promised funds; Frenkel' was hard at work trying to solve this problem. When Ioffe finally received the letters of credit in May, he completed a number of transactions and sent 2.4 million deutsche marks of instruments and machine tools, 60 000 DM of chemical products and a spectrograph to the Leningrad Physico–Technical Institute.

In June Ioffe journeyed to England and met with Rutherford, William and Lawrence Bragg and others, and finally met up with Kapitsa. Once again, however, uncertainties over letters of credit dogged his every step. Ioffe expressed particular displeasure at how the government had delayed funds. Nonetheless, Ioffe secured in Britain another eight boxes of books and instruments, including five Coolidge tubes and kenetrons (a type of vacuum tube) costing £7738 sterling. His meetings with Rutherford and the Braggs were particularly successful; among other things, they helped him to arrange publication of an article in *Nature*. In July he returned to Leningrad by way of Berlin, Hamburg and Leiden.

During his second trip abroad, in 1922, Ioffe succeeded in procuring another 100 boxes of equipment, including two powerful 200-kV transformers, string galvanometers and a Siemens oscillograph. While in Berlin, Ioffe met with Karl Wagner, Arnold Sommerfeld, Wilhelm Wien, Kasimir Fajans and Walter Gerlach. A meeting with physicists in Göttingen proved fruitful to Ioffe's efforts to

Abram Feodorovich Ioffe and his wife, Anna Vassilievna Ioffe. Ioffe, the dean of Soviet experimental physics in the 1920s and 1930s, was instrumental in reestablishing contacts with Western scholars after World War I. He founded the Leningrad Physico-Technical Institute, which became the leading Soviet physics institute between the two world wars. (Photograph from Peter H. Plesch.)

reestablish normal scientific exchanges with his European colleagues: The German physicists agreed to send to the Leningrad Scholars' Club several copies of reprints of all German articles in the main physics journals dating back to 1914, and one German scientist agreed to organize a center for the selection of individual reprints of scientific work to send to Russia.

By 1924 foreign contacts were an integral part of Soviet research and an indispensable source of scientific equipment. In that year the Optical Institute's Rozhdestvenskii traveled with $80 000 to Europe, where he purchased instruments that, as a colleague recalled, "opened the possibility of beginning systematic production of optical glass in [a Russian] factory." The Soviet Union opened a special customs point to receive the hundreds of boxes of scientific materials sent to the optical institute. During the 1923–24 academic year Ioffe, Aleksandr A. Chernyshev, Kapitsa and others represented the Leningrad Physico–Technical Institute abroad; the government had by then granted Ioffe's institute a monthly allowance of 1000 rubles for foreign purchases. The institute had also signed the first of a number of contracts with foreign firms for research and development. By 1926 one-fifth of all institute physicists had visited Western physics facilities, although younger scholars, who might have benefited the most from such visits, were excluded from foreign travel.

The Academy of Sciences played a major role in reestablishing international scientific contacts. Three of the 10 scholars sent abroad in 1920 were academicians, as were 8 of the 17 in 1922, 12 of the 25 in 1924 and 15 of the 44 in 1926. Such leading figures as academy president Aleksandr P. Karpinskii, geochemist Vladimir I. Vernadskii, Krylov, geologist Aleksandr E. Fersman and physiologist Ivan P. Pavlov traveled to the West in the first difficult years. The 200th-anniversary celebration of the Academy of Sciences, held in September 1925 and attended by over 120 scientists from Austria, England, Germany, Holland, India, Spain, Italy, the US and France, was especially important to the normalization of scientific ties; academy scientists and government officials used the event to make professional contacts with foreign researchers more acceptable from a political, economic and ideological standpoint.

In the 1927–28 academic year there were almost 400 foreign trips under government auspices, and 11 of these were to the United States. By the next year, however, the number of trips fell to 140. Physicists at Ioffe's institute in particular seem to have benefited from the normalized contacts. Among the almost 40 Soviet physicists who visited Europe and the US were Iurii A. Krutkov, Landau, Vladimir A. Fok, Iurii B. Rumer, Kapitsa, Frenkel' and Ioffe himself.

In the mid-1920s Soviet scholars participated more and more frequently in the international scientific arena, studied at German, American and British universities, and attended such prestigious gatherings as the international Solvay meetings in Brussels. These activities had a positive impact on Russian physicists in three ways. First, Russian physicists often based their ideas for the creation and organization of new research institutes on such examples of university and industrial physics laboratories

as the Physikalisch–Technische Reichsanstalt, MIT, Harvard, Berkeley, Caltech, General Electric and Westinghouse. Second, contacts with Western firms resulted in research contracts that brought in hard currency. Third, the contacts gave Soviet scholars the opportunity to exchange ideas with their Western counterparts, primarily through study sponsored by the International Education Board.

Frenkel' and the IEB

The IEB, which was funded by the Rockefeller Foundation, operated between 1923 and 1938, but was most active in the period 1923–28. Through the IEB Soviet scholars strengthened their ties with foreign scholars and brought back scientific instruments and ideas for their laboratories, as the experiences of Frenkel' and Ioffe demonstrate. Ehrenfest was instrumental in getting the IEB to support physicists. In March 1925 Augustus Trowbridge, IEB's director for Europe, met with Ehrenfest in Leiden about Ehrenfest's plans to get "a few of the more gifted young Russians into Western Europe." Ehrenfest recommended Krutkov and Frenkel' on the grounds that "one will be doing the most that one can for this branch of science in Russia at the present time since these men will go back with new ideas gained by researching for a year in the West." Trowbridge supported Frenkel' 's candidacy because of his impressive publication record.

In his application to the IEB, Frenkel' expressed his desire to work in Göttingen with Max Born on "the electronic theory of solid and liquid bodies" and "general electrodynamics in connection with the quantum theory" before traveling to Cambridge for a few months. He hoped to obtain results "regarding the nature and magnitude of attractive and especially repulsive forces between atoms, and, further, to extend the theory [to] liquid bodies, the liquid state of matter . . . being much nearer to the solid than the gaseous one." Frenkel' also proposed to "show that the electron must be considered as spatially unextended force centers." The IEB approved Frenkel' 's candidacy, funds for travel and a monthly allowance of $182 for Frenkel' and his wife.

When Frenkel' arrived in Berlin in November 1925, however, he found that Born had already departed to deliver a course of lectures at MIT. After consulting with Ehrenfest and Ioffe, Frenkel' left for Hamburg to study with Wolfgang Pauli and Otto Stern, making plans to return to Berlin to work with Albert Einstein and finally in the spring or summer to go to Göttingen with Born. He also consulted with Paul Langevin and Leon Brillouin in Paris. Frenkel' 's European sojourn had a twofold importance for Soviet physics. First, it served to cement ties with European theoreticians, to convince them of the high quality of research going on in the Soviet Union and to pave the way for such scholars as Landau, Petr I. Lukirskii, Viktor R. Bursian and George Gamow to follow him to their laboratories. Second, when Frenkel' returned to the Leningrad Physico–Technical Institute he was better prepared to provide first-rate leadership to a new generation of theoreticians in Leningrad: Fok, Georgii A. Grinberg and Dmitrii D. Ivanenko, and in the 1930s Matvei P. Bronshtein and Lev V. Rozenkevich, both of whom perished in Stalin's purges. Frenkel' went abroad

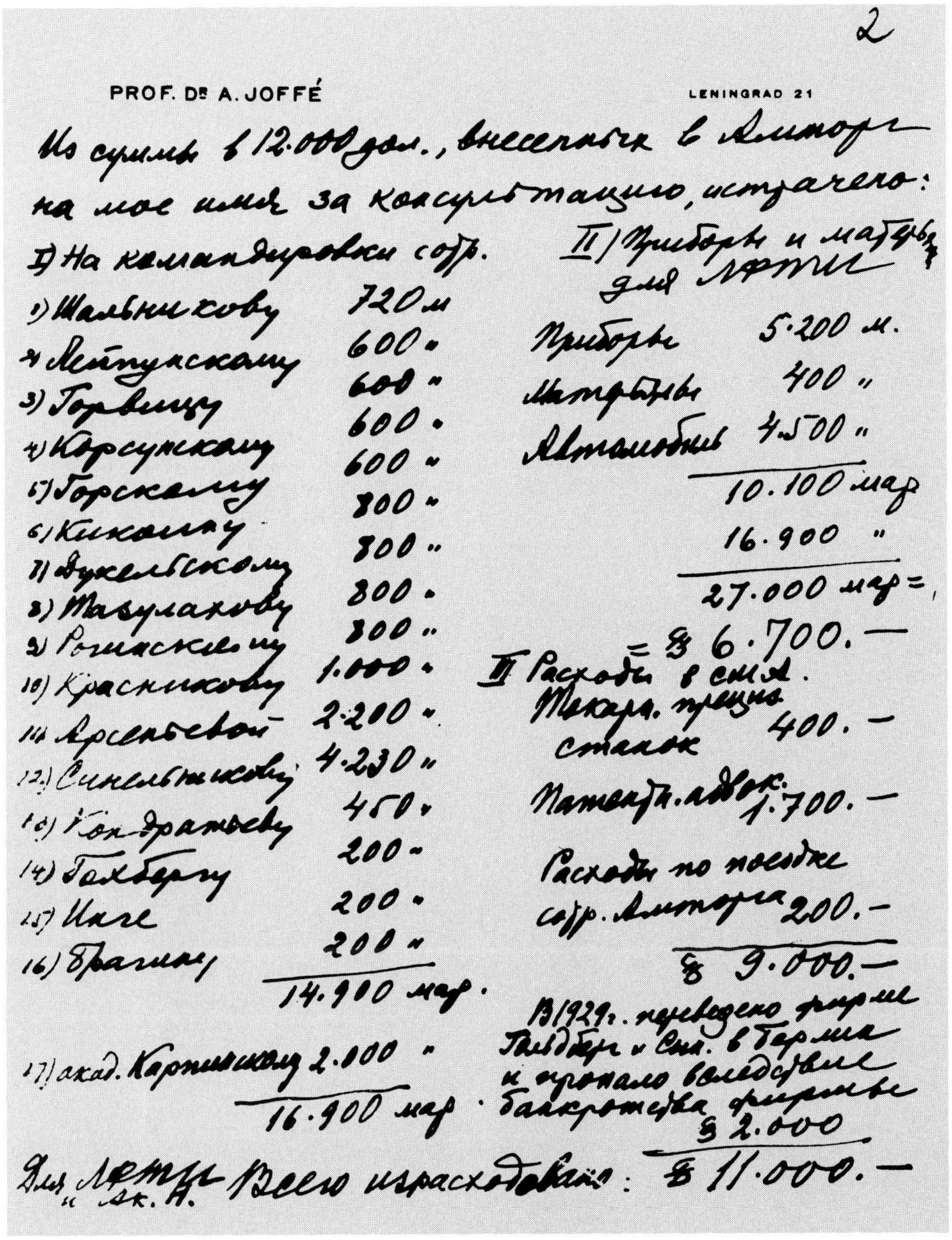

PROF. DR A. JOFFÉ   LENINGRAD 21

[Handwritten budget document in Russian]

Budget. This was Ioffe's plan for using the hard currency that he raised through contracts with Western firms. Listed are 17 scholars, mostly from his institute; they would use the money for foreign research travel.

again in 1930–31, primarily to the University of Minnesota, and upon his return contributed to the development of quantum mechanics in the USSR.

Ioffe at Berkeley

European scholars received Ioffe as an honored guest; they were relieved to see international scientific relations return to normal. His stories of what Soviet physics had achieved in isolation in less than a decade impressed them. Ioffe, for his part, gained confidence that his institute was moving along the right path, and welcomed the recognition he was getting for his work.

Ioffe traveled in the US and Europe several times during the 1920s, having first helped to reestablish regular scientific exchanges with European scholars in 1920. From November 1925 to February 1926 he visited Berlin, Paris and the United States. In the US he visited Columbia University, the Rockefeller Institute, the cities of Chicago, Madison, Boston, Kansas City and Pasadena, and en route visited GE and Westinghouse. These visits influenced his views on the organization of industrial research and development, especially in regard to their grand scale.

In January 1927 Ioffe went abroad for his most successful trip. He gave a series of lectures at Harvard and MIT and then traveled to Schenectady, Washington and Pittsburgh, where he participated in colloquia. The General Electric facilities in Schenectady especially drew his interest: Ioffe was impressed by the company's 400-kV Coolidge tubes with 40-cm spheres; he also discussed shears and displacements with Irving Langmuir. Ioffe traveled to Berkeley in February for a semester of teaching. He developed a close working relationship with Gilbert Lewis and Leonard Loeb and wrote *Physics of Crystals* during this time. The University of California awarded Ioffe an honorary PhD and paid him the compliment of offering him a professorship, a position Ioffe chose to decline.

After leaving the University of California, Ioffe stopped in New York and Boston, where he negotiated an agreement with Vannevar Bush and L. Marshall to develop and manufacture thin-sheet insulation with the potential backing of up to $50 million from T. J. Coolidge and J. P. Morgan. This was one of many new ventures for the newly established Raytheon Corporation, through which Ioffe would get a salary of $12 000, certain patent rights and royalties in exchange for development of the insulation. (The budget in the figure on page 59 shows how Ioffe planned to spend the anticipated income and salary.) Ioffe also negotiated with a consortium of German electrotechnical companies, including Siemens AG, to create a laboratory for the production of insulation, obtaining a contract that guaranteed him 10 million DM and 7 percent royalties. Ioffe used the funds received by the end of the summer of 1928 "for my collaborators and pupils being sent to Europe. All together there are 25 physicists from my institute abroad."

Although Ioffe and his institute failed to deliver on the Raytheon contract, Ioffe, Frenkel' and other physicists from the Leningrad Physico–Technical Institute had by the end of the 1920s succeeded in something more important: reestablishing scientific ties with European and American physicists. Many of the achievements of Soviet physicists in the international arena were the result of the successes of the Russian Association of Physicists on the domestic front.

The Russian Association of Physicists

During the period of the New Economic Policy in the mid-1920s, the Russian Association of Physicists grew in membership, national influence and reputation abroad. More than 600 individuals attended the association's fourth congress in Leningrad in 1924, where participants discussed the controversy over the wave and corpuscular theories of light, which was solved with the development of quantum mechanics. At the fifth congress, held in Moscow in 1926, discussions centered on the new physics—Louis DeBroglie's wave mechanics—and its implications for understanding the atomic and molecular properties of matter.

At Ioffe's suggestion, the next conference was held during a steamship trip down the Volga, without the press of the crowds but with freely held discussions on controversial questions of contemporary physics. (See the photograph on page 55.) The sixth congress, in 1928, began with a four-day meeting at the Moscow Scholars' Club, after which the attendees boarded a train to Nizhnyi Novgorod and then proceeded down the Volga on the steamship "Aleksei Rykov" to Tsaritsyn, stopping in the university towns of Nizhnyi, Kazan' and Saratov for meetings and popular lectures with local teachers and students. The general makeup and program of the sixth congress reflected a youthful and vital discipline and the growing international respect that Soviet physicists commanded. Vavilov described the international flavor: "Within the walls of a former seminary in Nizhnyi an English speech with American pronunciation on the thermodynamics of chemical processes was heard (Lewis); in the packed conference hall of Kazan' University the physics of crystals was discussed in a number of languages ... ; in Saratov Born drew matrix tables on a blackboard." Throughout it all Ioffe could be heard translating from Russian to English or German and back again. The steamboat trip symbolized the feelings of internationalism, corporate spirit and independence that Soviet physicists developed in the 1920s.

In an effort to ensure scientific priority and expand their participation in the international affairs of science, Soviet physicists submitted articles to a number of Western publications, particularly the major German physics journal, *Zeitschrift für Physik*. Before 1923 and the end of international scientific isolation, almost no Soviet physicists published in that journal. Then, after Ioffe and others reestablished Western contacts, a period of rapid and regular increase in the percentage of Russian articles published followed, reaching a peak of 16 percent in 1926 and averaging almost 12 percent for the period 1924–31. All in all, between 1920 and 1936, Soviet scholars wrote 8.5 percent of the 6962 articles published by the journal. With the appearance of a German-language Soviet journal, *Physikalische Zeitschrift der Sowjet Union*, in 1932, the number of articles published in *Zeitschrift für Physik* began to decline. *Physikalische Zeitschrift der Sowjet Union* was intended to communicate to the European community the ideas of Russian scientists, including those in the dialectical materialist philosophy of physics; to ensure Russian priority in discoveries; and, as Ioffe wrote in the foreword to the first volume, "to become the central publication organ of Soviet physics." The pressure for autarky, or self-sufficiency, in science also influenced Soviet physicists not to publish abroad. In protest against Nazism, Soviet physicists ceased to publish in the German physics journal altogether, as the figure on page 61 shows.

By the end of the 1920s Soviet physicists had become fully active in the international arena. They traveled abroad to purchase instruments, chemicals, books and journals and reequipped their laboratories to embark on research at the cutting edge of such areas as the physics of crystals. Under the leadership of Frenkel', Kapitsa, Semenov and Ioffe, they had reestablished scientific ties with the West. Russian physicists took advantage of sabbatical stays in American and British universities, engaged European and American physicists at international congresses and published widely in foreign journals. But policies introduced under Stalin in the late 1920s and 1930s were to reverse many of these achievements.

The Stalin era

During the rapid industrialization, forced collectivization and cultural revolution under Stalin, so-called bourgeois specialists—scientists, technologists and engineers—came increasingly under attack. "Cultural revolution" involved the advancement of workers into positions of responsibility on the basis of class origin or party membership rather than according to traditional definitions of merit, and also entailed "class war" directed against the specialists by an increasingly militant and proletarian Communist Party. Two show trials—the Shakhty and Industrial Party affairs—signaled an end to the autonomy of scientists. Additionally, Stalinist policies toward science required greater emphasis on the applicability of results, centralization of the administration of research and development, the introduction of long-range planning and increasing

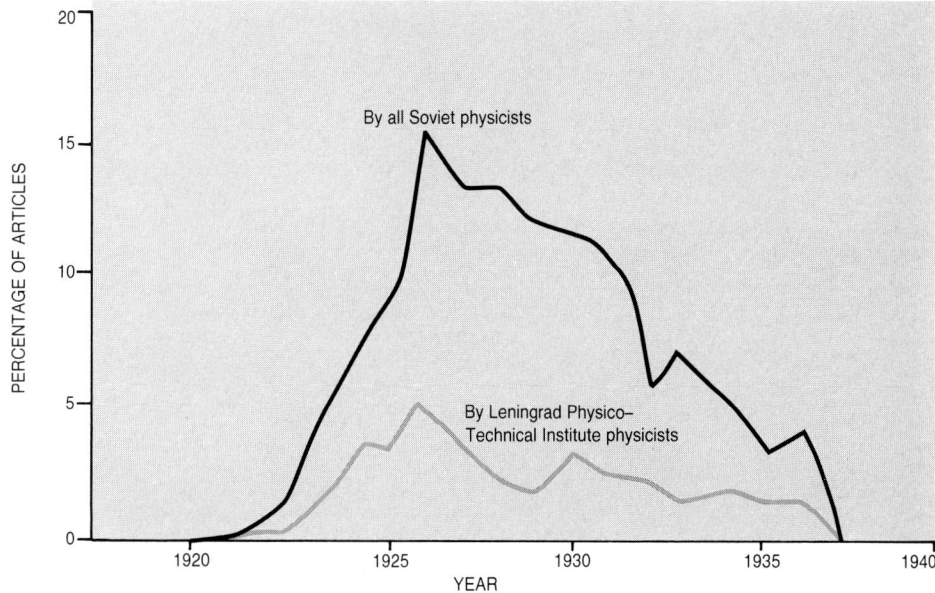

Articles by Soviet scholars in *Zeitschrift für Physik*, 1920–36, as a fraction of all articles in the journal.

autarky in science. As physicists left the sixth congress of the Russian Association of Physicists in 1928, they did not realize they would meet together only one more time as a national professional organization. The boat ride down the Volga had symbolized the independence of Soviet physicists. However, to an increasingly class-conscious party in Stalin's Russia, it signified physicists' elitism and detachment from government economic and political programs.

The pressure for autarky in the economy was mirrored in science, with the result that physicists were caught between countervailing forces. They had expended great effort reestablishing scientific ties with the West, knowing the importance of those ties in promoting currency in the selection of research topics and in verifying that work was on the right path. Now, however, for both political and philosophical reasons physicists were pressured to conduct research divorced from the West. The pressure was political, because physicists had to demonstrate that Soviet science could outdistance capitalist science, and philosophical, because they were required to reject a number of contemporary (and in many cases correct) theories of such "idealists" as Niels Bohr and Erwin Schrödinger. During the purges of the 1930s scholars had to protect themselves from unfounded charges of collusion with "enemies" of the Soviet Union. The academic councils of institutes voted to take responsibility for sending reprints to Western physicists, whereas before individuals had felt safe sending reprints themselves. Physicists felt obliged to indicate carefully and pointedly that requests from abroad were unsolicited. For Stalinist ideologues, success under these conditions would demonstrate the superiority of socialist science and secure the motherland from dependence on the West.

Soviet physicists were able to blunt the impact of autarky to a small degree. A few foreign scholars—Bohr, Langevin, Frédéric Joliot-Curie and Wolfgang Pauli, among others—managed to attend physics conferences held in the USSR throughout the 1930s and well into the 1950s. Of course, the institutes continued to subscribe to a large number of Western scientific journals. Largely, however, physicists had to abandon their foreign contacts. The number of Soviet scientists who traveled to the West dropped precipitously in the early 1930s, and although

exact numbers are not available, they probably could be counted on a few hands. In one well-known case, Kapitsa, working in Cambridge with Rutherford, returned to the USSR for his annual summer visit. Because the authorities wanted him to help establish low-temperature physics firmly as a field of research and to produce liquid oxygen for a number of applications, he was not allowed to return to his equipment and research in England. The number of physicists coming from Europe also declined. Only in the late 1950s, during de-Stalinization, would autarkic pressures abate to any degree and would Soviet physicists again participate in the international scientific enterprise as they had in the 1920s.

* * *

I would like to thank the International Research Exchanges Board and the Fulbright–Hays program for supporting the research that led to this article, and I would like to thank the referee for helpful comments. I used archival materials at the Leningrad Physico-Technical Institute, the manuscript division of the Saltykov-Shchedrin State Library in Leningrad, the Rockefeller Foundation International Educational Board, the University of California at Berkeley and MIT.

Bibliography

○ L. Badash, *Kapitza, Rutherford and the Kremlin*, Yale U. P., New Haven, Conn. (1985).

○ V. Ia. Frenkel', ed., *Erenfest–Ioffe: Nauchnaia perepiska [Scientific Correspondence]*, Nauka, Leningrad (1978).

○ V. Ia. Frenkel', *Ia. I. Frenkel'*, Nauka, Moscow–Leningrad (1966).

○ L. Graham, "The Formation of Soviet Research Institutes: A Combination of Revolutionary Innovation and International Borrowing," Social Studies Sci. **5**, 303 (1975).

○ A. F. Ioffe, *Dostizheniia fiziki [The Achievements of Physics]*, Gudok, Moscow (1928).

○ K. B. Ostrovitianov *et al.*, eds., *Organizatsiia sovetskoi nauki v pervye gody sovetskoi vlasti (1917–1925) [The Organization of Soviet Science in the First Years of Soviet Power]*, Nauka, Leningrad (1968), p. 375.

○ M. S. Sominskii, *A. F. Ioffe*, Akademiia Nauk SSSR, Moscow–Leningrad (1965).

○ S. E. Frish, A. I. Stozharov, eds., *Vospominaniia ob Akademike D. S. Rozhdestvenskom [Reminiscences of Rozhdestvenskii]*, Nauka, Leningrad (1976). ∎

Demonstration of fission at the Department of Terrestrial Magnetism of the Carnegie Institution, shortly after the announcement of the fission experiments at the Fifth Conference on Theoretical Physics in Washington. From left to right are Enrico Fermi, Niels Bohr and Léon Rosenfeld. (Photo courtesy of Carnegie Institution of Washington.)

Bringing the news of fission to America

The news of this discovery reached America in January 1939; notwithstanding communication problems, Niels Bohr succeeded in protecting the priority of Lise Meitner and Otto Frisch's interpretation of the experiment.

Roger H. Stuewer   PHYSICS TODAY/OCTOBER 1985

In January 1939 the news of the discovery of nuclear fission burst in America, sending physicists into their laboratories to try to confirm the startling new discovery. Some aspects of the story of how this news reached America are well known. Others, however, are not; they have remained hidden in private correspondence and other unpublished documents. By examining these materials in conjunction with the published literature, one can reconstruct the circumstances that converged to produce this historic event.

Appointment in the US

The story begins quietly in early 1938, when Niels Bohr began to finalize his plans to spend the second semester of the 1938–39 academic year as a visiting professor at the Institute for Advanced Study in Princeton. Bohr was last in the United States about a year earlier, in February of 1937, at the beginning of a six-month around-the-world trip, during which he lectured widely on his recently published[1] theory of the compound nucleus. Now, in early 1938, he was looking forward to an extended stay in Princeton, exploring a host of fundamental problems in quantum theory and nuclear physics with Albert Einstein, John von Neumann, Eugene Wigner and other eminent colleagues. His appointment was arranged by Oswald Veblen, professor in the Institute's School of Mathematics. By the end of April 1938, word of Bohr's prospective visit had reached John Wheeler, who was then on the faculty of the University of North Carolina in Chapel Hill. On 30 April, Wheeler included[2] the following remarks in a letter to Bohr:

Professor Veblen informs me that you will probably be at Princeton for half of the coming year. I shall be there permanently in the fall, and look forward very much to the opportunity of learning more from you of the fundamental principles from which one can hope to attack the problem of nuclear structure.

That Wheeler, too, would be in Princeton came as welcome news to Bohr, because the two had already established a warm working relationship when Wheeler, as a National Research Council Fellow, had spent the 1934–35 academic year in Copenhagen. Bohr, however, had already been considering another way of satisfying his need for a close collaborator: He intended to bring one with him to Princeton. The candidate he had in mind was Léon Rosenfeld.

Bohr first met Rosenfeld in April 1929, when Rosenfeld, as a 24-year-old postdoctoral student, attended the first small conference that Bohr arranged in Copenhagen.[3] Extended periods of intense collaboration followed and, now, in 1938, Rosenfeld was a full professor at his *alma mater*, the University of Liège. He was an expert on the problem of measurement in quantum electrodynamics, and it was on that problem that Bohr intended to work with him in Princeton.[4] Bohr probably raised the question of support for him through John von Neumann, because on 2 July 1938, von Neumann wrote[2] to Bohr from Budapest, telling Bohr that he had written to Veblen about Rosenfeld. Five weeks later, on 9 August, Bohr had his answer: von Neumann informed[2] Bohr by postcard that he had

just learned from Veblen that Rosenfeld's trip could be financed by a stipend from the Committee for the Relief of Belgium. Two or three months later, Bohr, his son Erik and Rosenfeld booked passage on the Swedish–American liner *Drottningholm*, scheduled to sail from Gothenburg, Sweden, on 7 January 1939.

In the fall of 1938 ominous clouds spanned the European political sky, and Bohr made his travel arrangements with great trepidation. Adolf Hitler had annexed Austria in March, and six months later, on 29 September, he and Neville Chamberlain had dismembered Czechoslovakia in Munich. Under these threatening conditions, international scientific meetings could no longer be held, and the eighth Solvay Conference, scheduled for the end of October in Brussels, was cancelled. Bohr had planned to hold another of his small conferences in Copenhagen just prior to the Solvay Conference; now he changed his plans, rescheduling for the last week in October. He hoped to attract at least a few people who were still willing and able to travel.

One who came was Enrico Fermi. The Italian racial laws had been promulgated in early September, affecting Fermi's wife Laura, and had forced Fermi to think about leaving Italy with his family. In view of this situation, the rules were broken and Fermi was informed confidentially by Bohr in Copenhagen that he was likely to

Roger H. Stuewer is professor of history of science and technology at the University of Minnesota.

Lise Meitner and Otto Hahn photographed in 1913 in the Kaiser Wilhelm Institute for Chemistry in Berlin-Dahlem. Their scientific collaboration lasted 30 years. (This and subsequent photos courtesy of AIP Niels Bohr Library.)

receive the Nobel Prize in Physics for 1938.[5] Confirmation came by the traditional telephone call from Stockholm on 10 November 1938, two weeks after Fermi had returned to Rome. On 6 December, he and his family left for Stockholm. Then, after the festivities, the Fermis traveled to Copenhagen to talk with Bohr. Leaving Copenhagen for Southampton, they boarded the Cunard White Star *Franconia* on 24 December and arrived in New York on 2 January 1939, where Fermi took up a position at Columbia University. In Copenhagen, it seems, Enrico and Laura Fermi made arrangements to meet Bohr and his son Erik when they, too, would arrive in New York with Rosenfeld, a mere two weeks later. By then, Wheeler also had made plans to be in New York.

Understanding fission

Meanwhile, another chain of events had been set into motion. Lise Meitner, imperiled in Berlin after the *Anschluss* of Austria in March 1938, was spirited into the Netherlands by Dirk Coster on 13 or 14 July, and from there went to Stockholm. Five months later, Otto Hahn and Fritz Strassmann, pursuing the researches they had begun with Meitner in Berlin, bombarded uranium with neutrons and found highly mystifying results—so mystifying that Hahn, still in his laboratory at 11:00 pm on Monday evening, 19 December 1938, decided to first reveal them by letter only to Meitner. He and Strassmann, Hahn wrote,[6] were coming "again and again to the frightful conclusion" that one of the products "behaves not like Ra, but rather like Ba." Replying[6] by return mail on 21 December, Meitner asked Hahn if he was "absolutely certain" about these results. Hahn was certain; that same day, even before receiving Meitner's reply, he had written[6] again, telling Meitner that "as chemists" he and Strassmann were forced to conclude that they were finding barium as a product of the reaction. Moreover, Hahn said, they could not "hush up" their results, "even if they are perhaps physically absurd." They submitted[7] the results the next day for publication in *Die Naturwissenschaften*. As "nu-

clear chemists related in a certain way to physics" they still could not "make this jump," they added. "Perhaps a series of rare coincidences might still have simulated our results."

Meitner, after receiving Hahn's second letter, left Stockholm on 23 December to travel to Kungälv, a small town just north of Gothenburg, to spend the holidays with friends and, as had been her custom, with her physicist nephew, Otto Robert Frisch. Frisch had been forced to leave Hamburg in 1933, and after a year in London had gone to Bohr's Institute in Copenhagen.[8] Traveling to Kungälv, he arrived just after Meitner had received yet another letter[6] from Hahn, dated 28 December. In that letter, Hahn had proposed a "new fantasy." He asked, "Would it be possible that uranium 239 bursts into a Ba and a Ma?"—Ma standing for the element masurium, today called technetium (Tc). "A Ba 138 and a Ma 101 would yield 239." The atomic numbers "of course do not work out"—Ba ($Z = 56$) plus Ma ($Z = 43$) did not yield U ($Z = 92$)—so "some neutrons would have to be transformed into protons.... Is that energetically possible?" asked Hahn.

Meitner replied[6] the following day, 29 December, telling Hahn that she and Frisch had "racked [their] brains" over his and Strassmann's "very exciting" results. Much later, Frisch recalled[9] that he and Meitner had discussed Hahn's letter first at breakfast in Meitner's hotel and then on a hike outdoors in the snow—Frisch on skis, Meitner walking. It occurred to Frisch that Hahn and Strassmann's results

might be explained on the basis of the liquid-drop model of the nucleus. They stopped to calculate on scraps of paper, and, as Frisch recalled[9] further:

> ... I worked out the way the electric charge of the nucleus would diminish the surface tension and found that it would be down to zero around $Z = 100$ and probably quite small for uranium. Lise Meitner worked out the energies that would be available from the mass defect in such a breakup.... It turned out that the electric repulsion of the fragments would give them about 200 MeV of energy and that the mass defect would indeed deliver that energy....

On 30 December 1938, Meitner informed[6] Hahn that she had just received a copy of his and Strassmann's manuscript for *Die Naturwissenschaften*, which Hahn had promised to send her. On 1 January 1939, she told[6] Hahn: "We have read and considered your paper very carefully, [and] *perhaps* it is indeed energetically possible that such a heavy nucleus bursts."

Meitner then returned to Stockholm and Frisch to Copenhagen. Arriving on New Year's Day 1939, Frisch sought Bohr out, catching him two days later. As Frisch explained[6] to Meitner in a letter on 3 January 1939:

> Only today was I able to speak with Bohr about the bursting uranium. The conversation lasted only five minutes, since Bohr immediately and in every respect was in agreement with us. He was only astonished that he had not thought earlier of this possibility, which

follows so directly from the present conceptions of nuclear structure. He was also completely in agreement with our view that this disintegration of a heavy nucleus into two large pieces is an almost classical process, which does not occur at all below a certain energy, but already occurs very easily a little above it. (One indeed has to require this in order to understand the great stability of natural uranium as compared to the very great instability of the (not so very much more energetic) compound nucleus.) Bohr still wants to consider this quantitatively this evening and to talk with me again about it tomorrow.

Much later, Frisch recalled[9] his meeting with Bohr in more dramatic terms (but erroneously placing it on the day Bohr was leaving for America). Frisch recalled that he had just begun to speak when Bohr burst out: "Oh, what fools we have been! We ought to have seen that before." That Bohr had not, however, is understandable in light of his past work. Although George Gamow had conceived[10] the liquid-drop model of the nucleus at the end of 1928, Bohr did not treat Gamow's model as seriously as he might have in succeeding years. He mentioned[11] it briefly and only once, in 1933. Furthermore, when Bohr proposed[1] his theory of the compound nucleus three years later, in 1936, he speculated that more and more energetic particles (for example, neutrons) when striking a heavy nucleus would simply dislodge more and more nucleons, eventually producing "an explosion of the whole nucleus." In the fall of 1937, Bohr still was much less inclined to view[12] a heavy nucleus as a liquid drop than as an elastic solid. Thus, to the extent that he still held these views on 3 January 1939, he would not have seen the possibility that Frisch and Meitner recognized—and he would have been all the more astonished that he had not.

That Bohr wanted to talk everything over once again with Frisch on the following day suited Frisch, because Frisch already had made plans to call Meitner on 5 January to begin composing a note to *Nature* over the telephone. By Sunday, 8 January, Frisch had completed a draft of this note. He sent this draft to Meitner that same day, along with a cover letter[6] in which he brought Meitner up to date on his further discussions with Bohr:

I wrote up a first draft on Friday [6 January] and on Bohr's request rode out to Carlsberg [Bohr's residence] still in the evening, where Bohr once again thoroughly discussed the matter with me. He let me recalculate my estimate of the surface tension, and he was in complete agreement with it; he had already hurriedly considered the electrical term, but had not realized it would be so large. Concerning the [formation of U^{239} by] resonance, he did not want to express himself directly, but seemed to see no difficulty with it. Later, I again considered this point a bit as it arises in the conclusion of the note; in any case Bohr did not take a position on this. Bohr only made several recommendations during the evening for a clearer formulation of several points; otherwise he was in agreement with everything. On the following morning [Saturday, 7 January], I then started to type up the draft and was able to take only two pages to Bohr at the train station (10:29 A.M.), where he put them in his pocket; he no longer had any time to read them.

Setting out to the US

By the time Bohr and his son Erik boarded their train to Gothenburg, to embark on the *Drottningholm* that same day—Saturday, 7 January 1939—Bohr had been thinking for no less than four days about Frisch and Meitner's interpretation of Hahn and Strassmann's experiments, and he was completely familiar with it.

Rosenfeld, who apparently traveled to Gothenburg independently to join the Bohrs there, has described[13] the

Fritz Strassmann.
The photo was taken in 1930.

Otto Robert Frisch, photographed in Hamburg in 1931.

circumstances attending their transatlantic voyage:

> As we were boarding the ship, Bohr told me he had just been handed a note by Frisch, containing his and Lise Meitner's conclusions; we should "try to understand it." We had bad weather through the whole crossing, and Bohr was rather miserable, all the time on the verge of seasickness. Nevertheless, we worked very steadfastly and before the American coast was in sight Bohr had got a full grasp of the new process and its main implications.

The "most puzzling point" to Bohr concerned the high probability of the new process as compared to other competing processes. Rosenfeld continued[13]:

> The answer turned out to be very simple, on the basis of Bohr's treatment of the excited compound nucleus as a system in thermodynamic equilibrium: it is just a consequence of the equipartition of energy between all the modes of motion of the system. The relative oscillation of two large fragments of the nucleus competes on equal terms with any other mode, such as the relative motion of a single neutron, leading to its emission.

The *Drottningholm*, after a nine-day voyage, docked at the Swedish-American Line's West 57th Street pier in New York at 1:00 pm on Monday, 16 January 1939. Enrico and Laura Fermi were there early. As Laura Fermi recalled,[5] even before the ship "came alongside the wharf, we recognized in a crowd the man we had come to meet, Professor Niels Bohr. He was standing by the rails of an upper deck, leaning forward, scanning the people on the dock." John Wheeler was also there. He recalled[14]: "I had my regular morning class on Monday the 16th and then went in on the train to meet the Drottningholm coming in that afternoon; I of course shook hands with those waiting for Bohr and with him and Rosenfeld when they came off. Bohr was staying in New York for a little while, but Rosenfeld went down with me on the train to Princeton."

In Copenhagen on 7 January, Frisch had left Bohr under the impression that after finishing the typing of his and Meitner's note, he would submit it immediately to *Nature* for publication. Bohr then had promised Frisch that he would protect Frisch and Meitner's priority by not saying anything to anyone in America about the new discovery until he had received word from Frisch that the note was actually in press. On board ship, therefore, Bohr conveyed the impression to Rosenfeld that Frisch and Meitner's note no doubt had already been submitted for publication. Unfortunately, Bohr failed to inform Rosenfeld of his promise of confidentiality to Frisch. Moreover, quite by chance, on the very day that Bohr and Rosenfeld arrived in New York, on 16 January, one of the regular Monday-evening meetings of Princeton's Physics Journal Club was

to take place, with Wheeler in charge. The inevitable occurred: As Rosenfeld recalled,[13] Wheeler "politely asked" him if he had anything to report, and Rosenfeld, "in spite of the fatigue of the voyage, . . . told them all about the problem we had struggled with during the journey."

Communication breakdown

The cat, therefore, jumped out of the bag—as Bohr learned, to his distress, when he himself arrived in Princeton shortly thereafter and found no news from Frisch awaiting him. He expected, however, that a letter from Frisch was imminent. As a result, immediately after his arrival in Princeton, he himself drafted a note outlining his deeper understanding of the new process that he had achieved on board ship with Rosenfeld. By 20 January, Bohr had still not received a letter from Frisch. He therefore decided to send his own note to Frisch, along with a cover letter[2] dated 20 January, asking Frisch to have his secretary, Betty Schultz, forward it to *Nature* "if, as I hope, Hahn's article has already been published, and your and your aunt's note has already been submitted to *Nature*." He was looking forward to hearing, he told Frisch, "about the latest news in this connection and how the experiments are proceeding at the institute, which I, despite the distance, follow in my thoughts." He added a "p.s." just prior to mailing his letter: "I have just seen Hahn and Strassmann's article in Naturwiss., which naturally

Léon Rosenfeld (right) talking to Walter Heitler. The photograph was taken during the 1934 conference at the Bohr Institute.

has caused much discussion here at the institute. ... "

The publication of Hahn and Strassmann's paper, of course, increased Bohr's distress still further. Still, however, he received no word from Frisch. On 24 January, no longer able to contain himself, he wrote[2] a second letter to Frisch, saying:

I still have not received any letter at all from the institute and sorely long to see the final version of your and your aunt's note to *Nature*, a copy of which you promised to send me. I therefore do not know whether you in your note reach the same conclusions about the splitting mechanism as those which are mentioned in my note and how far the latter contributes anything sufficiently new to be published.

A few lines later Bohr added:

As I mentioned in my last letter, the physicists here at the institute are very caught up in the whole question, and I already have seen preparations for experiments to detect radioactive matter of very short half-life, the appearance of which should be an immediate result of the new type of splitting of the nucleus. ... Working with Wheeler, I also have started a more thorough study of the different theoretical problems which the new splitting of the nucleus presents to us. Naturally, I am very interested in hearing more about what you yourself have been thinking in one direction or another, just

as I am excited to hear about all of the investigations at the institute.

What, in fact, *was* Frisch doing all of this time in Copenhagen? In his letter to Meitner of 8 January, in which he reported[6] the results of his discussions with Bohr, Frisch also had told Meitner:

Since Hahn and Strassmann's article appeared here yesterday, I discussed the entire matter somewhat, above all with [George] Placzek, who at the moment is very skeptical, but he of course always is. Early today he again flew back to Paris, and then will travel soon to America, to Bethe in Ithaca, where he has a position.

That discussion with Placzek, as Frisch later acknowledged,[9] prodded Frisch immediately into thinking about experimental confirmation of his and Meitner's interpretation. Assembling his apparatus, Frisch first observed the expected ionization pulses of the uranium fragments on Friday, 13 January, and he then confirmed his observations over the next three days. At that point he wrote up a report on his experiments, which he submitted to *Nature* along with his and Meitner's note. As he explained[6] in a letter to Meitner on 17 January:

Yesterday evening I finished both notes, and at about 5 A.M. took them to the airmail deposit box, so that they should be in London today in the afternoon. With that, however, my energy was exhausted, so that I no longer

wrote to you; rather, I do that now.

Unfortunately, Frisch at the same time did *not* write to Bohr in Princeton. Rather, he delayed five more days, until 22 January,[2] before writing the letter. Later, on 15 March, Frisch apologized to Bohr for this delay, explaining[2]:

This was partly due to a lack of imagination on my side, as I did not imagine that the appearance of Hahn and Strassmann's paper would raise such a run as it did. And then I was pretty tired after the experiment (I had been working long after midnight for several nights in track) and instead of sending you the manuscripts at once (the obvious thing to do) I kept them until I managed to write you a letter, which meant about six days delay. When I think it over now I can hardly find an excuse for my letting you without information as I did, but, you see, I did not think my experiment so terribly important (it seemed to me just additional evidence of a discovery already made) and the idea of cabling to you would have appeared unmodest to me.

When he finally did bring Bohr up to date, on 22 January, Frisch noted that he was currently planning various new experiments on "these 'fission' processes." He added[2] the following parenthetical remark on that historic term: "I wonder how you like this word; it was suggested by the [American] biochemist Dr. [William A.] Ar-

nold, who told me it was the usual term for the division of bacteria."

Conference on theoretical physics

As Frisch's letter of 22 January was being slowly transported to Princeton, crossing with Bohr's letters of 20 and 24 January to Copenhagen, Bohr's tension was increasing to the breaking point. He knew that in only two days the possibility of containing the news of the new discovery would evaporate completely. For, two months earlier, another completely independent chain of events had been set into motion whose culmination was imminent. On 30 November 1938, Merle Tuve, George Gamow and Edward Teller had drawn up[15] a proposal that the Carnegie Institution of Washington and George Washington University sponsor the Fifth Conference on Theoretical Physics in Washington, D.C., sometime during the period of 21–30 January 1939. Quite innocently and with due deliberation, they had chosen as the subject of this conference "Magnetic, electric, and mechanical properties of matter at very low temperatures." They had drawn up a preliminary list of ten participants whose expenses would be paid, and of twenty-four who would be invited without paid expenses. Bohr was on neither list. A few weeks later, however, the organizers got wind of Bohr's visit to Princeton—and of Fermi's transfer to Columbia University. Both, therefore, were included on the final list of 15 invited participants whose expenses would be paid. John R. Lapham, dean of George Washington's School of Engineering, sent[15] out formal letters of invitation to these people on 22 December 1938—on precisely the same day that Hahn and Strassmann in Berlin suomitted their paper for publication in *Die Naturwissenschaf-*

ten. On 6 January 1939, Lapham sent[15] out a second letter of invitation to 21 additional participants whose expenses would not be paid. The total number of participants would be increased by the attendance of about 20 local physicists. All, of course, had contributed to some area of low-temperature physics. The dates of the conference had now been fixed for 26–28 January 1939. On 23 January, J. A. Fleming, director of the Carnegie Institution's Department of Terrestrial Magnetism, sent its new president, Vannevar Bush, a list of all those who had been invited to attend, along with a notation indicating that William F. Giauque of Berkeley, Frederick G. Keyes and John C. Slater of MIT and Eugene P. Wigner of Princeton would be unable to come. Fleming also informed[15] Bush that "the first meeting will be held at the George Washington University in Room 105, Building C, 2029 G Street, N.W. at 2 p.m., January 26." (Today a plaque outside the lecture room commemorates this historic meeting.)

It was there—the building today is George Washington's Hall of Government—that the bombshell burst. Bohr knew that it was pointless to try to keep the new discovery secret any longer. If Hahn and Strassmann's article had arrived in Princeton on 20 January, it had certainly arrived elsewhere as well, its charge ready to explode. Moreover, Bohr knew that the news had already leaked a short distance northward. Isidor I. Rabi, as Wheeler recalled, actually had been present at Rosenfeld's discussion on 16 January and had carried the news of Hahn and Strassmann's discovery back to Columbia University.[4] Fermi himself, however, learned about it from Willis Lamb, who had been on a brief visit to Princeton.[16] Knowing, therefore, that

Hahn and Strassmann's discovery was no longer a secret at Columbia University, Bohr himself stopped off there on his way to Washington, evidently on 25 January, seeking to discuss it with Fermi. Instead he found Herbert L. Anderson.[17] After a brief interchange with Anderson, Bohr then left for Washington, where Fermi joined him. They conferred, and on 26 January they took the floor—before a single talk had been delivered on low-temperature physics. As stated in the report of the conference[15,18] that was submitted on 1 February by C. F. Squire, F. G. Brickwedde, Teller and Tuve:

Certainly the most exciting and important discussion was that concerning the disintegration of uranium of mass 239 into two particles each of whose mass is approximately half of the mother atom, with the release of 200,000,000 electron-volts of energy per disintegration. The production of barium by the neutron bombardment of uranium was discovered by Hahn and Strassmann at the Kaiser-Wilhelm Institute in Berlin about two months ago. The interpretation of these chemical experiments ... was suggested by Frisch of Copenhagen together with Miss Meitner, Professor Hahn's longtime partner who is now in Stockholm. ... Professors Bohr and Rosenfeld had arrived from Copenhagen the week previous with this news. ... Professors Bohr and Fermi discussed the excitation energy and probability of transition from a normal state of the uranium nucleus to the split state.

Given their advance information, the experimental team at Columbia University—Anderson, E. T. Booth, John Dunning, Fermi, Gynther Glasoe and

Participants in the Fifth Washington Conference on Theoretical Physics. First row, left to right: O. Stern, E. Fermi, J. A. Fleming, N. Bohr, F. London, H. C. Urey. Second row: F. G. Brickwedde, G. Breit, J. B. Silsbee, I. I. Rabi, G. E. Uhlenbeck, G. Gamow, E. Teller, M. Goeppert-Mayer, F. Bitter, H. A. Bethe, H. Grayson-Smith, J. H. Van Vleck, R. Jacobs, C. Starr, M. H. Hebb, C. F. Squire. Third row: H. Kuper, A. J. Mahan, R. D. Myers, R. B. Roberts, C. L. Critchfield, L. Baroff, E. Bohr, R. C. Meyer, K. F. Herzfeld, R. C. Lord Jr, D. R. Inglis, O. R. Wulf, P. Wang, E. A. Johnson, F. Mohler, R. B. Scott, E. H. Vestine, L. Rosenfeld, F. Seitz, G. H. Diecke, J. E. Mayer, J. H. Hibben, M. A. Tuve, H. M. O'Bryan, L. R. Hafstad, K. Cohen, H. J. Hoge, A. L. Sklar, F. D. Rossini. Missing from photo: N. Bjerrum, V. Bush, N. P. Heydenburg, R. D. Potter, A. E. Ruark.

F. G. Slack—was first off the mark.[19] Fermi came into his office after Bohr had left for Washington, and Anderson told him all about Hahn and Strassmann's discovery—only to learn that Fermi already knew about it from Lamb.[17] Rushing into the laboratory, still on 25 January, Anderson and his colleagues managed to detect the fission fragments. By then, however, Fermi had already left for Washington, and although Dunning telegraphed the news to him there, it appears that he did not receive Dunning's telegram and hence did not learn about the success of the Columbia experiments until the close of the Washington meeting.[17] (Anderson claims[17] that Fermi did receive Dunning's telegram, but Fermi himself does not mention this, and Bohr told[2] Frisch in a letter of 3 February that the first reports of the detection of the fission fragments came in on the last day of the conference.)

Meanwhile, R. D. Fowler and R. W. Dodson of Johns Hopkins University, evidently tipped off by one of their six colleagues attending the Washington meeting, swung into action. They confirmed the discovery on Saturday morning, 28 January.[20] They managed, in fact, to just beat out R. B. Roberts, R. C. Meyer and L. R. Hafstad of the Carnegie Institution's Department of Terrestrial Magnetism, all of whom, immediately upon hearing the news, rushed across town and into their laboratory, confirming the discovery that same day, on 28 January.[21] Owing to their proximity to the conference site, however, this group was the most fortunate one of all in one respect: As reported[22] in the *Science News Letter* of 11 February 1939, Roberts and his colleagues, in "a historic midnight experimental conference" on 28 January, demonstrated the existence of the fis-

sion fragments to Bohr and Fermi.

Luis Alvarez in Berkeley did not lose out by much. He caught an announcement of the discovery of fission in the *San Francisco Chronicle* while having his hair cut in the student union. Rushing out of the barber chair and back to the Radiation Laboratory, he first broke the news—dramatically—to graduate student Philip Abelson and then telegraphed Gamow in Washington for further details. He received them on the morning of 31 January, and that same afternoon he and G. K. Green observed[23] the predicted ionization pulses.

Publication

In the midst of all of this furious activity on the east and west coasts of the United States, Bohr returned to Princeton on Sunday, 29 January. He was still totally unaware of Frisch's experiments in Copenhagen: Even bombarding Frisch with telegrams had raised no sign of life. Finally, at long last, on 2 February, Frisch's letter[2] of 22 January, along with its two enclosed notes, arrived in Princeton. Bohr fairly jumped for joy, as is evident from his reply[2] to Frisch on the following day:

I need not say how extremely delighted I am by your most important discovery, on which I congratulate you most heartily.... The experiments of Hahn, together with your aunt's and your explanation have indeed raised quite a sensation not only among physicists, but in the daily press in America. Indeed, as you may have gathered from my telegrams and perhaps even, as I feared, from the Scandinavian press, there has been a rush in a number of American laboratories to compete in exploring the new field. On the last day

of the conference in Washington (January 26–28), where Rosenfeld and I were present, the first results of detection of high energy splitters were already reported from various sides. Unaware as I was, to my great regret, of your own discovery, and not in possession even of the final text of your and your aunt's note to *Nature*, I could only stress (which I did most energetically) to all concerned that no public account of any such results could legitimately appear without mentioning your and your aunt's original interpretation of Hahn's results. When Hahn's paper appeared, information about this could of course, for your own sake, not be withheld and was, in fact, the direct source of inspiration for all the different investigators in this country. When I came back to Princeton I learned from an incidental remark in a letter from [my son] Hans the first news of the success of your experiments. I at once telephoned this information to Washington and New York, and succeeded in obtaining a fair statement in a *Science Service* circular of January 30, of which I have sent a copy to my wife, but I could not prevent various misstatements in newspapers. This is of course regrettable but without any importance for the judgment of the scientific world, which here even more than in Denmark is accustomed to such happenings.

With Frisch and Meitner's, and Frisch's, notes at last in his possession, Bohr could make "a few corrections" in his own note to *Nature* and enclose "a new copy" for Frisch to pass on for publication. "Quite apart from the question how much or little new the

George Placzek, photographed in 1946. In 1934 Placzek used Bohr's Nobel prize medal to measure the absorption of slow neutrons by gold.

note contains," Bohr said, "I think that its appearance at the earliest possible opportunity will contribute essentially to clear up the confusion as regards the history of the discovery and its theoretical significance." Actually, it took a relatively long time for the notes to appear in print: Meitner and Frisch's, and Frisch's (both submitted on 16 January), appeared only on 11 and 18 February, respectively,[24] while Bohr's (dated 20 January, revised 3 February) appeared on 25 February.[25] Frisch, when he wrote once again to Bohr on 15 March, remarked[2] that these long delays occurred "probably on account of an accidental increase in the number of letters [received by *Nature*] and, perhaps, because we had not sufficiently stressed the importance of quick publication, when writing to the editor."

On 7 February 1939, Bohr took yet another opportunity to sketch[26] the history of the discovery and interpretation of nuclear fission in a letter to the editor of the *Physical Review*. The main purpose of Bohr's letter, however, was entirely different. George Placzek had visited Princeton a few days earlier, and stimulated by Placzek's penetrating questioning and inevitable skepticism, Bohr had conceived an ingenious argument that had led him to conclude that it was not the heavy uranium isotope U^{238} that is primarily responsible for fission, but rather the rare, light isotope U^{235}. At the end of his letter, Bohr remarked that he and Wheeler were currently engaged in a "closer discussion" of the mechanism of nuclear fission. That discussion occupied the entire balance of time—until May 1939—remaining to Bohr in Princeton. It remains one of the finest collaborative efforts in physics.[27] No

one did more to promote an accurate understanding of the history of nuclear fission, and of nuclear fission itself, than the man who brought the news of its discovery to America.

* * *

I am grateful to Aage Bohr for permission to quote from his father's correspondence; to Ulla Frisch for permission to quote from her husband's and Lise Meitner's correspondence; to Dietrich Hahn for permission to quote from his uncle's correspondence; and to John A. Wheeler for permission to quote from his correspondence. I am also grateful to Heinz H. Barschall for reading a draft of the manuscript and to Anne I. Goldman for translating Niels Bohr's letters of 20 and 24 January 1939, to Otto R. Frisch. This research was supported by grants from the American Council of Learned Societies, the Bush Foundation and the National Science Foundation, whose support I gratefully acknowledge.

References

1. N. Bohr, Nature **137**, 344 (1936); Naturwiss. **24**, 241 (1936).
2. Bohr Scientific Correspondence, Archives for the History of Quantum Physics, with repositories at the Bohr Institute, Copenhagen; the American Philosophical Society, Philadelpia; the AIP Center for History of Physics, New York; the University of Minnesota, Minneapolis; the University of California, Berkeley; the Science Museum, London; the Deutsches Museum, Munich; and the Accademia dei XL, Rome. No relevant letters between Bohr and Veblen, however, are extant in the Bohr Scientific Correspondence. That Veblen arranged Bohr's appointment is evident from other correspondence.
3. L. Rosenfeld, J. Jocular Phys. **2**, 7 (Institute of Theoretical Physics, Copenhagen, October 1945), reprinted in R. S. Cohen, J. J. Stachel, eds., *Selected Papers of Léon Rosenfeld*, Reidel, Boston (1979).
4. J. A. Wheeler, PHYSICS TODAY, November 1967, p. 50.
5. L. Fermi, *Atoms in the Family: My Life with Enrico Fermi*, Univ. Chicago P., Chicago (1954); R. Moore, *Niels Bohr: The Man, His Science, and the World They Changed*, MIT Press, Cambridge (1985).
6. Meitner Papers, Churchill College Archives, University of Cambridge. Parts of this correspondence are reprinted in F. Krafft, *Im Schatten der Sensation: Leben und Wirken von Fritz Strassmann*, Verlag-Chemie, Deerfield Beach, Fla. (1981).
7. O. Hahn, F. Strassmann, Naturwiss. **27**, 11 (1939).
8. O. R. Frisch, *What Little I Remember*, Cambridge U.P., New York (1979).
9. O. R. Frisch, PHYSICS TODAY, November 1967, p. 47.
10. G. Gamow, Proc. Roy. Soc. [A] **123**, 386 (1929).
11. See Bohr's remarks following Werner Heisenberg's paper in *Structure et Propriétés des Noyaux Atomiques: Rapports et Discussions du Septième Conseil de Physique* (Institut International de Physique Solvay) Gauthier-Villars, Paris (1934).
12. N. Bohr, F. Kalckar, *Det Kgl. Danske Videns. Sels. Math.-fys. Med.* **14**, No. 10, pp. 9 and 14 (1937). See also R. H. Stuewer, "Niels Bohr and Nuclear Physics," in A. P. French, P. Kennedy, eds., *Niels Bohr: A Centenary Volume*, Harvard U.P., Cambridge, Mass. (1985).
13. L. Rosenfeld, "Nuclear Reminiscences," in F. Reines, ed., *Cosmology, Fusion and Other Matters: George Gamow Memorial Volume*, Colorado Assoc. U.P., Boulder (1972).
14. R. H. Stuewer, ed., *Nuclear Physics in Retrospect: Proceedings of a Symposium on the 1930s*, Univ. Minnesota P., Minneapolis (1979).
15. Tuve Papers, Library of Congress, Washington, D.C.
16. E. Fermi, PHYSICS TODAY, November 1955, p. 12.
17. H. L. Anderson, Bull. Atomic Sci., September 1974, p. 57.
18. C. F. Squire, F. G. Brickwedde, E. Teller, M. A. Tuve, Science **89**, 180 (1939).
19. H. L. Anderson, E. T. Booth, J. R. Dunning, E. Fermi, G. N. Glasoe, F. G. Slack, Phys. Rev. **55**, 511 (1939).
20. R. D. Fowler, R. W. Dodson, Phys. Rev. **55**, 417 (1939).
21. R. B. Roberts, R. C. Meyer, L. R. Hafstad, Phys. Rev. **55**, 416 (1939).
22. W. Davis, R. D. Potter, Science News Letter, 11 February 1939, p. 87; Science Supplement, 10 February 1939, p. 5. See also M. A. Tuve, Science **89**, 202 (1939).
23. G. K. Green, L. W. Alvarez, Phys. Rev. **55**, 417 (1939).
24. L. Meitner, O. R. Frisch, Nature **143**, 239 (1939); O. R. Frisch, Nature Supplement **143**, 276 (1939).
25. N. Bohr, Nature **143**, 330 (1939).
26. N. Bohr, Phys. Rev. **55**, 418 (1939).
27. N. Bohr, J. A. Wheeler, Phys. Rev. **56**, 426 (1939).   □

Reminiscences of the early days of fission

**News of the discovery of fission and the onset
of World War II affected the author's activities, taking him from Princeton
to Los Alamos by way of Lawrence, Kansas.**

Henry H. Barschall

PHYSICS TODAY/JUNE 1987

Roger Stuewer has described how Léon Rosenfeld brought the news that uranium undergoes fission to Princeton's Physics Journal Club on the evening of 16 January 1939—the day on which Niels Bohr and Rosenfeld had arrived in New York from Denmark. (See Stuewer's article in PHYSICS TODAY, October 1985, page 48). Bohr had first learned of the discovery of fission from Otto Frisch on 3 January 1939, and Rosenfeld's report was the first information received by physicists in the United States. I was in the audience and the news had an immediate impact on my own activities, and it continued to affect my work for the next six years. The following are some of my recollections of that period.

At the time I was a graduate student at Princeton working under Rudolf Ladenburg in fast-neutron physics. After leaving Germany, I had arrived at Princeton in the fall of 1937. Ladenburg, who is best remembered for his contributions to atomic spectroscopy, had left the Kaiser Wilhelm Institute in Berlin in 1931 to become the Brackett Professor of Physics at Princeton. A relative of mine, Otto Meyerhof, who had been a colleague of Ladenburg's in Berlin, persuaded him to arrange for my acceptance as a physics graduate student even though I had neither an undergraduate degree nor any documented qualifications. It was therefore natural that I should work under La-

denburg.

Soon after the discovery of the neutron, Ladenburg initiated a research program to study the interaction of fast neutrons with nuclei. For this purpose he had acquired in 1934 a 400-kV power supply, which was used to accelerate deuterons. The deuterons impinged on a D_2O ice target and produced fast neutrons. It was an impressive installation that required a lot of attention to operate properly. Fortunately, a more experienced and exceptionally able graduate student, Morton Kanner, taught me patiently all the tricks one had to know to produce and detect neutrons. Almost all the vacuum equipment and all the electronics were built in the department, because there were no commercial suppliers.

Finding and fixing leaks in the vacuum system of the accelerator was an almost daily task. The leak detector was a gas discharge that changed color when alcohol was sprayed on the leak. The proof that the vacuum system was tight enough was the absence of strong x rays coming from the acceleration tube. The x rays were observed on a fluorescent screen at the end of a dark box. Holding one's hand between the fluorescent screen and the acceleration tube allowed one to see the bones of the hand when the accelerator was operated with a poor vacuum.

Sealing the leaks was an art. The expert knew which kind of wax (black or red, hard or soft) was most effective for what kind of leak. Painting with glyptal, a then widely used alkyd resin, was our last resort. Red glyptal was

used when other researchers needed to be warned of the prior use of paint, clear glyptal when the use of paint was not to be advertised.

Our home-built electronics required constant attention. Because the amplifiers were very slow, they amplified 60-Hz pickup as well as microphonic noise. Although we could avoid talking or walking around while taking data, we had a noisy mechanical vacuum pump that occasionally caused problems even though we had placed it in a soundproof box. There were no electronic pulse height analyzers. Instead we used a mechanical galvanometer and recorded the deflections of a light beam on photographic film. After the film was developed, we spent day after day measuring the deflections of the light beam with a magnifying glass to obtain pulse height distributions. Another experimental difficulty was the high background count of our detectors, which was due to an accident that Ladenburg had had several years earlier. Suspecting that a radium–beryllium neutron source had a small leak, he had attempted to touch up the soldered joint that was intended to provide a gas-tight seal for the capsule enclosing the radium–beryllium mixture. In the process, the source had blown apart and not only was he sprayed with radium, but, as he was careful to work under a chemical hood, the radium was blown into the entire ventilating system of the building. Radon from the decay of radium spread into our counters, and alpha particles from radon and its daughter products pro-

Henry H. Barschall is professor of physics at the University of Wisconsin in Madison.

One hundred tons of high explosives placed on a 25-ft tower were fired at the Trinity test site on 7 May 1945.

duced background counts. We were more concerned with this background than with possible health effects. In fact, we had no survey instruments to determine the radiation dose either due to radioactivity or due to x rays or neutrons. There was no shielding between the neutron source and the experimenters, although there was some lead between the experimenters and the acceleration tube.

Demonstrating fission

After hearing Rosenfeld's report on 16 January, Kanner and I discussed, as we walked back to the Graduate College, where we were living, whether we should follow up on the exciting news we had just heard. We were both anxious to work on our PhD theses and concluded it was unwise to divert our activities in another direction. But our decision to continue our research was soon overruled when Ladenburg brought Bohr to the laboratory. Bohr wanted us to demonstrate the fission process by observing neutron-induced fission in an ionization chamber. C. C. Van Voorhis had a vacuum system in which he could sputter uranium metal, and he quickly prepared a thin uranium target. When we put it into an ionization chamber, the neutron-induced fission events could be easily observed. Actually this was the first observation of fast-neutron-induced fission. In his first experiments, Frisch had used thermal neutrons to observe fission in an ionization chamber.[1]

During the next few months we had frequent discussions with Bohr and

John Wheeler, who were working on the theory of fission. They suggested many more experiments than we could do to test various aspects of their theory. In his article Stuewer quotes Bohr in a letter written to Frisch on 24 January 1939:

The physicists here at the institute are very caught up in the whole question, and I already have seen preparations for experiments to detect radioactive matter of very short halflife, the appearance of which should be an immediate result of the new type of splitting of the nucleus.

This refers to our activity—initiated by Bohr—during the week following his arrival at Princeton. Louis Turner, a molecular spectroscopist, had become very interested in the new discovery and joined our effort to look for short-lived activities. We soon observed the activities Bohr was looking for, and we sent a note[2] to the *Physical Review* on 26 April 1939; it appeared on 15 May 1939—that is, in the remarkably short time of 19 days. When I became an editor of the *Physical Review* in 1972, I wished I could achieve such short publication times, but in spite of all the modern facilities, it takes almost as many weeks to publish a paper now as it took days 50 years ago. The observation of the short-lived activities resulted in my first publication.

Our next project was motivated by Bohr's surprising prediction that fission induced by slow neutrons occurs in the rare isotope U^{235}, while fission induced by fast neutrons occurs pri-

marily in U^{238}. Although others could easily measure the fission cross section for slow neutrons, we were in a good position to measure the fission cross section for fast neutrons. Bohr and Wheeler urged us to perform such a measurement. Even today the measurement of a reaction cross section for fast neutrons is a difficult task, and often measurements performed at different laboratories differ by large factors. The principal difficulty is the determination of the neutron fluence. Ladenburg and his coworkers had previously measured the neutron output of the source, so that we could estimate the fission cross section, but we were not really sure that our results would give more than the order of magnitude. At a neutron energy of 2.5 MeV we found[3] a fission cross section for thorium of 0.1×10^{-24} cm^2, and for uranium a fission cross section of 0.5×10^{-24} cm^2, with an uncertainty of $\pm 25\%$. Recent measurements of these cross sections give 0.12 barn for thorium and 0.54 barn for uranium, values that agree better with the 1939 experiments than I would have expected.

The measurement of the fast-neutron fission cross section was quoted by Bohr and Wheeler[4] in their famous article on the mechanism of nuclear fission as proof that U^{238} is responsible for fast-neutron fission. They used the fact that the cross sections did not vary appreciably for neutron energies between 2.1 and 3.1 MeV to estimate the energy that had to be supplied to induce fission in Th^{232} and U^{238} nuclei.

Our last effort in studying fission was a measurement of the total kinetic energy of the fission fragments. Others had measured the energy of individual fission fragments; we wanted to measure the total energy by observing both fragments simultaneously. This required a uranium foil thin enough that the fragments would lose only a small amount of energy in passing through it. Van Voorhis prepared such a foil for us, and we measured the energy distribution of the fragments in an ionization chamber. The results[5] were consistent with the calculations Bohr and Wheeler had carried out and agree with currently accepted values.

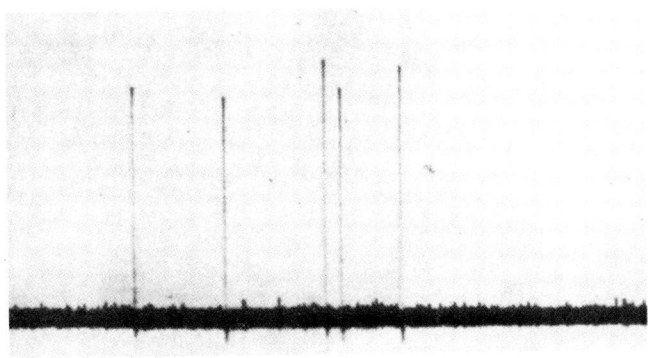

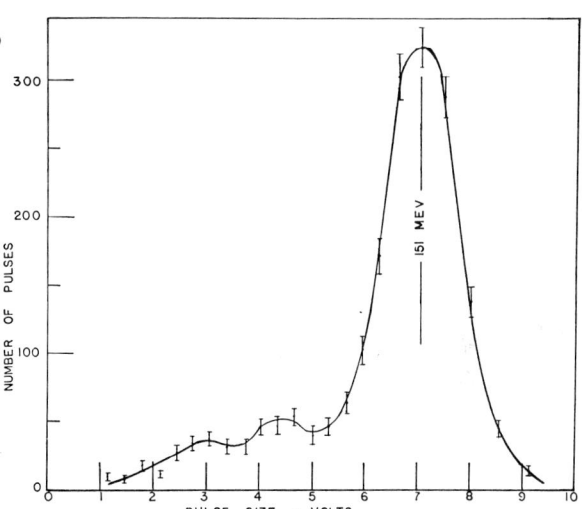

The discovery of fission attracted the interest of the press, where speculations about the possible military applications of the new discovery made headlines. Hence our experiments brought us in contact with newspaper reporters, an experience that appeared glamorous at the time.

Wartime research

Kanner and I had originally been reluctant to get involved in experiments on fission, because we wanted to work on our theses. In view of the results we had obtained, Ladenburg agreed to let Kanner use the fission experiments as his thesis. He in turn agreed to stay on for another year to help me with my thesis, as it was almost impossible for one person alone to operate the accelerator and take data. I completed my thesis research—measurements of the angular distributions of neutrons scattered by hydrogen, deuterium and helium—during my third year at Princeton.[6] Although Ladenburg was my major professor, the ideas and the understanding were largely provided by Wheeler.

In 1940, when I got my degree, some members of the Princeton faculty had moved to the MIT Radiation Laboratory. I was invited to stay on at Princeton for a year as an instructor, an opportunity I welcomed because there were few jobs for physicists available. During the 1940–41 academic year I continued experiments with fast neutrons and measured elastic and inelastic scattering by several intermediate and heavy elements. By 1941 questions were raised about whether, because of the war, my measurements should be published and whether as a recent immigrant from Germany I should be allowed to perform such measurements. Ladenburg finally informed me reluctantly that I could no longer work in his laboratory. In the meantime I was trying to find another job. One of the few openings was at the University of Kansas. James D. Stranathan, the chairman of the Kansas physics department, interviewed many applicants during the 1941 Washington APS meeting. I did not consider my chances good of being offered what appeared to be

Fragments from uranium fission induced by fast neutrons are indicated by sharp (150-MeV) pulses in the photographic record of galvanometer deflections (above). The graph below shows the pulse-height distribution. The large peak is due to pairs of fragments, the two smaller peaks are due to single fragments.[4]

one of the very few open positions. Fortunately Stranathan was just writing a textbook on modern physics[7] in which he quoted my work on fission, and I think that was the reason he offered me the position. I moved to Lawrence, Kansas, in the fall of 1941. After the attack at Pearl Harbor I found myself under the restrictions applicable to enemy aliens, the most unpleasant of which was that I was not allowed to leave the city limits of Lawrence without special permission from the US Attorney.

Although hardly any experimental equipment was available in the Kansas physics department, I was able to do some work on charged-particle detectors with a couple of graduate students and a colleague.

At the end of October 1942 I received an unexpected letter from Wheeler on University of Chicago—Metallurgical Laboratory stationery. He wrote:

I am writing to ask if there is enough possibility of your being able to join this laboratory to make it worthwhile for us to make you a formal offer. . . . There might be a little delay in getting your clearance through, but we have done such things before.

I responded:

Your letter came as a great surprise to me, since I did not know

that I was eligible to do defense work. As you can realize, I am most anxious to do anything which might be of direct value to the war effort.

On 6 November Wheeler responded that he had referred my letter to John Manley and that further communications would come from Manley or Arthur Compton. Shortly thereafter Manley wrote that he had asked the Internal Security District to send me an Alien Questionnaire, which I returned to Manley on 16 November 1942. Although various friends told me that War Department investigators were asking them about me, there was no indication that my clearance would be approved. In fact, the director of personnel at the Metallurgical Laboratory considered approval unlikely. In the meantime Henry Smyth, the chairman of the Princeton physics department, as well as Ladenburg tried to persuade me to return to Princeton to teach in a Navy program.

In view of my problems as an enemy alien I suggested that the University of Chicago might ask the War Department to help me become a US citizen. Naturalization was difficult for an enemy alien in wartime, but Sam Allison at the Metallurgical Laboratory immediately agreed to help. A letter to the Immigration and Naturalization

Building Z at Los Alamos, which housed a Cockcroft–Walton accelerator that had been moved from the University of Illinois, is shown in this 1944 photo. The accelerator produced neutrons of energies around 2.5 MeV, not far from the energies of neutrons produced in fission.

Service was written on University of California stationery and signed by the University Representative, University War Council. On 25 May 1943 Manley wrote that he had made contact with the War Department to expedite both my clearance and my naturalization. A confusing development was that Manley's letter came from Santa Fe, New Mexico, the supporting letter came from Berkeley and the rest of the correspondence was from Chicago. In any event, the intervention by the War Department resulted in immediate action by the Immigration and Naturalization Service. They agreed to hold a special hearing in Lawrence if I could persuade the judge of the District Court to interrupt his summer vacation. After a colleague who knew the judge arranged this, I was admitted to citizenship on 3 July 1943. The following three months were chaotic. By that time the students at the university were mostly soldiers and sailors, all of whom had to take physics, so that I had to give lectures several times a day. The university had to provide this instruction and was unwilling to let me leave. In the meantime research teams at both Chicago and Los Alamos wanted me. I wrote to Wheeler for advice. His answer, dated 9 July 1943, came from Wilmington, Delaware, on Du Pont stationery with the notation Explosives Department—TNX, another confusing development. He wrote:

> You ask whether I would recommend that you go to Chicago or Santa Fe. This is a very difficult question to answer. . . . The experimental instruments available for use are of a quite new type at Chicago. If you went there, you

would naturally work with Fermi. . . . It is an easy transition from Chicago to Santa Fe, but once one is at Sante Fe it is difficult to be released to go back to Chicago.

I interpreted Wheeler's letter as advising me to go to Chicago. On the other hand, he had originally suggested that I work with Manley, who had moved to Los Alamos. Manley had taken the initiative that had resulted in my naturalization; hence I was inclined to want to work with him even though the prospect of working with Enrico Fermi was difficult to ignore.

The first attempt to get me released from Kansas was made on 3 July, the date of my naturalization, by Arthur L. Hughes, who was then personnel director at Los Alamos and a well-known physicist. He wrote again on 13 July:

> We can ask the highest official in Washington connected with our project to apply pressure to your chancellor. We seldom use this procedure.

On 19 July Hughes wrote once again. This time his stationery bore the heading "US Engineer Project 'Y,'" and he now signed using the title "Assistant Director." He wrote:

> We have written to Dr. James B. Conant in Washington asking for his assistance in getting an early release for you.

On 31 July Hughes wrote:

> Dr. Conant is going to talk to General [Leslie] Groves who in turn would get the office of the Secretary of War to act.

On 17 August Hughes wrote (this time the letter was classified "Restricted"):

> Dr. Conant tried to get your Chancellor to release you but was com-

pletely unsuccessful. He suggested as a last resort that we call on General Groves to request the Secretary of War to ask for your release. . . . We have adopted this extreme approach in the case of only one other man.

Groves told me later that this person was Norman Ramsey.

On 30 August Hughes wired me, "After discussing your case with General Groves' office I am at liberty to ask you to come here as soon as possible," but on 6 September he wired, "On General Groves' instruction do not move from Kansas until you get signals from his office." On 12 September Groves wrote:

> I have had several telephone conversations with Chancellor [Deane] Malott, and in the course of these have promised him that no action would be taken by me until after he had received a letter from the Secretary of War.

My reaction to this letter was that surely in the middle of the war the Secretary of War had more important tasks than to write a letter on behalf of a recently naturalized instructor at the University of Kansas and that the chancellor had succeeded in talking the general into accepting a condition that he could not meet, but I was wrong. The letter from the Secretary of War really arrived, and a few days later I headed west.

Los Alamos

I had received another Restricted document ("Within the meaning of the Espionage Act, the contents of this document . . .") that explained living and employment conditions at Los Alamos: "Rent for furnished, equipped single rooms including utilities is $13 a month. Room service is $2 extra a month." The salary of a PhD with three years' experience was $355 a month, which was far more than my salary in Kansas. My instructions were to report to 109 East Palace Avenue in Santa Fe. This turned out to be a small office in the back of a one-story adobe building. Here I received a pass to Los Alamos and instructions on how to get there. I found the drive to the mesa on which Los Alamos is

Loudspeaker used to record the arrival of the sound wave and of the shock wave from the first nuclear explosion. Ten such loudspeakers were used to measure the shock wave's velocity.

located frightening. The road was narrow, with sharp switchbacks. I encountered large graders that were being used to widen the road, and I had to back down around the switchbacks. In addition, it was a hot day and my car's engine vapor-locked at the high altitude so that I had trouble maneuvering. I finally reached the top of the mesa, where armed military police inspected my pass.

At Los Alamos I was assigned, as I had expected, to Manley's group. He had brought to Los Alamos from the University of Illinois an accelerator of the same type as the one I had used at Princeton, and the measurements in progress[8] were very similar to those I had performed during my last year at Princeton. The group's goal was to find a suitable material to serve as a neutron reflector for a nuclear explosive. I had vaguely suspected that something like this was the reason I could not continue my work at Princeton, but I had not guessed the full story. I soon

learned what the new type of instrument at Chicago was that Wheeler had written about: Fermi had observed the first chain reaction ten months before I arrived at Los Alamos.

Manley's group was hardworking and congenial. Some group members whom I met on the day of my arrival have remained my friends ever since. We initially used neutrons of the same energy that I had used at Princeton, but we soon extended the measurements to lower energies. For these experiments we used electrostatic accelerators that had been built under Ray Herb's direction at the University of Wisconsin and moved to Los Alamos. I had heard about these machines, and the possibility of varying the energy easily over a wide range made them unique tools.

The cooperative spirit at Los Alamos made for an unusually effective operation. Data were analyzed by the theoretical group as soon as they were obtained. Viki Weisskopf did most of the analysis, but other theorists were also involved. The electronic equipment had improved enormously, largely because of advances made at the MIT Radiation Laboratory. There was always a staff of electronics technicians available to take care of all problems. Many of these technicians were young draftees of exceptional ability, and some of them became distinguished physicists.

When samples of U^{235} and Pu^{239} became available, we were assigned to study the properties of fissile nuclides. In particular, we measured fast-neutron multiplications of subcritical amounts of these materials.

The design of the nuclear explosive was completed toward the end of 1944. Early in 1945 Manley was given the task of performing blast measurements at the first bomb test, which was scheduled for the middle of the year. None of us had any experience with such measurements, and we spent much time learning about the new field—especially from British experts who had come to help us. While techniques for blast measurements of conventional explosions were well developed, the measurements we were asked to perform had special difficul-

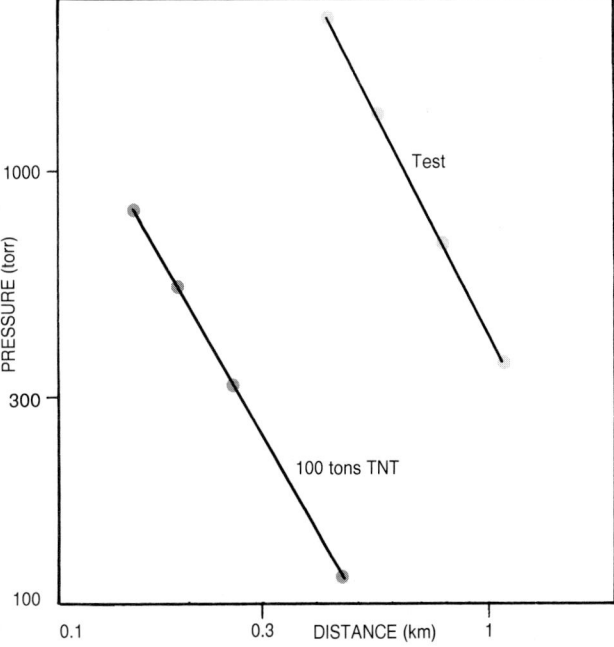

Peak gauge pressure as a function of distance from the explosive, measured for the firing of 100 tons of high explosives and for the nuclear explosion. Comparison of the two yields the TNT explosive equivalent of the nuclear device.

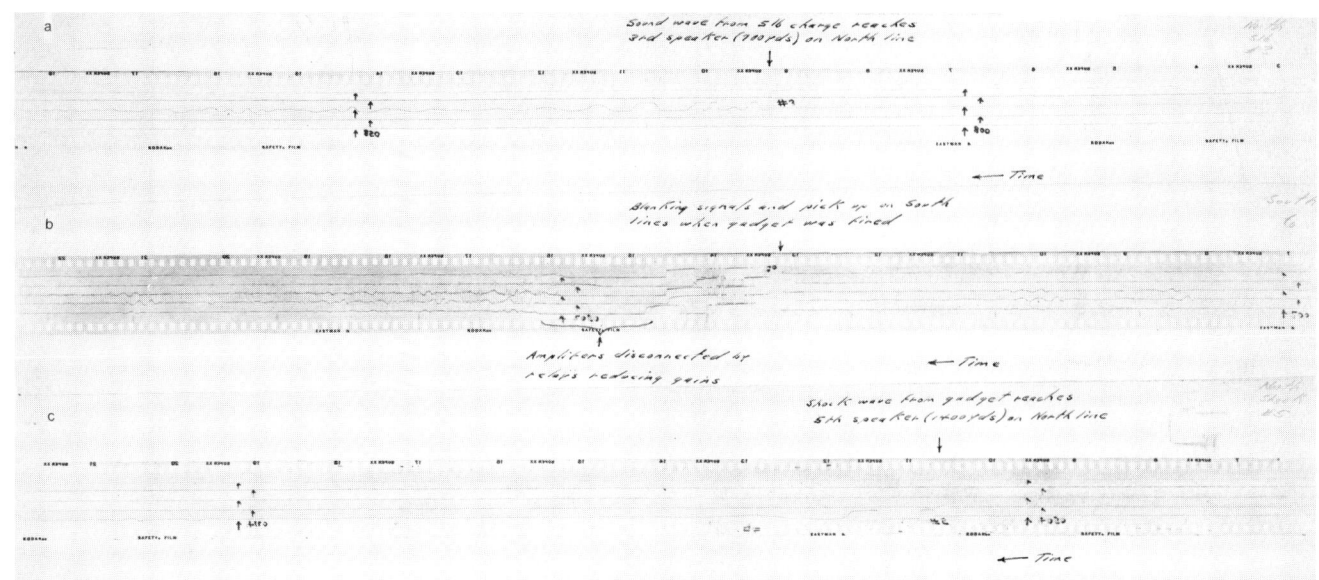

Oscilloscope record of the arrival of sound and shock waves from the Trinity explosion. Each line records signals from a different loudspeaker. **a:** Arrival of the sound wave from a 5-lb calibration explosion, set off just before the nuclear explosion. **b:** The electromagnetic pulse from the nuclear explosion is picked up as noise on the cables from the speakers and a mechanical relay reduces the gain of the amplifiers (arrows). **c:** The shock wave from the nuclear explosion arrives at the fifth north detector (arrow).

ties. To provide data in the event of a very low explosive yield, our instruments had to be capable of giving results for energy releases differing by a factor of at least 100. Furthermore, the nuclear explosion was expected to produce very strong electromagnetic signals that would disable most electronic equipment, and intense thermal radiation that was likely to burn flammable materials; of course, these radiations would arrive at the blast detectors long before the shock wave. Another problem was that previous blast measurements had usually been performed near sea level, while our test site was at high altitude.

My assignment was to measure the velocity of propagation of the shock wave as a function of distance from the explosion. The ratio of the velocity of the shock wave to the sound velocity is related to the peak pressure of the shock wave by the Rankine–Hugoniot equation. We used loudspeakers as detectors for the arrival of the shock wave. The motions of the speaker coil in the magnetic field would induce a large signal that could be transmitted to a shelter, where a movie camera would photograph the signal as displayed on an oscilloscope screen. Five loudspeakers were placed in a line at different distances from ground zero, so that four average velocities could be measured. Two such strings of speakers were arranged in opposite directions from the tower on which the explosive would be detonated.

An accurate knowledge of the velocity of sound at the time of the explosion

was necessary to deduce the peak pressure. Because the sound velocity depends on temperature and wind velocity, which vary with time, we wanted to measure the sound velocity immediately before the nuclear explosion. To accomplish this we decided to set off a small (5-lb, conventional) explosion near the nuclear explosion, a request that was approved only reluctantly. As soon as the signal from the small charge was recorded, the sensitivity of the system would be reduced.

We tested our detection and analysis equipment on 7 May 1945 by setting off 100 tons of high explosives on a platform with the center of the explosive about 10 meters above the ground. All the equipment worked satisfactorily. The nuclear test, originally scheduled for 4 July 1945, had to be postponed until the 15th. After various rehearsals and dry runs on the preceding days, everybody was ready on the evening of 15 July. For the benefit of optical measurements the test was planned to be done during the night, but storms on the evening of 15 July and in the early hours of 16 July led to a long wait. The weather cleared up around 5 am, and the explosion occurred at 5:30 am. I was watching the oscilloscope in a shelter 10 000 yards from the tower, and observed first the arrival of the signals from the explosion of the 5-lb charge. Then the shelter suddenly lit up as the nuclear explosion occurred. The arrival of the shock wave at the loudspeakers could be seen on the oscilloscope. About 20 seconds later the shock wave hit our

shelter.

We sent the film back to Los Alamos for developing and analysis. The data yielded an energy release equivalent to $10\,000 \pm 1000$ tons of TNT, a value close to expectations.[9]

With the completion of this measurement my activities initiated by the news of the discovery of fission ended, and I returned to research in fast-neutron physics. After the war the "Long Tank," the electrostatic accelerator that we used at Los Alamos, was returned to Wisconsin. It was the ideal tool for extending my research in neutron physics to other neutron energies. Hence I was delighted to be invited to join the Wisconsin faculty and to be able to continue my research there.

References

1. O. R. Frisch, Nature **143**, 276 (1939).
2. H. H. Barschall, W. T. Harris, M. H. Kanner, L. A. Turner, Phys. Rev. **55**, 989 (1939).
3. R. Ladenburg, M. H. Kanner, H. H. Barschall, C. C. Van Voorhis, Phys. Rev. **56**, 168 (1939).
4. N. Bohr, J. A. Wheeler, Phys. Rev. **56**, 426 (1939).
5. M. H. Kanner, H. H. Barschall, Phys. Rev. **57**, 372 (1939).
6. H. H. Barschall, M. H. Kanner, Phys. Rev. **58**, 590 (1940).
7. J. D. Stranathan, *The Particles of Modern Physics*, Blakiston, Philadelphia (1942).
8. H. H. Barschall, J. H. Manley, V. F. Weisskopf, Phys. Rev. **72**, 875 (1947).
9. K. T. Bainbridge, Los Alamos report LA-6300-H (1976).

Scientists with a secret

While the Nazi war machine was gearing up, a few physicists
realized that a fission chain reaction was feasible—would they be able
to get all groups to agree to hold back publication?

Spencer R. Weart

PHYSICS TODAY/FEBRUARY 1976

What are physicists to do if they make a discovery that promises to transform industry but also threatens to revolutionize warfare? Should they investigate the phenomenon within their traditions of free and open inquiry or keep the deadly secret to themselves? This is the dilemma that was faced by several groups of physicists who studied uranium fission in 1939 and 1940. In the spring of 1939 one group, foreseeing the unprecedented power of nuclear weapons, made a concerted attempt to restrict knowledge of chain reactions. But it was not until over a year later that censorship—imposed by the community of physicists on itself—became fairly complete.

Any attempt to keep a secret must by its very nature follow a course that is difficult to observe, creating confusion and misunderstanding. But this course, which the participants could not see clearly at the time, can now be pieced together from collections of papers made available to researchers, supplemented by oral history interviews conducted by the Center for History of Physics of the American Institute of Physics.

Fears of disaster

The first arguments over nuclear secrecy revolved around the unlikely figure of Leo Szilard. A short, round, exuberant Hungarian, Szilard in 1939 had

Spencer R. Weart is the director of the Center for History of Physics, American Institute of Physics, New York.

neither a job nor a home. But he was uniquely qualified to face the issues of nuclear energy and secrecy because for over five years he—and he alone—had been concentrating on these problems.

Since 1933 Szilard, then recently arrived in England to escape the Nazi persecution of Jews, had wondered if there was a way to release the energy that physicists knew to be bound up in nuclei.[1] The answer came with his realization that if one could bombard some element with a particle (say, a neutron) and make it radioactive in such a way that it emitted two particles, a chain reaction of awesome power might be induced. The possibility seemed much closer the next year, when Frédéric Joliot and his wife Irène Curie, working at the Radium Institute in Paris, discovered that, with alpha particles, one could indeed make nuclei radioactive artificially. Szilard decided to devote himself to nuclear physics and set out to search for some type of nucleus in which a chain reaction might be sustained.

From the start Szilard feared the consequences of his work. He attempted to gain some control by the only means then available to a scientist who wanted to restrict the use made of his work: He took out a patent on his ideas. Furthermore, he persuaded the British government to declare the patent secret; there was a small but real possibility, he warned them, of constructing "explosive bodies ... very many thousand times more powerful than ordinary bombs."[2] Meanwhile Szilard brashly

tried to alert his colleagues in Britain. His ideas, he told one professor in 1935, could cause an industrial revolution but might cause a disaster first. It would be necessary to bring about something like a conspiracy of the scientists working in the general field. In a letter to F. A. Lindemann, the head of physics at Oxford, he offered a mechanism to ensure secrecy—an agreement to make experimental results in the dangerous zone available only to those working in nuclear physics in England, America and perhaps one or two other countries, while otherwise keeping quiet.[3]

Szilard foresaw only too well the likely reaction to his efforts: "Unfortunately it will appear to many people premature to take some action until it will be too late to take any action."[3] And indeed the leading physicists in Britain were cool to Szilard's obstreperous advice. They thought his proposed chain reaction entirely unworkable (as was in fact the case for the mechanisms Szilard was then considering). They were suspicious when he sought to patent his ideas, suspecting that he was seeking pecuniary return, a motive incompatible with British traditions of disinterested science. Finally, they found the idea of scientific secrecy entirely alien. Even those scientists who felt most keenly the responsibility of scientists for the consequences of their discoveries traditionally felt that secrecy is abhorrent and that interference with the normal process of open criticism would not only impede scientific progress but pervert it.[4,5]

Szilard went on to study various elements for a possible chain-reaction mechanism; he had not quite reached uranium when he learned that Otto Hahn, Fritz Strassmann, Otto Frisch and Lise Meitner had discovered uranium fission. When Szilard heard of this in January 1939 in New York, where he had moved to escape the war that appeared ever more imminent in Europe, he discussed his concern with scientists at Columbia University.

Private messages

The leading nuclear physicist there was Enrico Fermi, who had fled Italy because Fascist race laws affected his Jewish wife, and who had arrived in New York scarcely three weeks ahead of the news of the discovery of fission. Like Szilard and other physicists, Fermi quickly recognized the possibilities this discovery opened. According to one account, he made a rough calculation of the size of the hole a kilogram of uranium would make in Manhattan Island if it underwent an explosive chain reaction.[6] However, he soon concluded that this would never happen: When a uranium nucleus was struck by a neutron and split in two, it seemed unlikely that it would release enough neutrons to sustain a chain reaction. When Szilard approached Fermi about the need to keep fission work secret, Fermi's response was direct: "Nuts!"

> From the very beginning [*Szilard recalled*] the line was drawn; the difference between Fermi's position throughout this and mine was marked on the first day we talked about it. We both wanted to be conservative, but Fermi thought that the conservative thing was to play down the possibility that this [chain reaction] might happen, and I thought the conservative thing was to assume that it would happen and take all the necessary precautions.[1]

Rebuffed by Fermi, Szilard remained alert for a way to control events. At about this time, late January, a telegram arrived at Columbia, addressed from Hans Halban, a physicist in Paris, to his colleague George Placzek. As Szilard recalled it long after, the telegram was opened by a secretary by mistake, and Szilard learned the contents: "JOLIOT'S EXPERIMENTS SECRET." Placzek had just come from a visit in Paris, and Szilard assumed that Placzek had learned of an experiment Joliot was doing; apparently Joliot had now decided to keep the experiment quiet for the time being. Szilard had little doubt what experiment would be so important as to require secrecy.

What Szilard felt was involved here was the sort of secrecy that had been traditional in science for centuries—the caution of the scientist who holds back his results until he is ready to publish

them, so they will not be broadcast in a distorted form and so that others will not take advantage of a hint to beat him to the next result. This was quite different from the sort of secrecy Szilard had in mind. There was some misunderstanding here, for Joliot did not actually begin fission experiments until late January, after Placzek had left Paris, and it is not clear what Halban and Placzek were corresponding about. But Szilard now believed (correctly as it happened) that Joliot's group was working on fission, and decided to send him a letter.

The only reason he was writing, Szilard said, was that there was a remote possibility that he would be sending a cable after some weeks, and the letter was to explain what his cable would be about. Some scientists in New York were concerned about the possibility that neutrons would be liberated in fission. Obviously, if more than one neutron would be liberated, a sort of chain reaction would be possible. In certain circumstances this might then lead to the construction of bombs which would be extremely dangerous in general and particularly in the hands of certain governments. Perhaps steps should be taken to prevent anything on this subject from being published. No definite conclusions had been reached, but if and when any steps were agreed on, Szilard would cable Joliot. Meanwhile Fermi was doing experiments to see whether the danger was real, and these would perhaps be the first to give reliable results. But if Joliot got definite results sooner, Szilard would be glad to have the uncertainty ended. Also, if Joliot felt that secrecy should be imposed, his opinions would be given very serious consideration.[3]

Neither Joliot nor his close collaborators Halban and Lew Kowarski responded. The letter was obviously a purely personal venture, and this impression must have been reinforced by a letter Fermi sent Joliot two days later. On 4 February 1939 Fermi wrote that he was then engaged in trying to understand what was going on in uranium fission—as was, he thought, every nuclear physics laboratory. After thus informing Joliot's team that they had competition, Fermi went on to ask help for another Italian refugee scientist and closed without saying a word about keeping secrets.[7] There was every reason to believe that Fermi would publish first if the French held back their own results.

Even as a personal request Szilard's letter made little impression on the French, for it stated that it was only meant to help them understand a cable that might follow. Weeks passed, no cable appeared, and the French, as Kowarski recalled, "considered that probably the whole idea was abandoned. We simply published."[8]

This publication, the first result of the joint efforts of Halban, Joliot and Kowarski, contained important news: Neutrons were indeed liberated when a uranium nucleus fissioned.[9] The experiment was of a kind that would only have been done in a few places, requiring ingenuity, a powerful source of radioactivity and an interest in chain reactions. It had not been easy to detect the few neutrons produced in fission amidst the flood of neutrons that had been required to provoke some fissions in the first place, nor had it been obvious that these neutrons were important. Although the French, like Fermi, believed scientists everywhere were

---

The many "secrets" of the atomic bomb

There was no single discovery that showed how atomic bombs could be built, but a combination of discoveries made at various times. Here is a partial list:

Published discoveries

1934 Artificial radioactivity can be produced with alpha particles (Joliot and Curie, *France*) or neutrons (Fermi, *Italy*).

December 1938 Neutrons can cause uranium to fission (Hahn and Strassmann, *Germany*, Frisch, *Denmark* and Meitner, *Sweden*).

March 1939
▸ Neutrons are produced during fission (Anderson, Fermi and Hanstein, *US*; Szilard and Zinn, *US*; Halban, Joliot and Kowarski, *France*).
▸ Two or three neutrons are emitted per fission (same groups).
▸ U^{235} is the fissionable isotope of uranium (Bohr and Wheeler, *US*).

Unpublished discoveries

June 1939–February 1940 A self-sustaining nuclear reactor can be built if a suitable moderator can be found (Szilard, *US*; Halban, Joliot, Kowarski and Perrin, *France*; Heisenberg, *Germany*; various groups, *USSR*).

Spring 1940
▸ Carbon is a suitable moderator for a nuclear reactor (Anderson and Fermi, *US*).
▸ Nuclear reactors can be used to produce a fissionable element, plutonium (Turner, *US*)—from this resulted the bomb that devastated Nagasaki.
▸ It is possible to isolate sufficient U^{235} to make an explosive critical mass (Frisch and Peierls, *UK*)—from this resulted the bomb that devastated Hiroshima.

hard at work on the question, there was in fact only one other group then carrying on a similar experiment—the group at Columbia.

Chain reaction—and invasion

By mid-March Fermi and Szilard, working with Herbert Anderson, Walter Zinn and others, had done their own experiments and independently learned the distressing news that neutrons were produced in fission. This was still far from proving that a chain reaction was possible, for that would depend on the precise number of neutrons emitted in each fission, a thing still more difficult to measure. The group estimated that there were about two neutrons per fission, which made it appear only barely possible that a chain reaction could be sustained (the true value is about 2.5 neutrons per fission).

On 15 March, as the Columbia physicists finished writing up their experiments for publication, German troops invaded the remnant of Czechoslovakia that had survived the Munich agreement. With this action, many felt, Hitler crossed his Rubicon, subjecting for the first time a non-German people and giving a clear signal that world war was inevitable. Despite their concern over this, the physicists sent their papers to the *Physical Review* the next day.

Szilard was not satisfied, and three days later he met with Fermi and with another Hungarian refugee physicist, Edward Teller. As Szilard recalled the meeting, he and Teller pressed for keeping their work secret, but Fermi was repelled by this idea, holding that publication was basic to scientific morality. "But after a long discussion, Fermi took the position that after all this is a democracy; if the majority was against publication he would abide by the wish of the majority . . ."[1] Fermi therefore arranged to ask the *Physical Review* to delay the publication indefinitely.

Szilard was now on the point of cabling Joliot, but before he did so he heard of the French team's note, just published in *Nature,* which revealed that some neutrons are emitted in fission. Fermi felt that there was now no secret to keep, so that there was no longer any sense in refusing to publish. Szilard denied this (the crucial number of neutrons emitted per fission was not yet published), and argued that "If we persisted in not publishing, Joliot would have to come around; otherwise, he would be at a disadvantage, because we could know his results and he would not know our results." Fermi was not convinced but, determined to be fair, he reluctantly agreed to put the matter before George Pegram, administrative patron of the Columbia group and a respected physicist. Pegram delayed his

SZILARD

KOWARSKI, HALBAN AND JOLIOT

decision for some time. Szilard's arguments were forceful, but others at Columbia replied that an attempt to restrict publication was both futile and an undesirable breach of scientific custom.[1,3]

Warnings

While Pegram deliberated, Szilard and his friends were determined to waste no time. Several of them talked the matter over, among them Victor Weisskopf, an emigré Austrian physicist. "We were very much afraid of the Nazis," Weisskopf recalled. "We knew this was a hopeless thing but we thought we had to try . . . And then the question was . . . how do we get to Joliot." As Weisskopf said in a recent interview, he had met Joliot's collaborator Halban years earlier and the two had become close personal friends, so Szilard and Weisskopf drafted a telegram to Halban, which Weisskopf signed. The telegram asked Halban to advise Joliot that papers on neutron emission had already been sent to *the Physical Review,* but that the authors had agreed to delay publication for the reasons indicated in Szilard's letter to Joliot of 2 February. The telegram continued:

NEWS FROM JOLIOT WHETHER HE IS WILLING SIMILARLY TO DELAY PUBLICATION OF RESULTS UNTIL FURTHER NOTICE WOULD BE WELCOME STOP IT IS SUGGESTED THAT PAPERS BE SENT TO PERIODICALS AS USUAL BUT PRINTING BE DELAYED UNTIL IT IS CERTAIN THAT NO HARMFUL CONSEQUENCES TO BE FEARED STOP RESULTS WOULD BE COMMUNICATED IN MANUSCRIPTS TO COOPERATING LABORATORIES IN AMERICA ENGLAND FRANCE AND DENMARK . . .[7]

The proposed scheme was similar to the one Szilard had conceived in 1935, with the additional idea that papers should be sent to journals, not for publication but to certify priority of discovery.

At the same time Weisskopf also cabled P.M.S. Blackett, a leading British physicist, asking whether it would be possible for *Nature* and the Royal Society's *Proceedings* to cooperate in delaying publication of fission research. Meanwhile another of Szilard's Hungarian physicist friends, Eugene Wigner, wrote P.A.M. Dirac and asked him to support Blackett. The matter was rather urgent, Wigner said; although American scientists were willing to cooperate, they realized that their interests might be prejudiced if scientists in other nations published results and they did not.[3,10] Blackett and another prominent physicist, John Cockcroft, promptly replied that they would support the secrecy plan. *Nature* and the Royal Society were expected to cooperate.[3]

FERMI

Szilard, Teller, Weisskopf and Wigner also talked the problem over with Niels Bohr, who was visiting the United States. Bohr doubted very much that fission could be used to cause a devastating explosion. And he thought that at any rate it would be difficult if not impossible to keep truly important results secret from military experts—the matter was already public. Nevertheless he agreed to go along with the attempt and drafted a letter to his Institute in Denmark (which apparently he did not immediately mail):

The Columbia group is busy organizing cooperation among all the physics laboratories outside the dictatorship countries, to keep possible results from being used in a catastrophic way in a war situation, and I must therefore ask you, if work along these lines is going on in Copenhagen, to wait before you publish anything until you have cabled me about the results and received an answer.[11]

But the conspirators still had to win the agreement of other American laboratories.

The most immediate problem was a group headed by Richard Roberts working under Merle Tuve at the Carnegie Institution in Washington, DC. They too had recently seen some neutrons released from uranium. But the neutrons they saw were emitted over a period of some seconds after fission occurred: These were not the true fission neutrons, but occasional neutrons produced as a side effect of the radioactivity of the fission fragments.[12,13] The development was announced in a news release of Science Service dated 24 February, written by Robert D. Potter, a science writer who kept in touch with the Columbia physicists and was infected with their excitement over chain reactions. Potter headlined the possibility of an explosive chain reaction propagated by neutrons. He carefully noted that Roberts's delayed neutrons might not be enough to sustain a chain reaction—in fact they are not—but he quoted Fermi as saying that the possibility of a chain reaction was certainly present.[14]

Szilard and his friends quickly approached the Washington group, who promised to cooperate in withholding future publications. The proposal was spread further within the United States by word of mouth and letter. Maurice Goldhaber of the University of Illinois was included and Ernest Lawrence of Berkeley was probably informed of the matter when he visited New York on 3 April.[15] John Tate, editor of the *Physical Review,* was brought in, for nearly all important physics papers in the United States passed through Tate's office; anyone else who showed an interest in fission neutrons could thus be put in touch with the conspirators. The attempt to restrict the circulation of information to physicists outside the dictatorships was well begun. It lacked chiefly the acquiescence of the French.

The French reply

The French knew what Weisskopf's telegram implied, for they were as alarmed as he by Hitler's march towards world war. However, like Bohr and Fermi, the French believed an atomic bomb was not likely to be built for many years, if ever. In this they were entirely correct, so far as atomic bombs were then conceived—masses of tons of natural uranium. Nobody had yet seriously considered the likelihood of isolating a substantial quantity of the rare fissionable isotope U^{235}, still less of the undiscovered element plutonium; and these two substances are the only ones that could in fact be used for a nuclear weapon. Unaware of these possibilities, Joliot and his collaborators thought that industrial nuclear power from nuclear reactors was a much more immediate prospect than weaponry.

It was up to Joliot, as head of the team, to answer Weisskopf's telegram, but he discussed it at length with his colleagues. Thinking back, they recalled a number of factors that entered their decision.[8,16] For one thing, Joliot believed strongly in the international fellowship of scientists, and in principle had little sympathy with secrecy.[17] For another, if he and his colleagues failed to publish, they might well be eclipsed by those who did. For they could scarcely believe that everyone would adhere to an unprecedented pact, a pact pushed forward, so far as they knew, only by two Central European refugees on the outskirts of the Columbia scientific community. (Had Fermi, Bohr or a leading American scientist written them about the scheme, the French might have found it more plausible.) And if they failed to be first to publish discoveries, the French might have trouble getting the money they would need to pursue the development of industrial nuclear energy. Finally, even if all the laboratories joined and stuck by the agreement, there would remain a powerful objection, the same one noted by Fermi and Bohr. It was scarcely likely that copies of papers circulated privately around America, France, Britain and Denmark could be kept out of

BREIT

Germany and the Soviet Union; moreover, German and Soviet scientists were surely aware of the importance of fission chain reactions.

Ideas of fission power and weapons had begun to show up in the popular press. The French were aware of at least some of the sensational news stories that emanated from the United States. The French were not in close touch with what was happening there, but it is very likely that they had seen a copy of a Science Service news release of 16 March, which summarized their own report, published in *Nature* on that date, of neutrons resulting from fission. Presumably they were not pleased to read that they had apparently been beaten to the discovery: Their result, the release said, "is comparable with, and a confirmation of, the announcement (Science Service, 24 February 1939) that scientists at the Carnegie Institution . . . had been able to observe the same important reaction in atomic transmutation."[18] This was an error, but it made it seem that the most important facts were already leaking out in America.

For all these reasons, the team cabled Weisskopf a discouraging reply around 5 April.

SZILARD LETTER RECEIVED BUT NOT PROMISED CABLE STOP PROPOSITION OF MARCH 31 VERY REASONABLE BUT COMES TOO LATE STOP LEARNED LAST WEEK THAT SCIENCE SERVICE HAD INFORMED AMERICAN PRESS FEBRUARY 24 ABOUT ROBERTS WORK STOP LETTER FOLLOWS

JOLIOT HALBAN KOWARSKI[3]

Szilard was well informed on the work of Roberts's group through their publications and through letters from Teller, who had visited them various times, and on the next day, Weisskopf having left New York, Szilard answered on his behalf. Roberts's paper, he noted, concerned delayed neutron emission, which was harmless. But the group had been approached and had promised to cooperate. The American group had delayed publishing papers; were the French inclined to delay their papers too, or did they think everything should be published?

That same day the French sent their final answer:

QUESTION STUDIED MY OPINION IS TO PUBLISH NOW REGARDS JOLIOT.[3]

The scheme fails

This reply, along with the preceding French publication of the fact that fission does produce some neutrons, doomed the attempt to restrict publication. Pegram, who was not aware how much progress Szilard and his friends had made aside from the French, after some days of deliberation decided that any attempt to impose secrecy was hopeless. Szilard was forced to give in. The Columbia scientists asked the *Physical Review* to print their papers.[19]

On April 7, the day of the final exchange of cables with Szilard, the French sent *Nature* the results of experiments and calculations that estimated the number of neutrons emitted per fission at between three and four. The report was duly published on 22 April 1939. This note convinced many physicists that uranium chain reactions were a real possibility. In Britain, George P. Thomson decided to warn his government of the dangerous prospects and meanwhile to begin experimenting with uranium.[20] In Germany, Georg Joos wrote a letter to the Reich Ministry of Education; independently and simultaneously, Paul Harteck and Wilhelm Groth wrote a joint letter to the War Office.[21] News of the French work may also have played a role in the start-up of Soviet nuclear energy research, perhaps provoking the letters on uranium which I.V. Kurchatov and others sent the Soviet Academy of Sciences about this time.[22] Thus in Britain, Germany and perhaps the Soviet Union, publication of the French results precipitated officially-supported programs of research into nuclear energy. The effort of Szilard and his friends, after coming within an inch of success, had failed disastrously.

Nevertheless, by the end of 1939 a blanket of secrecy had settled over fission research in certain countries. After war broke out in September, scientists in France, Germany and Britain withheld publication on fission and any other subject remotely of military interest. But in the United States, the Soviet Union and other neutral countries, publication was scarcely impeded.

US government: Do it yourself

Szilard continued to work on the problem. With Albert Einstein he set in motion a chain of events that led to the formation of an official government committee, under Lyman J. Briggs, which was supposed to support and coordinate fission work.[23] From the beginning Szilard hoped that the committee would also do something about secrecy. When he took up the matter with Briggs he added another element to his by now increasingly well developed scheme. Presumably to counter objections he had faced from younger men at Columbia, he wrote:

> For a physicist, who has not yet made a name for himself, refraining from publication means a sacrifice which he should not be asked to make without being offered some compensation. Some addition to the salary which he is normally drawing from the university might therefore be desirable and might require the creation of some special fund.[3]

But the Briggs committee remained all but inactive, leaving everything up to the physicists. As late as 27 April 1940, when the committee held one of its rare meetings, the only response Szilard could get was a suggestion from Admiral Harold Bowen, present as an observer, that the scientists working on uranium might get together and impose upon themselves whatever censorship they felt necessary. The government itself would do nothing.[3]

Szilard had already taken the single step that was entirely within his power: He withheld from publication a paper of his own. This paper, completed in February 1940, contained elaborate calculations of the characteristics of a nuclear reactor and concluded that there was a strong possibility of making one work. Had the article been published, it surely would have been a great stimulus to nuclear reactor work in various countries. But when Szilard sent it to the *Physical Review* he requested that printing be delayed until further notice.[2] For a second specimen of a withheld paper, in late April Szilard persuaded Herbert Anderson, a graduate student who had worked closely with Fermi on fission from the beginning, to hold back his doctoral thesis on neutron absorption in uranium, which was then already in proof.[24,25]

Anderson and Fermi had meanwhile been measuring the neutron-absorption cross section of carbon: This difficult-to-determine quantity was central to the question of whether or not a nuclear reactor could be built, for carbon seemed the only feasible moderator, and even carbon could be used only if it absorbed virtually no neutrons. This turned out to be the case: The cross section was extremely small. Szilard now approached Fermi and suggested that the value for the cross section should not be published. "At this point," Szilard recalled, "Fermi really lost his temper; he really thought that this was absurd." But while Fermi stuck by his principles, Pegram had second thoughts and finally asked Fermi to keep his work secret.[1]

This decision came late, but still in time: If the value for the carbon cross section had been published, the course of World War II might conceivably have been changed. For German scientists, using experiments they carried out later in 1940, wrongly concluded that carbon had a substantial neutron-absorption cross section. From that point on they abandoned carbon as a moderator and attempted to use the extremely rare isotope deuterium, which they never managed to get enough of.[21,26] Soviet scientists too at first did not seriously consider carbon as a moderator.[27] The French scientists were also committed to deuterium. They escaped to England when France fell to the Germans, and thereafter the British followed their lead in matters of reactors, regarding carbon as an unlikely choice. Anderson and Fermi's work could have put all these groups on a different track.

Prescription for a bomb

This was not the only hole in the dike that had to be plugged. In late May, Louis Turner at Princeton sent Szilard a copy of a paper on "Atomic Energy from U^{238}." In this paper Turner pointed out that if U^{238} were bombarded by neutrons, as would happen in a nuclear reactor, a series of steps would give rise to a new element. This he predicted to be fissionable—it was the element later named plutonium. Although Turner had not realized it, he had written the prescription for the easiest route to building an atomic bomb.

Szilard wrote back at once to say that his own paper was secret, implying that there was an official move underway to withhold papers. He persuaded Turner to write the *Physical Review* and delay publication.[3] It was well he did so: Turner's paper could have been an essential clue for the Germans and others. Meanwhile Szilard approached Harold Urey and asked him to try to set up a committee to regulate fission publications.

Before much progress had been made, the 15 June issue of the *Physical Review* appeared, containing a letter from Edwin McMillan and Philip Abelson at Berkeley. They had observed the production of neptunium when ura-

Two history-making releases from Science Service, as reprinted in *Science News Letter*. After reading an erroneous statement in the later (lower) article, which said that their results had already been published in America, the French team rejected Szilard's request for secrecy.

nium was bombarded with neutrons. This was the first and most essential step of the process that Turner had predicted should lead to plutonium. But Abelson and McMillan had simply failed to see the connection between their work on transuranic elements and the fission problem.[15,28]

This publication brought down a flurry of protest, which helped to settle the secrecy issue. From as far as Britain, scientists interested in fission protested the publication of such revealing information. But the most important news came from Gregory Breit at the University of Wisconsin. Breit had known Szilard and Wigner for years, and was awakened to the secrecy problem through long conversations with them. Around the beginning of June Breit found a way to circumvent the problems Szilard and others were running into. Recently named to the National Academy of Sciences, he had been put in the Division of Physical Sciences of the Academy's National Research Council. At a committee meeting he spoke up in favor of censorship. There was some skepticism, Breit later recalled, but a committee on publications was appointed to consider the problem. Breit was made chairman of a subcommittee concerned specifically with uranium. Acting on his own initiative, he immediately began writing letters to journal editors, proposing a voluntary plan under which papers relating to fission would be submitted to his committee before publication. Sensitive papers would be circulated only to a limited number of workers. Breit added that he expected ultimately to publish the papers in book form or otherwise, with a statement of the original date of the paper and with a suitable acknowledgment of the public spirit of the authors.[15]

There were some raised eyebrows, but the editors of scientific journals and other leading scientists agreed to the plan. "As recently as six months ago," Lawrence wrote Breit, "I should have been opposed to any such procedure, but I feel now that we are in many respects essentially on a war basis."[15] German troops were pursuing the remnants of the defeated French army, and none could doubt that the international situation was desperate.

Better than never

Within a few weeks Breit, who swiftly set up close communications with Fermi, Urey, Wigner and others involved in parallel efforts at secrecy, had imposed total censorship on American fission research. After passing the papers around by mail for comment, Breit's committee let some through as innocuous; other they withheld from publication.[25] Because of this procedure, carried out entirely by physicists

with no government participation, long before the United States went to war it was keeping vital scientific information within its own borders.

The extraordinary coincidence that history's most dangerous scientific secret appeared at the moment history's greatest war began made possible this unique case of scientific self-censorship. It was imposed against the grain—even some of the conspirators, like Szilard and Teller, would later argue strongly for the advantages of open publication. But it is worth noting that if self-censorship is difficult, under sufficiently deadly circumstances it can be achieved, and that if it may seem to come late, late may be far better than never.

* * *

I wish to thank first of all Gertrud Weiss Szilard, who kindly gave me permission to use the Szilard Papers and to publish the excerpts above. Thanks are also due to Hélène Langevin, who kindly made available the Joliot-Curie Papers; to Monique Bordry, who gave invaluable assistance in their use; to Gregory Breit, Otto Frisch, Victor Weisskopf, and particularly Lew Kowarski, who answered the questions I posed them, and to Charles Weiner, who assembled interviews and other materials at the Center for History of Physics of the American Institute of Physics. Further details and documentation on the subject of this article will be published in Volume 2 of The Collected Works of Leo Szilard (reference 1 below) and in a book I am preparing on French scientists and nuclear energy. All translations are my own except for the Bohr letter, for assistance with which (and for much else) I thank John Heilbron.

References

1. L. Szilard, "Reminiscences," *The Intellectual Migration, Europe and America, 1930–1960* (D. Fleming and B. Bailyn, eds.) Harvard U. P., Cambridge, Mass. A revised and expanded version will be in *The Collected Works of Leo Szilard*, volume 2 (G. W. Szilard, H. Hawkins, S. Weart, eds.), MIT Press, to be published.

2. *The Collected Works of Leo Szilard, Volume 1, Scientific Papers* (B. T. Feld, G. W. Szilard, eds.), MIT Press, Cambridge, Mass. (1972).

3. Szilard papers, La Jolla, Calif.

4. Bainbridge collection, American Institute of Physics, New York.

5. J. D. Bernal, *The Social Function of Science,* Routledge & Kegan Paul, London (1939), pages 150, 182.

6. Pegram collection, Columbia Univ. Library.

7. Joliot-Curie papers, Radium Institute, Paris.

8. Testimony of L. Kowarski before the US Atomic Energy Commission's Patent Compensation Board, Docket 18, 16 March 1967, Energy Research and Development Administration, Germantown, Md.

9. H. von Halban, F. Joliot, L. Kowarski, Nature **143,** 470 (1939); *The Discovery of Nuclear Fission: A Documentary His-*

tory (H. Graetzer, L. Anderson, eds.), Van Nostrand Reinhold, N. Y. (1971).

10. Copies are in ref. 3; the original is in Dirac papers, Churchill College, Cambridge, UK.

11. Bohr Scientific Correspondence (copies are held at the American Institute of Physics, New York; American Philosophical Society Library, Philadelphia; Bancroft Library, Berkeley, and Niels Bohr Institute, Copenhagen).

12. R. B. Roberts, R. C. Meyer, P. Wang, Phys. Rev. **55,** 510 (1939).

13. R. Roberts, L. R. Hafsted, R. C. Meyer, P. Wang, Phys. Rev. **55,** 664 (1939).

14. Science Service, 24 Feb. 1939; reprinted in Science News Letter, 11 March 1939, page 140.

15. Lawrence papers, Bancroft Library, Berkeley, Calif.

16. R. Clark, *The Birth of the Bomb: The Untold Story of Britain's Part in the Weapon that Changed the World,* Horizon, New York (1961); B. Goldschmidt, *Les Rivalités Atomiques 1939–1966,* Fayard, Paris (1967), page 27; interview of Kowarski by Weiner, American Institute of Physics.

17. F. Joliot-Curie, *Textes Choisis,* Editions sociales, Paris (1959), page 154.

18. Science Service, 16 March 1939; reprinted in Science News Letter, 1 April 1939, page 196.

19. R. B. Anderson, E. Fermi, H. B. Hanstein, Phys. Rev. **55,** 797 (1939); L. Szilard, W. H. Zinn, Phys. Rev. **55,** 799 (1939).

20. M. Gowing, *Britain and Atomic Energy, 1939–1945,* St. Martin's Press, New York (1964), page 34.

21. D. Irving, *The Virus House: Germany's Atomic Research and Allied Counter-Measures,* William Kimber, London (1967), page 32.

22. I. N. Golovin, *I. V. Kurchatov: A Socialist-Realist Biography of the Soviet Nuclear Scientist* (H. Dougherty, transl.), Selbstverlag Press, Bloomington, Ind. (1968), page 31.

23. R. G. Hewlett, O. E. Anderson Jr, *The New World: A History of the United States Atomic Energy Commission, volume 1: 1939–1946,* Pennsylvania State U. P., University Park, Pa. (1962), page 16; Briggs Committee correspondence, Atomic Energy Papers, Office of Scientific Research and Development, National Archives, Washington, DC.

24. E. Fermi, *Collected Papers, volume 2: United States 1939–1954,* (E. Segrè et al, eds.) University of Chicago Press, Chicago (1965), page 31.

25. Samuel A. Goudsmit collection, Library of Congress, Washington, DC.

26. W. Bothe, P. Jensen, "Die Absorption thermischer Neutronen in Elektrographit," 20 Jan. 1941, captured German report G-71, Technical Information Service.

27. Bulletin de l'Académie des Sciences de l'URSS, Ser. Phys. **5,** 555 (1941); a translation by E. Rabinowitch, Report CP-3021, is available from Technical Information Service, Oak Ridge, Tenn.

28. E. McMillan, P. H. Abelson, Phys. Rev. **57,** 1185 (1940). □

Leo Szilard: Giving peace a chance in the nuclear age

An honorary doctorate from Brandeis University provides a fitting epitaph: 'A prophet ahead of his time, yet a victim of its maladies.'

Barton J. Bernstein

PHYSICS TODAY/SEPTEMBER 1987

The idea came to him one day in September 1933 as he crossed Southampton Row near the British Museum in London. What Leo Szilard imagined so brilliantly that day was a transmutation of chemical elements in a nuclear chain reaction that could someday produce enormous explosive power. Szilard's concept unrolled in his mind five years before fission was discovered, nine years before the first self-sustaining reaction was achieved and 12 years before the atomic bomb was dropped.

To keep the idea out of the hands of Hitler's scientists, Szilard, who had emigrated earlier that year from Germany, assigned the patent rights to the British Admiralty to ensure its secrecy. In 1938, Otto Hahn and Fritz Strassmann split the uranium atom. Eight months later, fearing that German scientists were already working on an atomic bomb, Szilard drafted a letter for Albert Einstein to sign and send to President Franklin D. Roosevelt. The intent of the letter was to alert the President to the potential development of nuclear arms.

"If the A-bomb project could have been run on ideas alone," said Eugene Wigner, like Szilard a theoretical physicist born and educated in Hungary, "no one but Leo would have been needed." Szilard possessed such originality that he evoked superlatives—and possibly even hyberbole. Szilard also was capable of arousing annoyance and anger. This was often the reaction

Barton J. Bernstein is professor of history and Mellon Professor of Interdisciplinary Studies, as well as director of the international relations program, at Stanford University. Among the books he has written are *The Politics and Policies of the Truman Administration* (1971) and *The Atomic Bomb* (1976). This article is adapted from his introduction to *Toward a Livable World: Leo Szilard and the Crusade for Nuclear Arms Control*, edited by Helen S. Hawkins, G. Allen Greb and Gertrud Weiss Szilard (MIT Press, Cambridge, Mass., 1987).

of military brass hats and science administrators to his prodding and provoking manner. At times he offered so much unsolicited advice that General Leslie R. Groves, who headed the Manhattan Project during World War II, once wanted to imprison Szilard as a security risk. Groves's plan was thwarted by no less a figure than Henry L. Stimson, Secretary of War. In 1945, as the war was ending, Szilard infuriated some military and political officials by publicly opposing the use of nuclear bombs on Japan and resisting the course of postwar US policy, which, he argued, would lead inexorably to a nuclear arms race with the Soviet Union.

Even before the Trinity test at Alamogordo, Szilard had pleaded with Washington for international control of all atomic materials. Once the war was won, he devoted the remaining 19 years of his life to the cause of perfecting a "livable world"—one in which nuclear weapons would be controlled and ultimately eliminated.

By nature Szilard was restless, relentless, charming, eccentric and frequently bursting with enthusiasm about a new cause. An idealist, committed to improving the human condition in the shadow of the nuclear age, Szilard was a kind of one-man lobby for peace.

At various times in his last years he championed such concepts as world government, total disarmament and arms control. A pragmatist, he yielded to events by devising short-run solutions, including his preparation of rules for limited nuclear war and for reciprocal nuclear destruction of cities—a sort of early version of mutually assured destruction, known familiarly by the acronym MAD.

During the cold war of the 1950s Szilard helped promote the Pugwash movement, which brought together leading scientists from East and West to seek solutions to the problems

brought on by nuclear weapons. The Pugwash Conferences, financed by US industrialist Cyrus Eaton and named for the Canadian town where he had a summer home, began in July 1957, a month before the Soviet Union tested its ICBMs and three months before it launched the first Sputnik satellite. Although Szilard continued as a dedicated participant in the Pugwash movement, he believed that only smaller, informal meetings between Soviet and Western scientists, engaging in more open exchanges, could be truly productive.

Szilard had higher aims. He attempted to negotiate directly with Soviet Premier Nikita Khrushchev in the early 1960s. During the period, Szilard urged that a telephone hot line be installed between the Kremlin and White House and founded the Council for a Livable World, which he expected to lobby against the arms race and to support political candidates who endorsed that goal.

The superiority of scientists

Most of Szilard's projects contained goodly elements of intellectual elitism, principally of a scientific nature. Periodically he devised various schemes to create groups of the best and brightest to guide the nation out of the arms race or into a more stable nuclear balance. He believed in the superiority of scientists, extolled them for their capacity for objectivity and believed himself among the most preeminent.

He was a man of dazzling intellect, playful, abrasive at times, and obsessive. He delighted in twitting Army security officials and in operating somewhat outside established channels. He once wrote on a lengthy security form during World War II that his hobby was "baiting the brass." It did not endear him to the military.

He sought to push and prod Presidents, Secretaries of State and others at the pinnacles of power to endorse his

The arms race began with the test of Joe I, the first Soviet atomic bomb, announced by President Truman on 23 September 1949. Thereafter Leo Szilard (at left) increased his activity to control the weapon he did much to create.

ideas. Contemptuous of physical labor, he had angered Enrico Fermi so often by his reluctance to do his share in the lab at Columbia University in 1939 that Fermi refused to work with him after their famous fission experiment was completed.

Like radium giving off luminescence, Szilard effused new ideas. He helped pioneer information theory. With Einstein, he designed a liquid-metal pump refrigerator. ("It howled like a jackal," a friend recalled, then volunteered that similar pumps were later used in breeder reactors.) Szilard probably was the first to conceive of the cyclotron and the electron microscope. He also shared a key patent for nuclear reactors. "Had he pushed through to success all his inventions," said Denis Gabor, a Hungarian physicist who became a Nobel laureate and Szilard's

longtime friend, "we would now talk of him as the Edison of the 20th century."

Unfortunately for science, Szilard published surprisingly little—only 29 papers, not much more than an assistant professor up for tenure today at a major American university. "He was as generous with his ideas as a Maori chief with his wife," claimed Jacques Monod, a Nobel Prize-winning biologist. Though a self-declared "man of ideas," Szilard lacked the personal discipline to pursue them to completion. He often dropped interesting concepts just as others were beginning to explore them. James Franck, a physicist on the Manhattan Project, once suggested only partly in jest that Szilard should be kept in a freezer and pulled out when great ideas were needed. Wigner, another close friend and Nobel laureate, called him "the most

imaginative man . . . I ever knew," and said that "no one [had] more independence of thought and opinion."

Szilard was also a man who believed steadily (his brother would say "stubbornly") in rationality and devised elaborate political schemes that denied emotions, mistrusted passions and often ignored much of the troubling stuff of national culture—habits, inclinations and patterns of behavior. Recognizing this shortcoming in Szilard, Einstein said in 1930 about the then 32-year-old Szilard, "he tends to overestimate the role of rational thought in human life."

Perhaps it was the strain of self-imposed rationality that led Szilard to be puckish, impish, even childlike. He once admitted that he became a scientist because "in some ways I remained a child."

During the last five years before his death in 1964, Szilard devoted himself mostly to political activities. He completed "How to live with the bomb and survive," a paper he wrote and rewrote over a few years, and published it first in the *Bulletin of the Atomic Scientists* in 1960. A year later his masterpiece of science fiction, *The Voice of the Dolphins*, came out to critical acclaim. He conducted negotiations with Krushchev, took up residence in Washington to influence policy during the Kennedy Administration and gave a series of college lectures that launched the Council for a Livable World.

Szilard became seriously ill with bladder cancer in autumn 1959. At first he delayed entering a hospital and postponed his x-ray treatments until he could finish two scientific papers on cell regulation of enzymes and on antibodies. He rejected major surgery, later explaining dispassionately that surgery did not represent a good cost–benefit choice. If the proposed surgery had been likely to give him ten more years at the price of a few months of great discomfort, he would have done it "But the chances were not good," he said. So, he explained, he chose radiation treatments instead, "which certainly will not save my life but which gave me some hope that I will be able to work for some time." Ironically, after his death by a heart attack, the autopsy showed that the cancer had completely disappeared.

While he was a cancer patient in Memorial Hospital in New York City, Szilard delighted in entertaining visitors, often shocking them with his whirlwind activity, and in directing his physicians. "These radiologists don't know x rays," he asserted with intentional exaggeration. "I find myself having to give a course in radiology to these fellows. Anyway, I'm the chief consultant on my own case. It's quite fascinating." Exerting such control over his treatment, in which he was assisted by his wife, Gertrud (Trude), a practicing physician, was essential to Szilard, who normally rejected open dependence on others.

Despite the radiation treatments, Szilard believed that his bladder cancer would recur. "My chances are anything but good," he told a *Life* magazine reporter. "Say, six months to a year. I have plenty to occupy me in whatever time is left."

In the hospital he was finishing "How to live with the bomb" and parts of *Dolphins*, dictating his memoirs, writing letters to newspapers, directing some informal lobbying efforts, conducting television interviews, planning television debates on the arms race and world politics with Edward Teller, an-

Pugwash
Conferences were sponsored during the early years by Cyrus Eaton (at right), the Cleveland industrialist who made available his estate in Nova Scotia, near the little town of Pugwash, after the US government refused to grant visas for the first meeting to scientists from Eastern Europe. Szilard and Bertrand Russell were among those who backed Pugwash as a way of reducing superpower tensions.

other Hungarian emigré, and entertaining a flow of visitors. He handled all this with rumpled efficiency, dressed in a bathrobe and surrounded by a scatter of notepads, papers, letters and a tape recorder. When asked how he could conduct his affairs amid such confusion, he replied: "This hardly seems abnormal. I guess it's because I have spent so much of my time living in the rooms of hotels and faculty clubs."

Unsolicited advice

He also gave unsolicited advice to politicians. In April 1960, for instance, after noting that Hubert Humphrey's campaign for the Democratic Party's Presidential nomination was broke, Szilard proposed that John F. Kennedy should contribute $10 000 to help his rival's campaign and inform Humphrey that he (Kennedy) would consider it unfair to win the primaries if his principal competitor ran out of money. Szilard suggested that Kennedy should not publicize the offer, and that it would be better just to let the story leak out. Kennedy sent him a polite, perfunctory note, saying he welcomed the comments "and will certainly keep them in mind." Szilard had clearly misjudged Kennedy and American politics, revealing what some would decry as innocence and naiveté.

Szilard sometimes sneaked out of the hospital to attend meetings. In May 1960, together with Wigner, he re-

ceived the Atoms for Peace Prize, worth $37 500 to each of them. At the ceremony, Szilard and Wigner, despite their longtime differences on the Soviet threat and the need for more weapons, made common cause in criticizing the quest for a test ban treaty. Each thought that the proposed system of inspection points would, in Szilard's words, "lead to friction." Szilard preferred his solution: Assemble the world's atomic scientists to work on methods of detection and offer a $1 million reward to report any violations.

By early 1960, after many false starts and discarded revisions, Szilard's early thinking on "How to live with the bomb and survive" finally came together. It was a curious essay—perhaps a quintessential Szilard piece—for it was an amalgam of the prescient, the hopeful, the pessimistic and the overly rational. In it, he sketched some rules for the emerging nuclear stalemate he foresaw. As the United States and the Soviet Union each developed mobile ICBMs and thus virtually invulnerable second-strike capacities, Szilard thought it might be possible to stipulate ways of coexisting with the bomb. His overly mechanical and excessively rational solution called for each nation to define a "permissible threat" and even to work out guidelines for limiting nuclear war—with such a war, if it broke out, to be waged against evacuated cities in the enemy's country. Like

Fourth anniversary of the first nuclear pile was commemorated on 2 December 1946 when some members of Enrico Fermi's triumphant team posed for a photograph on the steps of Eckhart Hall at the University of Chicago. In front row (from left): Fermi, Walter Zinn, Albert Wattenberg and Herbert Anderson; middle row: Harold Agnew, William Sturm, Harold Lichtenberger, Leona W. Marshall Libby and Szilard (in raincoat); back row: Norman Hilberry, Samuel Allison, Thomas Brill, Robert Nobles, Warren Nyer and Marvin Wilkening.

many other strategists of the era, he wrote with the dispassion of a chess player, assuming that national leaders would not panic in crisis, accidents would not occur and communications would not break down.

Szilard grafted ideas from this essay into the title story of *The Voice of the Dolphins*. Written ostensibly after peace had been achieved in the last decades of the 20th century, the story emphasizes the wasteful cost of armaments, the dangers of an antiballistic missile defense system, the desirability of a no-first-use policy, the need to limit nuclear threats and to restrict nuclear retaliation to equivalent damage (a city for a city), the liability of allies, and various schemes for guaranteeing that nations would not cheat on arms control or disarmament agreements.

At the time, some of his proposals and perceptions were rather bold—his awareness of the dangers of an ABM system, his forecast of a stalemate in the nuclear arms race, his anticipation of the use of mobile ICBMs, his plea for no first use and his emphasis on avoiding an overwhelming counterforce capability because of the instability it created. Some of his other ideas, though characteristically ingenious, were not especially helpful—such as stressing that each nation, under an agreement, always had an interest in proving that it was not cheating and that citizens could be persuaded to

monitor honestly whether their own government was cheating.

In the title story, the dolphins provide the funding and even the advice that lead to disarmament and world peace. The story—in characteristic Szilard fashion—describes some negotiations and events, including narrowly averted wars, in elaborate detail but is troublingly vague in other respects. Often his narrative glides to happy events and conclusions, too frequently without adequate explanation.

Whimsy and hope

But, as he would have admitted, the tale expresses hope; it provides a guide, not a blueprint. A Soviet–American Biological Research Institute in Vienna studies dolphins, who love liver paste, help scientists win Nobel Prizes and lead them to discover a valuable substance that limits female fertility, thus checking the problem of soaring population—an issue that deeply troubled Szilard. The institute, made wealthy by this antifertility product, sponsors a television program to clarify political views and even devises ways to buy off politicians blocking peace initiatives. It is a tale of whimsy and hope, and the skillful avoidance of disaster. The story is, in important ways, similar to many Western utopian tales in which an act of will checks the otherwise inevitable slide toward cataclysm—in this case nuclear holocaust.

The fiction reveals much about Szilard: his cynicism about politics, his great respect for scientists, his neglect of human psychology, his delight in details and his fascination with convoluted plots, full of mystery and some genial deception. Indeed, at the end, the narrator admits that possibly the dolphins had not played any role. Could it be that the scientists had guided the world to peace? The story is, then, a powerful allegory, emphasizing the force of rationality, the forte of scientists.

Throughout the story, Szilard's great faith in scientists dominates. He laments "that scientists [are] on tap but not on top" in Washington. He declares that political issues are often complex, but that they are rarely anywhere as deep as scientific problems, particularly those in physics that had been solved in the first half of this century. He emphasizes that scientists, unlike politicians, seek the truth, and thus a critic need not ask why scientists take certain positions but only whether or not the positions are correct. And finally, in a burst of playfully expressed elitism, he asks: If in a democracy "one moron is as good as one genius, is it necessary to go one step further and hold that two morons are better than one genius?"

Most of the reviews were favorable, though Szilard grumbled that he could not prod the *New York Times Book*

Review to review it. So, he busily
devised his own advertisements to promote the book. Some old friends sent
him glowing tributes. Michael Polanyi, a Hungarian-born chemist who
had known Szilard since the 1920s,
suggested in a prescient letter: "Maybe . . . you will be remembered by these
light-hearted fancies long after your
contributions to science will have
joined the melting pot of anonymity."

The book brought Szilard some fame
and some money. It gave him a new
platform for his ideas, and—to his
delight—it was translated into Russian
in the Soviet Union. Szilard himself
called it to Khrushchev's attention
during their meeting in October 1960
and later gave him a copy of the slim
volume.

The meeting with Khrushchev grew
out of Szilard's efforts after 1959 to
open relations with the Soviet leader.
In September of that year Szilard sent
Khrushchev an advance copy of "How
to live with the bomb," and then in the
summer of 1960 he began urging the
Soviet premier to support informal
meetings of Soviets and Americans—
mostly scientists—to discuss world security issues. In one letter to Khrushchev, Szilard included his telephone
and room numbers at Memorial Hospital and invited the premier to call on
him during an impending visit to New
York City that fall for the United
Nations session. "I have given some
thought to the problem of what it would
take to avoid war between America and
Russia," Szilard wrote, "and that perhaps it might interest you to hear what
I might be able to say on the subject."

On 5 October, at Khrushchev's invitation, Szilard briefly left the hospital
and met with him for two hours.
Judging from Szilard's notes, it was a
friendly session. Szilard, ever playful,
gave Khrushchev a Schick injector
razor, showed him how to change
blades and promised to supply more
blades, he said, "as long as there is no
war." Khrushchev replied that no one
would have time to shave if war broke
out. On the Kennedy–Nixon Presidential campaign, Szilard mischievously
chided Khrushchev, saying he was
distressed that the Soviet premier had
emphasized only his disagreements
with the candidates.

They briefly discussed Szilard's idea
that prominent American citizens
would put together a manuscript on the
arms race, send the draft to Khrushchev, get his comments and then publish it in what they hoped would be "a
lively and interesting book." More
importantly, they also talked about
Szilard's concept of regional international police forces, as well as ways of
solving the Berlin conflict, hopes for

Battling cancer, Szilard, photographed here with his wife, Gertrud,
a physician, continued to dictate memos, letters, manifestos and
sometimes orders to his doctors.

ongoing private Soviet–American discussions and a plan for a Soviet–
American hot line. The meeting with
Khrushchev inspired Szilard and nour-
ished his hopes for Soviet–American
cooperation.

A market for wisdom

As 1961 started and Kennedy entered the White House, Szilard moved
to Washington, hoping, as he puckishly
phrased it, that he could "find a market
for [his] wisdom." He settled with his
wife into two hotel rooms at the Du-
Pont Plaza, which he quickly cluttered
with papers and files. He took to
spending hours in the hotel lobby. "I
can work happily in the lobby," he said.
"I have never owned a house, and don't
feel the need of owning one." In the
lobby, he would write, read his mail,
meet reporters and friends, and make
phone calls.

But his influence with the Kennedy
Administration proved minimal. Despite his early hopes for it, Szilard
found himself sharply criticizing the
Administration's ventures, especially
the abortive Bay of Pigs invasion and
the short-lived campaign to build bomb
shelters, which, he wrote Kennedy,
could be interpreted as girding for a US
first-strike policy.

Building on his own earlier ideas,
Szilard offered elaborate, detailed
strategies on how to move to a less
dangerous world. He called for intermediate stages of force reduction with
varying totals for different weapons—
planes, fixed-based ICBMs, submarine-

based missiles and land-based mobile
missiles. At a time when many of his
usual political allies opposed nuclear
testing, Szilard argued that some testing—especially to develop mobile missiles, which could be virtually invulnerable—might reduce public fear and
increase the likelihood of an arms
control agreement. At a time when the
US had about a 4:1 superiority in
ICBMs, he warned that a continuing
American build-up would sour any
chances for an arms control agreement.

He continued to propose ways to
prevent nuclear war. He cited the
dangers of antimissile defenses and
suggested that such systems could lead
to a spiraling increase in offensive
weapons systems. He recommended, as
others had earlier, an effort to establish
a "minimum deterrence"—specifically,
allowing enough invulnerably based
weapons on each side to destroy some of
the other nation's major cities, thereby
barring either side from initiating an
all-out nuclear war without also committing predictable suicide. Such a
scenario, he suggested, might require
only about 12 Soviet missiles to devastate key American cities and possibly
40 US weapons to destroy a comparable
number of Soviet cities.

Szilard argued that it was prudent to
insist on an agreement calling for
verifiable inspections. He believed the
Soviets would probably accept inspections as a condition to halt the arms
race and ensure security, as well as to
save money that otherwise might go for
weapons. Szilard's support for mini-

Eugene Rabinowitch, a Met Lab colleague of Szilard's and later a professor of biochemistry at the University of Illinois, shared Szilard's fears for the future and helped found the *Bulletin of the Atomic Scientists.*

mal deterrence did not mean he was abandoning his quest for arms control, as was the case for many others, but rather was intended as an early step along the long road to disarmament.

Looking around for ways to establish arms control and ultimately to end the nuclear arms race, Szilard frequently returned to one of his old ideas, first offered in 1930, for a *Bund*—that is, small groups of specialized intellectuals who could function as "think tanks." To make this happen, he negotiated with the Ford Foundation in fall 1961 to establish a National Society of Fellows, drawn partly from the Administration, to influence and educate high government officials. In this venture, he sought backing from, among others, Henry Kissinger, then a Harvard government professor best known for his support of limited nuclear war, and Joseph Rauh, a leader in Americans for Democratic Action and a former New Dealer. The Ford Foundation turned down Szilard's proposal and the venture died.

During this same period Szilard was trying to solve the Berlin crisis. After the Soviets erected the Berlin wall in August 1961 and Kennedy called up reserve forces, Szilard offered his services to the White House for private diplomacy. He wanted "to hop a plane and fly to Moscow" to offer Khrushchev a package proposal that East Germany would move its capital from East Berlin and that West Berlin would become a

free city. "I had a rather good conversation with him [Khrushchev] about this point in October" of 1960, Szilard said. The Kennedy Administration was not interested in such a scheme.

Such rebuffs did not lessen Szilard's enthusiasm for new ventures. In the autumn and winter of 1961, beginning at the Harvard Law School Forum on 17 November, Szilard visited eight campuses, where he gave a speech ("Are we on the road to war?") that led to the formation of the Council for a Livable World. For some who heard him then, his words were an inspiring call for action.

At Harvard, Szilard began his lecture in a prankish way: "I am here under false pretenses, and since I am about to be found out, I might as well confess at once, and throw myself upon your mercy. I am not here to deliver the kind of lecture which you may expect from me. I came here in order to invite those of you who are adventurous to participate in an experiment that might show I am all wrong. And, it might well be that something of a more serious nature is at issue also."

He then declared that the chances were "slim" of getting through the next two years without war. He argued that "the problems which the bomb poses to the world cannot be solved except by abolishing war." Stressing that arms control efforts had failed so far, he asserted that the Soviets were interested "in far-reaching disarmament," but

that he did not believe that any meaningful agreement was imminent. Instead, he suggested, the US should take some modest unilateral steps: declare a no-first-use policy, agree to use the bomb only if nuclear war erupted, move toward minimum deterrence and certainly refrain from developing counterforce superiority.

These ideas, as he knew, were not original. Even so, he evoked enthusiasm because he offered alternatives to despair, inaction and, possibly, war.

He also proposed the creation of a council, directed largely by scientists, to guide citizens to donate about 2% of their income to designated Congressional candidates. Implicit in his concept was his belief in scientific elitism. Szilard hoped that this peace lobby (originally called the Council for Abolishing War) would liberate the best impulses of the Kennedy Administration. In the next few years, the Council for a Livable World helped elect such Senate candidates as George McGovern of South Dakota, Joseph Clark of Pennsylvania, Frank Church of Idaho, Wayne Morse of Oregon and Jacob Javits of New York.

Judged against the political radicalism of the later 1960s, the council was moderate. It was rooted in mainstream liberal ideology and the traditional two-party system. Szilard himself was distrustful of the ideas of participatory democracy, of an assault on established leadership and authority, and of plans for transforming America. In this connection, he was offended by the Southern black sit-in movement and the sympathetic Northern white picketing and boycott movement. Unlike the emerging "new left," he believed in authority, hierarchy and the wisdom of the intelligentsia, especially if this group contained scientists. He respected property and rejected radical and leftist theories about concentrated power, class-based society and a military–industrial complex.

Even so, his ideas seemed politically daring in 1961. They were unacceptable to the Kennedy–Johnson–Humphrey wing of the Democratic Party. One Navy official told American intelligence people that the council was "subversive and Communist inspired." In 1961, many liberal academics, including most scientists, feared the taint of associating with the council and with Szilard's program. Thus, at Harvard in November 1961, when Szilard spoke, a prominent biologist, soon to win a Nobel Prize, avoided taking part lest, as he explained to a friend, it jeopardize his chance of becoming science adviser to a future President.

After the 1961 Harvard meeting, Szilard, tired and irritable from his lecture, had no desire to relate personally with the young people who had helped him. He had little understanding that they might want to hear a few kind words. Szilard was distant. Students had enlisted in his crusade, he believed, because of the merits of his analysis, not because of the charisma of his personality, and so his personal attention was irrelevant and unnecessary. Yet when they asked questions about politics or nuclear strategy, he was willing to spend time explaining, parrying, listening—at all times responding sympathetically to the younger people as near-equals.

By 1962, while still promoting the council, Szilard was also trying, once again, to set up a Soviet–American project for informal meetings. This "angels project," as he playfully called it, would bring together US government consultants and junior officials with their Soviet counterparts. The Americans, he explained, should be on the side of the angels and "would be willing to give up, if necessary, certain temporary advantages . . . for the sake of ending the arms race." Between August 1962 and June 1963, a period punctuated most notably by his panicky flight to Geneva during the Cuban missile crisis in October 1962, he maneuvered to arrange such a meeting. But the opposition of William C. Foster, head of Kennedy's recently established Arms Control and Disarmament Agency, helped scuttle the venture. At the outset, Foster barred agency members and advisers from taking part, and then the Soviets backed out in August 1963.

Hope and despair

For Szilard, his last year in Washington was a mixture of hope and despair—the collapse of the angels project, the failure of some similar ventures, the growth of the council, his periodic bursts of enthusiasm for arms control, his descents into occasional despondency during and after the Cuban missile crisis and his anxiety that the Kennedy and Johnson Administrations would deepen their commitment to the Vietnam War. "Starting with the Cuban missile crisis, last October," he wrote in 1963 to an older friend who was abandoning the US to live in Geneva, "I have been getting more and more convinced that the country will come to grief. If I were to stay in Washington until the bombs begin to fall and were to perish . . . I would consider myself, on my deathbed, not a hero but a fool."

In February 1964, briefly pessimistic, he said, "I myself shall make no further attempts to engage the Russians in 'private discussions' on the subject of

Atoms for Peace Award was presented to Szilard and another Hungarian-born physicist, Eugene Wigner (left background), at a White House ceremony in 1960 by President Eisenhower's science adviser, James R. Killian Jr.

arms control." He felt betrayed by the liberal Kennedy–Johnson Administration and defeated by the euphoria that followed the limited test ban treaty of 1963. In February 1964 he moved to La Jolla, California, hoping to continue his work in biophysics, on aging and memory, and to inspire other scientists at the recently formed Salk Institute, where he had become a permanent fellow. At the same time he continued advising the council and looking for ways to control the bomb and move toward disarmament.

His wife as well as some old friends thought he was achieving a level of contentment in La Jolla that he had never known before. Some sensed a general softening in Szilard in his final years. "You are warmer [since your hospitalization] and more human than you ever were before," Polanyi told him. But the promise of comfortable years in La Jolla, with production work at the Salk Institute, ended abruptly on 30 May 1964, two months after Szilard's 66th birthday, when he died of a heart attack in his sleep.

Acknolwedging that Szilard might have finally found an inner peace in those last months, Teller wrote in a eulogy. "I cannot help but think of that legendary, restless figure, Dr. Faust, who in Goethe's tragedy dies at the very moment when at last he declares he is content."

To the end of his life, Szilard continued his crusade for peace. Always

exuberant about the power of rationality, he brought to his efforts a fierce energy and a fertile imagination. Having done all he could to create the atomic bomb and all he could within the law to prevent its use on populations in 1945, he spent his last 19 years courageously seeking arms control, disarmament and world peace.

In helping to shape the postwar dialogue in this quest, Szilard made *some* difference. His thinking about nuclear strategy was, admittedly, not as influential as that of, say, Bernard Brodie and Thomas Schelling, perhaps partly because Szilard did not operate near the corridors of power. Still, his actual efforts to change international politics—conceiving and participating in various Soviet–American informal discussions, frequently writing and speaking to an elite public, organizing a "peace" lobby and meeting with Khrushchev—may have contributed slightly to a thaw in the cold war.

Szilard, like those in SANE or connected with the *Bulletin of the Atomic Scientists*, groups with whom he shared much intellectually, could not successfully oppose the larger forces of the cold war. Nor, like those groups, was he able to analyze those forces deeply. He never rooted his concern about peace and the bomb in a probing analysis of national culture, history, economic factors and ideologies. Even had he done so, however, he no doubt would have failed to change policy significantly.

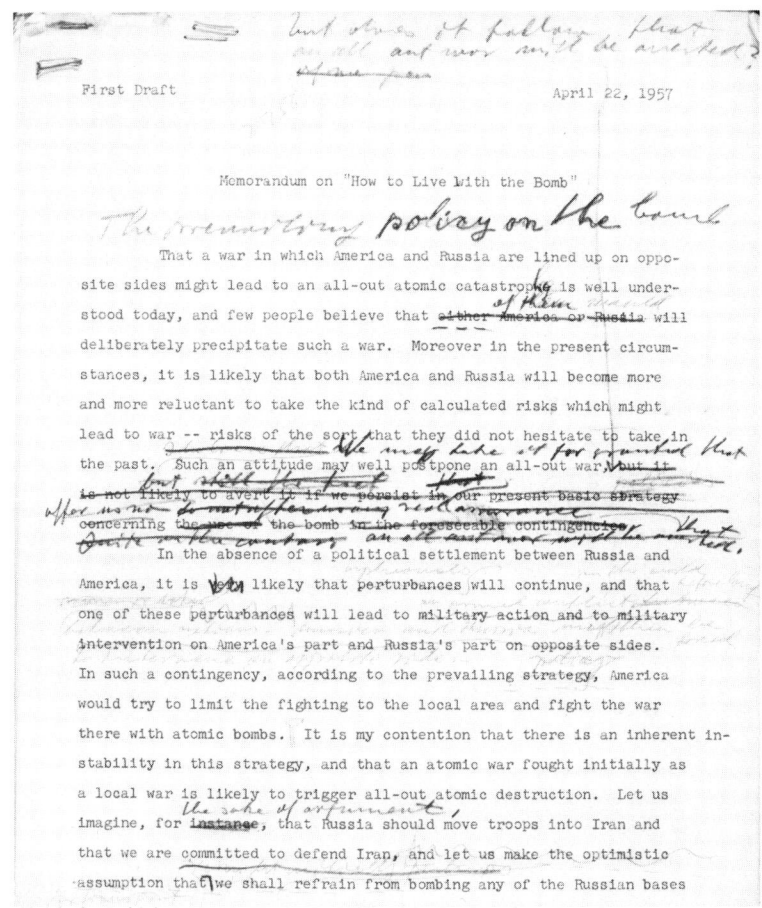

First draft of Szilard's manuscript on "How to live with the bomb" shows the author's scribbled emendations. The essay, considered quintessential Szilard for its curious amalgam of the prescient and the pessimistic, was eventually published in the *Bulletin of the Atomic Scientists* in 1960. (Courtesy of Mandeville Collection, University of California, San Diego.)

He kept faith with himself and his values. More than most other striking figures in science, he continued to think and act on an untrammeled conception of the moral responsibility of scientists. In the pursuit of peace and the effort to change national policy, he remained an outsider. Unlike J. Robert Oppenheimer or Teller, who served the government in the postwar years and became such bitter foes and symbols for such different positions, Szilard never excited hatred or rancor, nor was he intolerant. He did not seek positions on Federal advisory committees, for he valued the freewheeling ways that distance and independence could help guarantee. Strong-willed, quarrelsome and often unpredictable, Szilard would not have been a controllable adviser, and therefore he was not sought.

Yet, ironically, throughout his life Szilard believed ardently in the power of advice, in the need for wise men (especially scientists) to influence the government. Ultimately, in moving to create the Council for a Livable World, he shifted to emphasize electoral poli-

tics—but initially to gain a greater hearing for the wisdom he believed that he and his associates could offer. For him, electoral politics was not a substitute but a supplement to change policy by gaining access to government office-holders. They might listen, he believed, because the council could deliver votes.

Beyond the seemingly possible

Szilard's political ventures could sometimes miscarry and even offend possible domestic allies. In the mid-1950s, for instance, when he proposed that scholars band together to call for the resignation of Secretary of State John Foster Dulles, Szilard met quick rebuffs. And when he privately proposed a bipartisan Eisenhower–Stevenson ticket for 1956, he was told, correctly, that he did not understand US politics.

Perhaps it was this very misunderstanding of the American political system that inspired him to organize scientists to try to block the use of the atomic bomb against Japan and then to organize them to thwart formal mili-

tary control of postwar atomic energy. Such spirited imagination and political energy also led him to propose, unsuccessfully, settlements of the Quemoy–Matsu crisis in the mid-1950s and of the Berlin crisis in the early 1960s. Because he never worried about developing safe ideas or muting his moral obligations, he felt free to protest against many events that other scientists of his generation might privately decry or simply ignore.

Szilard acknowledged the need to reach beyond the boundaries of the seemingly possible. "Let your acts be directed towards a worthy goal," he once said, "but do not ask if they will reach it; they are to be models and examples, not means to an end."

Even political allies, like Eugene Rabinowitch, a biochemist who served as editor of the *Bulletin of the Atomic Scientists*, were sometimes inclined to chide Szilard for his boldness. "It is too easy to say," wrote Rabinowitch, "that some of his proposals are unrealistic or too cleverly contrived; but nobody can deny that they are ingenious, original and stimulating."

Szilard's sense of conscience, his intellectual boldness, his willingness to try new ideas, to unsettle conventions and to disregard old ways could be inspiring. He liked to be intellectually disruptive and original. He was willing to devise ideas, drop them and try others.

He was a man of moral vision who took on himself the great burdens of improving Soviet–American relations and of trying to save the human race from extinction in a nuclear holocaust. As Arthur Holly Compton, Szilard's World War II boss in the Manhattan Project, told him in 1960, "History will see you . . . as one who labored bravely to make of [our] age a condition of life under which men could enjoy an increasing degree of safety and mutual confidence, in spite of the threats of war." Szilard was a kind of moral hero who knew that total success could not be achieved, yet failure would mean disaster.

Perhaps the best brief testimonial to Szilard's creative brilliance and moral commitment was expressed by Brandeis University in October 1961, as it bestowed on him the degree of Doctor of Humane Letters: "Among the first to perceive the threat and the promise of nuclear energy . . . crusading indefatigably to help men understand how to live with themselves, and with their creations, in the atomic age. A prophet ahead of his time yet passionately part of it, a victim of its maladies, but demonstrating through his own courage, that they, too, may be conquered." □

The American connection to Soviet microelectronics

A former Soviet physicist solves a longstanding mystery about the identities of two Americans who disappeared during the Rosenberg spy case and engaged in technology transfer for the Kremlin

Mark Kuchment   PHYSICS TODAY/SEPTEMBER 1985

Mark Kuchment, a physicist educated in the Soviet Union, is now a science historian at Harvard's Russian Research Center and at the Fletcher School of Law and Diplomacy at Tufts University.

Ever since the cold war of the late 1940s, the US, joined at times by Europe, Canada and Japan, has sought to restrict sophisticated technology with military potential from reaching the Soviet Union and its Warsaw Pact allies. Despite such efforts, the US has not been completely successful in achieving the goal of limiting or eliminating the flow of militarily valuable data and products from West to East. The Soviet bloc has been able to acquire advanced technologies from the West in various ways—by espionage and entrepreneurship, as exemplified in Pentagon accounts of Vax 11/782 mini-computers and an array of microelectronics, seismographs and lasers shipped illegally to Warsaw Pact countries through a variety of real and phony companies in Europe and elsewhere. For all the horror stories of spying and smuggling, there have been virtually no instances of scientists and engineers defecting or emigrating to contribute to Soviet military R&D. That is why a case involving two American electrical engineers is so interesting. It is a conspicuous example of technology transfer that, as it happened, lifted a corner of the shroud of secrecy that long concealed military research in the Soviet Union.

About three years ago, while interviewing Soviet émigré scientists, I heard repeated accounts of the successful careers of two Americans—Filipp Georgievich Staros and Iozef Venia-minovich Berg, respectively chief designer and chief engineer of the principal electronics design bureau operating in Leningrad under the auspices of the military during the 1960s and 1970s. While the identities of the defectors could be established through the tales of these émigrés, the dramatic stories of Staros and Berg required some detective work to piece together their lives before their arrival in the USSR.

Staros, it turns out, was born Alfred Sarant, and Berg was Joel Barr. Both were associated in a Communist Party cell during the 1940s with Julius and Ethel Rosenberg, who were executed in 1953 after being convicted of passing US atomic bomb secrets to Soviet agents. Both were known as competent, but not particularly outstanding, electrical engineers. They arrived in the Soviet Union from Czechoslovakia

Leningrad Design Bureau located in the wing at left of the Dom Sovetov. The bureau housed the center for microelectronics and minicomputers in the Soviet Union during the 1960s. Filipp Staros ran the bureau from Room 1900-S, the corner office, with balcony, on the third floor.

at the end of 1955 or the start of 1956.

Here I will discuss in greater detail the career of Staros, who achieved a remarkable degree of celebrity in the USSR for an American.

According to his Soviet colleagues now residing in the West, Staros deserves considerable credit for developing a form of minicomputer used in automating steel mills, power stations and other industrial operations. Some even call Staros the father of microelectronics in the Soviet Union. His parenthood is said to have begun with a seminal paper delivered at a conference of scientists and engineers in November 1958, forcefully suggesting that microelectronics could lead the way toward advancing Soviet weaponry. It is widely accepted that it was Staros who first introduced the term *microelectronica* into Russian—though it wasn't legitimized until Volume 16 of the third edition of the *Great Soviet Encyclopedia* appeared in 1974. In the encyclopedia the word is defined as the field of computer technology dealing with integrated electronic assemblies, including semiconductors.

Staros's former coworkers agree that his ideas gained acceptance for the following reasons: The Soviet Union had encountered great difficulties in developing microelectronics and computers on its own and in introducing such technologies into production processes and weapons systems—though political and military leaders soon grasped the implications of these technologies for defense. So, whether Staros knew it at the time or not, his

appearance in the USSR was auspicious. For the men in the Kremlin he apparently embodied the "right stuff"—a combination of characteristics that the Soviets were convinced had led to scientific and technical advances in the US. He was both a well-trained researcher and a hard-driving manager.

Plugged into the Kremlin

Consider, then, these excerpts from three of my interviews with Soviet émigrés:

> Our director was outstanding. On top of being a good scientist and a strong personality, he emanated the aura of an American. In addition, he had high-level connections. He knew Dmitri Ustinov [the late Minister of Defense] and individuals from the Central Committee of the Communist Party and, I think, people from the KGB.

> Staros was invited several times to discuss his projects at meetings of the Military–Industrial Commission [the influential Voenno–Promyshlennaya Kommissiya or VPK].

> Everyone recognized him as not only a thorough professional but also a talented organizer.

There is no doubt that Staros's military connection was important. It provided him with access to the Soviet hierarchy, enabling him to obtain funds and facilities to carry out his research programs. Those dealing with

the Military–Industrial Commission, described rather loosely by Henry Kissinger[1] as "a party–state organization in charge of all the defense industries," often come face-to-face with the Soviet political establishment—members of the Central Committee of the Communist Party, deputy prime ministers, senior military officers and top scientists and bureaucrats in the State Committee on Science and Technology.

There also is no doubt that Staros's participation in the construction of the synchrotron at Cornell University in the late 1940s earned him a repute of sorts among Soviet scientists and bureaucrats. Observers of Soviet society such as Hans Rogger[2] and Kendall Bailes[3] claim that US technology and culture superseded those of Western Europe in the 1920s. But Soviet disillusionment with American society set in during the 1930s, due mainly to the Wall Street debacle in 1929 and the Great Depression, which Communists argued was dramatic proof that the predictions by Marx and Lenin of the West's economic collapse were indeed true. Ideology notwithstanding, Soviet enthusiasm for US technology was restored by the end of World War II. Following the death of Stalin in 1953 and especially during the period of détente, US science, technology and culture were held in high esteem.

Still, Soviet leaders were certain they could not depend upon the US or its allies for technological advances. Any sense of scientific and technical inferiority could be dispelled at home and abroad, it turned out, by such spectacular events as nuclear bombs and space flights. Despite these achievements, though, difficulties persisted in organizing research and in manufacturing products.

After the devastation and debilitation caused by World War II, here came Staros, well-trained and self-confident, apparently heralding a new era of automation and modernization for Soviet factories and industries, as well as for military systems and, possibly, office procedures. He attracted admirers in droves. He held out the possibility of making the right decisions about research programs. He also might show how to manage large research groups, consisting at times of unruly scientists. Such attributes coincided exactly with the Soviet ideal of an R&D operation, in which such outstanding scientists as Abram Ioffe, Mstislav Keldysh, Sergey Korolev and Igor Kurchatov had shown themselves to be successful project managers.

Staros's Soviet colleagues knew little of his past before he arrived from Prague, accompanied by his American wife, four children and another American engineer who called himself Iozef Berg. Some insisted that Khrushchev himself had brought Staros to the USSR. Eric Firdman, an émigré physicist, claims it was Pyotr Vasilievich Dementiev, then the Minister of the Aviation Industry. Others think Staros was recruited by Ustinov, who was Minister of Defense Industries at the time. No matter who was responsible for him coming to the USSR, Staros was treated well from the start. For one thing, his salary was 700 rubles per month—significantly greater than the 550 rubles a deputy minister of the USSR would have been paid each month. For another, within months of his arrival, he was appointed director of a newly established laboratory at a military research institute in Leningrad.

The choice of Leningrad by Soviet authorities appears to have been natural. Leningrad became the center for research on semiconductors after the Soviet Academy of Sciences set up its Physical–Technical Institute and its Institute of Semiconductors, both founded by Ioffe, then considered the country's most influential physicist.

A great transformation

The choice of Staros to head the institute came as a surprise. The somewhat mysterious origin of Staros is characterized in his official Soviet biography in a single sentence: "In 1941, graduated from a university in Toronto and started work as a researcher." Even so simple a statement turns out to be misleading—perhaps intentionally. No university in Toronto had any record of graduating a student named Staros—or of anyone with that name ever attending.

Assessing certain technological advances in the Soviet Union is difficult at best, but even harder when it involves military systems. Secrecy is paramount in such matters. The creators of such technologies are not usually identified, and then only when they have died. Thus, making known the identity of the designer of a Soviet computer would be a curious anomaly—and even more so when he happened to be an American who worked as an electrical engineer on Cornell's synchrotron in the 1940s and emigrated secretly through Mexico and Czechoslovakia to the Soviet Union in the midst of the US's most controversial spy case. But under the name of Filipp Staros, he attained great importance, though little prominence, as director of a renowned microelectronics institute in Leningrad during the 1960s. Moreover, in 1969 he was awarded a coveted State Prize.

After failing to trace Filipp Staros back to the US or Canada over nearly 18 months, I almost abandoned all hope of solving the riddle. In the summer of 1983, while in Europe on academic matters, I interviewed several émigrés from Czechoslovakia to seek more information about Staros. None of my contacts knew anything about him. The night of my return home to Cambridge, I relaxed with a copy of The New York Review of Books. Suddenly, there it was: a review of a book[4] about the Rosenbergs that mentioned their friends who had probably disappeared behind the Iron Curtain. I ran to Harvard Square, bought a copy of the book and found the name Sarant. All the pieces of the puzzle soon fell into place.

Alfred Sarant had received his BS in electrical engineering from The Cooper Union in New York City in 1941. During the war, he worked on communications systems at Forth Monmouth and Bell Laboratories. After the war he took part in constructing the synchrotron at Cornell's nuclear physics laboratory. I showed a snapshot of Sarant, obtained from his sister, to Philip Morrison, who had been at Cornell then. Morrison immediately identified Sarant as his next-door neighbor from 1947 to 1950 in Ithaca, New York. When Eric Firdman, who had worked under Staros in Leningrad, was shown the photograph, he hesitated; then, after a small mustache was drawn on the face, he insisted excitedly that it was Staros.

Of course, a snapshot is not convincing proof. Nor is the name Sarant took in the Soviet Union. Staros's Russian patronymic of Georgievich means son of George, and, indeed, the name of Sarant's father was George. The name Staros sounds Greek in origin. According to his sister, Electra Jayson, their father's full name was Epamenonda George Sarantopoulos, which was later changed to Nonda George Sarant. The family on both sides was Greek Orthodox. His former Soviet colleagues recalled that Staros claimed Greek ancestry and enjoyed Greek movies. Emigrés said Staros had told them he had four brothers. So did Sarant. Firdman and Sarant's sister independently gave identical, detailed descriptions of Sarant/Staros. Firdman recalled that Sarant boasted to his Soviet coworkers that he had once participated in building a synchrotron in the US, though he did not say where or when.

There are also several discrepancies in the known facts about Sarant/Staros. Sarant was born on 26 September 1918. Staros was born in 1917, according to the Soviet Encyclopedia, and, according to Firdman, his birth-

Alfred Sarant, shown here in a 1945 snapshot, has been identified by Soviet émigrés as the American electrical engineer who attained fame and fortune of sorts in the USSR under his adopted name of Filipp Staros.

day was observed on 24 February. When Sarant abandoned his wife and children in Ithaca, he left for Mexico on 9 August 1950 with Carol Dayton, the wife of a neighbor. In the USSR, Staros's American wife was called Anna. Staros's close friend was Joel Barr, who also disappeared in 1950. In the USSR, Staros's deputy was named Iozef Berg.

Why did Sarant leave the US and adopt a new identity? It is known that FBI agents interviewed Sarant in Ithaca on 18 July 1950 (a day after Julius Rosenberg was arrested), and accused him of keeping an apartment for espionage purposes at 65 Morton Street in Manhattan. FBI records indicate Sarant was an American Communist Party member until 1944 and that he and Julius Rosenberg belonged to the same cell. After his interrogation, Sarant was told he could visit relatives in New York City and, thereafter, he seems to have dropped from FBI files. In 1951, when asked about Sarant by Rosenberg's lawyers, US District Attorney Irving Saypol issued a statement: "There is insufficient evidence at the present time to warrant filing a complaint against Sarant . . . on any possible Federal charge."

Perhaps Sarant left the US in panic after being questioned by the FBI. The convictions of physicists Klaus Fuchs and Alan Nunn May, the arrest of the Rosenbergs, the escalation of the cold war, the opposition to Communism implicit in the McCarran–Walter immigration act and the imminent rise of McCarthyism all may have had something to do with Sarant's decision to defect. Possibly his decision was strengthened by his marital troubles and his love affair with Carol Dayton.

Beyond this, however, there is only speculation.

The importance of ideology

What is apparent from my interviews with émigrés and, more recently, with Americans who observed the careers of Sarant/Staros and Barr/Berg in Czechoslovakia and the Soviet Union is that both men were Communist ideologues. Both lived in Czechoslovakia between 1950 and 1956, working in electronics R&D for the military. One informat, Morton Nadler, who lived in Prague from 1948 to 1959 and worked for Antonin Svoboda, a member of the Czechoslovak Academy of Sciences and the country's leading computer scientist, met Staros and Berg just before they moved to the Soviet Union. They attempted to persuade Nadler to come with them and take part in the design and production of minicomputers used to control military and industrial systems. By that time Nadler was disenchanted with the Communist system and refused to move further East. For a while he exchanged letters with Staros, who bragged that he had first pick of the best and brightest graduate students in the USSR for a new laboratory and was inventing what he called "an eye in the sky," which Nadler interpreted as meaning an electonic spy satellite, though the first sputniks had not yet been launched into space. Nadler concluded from the lifestyles of Staros and Barr that they had "very good connections" and, because they did not socialize with other American expatriates in Czechoslovakia, that they probably had been spies in the U.S.

Later, in 1976, a US computer expert on a scientific exchange program with

the USSR bumped into Berg at a conference on electron microscopy in Tashkent, the capital of Soviet Central Asia. Berg claimed to have been born in South Africa and said he acquired his obviously New York accent from American friends in Johannesburg. During his stay in Leningrad, the American computer expert spent an evening with Berg and his family. Berg called for the scientist in his black Volga, a car generally reserved for important Soviet officials. He brought his guests to his cluttered six-room apartment, an unusually spacious housing arrangement anywhere in the Soviet Union. The guests were greeted by Berg's Czech-born wife, his daughter Vivian and a middle-aged woman by the name of Anna Staros, who had come to Leningrad from Vladivostok to visit her children. Judging by her command of the English language, the computer expert figured she was, like Iozef Berg, a native American. It was obvious from the conversation that she was the wife of Filipp Staros, then working for the Soviet Academy of Science in Vladivostok.

Berg appeared well informed about advanced computer technology in the US. The American scientist considered Berg's information to be possibly no more than two months behind the latest work. Berg asked questions about restricted information and classified technology, but when the American evaded the questions, Berg turned to other topics without showing any irritation or displeasure, though, the American recalled, the sidetrack had not passed unnoticed by Berg.

Once Barr and Berg became one person, it was easy to find out his American background. Barr had been born in New York City in 1916. He received a BS in engineering from City College of New York in 1938 and worked at Fort Monmouth at the same period as Sarant. He also worked at Western Electric and Sperry Gyroscope.

Some assume that the successful careers and special favors for Staros and Berg imply that they were Soviet spies when they lived in the US.

Joel Barr, like Sarant, disappeared soon after the Rosenbergs were arrested on charges of passing atomic secrets to the Soviet Union and turned up as a microelectronics expert, according to former colleagues, at the Leningrad Design Bureau during the 1950s and 1960s.

Considering the suspicious and conspiratorial nature of Soviet leaders, it is difficult to comprehend how else Staros and Barr could attain such positions of importance in the USSR. The most plausible explanation is that they provided knowledge and skills the Soviet Union then lacked. They excelled in microelectronics technology, and Sarant/Staros especially, by dint of his experience at Bell Labs, knew how to manage a research organization. Whatever the reasons, Staros became a respected member of the Soviet military R&D community, a rare privilege for a Russian scientist or engineer, let alone an expatriate American. To achieve this status, a Soviet scientist or engineer needs a second-class clearance from the KGB. Presumably, Staros and Berg had such security clearances.

When he left the US in 1950, Sarant had limited knowledge of US computer technology and microelectronics. Computers still used vacuum tubes. Perhaps the single development with the most far-reaching consequences for computers was the transistor, invented by William Shockley, John Bardeen and Walter Brattain at Bell Labs in 1947. Staros may have gained better understanding of computer design while working in Czechoslovakia, where he was in touch with Svoboda. Even so, his former Soviet colleagues say, Staros's opinion of Czech computer science was quite low.

Lab with a box number

Staros kept informed about developments in electronics mainly by reading US journals. By way of this form of technology transfer he was able to manage a series of spectacular successes in his early years in the Soviet Union. Firdman asserts that Staros "sped up work on airborne computers"—one of the major gaps in Soviet military technology in the 1960s. This helps explain why Staros may have been a favorite of Dementiev and Ustinov. With each achievement he was able to expand his lab, first into a design bureau and then into a combination design bureau and production plant. At the start, recalls Firdman, Staros had about a dozen people. When Firdman arrived in 1964, Staros's bureau was so secret it was known as *pochtovyi yashchik*—that is, a postal box number. It was called Post Office Box 155, Leningrad. But, says Firdman, it was known for its military research. The lab had more than 800 employees and some pilot production lines for semiconductors.

It was unquestioned that Staros had the support and confidence of the Soviet military and political oligarchy. One of the earliest signs came in 1958, when Staros presented a report to a conference of scientists and managers in the electronics industry. It advocated a major commitment to accelerate R&D in microelectronics. The military and political decision was to accept Staros's report. After 1960, Staros was accorded the title of chief designer. In 1967 he was awarded the additional title of Doctor of Technical Sciences, and in 1969 he received the State Prize. The citation, signed by Keldysh, the president of the Soviet Academy of Science, credited Staros with heading a "collective of specialists" who developed a small computer used to control production processes in the energy, metals, electronics and glass industries.

The computer was produced at Staros's Leningrad Design Bureau, located in a wing of the rococo Dom Sovetov, built in the style known derisively in the Soviet Union as Stalin Gothic. The machine was identified only as UM-1-NKh. Weighing 150 pounds, the 100-watt UM-1 contained 8000 transistors and more than 10 000 resistors and capacitors. At the time, its designer was said to be a certain Comrade Filippov. Not until Staros got the State Prize was the identity of Filippov publicly made known as Filipp Georgievich Staros.

The NKh in the computer's name formally stood for *Narodnoe Khozyaistvo* (State Economy), says Firdman, but the behind-the-bench joke was that it also stood for Nikita Khrushchev, considerd the godfather of Staros's design bureau. Khrushchev visited the lab in 1962 to see UM-1 and its 265-pount successor, named Electronica K-200.[5]

The Leningrad Design Bureau's K-200 attracted attention in the West. A process-control computer, using the first Soviet-made integrated circuits and capable of performing 40 000 operations per second, it was not considered an innovative departure by American and British reviewers, but its appearance was hailed as a well-engineered

ПРАВДА

Коллектив специалистов во главе с Ф Г Старосом разработал малогабаритную полупроводниковую управляющую машину и управляющие вычислительные комплексы, которые внедрены в металлургической, энергетической, стекольной и электронной промышленности.

Интересную работу в

Академик М. КЕЛДЫШ
Президент Академии наук
СССР.

State Prize for Staros and his Leningrad researchers was announced in *Pravda* on 9 November 1969 under the headline "New Squadron of Laureates." The citation was signed by Mstislav Keldysh, president of the Soviet Academy of Science.

machine, "surprisingly up to date."

Staros's influence had increased enormously in 1961 with the creation of a powerful bureaucracy, the State Committee of the Electronics Industry, which rose to the status of a ministry in 1965. The minister, Alexander Shokin, received his engineering degree in 1934 from Bauman Advanced Technical College in Moscow, one of the most prestigious engineering schools in the USSR. He worked for many years in the defense industry and, after World War II, became Deputy Minister of Radio Technology, which was responsible for producing electronic components for radar, communications equipment and computers. The components included various types of vacuum tubes, magnetrons, klystrons, transistors, semiconductors and integrated circuits. From its outset the State Committee of Electronics Industry was hailed as perhaps the haughtiest of the so-called "Eight Sisters"—the eight industrial ministries most important in military production. Obviously, as the head of the ministry, Shokin wielded great power. He also was under great pressure to come up with rapid developments equal in quality to those in the West. To do this, Shokin gave Staros the task of planning a semiconductor R&D facility just outside of Moscow.

With the full support of the Central Committee and the Council of Ministers, Staros created the Center for Microelectronics at Zelenograd, which has since become the Soviet Union's high-tech capital, a sort of state-run Silicon Valley.[6] The center was orga-

nized according to Staros's grand design. Firdman, now a computer specialist in the US, describes it:

All development of the center was undertaken by a group of five to ten people under the direction of Staros. Our project was not the result of wishful thinking. It was meticulously thought out. We were young and enthusiastic. Staros knew all the relevant people, enjoyed high authority and had carte blanche from Khrushchev....

The decisions to establish the Zelenograd center were all classified. It was the first of several more microelectronics laboratories that were soon set up in Riga, Minsk, Tallin, Erevan and Tbilisi. Firdman says the model for these centers were R&D labs at such US companies as IBM, Texas Instruments and Raytheon. The operation at the Zelenograd center had additional American idioms. As Firdman recalls it, Staros consumed dozens of US scientific journals every day. "Nobody could make an appointment with the boss without preparing himself by reading American scientific literature that dealt with the topic of discussion," says Firdman.

The Zelenograd center included several research institutes and design bureaus, a technical college (now the Institut Electronnoi Tekhniki) and a production plant. Staros was appointed Associate Director General of Research, a post he held concurrently with his job as chief of the Leningrad Design Bureau. The dual appointment

caused trouble for Staros. He was forced to stay in Leningrad to counter attacks by the local party authorities against his recruitment and research practices. The Leningrad bureaucrats objected to a foreigner, particularly an American, running a prestigious, top-secret lab. Moreover, they raged against Staros's habit of hiring on the basis of merit, often taking on Jews and nonparty members. To make matters worse, Staros resisted party attempts to force him to hire favored people. Under constant criticism particularly from Grigori Romanov, the second secretary of the Leningrad regional party, Staros spent less and less time at Zelenograd, which developed so successfully that his Soviet colleagues soon realized they had a ripe plum in their hands.

In desperation, Staros wrote to Khrushchev, detailing his grievances against the powerful Leningrad party and complaining about the lack of support from the Minister of the Electronics Industry, Shokin. Unfortunately, Staros's timing could not have been worse. He sent his letter in early October 1964. Khrushchev was overthrown a few days later, on 14 October.

In the aftermath, Staros's letter, dealing with electronics, was dutifully forwarded to none other but the minister involved, Shokin. The minister's action was predictable. Staros was summoned from Leningrad to Shokin's office. An émigré source gave an account of Shokin's words to Staros: "Filipp Georgievich, it seems to me you have the strange fantasy that you are

ИЗВЕСТИЯ СОВЕТОВ Н₁

Ф. Г. СТАРОС

Советская наука понесла тяжелую утрату. На 63-м году жизни скоропостижно скончался член президиума Дальневосточного отделения АН СССР, лауреат Государственной премии, доктор технических наук, профессор Филипп Георгиевич Старос.

Смерть вырвала из наших рядов неутомимого ученого, талантливого организатора, многие годы отдававшего все свои силы и яркий талант исследователя развитию советской науки и техники.

Возглавляя в течение 20 лет конструкторское бюро электронной промышленности, главный конструктор Филипп Георгиевич Старос внес большой вклад в становление и развитие отечественной микроэлектроники. Ему принадлежит ряд основополагающих идей, получивших признание и дальнейшее осуществление в работах ряда предприятий и организаций страны.

Последние годы Филипп Георгиевич Старос руководил коллективом ученых Дальневосточного отделения АН СССР, до конца оставаясь на переднем крае отечественной науки.

Светлая память о Филиппе Георгиевиче Старосе навсегда останется в наших сердцах.

Президиум АН СССР. Коллегия Министерства электронной промышленности СССР. Государственный комитет СССР по науке и технике. Дальневосточное отделение АН СССР.

Staros's obituary, lauding him as an "indefatigable scientist and talented organizer" at the Leningrad Design Bureau, appeared in *Izvestia* on 17 March 1979. Signed by a panoply of scientific and political organizations, the death notice said the memory of Staros "will remain in our hearts forever."

the founder of Soviet microelectronics. That is all wrong. The Communist Party created Soviet microelectronics, and the sooner you realize that fact the better it will be for you."

Staros's position at both Leningrad and Zelenograd clearly were in jeopardy. Early in 1965, he was removed from the Zelenograd associate directorship, but Shokin himself intervened to keep Staros at the top at the Leningrad Design Bureau. Then, in 1970, the Leningrad Party boss, Vasilyi Sergievich Tolstikov, was appointed ambassador to the People's Republic of China, and Romanov was elevated to the top of

the regional party. "Romanov tried to push our design bureau to merge with a big research and manufacturing organization named Positron, and Shokin was able to extricate Staros," recalls a former coworker at the Leningrad Design Bureau.

The Kremlin strikes back

Still, Staros undoubtedly realized that new masters were manipulating events he had no power to influence. In 1973, Romanov became an alternate member of the Politburo, which gave him absolute control of the Leningrad region. One of his first actions was to initiate a merger of Staros's design bureau with the larger research laboratory of the Svetlana radio manufacturing company. Staros faced another turning point, as he had at Cornell in 1950: He could stay on and endure the consequences or he could start life anew somewhere else. Thus, at the age of 60, Staros decided to accept the offer of running an electronics lab in the newly created Far East branch of the Soviet Academy in Vladivostok. To lure him to Vladivostok, 11 time zones from Leningrad, the prospect of Academy membership was dangled before him.

The move brought him neither any new possibilities for advanced research nor membership in the Academy. His name appeared on the list of prospective academicians several times, the last in 1979. It was bruited in Academy circles that he was supported by Keldysh and Defense Minister Ustinov, but the old boy network opposed him. He was admired by his workers but resented by his peers. Staros did not practice the research style so often favored by Soviet academicians: a systematic approach that requires patience and relative anonymity. Staros, by contrast, was impatient and egoistic.

At his death in March 1979, supposedly from a heart attack while riding in a taxi in Moscow, none of his computers was in mass production. His obituary appeared in *Izvestia* on 17 March, saying:

Soviet science has suffered a heavy loss. In the 63rd year of his

life, Professor Filipp Georgievich Staros, a member of the Presidium of the Far East Division of the Academy of Sciences of the USSR, a recipient of the State Prize and Doctor of Engineering, passed away suddenly. Death has torn from our ranks an indefatigable scientist and talented organizer, who for many years devoted all his efforts and brilliant talents to the development of Soviet scientific and technological research. As head of a design bureau in the electronics industry for 20 years, Chief Design Engineer Filipp Georgievich Staros made a great contribution to the formation and development of microelectronics for the fatherland. The memory of Filipp Georgievich will remain in our hearts forever.

The obituary was not signed by Staros's family, friends or colleagues, in keeping with Russian custom, but by the Soviet bureaucracy—the Academy of Science, the Ministry of the Electronics Industry and the State Committee on Science and Technology. This was an official expression of appreciation for an American expatriate who pioneered Soviet microelectronics in a national hour of need.

References

1. H. Kissinger, *The White House Years*, Little Brown, Boston (1979), p. 1233.
2. H. Rogger, Comparative Studies in Society and History **23**, 382 (1981).
3. K. Bailes, Comparative Studies in Society and History **23**, 421 (1981).
4. R. Radosh, J. Milton, *The Rosenberg File*, Holt Reinhart and Winston, New York (1983).
5. According to Sergey Khrushchev (son of Nikita Khrushchev) he attracted his father's attention from Staros. Personal conversation with Sergey N. Khrushchev at Harvard, Feb. 15, 1989.
6. For a description fo Zelenograd and its implications for Soviet defense, see L. Melvern, N. Anning. D. Hebditch, *Techno-Bandits, Houghton Mifflin*, Boston (1984).

CHANGING TIMES: SAKHAROV IN THE US ON HUMAN RIGHTS AND ARMS CONTROL

PHYSICS TODAY/FEBRUARY 1989

In this age of images and illusions, Andrei Dimitrievitch Sakharov impresses many people with his ability to lead a life of substance, not symbols. On his first visit to the US last November, sometimes speaking before as many as five groups a day in Washington, New York and Boston, Sakharov made those who heard him more aware of today's scary issues as he urged the release of political prisoners, the abolition of nuclear weapons, the protection of freedoms of speech, press, demonstration and travel for everyone and the reconciliation of "mankind's divisions [that] threaten it with destruction." It is for his tenacious position on the primacy of human values that he won the Nobel Prize in 1975 and the hearts and minds of people throughout the world.

Sakharov is a folk hero for the times. At home, an Armenian poet, Silva Kaputikyan, calls him "the conscience of the Soviet people." *US News and World Report* hailed him as "the most admired man of science since Einstein." In an editorial, *The New York Times* considered it "apt that Sakharov should be a guest in the US during election week" because few Soviet citizens "have argued so fearlessly for greater democracy . . . and more open society."

An unflinching dissenter

So it was characteristic of this unflinching dissenter that on his departure from Moscow he urged Soviet authorities to defend human rights and to improve conditions for patients in mental hospitals. It was just as natural that on arriving at Boston's Logan Airport on 6 November, he expressed his respect for the US and praise for its capacity of self-criticism as a "rare quality in the world," then launched into an appeal for a Soviet mathematician, Vasis Melanov, who remains in jail for protesting Sakharov's enforced exile in 1980 to the closed city of Gorki.

In the 1970s and 1980s the Kremlin forbade Sakharov to travel abroad because it regarded him as a security risk for his work in developing nuclear weapons during the 1940s and 1950s. Since General Secretary Gorbachev personally released him from exile in December 1986 and encour-

A historic meeting of Edward Teller (left) and Andrei Sakharov. The two men expressed both agreement and discord.

SUSAN STEINKAMP/US NEWS & WORLD REPORT

aged him to become reengaged in scientific and patriotic matters, Sakharov's influence on Soviet affairs has been extraordinary. One of Sakharov's first acts on returning to Moscow was to demand a mass amnesty for about 700 Soviet political prisoners. Within days the Kremlin set free some 140 of these. Sakharov believes there are now probably no more than 30 political dissidents in Soviet prisons, yet he has not let up on his demands for their release.

His stature among politicians and physicists in the West gives him great celebrity and credibility at home. Alone among dissidents, Sakharov has gained an audience with Gorbachev and other Soviet leaders. At the superpower summit last May, President Reagan invited Sakharov to a dinner with Gorbachev. Although all his medals, including three for the Order of Lenin, were stripped from him, Sakharov remained a member of the Soviet Academy of Sciences throughout his period of exile. Last October, when he was elected to the presidium of the Soviet academy (PHYSICS TODAY, January, page 61), he also became honorary chairman of Memorial, an unofficial anti-Stalinist group. The old guard struck back in

mid-January when the academy rejected the nomination of Sakharov by physicists for one of the organization's own specially reserved seats at the election on 26 March of the new Congress of People's Deputies. Undaunted by the rebuff, physicists from the Lebedev Physical Institute and other places, along with some 750 more citizens, held a hastily called rally a week later to champion his candidacy for an at-large seat representing Moscow.

Sakharov's election manifesto broadly resembles Gorbachev's domestic program. It calls for instituting the rule of law (not of arbitrary decisions by officials), strengthening the faltering economy and cleaning up the environment. During the meeting, Sakharov added some more planks to his platform, such as constructing nuclear power plants underground, curbing the power of the KGB and guaranteeing the freedoms of speech, press, travel and demonstration. His campaign seems to bestride a difficult path, championing the reforms of *perestroika* while at the same time faulting their insufficiency.

His most enthralling talk in the US was his first, delivered on 13 November to an audience of some 150 physicists, journalists, foreign diplomats, members of Congress and foundation leaders in the Great Hall of the National Academy of Sciences. The occasion also provided the first opportunity for Sakharov to sign the academy's members register since his election as a foreign associate in 1973. He was introduced in a moving tribute by Sidney Drell, deputy director of SLAC (see box below). Speaking in Russian, with an accompanying translation in English, Sakharov said he had listened to Drell's remarks "with a great deal of inner turbulence and distur-

bance." In 1983, he recalled, he had sent Drell an open letter, "Threat and Danger of a Thermonuclear War," in which he had expressed "all my anxieties about the present and future." That was the year of "the greatest pressure," he said. It was marked by a "libelous and abusive book, printed in 11 million copies, in which my name and the name of my wife were trampled in the mud. It also was the year in which four members of the Soviet Academy denounced me and my open letter. At the same time, I am moved never to forget the un-

equivocal support of this American academy."

As far back as 1973, Sakharov's leadership in defense of human values resulted in official denunciations and humiliations—though such colleagues as Pyotr Kapitza adamantly refused to sign any statement denouncing Sakharov. In 1980 he was forcibly removed from his modest apartment in Moscow to a guarded house in industrial Gorki for opposing the Soviet invasion of Afghanistan. That year the street in front of the Soviet embassy in Washington was

Celebrating Sakharov

For the 150 guests it was a memorable occasion. They had come to the Great Hall of the National Academy of Sciences on a rainswept Sunday night last 13 November to honor Andrei Sakharov on his first visit to the US. Elected a foreign associate in 1973, Sakharov at long last was able to sign the official registry of membership, which bears the signatures of members going back to 1863, when Congress selected the first 50 scientists. Sakharov received a three-minute standing ovation after he was introduced by **Sidney D. Drell,** *deputy director of SLAC and codirector of Stanford University's Center for International Security and Arms Control. Excerpts of Drell's moving tribute follow:*

Andrei, I always dared to hope—as did many of your friends and colleagues around the world—for this moment when you would be free to visit our shores and join us in this great Academy—which, since your election in 1973 as a Foreign Associate, is yours as well as ours. Still, as I look back over the arduous and at times tortuous path you had to travel to get here, this occasion seems to me to be as close to a miracle as I ever expect to witness. The recent changes in your country that have made possible your visit offer the further hope that our two great nations will embrace common principles of human dignity and mutual respect and that they will continue moving away from chilling confrontation toward constructive cooperation, the better to meet the challenges to the survival of humanity.

Twenty years ago Andrei Sakharov published his remarkable essay on "Progress, Coexistence, and Intellectual Freedom." The two basic theses which he developed in this essay are (1) the division of mankind threatens it with destruction and (2) intellectual freedom is essential to human society. His arguments remain as valid and compelling today as they were when they first appeared. This essay publicly marked Sakharov's emergence from the laboratory where he had worked as a scientist. It was soon followed by further writings and speeches of great impact, and Andrei became recognized not only as a scientific leader in search of nature's principles for the properties of matter, but also as a moral leader in search of ethical principles for a humanity striving for peace, for progress, and for basic human dignity.

From 1968 up to the present Andrei has continued to speak out—forcefully, courageously, persistently, and wisely on the main issues of our times. . . . He risked everything and sacrificed much in his support of prisoners of conscience and his opposition to oppression wherever it occurs in the world. In his devotion to truth and human dignity and his defense of the freedom of the human spirit Andrei has become, in the words of his 1975 Nobel Peace Prize citation, *"the spokesman for the conscience of mankind."*

Human history has been inspired and ennobled by the occasional occurrence of figures of indomitable courage. Each of us has our own personal honor roll of those rare individuals whose lives have become morality plays with the dimensions of an historical epic, the theme of which is the struggle between conscience and principle on the one hand and raw political power on the other. Andrei stands tall in my honor roll of those giants who have been driven to do battle for principle in the manner described so eloquently by the young lawyer, Gavin Stevens, in William Faulkner's *Intruder in the Dust:*

> Some things you must always be unable to bear. Some things you must never stop refusing to bear. Injustice and outrage and dishonor and shame. No matter how young you are or how old you have got. Not for kudos and not for cash: your picture in the paper nor money in the bank either. Just refuse to bear them. . . .

Andrei is most widely known for his courageous leadership in the defense of human principles that we hold dear and as the father of the Soviet hydrogen bomb. But you should also know that he is a great scientist whose brilliant career as a theoretical physicist is distinguished by seminal research contributions to fundamental physics, including the behavior of plasmas and the properties of elementary particles.

In 1950 Andrei, together with Academician Igor Tamm, an internationally honored and greatly admired former leader of Soviet physics and Andrei's teacher, wrote the pioneering paper in the controlled fusion effort in the Soviet Union. In this paper they introduced a confinement scheme for a hot plasma that is famous today under the name Tokamak. . . .

Andrei also made a contribution of crucial importance to our quest to understand the evolution of our universe following its physical beginnings in the "big bang" of 18 or so billion years ago. The problem he addressed is this: Physicists know that for each form of matter, there also occurs antimatter—for example, electrons and positrons, protons and antiprotons. Antimatter is a necessary consequence of joining the general principles of atomic theory—that is, the quantum theory—with Einstein's special theory of relativity. But we must wonder then what has happened to all the antimatter. In our universe—or all we can see of it as we peer far out into space to receive signals just arriving from distant events that occurred ten or more billions of years ago—why are the massive systems of stars and galaxies made almost exclusively of matter and not antimatter?

Andrei provided the clue for understanding this in 1968— the same year he published his original essay on "Progress,

renamed Sakharov Place to protest his exile.

It seemed appropriate that Sakharov's talk was delivered under an academy mural depicting Prometheus Bound and bearing a quotation from Aeschylus reading "Harken to the miseries that beset mankind." In his remarks, Sakharov firmly defended *perestroika*, but observed that many people wondered whether it might endanger the West by strengthening Soviet economic and military capabilities. "The threat of *perestroika* to our country and the world,"

he said, "does not lie in its success but in the possibility of its bloody failure.... That would be a calamity."

Nevertheless, he cautioned against unquestioning acceptance of every change occurring in the Soviet Union. "Beneath the slogan of 'Don't interfere with Gorbachev' is a passivity...a naiveté, an absence of thought.... We need to be realistic. I am now speaking in the West and saying that it is essential to assist *perestroika*, but with eyes wide open, with an understanding of the issues, not with naiveté."

Speaking haltingly, as if searching for the right words to express himself, Sakharov claimed that Drell's remarks were "exaggerations to a certain extent of my work in physics.... If one has made even a small contribution, it enters the general scholarly community and enables others to seize it and develop it in ways you may not have perceived. This is particularly true of my work in asymmetry...something I could not complete at the time I conceived it. But I take enormous satisfaction that it has entered the current of scholarly inves-

---

Coexistence, and Intellectual Freedom." His was the leap of imagination to see that the absence of antimatter can be explained rather elegantly by joining a recent experimental observation that there is a very tiny difference between the behavior of matter and of antimatter with several general postulates that separately had been made in other contexts. The most intriguing of these postulates is that the proton, the nucleus of the hydrogen atom—long believed to be a stable particle of nature—may in fact decay just like other forms of subnuclear matter, albeit very, very slowly. This bold hypothesis is currently being tested in laboratories around the world....

When the history books of the latter part of the 20th century are written they will tell that this was a time when mankind was first able to begin writing a history of the evolution of the universe following the big bang that is based on solid experimental data and theoretical concepts. And in that chapter of history as in other chapters of our times, Andrei's name will surely appear, this time as Andrei Sakharov, physicist.

Andrei's life in physics is clear evidence of the international character of science.... Science knows no boundaries, and efforts to create barriers—whether to keep new ideas within or to prevent new ones from entering from the outside—have universally proved harmful to progress. The great 19th century Russian playwright, Anton Chekhov, said it best, as follows:

There is no national science just as there is no national multiplication table; what is national is no longer science.

It is regrettable when, on occasion, governments need to be reminded of this basic fact. It may not be a law of nature, but it has proved to be a reliable rule of thumb, that national interests and true security are better served by keeping open the channels of communication of scientific achievements than by erecting barriers to stem the transfer of knowledge.

Just as good science knows no geographic or political boundaries, modern-day scientists have increasing difficulty in defining a boundary line between work in the laboratory and a concerned involvement in the practical applications of scientific progress. Sakharov himself is one of the most important examples of this involvement and of the serious difficulties, and on occasion the painful disillusionment, that a scientist or a scholar may encounter when he or she reaches out of the private shell of the laboratory or the study and participates in society.

Sakharov has written in an autobiographical essay published in 1974 that "I had no doubts as to the vital importance of creating a Soviet super-weapon—for our country and for the balance of power throughout the world," but tells of his concern for continuing bomb testing

throughout the following decade and of his involvement in a military-industrial complex "blind to everything except their jobs" and of his coming "to reflect in general terms on the problems of peace and mankind and, in particular, on the problems of a thermonuclear war and its aftermath." The involvement of scientists in war and weapons of death—as in other major issues of importance to the human condition—is in itself nothing new. Its distinguished honor roll of olden days includes such luminaries as Archimedes of Syracuse, Leonardo da Vinci, and Michelangelo. But never before have scientists dealt with weapons of absolute destruction, with weapons whose use could mean the end of civilization as we know it—if not of mankind itself. And never before has the gulf been so great between the scientific arguments—even the very language of science—and the political leaders whose decisions will shape the future.

The new fact that the fruits of our learning threaten the existence of all mankind presents an acutely heightened ethical dilemma to scientists. Our predicament is precarious because we have so little—if any—margin of safety. As much as any scientist I know, Andrei Sakharov has understood the special obligation of the scientific community to alert society to the implications of the products of scientific advances and to assist society in shaping the applications of these advances in beneficial directions.

Scientists who enter the political realm and participate in the public debate on the implications of scientific advances bear a special responsibility to speak accurately and responsibly on the technical challenges to society. Once again, Sakharov is a model for us all. He has spoken out courageously, passionately, and with outrage when appropriate on issues of social injustice and oppression; but, when speaking as a scientist on technical and factual issues, he has maintained the same high standards that we demand in our professional scientific lives. It is our obligation to do likewise....

By his actions, Andrei has been an inspiration to all of us. Constant in purpose, clear in vision, modest, and unflinching in his courage to speak out in circumstances of great personal danger, he has inspired support, admiration, and devotion from people of all stations and nations....

Andrei, I will close by asking all your friends here tonight to join me in a toast expressed in the words of your friend Lev Kopelev, author, compatriot, and known to many of us as the mathematician Rubin who appears in Alexander Solzhenitsyn's great novel, *The First Circle*. Kopelev's beautiful tribute is:

...the majesty of his spirit, the power of his intellect and the purity of his soul, his chivalrous courage and selfless kindness feed my faith in the future of Russia and mankind.

tigation and was snatched up by many others. As for demonstrating the instability of the proton, that may be much more difficult to prove than I had originally imagined. Nonetheless, what has come into question is the whole law of conservation of matter." As he went on describing his thoughts on physics and cosmology, his interpreter became confused and distraught. Detecting this, Sakharov admitted: "Now you see I'm a poor popularizer, especially compared with Drell. When he spoke he had your attention, but I am losing yours."

His point in turning to his work, he explained, was "to convey the whole drama of ideas—that once it begins it enters the scholarly exchange from one side of the ocean to the other, always developing a life of its own in the process. It is my hope that the interaction of ideas taking place in the scientific community will occur in all other human communities engaged in all kinds of activities.... The enormous responsibilities borne by the scientific community can be realized only through the most extensive international cooperation."

Examining the 'silent plagues'
Sakharov came to the US at the invitation of a newly formed organization with the portentous title of International Foundation for the Survival and Development of Humanity, of which he is a board member. Sakharov had become interested in the foundation during a conversation in February 1988 with Jerome Wiesner, MIT's president emeritus and President Kennedy's science adviser. On a visit to Sakharov's small flat in Moscow, Wiesner explained the origin and purpose of the proposed foundation—that the idea for it came from Yevgeniy P. Velikhov, a plasma physicist who is a vice president of the Soviet academy, during the International Forum for a Nuclear-Free World conducted in Moscow in February 1987. Discussing what such an organization might do, Velikhov and Wiesner decided it should examine the world's "silent plagues," such as hunger, desertification, global environmental pollution, international security and human rights deprivation.

Wiesner said the foundation had applied for permission to operate in the Soviet Union and a few Soviet academicians had agreed to serve on its board—namely Velikhov and Roald Z. Sagdeev, then director of the Institute of Space Research. Sakharov expressed his eagerness to join the board. Once a member of the foundation, he still faced the problem of getting a travel visa to attend

meetings outside the Soviet Union. That matter was resolved, with the help of Gorbachev, only last 5 October, when all foundation members were granted multiple visas to leave and enter the USSR for a two-year period. The Council of Ministers decree No. 1167 that provides for travel visas also allows the foundation the right to conduct meetings, publish documents and raise tax-free funds— the first time the Kremlin has allowed this to happen.

Foundation leaders believe they will need to raise about $2 million per year as well as 2 million rubles from voluntary contributions. Private philanthropy is uncommon in the Soviet Union, but the Chernobyl reactor fire stimulated the practice in 1986 when Soviet citizens contributed 500 thousand rubles for young victims. Several US foundations have donated to the international foundation, and Occidental Oil tycoon Armand Hammer has pledged $1 million.

So, at its organizational meeting, which ran for three days at the US academy last November, Sakharov presided over a human rights panel that included representatives of Helsinki Watch, Amnesty International and the US–Soviet Human Rights Commission. At one point he proposed that the foundation set up a human rights commission to investigate reported abuses and excesses. It turns out that Sakharov spent so much of his time in the US receiving awards, attending receptions and meeting VIPs that he had to ration his moments with foundation members, who include such well-known figures as Wiesner, Hammer, former Defense secretary Robert S. McNamara, former Notre Dame University president Theodore Hesburgh, Apple Computer president John Sculley and Princeton physicist Frank von Hippel. Six members came from the Soviet Union, including Sagdeev, Velikhov and Metropolitan Pitirim of the Russian Orthodox Church.

Entrée to the establishment
During the four days he spent in Washington, Sakharov met with President Reagan at the White House, drank cocktails with establishment figures at the Library of Congress, hobnobbed with an intellectual circle during dinner at the Smithsonian, discussed US politics during an entertaining evening at the home of Senator Edward M. Kennedy and dined among Washington's conservative elite with Ernest W. Lefever, president of a right-wing think tank, the Ethics and Public Policy Center. Although usually queasy about asso-

ciating with Soviets, the center invited Sakharov to speak at a banquet to honor Edward Teller as winner of the 1988 Shelby Cullom Davis Award for "integrity and courage" as "a patriot who has combined profound moral judgment with political wisdom."

After Sakharov agreed to attend Teller's dinner and possibly say a few words, he was urged by some US and Soviet friends to skip the event. But Sakharov insisted on going. Thus did Sakharov and Teller, two physicists whose work did so much to change the world, meet for the first time in the Washington Hilton Hotel on 16 November. Such a meeting would have been unthinkable almost any time since World War II when both men were working on nuclear weapons. It would have been impossible even two years ago while Sakharov was in exile. Indeed, it is difficult to imagine two men more different and yet with so many parallels in their lives.

So, when Sakharov and Teller encountered each other in a hotel suite for about 20 minutes before the dinner, they smiled warmly and shook hands eagerly. Then, dutifully obeying orders from magazine photographers to sit or stand before floodlamps and cameras while their pictures were taken, the two men chatted about personal and political matters—their respective states of health, the Chernobyl disaster and their shared enthusiasm for locating nuclear power reactors underground for added safety, and their vision of arms control for the superpowers.

Sakharov steered the conversation to SDI, stating his belief that deployment of a space-based ballistic missile defense would have the effect of destabilizing the balance of power if it could ever be made to function at all before the US and USSR went broke developing the system. Teller, for his part, emphasized that the technology of defense had not yet been given a sufficient chance to prove its value and that to abandon its promise prematurely would be a "terrible mistake."

After Sakharov was introduced in somewhat flamboyant language by conservative commentator William F. Buckley Jr, he received the most exhuberant standing ovation he got at any of his many US appearances. It seemed to take Sakharov by surprise, perhaps because the long applause came from an audience dressed up in black ties and long gowns.

The extemporaneous talks by both Sakharov and Teller were separated by more than an hour because Sakharov had decided to fly to Boston. Each extolled the other as a man of principles, conviction and ingenuity, and they commented on the parallel course of their physics careers in creating thermonuclear bombs. Both said they and their fellow scientists were convinced that work on the weapons was vital to their country's defense and necessary to maintain world peace—though Sakharov referred to the results as "a great tragedy." Both also agreed on the importance of maintaining open discussions, as Sakharov put it, "particularly when we disagree.... At least we will understand our different points of view and avoid confrontation." Teller was thankful that *glasnost* enabled Sakharov to see science and society outside the Soviet Union and "create a dialogue—one in which I would certainly like to participate."

In his talk, Sakharov said he and Teller agreed on some subjects, such as ensuring the safety of nuclear reactors, but he added that there were other topics of disagreement and cited SDI as a "grave error." He repeated what he has said to Teller in private, declaring that SDI would "destabilize the world situation" and that, "if deployed, even before the system were fully armed, there will be a tempta-

tion to destroy it [and] this in itself could trigger a nuclear war. The problem of SDI stands in the way of achieving a really profound arms control."

Teller emphasized his agreement with the Soviet physicist "except on one point"—SDI. "We must know what can be known," Teller asserted. Teller said he continues to work in nuclear energy, lasers and weapons defense, but Sakharov has been out of touch with those matters since his security clearance was revoked by the Kremlin 20 years ago. "He has not had the opportunity to work in the remarkable development of defensive systems in the Soviet Union [that] we have confidence in believing is years ahead of us."

'An atmosphere of trust'
Earlier in the week Sakharov had spoken on arms control on two occasions—on 15 November when he received the $50 000 Albert Einstein Foundation Peace Prize, established in 1979 by members of the Pugwash Conference, and in a question-and-answer session the day before at the Kennan Institute, named for George Kennan who is credited with initiat-

ing the US policy of Soviet containment after World War II. In accepting the Einstein Prize he characterized science as providing "a unified conception of the world at the same time that all of the evolutionary processes develop and are turbulent around us. For a man of science, this sense of unity of the entire world provides the kind of grounding and orientation that religion gives for those who have faith."

He detected that "an atmosphere of trust has begun to develop" between the US and Soviet Union after what he called "the beastly abuses of the Stalinist period." Measures are now needed to strengthen that trust, said Sakharov. He suggested that "the best thing for us, the Soviets, to do would be a unilateral reduction of military forces.... The reduction of that army would in no way jeopardize the security of the Soviet Union."

Less than a month later, Gorbachev seemed to adopt Sakharov's recommendation by proposing in a speech before the United Nations on 7 December that the Soviet Union would reduce its armed forces by 500 000 troops.

—IRWIN GOODWIN

HOW THE MILITARY RESPONDED TO THE LASER

A 'cash and crash' approach and interservice competition
led to a premature shift of emphasis from research
and exploration to development and scaling-up.

Robert W. Seidel

PHYSICS TODAY/OCTOBER 1988

"I feel as do others here that the LASER may be the biggest breakthrough in the weapons area since the atomic bomb."[1] This statement, made in 1962 in a letter by Major General August Schomburg, head of the Army Ordnance Missile Command, reflected an attitude that was pervasive in the military in the first years after the birth of the laser. According to one contemporary assessment, there was "scarcely an Air Force, Army [or] Navy agency that does not now support, or talk of sponsoring in the near future, some type of basic or applied research or experimental development with optical masers."[2]

The invention of the laser stimulated emissions of interest from the military far more coherent than those from any other potential user: The military supplied most of the funding for laser research and development; military need suggested many of the laser's applications; and the military was the principal market for the laser. As I will shortly explain, however, military laser programs underwent a premature shift in emphasis, from research and exploration to development and scaling-up (see figure 1). A number of factors conspired to bring about these developments:

▷ Interservice competition to develop devices suited to the missions of each branch
▷ Institutionalization of research programs in military as well as in contractor laboratories
▷ Adoption of the Manhattan Project and the wartime program to develop radar as models for military laser development, as Schomburg's remarks suggest.

Robert Seidel is a research historian at the Laser History Project, in Los Alamos, New Mexico.

The services responded to the invention of the laser according to their widely varied missions in national defense. This centrifugal tendency was only partly overcome by the DOD agency that was responsible for frontier research, the Advanced Research Projects Agency.[2] Because the transition from basic research to development and thence to advanced development and deployment was a profitable one for many of the companies performing laser research, there was a dynamic built into the system that favored this scaling-up. Moreover, the services sought to develop their own clientele of researchers through programs sponsored by the Office of Naval Research, the Army Research Office and the Air Force Office of Scientific Research. The interservice rivalries that underlay this competitive organization of research could hardly be overcome by the Advanced Research Projects Agency, especially when such an exciting technology as the laser was to be developed.

Stimulation of research

The military interest in quantum electronics did not begin with the laser, as historian of science Paul Forman has shown. That interest intensified greatly, however, after the laser was reduced to practice. By forcing a change from small to big science, from academic to in-house and contract laboratories and from open research to classified development, military interest in the laser transformed the nature of laser research and development. This transformation was accompanied by, and in turn accelerated, a shift in laser technology from solid-state to gas and chemical lasers.[3]

Military interest in the laser developed even before the first laser was built. After Charles Townes and Arthur

Airborne Laser Laboratory. Approximately 300 technical personnel at dozens of aerospace firms worked for 11 years to complete this demonstration platform for a high-energy laser weapon. In May 1988 the Airborne Laser Laboratory was moved from Kirtland Air Force Base in Albuquerque, New Mexico, to the Air Force museum at Wright–Patterson Air Force Base in Dayton, Ohio. **Figure 1**

Schawlow developed the theory of infrared and optical masers in 1958, Townes, who was then at Columbia University and a consultant to Bell Laboratories, thought enough of the military potential of the "optical maser" to offer it to John Wheeler of Princeton University, who was heading a project for the Institute for Defense Analysis and the Advanced Research Projects Agency. The project, as Wheeler explained at the time in a letter to Townes, sought "presently unappreciated ways in which science may be able to contribute vitally to national defense." Wheeler thought the laser "fell right in this ball park," and predicted that it would receive "the very serious attention of ARPA."[4]

That attention was forthcoming when Technical Research Group, a rising electronics firm with maser experience, bid for military support to develop the laser in 1959. TRG had hired Gordon Gould in 1958 from Columbia University, where, as a graduate student, he had written down the laser ideas for which he sought a patent in 1959.[4] Perhaps because of the interest aroused at ARPA by Wheeler's assessment, TRG received $1 million from ARPA, more than three times the amount TRG had sought. ARPA aimed not only to use the laser for radar and communications, as TRG proposed, but also to defend against the Russian missile threat, or "missile gap," with death rays.[5]

The laser offered a coherent, directed, concentrated beam of light that promised to realize an ancient dream, epitomized in Archimedes's idea to attack the Roman fleet at Syracuse by using mirrors and lenses to focus burning solar rays on ships at sea. Science fiction's preoccupation with burning "death rays" added modern sanction to the ancient dream.[6] The Soviet Union's large boosters, which lofted Sputnik and the first cosmonauts into space and might equally well launch warheads, provided suitable targets for the rays. The promise of beam weapons enhanced the services' interest in lasers and launched a number of industry and service research programs that transcended the interest in laser ranging, communication and detection.

Industrial defense organizations were quick to pursue beam-weapon applications. Within months of Theodore Maiman's invention of the ruby laser in 1960, his employer, Hughes Aircraft, had an in-house program to develop beam weapons, beginning with a "kill-a-rat" laser. Martin Marietta's advanced program division sought a cryogenic hydrogen "laser capable of beaming a million-degree ray to vaporize hostile space weapons." This "disintegrator ray is being designed for use at altitudes of 100 000 feet or farther out in space," reported *Electronics.* "It would be the size of a large army searchlight."[7]

The military services soon picked up on the work,

which was marketed aggressively by the firms involved. "Seldom has a development in technology fired the imagination of the scientific and technical community as has the development of the laser," the Air Force Office of Scientific Research reported in 1964, after finding more than 500 research groups studying the laser.[8] In the wake of Sputnik and a generation after the development of the atomic bomb, such a stimulus was eagerly awaited in the military, particularly in those laboratories that were its captives. Laser research and development was amplified by the dozens of in-house laboratories and hundreds of contractor laboratories to which the laser meant not only an exciting technical challenge, but the promise of profits. "Defense at the speed of light!" became a rallying cry for the military–industrial complex.

Pumping

The Defense Department's interest was important because of its position as the chief funder of scientific research and development in the United States since World War II.[9] Whether or not paying the piper means calling the tune in basic research, it is certainly true that in the competition for Federal dollars, military interests and priorities affect applied research. In defense contractor organizations such as Hughes Aircraft, where the laser was first built, basic research is certainly heavily influenced by the corporate interest in pleasing the customer. And Hughes, like many other defense firms, had only one real customer.

The rapid proliferation of laser types in the 1960s and 1970s made available a large number of candidates for advanced development and marketing to the services. Some, such as the neodymium glass and neodymium YAG lasers, the gas-dynamic and electron-beam sustained CO_2 lasers and the hydrogen fluoride chemical laser, became the foci of major military programs that scaled them to higher powers and higher energies.

The funding for the research that provided those lasers proved to be a significant fraction of the total national support for laser research and development. With most practical applications in the nondefense commercial sector too risky to fund, the large military commitment to lasers allowed industrial and university researchers to turn to the Department of Defense for research and development support.

Even after they won research and development contracts, industrial firms often funded their own research programs aimed at generating hardware and knowledge to market and to develop for other military purposes. The contractors supported this research until they could find a friendly program manager in one of the service research and development commands to fund it. Fundamental inventions tended to occur in these so-called "independent" research and development programs, so that firms retained proprietary rights in them.

As the services provided contracts for development as well as research, they were more attractive patrons of research than most. The DOD's official patron of ad-

vanced research, ARPA director Jack Ruina, strongly opposed Air Force and Army laser programs, fearing the "dilution of effort which parallel programs cause, in addition to the natural tendency of the industry to prefer service programs to ARPA programs. Service programs can lead to big development contracts."[10] ARPA's efforts to control this tendency were, however, overcome by the dynamic already established in the development of the laser.

The Army and the laser

The Army felt that it should participate in laser research "in order to protect its specific interests and requirements."[11] As early as 1960 the Army's Office of Ordnance Research asked that the TRG laser work sponsored by ARPA be declassified. It was interested in the laser beam "for the illumination of small targets for guided missiles equipped with optical homing devices, [such as] antitank weapons," which "would be relatively secure, since a coherent infrared source is essentially undetectable at any position outside the very narrow beam." The Army also coveted high-energy beam weapons for its antiballistic-missile and antipersonnel missions.

Research implied trained manpower. To "stimulate defense oriented research in modern optics," the Office of Ordnance Research donated $2 720 500 to 23 universities. The University of Rochester got the largest amount, $315 000, with other universities receiving from $35 000 to $200 000. The Army "recognized the adverse effects the continued shortage of trained optical research scientists could have on advanced and sophisticated defense programs." Supplying these universities with expanded optical facilities was "expected to provide inroads into advanced measurement in the reentry and midcourse ICBM flight regime." It also enhanced the supply of academic research.[12]

In its effort to develop a high-energy laser radiation weapon, the Army Missile Command Laboratories selected neodymium glass as the lasing medium because it could be produced in large and varied sizes, with high concentrations of ion doping and high optical quality, and was easily worked and shaped. Moreover, its 1.06-micron radiation was both invisible to the naked eye of the enemy and detectable with Army sensors. After scaling up these lasers to the point where the power and energy growth curves leveled off and further increases required "larger arrays and brute force increases in the size of devices," the Army researchers attempted to use them in laser blinding weapons such as a jeep-mounted system tested in 1968, but the devices offered repetition rates too low to be useful on the battlefield. The Army's efforts to scale up glass laser weapons for ABM uses led to devices such as the Big-X laser, which focused the output of several glass rods pumped by one of the world's largest induction power supplies. The transfer of the missile defense effort away from the Army Missile Command to the Army Ballistic Missile Agency in 1968 terminated efforts in this area. Between

US ARMY MISSILE COMMAND

Ground laser locator designators, which range and mark targets. These prototypes were made by International Laser Systems (left) and Hughes Aircraft Company (right). The devices were the result of seven years of research and three years of advanced development at the Army Missile Command Laboratories, Redstone Arsenal, Huntsville, Alabama. In 1980, after another six years of engineering development, both designators went into production. **Figure 2**

1962 and 1968, the Army spent $8.8 million on high-energy lasers, which they divided about equally between in-house and externally funded research.[13]

The Army Missile Command was more successful with laser target designators and guidance systems. In 1962, David Salonimer, an engineer at the Missile Command Laboratories at Redstone Arsenal in Huntsville, Alabama, developed the concept of using a pulsed Q-switched laser to guide artillery projectiles or missiles. Q switching, which kept the energy storage factor Q of the laser cavity low while inverting the population and then switched it to a high value to produce a very high rate of stimulated emission, provided pulses powerful enough to serve this purpose. Here neodymium glass lasers proved superior, too, because of their invisible beams, greater efficiencies and compatibility with highly sensitive sensors. A variety of systems were designed and some, such as the laser-guided bomb used in Vietnam and the Hellfire and Copperhead laser-guided tactical missiles, were successfully entered into the arsenal.[14] Figure 2 shows prototype laser designators, which range and mark targets.

The Army's greatest success was with laser rangefinding. After the invention of the laser, Army Ordnance set up a $100 000 emergency fund to study the laser, and formulated a $700 000 rangefinder program. The Army Electronics Command developed a rotating Q switch for rangefinder applications in 1962, while the Hughes Aircraft Company developed a Kerr-cell optical Q switch under Army contract. By the mid-1970s, the Army fielded the AN-GVS-5 hand-held rangefinder (figure 3) as well as tank laser rangefinders, and Hughes was doing a $50–$100 million a year business producing rangefinders.[15]

In the Army programs, work on tactical battlefield applications of the laser supplemented the more common focus on the ABM uses of the new device. The "down to earth" applications were more practical, but required technological innovations to provide greater eye safety for the troops using the lasers and to pulse and code the beam to insure reliable information and frustrate countermeasures. The "smart" weapons resulting from this work appeared in the Vietnam War less than a decade after the

invention of the laser and became the staple product of the industry for the military. The laser has continued to find its greatest military usefulness in such weapons.

The Air Force and the laser

Space applications dominated Air Force thinking about lasers, just as Earth-bound uses shaped Army plans. Frustrated by the civilian thrust of the space program after 1957, Air Force General Bernard Schriever convened a high-level scientific advisory committee to review the space program of the Air Force Research and Development Command. The committee, which was formed just before the 1960 election under Trevor Gardner, president of Hycon Manufacturing Company, underlined the threat posed by Sputnik, "which could have contained military intelligence and communication equipment or possibly a nuclear warhead," and had done "great damage to our image of world leadership." The committee urged Schriever's newly formed Air Force Systems Command to carry out a program of fundamental scientific investigation for space exploration and for the development of orbiting arms-control satellites. Spaceworthy systems of this sort were fundamental to the reception of the laser by Air Force laboratories.

Representative of Air Force responses to the need for space-based systems were the solar-powered laser, high-transmission-rate laser communications systems, and systems to identify and track objects in space.[16] As in the case of the Army's interest in laser-based defenses against ballistic missiles, the Air Force's focus on applications of the laser in space was at best premature and at worst unproductive of any practical devices.

The Air Force also had an air defense mission, and for the associated tactical needs worked to use lasers in rangefinding, precision delivery of weapons, navigation and location, reconnaissance and imaging, anti-tactical missiles and other countermeasures. The war in Vietnam presented technological opportunities for the laser, and the Air Force responded by setting up a quick-reaction program in 1965. Project 1559, as it was called, provided funds for translating research and development projects into weapons in six months to a year. After the Army suggested the concept of laser guidance for the precision delivery of weapons, the Air Force contracted with Texas Instruments to build the Paveway guided bomb. Although its advantages over conventional bombs were disputed, the laser-guided bomb won the hearts and minds of the service.[17]

The laser-guided bomb project also developed a laser target-illuminator system, Pave Arrow, for use by airborne forward air controllers in guiding F-100D strike aircraft to well-camouflaged and hidden targets in the Vietnamese jungle. Laser designator systems such as Pave Spike, Pave Tack and Pave Penny used television and infrared detectors to pick up visible and invisible laser radiation reflected from targets designated by ground or airborne reconnaissance forces and to allow weapons to home in on these targets.

The Air Force also investigated other, more direct applications of laser energy, such as damaging sensors used by the enemy. Like the Army's antipersonnel laser applications, these were not successful. However, as a sensor, the laser extended the range of human senses far beyond other optical devices such as telescopes and binoculars. Representative systems such as the KA98 imaging system used gallium arsenide lasers to scan target areas.[18]

The Air Force response to the laser was a wide-ranging exploration of its aerospace applications. It even created a career track that allowed Air Force officers to advance through the ranks in research and development. In-house research by Air Force officers distinguished the Air Force program from the other services. Beginning early in 1962 at the Air Force Weapons Laboratory at Kirtland Air Force Base in Albuquerque, New Mexico, these "blue suiters" conducted research on laser effects and laser damage, and ultimately they became active in developing laser devices. The Air Force Cambridge

Research Laboratory built the first Air Force laser and became a center of fundamental ruby and glass laser research.[19] Laser research and development, especially in high-power and high-energy systems and their components, became a *raison d'être* for these groups, especially when fundamental laser phenomena were classified.

Major General Donald L. Lamberson (figure 4), who worked on some of the early laser effects studies at the Air Force Weapons Laboratory as a lieutenant, developed the Airborne Laser Laboratory (figure 1) there in the 1970s as a colonel. This laser systems laboratory and demonstration prototype showed that one could install lasers in planes and use them against threats such as Sidewinder antiaircraft missiles and jet drones simulating cruise missiles.

For Lamberson and other military and civilian Air Force scientists, the laser became the focal point for the development of a new kind of scientific career. The Air Force Institute of Technology, at Wright–Patterson Air Force Base in Dayton, Ohio, developed a curriculum to train them for it; the Air Force extended tours of duty for researchers and hired them after they left the Air Force to staff its laboratories and continue their work; and the Air Force Weapons Laboratory created organizations such as the Laser Division, a "big science" facility to house the work.

Trained in Air Force schools and exposed to some of the largest laser devices in the world at the Air Force Weapons Laboratory or at captive contractor laboratories, these men had unparalleled opportunities to work on the frontiers of high-energy laser technology. Many of them became point men for the development of other laser systems as well as for the Strategic Defense Initiative. In 1982 Lamberson became deputy assistant secretary for directed-energy weapons to the under secretary of defense, "the focal point for the President's initiatives on defense

Hand-held laser rangefinder. The AN-GVS-5, built by RCA for the Army, had become standard military issue by the mid-1970s. **Figure 3**

against ballistic missiles."[20] Martin Stickley of the Air Force Cambridge Research Laboratory and Gregory Canavan of the Air Force Weapons Laboratory became directors, successively, of the inertial confinement fusion program at the Department of Energy. Peter Avizonis and Arthur Guenther headed large laser damage and weapons programs at the Air Force Weapons Laboratory. Although these men, unlike Lamberson, shed their Air Force blues, they remained with the Air Force as civilian researchers for many years, training new generations of technical officers.

Controlling laser research

Even as the first service programs proliferated in the wake of the invention of the laser, ARPA created a joint venture with the third service, the Navy, to contain the burgeoning research in lasers. Project Seaside was conceived by William Culver, whom Townes had brought to the Institute for Defense Analysis in 1961 from the Rand Corporation, where Culver had been a resident expert in optics. ARPA had asked IDA early in 1961 to review the service and contractor programs in radiation weapons.[21]

ARPA was concerned about the proliferation of these programs, and wished to pursue them "within the constraints of scientific sobriety and by organization and personnel of appropriate caliber." To that end, it asked IDA to determine whether or not projects were clearly devoted to achieving a feasible radiation weapon, appropriate in size. "If the project cannot be justified on a weapon or weapon technology basis," then ARPA wanted to know if it could be justified "as a means of supporting basic research, i.e. high caliber research . . . which contributes importantly to the total fund of knowledge in a specialized area." Albert Weinstein, the Pentagon's manager in the field, stated the goals of radiation weapons studies: "maintaining a sufficient knowledge and high sensitivity to new knowledge to assure that we would recognize a potential breakthrough in its very early stages" and "constraining the program within the boundaries of scientific sensibility and resisting the natural tendency to make this area a 'glamour wagon' upon which questionable programs would be loaded."[22]

By December Culver was ready to report that the wagon was even more glamorous than Weinstein dreamed: "Current Optical Maser developments," he wrote, "have led a number of people in government and industry to believe that it may be possible to generate and direct enough coherent optical power to make a useful radiation weapon." Since Maiman's invention, he explained, ten different types of lasers had been developed. Ruby lasers had emitted 10-megawatt peak powers and 50-joule pulses, burning holes through steel 0.1 inch thick. With 12 joules per pulse for every cubic centimeter of ruby, the only limit appeared to be the size of the crystal one could produce. Moreover, the invention of the neodymium glass laser, which could be made very large and which had per unit volume 50 times as many lasing ions as ruby, promised proportionately greater powers. Already service programs exceeded several million dollars per year, and they would become much larger if any of the ideas they were pursuing for generating high power proved practical. Although Culver believed that the chances for this were much less than 50–50, he recommended bringing together a group of prominent experts in masers to look at the problem. These people included

Major General Donald L. Lamberson. Lamberson led the Air Force high-energy laser program in the 1970s as a colonel and developed the Airborne Laser Laboratory (figure 1) before becoming deputy assistant for directed-energy weapons in the office of the under secretary of defense for research and engineering. **Figure 4**

Townes; Nicolaas Bloembergen of Harvard, the inventor of the three-level maser; Robert Kingston from Lincoln Laboratory, a solid-state maser expert; and Norman Kroll of Columbia University, a theoretical physicist.

The meeting convened at the end of 1961 and recommended that ARPA fund a high-priority research and development project in high-power laser techniques. The project, code-named Seaside, was set up under representatives of the office of the deputy secretary of defense for research and engineering, ARPA, IDA and ONR to build within a year and a half a scale model high-power device that would produce 10^3–10^4 joules of laser energy.[23] Project Seaside investigated laser kill mechanisms, scaled up the neodymium glass laser, funded projects to improve ruby and glass laser materials and built four high-power solid-state laser weapon prototypes. It focused Department of Defense laser research on the exploration of solid-state rather than liquid lasers. When the development of the carbon dioxide laser and the application of aerodynamic technology to it produced a superior device—the gas dynamic laser—the era of the solid-state antiballistic missile laser weapon gave way to the second phase of military laser development, which concentrated on flowing gas lasers.

In addition to running Seaside, ONR also organized the Department of Defense Conferences on Laser Technology, as well as a variety of smaller meetings on specialized laser topics, permitting researchers from the various services to share classified and unclassified research results. It became the focal agency for military interest in high-energy lasers, but did not, as Ruina intended, eclipse service programs.

In its coordination of high-energy laser programs, Seaside prefigured later projects such as ARPA's Eighth Card, a restricted-access program that sought to develop gas-dynamic carbon dioxide lasers in the late 1960s while unsuccessfully restraining service efforts in this area, and its successor, the Tri-Service Laser Program, which developed test beds of tactical laser weapons to acquaint all three services with the new technology. The laser team at ONR presaged later coordinating groups such as the High-Energy Laser Review Group, a Defense Department unit that sought to coordinate and control service laser research as it grew by leaps and bounds in the early 1970s. By concentrating in-house and contractor efforts on goals such as the development of antiballistic missiles, anti-antiship missiles and antiaircraft missiles, these groups paralleled the role played by the Special Group on Optical Masers, a unit of the Pentagon's Advisory Group on Electron Devices that coordinated low-energy laser research and tried to constrain laser development to agreed-upon goals and budgets.

Mixed results

Because of the enthusiastic response of the services to the laser, however, these controls were not sufficient to arrest the development of experimental weapons systems. The Airborne Laser Laboratory, which carried a gas-dynamic CO_2 laser, and the Mobile Test Unit, a tank that carried a CO_2 laser sustained by an electron beam, both went ahead and eventually disappointed expectations and broke budgets. The Navy's counterpart system, the Coastal Crusader, a ship that was to have housed a chemical laser, was halted by the office of the deputy secretary of defense for research and engineering and the High-Energy Laser Review Group, but the Navy ARPA chemical laser and the MIRACL system (figure 5) eventually were built by TRW and became the most powerful laser technology available.

All of these programs demanded increasing funds as they scaled devices to engineering proportions. In the late 1970s the high-energy laser research and development programs ran at about $150–200 million per year, and by 1980 they had consumed $1.5 billion, according to evidence presented to Congress. Low-energy laser research and development, which produced fieldable devices, absorbed a similar sum, and in 1980 the government's fraction of the laser market was about 60%, or $453 million. Clearly, the military dominated the laser industry.[24]

Despite the development of a number of scaled-up high-power and high-energy laser devices, the military programs were most successful in those areas where traditional research and development produced components and systems of more modest ambition. The glamour wagon attracted personnel, funding and interest from the services, stimulating growth in the industry and the laboratory, but has yet to travel very far. As in the case of the atomic bomb, a program eventually emerged to accelerate the development of this technology and consolidate the rival service programs under a single agency, the Strategic Defense Initiative Office. However, the earlier projects suggest that it was not a failure to scale up candidate laser devices nor an overemphasis on research that frustrated the search for laser weapons. Rather, the attempts to have the laser provide different devices for different applications corresponding to different missions concentrated efforts on development and early deployment, while the high power and high energy levels required for lethal effects in antimissile applications forced development of high-energy laser weapons that proved, for one reason or another, impractical. Unlike the atomic bomb, the laser did not scale into the "megaton" range very readily. In pursuing both the bomb-like and the radar-like aspects of the new technology, the Department of Defense may have chosen incompatible models of technological development.

References

1. Letter from Major General A. Schomburg to Lieutenant General J. H. Hinrichs, 16 January 1962, history office, US Army Missile Command, Redstone Arsenal, Huntsville, Ala.

2. Aviation Wk. Space Tech., 15 January 1962, p. 92.

3. P. Forman, Hist. Studies Phys. Bio. Sci. **18**, 149 (1987). R. W. Seidel, Hist. Studies Phys. Bio. Sci. **18**, 111 (1987).

4. C. Townes, A. L. Schawlow, Phys. Rev. **112**, 1940 (1958). M. Bertolotti, *Masers and Lasers: An Historical Approach*, Adam Hilger, Bristol (1983), p. 104. Letter from J. A. Wheeler to C. Townes, 30 September 1958, in response to letter from Townes to Wheeler, 22 July 1958, sending a copy of Townes's Air Force proposal for research on coherent infrared radiation: copy provided by Townes. TRG Inc, "Proposal to Study the Properties of Laser Devices," P-329 (12 December 1958), p. 93.

5. Interview with L. Goldmuntz by J. Bromberg, Laser History Project, 21 October 1983, p. 13. For the goal of the ARPA laser program, see Item Brief, ARPA program council case number 217, agenda item 9, meeting of 26 February 1960, ARPA order file 356, Defense Advanced Research Projects Agency, Arlington, Va.

6. On Archimedes's alleged use of burning mirrors, see the following: E. J. Dijksterhuis, *Archimedes*, Princeton U. P., New York (1987), p. 28. D. L. Simms, Tech. Culture **18**, 1 (1977). W. Knorr, ISIS **74**, 53 (1983). For a science fiction treatment, see H. G. Wells, *The War of the Worlds*, Octopus Books, New York (1985), ch. 5.

7. Electronics, 22 December 1961, p. 17.

8. Office of Aerospace Research, *US Air Force Achievements in Research*, US Air Force, Washington, D. C. (1965), p. 36.

9. National Science Board, *Science Indicators 1982*, National Science Foundation, Washington, D. C. (1983), p. 51.

10. A. Weinstein, "Memo for Col. Innes," 8 May 1962, ARPA order file 306, Defense Advanced Research Projects Agency, Arlington, Va.

11. Letter from H. Robl, director, physical sciences division, Office of Ordnance Research, to C. W. Cook, Institute for Defense Analysis, 12 September 1960, enclosing a document titled "LASER and IRASER Program" and dated 1 September 1960, ARPA order file 306, Defense Advanced Research Projects Agency, Arlington, Va.

12. Objective statement in research and technology work unit summary on contract DA-ARO(D)-31-124-G-699, "Optical Equipment," with Columbia University, Defense Technical Information Center, Alexandria, Va.

13. Untitled report on Army Missile Command laser program, history office, US Army Missile Command, Redstone Arsenal, Ala. (1978), pp. 1, 10, 12.

14. The Rocket, 16 August 1972, Redstone Arsenal, Ala. US Army Aviation Digest, January 1975, p. 24. G. Widenhofer, *History of Laser Designators*, 5 November 1982, unpublished document, US Army Missile Command advanced sensors directorate, Redstone Arsenal, Ala. G. Widenhofer, *Laser-Semiactive Guidance*, US Army Missile Command advanced sensors directorate, Redstone Arsenal, Ala. See also E. C. Jolliff, *History of the United States Army Missile Command, 1962–1977*, historical monograph project number DARCOM 84M, history office, US Army Missile Command, Redstone Arsenal, Ala. (20 July 1979), p. 169.

15. Letter from J. H. Hinrichs to A. Schomburg, 2 February 1962, history office, US Army Missile Command, Redstone Arsenal, Ala. C. S. Porter, Army Res. Dev. Mag. **5**, 10 (December 1964). G. F. Smith, J. Quant. Electron. **QE-20**, 581 (1984). F. G. Britton, Ordnance **47**, 533 (March–April 1963). R. C. Benson, R. O. Godwin, M. R. Mirarchi, "A Single-Pulse Ruby Laser for Ranging Applications," ARO report, December 1962, acquisition number AD 332-016, Defense Technical Information Center, Alexandria, Va. M. R. Mirarchi, R. C. Benson, R.

MIRACL, or mid-infrared advanced chemical laser. This powerful chemical laser was developed by TRW for the Navy's high-energy laser program, Sea Lite, to provide a defense against antiship missiles. It was turned over to the Strategic Defense Initiative after the Sea Lite program, like other service laser programs, was canceled. In 1985, during a lethality test at White Sands Missile Range, MIRACL destroyed a Titan I missile casing. **Figure 5**

Green, "Initial Testing of an Experimental Battery Operated Laser Range Finder," Army Electronics R&D Activity Report TR 2352, acquisition number AD 336-587, Defense Technical Information Center, Alexandria, Va. (April 1963). R. G. Buser, Natl. Defense **61**, 114 (September–October 1976). Testimony of G. F. Smith, vice president, Hughes Aircraft Co, US Senate Committee on Commerce, Science and Transportation, *Laser Technology—Development and Applications*, 96th Congress, 1st and 2nd sessions, hearings before the Subcommittee on Science, Technology and Space, serial no. 96-106, 12 and 14 December 1979; 8 and 12 January 1980, US GPO, Washington, D. C. (1980), p. 99.

16. "Report of the Air Force Space Study Committee," 20 March 1961, reprinted in *History of the Air Research and Development Command, 1 January–31 March, and Air Force Systems Command, 1 April–30 June 1961*, volume 3, supporting documents, document 15, Air Force History Center, Maxwell AFB, Montgomery, Ala. E. O. Dixon, "Investigation and Development of a Sun-Powered Laser Transmitter," report TDR-62-447AOER-2-03 on contract AF-33-616-8025, October 1962, acquisition number AD 292-146, Defense Technical Information Center, Alexandria, Va. D. G. Clute, "LARIAT and the Air Force Avionics Laboratory Electrooptical Research Facility," technical report AFAL-TR-66-158 on projects AF-6263 and AF-5244, Air Force Avionics Laboratory, Wright–Patterson AFB, Dayton, Ohio (May 1966).

17. For criticism of the laser-guided bomb's performance in Vietnam, see C. E. Hunt, "Guided Bombs—Problems of Utility and Cost," Air Force Operations Analysis Paper 71-2, Air Force History Center, Maxwell AFB, Montgomery, Ala. (February 1971). For a more positive evaluation, see D. L. Ockerman, "An Analysis of Laser Guided Bombs in South East Asia," air operations report 73/4, Headquarters, 7th Air Force, Thailand, Tactical Analysis Division, available at Air Force History Center, Maxwell AFB, Montgomery, Ala. (28 June 1973). See also L. E. McKenney, "Pave Spike/Introduction/Combat Evaluation," report on project TAC-71B-237T, Tactical Air Warfare Center (June 1973), acquisition number AD 526-308L, Defense Technical Information Center, Alexandria, Va. G. W. Abraham, "Short Trip 29, B-52/Laser Guided Bomb Feasibility Demonstration," final report, Strategic Air Command, Omaha, Neb. (July 1973), acquisition number AD 526-345, Defense Technical Information Center, Alexandria, Va.

18. "Laser Target Illuminator System, Portion of Laser Designa-

tor System (Shedlight 73S) Pave Arrow," Air Force Advanced Technology Laboratory research and development management report HAF R-16, Air Force History Center, Maxwell AFB, Montgomery, Ala. (26 June 1969). A. W. Blizzard Jr, E. R. Johnson, J. F. McCormack, "Pave Arrow Laser Target Designator System," final report, August–December 1968, July 1969, acquisition number AD 507-034L, Defense Technical Information Center, Alexandria, Va. W. C. Eppers Jr, Natl. Defense **61**, 210 (November–December 1976).

19. D. R. Jones, Air Univ. Rev., November–December 1963, p. 44.

20. *Biography: Major General Donald L. Lamberson*, US Air Force, Office of Public Affairs (November 1983). D. Kyrazis, "ALL Departure from KAFB," presentation to the ceremony marking the departure of the Airborne Laser Laboratory from Kirtland AFB to the Air Force museum at Wright–Patterson AFB (May 1988), available from Air Force Weapons Laboratory, Kirtland AFB, Albuquerque, New Mexico. See also D. Kyrazis, J. W. Spidle, *Getting Started: The Beginnings of Air Force High Energy Laser Work*, history office, Air Force Weapons Laboratory, Kirtland AFB, Albuquerque, New Mexico (September 1960).

21. W. H. Culver, Science **126**, 810 (1957). "High Power Radiation Weapons," task order T-14, 17 February 1961, ARPA order file 306, Defense Advanced Research Projects Agency, Arlington, Va.

22. Letter from J. M. Bridges, director, ARPA office of electronics, to K. Brueckner, Institute for Defense Analysis, 13 October 1961. Letter from A. Weinstein to E. Fubini, 21 December 1961, ARPA order file 306, Defense Advanced Research Projects Agency, Arlington, Va.

23. Letter from W. H. Culver to A. Weinstein, 15 December 1961, Culver Papers, Optelecom Co, Gaithersburg, Md. Memorandum from J. P. Ruina, director, ARPA, and E. G. Fubini, deputy director, office of the deputy secretary of defense for research and engineering, to H. Brown, director, defense research and engineering, 19 January 1962, ARPA order file 306, Defense Advanced Research Projects Agency, Arlington, Va. A. Weinstein, "Memorandum for Steering Group on High Power Laser (Project Seaside): Minutes of First Meeting," 5 February 1962, ARPA Project Seaside, file 5 February 1962–31 December 1962, ARPA order files, Defense Advanced Research Projects Agency, Arlington, Va.

24. R. W. Seidel, Hist. Studies Phys. Bio. Sci. **18**, 111 (1987). ■

THE PHYSICIST AS MAD SCIENTIST

Deep-rooted forces have created a stereotype
of scientists: sometimes noble, but sometimes cold-blooded,
domineering and a danger to humanity.

Spencer Weart

PHYSICS TODAY/JUNE 1988

A crazed scientist with a deadly "atomic robot" set out to enslave the human race in a recent Saturday morning cartoon show. He was no exception: Unstable scientists plotting to master and destroy can be found almost anywhere one looks in children's television and comics—and in a surprising amount of adult fiction as well. Probably no other profession is so consistently drawn upon for storybook villains. To many people, then, the words "nuclear physicist" bring to mind a weird and evil picture.

The mad scientist stereotype—so ominously significant for the public image of science and the recruitment of future scientists—can be understood best through history. This figure stems from an ancient heritage, which was reshaped in surprising ways during the first half of the 20th century. Watching that process at work will reveal the powers that flow through the imagery, giving it enduring popularity.

The imagery of nuclear science began to form at the start of the 20th century, soon after the discovery of radioactivity. Announcements from laboratories had a remarkable ability to impress the public. The first impressive fact was simply that radium compounds glowed perpetually in the dark; nothing like that had been seen before. Then came the news that radioactivity accompanied the transmutation of elements, and that this process released a quantity of energy that, atom for atom, vastly exceeded any known before. Meanwhile scientists announced that radioactivity affected living creatures. Pierre Curie killed a mouse with a dab of radium, and impressed reporters by remarking that he would not care to share a room with a kilogram of the element. In the

hands of criminals, he declared, it could be a great danger. However, Pierre and Marie Curie and other scientists followed up such warnings with reassurances that the problem was not radioactivity itself but only the chance of misuse. In the hands of experts who took proper precautions, the prodigious power of radioactivity would be all to the good. It could be used immediately to cure diseases, and perhaps someday to turn the wheels of industry.

These ideas, remarkable as they were, did not suffice for some publicists. When radium glowed in the dark, newspapers spoke as if the rays were a magical force. The ability of radioactivity to reduce some types of cancer led a few scientists to suggest it might cure almost any other disease, and journalists went on to speak of arresting the symptoms of old age or even granting immortality. Frederick Soddy, the chemist who along with Ernest Rutherford had discovered atomic transmutation, told the public that this discovery was the goal denied to ancient alchemists. Radioactivity might be the new philosophers' stone, he said, capable not only of creating limitless wealth but of acting as the elixir of life. Soon atomic scientists like Rutherford, Soddy and the Curies were hailed as new alchemists, indeed far greater than the old ones. It became common for admiring reporters to call them "wizards" of the laboratory. This was a plain hint about the historical traditions that bore on the way people saw scientists.

The scientist as sorcerer

The public image of the scientist partly evolved out of ideas about wizards. Here was an impressive figure, known to all from early childhood, reaching back through ancient sorcery legends to prehistoric shamans. It was a figure distrusted by the educated and the uneducated alike. Such an individual might release pestilence and other evils, as tribal witches supposedly did, or might unleash demons, as medieval sorcerers supposedly did, or might simply propagate heretical ideas, as some proto-

Spencer Weart directs the Center for History of Physics at the American Institute of Physics, in New York. This article is based in part on his recent book *Nuclear Fear: A History of Images* (Harvard U. P., Cambridge, Mass., 1988).

BRITISH LION

Nuclear scientist John Willingdon, in the 1950 thriller *Seven Days to Noon*, cracks under the strain of devising weapons. Running away with an atomic bomb in his satchel, he threatens to blow up London.

scientists in fact did. People have always feared the overweening individual whose evil thoughts or magical acts might endanger his neighbors.

The idea seemed to find confirmation in the lives of real people. Most notorious was Dr. Faust, a 16th-century itinerant rakehell who boasted that he controlled demonic powers. Pious churchmen mixed Faust's actual career together with legends of black magicians; their aim was to issue an impressive warning against the rising breed of humanist skeptics and scientists who were not guided by established Christian belief. Many other quasiscientists stirred up the same feelings. Of special note was Franz Mesmer, a physician and hypnotist who in the late 18th century attracted tens of thousands of disciples with his ability to work cures. He claimed that his powers were based on his discoveries in "magnetic" science, but medical authorities worried that his cult threatened the social order.

Through the 19th century both Faustian tales and Mesmeric cults grew in popularity. A stereotype was being refashioned. Popular authors from hack newspaper writers to Nathaniel Hawthorne modified the old tales of witch and sorcerer to create a new fictional figure, the Mesmeric "scientist" who endangered himself and those around him with a mixture of demonic and scientific powers. The stereotype was widely seen. For example, at the time radioactivity was discovered, America's best-selling book and most popular touring stage show featured

the scientific villain Svengali, dominating his victim with hypnotic rays.

Such figures were usually associated with surface characteristics of the legendary alchemist and sorcerer, such as laboratory glassware, but also with more essential features. Exactly like the traditional sorcerer, the new figure was remote from everyday life, even avoiding women's love in his obsessive devotion to the search for dangerous secrets; he boasted an impious pride, and he seized powers over life and death.

At first such evildoers threatened only a particular heroine, not the entire human race. But as technology altered entire landscapes, and sometimes for the worse, opinions shifted. The scientist–sorcerer evolved into a scientist–inventor, who brought dangers wholesale. Jules Verne did more than anyone to develop the new stereotype. A typical Verne novel, *For the Flag* (1896), told of an insane chemist who had invented a mighty explosive; on an island hideout he used the invention to destroy an attacking fleet and then to blow up the island and himself with it. The premises seemed plausible: Verne had apparently based his fictional character on a living inventor of explosives, who promptly sued for libel.

It was atomic physics, however, that offered for the first time the scientific possibility of doom on a universal scale. Around 1903 Rutherford and Soddy publicly speculated that radioactivity might be contagious, passing from atom to atom. Might a form of radioactivity be

Dr. Victor Frankenstein, in the classic 1931 movie. Mary Wollstonecraft Shelley's character became the archetype of the seeker after scientific truth whose overweening curiosity leads to disaster.

released in the laboratory that would spread outward in a chain reaction, converting the whole planet to gas? The press quickly brought this spectacular idea to the world's attention. It remained in public awareness as, from time to time, one or another eminent scientist repeated it. In 1921, for example, the chemist Walther Nernst told a German radio audience that humanity was like a primitive race living on an island of gunpowder, a race to whom no Prometheus had yet brought the perilous gift of fire. Hendrik A. Kramers's well-known 1923 textbook on atoms carried the thought a step further, wondering whether the novas seen in the sky might be outbursts of atomic energy "brought about perhaps by the 'super-wisdom' of the unlucky inhabitants themselves." By the 1930s even schoolchildren had heard about the risk that an unwary experimenter might blow up the world.

Still more chilling was the thought that someone might bring a cataclysm on purpose. Bomb-throwing anarchists had become notorious in the late 19th century, and the public connected their powerful new explosives with science. More than one novel featured the cold-blooded chemical expert who, in the name of liberty, plotted to make the world a shambles with his infernal devices. In Anatole France's *Penguin Island* of 1908 the evil chemists were replaced by physics-minded terrorists who destroyed entire cities with pocket-sized atomic explosives.

The dangerous scientist was usually shown as more or less mad. For example, a 1938 adventure novel, *The Doomsday Men*, featured a fanatic scientist who declared that life is only suffering, so that it would be best to end the world forthwith. He prepared to set off an atomic reaction that would peel the skin off the Earth "like an orange, only faster."

The unstable nuclear physicist was most convincingly portrayed in *Wings over Europe*, a play that enjoyed success with the critics, a modest run on the stage in

London and New York beginning in 1928, and frequent revivals among college theater groups during the 1930s. Center stage was held by a youthful scientist, brilliant but a bit unbalanced. He announced to the British Cabinet that he had found the secret of releasing atomic energy, the power to make a golden age. But the ministers only dithered about what transmutation might do to the gold standard and took an unwholesome interest in the prospects for new weapons. With his naive idealism trampled, the scientist lost his bearings, decided the world was wicked and got ready to blow it to dust.

Of course this was not the only way scientists were pictured. Throughout the first half of the 20th century the majority of writings showed scientists in a positive light, as noble geniuses working for the good of humanity. According to many enthusiastic magazine articles and advertisements, radioactivity and all other things scientific were far more likely to bring good than harm. Yet nuclear physics increasingly evoked fears along with hopes. Even in the new pulp science fiction magazines, which seemed firmly committed to a vision of scientific wonders, the heroic atomic inventor sometimes had to share space with an evil one. Pressure from somewhere was keeping an ugly side present in the public image of scientists.

The scientist as tyrant

The greatest blow to public confidence in progress had been the First World War. From submarine warfare to the production of explosives and poison gas, science and technology had obviously played a crucial role in the slaughter. To make sure that was obvious, after the war scientists hastened to inform the public of what they had done, and argued that without advanced science a nation must perish. The argument was designed to raise the funding and prestige of science, and it did that; yet it could also prompt the public less to admire scientists than to fear them. Imagery is influenced by real situations, and so

Origin of Dr. Doom. "Forbidden experiments" by a university science student lead to his transformation into an evil genius who seeks to become "the master of all mankind" in this comic book tale. (Selected panels from *Fantastic Four Annual*, 1964; reprinted in S. Lee, *Bring on the Bad Guys*, Simon and Schuster, New York, 1976.)

long as scientists devised weapons, the public had sound reason for concern.

While some felt new misgivings about how science might alter warfare, others insisted that science was not the problem but the solution. Many believed that by bringing universal prosperity—for example, by harnessing atomic energy—scientists would eventually remove the economic causes of war. Others exclaimed that new weapons, such as the atomic bombs that some already foresaw, would make war so dreadful that it would become unthinkable. Or scientists might reform the world through pure moral example. Had they not already formed their own international community, peaceful, cooperative and dedicated to reason?

A number of young scientists and science journalists, mostly on the political left, declared that the proper way to reshape society was to give a greater role to scientifically trained people—that is, to people like themselves. They insisted that science was languishing under the thumb of capitalists who used technology only to enrich themselves. Let affairs instead be efficiently organized by science-minded people, they said, and progress would be amazingly swift; atomic energy was only one of many wonders they promised. The play *Wings over Europe* carried the idea to its logical conclusion. At the end a group of scientists who had learned the secret of atomic energy came soaring over the world's cities in huge green airplanes, bearing atomic bombs, to enforce the will of a "League of United Scientists of the World."

Not everyone agreed that science should call the tune. In the 1920s and early 1930s, popular German writers argued that heartless scientific analysis was separating people from the intuitively grasped wholeness of life. Essays in American literary magazines declared that because science by definition had nothing to do with such ideas as beauty and morality and holiness, it was worse than useless in dealing with human problems. A leader of

the French Senate complained that technological breakthroughs had upset industry and brought on the Great Depression. A prominent British bishop suggested that there should be a ten-year moratorium on research to give society time to adjust.

It was an old debate, but world war and depression had raised it to a fever pitch. On one side, a few outspoken men proclaimed that the situation must be rectified by honoring scientists and engineers above others, reorganizing society to conform to their ideals. The people who were used to shaping society's ideals—clergymen, humanists, popular writers and the like—could well feel that their world was threatened. These were the groups who in fact spread talk about blasphemous scientists.

Atomic energy, universally acknowledged as the most powerful and mysterious of all scientific powers, was especially apt to be dragged into criticism of science. An example of critics who invoked the atom was Raymond B. Fosdick, an idealistic American lawyer. In a 1928 book, *The Old Savage in the New Civilization*, Fosdick repeated the warnings he had heard from scientists about the risk of atomic bombs and world doom. He explained that savage humanity, like a child playing with matches, could scarcely cope with the powers of technology. We needed to give less heed to scientists, and more to humanists who would boldly question society—people like himself.

The theme showed up in popular culture on a more

© 1943 LOEW'S INCORPORATED RE 1970 METRO-GOLDWYN-MAYER INC

AIP NIELS BOHR LIBRARY

Marie Curie in the romantic 1944 movie *Madame Curie* (above) and in reality (left). The movie showed her as a self-sacrificing heroine, utterly devoted to her work with mysterious and perilous radioactive substances. The real Curie, seen here in a famous photograph, cooperated with journalists who portrayed the scientist as an extraordinary being, remote from mundane concerns.

use of advanced science was becoming a key attribute of the totalitarian nightmare.

The scientist as monster

Alongside this social theme lurked a more personal one. This too became most obvious in popular movies, for example, the 1940 horror film *Dr. Cyclops*. The title character was a scientist who planned to improve the world with astonishing new radium rays, even though a colleague warned him, "You are tampering with powers reserved for God!" When a group of visitors discovered his secret, the scientist used his ray to shrink them to the size of mice; the rest of the story showed a war to the death between the scientist and his miniature victims. Here was no dictator of the world but a tyrant in a single house. The scientist's character—superficially well-meaning but in the end cruelly domineering—corresponded to a widespread human problem. From childhood conflicts with parents into adulthood, everyone had to deal with relationships with authorities. Often the authority seemed to be dangerously knowing, powerful, and heedless of the wishes of others. Indeed the secret desire to master and even to harm people was something everyone had detected in others, if not in themselves. These were threats that the stereotypical scientist, remote in his laboratory with his mysterious knowledge and powers, was especially apt to symbolize.

The emphasis on secrets points to another primitive theme. Tales about dangerous secrets are found among native peoples around the world, while Western culture has Adam and Eve, Prometheus, Bluebeard's wife, the

primitive level, in the full meaning of "primitive": archaic, archetypal and crude. The arrogant tyrant who sought to rule the world was a stereotype as old as civilization, and it was now increasingly associated with science. Especially memorable were American movie serials. Gene Autry descended into an underground atom-powered city to battle an evil prince armed with radium rays; Crash Corrigan overcame an undersea ruler who planned to enslave the world using "the atom—the most destructive force known to science"; and the incomparable Flash Gordon sabotaged an "atom furnace" to foil his archenemy Ming, who had boasted, "Radioactivity will make me Emperor of the Universe!" To be sure, Ming and his ilk might not always be scientists themselves. But the sorcerer's apprentice and a thousand more who came to grief by grasping after forbidden knowledge. This partly reflects universal experience, in which children are warned away from adult secrets, threatened with punishment if they look where they should not. It also reflects the plain fact that unknown things can indeed be dangerous. Here was another strand in the tangled web of anxieties that surrounded the public image of scientists, whose job is precisely to venture into the unknown.

The most primitive theme of all was the form that the danger often took: creation of a monster. The weird and dreadful creature had an old heritage, going back to sorcerers' golems and witches' familiars, but in 20th-century tales it was usually associated with science—and often nuclear science. For example, in 1931 after Arthur Compton announced that he might be on the way to releasing atomic energy, one citizen wrote an anxious letter to *The New York Times*. Might not the atom get out of hand, asked the writer, and "turn into a 'Golem' which could destroy man?" Fosdick put the new cliché more generally: Would all of technological civilization become "a Frankenstein monster that will slay its own maker?"

This monster was not a simple symbol. Scholars who looked into the many versions of *Frankenstein* noticed that the authors tended to conjoin the dreadful being with the scientist. The scientist in the first version, Mary Wollstonecraft Shelley's novel of 1817, himself cried that his creature was "my own spirit let loose from the grave." Not without reason did millions of people from the early 19th century on get confused over whether "Frankenstein" was the name of the scientist or the monster; the weird being put into action the scientist's own secret desire to master and punish.

The division of the dangerous character into two parts, a Dr. Jekyll and a Mr. Hyde, reflected a Western cultural tradition. In the 19th and 20th centuries particularly, many people spoke as if humans were split into halves, with a strict and rational side keeping precarious control over murky urges. This idea was central to stories where fictional scientists tried to suppress their emotions and concentrate on research, only to find their evil desires set loose in their creatures. Audiences brought up on this tradition, watching a Dr. Cyclops scoff at human feelings, were not surprised when he became uncontrollably tyrannical and murderous. That was the mad scientist's madness.

The dilemma was most clearly revealed in a 1936 movie. *The Invisible Ray* starred Boris Karloff in a switch from his famous role as Frankenstein's monster. Now he was the scientific genius Dr. Rukh, who built a radium-ray projector with the standard abilities to smash cities and work miraculous cures. But Rukh caught a sort of contagious radioactivity, so that he glowed in the dark and could kill with his touch. Meanwhile he was so dedicated to his work that he ignored his young wife, and she left him, whereupon he set forth on a murderous rampage. At the climax Rukh's mother smashed a vial of antidote he required to stay alive, and he burst into flames, consumed by radioactivity.

The studio claimed that all this showed scientific theory that might come true, but in fact there was not a scrap of reality in any of the images. The images came as usual from anxieties about "secrets we are not meant to probe" (as the mother had warned). Only now the punishment for probing was that the scientist himself became an uncanny creature, rejected by everyone he loved, dying in hellfire. The idea was summarized by a homicidal Professor Radium in a 1941 *Batman* comic that used the movie's ideas: "I have made myself a monster!" It was nothing less than the monster that any of us might

Mme JOLLIOT-CURIE est radieuse !

French political cartoon, "Madame Joliot-Curie is radiant." When the famous nuclear scientist joined the French government in 1936 as under secretary for science, a political cartoonist showed her dealing with her party's enemies in mad-scientist fashion.

fear we would become if our worst impulses took over. At the same time—for powerful symbols work on many levels at once—it was the dire punishment that traditionally threatens those who look too deeply into forbidden secrets.

The self-sacrificing scientist

Scientists of the 1930s did not analyze all the forces that were lining up against them, but they did take note of the attacks that were expressed in intellectual language. For example, Fosdick's critiques prompted a public counterattack by the dean of American atomic scientists, Robert Millikan. Science had done far more good than harm in his lifetime, he pointed out, and could be counted on to do far more. Going straight to the most potent image, Millikan scoffed at the idea of a scientist, "like a bad small boy," blowing up the world by accident.

Millikan and some others, such as Rutherford, tried to quiet not only doomsday fears but also utopian dreams about the powers of atomic science. That did not stop the newspaper reporters. After all, prominent physicists in England, Italy, France, Germany, the United States and the Soviet Union all said publicly that an astonishing atomic revolution might begin at any time. Some also continued to warn about dangers. For example, Frédéric Joliot, in his acceptance speech for the 1935 Nobel Prize he and his wife Irène Curie had won for the discovery of artificial transmutation, used the occasion to warn scientists to take care lest they make the planet explode like a nova. Yet Joliot also told the public of the wonders that atomic energy might bring, and he worked to raise funds for his laboratory so nuclear research could push ahead full speed. He never doubted that over the long run knowledge was better than ignorance.

Not even humanist critics at heart denied the value of science. For example, by 1939 Fosdick was a leader of the Rockefeller Foundation, and when Ernest Lawrence requested money for a giant cyclotron, Fosdick enthusiastically voted in favor. He saw the device as a tool of pure, disinterested science, an emblem of "the noblest expression of the human spirit." For the public at large, up to 1939 the image of science was still warped less in the direction of fear than of unreasonable aspirations. Although there were scientist villains in science fiction, they were outnumbered by scientist heroes by two or three to one. The ordinary mass fiction of magazines and books, when it showed scientists at all, portrayed them as clean-cut and useful characters. Nonfiction magazine and newspaper articles likewise had little but praise. It did not seem incongruous when Millikan told the public that scientists offered an example of right conduct that could improve religion and public life. After all, in his own life he devoted himself visibly and indefatigably to the search for truth and the advancement of civilization.

A popular 1944 movie, *Madame Curie*, captured the majority image of scientists as neatly as a butterfly on a pin. The young Pierre and Marie Curie sought no worldly power or material reward for themselves, but only the pure light of truth. In a central scene they bent over a dish containing their first radium, faces bathed in its soft glow, enraptured by the powers for good that they had discovered.

Yet the film also suggested that there was something odd about scientists. It emphasized how Marie worked herself to exhaustion with superhuman dedication, how she brushed aside the risk of cancer from the radium, seeking to master the mysterious forces of life and death. Indeed neither Marie nor Pierre Curie could easily show any human emotion. In the impressive closing scenes, when Marie learned that Pierre had died, for days she remained speechless and staring at nothing, frozen, unable to mourn. In historical fact, when the

real Marie's friends took her to her husband's corpse she kissed his cold face passionately, then clung to his body, had to be dragged from the room, and wept like any suffering human. However, that was not for the movie-going public. Why not?

The image of the real scientist had much in common with the stereotype of the evil one. To become a real scientist meant years of intense study and often frustrating research, requiring unusual abnegation. That was a main theme, for example, of Sinclair Lewis's *Arrowsmith*, a 1925 novel that nevertheless helped inspire more than one young reader to take the arduous path of science. Some scientists went on to claim that science could progress only by seeing the world in impersonal, emotionless terms. Few groups were as apt as scientists, then, to stand for suppression of human feelings. Of course, real scientists were often warm and gregarious types, much appreciated as colleagues and teachers. But to newspaper and magazine writers, the scientist was an odd character who ignored mundane concerns, risking his health and scorning riches (as scientists themselves said), an unworldly "wizard" who isolated himself in the pursuit of tremendous secrets. In the movies, Dr. Rukh and Dr. Curie both worked on radium day and night with enormous endurance and rigidly suppressed feelings. In short, the scientist was portrayed rather like the atom itself: prodigious energy locked up like a compressed spring.

Real nuclear scientists could become confused with the fictional ones. For example, in 1936 when Irène Curie entered French politics, lending her prestige to a party that promised to increase funding for research, a satirical cartoon showed her striding coldly among her party's enemies, blasting them with death rays from her fingertips. The bad scientist was one who let desire for power get out of hand; the good scientist was one who maintained self-control and renounced social ambition. Scientists tended to agree with that. Curie herself soon fled politics,

Who Decided to Build Atomic Bombs—Scientists or Politicians?

What could the public, who in a democracy are supposed to control crucial affairs through their representatives, have had to do with the decision to build atomic bombs? It is sometimes claimed that such decisions are determined by cold-blooded scientists, remote from ordinary people; that is one reason for misgivings about scientists. But in fact nonscientists of the early 1940s were not so ignorant and impotent as this claim suggests.

The most crucial decision came in 1941, after British physicists realized that atomic bombs could be built within a few years. The final decision on what to do rested with the government, for the problem worked its way up through committees to the desk of Winston Churchill. Churchill addressed the question with the same display of good cheer he had shown in the face of the German bombardment of Britain. "Although personally I am quite content with existing explosives," he told his chiefs of staff, "I feel we must not stand in the path of improvement." The British accordingly set up a large atomic project, and their resolve spread across the Atlantic, playing an important role in pushing the Americans to set up their own project.

Churchill held his position as wartime Prime Minister not least because he was one of the best-qualified statesmen in

the world to contemplate strategic bombardment. His taste for building and using new weapons was well known. Already in 1914, as First Lord of the Admiralty, he had ordered the first preemptive strike in the history of air warfare, a bomber raid on German zeppelins that were preparing to attack London. His colleagues in Parliament, and the public, could easily have predicted how he would view atomic bombs.

Churchill had indeed been publicly associated with precisely those weapons years earlier—in fiction. A novel published in 1932, Harold Nicolson's *Public Faces*, described a pacifist British Cabinet faced with the invention of atomic bombs in 1939. The novel's politicians recognized that such a weapon could devastate a city at one blow, and dithered over what to do. "It would be impious," exclaimed one trembling liberal, "for us to dabble in the Satanic potentialities." Nothing made the Cabinet so nervous as the thought that the man who was then the opposition leader—Churchill—might learn of the weapons. No politically aware person could doubt that he would insist that they be built, and used, to keep Britain on top. It was because of such views that in the real year 1941 Churchill was in a position to make the crucial decision.

The scientist among generals. In the 1964 movie *Fail Safe*, a physicist advised that the United States launch an all-out bombing attack on Russia. Scientists were the villains of this movie and the book on which it was based.

telling reporters that she preferred the calm of her laboratory.

While all the ideas and myths discussed above were generally known, that does not mean they seemed important to educated people in the 1930s. The most impressive images appeared chiefly in pulp magazine stories and horror movies, genres that no doubt affected youths but seemed of little concern to sensible adults. Nevertheless even the most bizarre tales had something to say about reality. Two years after Fosdick praised Lawrence's cyclotron as an emblem of the most pure disinterested research, parts of the device were converted into a pilot plant to make material for atomic bombs.

The scientist as spy

When the United States dropped atomic bombs on Hiroshima and Nagasaki, the news had no immediate impact on the shape of the stereotype of nuclear scientists. Rather, the stereotype already formed leaped into new prominence, its emotional force redoubled. In radio broadcasts within hours of the Hiroshima bombing, pundits spoke breathlessly of uncanny forces, cosmic secrets and doomsday; the name of Frankenstein was invoked everywhere from street corners to the US Senate. Previous talk about nuclear scientists, whether heroes or villains, had won little serious attention or credence. From August 1945 on, the whole tangle of imagery was an important part of the mental equipment of everyone within reach of a radio. This public response was not unreasonable, for the real existence of nuclear weapons gave fears a new validity.

Some groups were not satisfied to let feelings of awe and horror develop by themselves, but for their own purposes worked to shape public imagery. Nuclear scientists were first to launch a campaign, and with the glare of atomic bombs as a limelight, they had unprecedented visibility. Physicists were asked to attend countless private dinners, club gatherings and government meetings, to write articles for all sorts of magazines, to speak everywhere from the "Quiz Kids" radio show to the White House. They used the opportunity to inform the public of the astonishing prospects now near at hand, but that was not all. To encourage international control of nuclear weapons, some warned about the end of civilization or even the entire human race. Others spoke of how nuclear research could turn deserts into gardens—provided of course that the research of scientists like themselves was adequately supported.

To most people, all this meant the old idea of mastering cosmic forces. A sociologist who studied the 1946 Congressional hearings on nuclear energy concluded that even senators looked upon atomic scientists much the way earlier people looked upon wizards, as beings who had seized control over awesome, supernatural forces. Scientists had never before enjoyed such high prestige.

But the wizard was a master of forbidden secrets, and that old idea spread independently, indeed in spite of, the atomic scientists. In *The New York Times*'s index of the articles it published in the autumn of 1945, of all the articles about atomic energy, roughly two-thirds were largely about international or other "control"; of these, nearly half were largely about "secrecy." A collection of mid-1947 digests of American network radio comments on atomic energy shows the same preoccupations in roughly

Transformation of nuclear scientist into a raging, uncontrollable, mountain of muscle known as the Hulk. In the contemporary comic book series, Dr. Bruce Banner designs ``the most awesome weapon ever created by man: the incredible *gamma bomb*,'' and is caught in its explosion; its radiation turns him into a Jekyll-and-Hyde monster. (Top frame from *The Incredible Hulk* number 314, December 1985; middle and bottom frames from *The Rampaging Hulk* number 1, January 1977.)

the same proportions. Even President Truman thought that his government had exclusive ownership of a cosmic mystery. Countless newspaper columns, radio shows, magazine stories and entire novels were written about "the secret" of atomic bombs, as if it were a formula on a piece of paper in a safe somewhere.

Congress passed a law that would impose the death penalty for revealing atomic secrets; in principle, a scientist who told a friend even the results of pure research done privately at home might be legally executed. Spy stories in the news, such as the treason of Klaus Fuchs and the trial of Julius and Ethel Rosenberg, kept the pot boiling. The concern about loss of secrets harmed scientists more than anyone. Although not one American atomic scientist was ever shown to have turned traitor, no group was more closely inspected; physicists and mathematicians numbered more than half of the people who were identified as Communists in Congressional hearings. Hundreds of scientists were mercilessly pursued, often losing their jobs, some ending their lives in exile or suicide.

American scientists came to accept something so undemocratic that they would have found it unthinkable a decade earlier—a widespread peacetime system of guards and fences, locked safes, visitors making detailed inquiries about the personal lives of friends and plain spying. This was strongest in the new Atomic Energy Commission. By the end of the 1950s the government had investigated in detail some 150 000 people in connection with AEC employment. The system spread into many sections of government, industry and even the universities.

The same passion for controlling secrets beset every other nation that planned a nuclear energy program. In the Soviet Union, for example, fission research was put under the command of none other than the secret-police chief, Lavrenty Beria. Soviet nuclear laboratories were fenced off as a separate part of Beria's empire of prisons, with captive scientists working alongside free ones, although the latter had scarcely more choice. At these hidden installations, as at Los Alamos, fences and guards restricted the activities of everyone including the project leaders, but in Siberia it was plain that the fences were less to keep spies out than to keep scientists in.

Fiction drove the idea home. Many Soviet and American movies of the 1950s featured spies and traitors, while children's television shows such as "The Atom Squad" brought similar stories into the living room. Here was a new popular stereotype: the traitor who endangered people by grasping scientific secrets. And that meant atomic secrets more often than every other kind put together.

The stories became one more force promoting mythical imagery. For example, in each of three American films of the 1950s where spies tried to kidnap a man for his atomic secrets, the man in question had been in an accident that made him weirdly radioactive, like Karloff's glowing mad scientist. In yet another spy film the camera panned along a Los Alamos fence from a sign reading "Contaminated Area" to another reading "Restricted Area"—the peril from radioactivity merging with the peril from those who sought to know forbidden things.

Scientists had always been seen as queer, single-minded, powerful beings working outside normal society, and the stereotype was repeated in journalists' stories of atomic scientists. Whether dedicated and brilliant Manhattan Project workers or dedicated and brilliant traitors, they seemed only too inclined to secretly inflict violent change upon society. People put their finger on the emerging pattern when they likened the spy craze to a "witch hunt." Sometimes talk descended to a level even more primitive than the stereotype of an antisocial sorcerer. A woman wrote a United States Senator begging him to "try and stop those crazy scientists . . . from playing with those atomic bombs before we are all blown to bits. They act like children with a new toy." This was no new idea, but precisely what Millikan had called the "bad small boy" who endangered the world.

A sophisticated British film of 1950, *Seven Days to Noon*, featured such a scientist. Dr. John Willingdon, a childishly naive physicist who helped make atomic bombs, began brooding over them until his mind cracked. Stealing away a bomb in a satchel, Willingdon announced that he would blow himself up along with all London unless Britain promised to renounce nuclear weapons. Despite the crazed-genius stereotype, in its overall atmosphere the film resembled nothing so much as an ordinary police thriller. The mad scientist had stepped into the world of realism.

In the wake of the creation of atomic bombs, anxiety about nuclear scientists had become more than ever a condensed way of symbolizing the dangers of science and technology in general. It also became more than ever a symbol for the cruelest secrets of the heart: prying into forbidden secrets; the drive to master others; and the treacherous urge to destroy, like Willingdon, even one's own city.

These associations did not come of themselves but were forged by individual people. The ordinary citizen who repeated mythical themes was using familiar ideas in an attempt to grasp the news from Hiroshima. Others used imagery more deliberately to impress others, although the results were not always as intended. Talk about wondrous atomic powers and scientist–magicians could encourage people to follow the lead of scientists—or to fear them and bring them to heel.

By the end of the 1950s if not earlier, the second tendency had won. After all, it was reasonable to take the fearful hostility provoked by nuclear weapons and displace it onto those who might someday use the bombs, or who aided those who would use them or had used them already. The doubts raised by the outbreak of the First World War were driven home by the end of the Second, and were kept in the forefront by the unending threat contained in the cold war.

Does the history of this odd stereotype tell us anything about how to correct it? One thing is plain: The public's image of the scientist draws on deep and numerous roots, making it hard to manipulate. A particularly strong influence is the real situation: So long as people are bewildered and threatened by the advance of technology—above all, so long as there are many nuclear weapons poised to strike them—the public cannot be expected to view either technology or science with unmingled admiration. Therefore the surest way to improve the image of scientists is to make genuine improvements in the ways society puts science to work, beginning with weapons.

While scientists undertake that long task, they can also attack the stereotype directly, but only with care. Efforts to present scientists as dedicated wonder-workers have proven all too likely to advance the stereotype of the sorcerer. More modest approaches are better. First, there is no reason that scientists, like other groups who have been unfairly stereotyped, should not complain directly on every occasion when the mad scientist appears in the mass media. And second, work should continue to tell the history of real scientists—showing them as people working to improve civilization but not to seize personal control over it, people devoted to their research but not contemptuous of normal human feelings, people pursuing knowledge but not aiming to master cosmic secrets of utopia and doomsday. ∎

MAKING WEAPONS, TALKING PEACE

A nuclear physicist and adviser to four Presidents, the author reflects on the development of nuclear weapons, the creation of a new lab to rival Los Alamos and the negotiation of the elusive comprehensive test ban treaty.

Herbert F. York

PHYSICS TODAY/APRIL 1988

Herbert F. York is Director of the Institute on Global Conflict and Cooperation at the University of California, San Diego. This article is adapted from York's book, *Making Weapons, Talking Peace: A Physicist's Odyssey from Hiroshima to Geneva.* Copyright © 1987 by Herbert F. York. Reprinted by permission of Basic Books Inc, Publishers, New York.

My earliest memories of public affairs include the election of Herbert Hoover, the Great Depression, the election of Franklin Roosevelt, the rise of Adolf Hitler, the Japanese invasion of China, and the Italian invasion of Ethiopia—all of which took place while I was still in grade school. They continue with the Spanish civil war, which occurred when I was in high school. The European branch of World War II started in the same month that I entered college at the University of Rochester.

Soon after that Pearl Harbor came along, and my professors began to leave in order to contribute to the war effort. I eventually joined them by going off to Berkeley and the Manhattan Project.

When the war ended, I was demobilized, so to speak, along with tens of millions of others all over the world, and I began what I hoped and thought was going to be a normal peaceful career in pure science. It was not to be. After only three and a half years, major external events, including the explosion of the first Soviet atomic bomb and the Korean War, brought me back into the nuclear arms race, and my life has been largely caught up in it ever since. This is a chronicle of my involvement, beginning with my work on the Hiroshima bomb and ending with my service as Jimmy Carter's chief negotiator at the Comprehensive Test Ban talks in Geneva.

It begins in November 1942, just before my 21st birthday, when a recruiter from the Berkeley Radiation Laboratory found me at the University of Rochester, where I was a first-year graduate student in the department of physics. Sidney Barnes, a Rochester physics professor who had gone to Berkeley in early 1942, had given my name to a recruiter, and so when he came our way later that year, he was looking for me specifically.

I could not have remained a graduate student any longer even if I had wanted to. Like most of my contemporaries, I wanted to do my part to win the war. I flirted briefly with other war projects—at MIT, Columbia and McGill—but the lure of Berkeley was by far the

Giant 184-inch magnet, used at Ernest O. Lawrence's Radiation Laboratory in Berkeley during the Manhattan Project for extracting uranium-235 from natural uranium, is shown just before its conversion to peacetime use in the fall of 1945. The man with the hat in the center of the photograph is W. B. Reynolds, business manager of the lab. To his right is the author, and to his left is Lawrence. Three other future Nobel Prize winners are also pictured: Luis W. Alvarez is between Reynolds and York; Emilio Segrè is among the men standing on the lower pole of the magnet, second from right; and Edwin McMillan is the man with a mustache standing on the floor to the left of the coatless man in the front row.

strongest. I had recently started doing research using a special machine called a cyclotron, which Ernest O. Lawrence, the director of the Berkeley laboratory, had invented, and the idea of one day going there and joining him had been forming in the back of my mind for some time.

Lawrence's Berkeley Radiation Laboratory was the only one of the major institutions making up the Manhattan Project that was actually founded by its director. All the other units were created at the time by official fiat, and all the other directors were appointed by higher authority. In marked contrast, Lawrence's lab was already more than a decade old when it joined the project to make the bomb. This unique situation gave Lawrence an extra measure of leverage in his dealings with higher government and university authorities both then and after the war.

Like their colleagues everywhere, the physicists at Berkeley immediately realized that the discovery of fission in 1938 was of transcendent importance. Lawrence's younger coworkers quickly confirmed the new discoveries and expanded on them. Among other things, they discovered and determined the properties of plutonium, one of the two materials suitable for the making of nuclear bombs.

Lawrence followed these events very closely, but he devoted his personal energies to three other activities. One of these was the promotion of his plans for still bigger cyclotrons. Another was to follow developments in the expanding studies of uranium fission and to make recommendations about what should be done in the area. The third was the conversion of his laboratory from one engaged in peacetime research to one dedicated entirely to the uranium project. He pushed hard to focus attention on the possibility of building a bomb, rather than just on that of generating power for ship propulsion or similar purposes; he argued strongly that the production of both U-235 and plutonium should be pursued in parallel; and he very early on urged that the project be greatly expanded to bring in the large number of scientists and to build the institutions necessary to accomplish the task quickly.

Lawrence got the responsibility for "small-sample preparation, electromagnetic separation methods and certain experimentation [on plutonium]." In anticipation of this assignment Lawrence had already disassembled one of his smaller cyclotrons in order to use its magnet as the basic building block for a device capable of producing microgram quantities of U-235. In addition, he diverted the much larger magnet originally intended for his next and biggest cyclotron to the construction of a still larger device, later dubbed the Calutron, for producing multigram quantities of the same material.

In the summer of 1942 the great 184-inch magnet was turned on and experimental prototypes of the Calutron, Lawrence's gigantic mass spectrometer, were placed in operation between its pole pieces. By that time much of Lawrence's senior staff had left the laboratory to help launch other war-related projects. They were always sent off with his blessing and, usually, at his urging. As a result, a largely new and mostly even younger group had to be recruited to carry out the work at Berkeley.

Accordingly, my first assignment was as a member of the crew of the R-1 Calutron, one of two prototypes sitting between the pole pieces of the 184-inch magnet. The cochiefs were Frank Oppenheimer (Robert Oppenheimer's younger brother) and Fred Schmidt. Our task was to study the operating characteristics of the R-1 and, by using the Edisonian cut-and-try approach, to maximize its ability to extract U-235 in as pure a form as possible from natural uranium.

Secrecy at the Rad Lab

The details of all this came as a complete surprise to me, even though I had been quite certain even before my arrival in Berkeley that the Radiation Laboratory was engaged in some way in making a nuclear bomb. Nobody had told me that in so many words, but the general idea of nuclear energy and nuclear bombs had been in the air in the early 1940s. Given Lawrence's reputation in nuclear physics, it was easy to put two and two together and conclude that Berkeley was somehow involved in making uranium bombs.

As clearly as I can recall, the word *uranium* was breathed in my ear only once after my arrival. From then on, code words were used for everything of special relevance to the project. Uranium itself was called tuballoy, a code name I later learned had been invented by no less than Winston Churchill. The three natural isotopes of uranium were called W (U-224), X (U-235) and Y (U-238). Mixtures of isotopes especially enriched in U-235 were called R, and the residue of material depleted in U-235 was called Q. The object of the whole enterprise was to produce R as fast as possible and to have it be as rich in X as we could make it.

At one point I told Lawrence I could do a better job if I could know the trade-off between quality and quantity of uranium. After his next trip to Los Alamos, he handed me a small slip of paper on which was written the critical mass of enriched uranium (in arbitrary units) for four or five different levels of U-235 concentration (all, of course, in numbers only, without any of those forbidden words). Because of the compartmentalization of information, I was the only junior person in our part of the project who had this information. I did make use of it in designing product collectors in order to optimize the rate at which the plant approached the production of a critical mass.

Equipment installation for the hydrogen bomb tests at Bikini Atoll is inspected by the author in 1954.

Ernest O. Lawrence (left) and the author were
both present for the thermonuclear weapons tests
at Bikini Atoll in 1954.

After the war ended, I convinced myself that my efforts, while not at all profound or even particularly clever, had had a small effect on the outcome. I had helped either to make the Hiroshima bomb available a few days earlier or to make its explosion a little more powerful. I knew, of course, that hundreds of others could make similar statements about hundreds of other small contributions, but I was pleased with what I had been able to do.

I do not recall anyone preaching to me or others about the need for secrecy. The need was explained just once—but firmly. Thereafter the whole atmosphere of the time and place strongly reinforced it. To my recollection, following my first day I never again heard the word *uranium* either in a normal conversation or in a confidential aside. This custom became deeply ingrained in me and everyone I knew. As a result, after news of the bomb burst upon the public two and half years later, it was deeply shocking for me to read that forbidden word in the headlines and to hear people utter it out loud—with a certain awe, to be sure, but nonetheless as if it were just another, normal word. Hearing about the bomb was one of those things that caused a sudden, queasy feeling in the pit of my stomach.

In recent years, in classes and special lectures, I've had many occasions to describe to younger people the project, the bomb and its use. I've found that at the start a very wide gap separates us. The first thing most of my listeners learn about World War II is that we won it. That is, so to speak, the last thing I learned about it. The first thing they learn about the atomic bomb is that we dropped one on Hiroshima and another on Nagasaki. That is the last thing I learned about the project. For most people born after 1940, those events marked the beginning of the

nuclear arms race with the Soviets. For those of us in the project, they heralded the end of history's bloodiest war.

The Soviet bomb

Two especially important US authorities—Vannevar Bush and General Leslie Groves, who had directed the Manhattan Project—estimated that achieving the bomb would take the "backward" Soviets much longer than it had taken us—decades at the very least. Nearly everyone else—veterans of the US project, professional intelligence analysts, even Winston Churchill—estimated that four years would be enough for the Soviets to end the American monopoly. In 1947 the intelligence estimates said the first Soviet atomic bomb would come in another two years—that is, in 1949—but at that point the estimating process got stuck in a rut. As more time passed, the estimate continued to be "two more years" rather than "in 1949." Thus, on the eve of the actual event the latest estimate was still "two more years." In the context of this static prediction, the Soviet explosion was, literally speaking, a surprise.

It was also an exceedingly ominous event, one widely seen as bringing serious new dangers of a kind totally different from any we had ever faced before. The Soviet atomic bomb, combined with the projected acquisition of very long-range aircraft, promised the end forever of our historical invulnerability. Almost immediately, serious concern over the possibility of a devastating surprise attack on us rose within nuclear and military—especially Air Force—circles.

Lawrence's reaction to the Soviet atomic bomb was easily predictable. Seven years before, he had led the call for an active response to the possibility of a German atomic bomb, and now he reacted similarly to the reality of the Soviet bomb. A new nuclear threat seemed to call for a nuclear response, and so he actively sought ways to reinvolve his laboratory in its development. Shortly after the explosion in Central Asia, Lawrence and Luis Alvarez set out on a trip to Washington to explore the question of what the Rad Lab might do. They traveled east by way of Los Alamos in order first to visit with the laboratory director, Norris Bradbury, as well as with Edward Teller, then a visitor to Los Alamos on leave from the University of Chicago, and others who, they believed, had to be similarly concerned about the new turn of events.

Teller and the super

Teller told them that the hydrogen bomb was the proper answer to the new challenge. This type of bomb, often also called the superbomb because its power was estimated to be perhaps a thousand times that of an "ordinary" fission bomb, had preoccupied Teller since the earliest days of the Los Alamos project. He had wanted to push ahead with it even then, but he was frustrated in this desire by Robert Oppenheimer's insistence that all efforts be concentrated on the fission bomb. Oppenheimer believed that a fission bomb was much more likely to be produced in time to be of use during the war. Besides, it would be needed as the trigger for a hydrogen bomb in any event. Teller had only grudgingly accepted Oppenheimer's conclusions during the war, but ever since its end he had been trying to find a way to get serious work going on his pet project. The challenge of the Soviet bomb seemed to provide the impetus that was previously lacking, and he used every means and argument he could think of to exploit it.

At the time it was still not known exactly how a hydrogen bomb might be constructed, only that in principle a few barrels of liquid deuterium, perhaps laced with tritium, would produce a prodigious amount of explosive energy if it could be heated to a high enough temperature for a long enough time—that is, to a temperature of many tens of millions of degrees for a fraction of a microsecond, quite a long time for such extreme temperatures and pressures.

Teller had never been satisfied with the management of the American nuclear weapons program or felt that the total effort devoted to it was adequate. His conflict with Oppenheimer during the war was notorious, as was his refusal to participate in the postwar continuation of the work at Los Alamos unless the government and the laboratory would both commit themselves to a substantial nuclear test program. In the fall of 1949, Teller had fought the General Advisory Committee (GAC) of the Atomic Energy Commission over the issue of whether the US should initiate a high-priority program to develop and build superbombs. He was even disappointed when, on 31 January 1950, President Truman ordered the AEC to "continue its efforts on...the so-called hydrogen or superbomb." Truman's use of the word *continue* rather than *charge ahead* had led Teller to feel—quite mistakenly—that his cause had been lost. Even after the Los Alamos laboratory had accelerated and expanded its program in response to the President's directives, Teller continued to believe that the laboratory leaders—Bradbury and Carson Mark, among others—were not doing enough to recruit new people and were not diverting a sufficient fraction of the laboratory's effort from fission bomb development to the search for the super.

In mid-1951 all this strife finally came to a head. That May the George shot had proved that a mixture of deuterium and tritium really would explode if the right initial conditions could be achieved. It showed that a superbomb could be built, and Los Alamos set about reorganizing itself to do so in the shortest possible time. Bradbury, surely with Washington's concurrence, decided to put longtime senior laboratory staffers, and not Teller, in charge of the program. In Bradbury's judgment, Teller was no manager, and now that the basic idea was in hand, people who could manage and orchestrate a complex effort involving many different technologies had to be put in charge. More important, Teller's strident complaints about the laboratory management had so poisoned the atmosphere that it was simply impossible to give him the kind of autonomous position within the laboratory that he was demanding.

For Teller this was the last straw. Soon after these new organizational arrangements were announced, he left Los Alamos and returned to the University of Chicago. His memoir on the subject made it clear that one of his several reasons for leaving was his desire to press his campaign for a second laboratory.

Teller took his message to friends on the Congressional Joint Committee on Atomic Energy. There he found an especially sympathetic listener not only in the chairman, Senator Brian McMahon, but also in the chief staff aide, William L. Borden—the latter being the man who three years later formally accused Oppenheimer of being a Soviet agent.

While Teller was lobbying in the defense establishment and in Congress, he was also doing what he could within the Washington headquarters of the AEC itself. He found some support for his ideas from Gordon Dean, who had recently succeeded David Lilienthal as chairman of the AEC, and from Thomas Murray, one of the newer

commissioners. Following the usual procedures in such matters, they referred the idea of a second laboratory to the GAC for consideration at its December 1950 meeting and again at several later meetings. Except for Willard F. Libby, a University of Chicago chemistry professor and personal friend of Teller's, the GAC members took a dim view of the idea of a new laboratory and recommended against establishing one. They were influenced in part by Bradbury, who said it wasn't needed and would inevitably divert essential support away from Los Alamos. In addition, many of the members of the GAC had previously found themselves in protracted conflicts with Teller over related issues, and that probably influenced their attitudes toward his latest proposal. A year later, in December 1951, after the GAC had made its third or fourth negative recommendation on the matter to the AEC commissioners, Murray put in the call to Lawrence that in turn prompted a New Year's Day query to me about my views on the matter.

I did not have enough information at hand to reach a firm conclusion, so I went off on a series of trips to sound out the situation. I visited Los Alamos to learn the views of my friends and colleagues there, and then I traveled on to Chicago to have a thorough discussion with Teller. The responses came out as one would suppose: The Los Alamos people felt they could do all that was required, and Teller was convinced they could not. I went on to Washington, where I met with the same AEC and Air Force officials Teller had seen. Everyone in the Air Force seemed to think that not enough was currently being done to exploit the recent breakthrough and that new institutional arrangements were needed to set things right.

I also visited John Wheeler at Princeton University. He had long been involved in nuclear weapons and the search for the super, and he had recently set up a small group of theoretical physicists, based in the nearby Forrestal Center, to help Los Alamos carry out some of the key calculations needed to convert the invention by Teller and Stan Ulam from a set of sketches and theories into a workable weapons system. I found Wheeler very much in agreement with Teller.

A second lab at Livermore

My early reports to Lawrence on the results of my travels confirmed his own preliminary conclusions that a second laboratory was needed. Combining his own prejudgments with what Teller and I had told him, Lawrence informed his friends at the AEC that he would, if they wished, take on the task of establishing an additional weapons research center at Livermore, California, as a branch of the Radiation Laboratory and that he could staff it, at least at first, largely with people already on the payroll. This proposal changed the situation radically. It clearly meant much less initial expense and an immediate, if small, cadre of people ready to go to work as needed. As Oppenheimer later recalled, the GAC and the AEC very quickly "approved the second laboratory as now conceived because there was an existing installation, and it could be done gradually and without harm to Los Alamos."

Lawrence asked me to draw up plans for the new research center. In response I began to sketch out my ideas of how to go about it. After a few weeks Lawrence asked me if I thought I could run it. After only an overnight hesitation I told him it was worth a try, and he simply instructed me to do so. It really was that casual— no search committee or any of the other procedures to which we are now accustomed. In keeping with his standard style of operation, Lawrence gave me no new title, no immediate raise in salary or any other change in status.

Eisenhower's Science Advisory Committee, at one of its first meetings in December 1957, listens to York, at blackboard. At tables (clockwise from far left) are Albert Hill, Detlev Bronk, Edwin Land, I. I. Rabi, Robert Bacher, James R. Killian, James Fisk, Jerome Wiesner, Jerrold Zacharias (at far right, top table), Emanuel Piore, James Doolittle, Lloyd Berkner, Hans Bethe (on far right), Edward M. Purcell, Hugh Dryden, Alan Waterman and George B. Kistiakowsky. The four men in the back to the right of Zacharias are unidentified.

Finally, and in close accord with Lawrence's views of the matter, the AEC in June 1952 approved the establishment of a branch of the Berkeley laboratory at Livermore to assist in the thermonuclear weapons program by conducting diagnostic experiments during weapons tests and performing other, related research. The question of how soon, or even whether, the Livermore Laboratory would actually engage directly in weapons development was left open, however. Teller was extremely dissatisfied with the vagueness of the AEC's plans for the new laboratory.

Finally, after mulling the matter over, Teller, in the course of a well-lubricated reception held at the Claremont Hotel in Berkeley in early July to celebrate the launching of the new enterprise, suddenly announced to Lawrence, Gordon Dean and me that he would have nothing further to do with the plans for establishing a laboratory at Livermore. Lawrence was prepared to go ahead anyway, and he even suggested privately to me that we would probably be better off without Teller. However, at the insistence of Captain John T. Hayward (then deputy director of the AEC's Division of Military Applications), intense negotations were resumed among all concerned. Within days this led to a firm commitment by Dean that thermonuclear weapons development would be included in the Livermore program from the outset, as well as to a renewed commitment on the part of Teller to join the laboratory.

Mike the mighty

On 2 November 1952, the world's first large thermonuclear device, codenamed Mike, exploded on Elugelab Island at Eniwetok Atoll. Its yield was 10.5 megatons, fulfilling the prediction that the superbomb, if it could be made, would be a thousand times more powerful than the bomb dropped on Hiroshima. On the basis of ideas by Teller, Ulam and others, it was built and tested by the Los Alamos Laboratory. Because his relations with the Los Alamos leaders were so severely strained, Teller did not accompany the lab's team to observe it.

AEC authorities, on instructions from the White House, clamped an extraordinarily tight curtain of security over the whole operation. They intended to allow no post-test reports to be sent from the Pacific back to the laboratories or anywhere else until after there had been an on-the-spot analysis of what had happened. Even then, the first word would go directly to Washington only. The Task Force Command did, however, broadcast a coded signal that indicated the moment when the button had been pushed. Because of some experiments on long-range effects that we were doing at Livermore, we were given the means for decoding that message. The moment I received it at my office at the lab, I noted the time and telephoned Teller, then standing by at the Berkeley seismometer, to tell him just when "zero hour" had passed. He kept a very close watch on the seismometer, and at the appropriate time, some 14 minutes after zero hour, he saw the needle jump. He called me to say, "It's a boy!"

When I reflect back on that moment, as I sometimes do in preparing or giving lectures on the history of the nuclear era, a feeling of awe and foreboding always recurs. Even at the time, I thought of that moment and of that coded message as marking a real change in history—a moment when the course of the world suddenly shifted, from the path it had been on to a more dangerous one. Fission bombs, destructive as they might be, were thought of as limited in power. Now, it seemed, we had learned how to brush even those limits aside and to build bombs whose power was boundless.

Could the development of the H-bomb have been avoided? I do not see how. Even if there had been no other reasons—and there were others—the almost total lack of communications between Stalin's Russia and the rest of the world would alone have made its avoidance impossible as a practical political matter.

The birth of Mike led inevitably and immediately to a demand for developing practical versions of the new bomb. It settled, once and for all, Livermore's role in the program. There was obviously plenty of work ahead for both labs.

Our working philosophy, which I set out at the very beginning and which everyone readily accepted, called for always pushing at the technological extremes. We did not wait for higher government or military authorities to tell us what they wanted and only then seek to supply it. Instead, we set out from the start to construct nuclear explosive devices that had the smallest diameter, the lightest weight, the least investment in rare materials, the highest yield-to-weight ratio, or that otherwise carried the state of the art beyond the currently explored frontiers. We were completely confident that the military would find a use for our product after we proved it, and that did indeed usually turn out to be true.

In keeping with this philosophy, I at one point proposed to the AEC that we build and explode a bomb considerably bigger—more than 20 megatons—than any built before. According to procedures then in effect, the President personally approved or disapproved every test. In the case of this one, I was informed that when it was presented to President Eisenhower he said, "Absolutely not. They are already too big." Years later, Andrew Goodpaster, who had been the President's military assistant, told me it was during this period that Eisenhower concluded, "The whole thing is crazy. Something simply has to be done about it."

Controlling nuclear arms

Eisenhower had long been concerned about the nuclear arms race. He was convinced it was leading us somewhere we had better not go, and he looked for ways of getting it under control. The proposals presented in his "Atoms for Peace" speech had been an early stab at the general issue. Later he proposed "open skies" over Russia and America as a means for reassuring each country that no grand surprises were being prepared deep inside the other's territory. And ever since the "Bravo" test in the Pacific in 1954, when 23 Japanese fishermen suffered radioactive fallout on board their tuna trawler, Daigo Fukuryu Maru (Fortunate Dragon Number 5), Eisenhower had mulled over the possibility of a nuclear test ban both as a solution to the fallout problem and as a means for slowing down the arms race. Adlai Stevenson had raised the issue of fallout during the 1956 Presidential campaign in a way that Eisenhower took as a personal attack, and he backed off from his own tentative thrusts in order not to be seen as acquiescing to his opponent's charges. By mid-1957, however, he was again considering the possible merits of a nuclear test ban as a "first step" on the way to a more comprehensive arrangement. Premier Khrushchev apparently was thinking along similar lines independently. Thus, even before the furor over Sputnik had faded, Eisenhower turned to his new President's Science Advisory Committee to give him advice.

During my period of full-time service in the Pentagon,

first as the chief scientist of the Advanced Research Projects Agency and later as the director of defense research and engineering, I became more and more committed to the view that the nuclear arms race must somehow be brought under control and that a nuclear test ban would make a good first step. The general climate on the west side of the Potomac ranged from dubious to hostile. As a result, I found myself in the unusual and faintly amusing situation of being a member of a rather special minority in the national security establishment, one that included the commander-in-chief himself.

This situation had its dark side as well. On a memorable occasion in early 1960, a meeting took place between Secretary of Defense Thomas Gates, Chairman John McCone of the AEC, myself and one or two others. At one point the discussion turned to the question of whether the Soviets were cheating on the nuclear test moratorium, then still in effect. McCone had been arguing for some time that they were, and one of his purposes at this meeting was to persuade Gates to join him in putting pressure on the President to end his policy of continuing the de facto moratorium. I said that I had just gone over every shred of intelligence we had on this matter and had found no evidence whatsoever supporting such a claim. McCone replied that my saying this was tantamount to treason. I was flustered by that awful charge, but I reasserted my position. Although the meeting ended inconclusively, I have never forgotten it. I believe that history fully confirms I was right. When the Soviets unilaterally ended the moratorium, they did so with a huge bang, not a secretive whimper. They did deceive us about their preparations during the months they needed to get their extensive test series ready, but there remains to this day no reason to believe they conducted any clandestine tests before they suddenly did so openly.

A political missile gap

During my years at the Pentagon I maintained a relationship with the White House that was very unusual for someone in my position in the defense hierarchy. My calendar and the diary of George Kistiakowsky, the President's science adviser, show that I visited the White House and the old Executive Office Building about once a week over a three-year period. All of this was done with the full knowledge of the three Secretaries of Defense I worked for in succession: Neil McElroy, Tom Gates and Robert McNamara. I had a superb working relationship with all of them. They trusted me to deal directly with the White House on my own, and I behaved in accordance with that trust.

Among other things, I frequently acted as the de facto link between the Office of the Secretary of Defense on a number of matters. These included both intelligence about Soviet technical developments and the use of technology for producing such intelligence. The first included the Soviet nuclear program and missile program. The latter included such things as the programming of the U-2 "spy planes," the development of the new reconnaissance satellites and the monitoring of the nuclear test moratorium.

One of the issues that absorbed much of my attention throughout my years in Washington was the so-called missile gap. I first became involved with it a few months before Sputnik as a result of my service on the Gaither panel in the summer of 1957. That committee, headed by Rowan Gaither, a San Francisco attorney and friend of Lawrence, took an extreme view of the situation. It was not alone. That fall, after Sputnik, Secretary of Defense McElroy and many others began to talk about a coming "gap" favoring the Soviets, and the term *missile gap*

Five directors of Lawrence Livermore Laboratory assembled for this 1962 photograph, along with Edwin McMillan, who served as director of Lawrence Berkeley Laboratory. Left to right: the author, Harold Brown, Edward Teller, McMillan, John Foster and Kenneth Street.

entered political discourse with great force and frequency.

After I joined PSAC, and even more after I became a defense official, I learned how we obtained the intelligence we had, and I also became more sophisticated in interpreting it. A large part of what we knew then came from the flights of the U-2 over the USSR, particularly in the region of the missile test range near Tyuratam, in Kazakhstan. The more flights we made, the more we learned about the Soviet development program. And what we learned gave us pause. The Soviets had in fact successfully accomplished the development of a huge ICBM, several times as big as the Atlas and Titan we were working on. It was capable of delivering very large and powerful thermonuclear warheads to targets in our country in less than 30 minutes. But we saw no deployments of such rockets. The only launch facilities we found were those at Tyuratam. We continued to search and came up with nothing.

Many of those privy to these facts, including the President, began to suspect that the most probable explanation of why we found no rockets was that there were none, or at most only a few. Gates and I came to share this view. At the same time, important Democrats in the Congress, especially those eyeing the forthcoming 1960 Presidential elections, became steadily more strident in their insistence on the reality and importance of the missile gap. It was a difficult situation. We on the inside grew more convinced that there was no missile gap, but we couldn't prove it, not even to ourselves. The Soviet Union was, after all, the largest country on Earth, 7 million square miles in all, and the U-2 flights had covered only a tiny fraction of it. We were in effect trying to prove a negative on the basis of a very small sample.

A surprise "fire drill"

In the spring of 1960 I made a calculation of the probable outcome of an attack by Soviet ICBMs on our strategic forces at various times in the near future. I assumed total surprise and used only the official CIA estimates for the numbers, accuracy, destructive power and reliability of Soviet missiles. The estimates of Soviet capability were reduced from their earlier "missile gap" values, but they

remained formidable. I had doubts about some of the figures, but I set them aside for the purpose of this calculation.

Our retaliatory forces then consisted of long-range bombers and ICBMs based in North America, plus Thors and Jupiters based in Europe. Polaris missiles were not scheduled to enter the force until late 1961.

Given all these initial assumptions, my calculations showed that for a period beginning earlier in 1961 and lasting for many months the Soviets could hypothetically reduce our retaliatory forces to zero in a surprise attack. I didn't call it that, but the situation was identical to the one that in later years would be referred to as the "window of vulnerability."

I also worked out a straightforward means for mitigating the problem. If we were to speed up construction of the Ballistic Missile Early Warning System, move the deployment of the first two Polaris submarines forward six months and arrange to put some of our bombers on air alert, the problem indicated by my calculations would be solved. With those changes, at no time could a Soviet attack bring our retaliatory forces down to zero. All this, of course, assumed that the CIA's estimates were right. Years later we learned that the estimates of Soviet force I used were still too high. There was therefore no such "window," but I could not know that at the time.

I briefed the Secretary of Defense and others, including the Joint Chiefs of Staff. The chiefs took the calculations seriously and urged me to bring them to the attention of President Eisenhower and the National Security Council. It was arranged that I should do so at an NSC meeting scheduled for the morning of 5 May 1960, to be held as usual in the Cabinet Room.

The afternoon before the meeting someone on the White House staff called and told me that instead of going directly the the White House the next morning, I should report to the Pentagon helicopter pad and bring my briefing charts with me. I did as directed. Herbert ("Pete") Scoville, then deputy director of the CIA, met me at the pad. We flew off toward the Blue Ridge Mountains of Virginia and landed near the entrance to one of the President's special remote underground command posts. Shortly thereafter the President came in another helicopter and the secretaries of Defense and State in a third. Only after the meeting did I learn that the regular members of the NSC, excepting only the President himself, had not been notified of the change in location until 6 o'clock in the morning of the day it was held. It was a genuine "fire drill" in the sense that the principals were taken by surprise.

Despite the unusual nature of the meeting, it proceeded in straightforward fashion. I gave a brief report of my calculations, which those present took seriously. Out of that meeting eventually came a plan to accelerate the Ballistic Missile Early Warning System and Polaris, and to arrange to place a fraction of our B-52s on air alert at such time as a future estimate might confirm the need.

Ike's farewell address

President Eisenhower's farewell address is justly famed for its twin warnings about the "military–industrial complex" and the "scientific–technological elite." I was not surprised by his remarks, but, like many others, I wanted to know more about them. I had the opportunity to do so just a few years later. After leaving the presidency, Eisenhower spent his winters in Palm Desert, California, a town less than 100 miles from my home in La Jolla, and I called on him there on several occasions to pay my respects. Our conversation sometimes turned to the two warnings. I asked him to explain more fully what he meant, but he declined to do so, saying he didn't mean anything more detailed than what he had said at the time. I understood what he meant: The warnings were not the result of a methodical analysis; rather, they were the product of a remarkable intuition, whose power has generally been underestimated.

What, then, was the context of these remarks? What annoyed and irritated him? Whom are we to be wary of?

The context spanned the 40 months from the launching of Sputnik to the end of his administration. The people who irritated him were the hard-sell technologists who tried to exploit Sputnik and the missile gap psychosis it engendered. We were to be wary of accepting their claims, believing their analyses and buying their wares.

It seemed that the pursuit of expensive and complicated technology as an end in itself might become an accepted part of America's way of life. The Eisenhower administration, with the help of PSAC, was able to deal successfully and sensibly with most of the resulting rush of wild ideas, phony intelligence and hard sell.

At the beginning of the Eisenhower administration the authorities were hopeful that they could build air defenses capable of defending North America. We vigorously pursued the development of the means for doing so. These included radar warning networks, surface-to-air missiles of several kinds, and interceptor aircraft armed with a variety of antiair weapons. We also elaborated ideas and developed plans for vast civil defense programs. By the end of the Eisenhower administration, we concluded that we did not then know how to build a defense against missiles. We also decided, in effect, to stop extending our air defense system and to let what we had wither away. In sum, we decided to protect ourselves solely with the threat of nuclear retaliation. That policy has continued ever since.

Eisenhower considered the numbers of offensive weapons excessive, and believed that the quest for nuclear peace must be bolstered by diplomacy and negotiations of various kinds. He explored a number of possibilities—

"atoms for peace," eliminating the threat of surprise attack and a nuclear test ban. By the time he left office, the International Atomic Energy Agency had become a reality, and a nuclear test moratorium was in place. The goal of the IAEA was to prevent the proliferation of nuclear weapons while promoting the broadest possible use of nuclear power for electricity. The test moratorium was intended as a first step toward wider and more important forms of arms limitation. Both succeeded, though not to the extent we had hoped. The current limited test ban and the other limitations on weapons that have been elaborated since then all derived directly from those first, modest successes.

Thus the three major elements that make up the current nuclear situation all assumed their present shape in the Eisenhower administration. These are (1) the number and types of weapons making up our strategic nuclear forces, including the concept of the triad (bombers, missiles and submarines); (2) the abandonment of any serious attempt to build either active defenses or civil defense, and the concomitant acceptance of a policy of maintaining nuclear peace through the threat of massive retaliation; and (3) the initiation of a search for political measures, including arms control, as a means of reinforcing and extending the quest for peace. In each of these areas the Eisenhower administration saw revolutionary changes. Since then, all further developments have been evolutionary. In 1983 President Reagan initiated the Strategic Defense Initiative, quickly dubbed Star Wars, whose basic purpose was to reverse number 2 above and diminish the importance of number 3. If he were to achieve these ends, it would bring about a new strategic revolution, but I am certain he will not.

Creating an arms control agency

During the 1960 Presidential campaign, a Democratic party study committee had recommended the creation of an Arms Control and Disarmament Agency. John F. Kennedy incorporated the idea into his platform as a candidate. After his election he made good on his campaign promise. The Arms Control Act of 1961, among other things, called for the creation of the General Advisory Committee to work with both the President and the agency director. Because of the politically delicate nature of arms control and disarmament, Congress insisted that its members be appointed by the President with the advice and consent of the Senate. They feared that otherwise the committee and the process might somehow be taken over by "woolly-headed" types. Senators opposing arms control were afraid such people would "give away the store." Senators favoring arms control were afraid they would give the process a bad name and cause the public to draw back from it.

John McCloy, who had been Kennedy's special assistant for that subject, was named chairman of the GAC. I became one of the members, and so did some other longtime friends and colleagues, including George Kistiakowsky and I. I. Rabi. I was chosen because I was thought to be both an expert on nuclear weapons and interested in pursuing arms control as an essential element of national security policy. Kistiakowsky and Rabi had similar reputations in this regard.

Our committee brought together a wide spectrum of views. One example involved B-47 bombers. The US was decommissioning B-47s as fast as it added ICBMs to its arsenal. At first we mothballed the airplanes, but it soon became evident that we should simply destroy them. Someone proposed that we do so in a big public bonfire celebrating the event. Two of the GAC members, repre-

President Lyndon Johnson meets with his Science Advisory Committee in the Cabinet Room on 21 March 1966. Pictured clockwise from the President are the author, Charles Townes, Lewis Branscomb, Richard Garwin, Ivan Bennett, Kenneth Pitzer, Marvin Goldberger, Donald Hornig (chairman of the committee), an unidentified member, Franklin Long, an unidentified member, J. Robert Schrieffer, Melvin Calvin, an unidentified member, William McElroy, Sidney Drell, Philip Handler and George Pake.

senting opposite political extremes, opposed the idea.

"No. It would be phony and misleading," said Rabi, noting that we were replacing them with something better.

"No. We might need them again," said another member.

We also reviewed related programs in other agencies, particularly the Defense Department and the CIA. One was Project Vela, a program I had been instrumental in starting when I was in ARPA. It was in essence a collection of technical devices and activities whose purpose was to monitor nuclear testing in all environments, including underground and in outer space.

Among those instruments was the so-called Vela satellite, designed to detect nuclear explosions in both space and the atmosphere. In space it did so by detecting gamma rays emitted during nuclear explosions. As a happy by-product, the Vela satellite led to a whole new science—gamma-ray astronomy. In the atmosphere, the Vela satellite detected nuclear explosions by observing the intense light produced during the fireball stage of the explosion. Years later, in 1979, when I was in Geneva negotiating a test ban treaty with the British and the Soviets, one of these satellites detected a peculiar, intense light flash over the ocean south of Africa. The light signal was similar to, but not idential with, that emitted by a nuclear explosion. It proved to be unique in the 20-year history of the satellite. No other data corroborating a test at that time and place have ever turned up. Almost certainly it was a false alarm, but we could not know that when it was first reported. It became, for a while, the focus of much of our informal discussion in Geneva.

The two topics that interested me the most in those early days of ACDA were the nuclear test ban and the proposals to eliminate or limit antiballistic missiles or ABMs. Support for a test ban had been coming from two distinct groups: First, those who were primarily con-

cerned about radioactive fallout and its harmful effects on human health and genetics; second, those who were primarily concerned about the connection between nuclear testing and the nuclear arms race. By eliminating tests in the atmosphere, the Limited Test Ban Treaty of 1963 effectively satisfied the people in the first category. Those in the second continued to push for a complete ban, but they did not have enough weight to do so now that the environmentalists were satisfied and had dropped out. Government efforts to achieve a comprehensive test ban continued until 1981, including a period of two years in 1979–81, when I was the US chief negotiator. We came close at times during those 18 years, but we never quite made it.

In 1981 President Reagan determined that a comprehensive test ban was not in the best US interests. Since then certain public interest groups have continued to push for it, but within the government the matter has been dropped.

To my knowledge, the first government official to propose and seriously study an international agreement to limit ABMs was Jack Ruina of MIT, then director of ARPA. Ruina's idea originated from a half joking remark made by Jerome Wiesner, then Kennedy's science adviser. Jerry had said something to the effect that the only reason our people wanted to build an ABM was because the Soviets were building one. At the time the Soviets were installing an ABM system—named the Galosh by NATO authorities—around Moscow, and Khrushchev boasted about how good it was. He claimed it could shoot a fly out of the sky, and to prove his point he reminded us of the very annoying series of Soviet firsts in space.

Following up on Wiesner's remark. Ruina explored the possibility of a formal agreement to ban or limit such weapons. As Jack saw it, the situation in which no ABMs were deployed on either side had brought about "a curious and unprecedented stability," deriving from two factors:

First, the military balance was insensitive to the number and kind of offensive weapons in the arsenals of each country so long as these were invulnerable; second, the danger that either side would miscalculate the consequences of a nuclear attack was minimized. The introduction of ABMs by either or both sides would change that by introducing new, important, but incalculable changes in the strategic relationship. These uncertainties, in turn, could lead both to an arms race instability—that is, to an unrestrained series of attempts by each side to cope with the worst possible case presented by the other side—and to instability at a time of crisis that could raise the pressure to go first.

The idea of limiting defenses seemed strange—indeed, even perverse—not only to the Russians when they first heard about it but to most members of our own defense establishments. Eventually, however, many high officials in both the US and the USSR, including Secretary of Defense McNamara, accepted the idea that limiting ABMs could provide an effective damper on the arms race.

In January 1967 McNamara carried the debate about ABMs into the Cabinet Room. He arranged a meeting at which, in addition to President Johnson, there were present all past and current special assistants to the President for science and technology—James Killian, Kistiakowsky, Wiesner and Donald Hornig—as well as all past and current Defense Department research directors—myself, Harold Brown and John Foster. We were asked the simple question that must be faced after all the complicated *if*'s, *and*'s, and *but*'s have been discussed: "Will it work and should it be deployed?" The outside experts all gave the same answer: "No. There is no prospect of its defending our people against a Soviet missile attack." McNamara said he would speak for the current Pentagon officials. To no one's surprise, he agreed with us outsiders. No one there contradicted him. It was my impression that Harold did in fact agree with him and that Johnny and the Joint Chiefs did not, but none of them were invited to give their views during that meeting.

From PSAC to IDA and Jason

According to Wiesner, a proposal to reappoint me to PSAC for a four-year term beginning January 1964 was on Kennedy's desk when he was assassinated. In that, as in most other matters during his first year in office, Johnson was true to his promise to continue what Kennedy had started. I was pleased to accept. The invitation came only weeks after my resignation as Chancellor of the University of California at San Diego, and I looked forward to spending a substantial effort on this new assignment. It turned out to be very different from what I had anticipated. The relations between the President and PSAC, particularly with its chairman, Hornig, were very different from those that had prevailed in the Killian, Kistiakowsky and Wiesner eras. A golden age of science advice to the President had come and gone in the span of only seven years.

Two meetings of the committee with President Johnson illustrate this point perfectly. The first took place soon after I rejoined the group. The President was open, cordial and, above all, optimistic. He said, "You just tell me what it is I should do. Don't you worry about how to get it done. That's my job and I'll take care of it." Three years later, toward the end of his term (and mine), we had another such meeting. He started that one by saying, "You people just come in here and tell me what I ought to do. You never stop to think how hard it is for me to do it, and you never take the time to help me with it."

Ever since I left full-time employment in the Pentagon in 1961, I have maintained a close working relationship with two special national security organizations—the Aerospace Corporation and the Institute for Defense Analyses, known simply as IDA. Each works mainly for one primary customer: Aerospace for the Department of the Air Force and IDA for the Office of the Secretary of Defense, including the Office of the Joint Chiefs of Staff. One of IDA's offspring is named Jason, which didn't leave home for the first 14 years of its life. Since then Jason has been successively a division of the Stanford Research Institute and the Mitre Corporation. Jason is remarkable in many ways.

My involvement in starting Jason, combined with my continuing membership on the IDA board, made it natural for me to participate in many of Jason's activities, including its summer studies, even though I did not fit the

SERGEI KAPITZA

Paying off his bet to Peter Kapitza, who had wagered champagne that President Nixon would not finish a second term, York appeared in September 1974, a few weeks after the President's resignation, with a false headline: KAPITZA WINS BET, YORK HUMILIATED.

standard profile of a Jason member. Many of the chiefs and leaders of Jason have been friends from my early Berkeley days: Keith Brueckner, Marvin Goldberger, Harold Lewis and William Nierenberg. Many other Jasons, including Dick Garwin, Sid Drell and Freeman Dyson, are people with whom I have had other long working relationships. Jason pioneered the work in beam weapons of all kinds. It studied a wide variety of strategic defense issues, ranging from basing modes for the MX to the interaction between nuclear explosions and detection systems. More recently, it has reviewed essentially all of the technical questions relevant to President Reagan's Star Wars proposal.

Jason also did pioneering work in arms control. One of the earliest instances involved the Multiple Independently Targeted Reentry Vehicle. In the 1964 Jason summer study, a group chaired by Ruina and including Dyson and Murray Gell-Mann examined the possible impact of new technologies on national security. MIRV was among them. The group concluded that its introduction, combined with foreseeable improvements in accuracy, would create a situation in which striking first could confer—or seem to confer—a substantial, perhaps decisive, advantage. The group was right on target. These developments were precisely what led to the "window of vulnerability" debate of the late 1970s and early 1980s.

The most controversial of Jason's many projects involved the "electronic battlefield" in Vietnam. The basic idea, which from the start received strong support from McNamara, called for installing a variety of special sensors in the jungles of Vietnam. The sensors would detect and report the presence of people or vehicles in the jungles and swamps of that unfortunate country. The main purpose was to deny the protection of natural cover to attacking enemy soldiers or infiltrating guerilllas. Many organizations were involved in the concept, but Jason was central.

Jason's role in the project became known to the public. Because many students and professors were actively hostile to the war in Vietnam, and nearly all Jasons were college professors, a very dicey situation developed. Many Jasons, particularly those at Columbia University, were hassled and picketed by their students and colleagues. Others, including Drell, were prevented from speaking at European universities, even on subjects having no relation to defense work. A few of the Jasons became disaffected with the war and dropped out of the group. Others simply stopped participating.

I, too, felt pressure—from both friends and my own conscience—to resign from Jason. I believed that the war was a bad mistake, that our cause was hopeless and that by continuing to fight we would only prolong the misery and increase the death and destruction. I did not, however, in any sense condone the North Vietnamese actions, as did many others on campuses and elsewhere, nor did I think it was a moral issue except in the general sense that all wars, especially modern ones, involve important moral questions and deep ethical contradictions.

More important, and despite my almost total lack of empathy with the main action going on, I continued to believe that the security of the US—and thus of the West—was a most worthy goal. I still continue all my remaining relationships with the defense establishment, including Jason.

Many loyal and patriotic people did otherwise. Kistiakowsky decided it was too much for him. As a result, he publicly refused to have anything to do with the US defense establishment for the rest of his life. Some other colleagues also dropped out but without making any public statement. They simply ended their participation and declined any further invitations to give advice. In so behaving, they joined another group of veterans of defense science and technology—including Philip Morrison, Victor Weisskopf, Robert Wilson and many other Manhattan Project physicists who much earlier, in the first postwar years, decided they had done enough, or more than enough, of that kind of work.

On Carter's team

Late in 1975 I received an invitation to a small cocktail reception from a local Democratic group. The words "JIMMY CARTER" were written in large, bright red letters across the top of the sheet. I had never heard of him, and I did not bother to attend. A year later he was elected President. By that time I was aware he had been a member of the Trilateral Commission and, through that connection, had become acquainted with several of my old friends and colleagues, including Harold Brown and Cyrus Vance.

I showed up at Brown's office a day or two after he became Secretary of Defense. He was reading a thick briefing book presenting the insiders' view of the history of his office and the permanent staff's ideas about the Secretary's functions. He was more elated than I can recall ever seeing him. It was evident that he would throw himself into the job with an energy and intensity that few

others can bring to such responsibilities. I told him I wanted to help in whatever way I could. The result was another four years of full-time work on national security problems at the highest levels.

Both the form and the content of my contributions evolved over the four years. At first I worked as a direct consultant to Harold on issues having to do both with high-technology armaments and with arms control. At the end I served as chief US negotiator at the Comprehensive Test Ban talks in Geneva and otherwise dealt almost exclusively with arms control issues.

Since the bombing of Hiroshima all American presidents had actively sought means to contain, stop and reverse the nuclear arms race. Jimmy Carter differed from his predecessors not in kind but only in degree. He tried harder than they to explore the broadest set of possibilities. In one of modern history's all too common ironies, he accomplished the least. Events over which he had little control ultimately prevented him from adding very much to the limitations already worked out and put into place by the efforts of Eisenhower, Kennedy, Johnson, Nixon and Ford.

The principal White House study was directed by Carter's science adviser, Frank Press, a geophysicist from MIT. Press assembled a panel of experts including Bethe, Carson Mark, Wolfgang K. H. Panofsky, Ruina and me. In addition, the weapons laboratory directors, Harold Agnew of Los Alamos and Roger Batzel of Livermore, sat regularly with the panel. By that time, spring 1977, the argument over the utility of a test ban had come down to making a judgment about the relative value of two quite different factors, one of which weighed in on each side.

The main argument in favor of a test ban was that it was a necessary element of our nonproliferation policy. The Nonproliferation Treaty of 1970 called on the nuclear weapons powers to negotiate in good faith to end the nuclear arms race, and that was widely interpreted to require a comprehensive test ban as an early step. In 1975 the first quinquennial review of the nonproliferation regime focused attention on this point. To be sure, nuclear proliferation had proceeded much more slowly than had originally been expected, but this situation could change quickly. A change was especially likely if the superpowers continued to engage in "vertical proliferation," a phrase meaning the development and deployment of ever more varieties of nuclear weapons in their own arsenals.

The main argument against a test ban revolved around the issue of stockpile reliability. Nuclear weapons experts pointed out that these devices were built of both chemically active and radioactive materials, which steadily undergo changes that can adversely effect their performance. Occasional full-scale nuclear tests would be necessary in order to assure that old weapons still worked. Even rebuilt weapons would inevitably include small, supposedly harmless changes in their manufacture, and these, too, would have to be subjected to full-scale tests to assure performance. Opponents of a test ban also argued that we needed to continue testing in order to build safer and more secure bombs, to develop bombs properly optimized for new delivery systems and to learn more about weapons effects. Perhaps more important, continued testing was said to be needed in order to preserve a cadre of weapons design experts at the laboratories.

We studied the problem of stockpile reliability thoroughly and, except for the lab directors, decided that the nuclear establishment's worries were exaggerated. We concluded that regular inspections and nonnuclear tests of stockpiled bombs would uncover most such problems and provide solutions to them. Moreover, the laboratories could, if they tried, find ways around those that might

The author, who served as President Jimmy Carter's chief negotiator at the Comprehensive Test Ban talks in Geneva, is pictured with the former President in 1985.

remain. Agnew and Batzel disagreed. The in-house staffs in the Department of Energy and the Defense Nuclear Agency concurred with the laboratory directors, and the higher authorities in those agencies accepted their advice in the matter. The Joint Chiefs of Staff, whose nuclear arm is the Defense Nuclear Agency, also accepted the conclusions of the working-level experts immediately responsible for such matters. They really had no other choice.

Energy Secretary James Schlesinger also felt that President Carter was making a serious error in pushing ahead with a comprehensive test ban. Schlesinger had previously been chairman of the AEC, director of the CIA and Secretary of Defense, and his views were based on his experiences in those posts. In an attempt to dissuade Carter, he arranged to have the President meet with Agnew and Batzel so that they could explain to him why further tests were needed.

The intervention of the laboratory directors eventually caused a stir in the University of California—a stir that persisted for many years. The regents of the university managed the laboratories, and the directors were responsible to them through the university president. Most faculty members favored a test ban. In fact, roughly half of the faculty argued that the university should not be operating such labs. It was therefore no surprise that many faculty reacted negatively when they learned about what they regarded as unwarranted political intervention by persons who were, ostensibly, representing the university.

In sum, the Press panel reconfirmed Carter's intuitive view that a comprehensive test ban was in the national interest and could be adequately monitored. The strong opposition of the military and the nuclear establishment, however, made him realize that a test ban would be much more politically difficult to attain than a limitation on strategic forces.

Arms control has been a disappointment but not a failure. Large areas of the Earth, including Latin America and Antarctica, remain free of nuclear weapons at least partly because of deliberate diplomatic efforts to keep them that way. The most successful, and arguably

the most important, of all arms control policies have been those designed to limit the spread or proliferation of nuclear weapons to additional states.

Despite almost universal expectations to the contrary, there are still only five overt nuclear powers, and the last state to become one, China, did so in 1964, a full generation ago. In the quarter of a century since that happened, only four others—India, Israel, Pakistan and South Africa—have taken strong steps toward becoming nuclear powers, but none of them has yet built and deployed substantial, overt nuclear forces. This surprising but happy result must be credited to the combined antiproliferation policies and actions of the majority of the world's states.

After 30 years of actively working for a comprehensive test ban, I have been forced to conclude, as I did in the 1960s, that it will be politically possible and stable only in a world in which the great powers are clearly and forcefully moving away from their current dependence on nuclear weapons.

Some general conclusions

I crystallized my thinking about the correct US approach to national security early in 1961 as I prepared to meet with John McCloy, who had just been appointed by President Kennedy as his arms control and disarmament adviser and emissary. I organized my thoughts around three basic principles, each derived from my experiences of the last several years: (1) defense of the population is impossible in the nuclear era, (2) our national security dilemma has no technical solution and (3) our only real hope for the long run lies in working out a political solution.

The notion that we must "do something" radical soon about either the Soviet threat or the nuclear arms race or be doomed has been with us for more than 40 years. So far it has always proved to be wrong, and I expect it will remain so for the foreseeable future. The maintenance of an adequate balance of power, including the nuclear component, combined with classical diplomatic actions designed to control arms and preserve the peace, has bought us time. If we are wise enough, we will use it to find a way out of the grand nuclear dilemma. ∎

SOVIET SCIENTISTS REBEL, SAKHAROV AND SAGDEEV ELECTED TO NEW CONGRESS

PHYSICS TODAY/MAY 1989

An extraordinary spirit has swept Soviet science institutions in the aftermath of an attempt last winter by the academy's old-guard leadership to impose an unpopular slate of candidates on Soviet science.

In the new constitutional system adopted by the USSR last summer, Soviet voters and major Soviet institutions including the Communist Party currently are selecting 2250 delegates to a Congress of Deputies. The Congress's official purpose is to elect a two-chamber parliament called the Supreme Soviet and a president, who presumably will be Mikhail Gorbachev. The Congress also will meet once a year to consider constitutional questions, and it has the potential of becoming a constitutional convention somewhat comparable to the one that met in Philadelphia in 1787 and wrote the US Constitution. From this point of view, the question of who will be in a position to play Benjamin Franklin's role in the Soviet Congress is of more than passing interest.

On 26 March, 1500 delegates to the Congress representing territorial districts and the USSR's constituent republics were elected by voters in a general election; in contested districts where candidates failed to win a majority or in uncontested districts where the official candidate was struck off the ballot by a majority of voters, second-round elections are to be held. Meanwhile, the other 750 delegates are being selected by officially sanctioned organizations such as the party's youth league, unions, trade and professional groups, and the Academy of Sciences.

In an internal election held on 18 January, the Academy selected a slate of 23 candidates that did not include—to the dismay of many Soviet scientists—either Andrei Sakharov or Roald Sagdeev, even though they had received by far the largest number of nominations from scientific institutes and other academy sec-

Unprecedented demonstration of Soviet scientists took place outside the USSR Academy of Sciences Presidium on 2 February to protest the exclusion of Sakharov, Sagdeev and other advocates of democratization from the academy's slate of candidates for the Congress of Deputies.

tions. The result was especially surprising in view of the fact that Sakharov had been elected to the academy's presidium just three months earlier, after Sagdeev withdrew in his favor (see PHYSICS TODAY, January 1989, page 61). According to Bill Keller, the Moscow bureau chief for *The New York Times*, "The Academy of Sciences presidium evidently outdid all other organizations in limiting the choice it present[ed] to its members. . . ."

The election procedures that led to the exclusion of Sakharov and Sagdeev have been described in a special report by Paul Doty, director emeritus of the Center for Science and International Affairs at Harvard University, who was in the USSR at the time of the election. The election, Doty said, was "carried out by an

oddly composed group of full and corresponding members of the academy. Roughly, this group would contain the 45 regular voting members of the presidium, the 25 or so emeriti (over 75 years of age) and 6 heads of special regional sections of the academy or scientific centers. To this number (of about 80) were then added a much larger group numbering about 300 and composed of the full and corresponding members of the 19 departments that cover the major scientific specialties. This expansion was apparently agreed to as a step toward greater democratization. However, this group is made up predominantly of corresponding members who carry out much of the administrative work of the academy and it is the conservative bias of this group which allegedly led to the upset." Apparently some less conservative members were complacent and did not vote.

Grassroots rebellion

The rejection of Sakharov prompted more than 1000 rank-and-file scientists to demonstrate in the academy's courtyard on 2 February—an unprecedented manifestation of collective dissent by scientists. For a time Sakharov toyed with the idea of running for a territorial seat. He accepted at least two nominations, one for an at-large Moscow seat and one representing the Oktyabr district, where the Lebedev Institute is located. But then, on 15 February, Sakharov withdrew from the territorial races, saying that he felt "inextricably linked with the academy" and that he would "not run anywhere except for the Academy of Sciences."

During the following months, Soviet scientists organized a campaign to reject the Academy's slate of 23 candidates in its entirety. Instead, in a split decision reached in a dramatic election held at the academy on 20–21 March, some 1000 academy members and 500 representitives of scientific institutes elected eight delegates from the presidium's slate and rejected the other 15. Among the eight elected are five physicists: Yuri Ossipyan, director of the solid-state physics institute at Chernogolovka and president-elect of the International Union of Pure and Applied Physics; Zhores Alferov, head of a group at the Ioffe Institute that has done pioneering work on injection lasers; Andrei Gaponov–Grekhov, a radiophysicist in Gorki who ran unsuccessfully against Sakharov for the physicist vacancy on the presidium last October; Nikolai Karlov, also a radiophysicist; and Karl Rebane. The other three individuals elected to the Congress are jurist Sergei Aleksei, chemist Oleg Nefedov and mathematician Vladimir Platonov.

According to a vivid report issued by the Soviet news agency Tass, the voting lasted for three hours on Tuesday morning, 21 March. "Then for over seven hours eleven members of the electoral commission were counting votes without using calculating machinery. This took place at a round table on the premises of the Moscow Palace of Youth. The table was divided by a broad red band. Ten observers—representatives of electors from collectives of scientists—as well as numerous journalists were seated opposite to members of the commission."

The vote was preceded, on Monday 20 March, by a meeting of the academy's general assembly. "All who spoke were unanimous," Tass reported, "in the opinion that scientists—deputies in the supreme body of state authority—should not only pursue the interests of science but should, first and foremost, promote perestroika, the democratization of Soviet society, and the intellectualization of the process of making policy decisions."

Run-off election

On 10 April, Sakharov and Sagdeev were nominated by the academy's presidium for seats in the constitutional congress. Sakharov received 34 of 37 votes. In the second round of the academy's election, which was held 20 April Sakharov and Sagdeev were elected to the Congress of Deputies in what a US embassy official described as "virtually a clean sweep for the reformers."

Sakharov's election was almost a foregone conclusion, but Sagdeev's prospects were more uncertain and his victory says more about the emerging political balance. An outspoken advocate of democratization, competition and westernization in general, Sagdeev is extremely popular among scientists and science policy makers in the United States, and he has a strong following among the more liberal-minded Soviet scientists. But sometimes fellow scientists seem to resent him. In the election of new presidium members last October, he was accused by Kiril Kondratiev, an ally of academy president Guri Marchuk, of announcing scientific results in the West before he announced them in the USSR itself.

Sagdeev was very closely identified with the idea of a manned mission to Mars and with the Phobos mission, which ended sadly in March when it was announced that the USSR had lost contact with the second of two spacecraft sent to the Mars moon. (Contact with the first was lost last summer as the result of a command error.) Even though the loss may (or may not) be the fault of organizations and plans in which Sagdeev had no responsibility, his prestige was bound to suffer. Apparently his political philosophy and general eminence outweighed, in the eyes of academy voters, the Phobos news.

There were many startling upsets in the general elections held 26 March. Georgi Arbatov, the USSR's designated top US expert and a member of the academy's presidium, lost to the head of the Orthodox Church (but won as a member of the academy's slate in the second round); Boris Yeltsin, the former Moscow party chief who was demoted and denounced by Gorbachev two years ago, won in a landslide against the head of the country's limousine manufacturer. It is not true, however, that Marchuk was defeated. Contrary to some news reports in the US, the academy president holds one of the party-reserved seats in the Congress of Deputies.

—WILLIAM SWEET

SECTION 7

COMMENTARY ON THE ERA OF NUCLEAR WEAPONS

Pʜʏꜱɪᴄꜱ ᴛᴏᴅᴀʏ has printed articles of opinion by physicists with differing views on the nuclear arms race. These articles contain discussions of the strategic arms race with offensive weapons, the possibilities and problems of defensive weapons, and the recent shifts in the Soviet Union, spurred by the Soviet scientists. The articles consist of an interesting mix of personal reflections, policy analyses, ethical considerations, practical politics, and predictions of the future.

CONTENTS

A peril and a hope

With the discovery of fission mankind entrained cosmic forces
with human irrationality; we must face the problem of reducing the tens
of thousands of nuclear bombs ready to be released in seconds.

PHYSICS TODAY/JULY 1978

Forty years ago Otto Hahn and Fritz Strassmann discovered fission. The existence of this phenomenon could have been predicted before it was found if the theoretical physicists at that time had shown a little more inventiveness: It is a simple consequence of the facts that the Coulomb repulsion increases with the square of the number of protons and the nuclear attraction only with the first power. In 1938 enough was known about the nuclear force that it would not have been too hard to conclude that the nucleus must become unstable against a split in two parts around atomic number $Z = 90$.

Although the fission process itself was not foreseen, the possibility of a nuclear chain reaction was indeed thought of in the early 1930's by Leo Szilard. The implementation of such a reaction with fission would depend upon the number of neutrons released in the process. This number was not easily predictable; even today it would be hard to determine it theoretically from our knowledge of the nuclear forces. That this number is considerably larger than unity does not appear to be based upon any fundamental property of the nuclear forces. There are no very deep reasons; it might have been less than one, but it was not.

Cosmic fire

The fact that the number of neutrons emitted per fission is around two seems to be of very minor importance to Mother Nature. Apparently the only major roles of the fission process are to provide an upper limit to Z and to influence to some extent the abundance of fission products. Apart from this, we may forget about fission if we are interested only in the major features of our world. This is even more true about the chain reaction itself. Nature has not made much use of it; recently evidence was found that a natural chain reaction happened a billion years ago below the soil of Africa; but, to our knowledge, nuclear chain reactions have never played any role in the development of our Universe.

Do we humans count in this Universe? If we do, my previous statement is totally wrong. For humankind the existence of this chain reaction is of decisive importance, probably greater than we can fathom today. We know that fission can be used destructively as an explosive and constructively as a source of useful energy. It is therefore a very effective instrument of power, both political power and physical power.

This, in itself, is nothing new in the history of the natural sciences. New discoveries have led to new weapons, to new energy sources, and to countless applications from which a lot of good has emerged—as well as some results that have not been so good. (In the last 20 years it has been fashionable to emphasize the "not so good," but let us be objective and fair.)

There is indeed something different in the latest developments of physics, which I will call "the leap into the cosmos." Previously we were dealing mostly with processes similar to those occurring in our terrestrial environment. In the last few decades, however, we have taken a decisive step: We now deal with extraterrestrial phenomena. Nuclear physics and subnuclear physics deal with the excitation of quantum states that are beyond the reach of ordinary terrestrial energy exchanges: In general, nuclear reactions do not take place on Earth. Nuclear dynamics is dormant in our environment. It is of course true that natural radioactivity occurs on Earth, producing the heat in the depths of our planet, but these radioactive elements are the remnants of a very different age and of a different environment: They are the last embers of the cosmic explosion that produced terrestrial matter.

Today we physicists deal with cosmic processes in which many millions of electron volts per atom are exchanged rather than the few electron volts that are customary here on Earth. Of these cosmic processes, the fission chain reaction was one of the first to lead to major technological applications. Two hundred million electron volts per atom, twenty million times more than the most powerful chemical reaction. This is cosmic and not ordinary fire.

The first major application was a destructive one, which ended World War II by killing a quarter of a million people with two bombs. It is not surprising, therefore, that people are fearful and bewildered, and that they have misgivings even in regard to the more benign applications. The arm of technology grew by a factor of a million within the lifetime of one generation.

In 1940, however, we took little time to speculate on these questions. The discovery of fission came at a dark time in the history of mankind. Germany was in the grip of a collective mental disease of unparalleled virulence. The whole world was threatened by the expanding cancer of Nazism; the Germans discovered fission, and they might have used it if it had been usable at that time. Many of us physicists—those who were not yet too deeply involved in the development of radar, which saved England from being destroyed—worked hard to improve our

Victor F. Weisskopf is Institute Professor in the Massachusetts Institute of Technology Department of Physics and President of the American Academy of Arts and Sciences, and was formerly the Director General of CERN.

understanding of fission. Many of us hoped that the number of neutrons per fission would be low enough to prevent the making of a bomb. But it was not.

History takes a turn

A tremendous collective effort began. Within two years, the first nuclear chain reaction was produced under the leadership of Enrico Fermi, and within another three years, with the collaboration of many European physicists—in a truly international effort—a nuclear bomb was developed. On 16 July 1945 the first nuclear explosion was set off in a desert of southern New Mexico, at a place called Jornada del Muerte (a Spanish army perished there two hundred years ago). Human ingenuity had succeeded in the release of cosmic forces that were hidden and unknown 13 years earlier, when James Chadwick discovered the neutron.

Some of us saw this event, which, at our observation post, had the intensity of twenty midday suns: an expanding fire ball, white, and then yellow and orange, rising majestically into the sky, surrounded by a halo of blue light. The air was fluorescent with radioactive radiation.

Here it was: The laws of Nature did admit an explosive nuclear chain reaction. Because of some little detail in the equations of nuclear matter, the number of neutrons per fission was large enough. The history of mankind took a turn.

From then on, political developments, and no longer scientific ones, determined the course of events. Two bombs were exploded over Japanese cities; over 200 000 people lost their lives. Why two bombs? Why over populated areas? I am not sure I can answer these fateful questions. The bombs ended the war; they made an armed invasion unnecessary

US AIR FORCE

"**Fat Man,**" the plutonium bomb dropped on Nagasaki, Japan on 9 August 1945. Larger than "Little Boy," the uranium bomb detonated over Hiroshima, it had about the same yield, some 20 000 tons of TNT. The production of nuclear chain reactions gave Man control over cosmic forces.

and thus perhaps saved more people than they killed. But Japan may have given up anyway under the pressure of a near defeat and a Soviet declaration of war. Whatever the answer, the stigma has remained upon the United States, as the first and only country to have used nuclear bombs. I can not tell you whether the decisions were right or wrong—I don't know.

The war ended, and it was physics that had helped to win it, with radar and the atom bomb, as well as many other new gadgets. Physics and the physicists moved onto center stage in public life; the significance of basic physical science became generally recognized. New means of research, technical and financial, suddenly were available to physicists. The radar technology created sophisticated new methods for producing electromagnetic waves of all kinds. Maser and laser physics emerged from it, and with it a deeper understanding of the fundamental interactions of light and matter, leading to quantum electrodynamics, as well as of the structure of solids under various extreme conditions.

Nuclear and subnuclear research experienced an almost explosive development, with an ever growing array of cyclotrons, synchrotrons and linear accelerators. The cosmic scope of physical research widened and penetrated into new realms of phenomena:
▶ On the microscopic side, into the GeV region—with its mesons, hyperons, antiparticles, heavy electrons and quarks, entering the innermost structure of matter;
▶ On the macroscopic side, into plasma physics, space physics, a new cosmology and astrophysics—with its quasars and

pulsars, and the observation of the optical reverberation of the birth of the Universe, the famous three-degree radiation in space.

The physicists experienced a tremendous surge of support for their activities. This in turn led to an unparalleled expansion of their fields of endeavor and to a large range of new technical applications, both destructive and constructive, such as fusion bombs, rocketry, transistors, space travel, computers and many more. The physicists now moved to the center of public attention. It was their enthusiasm and self-confidence, spurred by their successes during and after the war, that brought them into contact with the great social and political problems of the times.

Many tried, and some succeeded, to bridge the "communication gap" between politicians and scientists. Attracted by great, optimistic ideas of how to establish a new order in the world they had helped to engender, they embraced the idea of international control of nuclear explosives. They hoped this would stop all future wars, and direct the applications of fission and fusion away from destructive bombs and towards an unlimited benign source of energy for all mankind. Some of them, perhaps, also experienced a reaction against their feeling of guilt for having been involved in the creation of a device that could annihilate all mankind.

Power—and anxiety

The political realities, however, were not conducive to the ideals of those who tried to unite the nations. Stalinism in the Soviet Union, and nationalism all over the world, including our own country, led

to a breakdown of attempts to lift nuclear matters beyond national sovereignty. Of course, neither the physicists nor the politicians had much experience in dealing with such complicated technical matters on an international scale. With the development of fusion bombs, the nuclear arms race began in earnest.

Among the benign implications of nuclear physics, there still remained the great promise of unlimited energy through the use of fission reactors and, perhaps later, through controlled fusion. But over the decades following World War II, doubts were raised even on that account. Because the yield per atom is so many million times higher than in any conventional way of producing power, the consequences of accidents caused by human error are much worse. A nuclear power station certainly can not explode like a bomb, but the possibility of accidents that could spread large amounts of radioactivity can not be completely excluded. Even if human ingenuity and care can keep the accident rate at an acceptable level, the public and some of the experts look with some anxiety at this "extraterrestrial" way of producing energy in our terrestrial abode. The fact that the public is acquainted with this cosmic force only through the experience of nuclear bombs strengthens this anxiety. Furthermore, the use of fission for power constantly creates new raw material for bombs, which, in a nationally and politically divided world, adds to the danger of further spread of nuclear weapons.

The dangers and the promises of nuclear-power generators are today in the center of discussion; many studies have been undertaken and more are under way. Emotion and vested interest unfortunately have led to a sharp division of opinions, and the arguments used on both sides are too often beyond the limit of dignified scientific discourse.

Apotheosis of irrationality

At the same time the nuclear arms race between the superpowers continues in an almost uncontrolled way. The Soviet Union and the United States assemble increasing numbers of bombs, and perfect their efficiency and their mode of delivery. More than 50 000 nuclear bombs are deployed and ready for use. Each country now has the capability of destroying the other many times over. Current science is totally unprepared to discuss intelligently, let alone to predict, the totality of horrors that would result from an all-out nuclear war. Consider only the effects of the vast amounts of radioactivity released upon our environment; these would be so devastating that the condition of life would be permanently and dangerously altered, without much hope of recovery. Even the detonation of a single weapon of modern design over a city would be a catastrophe unprecedented in human history. Yet two large countries keep building more and more of these horrendous means of annihilation, knowing well that any actual use of these devilish gadgets would mean certain destruction of a large part of the world—making it unfit for habitation, with little chance of a recovery of civilization. Why? Why? Why? Only because neither side knows where to stop, and both go on producing nuclear weapons intended for all sorts of imagined missions. Only because each party is under the grip of an unrealistic measure-countermeasure syndrome. It is the apotheosis of irrationality and antilogic; it is the triumph of craziness.

In comparison to this overwhelming threat the nuclear power controversy dwindles to picayune dimensions. What are the dangers of nuclear-power stations compared with the dangers of tens of thousands of bombs that can be released within seconds by a small group of human beings? What is the so-called "worst reactor accident" compared to nuclear war? The damages done by the former, which would come mainly from effects of radiation—serious as they are—are far less than the effects of a single bomb. Now think of the number of victims of a nuclear war and its irreparable effects on our environment, on our souls, alive or dead, and on our planet as a whole.

And the probabilities? Nobody really can estimate the probability of an all-out nuclear war, but one fact is clear: With all those bombs around, it is not zero. Nuclear power may be too risky, or it may not; I do not pretend to know the answer. But I *know* that tens of thousands of stored bombs are too risky.

Undoubtedly, it is extremely improbable that the US or the Soviet government will decide outright to set off a nuclear war, thereby annihilating both countries simultaneously. However, a nuclear clash may develop from local wars between smaller countries that have acquired, or will soon acquire, a few nuclear weapons, and may make use of them in desperation. Or it may develop from an escalation of the use of so-called "tactical" nuclear weapons, which are meant to be applied in defense against aggression by conventional weapons. The tens of thousands of nuclear bombs are a mighty tinderbox; they may explode whether it was directly intended or not.

Is there a way out?

I am only too well aware of the difficulties in the way of reducing that danger. I know also that the presence of these weapons is given credit for having prevented a world war for a longer period than ever before. But, as time passes, the weapons multiply and become more efficient; they are adapted to all kinds of purposes, such as the neutron bomb. They are therefore more likely to be used. The outbreak of a volcano becomes more violent after long pauses. Here we have

HIROSHIMA PEACE MEMORIAL MUSEUM

Nothing obscures the view south from the roof of a building around Yamaguchi cho. The city of Hiroshima has become a bare scorched field; Motoujina and a gas tank appear to be near at hand.

a man-made volcano: It could be removed.

I used the phrase "almost uncontrolled" in referring to the nuclear arms race. There have indeed been a few hopeful attempts, albeit small and tentative, that slowed it down a bit. One was the cessation of bomb tests above ground; the other was the first SALT agreement six years ago, which will, I hope, be expanded in the near future. They are not much, but they are two small steps in the right direction, and we can be proud that some of our colleagues were active in bringing them about.

The difficulties of going further are enormous. Some raise the fear that the other side will gain more, or might even dare to take the tremendous risk of a first strike regardless of the consequences. There are those who say that a free society can not compete in terms of nuclear armaments with an authoritarian system in any other way but through an all-out technological arms race. Voices are heard that we should not be too confident of negotiated contracts with our opponents; that we should rather assure a reasonable survival rate by a large effort of organized civil defense in order to make the population as ready as one can be for the great holocaust. I honor the intentions of those who advocate such measures but, to make them effective, if this is at all possible, would thoroughly change our way of life. I can not help sympathizing with the majority of our population, who do not want to live under a constant awareness of mankind's ultimate self-destruction.

We can not go on forever living under a continuous threat of annihilation. There must be ways and means to decrease the number of nuclear bombs.

Certainly this task can not be accomplished from today to tomorrow; it is bound to be an extremely difficult process.

The foremost problem

There is a hope and not only a peril in the nuclear development. The growing recognition of this awful threat may still change the attitudes of the sovereign states, to lead slowly to the recognition that ultimate military preparedness is much less safe than a reduction and eventual abolition of nuclear weapons. J. Robert Oppenheimer said in 1945, shortly after the end of the war:

"The point is that atomic weapons constitute a new field and new opportunities ... when people talk of the fact that this is not only a great peril, but a great hope, this is what they should mean. I do not think they should mean the unknown, though sure, value of industrial and scientific virtues of atomic energy, but rather the simple fact that, because it is a threat, because it is a peril and because it has special characteristics, there exists a possibility of realizing, of beginning to realize, those changes which are needed if there is to be any peace."

The task must be faced; it is the first and foremost problem of our time. Any one of us can and should play a role in this task, as a scientist and as a human being. The most important step is a new setting of priorities. The reduction and eventual abolition of nuclear weapons *must have absolute priority;* everything else must be subordinated to this goal. The consequences of nuclear war are irreparable, whereas the consequences of other setbacks in world politics can be corrected.

HAJIME MIYATAKE; HIROSHIMA PEACE MEMORIAL MUSEUM

A victim of the atom bomb exploded over Hiroshima. Over 200 000 people lost their lives in the two Japanese cities bombed in World War II. Today more than 50 000 nuclear bombs vastly more powerful than these are deployed and ready for use, and the inventories are growing every day. Weisskopf: "The reduction and abolition of nuclear weapons must have absolute priority."

our achievements in getting at the basic processes of life, the workings of DNA and RNA, and the tremendous developments in our knowledge of organized matter.

It is not the first time in history that human greatness and human folly grew side by side in the same period. Think of the Gothic cathedrals, together with the senseless and murderous crusades, 700 years ago. Think of the blossoming of art and philosophy during the Renaissance, along with the decimation of Europe's population during the religious wars, 500 years ago. Think of the music of Mozart and Beethoven, and of the slave ships plying the oceans, 150 years ago. Think of the greatest achievements of scientific thought, quantum mechanics and relativity theory, and the ascendancy of the murderous periods of Nazism, Fascism and other authoritarian regimes, 50 years ago. Finally, think of the great achievements of science today, together with the folly of the nuclear arms race.

The last folly is more serious than all the previous ones. We are dealing with cosmic forces. Our epoch may be the end of what has been a great age of mankind, great in spite of all the strife and wanton destruction. Our age has been great in its achievements in art, architecture, literature, music; great in its numerous social innovations, in spite of the fact that in most parts of the world social organization and the quality of human life leaves much to be desired; great in its medical successes, which have resulted in the doubling of the average age of man; great in its means of food production, communication and transportation—which makes a united world without hunger and want a possibility, if not a reality. And, last but not least, ours has been a great civilization because of the constant growth of our insights into the mysteries of Nature, the continuous opening up, leaf by leaf, of the blossoms of truth and wonder.

This change of priority is essential and can be achieved only by constant pressure of public opinion. Remember how effective public opinion was during the Vietnam War. In nuclear matters, the public is now interested only in the relatively unimportant issue of nuclear power. This must change. The issue of getting rid of nuclear armaments must receive much more public attention, support and pressure than it receives today. There is much too little discussion of these essential questions: Do we not already have enough bombs for deterrence? How

would the Soviet Union react to a restraint in development or to a cutback? We must find that right balance of risks most conducive to lowering the levels of nuclear armaments. What is needed is a combination of new technical ideas and common sense, based upon humane considerations.

Only when we see a chance of success in the abolition of nuclear armaments can we scientists be proud of the achievements we gained during the last decades: our leaps into the cosmos; our penetration into the innermost structure of matter;

The age of insight

If we do not succeed in abolishing the nuclear arms race and a nuclear war results, all these great steps will be brought to naught. The twentieth century would then be remembered as the time of preparation for the great catastrophe, and science would be seen as the main culprit and the main instrument of destruction. The twentieth century ought to be remembered as the age in which mankind acquired its widest and deepest insights into the Universe, and learned to control its martial impulses.

Let us hope, strive and act so that it will.

* * *

This article is an adaptation of a talk given 24 April 1978 at the Washington, D.C. meeting of The American Physical Society, as part of a commemorative session marking the fortieth anniversary of the discovery of nuclear fission.

□

US STATE DEPARTMENT

To peace! At the SALT talks in Vladivostok, November 1974, Soviet premier Leonid Brezhnev and US President Gerald Ford raise their glasses in a toast. What is needed to find the "right balance of risks" are "new technical ideas and common sense based upon humane considerations."

The social responsibility of scientists

The scientific community, which enjoys freedoms and benefits from its activities, has a duty to inform the public fully about the impact of scientific developments.

Andrei Sakharov   <u>PHYSICS TODAY/JUNE 1981</u>

Because of the international nature of our profession, scientists form the one real worldwide community which exists today. There is no doubt about this with respect to the substance of science: Schrödinger's equation and the formula $E = mc^2$ are equally valid on all continents. But the integration of the scientific community has inevitably progressed beyond narrow professional interests and now embraces a broad range of universal issues, including ethical questions. And I believe this trend should and will continue.

Scientists, engineers and other specialists derive from their professional knowledge and the advantages of their occupations a broad and deep understanding of the potential benefits—but also the risks—entailed in the application of science and technology. They also develop an awareness of the positive and negative tendencies of progress generally, and its possible consequences.

Colossal opportunities exist for the application of recent advances in physics, chemistry and biochemistry; technology and engineering; computer science; medicine and genetics; physiology and hygiene; microbiology (including industrial microbiology); industrial and agricultural management techniques; psychology; and other exact and social sciences. And we can anticipate more achievements to come. We all share the responsibility to work for the full realization of the results of scientific research in a world where most people's lives have become more difficult, where so many are threatened by hunger, premature illness and untimely death.

But scientists and scholars cannot fail to think about the dangers stemming from uncontrolled progress, form unregulated industrial development and especially from military applications of scientific achievements. There has been public discussion of topics related to scientific progress: nuclear power; the population explosion; genetic engineering; regulation of industry to protect the environment; protection of air quality, of flora and fauna, and of rivers, lakes, seas and oceans; the impact of mass media. Unfortunately, despite the urgent and serious nature of the issues at stake, such discussions are often uninformed, prejudiced or politicized, and sometimes simply dishonest. Experts, therefore, are under an obligation to subject these problems to unbiased and searching examination, making all socially significant information available to the public in direct, first-hand form, and not just in filtered versions. The discussion of nuclear power, a subject of prime importance,

Andrei Sakharov is a member of the Academy of Sciences of the USSR; he received the Nobel Peace Prize in 1975.

is an instructive example. I have expressed elsewhere my opinion that the dangers of nuclear power have been exaggerated in the West, and that such distortion is harmful.

With some important exceptions (primarily affecting totalitarian countries), scientists are not only better informed than the average person, but also strive for and enjoy more independence and freedom. Freedom, however, always entails responsibility. Scientists and other experts already influence or have the capacity to influence public opinion and their governments. (That influence should not be exaggerated, but it is substantial.) My view of the situation of scientists in the contemporary world has convinced me that they have special professional and social responsibilities. It is often difficult to separate one from the other—the communication of information, the popularization of scientific knowledge, and the publication of endorsements or warnings are examples of activities with both professional and social aspects.

Similar complications arise when scientists become involved in questions of disarmament: in developing strategy for or participating in international negotiations; in advancing proposals or issuing appeals to governments or to the public; and in alerting them to dangers. Disarmament is a separate, critically important issue which requires a profound, thorough and scientifically daring approach. I realize that a more detailed treatment is needed, but now I will simply outline a few ideas. I consider disarmament necessary and possible only on the basis of strategic parity. Additional agreements covering all kinds of weapons of mass destruction are needed. After strategic parity in conventional arms has been achieved, a parity which takes account of all the political, psychological and geographical factors involved, and if totalitarian expansion is brought to an end, then agreements should be reached prohibiting the first use of nuclear weapons, and later, banning such weapons.

Another subject closely connected to questions of peace, trust and understanding among countries is the international defense of human rights. Freedom of opinion, freedom to exchange information and freedom of movement are necessary for true accountability of the authorities which in turn prevents abuses of power in domestic and international matters. I believe that such accountability would make impossible tragic mistakes such as the Soviet invasion of Afghanistan and would inhibit manifestations of an expansionist foreign policy and acts of internal repression.

The unrestricted sale of newspapers, magazines and books published abroad would be a major step toward

Comments in honor of Andrei Sakharov

I am pleased to add my congratulations to Academician Andrei Sakharov on the occasion of his 60th birthday. Sakharov is one of the true spiritual heroes of our time. An outstanding scientist whose position ensured him all the security and comfort he might desire, he was willing to risk all to speak out on behalf of human rights and freedom. He persisted in this mission even after being subjected to increasingly harsh penalties.

Mr. Sakharov is a Russian patriot in the best sense of the word because he perceived his people's greatness to lie not in militarism and conquests abroad but in building a free and lawful society at home. His principled declarations on behalf of freedom and peace reinforce our belief in these ideals. We hope and pray that his exile will be ended and that he will enjoy a long and creative life on behalf of science and humanity.

Ronald Reagan
President

The banishment of Andrei Sakharov from Moscow to Gorky reminds one of the isolation of Robert Oppenheimer by denial of his security clearance in the nineteen fifties.

The results of these actions should be instructive to the men in the Kremlin. Oppenheimer became an even greater celebrity than before, and as a consequence the US government became in international opinion a country where outspoken criticism, even by eminent scientists, was suppressed, intimidated and discouraged. He became a martyr to the cause of peace and free expression of critical opinion, and the US lost a great deal of its lustre as the land of the free and the home of the brave.

The men in Kremlin likewise failed to understand what a great national asset they had in Sakharov, a great scientist, patriot and humanitarian. From his small apartment and with very limited means, these issued a constant stream of constructive criticism of actions of the rulers of the country he loved so dearly. To the outside world he showed that the Soviet Union, oppressive as it appeared, still allowed a small flame of freedom of thought and expression to light the pervading darkness. Now that tiny island of freedom in that small apartment in Moscow is suppressed and the intellectual map of the Soviet Union is uniformly black.

What have the policy makers in the Kremlin gained from this brutal act? The answer is the universal condemnation of the world outside and the loss of a constructive critic within. They lost the service of a great man who brought respect for the moral quality of the Soviet people and Soviet science. It also raises a question in my mind: Is the Kremlin so weak and so insecure that they dare not tolerate one small voice from one small apartment in Moscow?

Perhaps that voice is a mirror for them to see themselves in all their meanness and moral bankruptcy.

The World has given Sakharov the Nobel Peace Prize and the President of the United States has given Oppenheimer the Fermi prize. How long will it take the Politburo to make amends to one of the Soviet Union's great and courageous scientists and humanitarians?

I. I. Rabi
Columbia University

Since being exiled to Gorky a year ago, on 22 January 1980, Andrei Sakharov has been totally isolated from his friends and colleagues. It is more than the injustice of his confinement and the indignities of his present living circumstances, under constant watch, that are so outrageous. Andrei is a scientist, and perhaps a fellow scientist can best appreciate how serious and distressing it is to lose the stimulation of personal discussions and seminars and the access to scientific writings of colleagues. These are the life blood of theoretical physics; denied them, Sakharov's scientific career is in serious jeopardy. We should not, and will not, forget the very unhappy conditions of Sakharov's life at present. His letters from Gorky record a pattern that is both psychologically stressful and physically precarious.

Sakharov is a brilliant physicist, best known in the West as the father of the Soviet hydrogen bomb. In 1951 he published, with Academician Igor Tamm, the pioneering paper in the Soviet controlled-fusion effort. He has also made important contributions to the studies of gravitational and elementary-particle phenomena. In 1953 he was elected, at the age of 32, the youngest full member of the USSR Academy of Sciences.

In the years from 1953 to 1968 his social and political views underwent a major evolution. His 1968 essay "Progress, Coexistence and Intellectual Freedom" argued that the division of the World into opposing camps threatens it with destruction and that intellectual freedom is essential to human society. This essay publicly marked Sakharov's transition from a scientist in search of Nature's principles for the structure of matter to a moral leader in search of ethical principles for a humanity in quest of peace, progress, and basic freedoms. He has forcefully reiterated and developed these same basic arguments on a number of occasions in the intervening thirteen years.

Sakharov was a cofounder of the Committee on Human Rights in Moscow in 1970 with Valery Chalidze, who is now in the US, and Andrei Tverdokhlebov, who was later sentenced to a Siberian exile. In 1973 he took the courageous step of making worldwide public appeals for support for dissidents forcibly committed to

effective freedom of information in totalitarian countries. Perhaps even more significant would be the abolition of censorship, which should concern first of all the scientists and intelligentsia of totalitarian countries. It is important to demand a halt to the jamming of foreign broadcasts that deprives millions of access to the uncensored information needed to form an independent judgment of events. (Jamming was resumed in the USSR in August 1980 after a seven year interval.)

I am convinced that support of Amnesty International's call for a general, worldwide amnesty for prisoners of conscience is of special importance. The political amnesties proclaimed by a number of countries in recent years have helped to improve the atmosphere. An amnesty for prisoners of conscience in the USSR, in Eastern Europe, and in all other countries where political prisoners or prisoners of conscience are detained would not only be of major humanitarian significance but could also enhance international confidence and security.

The worldwide character of the scientific community assumes particular importance when dealing with such problems. By its international defense of persecuted scientists and of all persons whose rights have been violated, the scientific community confirms its international mandate, which is so essential for successful scientific work and for service to society.

Western scientists are familiar with the names of many Soviet colleagues who have been subjected to unlawful repressions. (I shall confine my discussion to the Soviet Union since I am better informed about it, but serious human rights violations occur in other countries including Eastern European countries.) The individuals I mention have neither advocated nor used violence since they consider publicity the only acceptable, effective and non-pernicious way of defending human rights. Thus, they are all prisoners of conscience as defined by Amnesty International. Their stories have much else in common. Their trials were conducted in flagrant violation of statutory procedures and in defiance of elementary common sense. My friend Sergei Kovalev was convicted in 1975 in the absence of the defendant and counsel, that is, with no possibility whatsoever for a defense. He was sentenced to seven years labor camp and three years internal exile for anti-Soviet agitation and propaganda allegedly contained in the *samizdat* news magazine *A Chronicle of Current Events*, but there was no examination of the substance of the charge.

Comparable breaches of law marked the trials of Yury Orlov, the founder of the Moscow Helsinki Group, and of other members of the Helsinki Groups and associated committees: Victor Nekipelov, Leonard Ternovsky, My-

psychiatric hospitals. This led to his first public warning by Soviet officials and to his public chastisement by some of his own academic colleagues for activities "hostile to the Soviet Union." The petty harassment which was initiated at this point by Soviet authorities began taking its toll physically on Sakharov, who suffers from serious heart problems, and on his wife, Yelena, also afflicted with serious health problems.

It was shortly after this that I first met Sakharov, when I attended a small working physics seminar organized by the Soviet Academy of Sciences in the summer of 1974 in Moscow. We shared scientific interests as well as our mutual concerns about the impact of science on the human condition, particularly in the field of nuclear weapons and their control. I found him a gentle colleague with whom I formed a warm bond of friendship. Sakharov invited me to his small, crowded, but humanly warm Moscow apartment for supper with several family members. As it turned out, this was to be his last supper for some time because it marked the start of a hunger strike which coincided with his delivery of a letter of protest to the heads of both the US and Soviet governments, just then starting summit talks in Moscow. His letter protested the restrictions on emigration for many of the ethnic minorities in the Soviet Union, and this hunger strike was his personal way of focusing the world's conscience and attention on this issue.

Sakharov achieved a pinnacle of reverence, respect, and recognition in the fall of 1975 when he was awarded the Nobel Peace Prize. As the Nobel citation so fittingly concluded:

Sakharov's love of truth and strong belief in the inviolability of the human being, his fight against violence and brutality, his courageous defense of the freedom of the spirit, his unselfishness and strong humanitarian convictions have turned him into the spokesman for the conscience of mankind, which the world so sorely needs today.

Characteristically, Andrei responded in his prize lecture by rededicating himself and calling attention to his countrymen who were prisoners of conscience with whom he wished to share the honor of the Nobel award, and, indeed, to prisoners of conscience everywhere:

Granting the award to a person who defends political and civil rights against illegal and arbitrary actions means an affirmation of principles which play such an important role in determining the future of mankind. For hundreds of people, known or unknown to me, many of whom pay a high price for the defense of these same principles—the price being loss of freedom, unemployment, poverty, persecution, exile from one's country—your decision was a great personal joy and gift.

These words were read by his wife, as Sakharov was unable to attend the ceremonies at which the award was presented for the same reason that he was unable to participate in the celebration of his sixtieth birthday.

In fact, while the Nobel ceremonies were in progress, Sakharov himself was in Vilna, Lithuania, in a vigil at the trial of a close friend and fellow leader of the human rights movements, the brave and then painfully ill biologist, Sergei Kovalev. And on that occasion, even as he was calling attention to a serious violation of human rights, he emphasized his concerns about survival in our nuclear armed world. I quote from Sakharov's Vilna statement:

It is absolutely unacceptable—even for a goal as important as respect for human rights—to make conduct in that area a precondition for disarmament negotiations. Disarmament must have first priority.

Sakharov is paying a high price today —and has paid for the last ten years—for having the courage to speak freely and courageously on fundamental issues that challenge the dogmas of the closed society in which he lives. In 1973 he told the Swedish journalist Olle Stenholm that

There is a need to create ideals even when you can't see any route by which to achieve them, because if there are no ideals, then there can be no hope, and then one would be completely in the dark. . . .

And in his book *My Country and the World*, published in 1975, he wrote

. . .The struggle for greater humanity in places of imprisonment and for human rights in general is not only the moral duty of honest persons throughout the world but constitutes a direct defense of human rights in their own countries.

Valentin Turchin, a former close associate in the human rights movement in Moscow and who, very happily, is now in the West, has noted the morality that compels Sakharov to speak out against injustice, near and far, great and small. Turchin has called him the "classical example of a prophet whose actions in defense of human rights arise out of his heart and soul." Turchin's words call to mind those of William Faulkner in *Intruder in the Dust*:

Some things you must always be unable to bear. Some things you must never stop refusing to bear. Injustice and outrage and dishonor and shame. No matter how young you are or how old you have got. Not for kudos and not for cash: your picture in the paper nor money in the bank either. Just refuse to bear them.

Sidney Drell
Stanford Linear Accelerator Center

kola Rudenko, Alexander Podrabinek (and his brother Kirill), Gleb Yakunin, Vladimir Slepak, Malva Landa, Robert Nazarian, Eduard Arutyunian, Vyacheslav Bakhim, Oles Berdnik, Oksana Meshko, Mykola Matusevich and his wife, and Miroslav Marinovich. Tatiana Osipova, Irina Grivnina and Felix Serebrov have been imprisoned pending trial. (On 2 April, Osipova was sentenced to five years labor camp and five years internal exile) Yury Orlov's lawyer missed part of the trial proceedings when he was locked up forcibly in chambers adjoining the courtroom. Orlov's wife was frisked in a crude way and her clothing ripped during a search for written notes or a tape recorder, all from fear that the court's grotesque secrets might be revealed.

In the labor camps, prisoners of conscience suffer cruel treatment: arbitrary confinement in punishment cells, torture by cold and hunger, infrequent family visits subject to capricious cancellation, and similar restrictions on correspondence.

The political prisoners share all the rigors of the Soviet penal regimen for common criminals while suffering the added strain of pressure to "embark on the path of reform," that is, to renounce their beliefs. I would like to remind you that not once has any international organization, such as the Red Cross or a lawyer's association, been able to visit Soviet labor camps.

Political prisoners are often rearrested, and monstrous sentences imposed. Ornithologist Mart Niklus, poet Vasily Stus, physics teacher Oleksei Tikhy, lawyer Levko Lukyanenko, philologist Viktoras Petkus and Balys Gajauskas have all received sentences of ten years labor camp and five years internal exile as recidivists. A new trial is expected for Paruir Airikian, who is still in labor camp. Within the last few days I have been shocked by the fifth (!) arrest of my friend Anatoly Marchenko, a worker and author of two talented and important books: *My Testimony* and *From*

Tarusa to Siberia. Imprisoned religious believers include Rostislav Galetsky, Bishop Nikolai Goretoi, Alexander Ogorodnikov, and Boris Perchatkin. Imprisoned workers include Yury Grimm and Mikhail Kukobaka. Alexei Murzhenko and Yury Fedorov are still imprisoned. I shall name only a few scientists deprived of their freedom; many others could be added to the list: Anatoly Shcharansky, the young computer scientist now famous around the world; mathematicians Tatiana Velikanova, Alexander Lavut, Alexander Bolonkin and Vazif Meilanov; computer scientist Victor Brailovsky; economist Ida Nudel; engineers Reshat Dzhemilev and Antanas Terleckas; physicists Rolan Kadiyev, Iosif Zisels and Iosif Dyadkin; chemists Valery Abramkin and Juri Kukk; philologists Igor Ogurtsov and Mustafa Dzhemilev; and Vladimir Balakhonov. (I have only recently received word of the tragic death of Juri Kukk in a labor camp.)

A common violation of human rights, and one which especially affects scientists, is denial of permission to emigrate. The names of many "refuseniks" are known to the West.

I was banished without a trial to Gorky more than a year ago and placed under a regimen of almost total isolation. A few days ago the KGB stole my manuscripts and notebooks which contained extracts from scientific books and journals. This is a new attempt to deprive me of any opportunity for intellectual activity, even in my solitude, and to rob me of my memory. For more than three years Elizaveta Alexeyeva, my son's fiancee, has been arbitrarily prevented from leaving the Soviet Union. I have mentioned my own situation because of the absence of any legal basis for the actions taken and because the detention of Elizaveta is undisguised blackmail directed against me. She is a hostage of the state.

I appeal to scientists everywhere to defend those who

have been repressed. I believe that to protect innocent persons it is permissible and, in many cases, necessary to adopt extraordinary measures such as an interruption of scientific contacts or other types of boycotts. I urge the use, as well, of all the possibilities of publicity and of diplomacy. In addressing the Soviet leaders, it is important to take into account that they do not know about—and probably do not want to know about—most letters and appeals directed to them. Therefore, personal interventions by Western officials who meet with their Soviet counterparts have particular significance. Western scientists should use their influence to press for such interventions.

I hope that carefully thought out and organized actions in defense of victims of repression will ease their lot and add strength, authority and energy to the international scientific community.

I have titled this letter "The Responsibility of Scientists." Tatiana Velikanova, Yury Orlov Sergei Kovalev and many others have decided this question for themselves by taking the path of active, self-sacrificing struggle for human rights and for an open society. Their sacrifices are enormous, but they are not in vain. These individuals are improving the ethical image of our world.

Many of their colleagues who live in totalitarian countries but who have not found within themselves the strength for such struggle, do try to fulfill honestly their professional responsibilities. It is, in fact, essential to work at one's profession. But has not the time come for those scientists, who often exhibit their perception and nonconformity when with close friends, to demonstrate their sense of responsibility in some fashion which has more social significance, and to take a more public stand, at least on issues such as the defense of their persecuted colleagues and control over the faithful execution of domestic laws and the performance of international obligations? Every true scientist should undoubtedly muster sufficient courage and integrity to resist the temptation and the habit of conformity. Unfortunately, we are familiar with too many counterexamples in the Soviet Union, sometimes using the excuse of protecting one's laboratory or institute (usually just a pretext), sometimes for the sake of one's career, sometimes for the sake of foreign travel (a major lure in a closed country such as ours). And was it not shameful for Yury Orlov's colleagues to expel him secretly from the Armenian Academy of Sciences while other colleagues in the USSR Academy of Sciences shut their eyes to the expulsion and also to his physical condition? (He is close to death.) Many active and passive accomplices in such affairs may themselves someday attract the growing appetite of Moloch. Nothing good can come of this. Better to avert it.

Western scientists face no threat of prison or labor camp for public stands; they cannot be bribed by an offer of foreign travel to forsake such activity. But this in no way diminishes their reponsibility. Some Western intellectuals warn against social involvement as a form of politics. But I am not speaking about a struggle for power. This is not politics. It is a struggle to preserve peace and those ethical values which have been developed as our civilization evolved. By their example and by their fate, prisoners of conscience affirm that the defense of justice, the international defense of individual victims of violence, the defense of mankind's lasting interests are the responsibility of every scientist.

An autobiographical note

I was born on 21 May 1921, in Moscow. My father was a well-known physics teacher and the author of textbooks and popular science books. My childhood was spent in a large communal apartment where most rooms were occupied by our relatives with only a few outsiders mixed in. Our home preserved the traditional atmosphere of a numerous and close family—respect for hard work and ability, mutual aid,

---

Sakharov's 60th birthday

Andrei Sakharov was 60 years old on 21 May. To honor the occasion, the New York Academy of Sciences, The American Institute of Physics and The American Physical Society sponsored an international conference, held in New York City on 1–2 May.

Sakharov prepared the accompanying article for presentation at the conference. Although it is traditional for the honored scientist to be present at such celebrations, to receive the kudos in person, Sakharov was of course not able to attend. However, a film was shown at the conference of Sakharov reading the article, in Russian, at his apartment in Gorky, at the end of March 1981. The autobiographical notes were prepared for Russian readers of *samizdat*, the privately circulated, unofficial publication system in the USSR. (The English translations are courtesy of Khronika Press.) The photographs that illustrate this article were taken by Jeri Laber of the U.S. Helsinki Watch in Moscow, September 1971.

The accompanying contributions by Ronald Reagan, I. I. Rabi and Sidney Drell are based on presentations at the conference.

Sakharov's exile continues to become more repressive. According to his latest report, his diaries (scientific and personal), correspondence, scientific notebooks and manuscripts of his autobiography were stolen from his apartment, presumably by—or at least with sanction from—the KGB. Tanya Yankelovich, Sakharov's stepdaughter, said that he is allowed to receive reprints, but he is not allowed to see colleagues nor to visit libraries, since last June.

At the conference, Antonio Zichichi announced that a thousand European scientists have agreed that all their scientific papers will say on the front page: Dedicated to Andrei Sakharov on his 60th birthday. There will also be a concert in Milan, Italy, to raise funds for his defense, and a conference in Rome in the fall in his honor.

The New York conference included a concert, informal discussions, and formal sessions on:
► Sakharov's contributions to science, with lectures by John Wheeler of the University of Texas, Val Fitch of Princeton University and Harold Furth of the Princeton Plasma Physics Laboratory.
► Issues of war and peace, with talks by Herbert York, of the University of California at San Diego; Stanislaw Ulam, of Los Alamos; and McGeorge Bundy, of New York University.
► Human rights and justice, with contributions from Sir Karl Popper, of the University of London, Bayard Rustin of the A. Philip Randolph Institute, Harrison Salisbury, of the New York Times, Philip Handler, of the NAS, and Ernest Nagel, of Columbia.

---

love for literature and science. My father played the piano well; his favorites where Chopin, Grieg, Beethoven and Scriabin. During the Civil War he earned a living by playing the piano in a silent movie theatre. I recall with particular fondness Maria Petrovna, my grandmother and the soul of our family, who died before World War II at the age of 79. Family influences were especially strong in my case because I received my early schooling at home and then had difficulty relating to my own age group.

After graduating from high school with honors in 1938, I enrolled in the Physics Department of Moscow University. When war began, our classes were evacuated to Ashkhabad, where I graduated with honors in 1942. That summer I was assigned work for several weeks in Kovrov, and then I was employed on a logging operation in a remote settlement near Melekess. My first vivid impression of the life of workers and peasants dates from that difficult summer of 1942. In September I was sent to a large arms factory on the Volga, where I worked as an engineer until 1945.

I developed several inventions to improve inspection procedures at that factory. (In my university years I did not manage to engage in original scientific work.) While still at the factory in 1944, I wrote several articles on theoretical physics which I sent to Moscow for review. Those first articles have never been published, but they gave me the confidence in my powers which is essential for a scientist.

In 1945 I became a graduate student at the Lebedev Institute of Physics. My advisor, the outstanding theoretical physicist, Igor Tamm, who later became a member of the Academy of Sciences and a Nobel laureate, greatly influenced my career. In 1948 I was included in Tamm's research group which developed a thermonuclear weapon. I spent the next twenty years continuously working in conditions of extraordinary tension and secrecy, at first in Moscow and then in a special research center. We were all convinced of the vital importance of our work for establishing a worldwide military equilibrium, and we were attracted by its scope.

In 1950 I collaborated with Igor Tamm in some of the first research on controlled thermonuclear reactions. We proposed principles for the magnetic thermal isolation of plasmas. I also suggested as an immediate technical objective the use of a thermonuclear reactor to produce fissionable materials as fuel for atomic power plants. Research on controlled thermonuclear reactions is now receiving priority elsewhere. The Tokamak system, which is under intensive study in many countries, is most closely related to our early ideas.

In 1952 I initiated experimental work on magnetic–explosive generators (devices to transform the energy of a chemical or nuclear explosion into the energy of a magnetic field). A record magnetic field of 25 megagauss was achieved during these experiments in 1964.

In 1953 I was elected a member of the USSR Academy of Sciences.

My social and political views underwent a major evolution over the fifteen years from 1953 to 1968. In particular, my role in the development of thermonuclear weapons from 1953 to 1962 and in the preparation and execution of thermonuclear tests, led to an increased awareness of the moral problems engendered by such activities. In the late 1950s I began a campaign to halt or to limit the testing of nuclear weapons. This brought me into conflict first with Nikita Khrushchev in 1961, and then with the Minister of Medium Machine Building, Efim Slavsky, in 1962. (This is the Ministry responsible for nuclear weapons and industry in the USSR.) I helped to promote the 1963 Moscow Treaty Banning Nuclear Weapon Tests in the Atmosphere, in Outer Space and Under Water. From 1964 when I spoke out on problems of biology (at the Academy of Sciences, during a debate on the election of one of Trofim Lysenko's associates), and especially from 1967, I have been interested in an ever-expanding circle of questions. In 1967 I joined the Committee for Lake Baikal, which was organized to protect one of the purest lakes in the world from industrial pollution. My first appeals for victims of repression date from 1966–67.

The time came in 1968 for the more detailed, public and candid statement of my views contained in the essay "Progress Coexistence and Intellectual Freedom." These same ideas were echoed seven years later in the title of my Nobel lecture: "Peace, Progress and Human Rights." I consider the themes of fundamental importance and closely interconnected. My 1968 essay was a turning point in my life. It quickly gained world-wide publicity. (It was published in English by *The New York Times*.) The Soviet press was silent for some time, and then began to refer to the essay very negatively. Many critics, even sympathetic ones, considered my ideas naive and impractical. But thirteen years later, it seems to me that these ideas foreshadowed important new directions in World and Soviet politics.

After 1970, the defense of human rights and of victims of political repression became my first concern. My collaboration with physicists Valery Chalidze and Andrei Tverdokhlebov, and later with the mathematician Igor Shafarevich and geophysicist Grigory Podyapolsky, on the Moscow Human Rights Committee was one expression of that concern. (Podyapolsky's untimely death in March 1976 was a tragedy.)

After my essay was published abroad in July 1968, I was barred from secret work and excommunicated from many privileges of the Soviet establishment. The pressure on me, my family and friends increased in 1972, but as I came to learn more about the spreading repressions, I felt obliged to speak out in defense of some victim almost daily. In recent years I have continued to speak out as well on peace and disarmament, on freedom of contacts, movement, information and opinion, against capital punishment, on protection of the environment, and on nuclear power plants.

In 1975 I was awarded the Nobel Peace Prize. This was a great honor for me as well as recognition for the entire human rights movement in the USSR. In January 1980 I was deprived of all my official Soviet awards (the order of Lenin, three times Hero of Socialist Labor, the Lenin Prize, the State Prize) and banished to Gorky where I am virtually isolated and watched day and night by a policeman at my door. The regime's action lacks any legal basis. It is one more example of the intensified political repression gripping our country in recent years.

Since the summer of 1969 I have been a senior scientist at the Academy of Sciences' Institute of Physics. My current scientific interests are elementary particles, gravitation and cosmology.

I am not a professional politician. Perhaps that is why I am always bothered by questions concerning the usefulness and eventual results of my actions. I am inclined to believe that moral criteria together with uninhibited thought provide the only possible compass for these complex and contradictory problems. I shall refrain from specific predictions, but today as always I believe in the power of reason and the human spirit.   □

Sakharov speaks from exile on nuclear-arms issues

PHYSICS TODAY/AUGUST 1983

"What you say and write about the appalling dangers of nuclear war is very close to my heart and has disturbed me profoundly for many years now," begins Andrei Sakharov in an open 7000-word letter to Sidney Drell (SLAC). The complete text of the letter, translated by Richard Lourie and Efrem Yankelevich, was published in *Foreign Affairs* in June (Summer, page 1001). Sakharov, exiled in Gorky, was responding to Drell's recent public comments on nuclear war and nuclear weapons. Drell is a member of Jason—a group of top physicists who regularly consult for the Defense Department and Federal agencies—a longtime adviser to the government on issues of national security and arms control, and an outspoken supporter of arms control. (For a comprehensive presentation of his views, see "Facing the Threat of Nuclear Weapons," recently published by the University of Washington Press.) In particular, Sakharov addressed issues explored by Drell in his testimony in September before a House subcommittee about the technical capabilities of nuclear weapons and the resulting consequences of a nuclear confrontation (before the Subcommittee on Oversight and Investigations of the House Committee on Science and Technology) and to the general public (at Grace Cathedral in San Francisco in October) about how technological improvements to nuclear weapons lend urgency to disarmament talks.

Sakharov spends the first third of his letter agreeing with Drell about the horrors of nuclear war. In fact, his discussion of the direct and indirect effects of nuclear war adds weight to Drell's comments. Sakharov also refers, for example, to the recent estimate of the Royal Swedish Academy, "according to which an attack on the principal cities of the Northern Hemisphere by 5000 warheads with a total power of 2000 megatons will kill 750 million people as a result of the shock wave alone." He then adds to that estimate. The number of deaths would be greater, he says, because the overall number of weapons owned by the five nuclear powers is three to four times

Andrei Sakharov, physicist and winner of the Nobel Peace Prize in 1975, shown here in his Moscow apartment before he was sent into internal exile in Gorky.

more than the total used by the Academy for this estimate, and the effects of both thermal radiation and fallout would also be greater. So, he says, "it should be said that all-out nuclear war would mean the destruction of contemporary civilization, hurl man back centuries, cause the death of hundreds of millions or billions of people, and, with a certain degree of probability, would cause man to be destroyed as a biological species and could even cause the annihilation of life on Earth. Clearly, it is meaningless to speak of victory in a large nuclear war which is collective suicide."

Strategy. Because the use of nuclear weapons is suicidal, and because it is not possible to limit their use once the "nuclear threshold" has been crossed, *"Nuclear weapons only make sense as a means of deterring nuclear aggression by a potential enemy,* i.e., a nuclear war cannot be planned with the aim of winning it," Sakharov says. If the use of nuclear weapons is unacceptable, they cannot act as a deterrent for conventional weapons, which leads Sakharov to conclude that strategic parity of conventional weapons must be

restored, a conclusion Drell also draws.

What is meant by a deterrent, however? Sakharov asks, is it possible to "simply limit oneself to the criterion of achieving a reliable deterrent—when that criterion is understood to mean an arsenal sufficient to deal a devastating blow in response?" Drell has said that the US has enough secure nuclear arms available on bombers and submarines, as well as ongoing modernization programs to assure the security and effectiveness of these systems, that a window of vulnerability is not created by the Soviet numerical advantage in silo-based missiles. This, and the absence of a survivable land-basing scheme for them, leads Drell to conclude that neither the development nor the deployment of the MX is needed.

Sakharov disagrees. We must consider specific scenarios, he says. For example, will a country retaliate with nuclear weapons after being devastated? What advantage would they gain? If they will not retaliate, doesn't this scenario encourage an aggressor to assume or hope for capitulation? To avoid such potential advantages that promote the fighting of a nuclear war,

Sakharov believes that parity in each variant of nuclear arms must also be restored.

"Of course I realize that in attempting not to lag behind a potential enemy in any way, we condemn ourselves to an arms race that is tragic in a world with so many critical problems admitting of no delay. But the main danger is slipping into an all-out nuclear war. *If* the probability of such an outcome could be reduced at the cost of another ten or fifteen years of the arms race, then perhaps that price must be paid, while at the same time, diplomatic, economic, ideological, political, cultural, and social efforts are made to prevent a war," he says.

In our hopes and efforts for peace Sakharov reminds us not to lose sight of the complexity of the "specific political, military, and strategic realities of the present day." According to Sakharov, the practical problem of getting objective information about these realities is complicated by pro-Soviet propaganda, including pro-Soviet elements in mass media in the West. In line with his belief that a balance in conventional arms is needed to effect a reduction in nuclear arms, Sakharov cites the resistance to President Carter's attempt to reinstate the draft as one instance of public opinion gone awry due to insufficient information.

Disarmament talks. To achieve the goal of reducing the number of missiles, including "not moving the missiles behind the Urals but *destroying* them," there must first be a fair assessment of the quality, not just the quantity of the missiles. In fact, Sakharov endorses a counting scheme proposed by Drell, which uses the aggregate total of launchers plus warheads to assess nuclear strength. Such factors as accuracy, range, and degree of vulnerability have to be taken into account at the disarmament talks. Thus, he says, "One also must not consider powerful

Soviet missiles, with mobile launchers and several warheads, as being equal to the now-existing Pershing I, the British and French missiles, or the bombs on short-range bombers, as the Soviet side sometimes attempts to do for purposes of propaganda." Similarly, as the Soviets have an advantage in silo-based missiles, Sakharov suggests that, "Perhaps talks about the limitation and reduction of these most destructive missiles could become easier if the United States were to have MX missiles, albeit only potentially (indeed, that would be best of all)."

In addition, he says, "Much is written about the possibility of developing ABM systems using super-powerful lasers, accelerated particle beams, and so forth. But the creation of an effective defense against missiles along these lines seems highly doubtful to me." (See page 17, this issue.) Thus the specific and substantial military capabilities of large silo-based missiles must be considered. Sakharov says that one large rocket can carry a charge of up to 15–25 megatons. If used on a city, such a charge is capable of totally destroying dwellings in a 250–400 km^2 area, and of creating thermal radiation effects in a 300–500 km^2 area and radioactive fallout over an area 500–1000 km long by 50–100 km wide. These rockets can also accommodate multiple reentry vehicles. As an example, Sakharov considers an attack on Soviet launch sites by the 100 MX missiles proposed by the Reagan Administration for the first round of deployment. These missiles could carry 1000 600-kiloton warheads. Sakharov refers to American data that take into account both accuracy and the known hardness of Soviet launch sites, and lead to the determination that there is a 60% probability of destroying one launch site. Thus during an attack on 500 Soviet sites, with two warheads for each site, he calculates that "only" 80 missiles would remain. This ability, to destroy three

to four times more enemy missiles than are used, is destabilizing. Eliminating them is thus, for Sakharov, the top priority for arms talks. As the Soviets will not give up their advantage voluntarily, the West must come to the arms talks with something to give up. Thus he says "If it is necessary to spend a few billion dollars on MX missiles to alter this situation, then perhaps this is what the West must do."

Social and political problems, however, not technology, precipitate wars, whether conventional or nuclear, he says. The "relentless expansion of the Soviet sphere of influence," and the exploitation of developing countries both by the Soviets and the West are sources of concern for Sakharov. He notes the Soviet invasion of Afghanistan, not only for the cruelty of the confrontation itself and the implied danger of escalation to global war, but also as a "fundamental reason that the SALT II agreement was not ratified." Peace is connected to openness in society and to human rights. "Citizens have the right to control their national leaders' decision-making in matters on which the fate of the world depends. But we don't even know how, or by whom, the decision to invade Afghanistan was made!" he says. Even factual information is not freely accessible in the Soviet Union and many citizens have been incarcerated for transmitting information. Sakharov cites the plight of Anatoly Shcharansky, in Chistopol Prison, and Yuri Orlov in a Perm Labor Camp, but he neglects to speak of himself. As of this writing, Andrei Sakharov and his wife, Yelena Bonner, had both suffered heart attacks. Sakharov was being denied permission to travel to Moscow for treatment. His wife refused hospitalization for her condition and returned to Gorky because she felt that her husband could not be left alone. To date, no arrangements have been made for them to be hospitalized together in Moscow. —JC

Physics and the military

PHYSICS TODAY/OCTOBER 1984

Charles Schwartz

This is a time when militarism dominates the planet. The two major powers are locked in a nuclear arms race with new weapons that appear more designed for fighting a nuclear war than for avoiding one. Arms control, not very successful in the past, is now barely more than a charade. Science and technology provide the leading edge in this mad race.

While we in the "advanced" nations sit in fearful anticipation of a coming nuclear war, millions elsewhere experience the immediate force of conventional war and military repression sponsored by the superpowers in their global reach; here, again, the militarists make full use of modern science. For many people outside of the US and the USSR there is little difference between these two supepowers in how they use their military might. The Europeans protest against SS-20 missiles as well as against Cruise and Pershing missiles; and how does one distinguish between their actions in Grenada, Afghanistan, Nicaragua, Poland, Chile, and so on?

Where does one look for a political force that might be able to turn these bellicose policies around? The usual choice offered by elections seems quite inadequate: previous administrations have contributed greatly to our present predicament. I think the opposition must come from a viable grassroots movement against the nuclear arms race and against excessive militarism in general. The campaign for a nuclear weapons freeze and the Caholic Bishops' Pastoral Letter are significant elements of this movement, and so are the peace marches and demonstrations carried out at the Livermore laboratory and elsewhere. The outcome of this political struggle is still undecided. The central question for each scientist and for every institution of science is: Which side are you on? There is not much space for neutrality when the stakes involve the possible destruction of our civilization.

© PETER BONO 1984

Charles Schwartz is professor of physics at the University of California, Berkeley.

Most of the scientists working in the weapons laboratories and the defense industries continue their tasks: patriotism, economic pressure and team loyalty are potent forces. Some, in order to deal with the emotional pain of developing weapons of mass destruction, find comfort in such rationales as, "We must modernize our systems to maintain a surivable deterrent," or "I only work on defensive systems, not offensive ones," or "It is better that a liberal like me be involved in this rather than leaving it all to narrow-minded reactionaries."

But what about academic scientists? Most say that they have nothing to do with weapons development, and most of their research money comes not from the Defense Department but from NSF or DOE and is for pure research only. There are a number of academic scien-

tists who have been active in support of the peace movement—this as a purely individual political activity—while others choose, or rather have been chosen, to work as advisers and consultants to the Pentagon.

For many, this is the end of the story: We in university physics departments are seen as insulated from the military; there may be a problem, but we are not a part of it. My purpose now is to dig deeper into this question: I have identified five aspects of our connection to the military.

▶ Recently I have heard increasing numbers of unhappy physics undergraduate students report on their search for a job outside the university. Most all of the jobs available are involved with weapons development, something these students do not want to do. There has been a strong shift in priorities of the federal budget for R&D—from 50% dedicated to military programs a few years ago to 70% today—and this shift must have a large impact on the new job openings our students will have to choose from. Yet reliable statistics on technical jobs, how many in the civilian sector versus how many in the military, are simply not available. I think we, as educators, have an obligation to collect such data and provide it for prospective students early enough so that they may make well informed choices about their future careers.

▶ Several areas of research conducted in our department are not so far removed from weapons projects. A few graduate students have encountered this dilemma, and told me of their chagrin upon realizing the potential military applications and the active military interest in research projects they were becoming involved with. Some faculty colleagues have also raised similar concerns regarding their own research work. I think we are all generally aware of the military interest in, and funding for, research involving lasers (many kinds), accelerators, new materials, radiation detectors, electronic devices, plasma physics and nonlinear dynamics. What we should do, as a minimum, is devote some collective effort to analyzing and publicizing the potential for military applications in all areas of research relevant to us for the purpose of providing clear warnings to those of our students and colleagues who might otherwise blunder unknowingly into areas they would wish to avoid.

▶ There is far too little effort to bring these troublesome questions about the end uses of our science into the regular educational curriculum; occasionally, those who try to raise such questions in our classrooms and academic rituals are told sternly that they are acting improperly. These issues are controversial and painful but trying to avoid them does not make them less of a reality. Indeed, our silence implies acquiescence to, if not support for, the military's plans for our science. I suggest that we need to become more willing to discuss and explore these questions wherever they arise in our work; I also think the Department of Defense could profitably establish an ongoing seminar, with invited experts from all over, to help us all deepen our understanding of the many facets of militarism and its relation to science.

▶ A number of universities, while having no direct weapons work on their campuses, are closely allied with special laboratories dedicated to military research and development. The prestige, the access and the cloak of legitimacy which universities thus extend to weapons work is greatly valued by the military; but this liaison makes many of us in academia feel badly abused. Nowhere is the situation more blatant than my own institution, tied as it is to Livermore and Los Alamos. I think this will be a live issue on campus in the coming year as our Regents prepare for negotiations on renewal of the contract for the nuclear weapons labs. Our faculty cannot claim ignorance of their purpose. The faculty is responsible for them.

▶ Now let us consider the most important contribution we make to the military and the hardest of all for us to deal with. Every year we teach our basic introductory physics courses to over a thousand students who are just beginning their careers in the physical sciences and engineering disciplines. This is the primary raw material on which the military-industrial complex feeds; and we, the faculty and graduate students who teach this basic science of physics, are thus an essential and irreplaceable part of the military production process. This is such an elementary fact, yet it is the one that I, and probably many others, have the greatest difficulty in confronting squarely. There is a great temptation to find some excuse that will let us off the hook. For example: "We only teach basic truths about nature; there is nothing good or evil about this; how others may use this knowledge is something we cannot control." But we *do know* that this knowledge we teach *will be used* to create ever more weapons, as things are going now, and the results are very frightening. We cannot, in honesty, deny our central part in this. What we might do by way of resistance is a very hard question to which I offer no answers now. None will be easy; but I think we need to start discussing this very question with one another in an ongoing way.

Some recent statements by leading science officials in Washington are quite candid on the subject of how basic science and the universities are needed as integral parts of the military system.

Reading the Pentagon budget statements over many years one sees, along with requests for R&D money assigned to particular weapons programs, the particular item called "science and technology base." This refers to efforts at supporting pure research and training of technical personnel which will have long term utility to the military. In Congressional testimony (1982)[1] Richard DeLauer, recent Undersecretary of Defense for Research and Engineering stated, "The DOD depends on the university community to provide scientific and technical personnel to DOD, to do basic and applied research and to provide expert consultants and independent advice." Among the specific new efforts directed towards the universities DeLauer mentioned: funds "to upgrade selected equipment in the universities where it will add to their capability for research in the high-leverage technologies needed by DOD"; also a new program of "graduate fellowships in selected technologies"; and an "apprenticeship program" to "motivate talented high school youngsters to enter technical fields."

Direct Defense Department funding is only a small fraction of the overall research support that universities receive from the federal government (with exceptions to this rule in particular fields, like computer science). However, as acknowledged in a recent report of the Defense Science Board (see page 256 of reference 1) the different agencies (such as DOD, DOE, NSF and so on) share the funding of basic research in a mutually supporting way: "Research and development in universities is supported by many sponsors, each relying on complementary funding from the other sponsors to leverage its own expenditures."

An example of this interagency cooperation is found in x-ray laser research, which the Defense Department's Advanced Research Projects Agency supported until the late 1970's. Then, in his FY 1978 report to Congress,[2] the Director of DARPA stated that although this project was "technically very successful" its "military impact was at best indirect [and] long range." Therefore, he said, "We have ... terminated the program and have recommended further that research in this area be funded by NSF."

The degree of federal coordination in research funding and the priority given to military programs has, I think, been especially strong under the present administration. The two highest science policy offices in Washington, outside of the Pentagon, have been given to people from the Los Alamos nuclear weapons laboratory: George Keyworth, science advisor to the President and director of the Office of Science and Technology Policy; and Edward Knapp, director of the National Science Foundation.

For those who still cling to the illusion that most of what we academic physicists do is separate and insulated from military objectives, I recommend reading the article by Keyworth, published in the 6 April, 1984, issue of *Science*. He describes the importance of nondefense, basic research as clearly integrated into a larger picture in three ways: 1) "research grants to universities ... permit the training of tens of thousands of graduate students ... This new talent will be responsible for maintaining American technological leadeship." 2) "basic research . . . provides the new knowledge that drives our ecomonic growth, improves our quality of life, and underlies our national defense." 3) "well-chosen basic research projects can stimulate productive partnerships . . . that will . . . . speed the application of new knowledge to our increasingly technological defense needs."

The sugestions I made earlier for constructive responses to the five situations described, although all academic activities, nevertheless have a political and moral aspect in the context of opposition to militarism. They represent a minimum of our exercise of a sense of social responsibility in the practice and teaching of our science. Yet we might also find the need to do more than just study the issues and provide information to others; we may find the urgent need to actively separate ourselves from the plans that our government has made to use us. This is a threatening idea to all of us, and one that many will try to avoid. We need to take a moment to pause, stand back, take a good look at where we are headed, and ask ourselves, How much farther can we afford to coast? When we have reached the very brink then it will be too late to apply the brakes.

Let me quote Victor Weisskopf. He was my thesis supervisor thirty years ago at MIT. He has been very outspoken lately about the dangers of the nuclear arms race; and this is what he said in a recent speech at Los Alamos entitled, "We Meant So Well".[3]

We must never be fatalistic about the inevitabiity of nuclear war. There *are* ways to avoid the holocaust and we must never cease to search for them If we don't succeed, our century will be remembered by the unfortunate survivors as the time of preparation for the great catastrophe, and science will be seen as the main culprit. Our century ought to be remembered as the age in which humankind acquired its deepest insights into the universe and learned to control its martial impulses.

Let us hope, strive and act so that it will.

References

1. *Hearings on Military Posture, etc.,* Committee on Armed Services, House of Representatives, 97th Congress, 2nd session Part 5 (2–30 March, 1982) pp. 83–85.

2. *DARPA FY 1978 R&D Program,* George H. Heilmeier, before subcommittee of the House Armed Services Committee (February 1977) pp 111–19.

3. *Bulletin of the Atomic Scientists,* August/September 1983, p. 26. □

Science, technology and the arms race

The nuclear arms buildup continues because the superpowers use nuclear weapons primarily as political tools and fail to relate to their real potential for mass destruction.

PHYSICS TODAY/JUNE 1981

Wolfgang K. H. Panofsky

The arms race, in particular the nuclear weapons competition between the Soviet Union and the US, threatens the very existence of man's civilization. One must admit, albeit reluctantly, that the nuclear balance between these two powers deserves at least partial credit for the absence of all-out hostilities for the longest period in recent history. But even in the absence of total conflict, the increase of potential devastation and the growth in the number of states possessing nuclear weapons have created tensions that increasingly overshadow all other concerns of mankind. Many hold science, which led to the release of nuclear energy, responsible for this evolution. But others look to science for the tools to reverse this threat to all humanity.

What has been the influence of science and technology on the arms race in the 35 years since Alamagordo and Hiroshima? I feel that the arms race continues to persist because nuclear policy makers tend simultaneously (1) to overestimate what science and technology can predict in some areas and (2) to ignore scientific and technical realities in other areas.

On the one hand the apparent acceptance by policy makers of the concept of limited nuclear war presupposes that we can, on the basis of scientific and technical knowledge, predict the effects of exchanges of nuclear weapons at various levels. The fact is such predictions are extremely uncertain and no one can give assurance that a limited conflict will stay limited. *Once nuclear war is initiated by any power, under any doctrine, in any theater, or for any*

Wolfgang Panofsky is director of the Stanford Linear Accelerator Center. Drawings by Robert Osborn from *Missile Madness* by Herbert Scoville and Robert Osborn, published by Houghton Mifflin, Boston. © 1970 by Herbert Scoville and Robert Osborne.

strategic or tactical purpose, the outcome will involve truly massive casualties and devastation, leading to effects on the future of mankind that are essentially uncalculable.

On the other hand policy makers seem to subjugate the known, scientifically determined capabilities of nuclear weapons to political objectives in deciding on how large nuclear arsenals should be. Instead the nuclear arms race is driven by the *perception* of nuclear weapons as symbols of power rather than by *physical* realities. Unfortunately, perceptions tend to become *political* realities, and thus nuclear weapons become symbols of power and strength and trading objects in arms control negotiations. Such a shift of the role of nuclear weapons from threats of massive death and destruction to political tools has denied each nation a rationale for defining what nuclear weaponry is sufficient for its needs.

If these facts were emphasized more often and more persuasively, then world leaders might begin to realize that such controversies as employment or non-employment of specialized weapons such as the neutron bomb, or the exact numbers of weapons systems permitted under arms control agreement, lack technical or military significance.

We are led finally to the profoundly pessimistic conclusion that arms control agreements have not been successful in limiting the destructive options of war, in particular those involving nuclear weapons, since so many other forces and processes (such as those above) have in fact amplified the nuclear arms race. This trend must be reversed or else the future will indeed be dim.

Many remedies for this condition have been discussed and will continue to be discussed. Leading the list of remedies must be a conscientious policy decision by the nuclear nations to govern their policies in regard to nuclear weapons by true technical and military circumstances, and to minimize their justification as symbols of prestige and power.

More specific suggested measures following from such a policy are summarized in the box. The hope that measures such as those listed might lead to a halt and even reversal in the growing accumulation of nuclear arms is based on the conclusion that a distorted perception over the function of nuclear weapons lies at the heart of the arms race. In particular, policy makers must acknowledge that nuclear weapons cannot compensate for deficiencies in conventional military power. *Only a continuing and insistent reminder about the technical realities can form the basis for agreements and unilateral acts leading to lowered limits on nuclear weapons deemed sufficient for national and international security. The present course is set for disaster.*

Science and military devices

The relation of science to war and preparation for war has been what I might call a love/hate relationship throughout recorded history. Let me remind you of the story of Archimedes: In the defense of Syracuse against the Romans, Archimedes reputedly devised numerous instruments of war that contributed materially to delaying Roman entry into that besieged city for many years. When the conquest of Syracuse finally took place in 212 B.C., Archimedes was engaged in "pure" science, drawing circles in the sand in studying basic geometry. When a Roman soldier approached him, Archimedes was reported to have uttered as his last words, "Don't disturb my circles." This story illustrates that scientists wish to be recognized by all, friend and foe alike, for their contributions to pure science, irrespective of whether they also play an important and at times even decisive role in providing armaments to their home country. A vexing conflict between two moral responsibilities of the scientist in relation to armaments emerges: his duty to his fatherland, on the one hand, and his responsibility to prevent the abuse of lethal weapons, on the other.

This dilemma is rooted directly in the nature of the scientific process. New science and technology evolve through a long chain: Basic scientific discoveries eventually lead to the development of specific new devices, and finally to their testing, production, distribution and use.

Each of these steps offers different needs and opportunities for control by human institutions. Most scientists and non-scientists would agree that *basic* science motivated solely by the desire for improved understanding of nature should remain unfettered. I would maintain that, however fallible humans and their institutions may be, decisions on how to live harmoniously with nature and its products will be made more wisely with understanding of the workings of nature than without. Yet once basic new revelations are apparent, applying restraints to their potential fruits, lest they be abused for asocial ends, is difficult. In regard specifically to the military arms control problem, one finds that controls are very difficult to apply to early phases of the development chain. As a practical matter, particularly if one takes into account the problems of "verification" (that is, policing of an arms control agreement) one finds that the areas one can primarily control are the test phase of the development and the final deployment.

It has become trite to say that the rate of progress in arms control has been so small that technological advances in military systems have outpaced the gains of arms control. Yet I would argue that if technological and purely military factors had been the only ones determining the buildup of arms, then arms-control efforts might well have stemmed and possibly reversed the nuclear arms race. In truth it has been largely political factors that have made it infeasible thus far for arms-control efforts to prevail over the arms buildup.

Other factors in the arms race

Evolution of the tools of war—as facilitated by advances in science and technology—has aggravated the action and reaction cycles between nations and has led the competition between offensive and defensive technology systems through successive stages. Yet while these cycles can be clearly identified, it appears that the nature of the arms competition is in fact *not* dominated by such direct causal factors. Examination of this history of the arms race suggests that non-scientific factors have preempted what might be logically predicated on technical or even military grounds. For instance, if one observes the arms competition between the US and the Soviet Union, one finds very few instances of clear action-reaction processes. Apparently other forces drive the decisions by both sides to proceed with arms acquisitions. What are these forces?

Let me describe a few:

▶ It appears clear that in all countries—capitalist and socialist alike—there are strong *institutional pressures* to augment armaments, which are exerted by the institutions charged with developing and producing or managing and using arms. That equivalent institutions within different social systems apply equivalent pressures for arms expansion does not mean that the sources for military expansion are identical, but it makes it indeed difficult to attribute arms growth to single factors such as the profit motive, communist or imperialist expansionism, inter-service rivalry, and so on. We are dealing here with the institutional inertia or other manifestations of historical persistence that are inherent in any highly organized human activity. Institutions always find it difficult to produce only a fixed and limited amount of any one commodity; the producer always finds reasons why more of what he can produce is needed.

▶ A second driving force in the arms race is what I will call *mirror imaging*. Rather than react to a potential oppo-

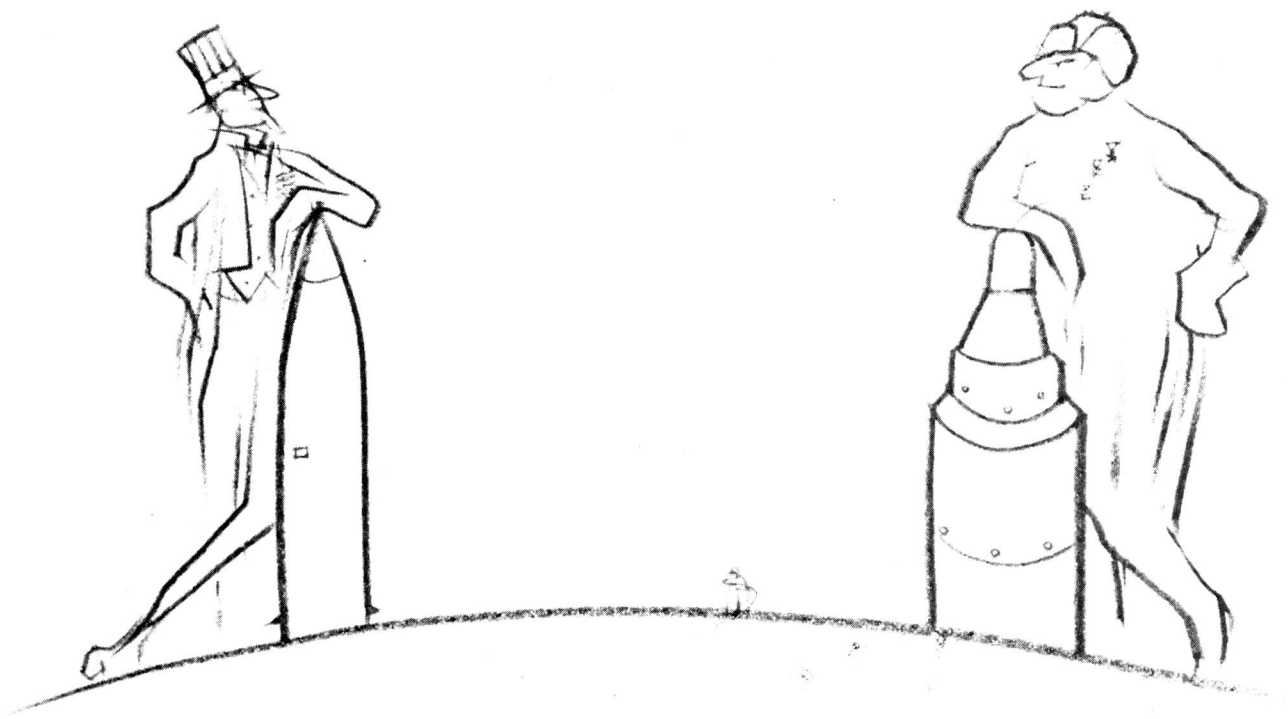

nent's new military system by implementing specific countermeasures, a nation frequently reacts by saying "We must have it too," and tends to be relatively uncritical with regard to the technical merit of the item in question. For instance, the public rationale given, among other factors, for the US to develop an improved hard-silo killer for its missile arsenal is that the Soviets have developed enough striking power to threaten preemptively the US Minuteman silos; yet the public rationale for the US "silo killer" missiles disclaims any intent for a "preemptive" or "first" strike. A "mirror image" response aims to silence the implied threat that the Soviets would receive *political* advantage and bargaining leverage out of these weapons, unless the US could field a matching threat. Conversely, one of the motives for some Soviet deployments appears to be to make "mirror images" of US moves rather than to take countermeasures. For example, the Soviets continue to upgrade the technological complexity of many of their military systems, following the US lead. Yet in so doing they at times inherit maintenance and reliability problems that have so frequently plagued the US systems in the past.

Very new and frequently spectacular technology is particularly sensitive to this aspect of the arms race. For example, the militarily highly dubious development of particle-beam weapons is largely promoted by the argument that "The other side is doing it," rather than by valid military-technical arguments. The difficulty here is that the political prestige engendered by a perceived technical-military "break-

through" is simply too valuable a commodity for decision makers to forego, even if the true military significance is minor or non-existent.

▶ Possibly the most serious driving factor in the arms race is *the asymmetry of perception*, resulting in uncertainty and fear. Each side in an adversary situation tends to feel insecure and threatened by the technical developments of the opponent, particularly if the technical details are poorly known and if the doctrine under which they are developed is ambiguous. For instance, the claim that the opponent is not only striving for but has in fact attained nuclear superiority rather than parity is being heard currently *both* from US and Soviet spokesmen. These spokesmen may in fact be sincere, yet both cannot be correct. In fact most discussions on superiority tend to be based on simplistic numerology. Factors that are difficult to quantify such as the reliability of allies, asymmetries in geography, access to ocean areas, the length and number of boundaries shared with potential adversaries, make "superiority" quality difficult to users. As a result attributions of "superiority" are based on *perceptions* of potential adversaries rather than an *objective reality*.

Given the asymmetry of information and interpretation it is understandable that both sides in the current arms buildup feel genuinely threatened. Political leaders at times react to these perceptions by "sabre-rattling," that is, by proclaiming their military prowess "second to none" to reassure their constituencies. At other times, in particular when military appropriations or other allocations of resources are at

stake, political leaders emphasize the "gaps" in the armor of their country, the areas of perceived deficiencies and weakness. The perceived, not the real, military situation strongly affects decisions on the acquisition of military weapons.

▶ The development and deployment of arms generally requires very long and frequently uncertain lead times—well over a decade elapses between development for most systems and their completion and final deployment. Thus military needs should be based on a combination of judgment of the opponent's intentions, evaluation of his current status, and on intelligence projections from that base. The validity of these projections and their accuracy are not only impaired by the basic uncertainty in foreseeing future developments and further intentions starting from known current conditions, but also by the ignorance imposed by the *secrecy* that shrouds these conditions. Thus it is not surprising that judgments as to future defense needs can span a very wide range between "best" and "worst" case projections. This gap is simplified by secrecy, and thus secrecy becomes a major factor in driving the arms race. I challenge national leaders, in particular those of the Soviet Union, to examine critically whether extreme secrecy on technical developments is really in the best security interest of their country, quite apart from the damage such secrecy does to basic human values, and incidentally, to the feasibility of negotiated, verifiable arms control.

▶ The complaint by a technical person that for the most part political and perceptual rather than technical fac-

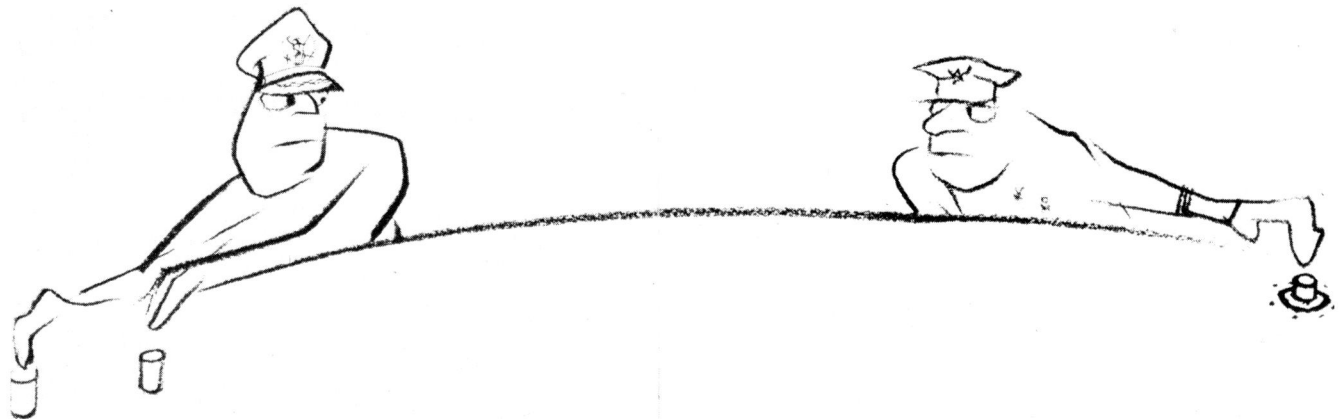

tors have driven the arms race must in fairness be balanced by the existence of what has been designated as the *technological imperative*. While a vast number of technical developments are on balance beneficial, there may be a rational assessment that a certain new development, if generally adopted, might result in decreased security for all; the strategically destabilizing influence of high accuracy in ballistic missiles is an example. Yet prohibiting or impeding the evolution of such a "counterproductive" technology under current conditions is generally infeasible. A development of what J. Robert Oppenheimer called a "technically sweet" military technology is hard to stop, either unilaterally or through negotiations. Against the unilateral, decision, an argument based on the question "What will we do if the other side gets it first?" tends to prevail. In a negotiation the difficulty of policing the evolution of a new technology in its infancy is usually a block to agreement, unless very low standards of verification are acceptable, or unless the openness of the societies involved leads automatically to public exposure.

Thus new technology tends to generate its own momentum and impedes unilateral or negotiated control.

▶ There has been much speculation about the escalating role of *arms-control negotiations* in the weapons buildup. I will not discuss this subject here in detail. I would like to conclude, however, that the arms-control efforts during the last decade have on balance led to positive results. In particular the SALT I agreement has increased world security, and so has the Nuclear Non-Proliferation Treaty. The various achievements in less central areas of arms control have certainly also made net positive contributions. In addition, the very existence of arms-control discussions, even those not leading to positive results, has maintained channels of communication and has permitted exchange of data, interactions which, in the absence of such talks, would hardly have been possible.

Despite this net positive assessment of arms-control efforts, one should also acknowledge the negative by-products of arms-control negotiations. Linking these negotiations to domestic politics, and in particular the extensive ratification discussions in the US, have given undue political importance to the detailed numbers of military systems, especially those employing nuclear weapons, their actual technical performance does not deserve. No analyst could reliably predict how the outcome of a military conflict would depend on the precise number of nuclear weapons deployed. Yet in the debate as to who has gained or lost in an arms-control agreement the detailed numbers gain undue significance.

It is also true that arms control agreements tend to leave "uncashed bargaining chips" on the table. When a key military system has played a politically important role in an arms-control negotiation, it is then politically very difficult to abandon the system after the negotiation—unless eliminating it was part of the agreement reached. For example, it was a matter of considerable difficulty for the US to discontinue deploying the ABM forces permitted under SALT I, however limited their strategic importance may be. The Soviets have not as yet succeeded in doing so, and are in fact putting their ABM deployment around Moscow through a new round of modernization, notwithstanding the limited potential effectiveness of this system.

The domestic debates in the US and presumably also in the Soviet Union surrounding the arms-control negotiations also help drive the military buildup in diverse ways: For instance, during the ratification process in the US Senate, the price for a favorable vote by some senators was escalation in defense expenditures. Similarly others insist that an arms buildup is essential to make it possible to "bargain from strength." Thus for many years the Soviet Union showed no interest in nuclear arms control until a very substantial nuclear arsenal was at hand.

After the Soviet Union deployed the mobile and accurate SS-20 medium-range ballistic missiles targeted against Western Europe, the US proposed to station, on the soil of members of NATO, a number of missile systems that can reach the Soviet Union. However, members of NATO were willing to accept this more advanced "Theatre Nuclear Force" only if the US agreed to pursue negotiations with the USSR on limiting these forces.

These examples indicate how interwoven arms control and weapons buildups have become: We are hearing "no arms control unless we first arm," but also "no new arms unless we move to new arms control." While couplings of this kind may represent current political reality, it breeds cynicism among the non-nuclear nations about whether the efforts of the Soviet Union and the Western allies to reach meaningful and incisive arms control are genuine. On technical grounds there is no basis for such linkage: A good arms control agreement increases the security of all participants, irrespective of buildup in areas not controlled by the agreement.

All the above factors, combined with political developments, have led both the Soviet Union and the United States to escalate their nuclear arsenals to the potential of worldwide destruction. I should like to add, however, that the *rate* of growth of military expenditures of less developed countries is actually higher than those of the two superpowers, and that the fraction of GNP dedicated by some smaller nations to military pursuits is higher than that of the United States, or even that of the Soviet Union.

The two superpowers have at times, driven by the non-scientific, non-technical factors I have listed, procured large and expensive systems that are of only marginal value to their security. The MIRV buildup by the US and then by the Soviet Union is an excellent example. Most would agree that the security of both countries would be greater if neither had deployed MIRVs. Originally the US decided to develop and deploy them primarily as a penetration aid against Soviet ABM; yet it is well known that Soviet ABM evolved at a rather slow and irregular pace. Economic arguments (cost per reentry vehicle) also played a role. Even when it became evident that Soviet ABM was not a very effective countermeasure against US missiles, US MIRV deployment continued. The US land-based MIRV missile buildup was then matched and exceeded by the Soviet Union, even though the need for MIRV to penetrate US defenses was clearly absent. Whether the Soviet motive in building up MIRV forces was to "mirror image" the US buildup, or was to threaten preemptive attack against US missile silos, or was simply to lengthen the potential target list, we can only surmise.

There is no doubt *that had arms control efforts involving the Soviet Union and the US led to more incisive results at earlier times, the security of both nations and of the international community would now be greater.*

There is a further factor that has militated against unilateral or success-

fully negotiated steps to limit deployment of strategic weapons; this is the vacillation and ambiguity of stated doctrine by the US and the lack of clarity in the available statements of Soviet doctrine.

Limited conflict

All declarations of US strategic doctrine give highest priority to *deterring* the start of *nuclear* war by an opponent by demonstrating to the potential initiator that he faces an "unacceptable" outcome. This highest priority goal has been supplemented during various US administrations by listing additional objectives. Unfortunately, once strategic doctrine wanders beyond the limited goal of deterring nuclear war it faces severe internal inconsistencies. For instance, US doctrine tends to include provisions aimed at limiting damage, or assuring that US damage be less than that suffered by the adversary, *should deterrence fail.* Yet a damage-limiting posture tends to contravene deterrence: One nation's measures to limit damage to itself in nuclear war generate the perception by its opponent that it has become more difficult to deter from starting nuclear conflict. Similarly, extending the doctrine on the use of nuclear weapons to try to deter not only nuclear war but *war at all levels* projects the image that nuclear retaliation is threatened against even minor, non-nuclear incursion.

Such a threat is hardly credible in the face of today's nuclear balance. As a matter of history, military intervention, not involving nuclear weapons either by nonnuclear nations or the superpowers have not been deterred. The broadening of the doctrine raises the perceived requirements for nuclear arms, and thus contributes to the arms race.

On the Soviet side, the available literature is even less definite as to the priority given to deterrence in strategic doctrine; there are extensive discussions of extended nuclear warfighting sequences. Soviet leaders have made it clear that they are well acquainted with the potential horrors of nuclear war, and both Soviet and US leaders continue to express great doubts whether a limited nuclear war could ever remain limited. Yet the actual development and deployment of hardware on both sides can only have a "rational" basis if its use is anticipated in connection with various limited conflicts, and if limited conflicts can remain confined.

Recently in the US *limited* use of nuclear weapons has been emphasized more frequently in policy declarations. Part of the basis of this emphasis is "mirror imaging," citing selected Soviet sources on nuclear doctrine, and Soviet weapons deployments. The proponents of the policy of limited nuclear war present the "straw man" that under past doctrine US leaders would have no choice but to respond to a nuclear attack of some kind either by not using nuclear weapons at all or by using them in a full-scale counterattack. But this has never been the real choice.

What might be done to limit nuclear-weapon growth

The policies of nuclear nations should be revised to relate to the physical realities of nuclear weapons rather than regard them as symbols in international power politics. The following specific measures would help implement such a change in policy:

Sufficiency. Each nuclear nation should analyze its nuclear weapons needs under a strategy solely dedicated to deter nuclear attacks by others under a variety of assumptions. Such deterrence should retain the option to counter-attack a spectrum of military and economic targets with varying degrees of severity, but *not* to reply *in kind* to all possible modes of attack the initiator of nuclear war might choose. Sites of limited economic or military value, and those targets that cannot be successfully attacked should be eliminated from the list of possible objectives of nuclear attack. This prescription should result in specifying the number of survivable nuclear weapons "sufficient" for national security.

Arms control. Concurrent with such reexamination of policy and drastically downward revision of required number and kind of weapons, arms control negotiations must be given higher priority than they have been given in the past, or even highest priority in the discussion among nations. This implies that the linkage

between arms control issues and other topics of controversy between nations must be minimized. Those arms control measures should be designed to improve the security of *both* sides, irrespective of "linkage."

Unilateral action. Based on internal examination, each nuclear nation should consider unilateral steps in decreasing nuclear armament. Candidate steps would include cutting off or reducing the production of weapons-grade, fissionable material and eliminating clearly redundant weapons and those weapons susceptible to first-strike attack–and therefore only useful for a first strike themselves.

Technical nature of nuclear weapons. Because the current generation of decision makers in most countries lacks personal acquaintance with nuclear weapons and their effects, a renewed effort should be made within each country and internationally to disseminate fully and publicly technical facts on the nature and destructive effects of nuclear weapons. Such studies should not only emphasize what is known, but also what is unknown. International communication and exchanges among scientists, including those who have contacts with military activities, should be greatly amplified.

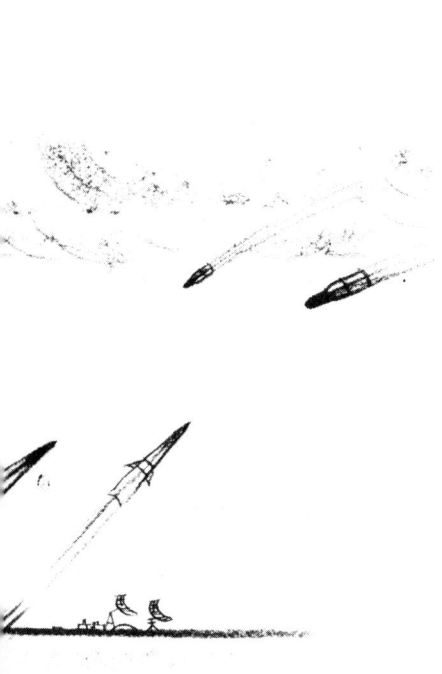

buildup of nuclear weapons is largely insensitive to announced military doctrine. This conclusion is not only based on the often-stated belief, which I share, that it is unlikely that "limited" war will stay limited, but also on the actual physical and technical effects of so-called limited exchanges.

Most analyses of limited nuclear war tend to be grossly simplistic in largely ignoring the *collateral effects* of such exchanges. As an example, many studies on the use of nuclear weapons in theater conflict do not include the effects of targeting errors or intelligence failures in indentifying and locating targets. One must remember that when a CEP (Circular Error Probable) is assigned to a weapons system, this means that one-half the rounds will impact outside this error circle, and some of those impact points just might be highly populated areas. Analyses also tend to ignore the fact that in all past wars in Europe a large fraction of the population, and in particular the urban population, did not remain in cities but traveled along roads to other areas. Yet refugee casualties are rarely included in theater nuclear war considerations. In addition, although some military doctrines discount the value of occupying cities by troops in localized warfare, the taking of cities has been in fact a major political objective in past wars. Therefore, while the assumption is often made in analyses of limited warfare that enemy columns will avoid centers of population, this is apt to be false. On top of all this, civil defense measures and evacuation, so much discussed publicly in the US and implemented on a considerable scale in the Soviet Union, are apt to be highly ineffective—simplistic calculations to the contrary notwithstanding.

All these considerations indicate that the vast volume of discussions and literature on limited uses of nuclear weapons tends to disregard physical reality. No national leader should assume that he can fine-tune the effects of nuclear weapons in an actual conflict. This point can possibly best be illustrated by the so-called neutron bomb, or more correctly, enhanced radiation warhead, which has played such a large *political* role in Europe. The neutron bomb is a specialized device that minimizes blast while maximizing lethal effects from radiation. In this role it is advertised to be a powerful weapon against the crews of enemy tanks, while minimizing collateral damage to friendly populations, structures, and economic assets of the country where the device is used. Yet the actual physical differences between the effects from the neutron bomb and other nuclear weapons are minimal. Roughly speaking, the 1-kiloton version of the US Lance Enhanced Radi-

ation Warhead produces as much neutron radiation as a "normal" 10-kiloton device would. The blast effect is still equivalent to 1000 tons of TNT, much larger than any blockbuster of World War II. The lethal radius of the neutrons depends significantly on intervening structures, and there is a wide gap between a lethal radiation dose and an instantaneous disabling dose, the latter being about 20 times larger. All these physical factors together with the problems cited above (that is, targeting errors, intelligence failures, and the presence of refugees in battle zones) indicate quite clearly that the "highly specialized" enhanced-radiation warheads are not all that specialized. The "enhanced radiation" of the neutron bomb is unlikely to be exploitable.

The controversy over whether the neutron bomb will lower the nuclear threshold and thus make nuclear war more likely has substance only if governments assume that the neutron bomb will produce consequences significantly different from those produced by ordinary nuclear weapons. Such an assumption is fallacious in view of the physical facts.

Nuclear weapon defense

The escalation in numbers of strategic nuclear weapons by the superpowers and the decrease in stability of the strategic deterrent due to the vulnerability of land-based missile silos to accurate enemy MIRVs, have revived discussion of the need for widely or locally deployed anti-missile defenses. The Soviet Union has carried out more extensive recent development programs than has the US, and the Moscow ABM installation is in the process of being replaced with a new system. While current activities are consonant with the provision of the SALT I Treaty, expanded deployments beyond these limits would invalidate the most significant arms control agreement yet enacted by the Soviet Union and the US.

One firm technical conclusion in the nuclear age is that a nuclear military conflict is apt to be "offense-dominated" rather than "defense-dominated." Because a single penetrating and exploding nuclear warhead does such an enormous amount of damage, the standards of interception required for defensive systems to be effective must be very high. Formal exchange calculations which compare the cost of defenses against nuclear weapons, be they air defenses or ballistic missile defenses, with the incremental cost required for the offense to cancel the effect of the defense, always show that defensive costs are higher, if casualties are to be held to moderate levels. There could be specialized exceptions to this general conclusion. For example, de-

---

The US has never been limited to the so-called MAD (Mutual Assured Destruction) strategy of all or none nuclear response. Rather, US has always had the option to launch parts of the nuclear force against a fraction of the listed targets, although the ability to implement such a "flexible" response can be and is being improved. Nevertheless, many US leaders and particularly former Secretary of Defense Harold Brown, have emphasized that they do not believe that a limited conflict will remain limited. At the same time current doctrine is that the deterrent purpose of US weapons is to convince an adversary that "first use" of nuclear weapons in any situation, for any purpose, limited or comprehensive, cannot lead to an outcome favorable or acceptable to the initiator. Rigidly interpreted, such a doctrine might imply that the US should be prepared to fight to "victory" *any* nuclear war fought for any "limited" goal set by the opponent. More liberally interpreted, this doctrine deviates very little from earlier declarations on "flexible response."

Although currently stated US doctrine implies in fact little significant change in number and kind of weaponry, the frequent public reformulations of strategy give the false impression of truly significant changes in doctrine. This in turn gives the impression that consequently significant changes in future military weapons may result from the changed doctrine—an impression that is largely false.

Once nuclear weapons are used at all in war, the ultimate impact is largely uncalculable; therefore the threat faced by mankind due to the advent and

USA

fense of very hard point targets such as missile silos may appear, under certain circumstances, to be militarily effective. However, the conclusion remains valid that, at least in reference to all currently designed systems, ABM is a poor investment for any large-scale strategic exchange, whether its targets are economic or military, and is an escalatory component of the arms race.

This negative *technical* conclusion on the value of anti-nuclear defenses is *politically* difficult to accept. As a matter of psychological and political necessity, it is unpalatable to leave the homeland undefended; yet the relative impotence of defense is a de facto truth in the nuclear age. Once there is clear recognition that nuclear strategic balance is offense-dominated, an escalating and self-defeating offense-defense competition can be avoided.

Sufficiency

There are now about 50 000 warheads stockpiled worldwide; most of these are more powerful than the two that together killed one-quarter million people in Japan. All studies continue to show that if a large fraction of these were used in war, several hundred million people would die and immense suffering would follow. Such studies make a variety of assumptions about targets, shelter, or effectiveness of evacuation. Yet they tend to consider only the "prompt" effects of nuclear weapons–blast and early radiation, combined with radioactive fallout. The casualties induced by delayed consequences—the effect of fires, food shortage, absence or maldistribution of medical care, societal breakdown, epidemics—are not included; they are omitted as "too difficult to calculate." Some of the long-range effects—ecological imbalances, depletion of the

ozone layer, synergistic effects, and the genetic burdens—have been studied, but it is generally agreed that the unknown exceeds the known.

How, in the face of these horrendous facts, have a series of "rational" decisions led to the status quo? What reasoning has led the superpowers to conclude that their nuclear weapons stockpiles and at times even their supplies of weapons-grade nuclear raw materials are insufficient?

In the interest of the future of mankind and kinds of practices and forces that have driven the armaments in the world to their current level must not continue.

Since most of the factors discussed previously that drive the arms race are fundamentally not directly related to science and technology, remedies must be sought largely in the political arena. I would like to defend here the thesis that if the technical and scientific nature of weaponry, and particularly of nuclear weaponry, were more clearly recognized and considered by the political decision-makers, then the largely non-technical factors which now drive the arms race would be greatly diminished.

After the first nuclear weapon was detonated in 1945 at Alamagordo, many observers expressed the sentiment that this new weapon would make further wars impossible, and that nations would not build up stockpiles of these weapons. These prophecies have proven false. *No strategic doctrine adopted officially by any power has provided a quantitative interpretation of the word popularized by Kissinger, "sufficiency." No government has answered the question "When is enough enough?" applied to nuclear weapons.*

In principle, if there were an accepted answer to the quest for an interpre-

tation of sufficiency, then limitation of nuclear arms could be achieved *both* through negotiated arms control *and* through unilateral action. Yet we have not seen a halt in the growth of nuclear stockpiles or the number of deployed nuclear weapons. Fifty thousand nuclear weapons worldwide are still not "sufficient." I attribute this failure largely to disregard by government leaders of generally accepted technical facts on nuclear weapons.

Too often the radically different nature of nuclear weapons and nonnuclear arms is ignored, or at least minimized by some. Conventional wisdom applied to nonnuclear weapons is that you require many more bullets than targets. For ordinary "bullets" this statement is self-evident, because a large fraction of available bullets would in fact not actually be fired in war; of those actually used, a large fraction would miss their intended targets. However, applied to nuclear weapons this homily should be far less persuasive. A nuclear weapon that misses its assigned target still does an enormous amount of damage. All parties acknowledge that the principal aim of nuclear weapons is to *deter* nuclear war; the importance of stockpiled weapons beyond those which might actually be used in a conflict is therefore a matter of *political* leverage rather than *military* utility. It is this political role of nuclear weapons, rather than their actual potential utility in warfare, which has prevented the nations of the world from defining nuclear "sufficiency."

* * *

Based on a paper presented at the Colloquium on Science Disarmament sponsored by the Institut Francais des Relations Internationales, 16 January 1981.   □

USSR

Pontifical academy urges no first use

PHYSICS TODAY/DECEMBER 1982

"All disputes that we are concerned with today, including political, economical, ideological, or religious ones, are small compared to the hazards of nuclear war." So states, in part, the preamble to a declaration on the prevention of nuclear war presented on 24 September to Pope John Paul II for his signature. In this document a group of world leaders in science, working with the enthusiastic support and endorsement of the Pope, have unanimously recommended that all nations agree to accept the no-first-use doctrine for nuclear weapons and to cease developing nuclear weapons. The statement points out the dangers of nuclear warfare involving increasingly sophisticated weaponry, stresses the absolute necessity of taking steps to reverse the trends leading to nuclear-arms escalation, and calls for scientists and national leaders to use their moral and ethical judgment as well as their intelligence in the interests of peace.

The declaration is one result of the efforts of a permanent working group—a group convened to consider the problem of nuclear war—set up in the fall of 1979 by the Pontifical Academy of Sciences, under the leadership of the current president of the Academy, Carlos Chagas, a Brazilian medical biologist. This group's efforts culminated in an historical gathering at the Vatican on 23–24 September, at which the heads of the scientific academies from 35 countries met with other members of the Pontifical Academy to sign this declaration. Among the signers were Sir Andrew Huxley, president of the Royal Society, E. Velikhov, vice president of the Soviet Academy of Science, Frank Press, president of the National Academy of Sciences, and such prominent physicists as Victor Weisskopf, Charles Townes, Carl Friedrich von Weizsacker and Louis Leprince–Ringuet.

The Pontifical Academy is the only existing international scientific academy and acts in an advisory capacity to the Vatican. Currently there are 70 members representing many countries, selected because of their scientific prominence and their interest in science and society issues. The Academy holds plenary sessions every two years and organizes study weeks on topics of special interest. Such topics have included the origin of the universe, particle physics, technical innovation in agriculture, and nuclear and nonnuclear sources of energy. The Vatican Press publishes proceedings from these sessions, which generally include extensive international participation by members and nonmembers.

Since its inception, the working group on nuclear war has provided the Pope with the facts about the dangers of nuclear war, and it has stimulated the delegations sent from the Pontifical Academy to inform world leaders about the medical effects of nuclear war (PHYSICS TODAY, April, page 57). Preliminary drafts of the declaration on nuclear-war prevention were prepared at meetings held in London in March and in Rome this June, and were sent to the academy heads with an invitation to attend the September meeting at which the statement was signed.

The declaration begins with a description of the current world situation and the danger of nuclear war. It states, "Mankind is confronted today with a threat unprecedented in history, arising from the massive and competitive accumulation of nuclear weapons." The statement goes on to explain that present stockpiles of nuclear weapons are capable of killing "hundreds of millions of people" immediately and untold millions more from after-effects. Against this threat it says, "Science can offer the world no real defense against the consequences of nuclear war." It states further that "there is no prospect" of preventing the destruction of the political, economical and cultural base upon which society depends.

The declaration expresses further concern because "the world situation has deteriorated." It notes the breakdown in communications between countries, and the existence of "serious inequalities" and partisan interests capable of providing the impetus for war.

The statement is concerned not only with the build-up of nuclear weapons, but also with the increase in capabilities of chemical, biological and conventional weapons, as they are steadily improved with the results of R&D. It states that, "It is therefore to be expected that also the means of non-nuclear war, as horrible as they already are, will become more destructive if nothing is done to prevent it." In view of this danger, "It is the duty of scientists to help prevent the perversion of their achievements and to stress that the future of mankind depends upon the acceptance by all nations of moral principles transcending all other considerations. Recognizing the natural rights of man to survive and to live in dignity, science must be used to assist mankind towards a life of fulfillment and peace."

The declaration then goes on to enumerate the key dangers involving nuclear weapons and their potential use, and to ask all nations to agree to a series of steps designed to prevent a nuclear conflict. It declares that:
▶ Due to the fundamental difference between nuclear and conventional weapons, nuclear weapons should not be "an acceptable instrument of warfare."
▶ There should "be no armed conflict between nuclear powers" because of the risk of involving nuclear weapons.
▶ Any use of force to settle international conflicts always "entails the risk of military confrontation of nuclear powers."
▶ The continued proliferation of nuclear weapons increases the risk of nuclear war.
▶ "The current arms race increases the risk of nuclear war," and so it must be stopped, weapons development must be curtailed, and current stockpiles must be reduced in a move toward complete nuclear disarmament.

To reduce the risk of nuclear confrontation and to encourage nations to settle disputes by peaceful means, it calls for all nations to agree:
▶ Never to "be the first to use nuclear weapons."
▶ To seek an end to a conflict immediately if nuclear weapons are used.
▶ Not to use force or the threat of force to violate the political independence of any state.
▶ To renew attempts to reach verifiable nuclear-arms-control agreements.
▶ To find effective ways of preventing the proliferation of nuclear weapons.
▶ To do everything practical to reduce the likelihood of nuclear war by accident or error.
▶ To continue present arms agreements while negotiating for "broader and more effective agreements."

The declaration ends by appealing to national leaders, scientists, religious leaders and people in general to each assume responsibility for preventing nuclear war. In particular it appealed to scientists "to use their creativity for the betterment of human life and to apply their ingenuity to exploring means of avoiding nuclear war and developing practical methods of arms control." —JC

Teaching about the arms race

Dietrich Schroeer

If progress is to be made toward peace, public education on arms-race issues is crucial both in the short run and for the long term. Because technical information is invariably essential to an understanding of issues, physicists can make a major contribution. However, educational efforts on the arms race face great difficulties, which we must first recognize if we are to overcome them:

▶ Teaching about the arms race has in the past been a cyclical activity, rising and falling in parallel with public

Dietrich Schroeer is professor of physics at the University of North Carolina, Chapel Hill, North Carolina.

concern.[1] These cycles are not all bad; the arms race is a real problem, not just an academic one, and hence it must be addressed whenever it surfaces in the political arena. But cycles tend to inject an aura of crisis into discussions, providing short-term motivation at the cost of credibility and efficiency in the long run. The longer-term approach has more educational implications. It means that arms-race education must be general enough to be useful for analyzing unknown future issues. The danger with short-term education is that by the time current educational efforts have roused public support for, say, a new round of arms control, the political need and opportunity may have passed.

▶ The arms race presents interdisciplinary problems. Physicists may be unaware of contributions other disciplines can make to this issue and insensitive to their own biases.

▶ There is little encouragement for physicists undertaking arms-race educational efforts. Peer recognition for either research or teaching in the area is sparse. There is little financial support for this kind of work. The DOD looks for technical research of direct use to its missions, the Arms Control and Disarmament Agency is in disfavor within the government, the National Science Foundation has difficulty supporting anything involving the social sciences or science education and the Ford Foundation also appears to be

PHYSICS TODAY/MARCH 1983

What physicists can do as individuals to help provide their students and the public with the technical information needed to understand the issues involved in the nuclear cold war.

decreasing its support for these activities.

Nonetheless, we cannot afford to give up. In this article my aim is to encourage teaching efforts directed toward controlling the arms race. I will review, in an impressionistic manner, some relevant technologically-based education to which physicists can make professional contributions. I will describe a wide range of educational efforts, from PhD research to education of the general public. Finally, I will analyze the role of physicists and their organizations in arms-race education.

Arms-race issues are intrinsically contentious, leading even to disagreements on how they should be confronted. I come to the subject with an

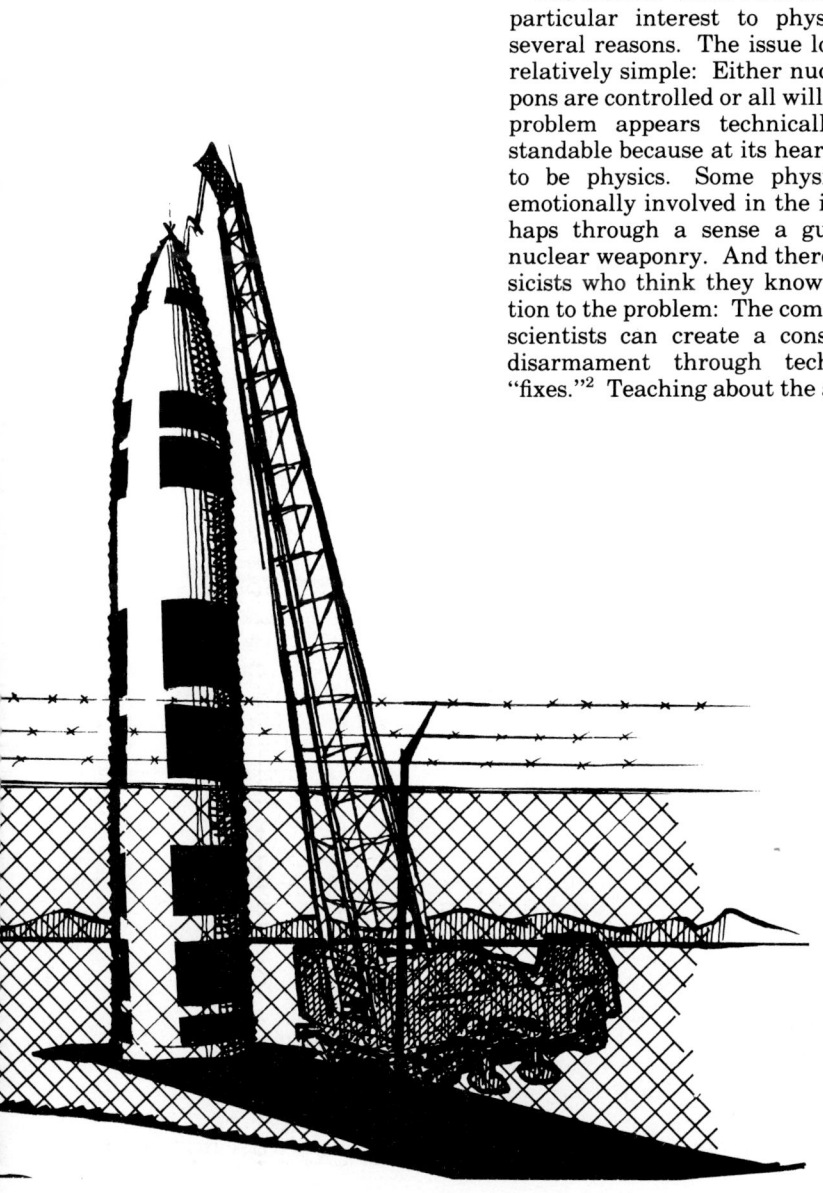

academic perspective: I am skeptical of broad public education in a crisis atmosphere. I favor teaching over the long-term to raise the competence of the public, policy makers and scientists; I am cautious about trusting technical expertise too far, believing we must increase the political sensitivity of scientists as much as the technical understanding of the public. Therefore, in this review I will focus on long-term arms-race education. There are many others who feel that political changes must be achieved in the short run as well. They emphasize a strong effort by technical experts to persuade the public on the grass-roots level. But everyone agrees: The nuclear arms race must be controlled.

Approaching the problem

The nuclear arms race has been of particular interest to physicists for several reasons. The issue looks to be relatively simple: Either nuclear weapons are controlled or all will end. The problem appears technically understandable because at its heart it seems to be physics. Some physicists are emotionally involved in the issue, perhaps through a sense a guilt about nuclear weaponry. And there are physicists who think they know the solution to the problem: The community of scientists can create a consensus on disarmament through technological "fixes."[2] Teaching about the arms race

is motivated by both these technical and emotional interests.

Difficulties for scientists. Technical interest and emotional commitment, however, should not blind physicists to the fact that working on the arms race requires efforts that are likely to be professionally unrewarded and that demand critical self-analysis to recognize biases toward overquantification and technological fixes.

The difficulties are manifold. The arms race makes contact with a very real world of cost–benefit engineering, secrecy and politics. Engineering is somehow unsatisfying to a scientist, because it aims at a product rather than at knowledge.[3] Industrial and military secrecy attacks the scientific norms[4] and has Machiavellian overtones.[5]

Physics training is too narrow to allow easy evaluation of social issues in a political context. To deal with the effects of nuclear war one must understand subtleties not only of physics, but of chemistry, geology, geography, biology, meterology and psychology. And one must be able to integrate understandings of sociology, political science, international relations, economics, law and ethics.

Physicists are well-known for their interest in quantifying everything and their ability to do order-of-magnitude estimates that can then be turned into precise calculations. In social issues this approach can provide a starting point for discussion, but can also raise problems through overquantification. In carrying out analyses of policy options, physicists are likely to prefer feasible alternatives to those more desirable but less feasible.[6] They may prefer technological fixes for social issues.[2]

Working on the arms race means direct contact with the general public.[7] Science is a communal effort for which the rewards ordinarily come from inside the scientific community, from scientific peers. There have been suggestions that such activities as writing popular books may in fact harm a scientist's standing within the scientific community,[7] that "visible scientists" face[8] some difficulties from their peers. As long as science is a peer-oriented activity, public education is likely to be seen as a second-class activity, with few professional rewards.

Social issues such as the arms race often surface as crises. During crises people demand prompt solutions—not because solutions exist, but because they must be made to exist. This style of action is foreign to physicists who are used to performing long-term research.

The current interest. If the current interest in the arms race is likely to be only temporary, it may not seem worthwhile to become professionally involved. The interest of scientists in the arms race has been cyclical over roughly ten-year periods. Within the scientific community, events trigger an interest aimed at a political goal. Once the goal has been reached—as in the case of

The oil shortage has become an oil glut, and the funding for energy work is vanishing. Physicists who developed energy courses in the 1970s are now developing arms-race courses. The Union of Concerned Scientists has gone from concerns with nuclear power reactors to organizing nuclear-war convocations. Students, who a few years ago were likely to be in general-survey courses devoted to science-and-society or energy, are now signing up for arms-

come available, this renewed interest and activity will likely be short lived.

Education channels

Education in arms-race issues runs the gamut in sophistication from PhD research to the most public exposure in the mass media. At this point I will review a sampling of educational efforts that are particularly relevant to physicists because they focus on technological aspects of the arms race.

Advisory education. Organizations that inform public-policy makers are educational in the sense that their advanced research efforts educate through publication. The Brookings Institution, the Hoover Institute, the Hudson Institute, the Institute for Policy Analysis and others are well recognized think tanks. But a technological focus is rare. We list here five major organizations whose work involves significant technological components:

> If I could ask two new departures of our great educational system as it prepares young people for the next century, I would ask that they be given a thorough grounding in the physics and politics of nuclear weapons.
>
> —Thomas J. Watson Jr

the nonproliferation treaty—or has become impossible, retrenchment takes place. Some of the scientists go back to doing "real" science, and some social activists go on to other socio-scientific issues that are more current, such as the energy crisis. A decade later, new events trigger renewed interest. In 1946 the interest was that of control of nuclear weapons, the middle 1950s presented the problem of radioactive fallout from atmospheric nuclear weapons tests and the late 1960s had the Vietnam war and the ABM debate.

The current concern over the arms race is the result of several separate events leading to the perception that nuclear-war fighting is being advanced as more "thinkable." At the same time, tension between the superpowers has increased and economic costs of the US military posture are escalating rapidly. The response of the scientific community has been to reactivate the old networks of scientists. The November 1981 Convocation of the Union of Concerned Scientists[9] tried to reactivate the network of scientists established during the ABM debate. The Federation of Atomic Scientists has started an arms-race educational effort by scientists.[10] The Forum on Physics and Society of The American Physical Society held a short course on the Arms Race in January of 1982 with heavy emphasis on technical aspects.[11] The American Association for the Advancement of Science has a new committee on the arms race,[12] and so does the National Academy of Sciences under the direction of Marvin Goldberger. Essays and editorials have appeared in relevant science journals.[13,14]

A cynic might say that the current interest in the arms race comes from the fact that the energy crisis has temporarily receded as a critical issue.

race courses. Editorial comments on oil shortages are being replaced by cartoons on the MX missile systems.

But the rising interest in the arms race cannot really be ascribed to disenchantment with the energy crisis. The current political situation makes the arms race an important and vital issue; scientists, science teachers, students and the public are each responding to the importance of this situation. However, physicists should be careful to distinguish the long-term importance of anti-arms-race work from shorter-term interest induced by current politics. Even though at the moment there is great interest, anti-arms-race efforts should not totally replace energy-policy work. Nor should they be restricted to the current issue of nuclear war fighting. Specific goals of a nuclear freeze should be distinguished from continuing efforts to produce a public educated to cope with the long-term threat of nuclear war.

The latter educational effort is much harder and uncertain. There is the danger that the current student interest may wane in a few years. For long-term survival, arms-race courses should aim at developing a continuing constituency as well as responding to the short-term excitement. In that sense the present situation is unpromising. Although the interest is growing, the funding for anti-arms-race activities is drying up; Federal and state support of academics is shrinking and even foundation support (for example, the Ford Foundation) is in danger of disappearing. The interest in the arms race seems perversely to be in an inverse relationship to the available support. A good measure of the success of arms-race education may well be its ability to institutionalize itself. Unless funding and professional rewards be-

▶ **Rand:** The analyses carried out by the Rand Corporation can be quite technical in nature. Its conclusions are influential with the policy makers in its narrow clientele, the Department of Defense, but some of its published material is more broadly useful.

▶ **NAS:** The National Research Council of the National Academy of Sciences does technical reviews of social issues at the request of the Federal government. These reports are released to the public and can have considerable influence both on policy makers and on the public.[15] Characteristic of these studies, and of the potential for controversy, is the NAS–NRC report of 1975 on the world-wide effects of nuclear war.[16] That study restricted itself to very narrow global and quantifiable questions. The position of the NAS to be "scientific," that is, to consider in this study only physical phenomena susceptible to quantitative analyses, prevented it from carrying out a true educational function in this instance.

▶ **OTA:** A somewhat more egalitarian analytic institution is the Office of Technology Assessment. It was established in 1972 to carry out policy analyses of technologies for Congress. After some early troubles[17] it has carried out quite good technical reviews. Perhaps because of the more populist nature of its clientele (the Congress), its reports are generally useful as educational material; its analysis in 1979 of the effects of nuclear war[18] included some excellent tutorial material.

▶ **IISS:** The International Institute of Strategic Studies in London is less technically oriented than the above organizations. But its annual *Strategic Survey* and its report *The Military Balance*[19] include authoritative reviews of strategic technology and the military balance. Some of its Adelphi monographs are fairly technical analyses of policy alternatives.

▶ SIPRI: The Stockholm International Peace Research Institute similarly publishes a *Yearbook*[20] which keeps track of world armaments and focuses essays on the most recent technological developments. Other SIPRI books cover topics such as chemical warfare, tactical nuclear weaponry and so on. The technological analyses are deeper than most, even if still superficial from a physicist's viewpoint.

PhD research. Among the very few institutions where advanced work in physics can be done in the area of arms-race research, the Program in Science and Technology for International Security in the department of physics at the Massachusetts Institute of Technology under the direction of physicists Bernard T. Feld and Kosta Tsipis[21] is outstanding. The physics program at MIT is involved in arms-race education on three levels. It trains undergraduate and graduate students, it prepares technical reports for various groups and it produces an expert faculty. For example, Steven Fetter was a junior when he participated in research at MIT on catastrophic releases

of radioactivity which was reported in a *Scientific American* article he co-authored with Tsipis.[22] Consider the MIT analysis of particle-beam weapons. A week-long study in 1978 assessed the technical feasibility of such a weapons system and resulted in a detailed technical report.[23] A version of this report was then published as a technical peer-reviewed article in *Nature*.[24] Finally a more generally accessible article appeared in *Scientific American*.[25] The authors also undertook extensive public education about this issue through speaking engagements, such as at APS meetings and in many public forums. This is technical and socially-responsible education at its best. It provides a source of technical information and develops a cadre of independent arms-race experts outside the DOD. But such efforts are inevitably rare. Advanced research requires a funding source. DOD does not support independent, academic public review of its own policies. The program at MIT is fortunate to be funded in part by the Ford Foundation. (The Ford Foundation has significantly funded academic programs on arms control and international security research at the University

of Aberdeen in Scotland, the University of Lancaster in Great Britain and in the US at Caltech, Carnegie–Mellon, Harvard, Indiana University, MIT, UCLA and the University of California at San Diego.)

Graduate work in public-policy analysis. Analyses of arms-race issues with a technical focus are rare; usually technology is seen as only one factor influencing public-policy choices. Many political-science departments consider problems of the arms race, but to them the technology may be relatively peripheral. Physicists need to understand this viewpoint.

Harvard University's John F. Kennedy School of Government is typical of this policy approach. This school trains future government leaders and provides some understanding of government to students training for other professions. Its educational degree program include Master's or PhD degrees in public policy, public administration, or in political economy and government.[26] These programs deal with issues that often derive from the research interests of the faculty. The Center for Science and International Affairs within this school has the strongest interests in arms-race issues. Funded by the Ford Foundation, this Center brings together a prominent science- and technology-based faculty including Harvey Brooks, Albert Carnesale, Paul Doty and Michael Nacht. Students take courses not only at the Kennedy School, but also at other Harvard departments and at neighboring institutions such as MIT. The faculty educates itself in the interdisciplinary Harvard Arms Control Seminar. And the public is reached by conferences, workshops and published articles in the Center's own *International Security* journal.

Undergraduate programs. If one includes ROTC and international-studies departments, there are actually undergraduate degree programs in US universities and colleges nominally related to arms-race issues. But these programs tend neither to focus on the arms race itself nor to have much technological emphasis. The Stanford University program in arms control and disarmament is one of the very few undergraduate degree programs focusing on the arms race.[27] With 1000 students over 12 years, this is a major effort. Its faculty is drawn from other departments throughout the university, and from adjunct institutions such as the Stanford Linear Accelerator Center (for example, Sidney Drell and Wolfgang Panofsky).

The Stanford arms-control program has produced extensive teaching materials, including a textbook.[28] Undergraduates can become involved in advanced research and often go on to

---

Extrapolate from the Hiroshima and Nagaski A-bomb explosions, show that a 1-Mt H-bomb is lethal in cities over about 30 mi². Pretend you are a MAD Russian military planner and target 50 1-Mt H-bombs onto US Metropolitan areas with the objective of maximizing civilian casualties (take the population data from the US *Statistical Abstracts*.) Discuss your calculations.

Targets

City	Pop. density	Area	Bombs	Deaths
New York	26,000/mi²	300 mi²	10	7.8 M
San Francisco	15,700/mi²	45 mi²	1	0.5 M
Philadelphia	15,100/mi²	129 mi²	4	1.8 M
Chicago	15,100/mi²	222 mi²	7	3.2 M
Boston	14,000/mi²	46 mi²	1	0.4 M
Washington	12,300/mi²	61 mi²	2	0.7 M
First 25 1-Mt bombs			25	14.4 M
Newark	16,300/mi²	24 mi²	1	0.4 M
Baltimore	11,600/mi²	78 mi²	2	0.7 M
Detroit	11,000/mi²	138 mi²	4	1.3 M
St. Louis	10,200/mi²	61 mi²	2	0.6 M
Cleveland	9,900/mi²	76 mi²	2	0.6 M
Pittsburgh	9,400/mi²	55 mi²	2	0.5 M
Milwaukee	7,600/mi²	95 mi²	3	0.7 M
San Francisco	15,700/mi²	(15 mi²)	1	0.2 M
Minneapolis	7,900/mi²	55 mi²	2	0.4 M
Long Beach	7,400/mi²	49 mi²	1	0.2 M
Seattle	6,400/mi²	84 mi²	2	0.4 M
Los Angeles	6,100/mi²	464 mi²	3	0.5 M
Second 25 1-Mt bombs			25	6.6 M
TOTAL of 50 1-Mt bombs			50	21.0 M

Student assignment in the author's course on science, technology and the arms race. Table gives results of assignment showing the saturation effects of nuclear attacks which helps the student to understand the concept of mutual assured destruction.

graduate work. Science and technology enter indirectly, in the form of science faculty who teach in their "spare" time; in some cases faculty members are inspired by active consulting work with the government. Here, too, funding is by the Ford Foundation.

"Enrichment" in physics courses. Physicists active in arms-race issues face the problems of interdisciplinarity. But they know physics, are able to do order-of-magnitude calculations and have some feeling for technology. They can exploit these abilities by "enriching" regular physics courses with examples related to the arms race.

There are many obvious examples. Explosions of nuclear weapons can be used to illustrate nuclear physics (fission, fusion, chain reactions), thermodynamics (fireball, shock waves), plasma physics (ionization phenomena) and atomic physics (x-ray production). Delivery systems involve mechanics (the Bernoulli effect on aircraft wings, Newton's laws for rocket engines, inertial-guidance gyroscopes), electricity and magnetism (communications, electronic countermeasures, computers) and acoustics (sonar). Arms control is related to geophysics (seismic detection of underground nuclear tests), optics (photographic and infrared reconnaissance), wave phenomena (radar systems) and so on.

logical knowledge. A good source of examples of this kind in an educational context is the series of science-and-society tests[31] published by David Hafemeister in the *American Journal of Physics*. Some technical references can be found in a bibliography[32] on "Physics and the Nuclear Arms Race."

Arms-race courses. The early 1970s saw the rise of physics-and-society courses in universities and colleges as part of the general-education curriculum.[33] These were often very general, dealing with questions of scientific responsibility and social control. Some of these socially-conscious education efforts are now shifting to consider the arms race.

The current concern generally is to give students a feeling for the technology involved in the arms race, and how it relates to the political situation. Courses for nonscience majors are a response to this concern. These courses tend to be descriptive in the technical area and focus on political aspects (see reference 10 for lists of these courses and copies of syllabuses). Lester Paldy at the State University of New York at Stony Brook and Alvin Saperstein of Wayne State University, for example, teach courses of this type; Paldy uses guest speakers, and Saperstein offers a very large course with team teaching. Since 1976 I have been teaching a somewhat more technical course enti-

ing graduate study in the area and going on to professional work in international-security agencies. However, we should be satisfied even if we do not produce activists, as long as we instill knowledge and analytical skills.

Educating physicists. One part of contemporary arms-race education should be to educate physicists beyond their professional limits. Physicists who want to teach or work on the arms race should themselves learn more and learn more broadly. Some efforts are in progress to offer these kinds of educational opportunities. The American Physical Society, through its Forum on Physics and Society, together with the American Association of Physics Teachers, is developing a second short course on the arms race for the April 1983 meeting in Baltimore.[11] The AAPT, through the *American Journal of Physics*, is providing articles[31] and bibliographies.[32] The American Institute of Physics is publishing this issue of PHYSICS TODAY concentrating on nuclear-arms education. And *Scientific American* is continuing to publish informative and educational articles on arms-race issues,[34] aimed at scientific literates.

To some extent the education of physicists suffers because good technical articles on the arms race are not readily available. It is a rare event to come across an article that goes beyond the level of *Scientific American*—say an article in *Science* analyzing nuclear attacks,[35] or one in *Nature* discussing particle-beam weapons.[24] There are no journals that readily serve as outlets for good technical analyses of arms-race technologies.

The lack of good technical literature may not be entirely the fault of the physicists. But physicists also do not read important literature relevant to the arms race that *is* available. Arms-race physicists should be aware[32] of books such as the arms-race text by Albert Legault and George Lindsey, *Dynamics of The Nuclear Balance*,[36] of technical reviews such as L. W. Ricketts, J. L. Bridges and J. Millette's *EMP Radiation and Protective Techniques*,[37] of the best social-science analyses such as Karl Spielmann's *Analyzing Soviet Strategic Arms Decisions*,[38] or of the models of the arms race of Lewis Richardson.[39] In other words, physicists should educate themselves to approach arms-race issues with all the best professional knowledge available. The professional societies in physics could make this kinds of education available. Self-educational efforts such as the *ad hoc* committee on arms-race studies of the APS Forum on Physics and Society may help in these efforts.[40]

Often the problems of political choice have become buried in debates among experts over highly technical alternatives.

—Harvey Brooks

Arms-race examples can be incorporated into physics courses on many different levels. In liberal-arts courses the chain reaction can be shown as a demonstration substituting ping-pong balls for neutrons, and the thermonuclear fusion bomb can be discussed qualitatively as in Howard Morland's article in the *Progressive*.[29] Fission and fusion mass-defect calculations and the exponential growth of chain reactions can be done in sophomore physics courses. Students in nuclear-physics courses can calculate the shapes of critical masses or the energy dependence of fusion cross sections or do computer Monte Carlo calculations of critical masses. And advanced plasma-physics courses can explore the relations of H-bombs to energy production by inertial-confinement fusion using microballoons.[30]

There is sufficient unclassified material to allow these discussions to be based on good physics and solid techno-

tled "Science, Technology and the Arms Race" to acquaint non-scientists with the science and technology of the arms race, including order-of-magnitude calculations; the course covers politics mostly as a consequence of weaponry (see figure).

Ideally, the calculations will explain political alternatives without determining the political choices. The objective of the calculations is to let the students make judgments based on their own understanding, not on the basis of what experts tell them. Such courses can be quite successful; enrollments are growing. The students become sensitized to the issues; a few learn to think about MAD as they sit in the Minuteman launch-control centers. And some students become deeply involved in the topic, going on to give public lectures on the subject, attending a Student Pugwash Conference or the International School on Disarmament and Research on Conflict, pursu-

Public education

In spite of the technical, professional

and disciplinary difficulties I have outlined much public education is taking place. In the short term, these efforts may in fact be the most useful in achieving actual policy changes. The recent rise in the quantity and intensity of public education on the nuclear arms race is phenomenal.[1] Because it is inspired by current events, this education tends to focus on political aspects of arms-race issues, and in the public forum scientists often participate in a politically inspired manner. But there are some technically-oriented aspects to this public education. Here I will give a feeling for styles of doing public education on a technical level that physicists could undertake in their professional capacities. I will say nothing about politically oriented efforts. For a guide on such activities see reference 41.

Publications. One of the major driving forces behind the current concerns has been the first-time involvement of physicians. Part of their success was the introduction of new technical information, namely detailed descriptions of the medical consequences of nuclear war. In a similar style, physicists could do more education specifically related to their expertise. There is a great need for proper technical analyses of

verification, of antisubmarine warfare, of space-based weapons and so on. These analyses need to be reported on all levels of sophistication, including direct reports to the general public. The present tendency, fostered by the publishers (and the buying public), is toward analyses inspired and influenced by political viewpoints. More apolitical studies useful to the public are needed. Such studies should involve physicists from all parts of the political spectrum, united to analyze scientific and technological possibilities dispassionately.

The media. The broadcast media, including television, are now quite interested in the arms race. Past efforts in this area rarely dealt with technology.[42] The NOVA series of public television for many years was the only major tv outlet that aired any technical programs. Now the commerical networks have become involved. Symptomatic of this renewed public interest in the arms race have been television network programs on the arms race such as ABC's "Second to None" in 1979, CBS's "The Defense of the United States" in 1980, the showing by PBS in 1981 of the BBC's "Nuclear Nightmares" and NBC's "Facing up to the Bomb" in 1982.

Physicists could encourage and help produce more of these programs. For example, "The War Game" was produced[3] for the BBC in the 1960s to depict the aftermath of nuclear war; it was never shown. This program portrays superbly the human consequences of a nuclear war. However, it needs updating. Physicists could unite with other professionals such as physicians, social scientists and humanists and take the OTA report on nuclear war[18] as the basis for a new version of "The War Game."

Public programs. The most active part of public education on the arms race in the last few years has been the direct grass-roots approach of political activism. The results of this work have included several efforts involving scientists:

▶ UCS: The Union of Concerned Scientists[9] is a scientist-based organization that looks at socio-technological issues such as the safety and desirability of nuclear power reactors. Under the direction of Howard C. Ris Jr, it has developed a nuclear-arms program. It has organized two nationwide convocations: in November 1981 on the threat of nuclear war, and in November 1982 on solutions to the nuclear arms race. These day-long convocations perform a public education function by informing the public about the nuclear arms race. The UCS activities *per se* are educational, frequently operating under the guidance of scientists. But obviously the motivation is social concern, and the goal is to inform the public toward political action.

▶ FAS: The Federation of Atomic Scientists has been concerned about nuclear arms since 1946. Its response to the current interest in the arms race has been to establish a nuclear-war education project, under the co-chairmanship of John Harris and Eric Markusen.[10] It is building up a network of people, particularly scientists, interested in or actively teaching arms-race subjects. The FAS project is not only collecting course outlines but is actively developing teaching materials, as, for instance, Barry Casper of Carleton College is doing. Its newsletter *Countdown* contains listings of faculty, courses and educationally relevant organizations. The FAS umbrella encourages these efforts to provide technical information, but arms control is the ultimate goal.

▶ Ground Zero: The Ground Zero program was formed under the direction of Roger Molander,[44] a staff member of the National Security Council from 1974 to 1981. Its goal is a grass-roots nationwide campaign to educate Americans about the threat of nuclear war. Its first week-long program in April of 1982 involved a million people in 350 colleges and 1200 high schools.

Another approach to education

The organization Ground Zero has developed Firebreaks: A War/Peace Game, and it is inviting citizens to play the game on a nation-wide basis during April. Ground Zero is offering the game as this year's educational event following last year's Ground Zero Week, during which seminars, lectures and discussions were held. In the game, players assume the role of experts with the task of resolving a mock international crisis. Participants have a chance to explore first-hand a sequence of events that could lead to nuclear war and to examine the tools—or "firebreaks"—available to prevent it.

For information about the group Game Kit ($10; for high school groups $5) and the schedule for the nationwide event write to: Firebreaks: A War/Peace Game, Ground Zero, Suite 400, 806 Fifteenth Street N. W., Washington, DC 20005.

Molander's inside knowledge of the arms-control bureaucracy makes this an activist program, even if it acts in an educational manner. And the involvement of scientists is somewhat lower than in the two previous programs.

We could mention other public programs such as the Freeze campaign or the Student Pugwash conferences, all of which have educational components involving scientists to some extent. (See *Countdown*[10] for descriptions of some of these.) Each of these activities has its own style, its own balancing of technical with political considerations and its own particular accomplishments in public education. The main difficulty of this and other public-policy issues is the confounding of expertise with special pleading. Is it more important to communicate information to the public, or to guide the debate toward a desired political goal?

References

The following extensive list of references is included to provide prospective teachers with materials they and their student may find useful in discussing the arms race.

1. J. Kavlin, "A talk with Louis Harris," Bull. At. Sci. **38**, 1 July 1982, page 3.
2. R. Gilpin, *American Scientist and Nuclear Weapons Policy*, Princeton U. P., Princeton (1962), pages 51, 309.
3. W. Kornhauser, *Scientist in Industry: Conflict and Accomodation*, U. California P., Berkeley (1962).

4. R. K. Merton, "The Normative Structure of Science," in *The Sociology of Science*, U. Chicago P., Chicago (1983), page 267.
5. A. de Volpi, G. E. Marsh, T. A. Postol, G. S. Stanford, *Born Secret: The H Bomb, the Progressive Case and National Security*, Pergamon, New York (1981).
6. A. M. Weinberg, *Relections on Big Science*, MIT, Cambridge (1967).
7. W. O. Hagstrom, *The Scientific Commu-*

nity, Southern Illinois U. P., Carbondale (1965), page 34.
8. R. Goodell, *The Visible Scientists*, Little, Brown, Boston (1977).
9. For information write the Nuclear Arms Program, Union of Concerned Scientists, 1384 Massachusetts Ave., Cambridge, MA 02138.
10. The FAS Nuclear War Education Project publishes the newsletter *Countdown*. For information write the project at FAS, 307 Massachusetts Ave. NE,

Washington DC 20002. The UCS has published the book *Beyond the Nuclear Freeze: The Road to Nuclear Sanity*, D. Ford, H. Kendall, S. Nadis, Beacon, Boston (1982).
11. A second short course on the arms race will be held 17 April 1983, in conjunction with the Spring meeting of The American Physical Society, in Baltimore. For details write the author.
12. For information about the Committee on Science, Arms Control and National

Security of the AAAS write R. Scribner, AAAS, 1776 Massachusetts Ave. NE, Washington DC 20036.
13. H. L. Davis, PHYSICS TODAY **35**, June 1982, page 112.
14. J. S. Rigden, "Editorial: Physics and the arms race." Am. J. Phys. **48**, 177 (1980).
15. P. M. Boffey, *The Brain Bank of America*, McGraw–Hill, New York (1975), reviewed in Science **188**, 1094 (1975).
16. *Long-term World-Wide Effects of Multiple Nuclear-Weapons Detonations*, National Academy of Science, (1975); see also Science **190**, 248 (1975).
17. B. M. Casper, "The rhetoric and reality of Congressional technology assessment," Bull. At. Sci. **34**, February 1978, page 20.
18. *The Effects of Nuclear War*, Office of Technology Assessment (1979).
19. *Strategic Survey: 1982*, and *The Military Balance: 1982–1983*, International Institute for Strategic Studies, London (1982).
20. Stockholm International Peace Research Institute, *World Armaments and Disarmament: SIPRI Yearbook 1981/82*, Crane Russak, New York (1981).
21. For further information write to K. Tsipis, director, Program in Science and Technology for International Security, Department of Physics, MIT, Cambridge, MA 02138.
22. S. A. Fetter, K. Tsipis, "Catastrophic releases of radioactivity," Sci. Am. **244**, April 1981, page 41.
23. G. Bekefi, B. T. Feld, J. Parmentola, K. Tsipis, *Particle Beam Weapons*, Report #4 of the Program in Science Technology for International Security, MIT (1978).
24. G. Bekefi, B. T. Feld, J. Parmentola, K. Tsipis, "Particle-beam weapons—a technical assessment," Nature **284**, 219 (1980).
25. J. Parmentola, K. Tsipis, "Particle-beam weapons," Sci. Am. **240**, April 1979, page 54.

Scientists giving political advice, especially to the public [ought] ... to moderate appeals to passion ... [and] discontinue any attempts to "save mankind by frightening men to rationality."

—R. Gilpin

What can physicists do?

What can and should physicists do to help improve and expand arms-race education? They can take advantage of the present interest in the topic and try to satisfy this interest.[1] Any immediate results that can be obtained would be important both to improve the level of the present political debate (helping—some would say—to save our civilization) and to attract outside support to underwrite more work on the subject. But physicists should also work toward the longer-term goal of an informed public that stays interested in the arms race and can understand future issues as they arise. Efforts directed to this long-term goal should include the following:

▶ Research in technical aspects of the arms race should be fostered to provide credible analyses of important present and future technologies. A research agenda needs to be set. Perhaps the arms-race studies being sponsored by the APS Forum on Physics and Society will satisfy this need.[40]

▶ Funding for research has to be identified, actively solicited and created.

▶ Publication outlets for technical analyses must be developed.

▶ Technical graduate research programs should be fostered within existing science departments, legitimizing degree work in arms-control areas; student work on these topics provides new knowledge, future workers and professionalism.

▶ More technological considerations could be injected into already existing international-security-studies programs (often housed within political science departments). And arms-race teaching and research centers, such as the proposed Peace Academy, might be established.

▶ The teaching of arms-race subjects should be "regularized." Seminars should be turned into permanent courses with departmental support, allocated faculty positions and regular listings in college course catalogs.

We need to create a new generation of physicists interested in arms-race matters. The old-timers from the Manhattan District and PSAC days have done yeoman duty as advisors to presidents, Congress and the DOD. But not enough replacements have been trained; not enough permanent institutions have been built to develop and support a new crop of arms-race physicists. We sorely need an invisible college of independent arms-race scientists who are able to perform public-education functions, are specialists enough to be credible as technical experts and yet are humble enough to recognize the limitation imposed on them by that expertise. Physicists are needed who are willing to learn as well as educate.

26. For further information write to the John F. Kennedy School of Government Harvard University, 79 Boylston Street, Cambridge MA 02138.

27. For further information write to John W. Lewis, Arms Control and Disarmament Program, Stanford University, Stanford CA 94305.

28. J. H. Barton, L. D. Weiler, eds., *International Arms Control: Issues and Agreements*, Stanford U. P., Stanford (1976); revised edition is in preparation.

29. H. Morland, "The H-bomb secret," The Progressive **43**, November 1979, page 14.

30. F. Winterberg, *The Physical Principles of Thermonuclear Explosive Devices*, Fusion Energy Foundation, 888 Seventh Ave., New York, NY 10019 (1981).

31. D. Hafemeister, "Science and society test for physicists: the arms race," Am. J. Phys. **41**, 1191 (1973); "Science and society test V: nuclear proliferation," Am J. Phys. **48**, 112 (1980).

32. D. Schroeer, J. Dowling, "Resource letter PNAR1: Physics and the nuclear arms race," Am. J. Phys. **50**, 786 (1982).

33. C. D. Spencer, D. Schroeer, "Teaching a physics and society course," Am. J. Phys. **44**, 135 (1976).

34. H. F. York, ed., *Arms Control: Readings from Scientific American*, Freeman, San Francisco (1973); B. M. Russett, B. G. Blair, *Progress in Arms Control? Readings from Scientific American*, Freeman, San Francisco (1979).

35. K. Tsipis, "Physics and calculus of countercity and counterforce nuclear attacks," Science **187**, 393 (1975).

36. A. Legault, G. Lindsey, *The Dynamics of the Nuclear Balance*, Cornell U. P., Ithaca (1976).

37. L. W. Ricketts, J. E. Bridges, J. Milette, *EMP Radiation and Protective Techniques*, Wiley, New York (1976).

38. K. F. Spielmann, *Analyzing Soviet Strategic Arms Decisions*, Westview, Boulder (1978).

39. A. Raraport, "Lewis F. Richardson's mathematical theory of war," J. Conflict Res. **1**, 249 (1957).

40. For information about possible arms-race related research projects under the sponsorship of the Forum of Physics and Society of the American Physical Society, contact Leo Sartori, Physics Department, University of Nebraska, Lincoln NE 68588.

41. J. Primack, F. von Hippel, *Advice and Dissent: Scientists in the Political Arena*, Basic, New York, (1974), particularly Part V on "Public Interest Science," page 239.

42. J. Dowling, *War Peace Film Guide*, World Without War Publications, 67 E. Madison, Suite 1417, Chicago IL 60603 (1980).

43. For a complete listing of films and TV programs related to the arms race, write John Dowling, Department of Physics, Mansfield State College, Mansfield PA 16933.

44. For information write Ground Zero, 806 15th Street, N.W., Suite 421, Washington DC 20005.

45. T. J. Watson Jr, "America's advantage is that we educate for the future," Brown Alumni Monthly **83**, October 1982, page 32. □

SCIENCE AND SCIENTISTS FOR A NUCLEAR-WEAPON-FREE WORLD

Soviet scientists persuaded the USSR's political leadership to downplay work on missile defenses during the 1970s and 1980s. Current cooperative research by Soviet and US scientists is laying the foundation for the elimination, with verification, of offensive weaponry.

Evgeny P. Velikhov

PHYSICS TODAY/NOVEMBER 1989

The scientific communities of the USSR and the US have been active both in developing new weapons and in attempting to curb the arms race, or at least in trying to prevent it from leading to the ultimate catastrophe. Scientists on the two sides have worked sometimes independently and sometimes with a degree of interaction, direct or indirect.

As stockpiles of arms were built up, the real or imaginary successes of one side provoked the other side and often were used—sometimes for want of better arguments—to justify development of new weapons systems. But a dialogue among scientists seeking to comprehend the situation and to stop the slide down to a catastrophe has never ceased.

Naturally, in years of political stagnation, it could only be hoped that the dialogue among scientists would see better days. But even during the worst years of the cold war, the dialogue had a considerable impact on public opinion, providing it with a scientific basis. During the last few years of accelerating political developments, the dialogue has had great difficulty keeping pace with events.

From our perspective on the Soviet side, we must begin by giving due credit to Soviet science and scientists for responding effectively to the challenge from a most powerful rival who started far ahead. Soviet science provided the means of ending the US nuclear monopoly, which we considered potentially fatal to us, and of establishing strategic parity. Thus, as we see it, Soviet science helped pave the way for a policy of bilateral nuclear disarmament.

Now the question is, What can science do to make the world free of nuclear weapons by the start of the third mil-

Evgeny P. Velikhov is a vice president of the USSR Academy of Sciences, director of the Kurchatov Institute, and a member of the Central Committee, the Supreme Soviet and the presidium of the Academy of Sciences. This article is adapted from a version that originally appeared in *International Affairs,* which is published by the Soviet Foreign Ministry.

lenium? There are only 12 years to go, and the idea of disarmament still encounters powerful opposition.

Soviet ABM debate

Once again, the issue of missile defenses is being spiritedly debated. Soviet scientists have had a long, active involvement in discussions of this issue.

Even at an early stage of the debate, in the late 1960s, the idea emerged of using lasers or charged-particle beams to hit warheads as they approach their targets. It was obvious from the outset, however, that the undertaking was almost certainly futile: Compact reentry vehicles, specially designed to withstand the high temperatures of reentry from space, are hard to locate and lock on to for enough time to administer fatal doses of energy. The propagation of beams of the required power through the atmosphere is itself a great problem.

The main opposition to Soviet beam weapons proponents came from Academician Lev Artsimovich, who was the leader of the Soviet magnetic fusion program. Work was launched on directed-energy weaponry, but soon the difficulties proved insurmountable. At the insistence of Academician Yuli Khariton, deputy director of the Kurchatov Institute's nuclear weapons program, an honest and principled critique was written and handed over to the government, which decided to stop the program around the time the antiballistic-missile treaty was concluded in 1972.

Today, a reminder of that work is the collection of empty structures on the testing ground near Lake Balkhash—the so-called Sary Shagan site—which US officials and other Americans often enquire about (see the box on pages 34 and 35).

Another reminder of the interest in directed-energy weaponry at that time is the amendment to the 1972 ABM Treaty allowing the testing of ground-based missile-defense systems based on "other physical principles." The United States finally agreed to the amendment because of corresponding work in its labs on directed-energy weapons.

In the early 1970s, enthusiasm about space technolo-

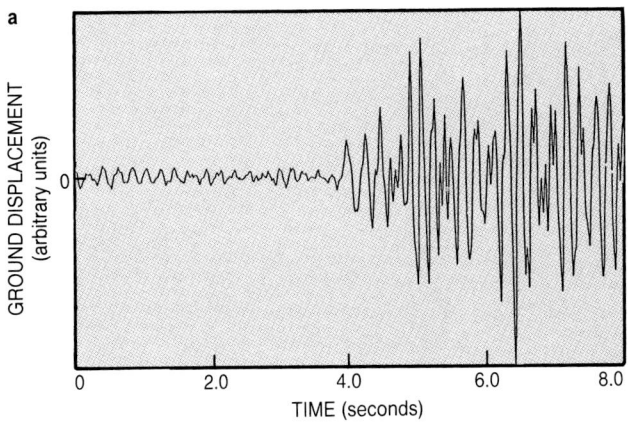

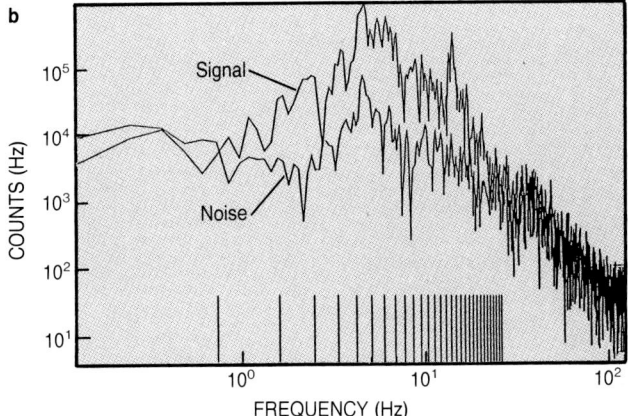

Results from nuclear explosive monitoring program conducted by the Natural Resources Defense Council and the Soviet Academy of Sciences. **a:** Recording of the vertical component of ground motion caused by a 10-ton chemical explosion at a distance of 650 km from the USSR's Kazakhstan test site. **b:** Spectra of 4 seconds of signal and noise from the same 10-ton chemical explosion. Spectra are uncorrected for system response. The vertical lines on the frequency scale at the bottom indicate (equally spaced) frequency samples in the spectral range where the signal-to-noise ratio is greater than unity. This band is from about 0.7 to 28 Hz.

gy gave rise to a Soviet version of the "Star Wars" concept. The idea of space-based neutral-particle-beam weapons was first proposed in 1969 by Academician Gersh Budker, a leading designer of colliding-beam accelerators. But this proposal was defeated within the Soviet Academy of Sciences by criticism from Artsimovich and Academician Boris Konstantinov. This debate vaccinated Soviet scientific opinion against the infectious ideas associated with directed-energy weaponry and prepared the way for the discussion that followed President Reagan's March 1983 Star Wars speech.

Second vaccination

Around 1980, well before Reagan's Strategic Defense Initiative, the idea of effective missile defenses received fresh impetus in the USSR from a proposal by Academician V. N. Chalomey, a designer of booster rockets for the Soviet missile and space programs and vice president for pure physics in the Soviet Academy of Sciences. Chalomey proposed the creation of a space-based defense using interceptor missiles, a concept very similar to the US first-phase scheme that the Reagan Administration proposed during its last year.

Chalomey made his proposal directly to General Secretary Leonid Brezhnev, and therefore the discussion proceeded in a very tense atmosphere at a very high level. A review commission was set up under the chairmanship of Vitali Shabanov, who was deputy minister of defense. Owing to the principled stand of a number of scientists and military experts, the heated debate resulted in a correct decision: The proposal was turned down.

Just imagine what would have happened had we agreed to begin work on the system. Apart from the fantastic expenditures, we would have given an excellent trump card to the Reagan Administration and to US cold warriors. We now most certainly would have a host of

costly and potentially dangerous weapons up in space, yielding no greater mutual security, and their removal would be very difficult.

The second "vaccination" occasioned by the Chalomey proposal enabled us to respond to Reagan's speech very rapidly and energetically. Less than a month after the speech, the members of the Soviet Academy of Sciences held a discussion of the Strategic Defense Initiative at the academy. This discussion was not an easy one—because of secrecy restrictions and the absence of a government position on the matter, we could not express all the arguments. But the resulting document was good, and I would subscribe to it even today. The report was published as a book in 1986,[1] and it was cited by US Star Wars opponents during the Congressional debate on SDI.

Influenced to some extent by the recommendations of Soviet scientists and by the opinion of the world scientific community, the Soviet leadership formulated the idea of "asymmetric" but effective responses to possible US space-based defenses. This was a level-headed decision, and I am sure that it helped sober US public opinion and ease the way for decisions by Congress to limit the SDI budget and protect the integrity of the ABM Treaty. It represented a break in the positive-feedback cycle that has sent the arms race spiraling upward.

Limiting tests

In 1983, before Reagan made his speech, we discussed with our counterparts on the Committee on Science and Arms Control of the US National Academy of Sciences the desirability of an agreement on nondeployment of weapons in outer space. Later, Soviet scientists discussed with Marshal Sergei Akhromeyev, chief of the general staff of the Soviet Armed Forces, the advisability of a Soviet unilateral moratorium on the testing and deployment of weapons in space. A draft treaty was submitted to the

Visit to a Laser Facility at the Soviet ABM Test Site

On 8 July, Evgeny Velikhov took a delegation of ten Americans on a visit to the Soviet testing facility near Sary Shagan on the western shore of Lake Balkhash in Kazakhstan. Sary Shagan is the USSR's main test center for air defense and antiballistic missile defense. The members of the delegation were Congressmen Robert Carr, Jim Olin and John Spratt; reporters Bill Keller of The New York Times *and Jeffrey Smith of* The Washington Post; *Chris Paine, an arms control aide to Senator Edward Kennedy; John Adams, Thomas Cochran and Jacob Sherr of the Natural Resources Defense Council, which organized the delegation; and physicist Frank von Hippel of Princeton University. Von Hippel's report on the visit follows:*

The 1985 joint report of the departments of Defense and State, "Soviet Strategic Defense Programs," showed an artist's rendering of the Sary Shagan facility with a powerful laser beam shining vertically from it (see the diagram below). The caption stated that "the directed-energy R&D site at Sary Shagan proving ground includes ground-based lasers that could be used in an antisatellite role today and possibly a ballistic missile defense role in the future." The 1984 and 1985 editions of the DOD report "Soviet Military Power" showed the same rendering with similar statements.

Unfortunately, the final confirmation of the Sary Shagan visit only occurred after the NRDC group had arrived in the USSR, and there were no US laser experts along. And because most of the day was spent in transit by air and bus, the actual site visit only lasted about three hours. But members of the group were able to make videotapes and to take over 100 photos of the equipment at the facility, and these have been examined by US government and independent experts. US intelligence experts have confirmed, from the external photos of the facility, that it is indeed the one represented in the DOD publications.

The big surprise for both the delegation and for most of those involved in the debate over US antisatellite and missile-defense policy was that we found no high-powered lasers. Judging from the DOD's representations of the facility, we expected a laser comparable to the 2-megawatt deuterium fluoride chemical laser that the SDI program has built at the US White Sands ABM test site. Instead we saw:

▷ A set of 19 pulsed ruby lasers with a total average output of about 100 watts, capable of emitting 30-nanosecond pulses at 10 Hz (see the photograph below).

▷ A CO_2 pulsed laser with a stated power of 20 kilowatts (with only 5–10% of this output actually making it through a long beam line to a beam director).

Of course, Sary Shagan is not the only facility at which the Soviet Union conducts research with lasers that may potentially be relevant to a Soviet antisatellite program. Indeed, even at Sary Shagan, CIA analysts told us after our return, there may be a "facility B" at which laser research is done, and there is a facility in the mountains above Dushanbe, near the Soviet–Afghanistan border, that has excited considerable speculation. These facilities are now near the top of the lists of those in the US urging the USSR on to more military *glasnost.*

Ruby and CO_2 lasers

The figure at the bottom of this page shows a representation of the parts of the facility that we saw. There are two buildings. The large building at the right contains the ruby lasers and has the beam director at its near end. The smaller building at the left contains the CO_2 laser. The low tunnel structure connecting the two carries the CO_2 laser beam to the basement of the large building, where it is deflected upward and then to the beam director along the same beam line as the ruby laser beam.

We followed the ruby laser beam line from the laser room to the beam director. In between is a sensor room,

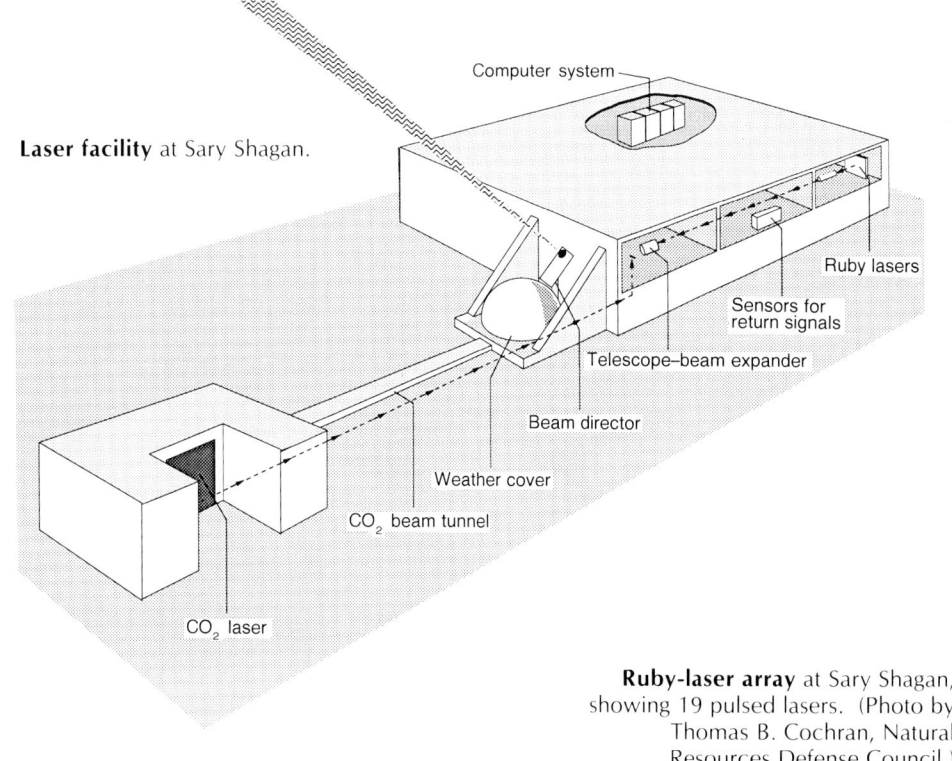

Laser facility at Sary Shagan.

Computer system

Ruby lasers

Sensors for return signals

Telescope–beam expander

Beam director

Weather cover

CO_2 beam tunnel

CO_2 laser

Ruby-laser array at Sary Shagan, showing 19 pulsed lasers. (Photo by Thomas B. Cochran, Natural Resources Defense Council.)

where the return signals are detected, and a combined beam expander and telescope. This device is simply a reflector telescope. When it is acting as a beam expander, the ruby-laser beam enters from behind through a hole in the center of the primary mirror, is reflected from the secondary mirror back onto the front of the primary mirror and then goes to the beam director. In this process, the beam is expanded to a diameter of about 1 meter. When the return signals come from the beam director, the process works in reverse and the photons are concentrated into a small beam that passes through the primary mirror to the sensor room behind it.

The 45°, 30-cm-diameter mirror that sits in the expanded beam (shown to the left of the telescope's secondary mirror in the figure at the bottom of this page) serves to redirect the vertical CO_2 laser beam coming up from the basement to the same beam director. This 45° mirror, which provided the only possible route to the beam director from any hidden laser, was uncooled and appeared to be gold plated like the primary mirror. Polished gold has a reflectivity of about 99% from the 10.6-micron CO_2 wavelength down to the red end of the visible. When the beam director is uncovered, outside air (and dust) flows directly into the telescope. (A particle of dust on a mirror would be heated to high temperature by a high-power laser beam and would burn through the mirror's reflective coating.)

The CO_2 laser is closed-cycle, with the gas being pre-ionized by a 250-kV pulsed electron beam and the excitation energy provided by a lower-voltage electric discharge. The general dimensions of this laser are similar to those of a 25-kW laser sold commercially by United Technologies Industrial Lasers.

Apparent significance

According to the engineers at the facility, it is used several times a week to track and range aircraft. A few attempts have been made to track a low-flying satellite equipped with a retroreflector—the last time in August 1988—but stable tracking was not achieved.

Clearly the Sary Shagan facility would *not* be useful as an antisatellite installation. Even a diffraction-limited 20-kW CO_2 laser beam with an initial diameter of 0.3 meters would have an intensity at orbital altitude much less than that of sunlight.

About a kilometer away from the laser facility described above, the delegation visited a huge empty room, most of which is underground. (See the photograph below.) This structure, which appeared to be about 70 meters long, 30 meters wide and 10 meters high, has very heavy walls and doors. We were told that one idea for the energy source for the high-powered laser, which was to have been installed in this room, was to create intense bursts of electromagnetic energy by imploding magnetic coils with conventional explosives. A very heavy, low wall that lies adjacent to the underground room was meant to protect its roof from the blast of such explosions.

—Frank von Hippel

Abandoned vault, originally intended for tests of a high-powered laser. (Photo by Congressman Robert Carr.)

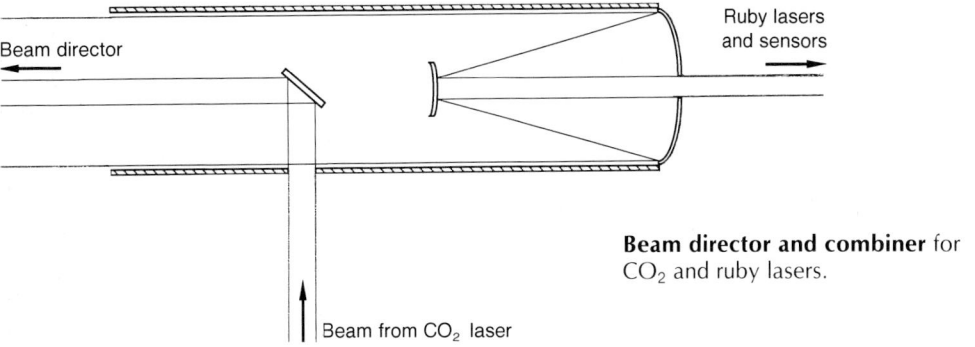

Beam director

Ruby lasers and sensors

Beam from CO_2 laser

Beam director and combiner for CO_2 and ruby lasers.

United Nations by Andrei Gromyko. In August 1983, the Soviet Union declared a unilateral moratorium on weapons testing in space, which remains in force to this day. Although the treaty and the moratorium were both rejected by the Reagan Administration, the US Congress responded by withholding funds for the testing of anti-satellite weapons against targets in space beginning in 1985, and last year the Pentagon abandoned development of the ASAT interceptor that was to have been deployed on F-15 fighter aircraft. [This year, however, the three US military services requested funds for development of more advanced systems, after Congress lifted its ban on tests in space—see PHYSICS TODAY, April 1989, page 59.]

In the spring of 1983, we also discussed with the NAS committee the advisability of signing a comprehensive nuclear weapons test-ban treaty. At the time, the main argument against such a treaty was the alleged difficulty of verification. We began discussing this problem with US scientists in 1986, during a unilateral Soviet moratorium on underground testing. Our hope was that the moratorium would help change US and world opinion about the sincerity of Soviet nuclear disarmament proposals.

In May 1986, we reached an agreement with the Natural Resources Defense Council, an independent organization in the United States that works primarily on environmental and arms control matters, providing for seismic-monitoring experiments around the Soviet nuclear test site in Kazakhstan and at the US test site in Nevada (see PHYSICS TODAY, November 1987, page 83, July 1986, page 63, and August 1986, page 57). The necessary permits were requested from the two governments. We on the Soviet side were so sure about our government's positive response that the US scientists were allowed to begin monitoring a week before final permission was received, and the first seismogram was obtained in July. The work was organized with unprecedented speed—in part because, on the US side, it was financed from private funds and the usual red tape was reduced to a minimum.

It took the Reagan Administration two years to allow similar on-site monitoring in Nevada. Permission was obtained only after Defense Secretary Caspar Weinberger and DOD Assistant Secretary Richard Perle resigned and the talks on ending tests resumed. This project has demonstrated[2] that the signal of a 10-ton chemical explosion can be reliably detected against the background seismic noise at a distance of 650 km in the USSR and at 300 km in the United States. (See the figure on page 33.)

Since 1986, the US House of Representatives has passed an annual amendment that would cut off funding for US underground tests with yields over 1 kiloton if the Soviet Union halts such tests. The Soviet government has declared that if such a limitation is adopted, it will observe it too. Regrettably, the US Senate has not agreed to this amendment.

The US government has come up with a number of other arguments against the complete cessation of tests and speaks of the need to continue testing to ensure the reliability of US nuclear warheads. But as has been stated by a number of Soviet and US scientists, among them some scientists at the Lawrence Livermore National Laboratory, reliability can be checked without actually exploding high-yield nuclear weapons.[3]

Among the real reasons for the resistance to a ban is a desire on the part of the US to continue the development of new types of weapons, above all the so-called third-generation weapons such as hydrogen-bomb-pumped x-ray lasers and optimized warheads for new weapons systems. It is exactly the potentially destabilizing impact of these new weapons that motivates our interest in a test ban.

Another area of cooperative activities between Soviet and US scientists is a joint project on the verification of future arms control agreements, which was organized by the Committee of Soviet Scientists for Peace and Against the Nuclear Threat and by the Federation of American Scientists. One problem being studied in this project is the detection of nuclear weapons in space-launch payloads, ships or submarines. The scientists are investigating the limits of techniques involving the detection of warheads by their gamma and neutron emissions. Naturally, in the case of space-launch payloads, it would be much easier to inspect a payload at the launch site than when it is already in space.

The Soviet government has proposed to the US government that such work, including experiments at launch sites and on board ships, be conducted on a government-to-government basis. So far, the official response has been negative. Therefore we have decided to carry out such experiments on a nongovernmental basis in the hope that governmental organizations will join us when the political situation is ripe. The verification measures in the treaty on the elimination of intermediate- and shorter-range missiles are important steps in this direction.

Strategic arms reduction

The strategic arms reduction talks have imparted urgency to the question of verifying the destruction of nuclear warheads and assuring that their fissile materials are used only for safeguarded, nonweapons purposes. The joint CSS–FAS project includes work on this problem and on the verification of an agreement to end the production of fissile materials for weapons, for which the problems in the US and Soviet materials-production complexes have opened a window of opportunity.

Moving toward a nuclear-weapon-free world also requires study of intermediate states, in which destabilizing situations might prompt a renewed arms race. As part of a project to explore the route to a nuclear-weapon-free world, the CSS has issued a report on the stability of the strategic balance if nuclear arsenals were reduced to just 5% of current levels. The report shows that conditions can be found under which the balance will be stable at such low levels. Unfortunately, however, this transition requires the development of a new single-warhead missile such as the proposed US Midgetman, which calls for a hardened carrier, or the Soviet SS-25. This problem should be discussed between the US and Soviet governments now, because open discussion of military plans will increase mutual trust.

We are just setting off on a road leading away from an irrational world that is irresponsible to past and future generations toward a world based on reason and mutual security. The political process has started, and scientists must mobilize to provide the necessary scientific and technological backing for the journey.

References

1. E. Velikhov, R. Sagdeev, A. Kokoshin, eds., *Weaponry in Space: The Dilemma of Security*, Mir, Moscow (1986).
2. H. K. Given, N. T. Tarasov, V. Zhuravlev, F. L. Vernon, J. Berger, I. L. Nersesov, "High-Frequency Seismic Observations in Eastern Kazakhstan, USSR, with Emphasis on Chemical Explosion Experiments," to appear in J. Geophys. Res.
3. R. E. Kidder, "Maintaining the US Stockpile of Nuclear Weapons During a Low-Threshold or Comprehensive Test Ban," report no. UCRL-53820, Lawrence Livermore National Laboratory, Livermore, Calif. (October 1987). For a rebuttal, see G. H. Miller, P. S. Brown, C. T. Alonso, "Report to Congress on Stockpile Reliability, Weapon Remanufacture and the Role of Nuclear Testing," report no. UCRL-53822, Lawrence Livermore National Laboratory, Livermore, Calif. (October 1987). ∎

Physicists sign appeal for nuclear freeze

PHYSICS TODAY/JANUARY 1984

By late November 15 000 physicists in 44 countries had signed the following appeal for a nuclear freeze:

We call for an agreement to halt the testing, production and deployment of nuclear weapons and nuclear weapons delivery systems. Meanwhile, no further nuclear weapons or delivery systems should be deployed anywhere.

Over half of the living winners of the Nobel prize in physics were among the signers.

In mid-November groups of physicists presented the call to representatives of national governments and to international organizations. The presentations, like the circulation of the call, were conducted without any formal organization. On 18 November Philip W. Anderson (Princeton University), James W. Cronin (University of Chicago) and Robert Serber (Columbia University) met with Javier Pérez de Cuéllar, Secretary General of the United Nations, to present the call. (Sheldon Glashow of Harvard University was prevented by illness from participating.) Pérez de Cuéllar gave the group a sympathetic reception, according to Serber, and welcomed the appeal. He assured the physicists that he shares their goals and is doing all he can to bring about disarmament, Serber told us.

Presentations were also made to officials of the governments of Finland, France, Italy, Japan (to the prime minister), Spain and West Germany, according to Rolf Hagedorn (CERN), one of the initiators of the petition. He told us that receptions for the most part were polite but negative.

In the US, signers made attempts to present the petition to President Ronald Reagan, Vice President George Bush, the Office of Science and Technology Policy, House majority leader Thomas P. O'Neill (D-Mass.), and Senate majority leader Howard Baker (R-Tenn.), without any success at this writing. Hagedorn did not know of any attempts to present the call in the USSR, where over 750 physicists signed the statement.

The idea for the call arose during a conversation at CERN in the summer of 1982. Daniele Amati (CERN), Nina Byers (UCLA), Rolf Hagedorn (CERN), Jack Steinberger (CERN), Victor Weisskopf (MIT) and Christophe Wetterich (CERN) discussed what they could do about the nuclear arms race. They wrote the appeal and sent it to 120 well-known physicists asking for their endorsements. Almost 80 responded positively. The organizers then found collaborators in many countries and, in the spring of 1983, sent out copies of the call with over 80 signatures of prominent physicists. From that time the gathering of signatures has proceeded—and still proceeds—informally, as physicists pass it on to each other.

In a covering letter circulated with the petition, the organizers state that the nuclear arms race is accelerating. The increased precision of missiles may invite first-strike use and launch on warning; new tactical nuclear weapons may lower nuclear thresholds. The letter continues to explain that the appeal has been circulated among physicists because it arose in a discussion among physicists. It was felt that an appeal from physicists all over the world, across political and national boundaries, might be a constructive contribution to efforts to curb the arms race. The organizers also felt it was appropriate for physicists to present an appeal because physicists have been instrumental in the invention of nuclear weapons and are still directly involved with their production and development. —DG

Petition presented at UN. From left: Jan Martinsen (UN under-secretary-general for disarmament affairs), Philip Anderson (Princeton), Javier Pérez de Cuéllar (UN secretary-general), Robert Serber (Columbia), Sidney Katz (coordinator of the presentation from the Center for Defense Information) and James Cronin (University of Chicago).

Role of strategic space forces in a non-nuclear world

Freeman J. Dyson

PHYSICS TODAY/JUNE 1984

Unfortunately for humanity, the purging of nuclear weapons from the Earth is too big a job for technology to do alone. The dream of the omnipotent celestial laser-beam patrol fails on technological grounds. The space-based antimissile system has many technical weaknesses. Even if death-ray weapons could be aimed and focused with perfect accuracy and could deliver a sufficient concentration of energy to destroy a missile, they would still be at a great disadvantage in terms of vulnerability. A missile is vulnerable to death-ray attack only for a few minutes while it is in flight; a death-ray machine in space is vulnerable all the time. A death-ray machine is a large and delicate piece of apparatus; a single pebble colliding with it at orbital velocity would have a good chance of putting it out of action. The same technology that allows us to aim the death ray with the necessary precision also allows us to aim the pebble. When the experts play their little games, with ground-based missiles on one side and death-ray machines in space on the other, the ground-based side almost always wins. It is of course possible to adjust the numbers of weapons on the two sides so as to make it an even game, but then the cost of the space weapons is outrageously high compared with the cost of the ground weapons. These imaginary battles ignore many aspects of the real world, but they lead us to a clear conclusion: So long as large land-based or sea-based missile forces exist and are not subject to severe political constraints, there is no technological magic by which space-based weaponry can disarm them. The extension of the technological arms race into space cannot by itself make ground-based missiles obsolete or ineffective. For this reason I call the future in which space weapons proliferate without end the technical-follies future. It is an extension into the future of the same folly which gave us the MX. It is a future of double folly, the small-scale folly of

Freeman J. Dyson is professor of physics at the Institute for Advanced Study, Princeton, New Jersey.

militarily useless weapons, and the large-scale folly of unattainable strategic objectives.

The arms controllers' future makes space a peaceful sanctuary and leaves us to deal as best we can with our strategic problems on the ground. The technical-follies future makes space a battleground and does nothing to make the problems on the ground more tractable. But there is a third possible future, a future in which nuclear weapons are legally banned from the Earth and from space, and in which the resources of non-nuclear technology are used in an energetic fashion to help make the ban effective. This third future I call the defense-dominated future. In the defense-dominated future, weapons of mass destruction are disarmed, not by defensive technology alone, but by legal and political restraints strengthened by the active intervention of technology.

So long as we maintain overwhelmingly destructive nuclear forces on Earth, we would be wise to keep space disarmed so far as possible. But if we can ever achieve drastic disarmament on Earth, a deployment of appropriately designed space weaponry may help us to push the negotiated reduction of nuclear arsenals all the way to zero. These are the premises of the defense-dominated future: The Earth becomes a non-nuclear sanctuary stabilized by substantial military forces in space; space forces are specifically designed to allay fears and to diminish incentive for secret or open nuclear rearmament. To achieve these purposes, the space forces would not need

to attempt the almost impossible task of nullifying a full-scale onslaught of the present-day Soviet or American missile forces. It would be sufficient for the space forces to be capable of nullifying much smaller threats. The smaller threats of a non-nuclear world would be either residual nuclear forces concealed by a country secretly violating a disarmament treaty, or embryonic nuclear forces deployed by a country openly abrogating the treaty, or forces belonging to smaller countries which had never acceded to the treaty. Space forces which could defeat these smaller threats are not beyond the realm of technical possibility. Such forces would not by themselves remove all danger of breakdown of the non-nuclear regime, but they would powerfully strengthen the political and institutional structures on which the durability of the regime would depend.

What kind of space forces would the defense-dominated future require? Certainly not space battleships, and probably not death-ray generators or high-energy lasers. One of the primary requirements for an effective space force is to be itself inconspicuous and invulnerable. The most likely shape for the space force would be a multitude of small vehicles, scattered in orbits around the Earth, carrying telescopes and sensors of various kinds. The purpose of these vehicles would be to collect accurate and timely information. In a defensive battle, information is more important than exotic kill-mechanisms. If the defense has adequate information, it can relay the information to small non-nuclear interceptors, either ground-launched or in orbit, which can use their own sensors to home onto a flying missile and kill it by direct impact. The idea of a space force of this kind is not new. It was proposed in the 1950s and given the name BAMBI (Ballistic Missile Boost Intercept). It was then rejected as technically impracticable and prohibitively expensive. During the subsequent twenty years, sensors and microcomputers have become enormously more capable and also cheaper. If the world ever decides to move along the road toward the defense-dominated future, it is possible that space forces of the BAMBI type can be built at reasonable cost, and that they can be effective enough to help stabilize the world against backsliding into nuclear terror.

In the defense-dominated future as I have described it, the space forces play a modest role, patrolling the Earth inconspicuously and serving as an adjunct to earthbound political arrangements. These forces could probably operate most efficiently without a single air-force officer in orbit. It is a far cry from the Deep Space Bombardment Force, or from the Galactic Empire space force we saw in the *Star Wars* film. And that is all to the good.... Space forces, like air forces, should be firmly harnessed to the strategic needs of earthbound humanity.

In the end, the goals of the arms controllers' future and the defense-dominated future are the same, and only the means are different. The goal of both futures is a stable world with a minimum of nuclear armament. The arms controllers' future chooses first to disarm space and to leave nuclear offensive forces on the ground intact, in the hope that a stable regime of nuclear deterrence will allow gradual steps toward disarmament. The defense-dominated future chooses first to disarm nuclear forces on the ground and to let space forces grow, in the hope that a disarmed world will settle down more comfortably if it has space forces providing substantial protection against the risks of surprise attack. Both futures, if they fulfill their promise, converge to a common end. The real future, if we are wise, will probably lie somewhere in the middle between the arms controllers' and the defense-dominated extremes. If we are unwise, the technical-follies future is there, waiting for us to stumble into it.

★ ★ ★

From the book "Weapons and Hope" by Freeman J. Dyson. © *1984 by Freeman J. Dyson. A Cornelia and Michael Bessie Book. Reprinted with permission of Harper & Row, Publishers, Inc.* □

Some physicists speak out in favor of Star Wars research

PHYSICS TODAY/JANUARY 1986

The physicists who wrote and are circulating petitions opposing Star Wars have sometimes made the claim that it is hard to find any physicist willing to defend the general concept of a leakproof missile-defense system. While the claim sounds implausible, the anti-SDI petitioners are not alone in advancing it. Physicist and science writer Jeremy Bernstein, reviewing William J. Broad's book *Star Warriors* in the *New York Times Book Review* recently claimed that Broad was "not able to find a single scientist" who would say that a leakproof nuclear umbrella could be built.

At another extreme, Lieutenant General James A. Abrahamson, the chief of the SDI program, has claimed that opposition to Star Wars among scientists is confined to "a few diehards." Roughly 2500 members of science faculties had signed petitions opposing Star Wars by the beginning of November, but Administration officials sometimes argue that the anti-SDI petitioners have no real involvement in SDI work and that their opposition is therefore shallow and irrelevant.

Whatever one may say, it was at least clear by the end of 1985 that some scientists were willing to speak up in favor of pursuing research vigorously within the framework of the SDI program.

Software issue. Last fall, when some computer scientists claimed that it would be intrinsically impossible to construct a workable data-processing system for missile defenses, Charles Seitz of Caltech, Solomon Buchsbaum of AT&T Bell Labs and Danny Cohen of the University of Southern California, among others, made strong public statements to the contrary.

In testimony to a Senate Armed Services subcommittee on 3 December, Buchsbaum said that a large system could compensate for errors. "The network as a whole is more reliable than its individual components," Buchsbaum said, alluding to experience with the telephone system. Buchsbaum is executive vice president, customer planning, at AT&T Bell Labs, and is chairman of the White House

BUCHSBAUM

Science Council. He also is on the SDI advisory committee and is a senior consultant to the Defense Science Board.

Buchsbaum does not claim that a perfectly leakproof missile-defense system can be built or that offensive missiles can be rendered "impotent and obsolete," as the President put it in his March 1983 speech. (Neither, for that matter, do SDI officials claim a leakproof defense can be had.) But Buchsbaum does think it will be possible to design a system that would be "reliable, robust and resilient." In his Senate testimony, Buchsbaum said he recognized that SDI "faces enormous challenges and problems which I do not minimize. However, a vision of the world in which the two superpowers have agreed to constrain their respective nuclear forces to levels much lower than today's and, at the same time, have also agreed to protect themselves and their allies against nuclear attack with defensive systems—a protective shield—of reasonable effectiveness is an attractive one."

On the specific question of whether one could build an error-free control system containing tens of millions of lines of software, Buchsbaum said that

"this is the wrong question." By compartmentalizing crucial functions and by building redundancy into the system, one could design a system that would be—like the telecommunications network—much more reliable than its components, Buchsbaum said.

Danny Cohen, a computer scientist who headed a 1985 panel that evaluated potential software for SDI, said at the same hearings: "There are those who claim they cannot produce adequate software. We agree that they cannot. There are experts who claim they can. We agree with them." Cohen argued that an adequate system could be designed by relying on autonomous redundant subsystems with different program codes.

Charles Seitz, a computer scientist who also served on the SDI software panel, told PHYSICS TODAY that he considers most scientific objections to Star Wars ill-founded. In particular, he thinks that most computer scientists do not agree with David Lorge Parnas of the University of Victoria in Victoria, British Columbia, who resigned from the software panel last summer, claiming that it would be impossible to design and build adequate software for a Star Wars system.

Seitz may be typical of many scientists who evaluate individual SDI components on their merits and who take a wait-and-see attitude toward missile-defense systems. He would not sign a petition either opposing or favoring Star Wars, he told PHYSICS TODAY, but he would sign a statement saying that adequate computer software could be designed for such a system.

Pro-SDI petition. To date, probably the strongest statement favoring Star Wars was adopted at an SDI seminar held in Washington on 9–10 November under the sponsorship of the Global Foundation. The statement was signed by several nonphysicists and the following physicists: Peter Auer (Cornell), R. V. Jones (University of Aberdeen, Scotland), Behram N. Kursunoglu (president and chairman of the board, Global Foundation Inc, and professor and director of the Center for Theoretical Studies at the University of Miami),

Carlo Salvetti (University of Milan), S. Fred Singer (George Mason University), Joseph Weber (University of Maryland), Alvin M. Weinberg (Institute for Energy Analysis, Oak Ridge) and Eugene P. Wigner (Princeton). The statement read in part:

We accept the concept of a strategic defense against nuclear missiles. We, therefore, support research to establish the feasibility of such a strategic defense. . . .

Defense, if sufficiently effective, could reduce the likelihood of a nuclear attack directed against strategic missiles and against cities and populations. . . .

We find defense morally preferable to the current strategy of naked offensive confrontation. . . .

The danger of the offensive standoff grows as increasing missile accuracy makes possible precise strikes against retaliatory forces, as nations other than the two superpowers acquire nuclear arms and as the possibility of an accidental launch increases. In our view, a successful defensive system could contribute greatly to nuclear stability.

The Global Foundation is based in Coral Gables, Florida, and is a nonprofit organization that describes itself as interested in global issues and frontier problems in science, such as the impact of physics on forefront medicine. Kursunoglu says that the views expressed in the Star Wars statement do not necessarily reflect the opinions of the trustees of the Global Foundation.

—WILLIAM SWEET

ARMS CONTROL PHYSICS: THE NEW SOVIET CONNECTION

Informal contacts established between US and Soviet physicists during an unpromising period in relations between the superpowers have taken on new significance since Gorbachev's emergence.

Frank von Hippel

PHYSICS TODAY/NOVEMBER 1989

US and Soviet scientists have been holding quiet discussions on matters like missile defense since the first International Pugwash Conference on Science and World Affairs in 1957 (see PHYSICS TODAY, September, page 81). Sometimes these discussions have constituted an informal "backchannel" between the governments for consideration of possible arms control initiatives. Thus in conversations between US and Soviet scientists in 1964, US scientists argued that ABM systems would not be effective against a determined adversary and that their deployment would stimulate an offense–defense arms race; they accordingly proposed a treaty to limit ABM systems. Two high-level Soviet scientists, Lev Artsimovich (who was head of the Soviet fusion program) and Mikhail Millionshchikov (who was vice president for applied physics and mathematics of the Soviet Academy of Sciences), subsequently helped bring their government around to this position, thereby contributing to the achievement in 1972 of the ABM Treaty.[1]

When cold war tensions have been at their worst, informal contacts between US and Soviet scientists have not always been welcomed in government circles. Yet as the history of the ABM Treaty illustrates, such discussions

Frank von Hippel, a physicist and researcher on the technical basis for arms control and on energy policy, is a professor of public and international affairs at Princeton University.

often have provided an opportunity to investigate new, experimental ideas that government agencies have been loath to explore for fear of reducing political maneuvering room. Such informal discussions also have been convenient forums in which to go beyond the issues of the day and to develop a basis for longer-range planning.

Therefore, despite misgivings, government officials who are interested in the possibility of progress in arms control negotiations have sometimes welcomed the type of research done within the framework of private initiatives. They see independent scientists as scouts mapping out technical territory that the armies of government technical experts will be able to secure very quickly once they are permitted to move forward.

During the early 1980s, a period of renewed cold war tension, a number of US scientists began cautiously to cultivate contacts with independent-minded Soviet counterparts such as Evgeny P. Velikhov, the Soviet Academy's current vice president for applied physics and mathematics, and Roald Sagdeev, until recently the head of the Soviet Institute for Space Research. With the emergence in the mid-1980s of Mikhail Gorbachev as leader of the USSR and the advent of *glasnost*, it has been possible to experiment with more open types of exchange, including joint research programs on the technical basis for new arms control policy initiatives. At the same time, some Soviet scientists have been catapulted into positions of top advisory authority, a situation that involves some risks but also great opportunities.

As some of the Soviet scientists have taken on greater

responsibilities, demands on their time have become voracious, limiting their availability for casual discussion. Despite their closeness to power, however, they have continued to operate with an activist style.

As a result of various fortuitous circumstances, I have been personally involved in a number of exchanges with these Soviet scientists, and my professional life has become more exciting as a result. I first met Velikhov at a meeting of the International Physicians for Social Responsibility in the summer of 1983. Earlier that year, a group of Soviet academicians had sent an open letter to the US scientific community asking whether, in light of President Reagan's "Star Wars" speech of March 1983, there had been a change in the professional consensus in the US regarding the feasibility of effective missile defenses. Only the Federation of American Scientists responded directly, and that response led to an invitation from Velikhov to visit the Soviet Union. Despite a partial boycott on bilateral scientific contacts, which the federation had joined mainly because of the Soviet government's treatment of Andrei Sakharov, the group decided to accept Velikhov's invitation. In November the federation sent a party to the USSR that included FAS President Jeremy Stone, John Pike of the FAS staff, John Holdren of the University of California (FAS vice chairman), Berkeley and myself (FAS chairman).

About this time, Velikhov also received an invitation from Senator Edward M. Kennedy to come to the United States to testify about the "nuclear winter" debate. Velikhov came with a party that included Sergei Kapitsa (Vavilov Institute of Physical Problems) and Vladimir V. Aleksandrov (Computing Center of the Academy of Sciences). While in the US Velikhov also visited Princeton and FAS headquarters in Washington, affording an opportunity for further conversations.

The Committee of Soviet Scientists

Velikhov met with the FAS and came to the Kennedy hearing as chairman of the Committee of Soviet Scientists for Peace and Against the Nuclear Threat. This committee, mostly made up of high-level members of the Soviet Academy of Sciences, was established in the spring of 1983, following Reagan's Star Wars speech. Velikhov had succeeded Artsimovich as head of the Soviet fusion program and Millionshchikov as the Soviet Academy's vice president for applied physics and mathematics.

In early 1983, Velikhov had already had discussions with Richard Garwin of IBM and others concerning the possibility of a ban on antisatellite weapons. These discussions took place in a meeting between the US National Academy's Committee on International Security and Arms Control and a counterpart group from the Soviet Academy of Sciences. The two groups had been meeting since June 1981. (Marvin Goldberger, at that time president of Caltech, was the first chairman of the US Academy's committee, and Academician N. N. Inozemtsev, director of the Institute of World Economics and International Relations, headed the Soviet Academy's group.)

These discussions helped persuade the Soviet government to declare, in August 1983, a unilateral moratorium on the testing of Soviet ASAT (antisatellite) systems. That same month the USSR introduced at the United Nations a draft ASAT treaty that owed a good deal to a model treaty that Garwin and Kurt Gottfried of Cornell had developed in cooperation with the Union of Concerned Scientists (see PHYSICS TODAY, November 1984, page 99).

Velikhov told me that the reason he decided to organize the Committee of Soviet Scientists was to educate a new generation of Soviet scientists, including himself, about nuclear arms control and to reopen the US–Soviet dialogue on strategic defense with the roles reversed. Now it would be the Soviet scientists who would try to convince the US government, with US scientists as intermediaries, that the pursuit of ballistic missile defenses would be counterproductive.

The Committee of Soviet Scientists therefore opened exchanges with scientists representing the full spectrum of US opinion on SDI—including, to the surprise of some of us, Edward Teller. (Velikhov and Teller had both attended conferences in July 1982 and July 1983 in Sicily on "how to avoid nuclear war" and "the technical basis for peace.")

As a result of such exchanges, its own studies and previous Soviet studies (see the article by Velikhov on page 32), the CSS became quite expert on the technical aspects of SDI, and a group under the leadership of Velikhov, Sagdeev and Andrei Kokoshin (at that time head of the division of military–political affairs of the Institute of US and Canadian Studies) wrote up its conclusions in a book which was printed in 1986 in both Russian and English.

The primary message of the book was that it would be feasible to neutralize space-based defenses with much less expensive countermeasures. The analyses presented were based on descriptions of proposed SDI weaponry that had been published in US publications, such as *Aviation Week and Space Technology*, and on back-of-the-envelope physics calculations—much in the style of Garwin and Hans Bethe. However, open publication of a technical discussion of possible future weapons systems was an unprecedented event in the Soviet Union and led to a number of letters from irate Soviet citizens demanding that the members of the CSS be prosecuted for revealing how the Soviet Union would neutralize the US Star Wars system. Fortunately, the Soviet government ignored these demands.

In the upshot, the direct impact of the CSS on the US debate over SDI was negligible. But the indirect impact was more important, because it helped persuade Gorbachev, who became General Secretary in early 1985, to announce that the Soviet Union would not compete with the United States in attempting to establish space-based defenses but would instead make an "asymmetric response" based on the types of countermeasures described by the Soviet scientists. This undermined the argument of American SDI proponents, who had said that if the United States did not go full speed ahead, the Soviets would deploy space-based defenses first.

In parallel with its critique of space-based defense, the Committee of Soviet Scientists also sponsored some studies on nuclear winter.[2] In part because of the limited capacities of Soviet computing facilities, these studies contributed more to the internationalization of scientific

GLORIA B. LUBKIN

Roald Sagdeev (left), a leading figure in the Committee of Soviet Scientists, and Richard Garwin of IBM, well known for his contributions to scientific research on arms control in the United States, chat while Sagdeev cooks pilaf Tatar style during a recent visit to the United States.

concern about nuclear winter than they did to the understanding of the phenomenon itself.

In-country seismic monitoring

For many years the US government had been on record as favoring a comprehensive nuclear test ban, but the Reagan Administration had backed away from this position. Gorbachev apparently hoped that world opinion would persuade the United States to join in a moratorium. In fact, although the Soviet moratorium was widely praised by advocates of arms control, the public response was far weaker than that to the 1958–61 moratorium, perhaps because moving testing underground had effectively ended the fallout that had so frightened the public.

When Gorbachev came into power in early 1985, *glasnost* and a personal relationship between Velikhov and Gorbachev gave the CSS the opportunity to undertake new ventures.

Its first new move was in direct support of Gorbachev's first arms control initiative: the Soviet unilateral moratorium on underground testing that began in August 1985.

In announcing the moratorium, Gorbachev presumably was inspired by the bilateral test moratorium of 1958–61, with which the US and Soviet governments signaled to each other and the rest of the world that they were seriously interested in ending all nuclear testing (see the box on page 43). That moratorium led to the Kennedy–Khrushchev Partial Test Ban Treaty of 1963, which ended nuclear testing in the atmosphere, in outer space and in the oceans, but not underground. Now Gorbachev wanted to complete the job by ending all underground testing of advanced nuclear weapons systems.

One of the major obstacles that had prevented Kennedy and Khrushchev from consummating a comprehensive test ban had been a claim by US weapons laboratories that the Soviets might be able to continue to develop new nuclear weapons with small underground tests, which seismic sensors beyond the borders of the USSR would be unable to distinguish from earthquakes. The 1958 International Conference of Experts sketched out a verification system involving a worldwide network of

seismic stations and on-site inspections at the locations of suspicious seismic events, but in the end, the system that the US government wanted was too intrusive for the Soviets, and the underground part of the ban had to be dropped.

Meetings with Shevardnadze, scientists

In September 1985, at a meeting in Copenhagen commemorating the centennial of the birth of Niels Bohr, Velikhov suggested to me that the Soviet government might be willing to let an outside group set up a seismic monitoring system in the USSR. I was already aware that one group, Parliamentarians Global Action, was interested in establishing a monitoring system for a bilateral US–Soviet moratorium. (Founded in 1979 by British and Canadian legislators, Parliamentarians Global Action had launched a peace initiative in May 1984—now known as the Five Continent Peace Initiative—with the objective of getting the two superpowers to reopen a constructive dialogue on arms control.) During the next several months, I learned that similar proposals were being made by Jack Evernden of the US Geological Survey and Thomas Cochran, a physicist with the Natural Resources Defense Council, an independent organization that does legal and technical work on environmental, energy and arms control policy.

The Parliamentarians were proposing to establish a monitoring system under the auspices of their peace initiative. Evernden was attempting to expand a scientific seismic monitoring agreement between the US Geological Survey and the Soviet Academy to provide information toward the design of a network to monitor a 1-kiloton threshold test ban. The NRDC was proposing a nongovernmental monitoring project to provide information about tests that both the US and the USSR keep secret.

In early 1986, Cochran wrote to Reagan and Gorbachev asking permission for private groups to establish seismic stations in the two countries for joint verification research. But after learning that the NRDC proposal was unacceptable to the Soviet foreign ministry, Cochran was ready to give up. At this point Jeremy Stone of FAS suggested trying to work through the Soviet Academy instead, and he invited Cochran to attend a meeting with

Academy representatives that was to take place at Airlie House in Warrenton, Virginia, in March 1986. At that meeting, which I also attended, the idea of nongovernmental monitoring met with a favorable reception from the Soviet delegation, which was headed by Sagdeev.

In April 1986, I accompanied a delegation from Parliamentarians Global Action on a visit to Soviet Foreign Minister Edward A. Shevardnadze. The delegation urged that the Soviet Union extend its testing moratorium in the hopes that the United States might still be persuaded to join.

To a physicist unfamiliar with the rituals of diplomacy, the meeting with Shevardnadze seemed very stylized. After we had shaken hands, we sat down, and tea and cookies were brought in. Then the spokesman for the Parliamentarians made his statement, and Shevardnadze responded politely. There were one or two additional polite exchanges; then the delegation got up, shook hands with Shevardnadze and left. It was not clear to me that anything had been accomplished, and so I suggested that we meet with Velikhov.

That meeting could not have been more different. The first thing Velikhov asked was, "Do you have any good ideas?" This led to a brainstorming session, at the end of which Velikhov and I agreed to organize a workshop the following month in Moscow. I agreed to bring representatives of the Western groups that were interested in setting up seismic stations in the USSR.

After the meeting, as we drove away, one of the Parliamentarians said to me in wonder: "You scientists can talk to each other!"

May 1986 workshop

By the time the workshop on test ban verification convened in May 1986 in Moscow, there were several reasonable proposals on the table. But the monitoring system suggested by the Parliamentarians would have required the Reagan Administration to join the Soviet test moratorium, and Evernden's plan also would have required US governmental support. Under the circumstances, the idea of doing the project on a nongovernmental basis carried the day, and the NRDC was able to make an immediate commitment to the project.

The final project incorporated elements from each of the three proposals. The design of the seismic monitoring system was based on that developed by Evernden; the scientific director of the project, Charles Archambeau of the University of Colorado at Boulder, had originally been recruited by the Parliamentarians for their project; and the NRDC undertook to raise funds for the US side of the project and to administer it.

The workshop led to an agreement between the Soviet Academy and the NRDC to set up three seismic stations around the principal US and Soviet underground test sites. Less than two months later, a team of seismologists recruited by Archambeau and led by Jon Berger and James Brune of the University of California's Scripps Institution of Oceanography were taking data in Kazakhstan, 100 kilometers from the Soviet test site. Cochran was in charge of the overall US effort.

This project had a rather big political impact, probably bigger than Gorbachev's unilateral moratorium on testing, because it vividly demonstrated, before the conclusion of the treaty eliminating intermediate-range missiles in Europe, the Soviet government's willingness to accept in-country monitoring of an arms control agreement. The project also had technical merit: It provided data that would be valuable to the design of an in-country monitoring network to verify a low-threshold underground test ban treaty.

International Scientists' Forum in Moscow

A key event in the developing exchanges between Western and Soviet scientists was the International Scientists' Forum on Drastic Reductions and Final Elimination of Nuclear Weapons, which Velikhov organized with some assistance from Western scientists, including myself, and which took place in Moscow in February 1987 (see PHYSICS TODAY, April 1987, page 67).

It was at this meeting that Sakharov reemerged as a public figure in the USSR, and it was here that he argued, in public for the first time, that the USSR should stop conditioning its agreement to a new strategic arms reductions treaty on a US promise not to deploy strategic defenses. Sakharov argued that the SDI program would collapse under its own weight.

There was a workshop at the forum on the idea of clarifying the gray areas of the ABM Treaty by imposing quantitative limits on parameters such as laser brightness. This idea was subsequently taken up by both the Soviet government and by Paul Nitze, who at that time was Reagan's specially designated arms control expert in the State Department. (Strobe Talbot, *Time* magazine's chief diplomatic correspondent, credits John Pike of FAS with implanting this idea in both governments.[3])

Also at the forum, a group of West European advocates of "non-offensive defense," who had been developing their ideas for several years in a series of Pugwash workshops, had the opportunity to present to Gorbachev's advisers their arguments that the best way to stabilize and shrink the conventional-weapons confrontation in Europe would be to preferentially eliminate offensive weaponry such as tanks. Subsequently, three of the West European analysts, physicist Anders Boserup (University of Copenhagen), economist Robert Nield (Cambridge University) and philosopher Albrecht von Mueller (Max Planck Society, Starnberg), and I were invited to write a letter to Gorbachev explaining the idea of non-offensive defense and its implications for arms control efforts.[4] These inputs probably helped provide the intellectual basis for Gorbachev's decision in late 1988 to unilaterally eliminate 5000 tanks from Soviet forces in Eastern Europe (about one-half the total number there) and another 5000 from the Western USSR.

Finally, it was decided at the forum to launch three new cooperative East–West ventures:

▷ An international journal, *Science and Global Security*, to publish the results of research pertaining to arms control, disarmament and the environment
▷ A five-year Cooperative Research Project on Arms Reductions under the joint auspices of the Federation of American Scientists and the Committee of Soviet Scientists
▷ An International Foundation for the Survival and Development of Humanity.

Joint journal on science and security

Although a number of American journals, including *Scientific American* and PHYSICS TODAY, publish occasional articles on the technical basis for arms control agreements, and the APS Forum on Physics and Society has published a number of useful collections of articles, there has been no journal with the publication of such work as its primary mission. *The Bulletin of the Atomic Scientists* was for many years devoted primarily to the publication of articles by scientists concerned about the arms race on the subject, but it has been used by them primarily to reach the public and to communicate with one another about political concerns, rather than to make detailed presentations of technical work.

With the creation of *Science and Global Security*, a

Test-Ban Technical Talks of 1958

At the end of 1957, after the USSR proposed a two- or three-year trilateral ban on nuclear weapons tests, President Eisenhower suggested, amid growing world concern about fallout from such tests, that Great Britain, the United States and the Soviet Union conduct "technical studies of the possibilities of verification and supervision." In early 1958, a panel convened by the President's science advisory committee and headed by Hans Bethe concluded that a ban could be policed and would be advantageous to the US, and in April that year Eisenhower suggested technical talks to Soviet leader Nikita Khrushchev. Khrushchev accepted, and in the summer of 1958 teams of British, American and Soviet scientists met in Geneva. This was the first instance of technical cooperation between US and Soviet scientists in furtherance of arms control objectives.

Despite cold war tensions and questions concerning the political independence and authority of the three teams, they were successful in reaching agreement on a proposed system of in-country monitoring stations. Leaders of the three nuclear weapons states thereupon agreed to diplomatic negotiations, which opened in an atmosphere of optimism at the end of October 1958. Immediately upon the beginning of the diplomatic negotiations, however, the US delegation presented new scientific evidence indicating that a larger system of in-country stations might be required. The US delegation also raised the question of whether a weapons test could be concealed ("decoupled") in a big hole. The Soviet side, whose scientific team had been operating under much tighter political supervision from the start, reacted with suspicion, finding it hard to believe that the United States would change its position in so important a matter solely on the basis of scientific findings. Both sides nevertheless declared testing moratoria, and the talks proceeded, only to break down in May 1960 with the shooting down of the U-2 piloted by Francis Gary Powers and Khrushchev's breaking up of the Paris summit conference.

Talks resumed during the Kennedy Administration, but by that time a history of unilateral test moratoria followed by frenzied testing on both sides had generated wide cynicism, and when a partial test-ban treaty finally was concluded in August 1963, it involved no mutual system of

'Molodets!'—"Bravo!" or "Fine fellow!"—Khrushchev exclaimed, greeting W. Averell Harriman the day after the conclusion of the partial test-ban treaty in 1963. Harriman struggled hard for a comprehensive ban and was a key figure in the negotiation of the partial ban.

in-country verification. The accord barred testing in the atmosphere, the oceans and space by the three parties to the agreement, but not underground testing, and in the following years the US and USSR continued to conduct such tests at increasing rates.

Could a comprehensive test-ban agreement have been achieved in 1963? The two sides seemed to some to be close, in that the USSR said it would permit three in-country challenge inspections, while the US insisted on six. Even, however, if the two sides could have reached agreement on the number of stations, it would have been hard to agree on what the inspectors should do.[1]

While agreeing by and large on the positive value and precedent-setting character of the Geneva technical talks, participants and scholars have drawn from them varying conclusions about the benefits and pitfalls of joint scientific research in matters of acute political sensitivity:

▷ James R. Killian Jr, president emeritus of MIT and science adviser to Eisenhower: "While unexpressed, it seemed clear that many political officers of our government hoped fervently that the whole test-ban controversy would be resolved by scientists; the political level looked for scientific certainty to minimize the difficulty of weighing other imponderables in the decision making. In the end, a limited test ban was achieved by diplomacy aided in essential ways by scientists."[2]

▷ Charles S. Maier, Harvard University: "In hindsight, it can be seen that the new scientific resources of the President provided a two-edged blade. Eisenhower himself clearly belonged among the supporters of a test-ban treaty. . . . Nonetheless, his style of administration and careful delegation of authority constrained his own

convictions. . . . This is not to argue that caution should have been abandoned, but the search for loopholes became all-consuming. Here the scientists' ingenuity hobbled the President even while it aided him. . . . The new scientific advisory panels could only multiply choices or frame them more intelligently; they could not resolve them."[3]

▷ Robert Gilpin, Princeton University: "In retrospect there can be no doubt that the Conference of Experts was unique in the annals of diplomacy. . . . [But] in the mistaken belief that one can separate the technical and political aspects of national policy, [the] American political leadership in the summer of 1958 assigned to a group of inexperienced private citizens the task of negotiating the first part of what might have been an extremely important arms control agreement. As should have been expected, the American scientists . . . fell into a number of regrettable errors. Yet, under the circumstances, it is surprising that there were no greater errors committed and that the the scientist-diplomats did as well as they did."[4]   —WILLIAM SWEET

References
1. G. T. Seaborg, *Kennedy, Khrushchev and the Test Ban*, U. Calif. P., Berkeley (1981), pp. 191–2.
2. J. R. Killian Jr, *Sputnik, Scientists and Eisenhower: A Memoir*, MIT P., Cambridge, Mass. (1977), pp. 160, 174.
3. C. S. Maier, introduction to G. B. Kistiakowsky, *A Scientist at the White House*, Harvard U. P., Cambridge, Mass. (1976), p. lii.
4. R. Gilpin, *American Scientists and Nuclear Weapons Policy*, Princeton U. P., Princeton, N. J. (1962), pp. 202, 218–19.

specialized journal for just that purpose now exists. Harold Feiveson, a Princeton physicist and political scientist who now works primarily on arms control, is the US editor, and the publisher in English is Gordon and Breach Science Publishers, which has a long-standing relationship with Sagdeev and a number of the other leading members of the Committee of Soviet Scientists. We expect that the publisher in Russian will be Nauka ("Science"), the publishing house of the Soviet Academy.

The first (double) issue of *Science and Global Security* appeared in English in October 1989 and contained articles on arrangements to verify nuclear weapon dismantlement, on the verification of limits on sea-launched cruise missiles and on space-reactor arms control. Future issues will also contain articles relating to the verification of arms control agreements, including an analysis of the results of the NRDC–Soviet Academy in-country seismic verification project.

The Cooperative Research Project

The FAS–CSS cooperative research project, of which Sagdeev (who succeeded Velikhov as chairman of the CSS) and I are codirectors, has thus far focused principally on approaches to verifying limits on nuclear warheads, a ban on nuclear reactors in Earth orbit and a ban on testing ground-based lasers as antisatellite weapons. On the US side, about 20 analysts, almost all of whom are university and national laboratory physicists, have contributed. On the Soviet side, the contributors have been six physicists from Sagdeev's group at the Space Research Institute and four analysts from Kokoshin's military–political affairs group at the US–Canadian Institute. Funding for the FAS group has been provided by the W. Alton Jones Foundation, the Carnegie Corporation and an anonymous philanthropist.

The focus of our work on the verification of limits on nuclear warheads is deliberately complementary to the traditional approach of limiting ballistic missiles and nuclear weapon "launchers"—the approach followed in the 1972 strategic arms limitation agreement (SALT), the 1979 SALT II treaty, the 1988 treaty eliminating intermediate-range missiles in Europe (the so-called INF Treaty) and the current strategic arms reduction negotiations (START). In the past, nuclear warheads were beyond the reach of arms control because they are small and cannot be counted from satellites. Now, with *glasnost*, it is possible to consider agreements requiring on-site verification.

Another reason it is now important to focus on verifying limits on warheads is that nuclear arms negotiations are concentrating for the first time on *reductions*. The INF Treaty eliminated a class of land-based nuclear missiles, and the START negotiations have the objective of reducing by approximately 50% the warheads currently deployed on strategic ballistic missiles. But neither the INF Treaty, as ratified, nor the START treaty, as it is being negotiated, include procedures for eliminating the warheads being retired. These warheads and the fissile materials they contain are therefore available for other uncontrolled weapons systems or for a sudden "breakout" from the constraints of arms reductions agreements.

One of the FAS–CSS papers therefore has examined possible arrangements for verifiably eliminating warheads and placing their fissile materials under safeguards without revealing warhead designs to the other side. Another explores the problem of verifying declarations of stockpiles of warheads (deployed and nondeployed) and stockpiles of weapons-usable fissile materials in weapons or available for their manufacture.

Two other papers have examined general approaches to the problem of verifying limits on long-range, nuclear-armed, sea-launched cruise missiles—one of the major unresolved issues in the START negotiations. All of these papers appeared in the first issue of *Science and Global Security* or will appear in *Reversing the Arms Race*, a book Sagdeev and I are editing for publication in English next year by Gordon and Breach.

A major joint FAS–CSS study relating to controls on nuclear warheads examined the applicability and limitations of passive radiation detection and of radiographic techniques for detecting nuclear warheads. This paper provided the theoretical basis on the US side for the recent Velikhov–NRDC project in which the gamma rays and neutron emissions from a Soviet cruise-missile warhead were measured on a Soviet cruiser through a three-inch-thick launch tube, at short range and at distances up to 70 meters, respectively (see the box on page 45).

Other papers have explored the application of similar techniques to the verification of limits on the number of warheads carried by a ballistic missile without actual removal of the nose cone. Two overview papers considered how nuclear balance would be achieved if each side's strategic arsenal was reduced from current levels of over 10 000 warheads to 2000 warheads or less.

A number of papers have also been written relating to the desirability and verifiability of a ban on nuclear reactors in Earth orbit. One of the indirect results of this effort was the uncovering of the fact, previously kept secret by both the US and Soviet governments, that gamma and positron emissions by Soviet nuclear reactors in Earth orbit have become a major problem for gamma-ray astronomers.[5]

In May 1989 FAS and CSS launched a joint assessment of a possible method of verifying limits on the potential brightness (measured in watts per steradian) of ground-based lasers being fired into space. One method would involve measurements at a distance of a kilometer from the laser of the light scattered out of the beam by atmospheric aerosols. The lead on this project is being taken by FAS under the direction of Ron Ruby, a physicist at the University of California, Santa Cruz.

The Soviet scientists in this cooperative project have repeatedly shown their independence from official Soviet positions as the following two examples will attest.
▷ In February 1988 the CSS brought to a joint workshop in Key West a manuscript that found no technical basis for Gorbachev's statement two months earlier that the Soviet Union had created "national means for verifying the presence of nuclear weapons on various naval ships . . . without conducting any on-the-spot inspection on board the vessels themselves."
▷ In May 1988 Sagdeev signed a joint CSS–FAS proposal to ban nuclear reactors in Earth orbit. It quickly became clear that this statement did not reflect Soviet policy when the Soviet space-reactor community counterattacked with a *Tass* article arguing the importance of space reactors and subsequently sent a delegation to the annual US space-reactor contractor in Albuquerque with offers to sell Soviet space reactors to the US. The leader of the Soviet delegation, Academician Nikolai N. Ponomarev-Stepnoi, when asked at the meeting about Sagdeev's opposition to space reactors in Earth orbit, said that he had met with Sagdeev before leaving Moscow and that Sagdeev "told me his opinion and I told him mine. And we were both so glad that we could tell each other our own opinion in our own country finally."

The International Foundation

At the January 1987 forum, Velikhov proposed the creation of a new foundation dedicated to fostering international cooperative solutions to the arms race,

Measurements of Radiation from a Soviet Warhead

One of the main issues that have blocked movement toward a Strategic Arms Reduction Treaty concerns limitations on long-range, nuclear-armed, sea-launched cruise missiles.

The Soviet government argues that unless the multiplication of this new type of long-range nuclear weapon is restricted, an agreement limiting other types of strategic nuclear weapons could be rendered meaningless. It claims that limits on nuclear cruise missiles could be adequately verified, and it has proposed a joint US–Soviet verification experiment on nuclear warhead detection on ships.

In contrast, the US government, which is ahead in the development of both the nuclear and nonnuclear variants of these missiles, feels no urgency about achieving a limitation agreement. It argues that the manufacture, storage and deployment of the small missiles can be so inconspicuous that verification of limits would be impossible. The US Navy has made it clear that it opposes any verification arrangement that would allow Soviet officials to inspect US ships or would reveal the presence of warheads to the general public. (Because the presence of nuclear warheads is a politically sensitive issue in many ports that naval vessels visit, the Navy neither confirms nor denies the presence of nuclear warheads on its ships.)

In an attempt to break this impasse the Soviet government, on 5 July, joined with the Natural Resources Defense Council in a demonstration of warhead detection on one of its ships, the guided-missile cruiser Slava. The demonstration occurred in the Black Sea off Yalta.

The warhead was easy to detect because the cruise-missile launcher was unshielded and in an exposed location above the deck. Even so, the demonstration of military *glasnost* was important because it was the first time that either government knowingly let anyone other than its own technicians close to one of its warheads with a high-resolution gamma detector—the instrument that the NRDC group brought along.

For this experiment, Thomas Cochran of the NRDC had recruited four other physicists: Steven Fetter of the University of Maryland, the lead author of the FAS–CSS study "Detecting Nuclear Warheads"; Lee Grodzins of MIT; Harvey Lynch of Stanford; and Martin Zucker of Brookhaven.

The NRDC group used a high-purity 150-cm^3 germanium crystal scintillation detector with a resolution of about 2 keV. The detector was placed on the launcher at a location directly over the warhead designated by the Soviet team (see cover). The gamma rays had to penetrate the launcher material, which was steel about 8 cm thick.

The background-subtracted gamma spectrum resulting from a total of about 20 minutes of observing time is shown in the figure below. It clearly shows peaks associated with the alpha decays of the fissile isotopes U^{235} (especially at 186 keV) and Pu^{239} (especially at 375 and 414 keV). It also shows gammas from the alpha decay of Am^{241}, itself a decay product of Pu^{241} (722 keV) and from Tl^{208}, a decay product of the artificial isotope U^{232} (especially at 2614 keV). The presence of U^{232} indicates that some of the uranium in the warhead must have been exposed to neutrons—implying that the Soviets must have recycled uranium from their production reactors for use in weapons.

The spectrum was surprising in two principal ways to the US experimenters, whose expectations were formed by calculations in the FAS–CSS report of the radiation from a number of highly simplified models of nuclear warheads. First, the number of 1001-keV gammas from the decay of U^{238} was unexpectedly low, indicating that all the uranium in the warhead is highly enriched (only about 4% U^{238}). And second, the observation of the low-energy gammas from the U^{235} decays indicates that there is virtually no heavy-metal shielding between the U^{235} and the detector, since the mean free path of these gammas is only about a millimeter in lead.

A group from the Soviet Academy's Institute of Earth Physics carried out measurements similar to those of the NRDC group on top of the launcher (with an instrument of somewhat poorer resolution), and a group from the academy's Geochemistry Institute attempted—unsuccessfully—to detect the warhead with a 0.25-m^2 array of sodium iodide counters from a landing ship passing along the side of the cruiser.

The most interesting Soviet experiment was done, however, by a group from the Kurchatov Institute's physics division, whose head, Academician Spartak Belyaev, was present. This group conducted neutron measurements from a helicopter using a set of He^3 proportional counters with a total area of 2.5 m^2. The helicopter detected statistically significant signals at distances of up to 70 meters (where the signal was about 1 neutron per second) from neutrons emitted as a result of the spontaneous fission of Pu^{240} in the warhead. The Kurchatov group apparently has worked with the Soviet Navy doing this sort of measurement before.

—Steven Fetter and Frank von Hippel

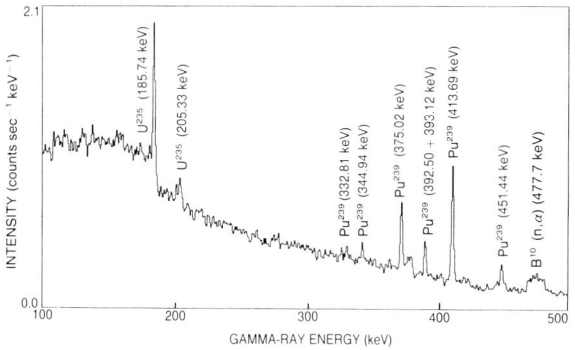
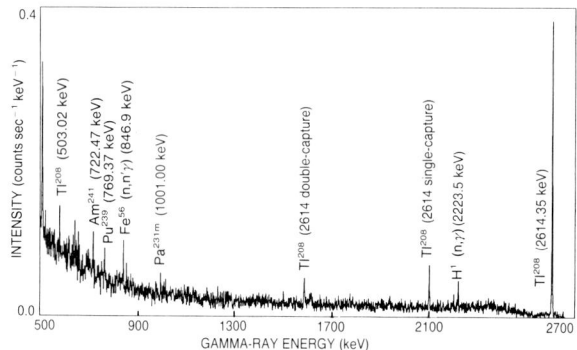

Gamma-ray signal recorded by germanium detector placed on top of cruise-missile launcher 3.4 meters from the lid for ten minutes. Background signal from the neighboring, empty launcher has been subtracted.

environmental degradation, underdevelopment, denial of human rights and lack of education. This proposal was inspired in part by Velikhov's exposure to US foundations such as the Carnegie Corporation and the MacArthur Foundation, which had supplied prompt support for the NRDC–Soviet Academy seismic monitoring project, despite the fact that it was not greeted with universal enthusiasm within the US government. The American who played the greatest part in organizing the foundation was Jerome Wiesner, science adviser to President Kennedy and president emeritus of MIT.

In January 1988, with start-up grants from a number of US foundations and the Soviet Peace Fund, the board of directors of the International Foundation for the Survival and Development of Humanity had its first meeting in Moscow. The name of the foundation, like that of the Committee of Soviet Scientists for Peace and Against the Nuclear Threat, is somewhat awkward sounding to Western ears. But it appears to be important to declare your purpose in naming an organization in the Soviet Union.

The Soviet government has given the foundation a unique charter that permits it, "for completion of its projects, . . .to create enterprises and organizations." Thus far, the foundation has used this authority to support a Soviet human rights group sponsored by Sakharov and to create a Soviet chapter of Greenpeace under the chairmanship of Alexei Yablokov, a leading Soviet ecologist.

An assessment

Cooperative and open research on the technical basis for arms control policy initiatives by independent US and Soviet scientists is a hopeful indication that the pressures driving the arms race may be dissipating. Such cooperation would not be permitted by either country if there were not an increasing consensus that the arms race is more of a threat to both countries than we are to each other.

Of course, the cooperation has a political impact—it strengthens the basis for further cooperation. Therefore, if one thinks that strengthening cooperative relationships between the US and USSR is dangerous because it undermines the will to invest in defenses, one will not welcome these initiatives.

It is often asked whether the types of joint projects discussed here are not better left to the governments. But governments don't like to sponsor research on questions relating to policy decisions that have not been made yet. Such research may undermine objections they may later wish to make to a policy option. Thus, for example, Colonel Ed Nawrocki, an assistant to former Assistant Secretary of Defense Richard Perle in the Reagan Administration, explained their office's opposition to the NRDC–Soviet Academy seismic-monitoring project as follows:

"The NRDC's goals were totally the opposite of our own. They went into this project to prove that a comprehensive test ban treaty is verifiable. [And we'd made verification the main public objection to a comprehensive test ban because] verification is such a 'showstopper,' as Perle is fond of saying. So the government didn't go much beyond verification as a reason why we shouldn't have a CTB. And the NRDC was out to undermine the verification argument against a CTB."[6]

The NRDC–Soviet Academy project has therefore been inconvenient to those who oppose a test ban because they think it is important to continue to develop new types of nuclear weapons, forcing them to "come out of the closet" and make their arguments before a not completely sympathetic public.

At this time, the future of the Committee of Soviet Scientists is uncertain. Both of its key scientific leaders, Velikhov and Sagdeev, have been caught up in the dramatic events associated with the effort to open up the USSR and reenergize its economy. Sagdeev has emerged, with Sakharov, as the leading advocate of reform in the Soviet Academy and its institutes (see PHYSICS TODAY, January 1989, page 61).

Velikhov has become a member of the Central Committee of the Communist Party of the USSR, the chairman of the armed forces subcommittee of the Supreme Soviet's Committee on Defense and State Security and the new director of the giant Kurchatov Institute of Atomic Energy.

In the meantime the CSS has received a charter that allows it to recruit a full-time staff and establish a small "think tank" with 17 full-time researchers working on the technical basis for disarmament. Whether, with its two original leaders so distracted, the CSS will successfully make the transition from an *ad hoc* organization of high-level and talented amateurs to a group whose work is done primarily by much less senior full-time professionals remains to be seen.

The next generation

A first step to ensure that there will be a next generation of Soviet physicists working on the technical basis for Soviet policy initiatives on global problems was taken early in September when the Committee of Soviet Scientists joined with the Moscow Physico-Technical Institute to sponsor an eight-day International School on Science and World Affairs. There were 24 Soviet physics students, ranging from second-year undergraduates to research associates; nine US postdocs with physics PhDs; one US graduate student; five physics graduate students from Imperial College, London; and two physics graduate students from Beijing. The US participants all are engaged in arms control research—at universities, with public-interest groups or as Congressional fellows. The faculty of the school also was diverse and lectured on global climate as well as arms control issues.

The US postdocs came away from the school quite impressed with the Soviet students. Some had obviously done considerable independent study in preparation for the school. The MacArthur Foundation sent the message that it would entertain applications from Soviet students for its predoctoral and postdoctoral fellowships in international security. A number of US-university arms control research groups had already expressed their willingness to host Soviet postdocs embarking on careers in that area. Obviously the door is wide open for further development of the US–Soviet "connection" in arms control research.

References

1. R. L. Garthoff, in *Ballistic Missile Defense*, A. B. Carter, D. N. Schwartz, eds., Brookings Institution, Washington, D. C. (1984), p. 298n.
2. V. Aleksandrov, G. Stenchikov, in *Proc. on Applied Mathematics*, Computing Center, USSR Academy of Sciences, Moscow (1983). G. S. Golitsyn, A. S. Ginsburg, *Possible Climatic Consequences of Nuclear War and Some Natural Analogues: A Scientific Investigation*, Committee of Soviet Scientists for Peace and Against the Nuclear Threat, Moscow (1984).
3. S. Talbot, *Master of the Game*, Norton, New York (1988), pp. 347–8.
4. *FAS Public Interest Report*, February 1988, p. 14.
5. J. R. Primack *et al.*, Science **244**, 407 (1989).
6. P. G. Schrag, *Listening for the Bomb: A Study in Nuclear Arms Control Verification Policy*, Westview, Boulder, Colo. (1989), p. 84.

APS and Academy members polled on SDI; physicists mobilize

PHYSICS TODAY/JUNE 1986

Physicists, as people who prize intellectual prowess, tend to be suspicious of mass public-opinion polls. Despite that or maybe because of it, they continue to sign petitions for or against the SDI program in great numbers, and when polled on the subject, they show a willingness to express their opinions at considerable length.

The Cornell–Illinois anti-SDI petition continues to gather signatures on university campuses (see PHYSICS TODAY, November, page 95). As of 13 May, when a press conference was held in Washington to publicize the latest results, 3700 faculty members and 2800 graduate students had pledged not to engage in SDI research. Majorities in 59 physics "research departments," as defined by petition organizers, had taken the pledge, according to David Wright of the University of Pennsylvania.

Signatories of the Cornell–Illinois petition include a large number of prominent physicists, ranging from Philip W. Anderson and Subrahmanyan Chandrasekhar to Carlo Rubbia and Steven Weinberg.

A new anti-SDI petition circulating at industrial and national laboratories is sponsored by about a dozen scientists, including physicists Anderson, Owen Chamberlain, Ernest D. Courant, Albert Crewe, Gerhart Friedlander, Pierre C. Hohenberg, J. Carson Mark, Edwin M. McMillan and Robert W. Wilson and computer scientists John Backus and Kenneth L. Thompson. It reads in part:

We, the undersigned scientists and engineers currently or formerly at government and industrial laboratories, wish to express our serious concerns about the Strategic Defense Initiative, commonly known as "Star Wars." Recent statements from the Administration give the erroneous impression that there is virtually unanimous support for this initiative from the scientific and technical community. In fact the SDI has grown into a major program without the technical and policy scrutiny appropriate to an undertaking of this magni-

SHURCLIFF

tude....

The stated goal of the SDI is developing the means to render nuclear weapons "impotent and obsolete." We believe that realization of this dream is not feasible in the foreseeable future. The more limited goal of developing partial defenses against ballistic missiles does not fundamentally alter the current policy of deterrence, yet it represents a significant escalation of the arms race and runs the serious risk of jeopardizing existing arms-control treaties and future negotiations....

We urge the Congress to heed these concerns and to limit the SDI to a scale appropriate to exploratory research, while assessing ... the program in comparison with alternative strategies for strengthening the overall security of the nation.

The aim of the final paragraph, says Hohenberg, a physicist at AT&T Bell Labs, is to "hold SDI to what the Administration says it is—a research program."

The laboratory letter, unlike the petition written at Cornell and the University of Illinois, is not a pledge to reject funding from SDI. Laboratory personnel typically have little control

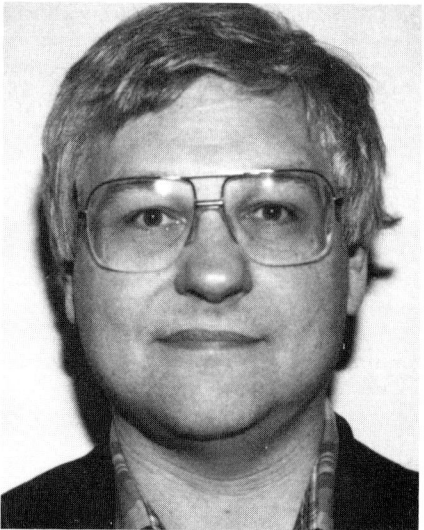

HOFFERT

over where their money comes from and in some cases taking a pledge might be tantamount to promising to resign.

Hohenberg says that "the letter is an attempt to redress the view propounded by some SDI officials that opponents to SDI are not in the mainstream of the scientific community" (see PHYSICS TODAY, January, page 79). The letter is a direct appeal to Congress, Hohenberg says, and its message is that "doubts are very much in order." The plan is to gather signatures for a few months and then send the letter directly to every member of Congress.

UCS poll. Because individual physicists might have many reasons not to sign statements on the Strategic Defense Initiative, petitions are open to the objection that they may not reflect the full range of opinion in the community. To correct for that possibility, the Union of Concerned Scientists hired Peter D. Hart Research Associates Inc to do a scientific survey. Hart Research conducted telephone interviews of roughly 25 minutes each with 549 physicists selected at random from the 1985 American Physical Society membership directory. UCS released the results in March.

The survey revealed that physicists oppose SDI by a ratio of nearly two to one, though many physicists saw merits in some aspects of SDI. By a margin of 54% to 29%, the physicists viewed SDI as a step in the wrong direction—even though pluralities approved of other weapons systems such as the Trident submarine, the cruise missile and the Stealth bomber. Opposition to SDI was strongest among those who said they knew the most about it. (Only 9% said they knew little or hardly anything about SDI.)

The physicists sampled opposed deployment of SDI systems by 62% to 23% and testing of SDI systems by 49% to 43%, but they favored basic laboratory research on SDI by 77% to 21%.

Two-thirds of them considered it improbable that SDI could provide a defense for the population of the country as a whole, but a plurality considered it likely that a missile-defense system could provide an effective point defense for hardened military targets such as missile silos.

The physicists agreed by a margin of 83% to 11% that "even if we develop a system that works against today's weapons, the Russians can develop effective countermeasures." They believed by a narrow margin that SDI might have some utility as a bargaining chip, but by a larger margin they considered that SDI would more likely escalate the arms race. "When asked to select the most important of four approaches the United States might pursue to reduce the threat of nuclear war, 61% chose 'negotiate new arms control agreements with the Soviet Union.' Arms control ranked substantially ahead of any other item," Hart reported to the Union of Concerned Scientists in March.

The attitude of American physicists toward arms control and SDI cannot be attributed to some kind of blind hostility to President Reagan. Only 28% of those polled described themselves as Republicans, but 47% approved of Reagan's general performance, while 50% disapproved.

General public. The physicists polled by Hart obviously tend to think that they are specially qualified to have opinions about SDI and specially well informed. Do their views about SDI differ materially from the views held by the general public?

According to a roundup of SDI surveys in the August–September issue of *Public Opinion* magazine, polls tend to indicate that a slight majority of the American public considers Star Wars a good idea and would like to see the United States develop a space-based defense against nuclear missiles. *Public Opinion* is published by the American Enterprise Institute in Washington and is often classified as a neoconservative publication.

"When an issue is complicated and the public not well informed," the editors of *Public Opinion* observe, "responses bounce all over, depending on which nerve the pollsters touch." *Public Opinion* found just one survey that asked Americans, without making any complicated explanations that might bias the results, whether they thought a Star Wars system could work: the CBS News/*New York Times* survey. When the survey asked this question in January 1985, 62% answered yes and 23% no. When the question was asked again in November 1985, 58% said yes and 27% no.

It seems apparent that physicists are in fact better informed about SDI than the general public, significantly more likely to consider the initiative undesirable and considerably more likely to see it as unrealistic.

Academy poll. The margins between informed or elite opinion and general opinion may be even bigger when it comes to members of the National Academy of Sciences. These individuals were polled this spring by William A. Shurcliff, a retired Harvard physicist living in Cambridge, Massachusetts. An optics specialist, Shurcliff was senior editor of the Smyth report on the Manhattan Project, senior author of the report on the first atomic-bomb tests at Bikini and a leader from 1967 to 1972 of the forces that persuaded Congress to ditch the civilian supersonic transport.

Between 16 March and 17 April, Shurcliff polled all 1505 US residents listed in the 1985 NAS directory. He received 530 replies (36%).

"By majorities of more than 20:1," Shurcliff reports, "the responding members declared that the proposed Star Wars program would not provide an effective shield, would not defeat a high-altitude attack, would not prevent delivery of A-bombs by other methods (for example, low-altitude delivery or smuggling) and would not protect our European allies. In overall attitude toward the Star Wars program, 20 members were for it and 461 against it."

Shurcliff made it clear in the opening sentence of his survey letter that he considered SDI "doomed to failure" and that he intended to use the results of the poll to persuade Congress to curtail the program, as he used a similar survey 17 years ago to rally opposition to the supersonic airliner. These facts, together with the self-selecting response mechanism, would seem to disqualify the poll from being considered scientific.

"We would never run something like that in the magazine," comments Victoria Sackett, deputy managing editor of *Public Opinion*, "because there are three things there [the known bias of the pollster, his intended use of the poll and the self-selecting sample] that color the response, though it is hard to say in what direction. If the results turned out to be similar to a carefully drawn probability sample, it would be by chance."

SDIO reaction. Considering the latest poll and survey results, it is scarcely surprising that some physicists thought they detected a certain estrangement between the SDI Organization and the physics community when APS met in Washington at the end of April. Lieutenant Colonel Simon (Pete) Worden, special assistant to SDI chief Lieutenant General James Abrahamson, showed up for a morning meeting on 30 April to describe SDI architecture, but he came late and left early because he was scheduled to give a talk at the War College. The SDI Organization went completely unrepresented in two press conferences on SDI held that afternoon. In the evening of the same day, Richard D. Bleach of the SDI Organization was to give a talk about technical-personnel requirements at the session on SDI and the physics community, but he canceled in the morning, and James Ionson did not show up to speak about progress in SDI-related physics.

Ionson, who is director for innovative science and technology with SDIO, says that he decided not to attend on his own initiative, for a combination of professional and personal reasons. He says that his office canceled his talk with APS, but APS has no record of the cancellation and the message did not get through to the session organizers. The external-affairs office of SDIO, Ionson says, had nothing to do with his decision and did not even know about the invitation. Ionson observes that it is not his job to do public relations for SDIO.

Bleach is reported to have been ordered by SDIO external affairs not to speak at the APS session. Denying reports that SDI officials were ordered not to show up for the APS sessions, Worden claims that the officials decided among themselves not to participate. The evening session "did not look like an opportunity to discuss the technical issues," Worden says. "It looked like an occasion to beat up on SDI."

Worden adds that the SDI Organization has taken note of the many physicists who have signed anti-SDI petitions and their "strong feelings," but he believes that only a small percentage of the people in the technical and engineering communities have signed. Very little SDI research is basic research, Worden notes.

Finally, Worden says that SDI officials are upset about the persistent "misconception" among physicists that the aim of SDI is a "perfect defense."

"We are looking for a better way to deter war and do arms control," Worden says.

A new pro-SDI group, the Science and Engineering Committee for a Secure World, has been formed. Its acting chairman is Frederick Seitz of Rockefeller University, and Martin I. Hoffert, who is chairman of the applied-science department at New York University, serves as spokesman for the group.

In a statement read by Hoffert to a Senate subcommittee on 9 May, the group said: "We are confident that there are thousands of scientists and engineers across America and elsewhere who agree with us that it is unscientific and unwise to hastily oppose the promising Strategic Defense Initiative at this early stage of its research and development, and who believe that the concept of developing a defensive system to protect our people from a nuclear attack makes good common and good moral sense."

Around 90 scientists—about half of them physicists—have signed on to the Science and Engineering Committee for a Secure World. Signatories include physicists Hans Mark (chancellor of the University of Texas), John A. Wheeler (University of Texas), Harold Agnew (former director of Los Alamos National Laboratory) and William Nierenberg (Scripps Institution for Oceanography).

—WILLIAM SWEET

SECTION 8 ———————————————————————
FOR FURTHER READING

The physics, technology, and history of the nuclear arms race are continually in motion as new technologies are advanced, and as political leaders and events change. For further reading, a list of additional references is listed below. More extensive reading lists, along with commentary, have been prepared by Dietrich Schroeer and can be found as Appendix B in the two AIP Conference Proceedings edited by Schroeer and Hafemeister (below).

William Burrows, *Deep Black*: *Space Espionage and National Security*, Random House, New York, 1986.

Ashton Carter, John Steinbrunner and Charles Zracket, *Managing Nuclear Operations*, Brookings, Washington, DC, 1987.

Thomas Cochran, William Arkin, Robert Norris and Milton Hoenig, *Nuclear Weapons Databook, Vol. I–IV*, Ballinger, Cambridge, MA, 1984–89.

Paul Craig and John Jungerman, *Nuclear Arms Race: Technology and Society*, McGraw-Hill, New York, NY, 1986.

Jonathan Dean, *Meeting Gorbachev's Challenge: How to Build Down the NATO-Warsaw Pact Confrontation*, St. Martin's Press, New York, NY, 1989.

Sidney Drell, Philip Farley and David Holloway, *The Reagan Strategic Defense Initiative: A Technical, Political, and Arms Control Assessment*, Ballinger, Cambridge, MA, 1985.

Gloria Duffy, *Compliance and the Future of Arms Control*, Stanford Press, Stanford, CA, 1988.

Lewis Dunn and Amy Gordon, eds., *Arms Control Verification and the New Role of On-Site Inspection*, Lexington Books, Lexington, MA, 1980.

Steve Fetter, *Toward a Comprehensive Test Ban*, Ballinger, Cambridge, MA, 1988.

Samuel Glasstone and Philip Dolan, *The Effects of Nuclear Weapons*, Departments of Energy and Defense, Washington, DC, 1977.

David Hafemeister and Dietrich Schroeer, eds., *Physics, Technology and the Nuclear Arms Race*, American Institute of Physics, New York, NY, 1983.

Richard Hewlett, *A History of the United States Atomic Energy Commission.* Vol. 1 with Oscar Anderson, *The New World, 1939–1946*, University Park: Pennsylvania State University Press, 1962

——Vol. 2 with Francis Duncan, *Atomic Shield, 1947–1952*, University Park: Pennsylvania State University Press, 1969.

——Vol. 3 with Jack Holl, *Atoms For Peace and War, 1953–1961: Eisenhower and the AEC*, Berkeley: University of California Press, 1989.

International Institute for Strategic Studies, *The Military Balance*, IISS, London, UK.

Fred Kaplan, *Wizards of Armageddon*, Simon & Schuster, New York, NY, 1983.

Allan Krass, Peter Boskma, Boelie Elzen and Wim Smit, *Uranium Enrichment and Nuclear Weapon Proliferation*, Taylor and Francis, New York, NY, 1983.

Michael Krepon and Mary Umberger, eds., *Verification and Compliance*, Balinger, Cambridge, MA, 1988.

Michael Krepon, *et al.*, eds., *Commercial Observation Satellites and International Security*, St. Martin's Press, New York, NY, 1990.

John Lamarsh, *Introduction to Nuclear Engineering*, Addison Wesley, Reading, MA, 1977.

Barbara Levi, Mark Sakitt and Arthur Hobson, eds., *The Future of Land-Based Strategic Missiles*, American Institute of Physics, New York, NY, 1989.

National Academy of Sciences, *Nuclear Arms Control: Background and Issues*, National Academy Press, Washington, DC, 1985.

Joseph Nye, *Nuclear Ethics*, Free Press, New York, NY, 1986.

Joseph Nye, Graham Allison and Albert Carnesale, *Fateful Visions: Avoiding Nuclear Catastrophe*, Ballinger, Cambridge, MA, 1988.

Office of Technology Assessment, Reports on SDI, ASATs, Nonproliferation, MX-Basing, Effects of Nuclear War, New Technologies for NATO, etc., Office Tech. Assessment, Washington, DC.

A. Barrie Pittock (Vol. I), Mark Harwell (Vol. II), *Environmental Consequences of Nuclear War, Vol. I–II*, Wiley, New York, NY, 1986.

Jeffrey Richelson, *The U.S. Intelligence Community*, Ballinger, Cambridge, MA, 1985.

Lawrence Scheinman, *The IAEA and World Nuclear Order*, Resources for the Future, Washington, DC, 1987.

Dietrich Schroeer, *Science, Technology and the Nuclear Arms Race*, Wiley, New York, NY, 1984.

Dietrich Schroeer and David Hafemeister, eds., *Nuclear Arms Technologies in the 1990s,* American Institute of Physics, New York, NY, 1988.

Richard Scribner, Ted Ralston and William Metz, *The Verification Challenge*, Birkhauser, Boston, 1985.

Glenn Seaborg, *Stemming the Tide: Arms Control in the Johnson Years*, Lexington Books, Lexington, MA, 1987.

Gerard Smith, *Doubletalk: The Story of SALT I*, Doubleday, Garden City, NY, 1980.

George Sutton and Donald Ross, *Rocket Propulsion Elements*, Wiley, New York, NY, 1976.

Kosta Tsipis, *Arsenal: Understanding Weapons in the Nuclear Age*, Simon & Schuster, New York, NY, 1983.

Kosta Tsipis, David Hafemeister and Penny Janeway, eds., *Arms Control Verification: The Technologies That Make It Possible*, Pergamon-Brasseys, Washington, DC, 1986.

Richard Rhodes, *The Making of the Atomic Bomb*, Simon & Schuster, New York, NY, 1988.

Leonard Spector, *Going Nuclear*, Ballinger, Cambridge, MA, 1987.

Strobe Talbott, *Endgame: The Inside Story of SALT II*, Knopf, New York, NY, 1979.

———*Deadly Gambits*, Knopf, New York, NY, 1984.

———*The Master of the Game*, Knopf, New York, NY, 1988.

Frank von Hippel and Ronald Sagdeev, eds., *Reversing the Arms Race*, Gordon and Breach, New York, NY, 1980.